Statistical Inference Based on Divergence Measures

STATISTICS: Textbooks and Monographs
D. B. Owen
Founding Editor, 1972–1991

STATISTICS: Textbooks and Monographs

Recent Titles

Asymptotics, Nonparametrics, and Time Series, *edited by Subir Ghosh*

Multivariate Analysis, Design of Experiments, and Survey Sampling, edited by Subir Ghosh

Statistical Process Monitoring and Control, *edited by Sung H. Park and G. Geoffrey Vining*

Statistics for the 21st Century: Methodologies for Applications of the Future, *edited by C. R. Rao and Gábor J. Székely*

Probability and Statistical Inference, *Nitis Mukhopadhyay*

Handbook of Stochastic Analysis and Applications, *edited by D. Kannan and V. Lakshmikantham*

Testing for Normality, *Henry C. Thode, Jr.*

Handbook of Applied Econometrics and Statistical Inference, *edited by Aman Ullah, Alan T. K. Wan, and Anoop Chaturvedi*

Visualizing Statistical Models and Concepts, *R. W. Farebrother and Michael Schyns*

Financial and Actuarial Statistics, *Dale Borowiak*

Nonparametric Statistical Inference, Fourth Edition, Revised and Expanded, *edited by Jean Dickinson Gibbons and Subhabrata Chakraborti*

Computer-Aided Econometrics, *edited by David EA. Giles*

The EM Algorithm and Related Statistical Models, *edited by Michiko Watanabe and Kazunori Yamaguchi*

Multivariate Statistical Analysis, Second Edition, Revised and Expanded, *Narayan C. Giri*

Computational Methods in Statistics and Econometrics, *Hisashi Tanizaki*

Applied Sequential Methodologies: Real-World Examples with Data Analysis, *edited by Nitis Mukhopadhyay, Sujay Datta, and Saibal Chattopadhyay*

Handbook of Beta Distribution and Its Applications, *edited by Richard Guarino and Saralees Nadarajah*

Item Response Theory: Parameter Estimation Techniques, Second Edition, *edited by Frank B. Baker and Seock-Ho Kim*

Statistical Methods in Computer Security, *William W. S. Chen*

Elementary Statistical Quality Control, Second Edition, *John T. Burr*

Data Analysis of Asymmetric Structures, *edited by Takayuki Saito and Hiroshi Yadohisa*

Mathematical Statistics with Applications, *Asha Seth Kapadia, Wenyaw Chan, and Lemuel Moyé*

Advances on Models, Characterizations and Applications, *N. Balakrishnan, I. G. Bairamov, and O. L. Gebizlioglu*

Survey Sampling: Theory and Methods, Second Edition, *Arijit Chaudhuri and Horst Stenger*

Statistical Design of Experiments with Engineering Applications, *Kamel Rekab and Muzaffar Shaikh*

Quality By Experimental Design, Third Edition, *Thomas B. Barker*

Handbook of Parallel Computing and Statistics, *Erricos John Kontoghiorghes*

Statistical Inference Based on Divergence Measures, *Leandro Pardo*

Statistical Inference Based on Divergence Measures

Leandro Pardo

Complutense University of Madrid
Spain

CRC Press
Taylor & Francis Group
Boca Raton London New York

CRC Press is an imprint of the
Taylor & Francis Group, an **informa** business
A CHAPMAN & HALL BOOK

Published in 2006 by
Chapman & Hall/CRC
Taylor & Francis Group
6000 Broken Sound Parkway NW, Suite 300
Boca Raton, FL 33487-2742

First issued in paperback 2020

ISBN 13: 978-0-367-57801-5 (pbk)
ISBN 13: 978-1-58488-600-6 (hbk)

Library of Congress Cataloging-in-Publication Data

Pardo, Leandro.
Statistical inference based on divergence measures / Leandro Pardo
p. cm. -- (Statistics, textbooks and monographs ; v. 185)
Includes bibliographical references and index.
ISBN 1-58488-600-5
1. Divergent series. 2. Entropy (Information theory) 3. Multivariate analysis. 4. Statistical hypothesis testing. 5. Asymptotic expansions. I. Title. II. Series.

QA295.P28 2005
519.5'4--dc22 2005049685

This book is dedicated to my wife, Marisa

Preface

The main purpose of this book is to present in a systematic way the solution to some classical problems of statistical inference, basically problems of estimation and hypotheses testing, on the basis of measures of entropy and divergence, with applications to multinomial (statistical analysis of categorical data) and general populations. The idea of using functionals of Information Theory, such as entropies or divergences, in statistical inference is not new. In fact, the so-called Statistical Information Theory has been the subject of much statistical research over the last forty years. Minimum divergence estimators or minimum distance estimators (see Parr, 1981) have been used successfully in models for continuous and discrete data. Divergence statistics, i.e., those ones obtained by replacing either one or both arguments in the measures of divergence by suitable estimators, have become a very good alternative to the classical likelihood ratio test in both continuous and discrete models, as well as to the classical Pearson–type statistic in discrete models. It is written as a textbook, although many methods and results are quite recent.

Information Theory was born in 1948, when Shannon published his famous paper "A mathematical theory of communication." Motivated by the problem of efficiently transmitting information over a noisy communication channel, he introduced a revolutionary new probabilistic way of thinking about communication and simultaneously created the first truly mathematical theory of entropy. In the cited paper, two new concepts were proposed and studied: the entropy, a measure of uncertainty of a random variable, and the mutual information. Verdú (1998), in his review paper, describes Information Theory as follows: "A unifying theory with profound intersections with Probability, Statistics, Computer Science, and other fields. Information Theory continues to set the stage for the development of communications, data storage and processing, and other information technologies." Many books have been written in relation to the subjects mentioned by Verdú, but the usage of tools arising from Information Theory in problems of estimation and testing has only been described by the book of Read and Cressie (1988), when analyzing categorical data. However, the interesting possibility of introducing alternative test statistics to the classical ones (like Wald, Rao or Likelihood ratio) in general populations is not yet found in any book, as far as I am concerned. This is an important contribution of this book to the field of Information Theory.

But the following interesting question arises: Where exactly can be situated the origin of the link between Information Theory and Statistics? Lindley (1956) tries to answer our question, with the following words with reference to the paper of Shannon (1948), "The first idea is that information is a statistical concept" and "The second idea springs from the first and implies that on the basis of the frequency distribution, there is an essentially unique function of the distribution which measures the amount of the information." This fact provided Kullback and Leibler (1951) the opportunity of introducing a measure of divergence, as a generalization of Shannon's entropy, called the Kullback-Leibler divergence. Kullback, later in 1959, wrote the essential book "Information Theory and Statistics." This book can be considered the beginning of Statistical Information Theory, although it has been necessary to wait a more few years for the statisticians to return to the problem.

The contents of the present book can be roughly separated in two parts. The first part is dedicated to make, from a statistical perspective, an overview of the most important measures of entropy and divergence introduced until now in the literature of Information Theory, as well as to study their properties, in order to justify their application in statistical inference. Special attention is paid to the families of ϕ-entropies as well as on the ϕ-divergence measures. This is the main target of Chapter 1. Chapter 2 is devoted to the study of the asymptotic behavior of measures of entropy, and the use of their asymptotic distributions to solve different statistical problems. An important fact studied in this chapter is the behavior of the entropy measures as diversity indexes. The second part of the book is dedicated to two important topics: statistical analysis of discrete multivariate data in Chapters 3, 4, 5, 6, 7 and 8, and testing in general populations in Chapter 9.

The statistical analysis of discrete multivariate data, arising from experiments where the outcome variables are the number of individuals classified into unique nonoverlapping categories, has received a great deal of attention in the statistical literature in the last forty years. The development of appropriate models for those kind of data is the common subject of hundreds of references. In these references, papers and books, the model is tested with the traditional Pearson goodness-of-fit test statistic or with the traditional loglikelihood ratio test statistic, and the unknown parameters are estimated using the maximum likelihood method. However, it is well known that this can give a poor approximation in many circumstances, see Read and Cressie (1988), and it is possible to get better results by considering general families of test statistics, as well as general families of estimators. We use the word "general" in the sense that these families contain as particular cases the Pearson and loglikelihood ratio test statistics, for testing, as well as the maximum likelihood estimator, for estimating. In Chapter 3, the problem of testing goodness-of-fit with simple null hypothesis is studied on the basis of the ϕ-divergence test statistics under different situations: Fixed number of classes, number of classes increasing to infinity, quantile characterization, dependent observations and misclassified data. The results obtained in this chapter are asymptotic and consequently valid just for large sample sizes. In Chapter 4, some methods to improve the accuracy of test statistics, in those situations where the sample size can not be assumed large, are presented. Chapter 5 is addressed to the study of a wide class of estimators suitable for discrete data, either when the underlaying distribution is discrete, or when it is continuous, but the observations are classified into groups: Minimum ϕ-divergence estimators. Their asymptotic properties are studied as well as their behavior under the set up of a mixture of normal populations. A new problem of estimation appears if we have some functions that constrain the unknown parameters. To solve this problem, the restricted minimum ϕ-divergence estimator is also introduced and studied in Chapter 5. These results will be used in Chapter 8, where the behavior of ϕ-divergences test statistics in contingency tables is discussed. Chapter 6 deals with the problem of goodness-of-fit with composite null hypothesis. For this problem, we consider ϕ-divergence test statistics in which the unknown parameters are estimated by minimum ϕ-divergence estimators. In addition to the classical problem, with fixed number of classes, the following nonstandard cases are also treated: ϕ-divergence test statistics when the unknown parameters are estimated by maximum likelihood estimator, ϕ-divergence test statistics with quantile characterizations, ϕ-divergence test statistics when parameters are estimated from an independent sample, ϕ-divergence test statistics with dependent observations and ϕ-divergence test statistics when there are some constraints on the parameters. Chapter 7 covers the important problem of testing in loglinear models by using ϕ-divergence test statistics. In this chapter, some of the most important results appeared in Cressie and Pardo (2000, 2002b), and Cressie et *al.* (2003) are presented. The properties of the minimum ϕ-divergence estimators in loglinear models are studied and a new family of test statistics based on them is introduced for the problems of testing goodness-of-fit and for testing a nested sequence of loglinear models. Pearson's and likelihood ratio test statistics are members of the new family of test statistics. This chapter finishes with a simulation study, in which a new test statistic, placed "between" Pearson's chi-square and likelihood ratio test statistics, emerged as a good choice, considering its valuable properties.

Chapter 8 presents a unified study of some classical problems in contingency tables using the ϕ-divergence test statistic as well as the minimum ϕ-divergence estimator. We consider the problems of independence, symmetry, marginal homogeneity and quasi-symmetry in a two-way contingency table and also the classical problem of homogeneity.

The domain of application of ϕ-divergence test statistics goes far beyond that of multinomial hypothesis testing. The extension of ϕ-divergence statistics to testing hypotheses in problems where random samples (one or several) obey distributional laws from parametric families has also given nice and interesting results in relation to the classical test statistics: likelihood ratio test, Wald test statistic or Rao statistic. This topic is considered and studied in Chapter 9.

The exercises and their solutions included in each chapter form a part of considerable importance of the book. They provide not only practice problems for students, but also some additional results as complementary materials to the main text.

I would like to express my gratitude to all the professors who revised parts of the manuscript and made some contributions. In particular, I would like to thank Professors Arjun Gupta, Nirian Martín, Isabel Molina, Domingo Morales, Truc Nguyen, Julio Angel Pardo, Maria del Carmen Pardo and Kostas Zografos. My gratitude, also, to Professor Juan Francisco Padial for his support in the technical development of the book.

Special thanks to Professor Arjun Gupta for his invitation to visit the Department of Mathematics and Statistics of Bowling Green State University as Distinguished Lukacs Professor. Part of the book was written during my stay there. My final acknowledgment is to my wife, Professor Maria Luisa Menéndez, who read many times the early drafts of the manuscript. She gave me valuable advice and suggested many improvements. Her enthusiasm sustained me during the period spent in writing the manuscript, and this book is dedicated to her.

Leandro Pardo

Madrid

Contents

1

Divergence Measures: Definition and Properties

1.1. Introduction

Let $\boldsymbol{X}$ be a random variable taking values on a sample space $\mathcal{X}$ (usually $\mathcal{X}$ will be a subset of $\mathbb{R}^n$, n-dimensional Euclidean space). Suppose that the distribution function F of $\boldsymbol{X}$ depends on a certain number of parameters, and suppose further that the functional form of F is known except perhaps for a finite number of these parameters; we denote by $\boldsymbol{\theta}$ the vector of unknown parameters associated with F. Let $(\mathcal{X}, \beta_{\mathcal{X}}, P_{\boldsymbol{\theta}})_{\boldsymbol{\theta}\in\Theta}$ be the statistical space associated with the random variable $\boldsymbol{X}$, where $\beta_{\mathcal{X}}$ is the σ-field of Borel subsets $A \subset \mathcal{X}$ and $\{P_{\boldsymbol{\theta}}\}_{\boldsymbol{\theta}\in\Theta}$ a family of probability distributions defined on the measurable space $(\mathcal{X}, \beta_{\mathcal{X}})$ with Θ an open subset of $\mathbb{R}^{M_0}$, $M_0 \geq 1$. In the following the support of the probability distribution $P_{\boldsymbol{\theta}}$ is denoted by $S_{\mathcal{X}}$.

We assume that the probability distributions $P_{\boldsymbol{\theta}}$ are absolutely continuous with respect to a σ-finite measure μ on $(\mathcal{X}, \beta_{\mathcal{X}})$. For simplicity μ is either the Lebesgue measure (i.e., satisfying the condition $P_{\boldsymbol{\theta}}(C) = 0$, whenever C has zero Lebesgue measure), or a counting measure (i.e., there exists a finite or countable set $S_{\mathcal{X}}$ with the property $P_{\boldsymbol{\theta}}(\mathcal{X}\text{-}S_{\mathcal{X}}) = 0$). In the following

$$f_{\boldsymbol{\theta}}(\boldsymbol{x}) = \frac{dP_{\boldsymbol{\theta}}}{d\mu}(\boldsymbol{x}) = \begin{cases} f_{\boldsymbol{\theta}}(\boldsymbol{x}) & \text{if } \mu \text{ is the Lebesgue measure,} \\ \Pr_{\boldsymbol{\theta}}(\boldsymbol{X} = \boldsymbol{x}) = p_{\boldsymbol{\theta}}(\boldsymbol{x}) & \text{if } \mu \text{ is a counting measure,} \\ & (\boldsymbol{x} \in S_{\mathcal{X}}) \end{cases}$$

denotes the family of probability density functions if μ is the Lebesgue measure, or the family of probability mass functions if μ is a counting measure. In the first case $\boldsymbol{X}$ is a random variable with absolutely continuous distribution and in the second case it is a discrete random variable with support $S_{\mathcal{X}}$.

Let h be a measurable function. Expectation of $h(\boldsymbol{X})$ is denoted by

$$E_{\boldsymbol{\theta}}\left[h(\boldsymbol{X})\right] = \begin{cases} \displaystyle\int_{\mathcal{X}} h(\boldsymbol{x}) f_{\boldsymbol{\theta}}(\boldsymbol{x}) d\boldsymbol{x} & \text{if } \mu \text{ is the Lebesgue measure,} \\ \displaystyle\sum_{\boldsymbol{x}\in S_{\mathcal{X}}} h(\boldsymbol{x}) p_{\boldsymbol{\theta}}(\boldsymbol{x}) & \text{if } \mu \text{ is a counting measure.} \end{cases}$$

Since Mahalanobis (1936) introduced the concept of distance between two probability distributions, several coefficients have been suggested in statistical literature to reflect the fact that some probability distributions are "closer together" than others and consequently that it may be "easier to distinguish" between a pair of distributions which are "far from each other" than between those which are closer. Such coefficients have been variously called measures of distance between two distributions (see Adhikari and Joshi, 1956), measures of separation (Rao, 1949, 1954), measures of discriminatory information (Chernoff, 1952, Kullback, 1959) and measures of variation-distance (Kolmogorov, 1963). Many of the currently used tests, such as the likelihood ratio, the chi-square, the score and Wald tests, can in fact be shown to be defined in terms of appropriate distance measures.

While the cited coefficients have not all been introduced for exactly the same purpose, they have the common property of increasing as the two distributions involved are "further from each other". In the following, a coefficient with this property will be called *divergence measure* between two probability distributions.

Before introducing the families of divergence measures that will be used in later chapters for studying different statistical problems, we consider two classical and important distances: the Kolmogorov and Lévy distances. Our aim is to illustrate the important role that distance measures play in Probability and Statistics.

Given two probability measures $P_{\boldsymbol{\theta}_1}$ and $P_{\boldsymbol{\theta}_2}$ with associated unidimensional distribution functions $F_{\boldsymbol{\theta}_1}$ and $F_{\boldsymbol{\theta}_2}$, respectively, the Kolmogorov distance, introduced by Kolmogorov (1933), between $F_{\boldsymbol{\theta}_1}$ and $F_{\boldsymbol{\theta}_2}$ (or between $P_{\boldsymbol{\theta}_1}$ and $P_{\boldsymbol{\theta}_2}$) is given by

$$K_1(F_{\boldsymbol{\theta}_1}, F_{\boldsymbol{\theta}_2}) = \sup_{x \in \mathbb{R}} \left| F_{\boldsymbol{\theta}_1}(x) - F_{\boldsymbol{\theta}_2}(x) \right|. \tag{1.1}$$

It is the well known Glivenko-Cantelli Theorem, based on the previous distance, which states that the empirical distribution function is a uniformly strongly consistent estimate of the true distribution function; i.e., given a random sample $X_1, ..., X_n$ from a population with distribution function $F_{\boldsymbol{\theta}_0}$, for any $\varepsilon > 0$ it holds

$$\lim_{n \to \infty} \Pr \left\{ K_1(F_n, F_{\boldsymbol{\theta}_0}) > \varepsilon \right\} = 0,$$

where F_n is the empirical distribution function, i.e.,

$$F_n(x) = \frac{1}{n} \sum_{i=1}^{n} I_{(-\infty, x]}(x_i),$$

and I_A is the indicator function of the set A.

On the other hand, the Lévy distance is

$$K_2(F_{\boldsymbol{\theta}_1}, F_{\boldsymbol{\theta}_2}) = \inf \left\{ \varepsilon > 0 : F_{\boldsymbol{\theta}_1}(x - \varepsilon) \leq F_{\boldsymbol{\theta}_2}(x) \leq F_{\boldsymbol{\theta}_1}(x + \varepsilon), \text{ for all } x \right\};$$

it assumes values on $[0, 1]$ and it is not easy to compute. It is interesting to note that convergence in the Lévy-metric implies weak convergence for distribution function in $\mathbb{R}$ (Lukacs, 1975, p. 71). It is shift invariant, but not scale invariant. This metric was introduced by Lévy (1925, pp. 199-200). Some other results about probability metrics, relationships between K_1 and K_2 as well as other interesting results can be seen in Gibbs and Su (2002).

1.2. Phi-divergence Measures between Two Probability Distributions: Definition and Properties

In this section we shall introduce different divergence measures; in all the cases it must be understood provided the integral exists.

Kullback-Leibler divergence measure, between the probability distributions $P_{\boldsymbol{\theta}_1}$ and $P_{\boldsymbol{\theta}_2}$, is

$$D_{Kull}(\boldsymbol{\theta}_1, \boldsymbol{\theta}_2) = \int_{\mathcal{X}} f_{\boldsymbol{\theta}_1}(\boldsymbol{x}) \log \frac{f_{\boldsymbol{\theta}_1}(\boldsymbol{x})}{f_{\boldsymbol{\theta}_2}(\boldsymbol{x})} d\mu(\boldsymbol{x}) = E_{\boldsymbol{\theta}_1} \left[\log \left(\frac{f_{\boldsymbol{\theta}_1}(\boldsymbol{X})}{f_{\boldsymbol{\theta}_2}(\boldsymbol{X})} \right) \right], \tag{1.2}$$

which was introduced and studied by Kullback and Leibler (1951) and Kullback (1959). Jeffreys (1946) used a symmetric version of (1.2),

$$J(\boldsymbol{\theta}_1,\boldsymbol{\theta}_2) = D_{Kull}(\boldsymbol{\theta}_1,\boldsymbol{\theta}_2) + D_{Kull}(\boldsymbol{\theta}_2,\boldsymbol{\theta}_1),$$

as a measure of divergence between two probability distributions. This divergence measure is also called J-divergence.

Rényi (1961) presented the first parametric generalization of (1.2),

$$\begin{aligned} D_r^1(\boldsymbol{\theta}_1,\boldsymbol{\theta}_2) &= \tfrac{1}{r-1}\log\int_{\mathcal{X}} f_{\boldsymbol{\theta}_1}(\boldsymbol{x})^r f_{\boldsymbol{\theta}_2}(\boldsymbol{x})^{1-r} d\mu(\boldsymbol{x}) \\ &= \tfrac{1}{r-1}\log E_{\boldsymbol{\theta}_1}\left[\left(\frac{f_{\boldsymbol{\theta}_1}(\boldsymbol{X})}{f_{\boldsymbol{\theta}_2}(\boldsymbol{X})}\right)^{r-1}\right], \qquad r>0,\ r\neq 1. \end{aligned}$$

Later, Liese and Vajda (1987) extended it for all $r \neq 1, 0$, by

$$\begin{aligned} D_r^1(\boldsymbol{\theta}_1,\boldsymbol{\theta}_2) &= \tfrac{1}{r(r-1)}\log\int_{\mathcal{X}} f_{\boldsymbol{\theta}_1}(\boldsymbol{x})^r f_{\boldsymbol{\theta}_2}(\boldsymbol{x})^{1-r} d\mu(\boldsymbol{x}) \\ &= \tfrac{1}{r(r-1)}\log E_{\boldsymbol{\theta}_1}\left[\left(\frac{f_{\boldsymbol{\theta}_1}(\boldsymbol{X})}{f_{\boldsymbol{\theta}_2}(\boldsymbol{X})}\right)^{r-1}\right], \qquad r\neq 0,1. \end{aligned} \tag{1.3}$$

In the following, expression (1.3) will be referred as Rényi divergence. The cases $r=1$ and $r=0$ are defined by

$$D_1^1(\boldsymbol{\theta}_1,\boldsymbol{\theta}_2) = \lim_{r\to 1} D_r^1(\boldsymbol{\theta}_1,\boldsymbol{\theta}_2) = D_{Kull}(\boldsymbol{\theta}_1,\boldsymbol{\theta}_2)$$

and

$$D_0^1(\boldsymbol{\theta}_1,\boldsymbol{\theta}_2) = \lim_{r\to 0} D_r^1(\boldsymbol{\theta}_1,\boldsymbol{\theta}_2) = D_{Kull}(\boldsymbol{\theta}_2,\boldsymbol{\theta}_1),$$

respectively. The divergence measure $D_{Kull}(\boldsymbol{\theta}_2,\boldsymbol{\theta}_1)$ is called the Minimum discrimination information between the probability distributions $P_{\boldsymbol{\theta}_1}$ and $P_{\boldsymbol{\theta}_2}$. Other two well known parametric generalizations of (1.2) are the one called r-order and s-degree divergence measure, and the other called 1-order and s-degree divergence measure. They were given by Sharma and Mittal (1977), by

$$\begin{aligned} D_r^s(\boldsymbol{\theta}_1,\boldsymbol{\theta}_2) &= \tfrac{1}{(s-1)}\left(\left(\int_{\mathcal{X}} f_{\boldsymbol{\theta}_1}(\boldsymbol{x})^r f_{\boldsymbol{\theta}_2}(\boldsymbol{x})^{1-r} d\mu(\boldsymbol{x})\right)^{\frac{s-1}{r-1}} - 1\right) \\ &= \tfrac{1}{(s-1)}\left(\left(E_{\boldsymbol{\theta}_1}\left[\left(\frac{f_{\boldsymbol{\theta}_1}(\boldsymbol{X})}{f_{\boldsymbol{\theta}_2}(\boldsymbol{X})}\right)^{r-1}\right]\right)^{\frac{s-1}{r-1}} - 1\right), \end{aligned} \tag{1.4}$$

for $r, s \neq 1$ and

$$\begin{aligned} D_1^s(\boldsymbol{\theta}_1,\boldsymbol{\theta}_2) &= \tfrac{1}{(s-1)}\left(\exp\left((s-1)\int_{\mathcal{X}} f_{\boldsymbol{\theta}_1}(\boldsymbol{x})\log\tfrac{f_{\boldsymbol{\theta}_1}(\boldsymbol{x})}{f_{\boldsymbol{\theta}_2}(\boldsymbol{x})}d\mu(\boldsymbol{x})\right)-1\right) \\ &= \tfrac{1}{(s-1)}\left(\exp\left((s-1)E_{\boldsymbol{\theta}_1}\left[\log\left(\tfrac{f_{\boldsymbol{\theta}_1}(\boldsymbol{X})}{f_{\boldsymbol{\theta}_2}(\boldsymbol{X})}\right)\right]\right)-1\right), \end{aligned} \quad (1.5)$$

for $s \neq 1$.

It can be easily shown that

i) $\lim_{s\to 1} D_r^s(\boldsymbol{\theta}_1,\boldsymbol{\theta}_2) = rD_r^1(\boldsymbol{\theta}_1,\boldsymbol{\theta}_2)$

ii) $\lim_{r\to 1} D_r^s(\boldsymbol{\theta}_1,\boldsymbol{\theta}_2) = D_1^s(\boldsymbol{\theta}_1,\boldsymbol{\theta}_2)$

iii) $\lim_{s\to 1} D_1^s(\boldsymbol{\theta}_1,\boldsymbol{\theta}_2) = D_{Kull}(\boldsymbol{\theta}_1,\boldsymbol{\theta}_2)$, $\lim_{r\to 1} D_r^1(\boldsymbol{\theta}_1,\boldsymbol{\theta}_2) = D_{Kull}(\boldsymbol{\theta}_1,\boldsymbol{\theta}_2)$.

The Kullback-Leibler divergence measure is the most famous special case of the ϕ-divergence family of divergence measures defined simultaneously by Csiszár (1963) and Ali and Silvey (1966).

Definition 1.1

The ϕ-divergence measure between the probability distributions $P_{\boldsymbol{\theta}_1}$ and $P_{\boldsymbol{\theta}_2}$ is defined by

$$\begin{aligned} D_\phi(P_{\boldsymbol{\theta}_1},P_{\boldsymbol{\theta}_2}) = D_\phi(\boldsymbol{\theta}_1,\boldsymbol{\theta}_2) &= \int_{\mathcal{X}} f_{\boldsymbol{\theta}_2}(\boldsymbol{x})\phi\left(\tfrac{f_{\boldsymbol{\theta}_1}(\boldsymbol{x})}{f_{\boldsymbol{\theta}_2}(\boldsymbol{x})}\right)d\mu(\boldsymbol{x}) \\ &= E_{\boldsymbol{\theta}_2}\left[\phi\left(\tfrac{f_{\boldsymbol{\theta}_1}(\boldsymbol{X})}{f_{\boldsymbol{\theta}_2}(\boldsymbol{X})}\right)\right], \quad \phi \in \Phi^* \end{aligned} \quad (1.6)$$

where Φ^* is the class of all convex functions $\phi(x)$, $x \geq 0$, such that at $x = 1, \phi(1) = 0$, at $x = 0$, $0\phi(0/0) = 0$ and $0\phi(p/0) = \lim_{u\to\infty}\phi(u)/u$.

Remark 1.1

Let $\phi \in \Phi^*$ be differentiable at $x = 1$, then the function

$$\psi(x) \equiv \phi(x) - \phi'(1)(x-1) \quad (1.7)$$

also belongs to Φ^* and has the additional property that $\psi'(1) = 0$. This property, *together with the convexity, implies that* $\psi(x) \geq 0$, *for any* $x \geq 0$. *Further,*

$$\begin{aligned} D_{\psi}(\boldsymbol{\theta}_1, \boldsymbol{\theta}_2) &= \int_{\mathcal{X}} f_{\boldsymbol{\theta}_2}(\boldsymbol{x}) \left(\phi\left(\frac{f_{\boldsymbol{\theta}_1}(\boldsymbol{x})}{f_{\boldsymbol{\theta}_2}(\boldsymbol{x})}\right) - \phi'(1)\left(\frac{f_{\boldsymbol{\theta}_1}(\boldsymbol{x})}{f_{\boldsymbol{\theta}_2}(\boldsymbol{x})} - 1\right)\right) d\mu(\boldsymbol{x}) \\ &= \int_{\mathcal{X}} f_{\boldsymbol{\theta}_2}(\boldsymbol{x}) \phi\left(\frac{f_{\boldsymbol{\theta}_1}(\boldsymbol{x})}{f_{\boldsymbol{\theta}_2}(\boldsymbol{x})}\right) d\mu(\boldsymbol{x}) \\ &= D_{\phi}(\boldsymbol{\theta}_1, \boldsymbol{\theta}_2). \end{aligned}$$

Since the two divergence measures coincide, we can consider the set Φ^* *to be equivalent to the set*

$$\Phi \equiv \Phi^* \cap \{\phi : \phi'(1) = 0\}.$$

Kullback-Leibler divergence measure is obtained for $\psi(x) = x \log x - x + 1$ or $\phi(x) = x \log x$. We can observe that $\psi(x) = \phi(x) - \phi'(1)(x-1)$. We shall denote by ϕ any function belonging to Φ or Φ^*. In the following table we present some important measures of divergence studied in the literature which are particular cases of the ϕ-divergence. More examples can be seen in Arndt (2001), Pardo, L. (1997a) and Vajda (1989, 1995).

ϕ-*function*	*Divergence*
$x \log x - x + 1$	Kullback-Leibler (1959)
$-\log x + x - 1$	Minimum Discrimination Information
$(x-1)\log x$	*J*-Divergence
$\frac{1}{2}(x-1)^2$	Pearson (1900), Kagan (1963)
$\frac{(x-1)^2}{(x+1)^2}$	Balakrishnan and Sanghvi (1968)
$\frac{-x^s + s(x-1) + 1}{1-s}, \ s \neq 1,$	Rathie and Kannappan (1972)
$\frac{1-x}{2} - \left(\frac{1+x^{-r}}{2}\right)^{-1/r}, \ r > 0$	Harmonic mean (Mathai and Rathie (1975))
$\frac{(1-x)^2}{2(a+(1-a)x)}, \quad 0 \leq a \leq 1$	Rukhin (1994)
$\frac{ax \log x - (ax+1-a)\log(ax+1-a)}{a(1-a)}, \ a \neq 0, 1$	Lin (1991)
$\frac{x^{\lambda+1} - x - \lambda(x-1)}{\lambda(\lambda+1)}, \ \lambda \neq 0, -1$	Cressie and Read (1984)
$\lvert 1 - x^a \rvert^{1/a}, \ 0 < a < 1$	Matusita (1964)
$\lvert 1 - x \rvert^a, \ a \geq 1$	χ − divergence of order a (Vajda 1973) Total Variation if $a = 1$ (Saks 1937)

From a statistical point of view, the most important family of ϕ-divergences is perhaps the family studied by Cressie and Read (1984): the power-divergence family, given by

$$I_\lambda(\boldsymbol{\theta}_1,\boldsymbol{\theta}_2) \equiv D_{\phi_{(\lambda)}}(\boldsymbol{\theta}_1,\boldsymbol{\theta}_2) = \frac{1}{\lambda(\lambda+1)}\left(\int_{\mathcal{X}} \frac{f_{\boldsymbol{\theta}_1}(\boldsymbol{x})^{\lambda+1}}{f_{\boldsymbol{\theta}_2}(\boldsymbol{x})^{\lambda}} d\mu(\boldsymbol{x}) - 1\right)$$
$$= \frac{1}{\lambda(\lambda+1)}\left(E_{\boldsymbol{\theta}_1}\left[\left(\frac{f_{\boldsymbol{\theta}_1}(\boldsymbol{X})}{f_{\boldsymbol{\theta}_2}(\boldsymbol{X})}\right)^{\lambda}\right] - 1\right),$$

for $-\infty < \lambda < \infty$. The power-divergence family is undefined for $\lambda = -1$ or $\lambda = 0$. However, if we define these cases by the continuous limits of $I_\lambda(\boldsymbol{\theta}_1,\boldsymbol{\theta}_2)$ as $\lambda \to -1$ and $\lambda \to 0$, then $I_\lambda(\boldsymbol{\theta}_1,\boldsymbol{\theta}_2)$ is continuous in λ. It is not difficult to establish that

$$\lim_{\lambda\to 0} I_\lambda(\boldsymbol{\theta}_1,\boldsymbol{\theta}_2) = D_{Kull}(\boldsymbol{\theta}_1,\boldsymbol{\theta}_2)$$

and

$$\lim_{\lambda\to -1} I_\lambda(\boldsymbol{\theta}_1,\boldsymbol{\theta}_2) = D_{Kull}(\boldsymbol{\theta}_2,\boldsymbol{\theta}_1).$$

We can observe that the power-divergence family is obtained from (1.6) with

$$\phi(x) = \begin{cases} \phi_{(\lambda)}(x) = & \frac{1}{\lambda(\lambda+1)}\left(x^{\lambda+1} - x - \lambda(x-1)\right);\ \lambda \neq 0, \lambda \neq -1, \\ \phi_{(0)}(x) = & \lim_{\lambda\to 0}\phi_{(\lambda)}(x) = x\log x - x + 1 \\ \phi_{(-1)}(x) = & \lim_{\lambda\to -1}\phi_{(\lambda)}(x) = -\log x + x - 1. \end{cases}$$

The power-divergence family was proposed independently by Liese and Vajda (1987) as a ϕ-divergence under the name I_a-divergence.

The power-divergence family, we shall refer to it in later chapters of this book, has been used by Cressie and Read, specially for discrete random variables with finite support (with multinomial data), to link the traditional test statistics through a single-valued parameter, and provides a way to consolidate and extend the current fragmented literature. As a by-product of their analysis, a new test statistic emerged "between" chi-square test statistic and the likelihood ratio test statistic that has some valuable properties. In the last years, many papers in the statistical literature have appeared using the power-divergence family to get competitive estimators as well as test statistics.

The divergence measures of Rényi and Sharma and Mittal given in (1.3) and (1.4), as well as the measure given by Bhattacharyya (1943)

$$B(\boldsymbol{\theta}_1,\boldsymbol{\theta}_2) = -\log\int_{\mathcal{X}}\sqrt{f_{\boldsymbol{\theta}_1}(\boldsymbol{x})f_{\boldsymbol{\theta}_2}(\boldsymbol{x})}d\mu(\boldsymbol{x}),$$

are not ϕ-divergences measures. However, such measures can be written in the following form:

$$D_{\phi}^{h}(\boldsymbol{\theta}_1, \boldsymbol{\theta}_2) = h\left(D_{\phi}(\boldsymbol{\theta}_1, \boldsymbol{\theta}_2)\right), \tag{1.8}$$

where h is a differentiable increasing real function mapping from

$$\left[0, \phi(0) + \lim_{t \to \infty} \frac{\phi(t)}{t}\right]$$

onto $[0, \infty)$; this condition will be justified in Proposition 1.1, with $h(0) = 0$, $h'(0) > 0$, and $\phi \in \Phi^*$. In the following table we list the functions h and ϕ that yield to the mentioned divergence measures:

Divergence	$h(x)$	$\phi(x)$
Rényi	$\frac{1}{r(r-1)} \log\left(r(r-1)x + 1\right); r \neq 0, 1$	$\frac{x^r - r(x-1) - 1}{r(r-1)}; r \neq 0, 1$
Sharma-Mittal	$\frac{1}{s-1}\left(\left(1 + r(r-1)x\right)^{\frac{s-1}{r-1}} - 1\right); s, r \neq 1$	$\frac{x^r - r(x-1) - 1}{r(r-1)}; r \neq 0, 1$
Bhattacharyya	$-\log(-x + 1)$	$-x^{1/2} + \frac{1}{2}(x + 1)$.

The new family of divergences (1.8), called the (h, ϕ)-divergence measures, has been introduced and studied in Menéndez et *al.* (1995). An interesting application of Bhattacharyya divergence in signal selection can be seen in Kailath (1967).

1.2.1. Basic Properties of the Phi-divergence Measures

In this Section we present some of the most important properties, from a statistical point of view, of the ϕ-divergence measures. A complete study of their properties can be found in Vajda (1989, 1995). It is reasonable to demand of a divergence the property of increasing when two distributions move apart. The first proposition is an immediate consequence of this idea, and it will be a basic tool in later chapters. In the following we assume the existence of the first derivative of ϕ at $x = 1$. This assumption is not necessary in order to establish the following results but with this condition some proofs will be easier. In Vajda (1995) proofs are given without the assumption of that restriction.

Proposition 1.1

Let $P_{\boldsymbol{\theta}_1}$ and $P_{\boldsymbol{\theta}_2}$ be two probability distributions and let $\phi \in \Phi^*$ be differentiable at $t = 1$. Then

$$0 \leq D_{\phi}(\boldsymbol{\theta}_1, \boldsymbol{\theta}_2) \leq \phi(0) + \lim_{r \to \infty} \frac{\phi(r)}{r},$$

where

$$D_\phi(\boldsymbol{\theta}_1, \boldsymbol{\theta}_2) = 0 \quad \textit{if } P_{\boldsymbol{\theta}_1} = P_{\boldsymbol{\theta}_2}, \tag{1.9}$$

and

$$D_\phi(\boldsymbol{\theta}_1, \boldsymbol{\theta}_2) = \phi(0) + \lim_{r\to\infty} \frac{\phi(r)}{r} \quad \textit{if } S_1 \cap S_2 = \emptyset. \tag{1.10}$$

If ϕ is also strictly convex at $t = 1$, then (1.9) holds if and only if $P_{\boldsymbol{\theta}_1} = P_{\boldsymbol{\theta}_2}$. If moreover,

$$\phi(0) + \lim_{r\to\infty} \frac{\phi(r)}{r} < \infty,$$

then (1.10) holds if and only if $S_1 \cap S_2 = \emptyset$, where S_i, $i = 1, 2$, is the support of the probability distribution $P_{\boldsymbol{\theta}_i}$, $i = 1, 2$.

Proof. Using the nonnegativity of the function ψ given in (1.7), we have $D_\psi(\boldsymbol{\theta}_1, \boldsymbol{\theta}_2) \geq 0$, but we know that $D_\phi(\boldsymbol{\theta}_1, \boldsymbol{\theta}_2) = D_\psi(\boldsymbol{\theta}_1, \boldsymbol{\theta}_2)$, then $D_\phi(\boldsymbol{\theta}_1, \boldsymbol{\theta}_2) \geq 0$.

It is known that for every convex function ϕ the following inequality holds

$$\phi(t) \leq \phi(0) + t \lim_{r\to\infty} \frac{\phi(r)}{r}, \quad (t \geq 0). \tag{1.11}$$

If ϕ is strictly convex at some $t_0 \in (0, \infty)$ then the inequality in (1.11) is strict for all $t > 0$. Using (1.11) we have

$$\begin{aligned} D_\phi(\boldsymbol{\theta}_1, \boldsymbol{\theta}_2) &\leq \int_{\mathcal{X}} f_{\boldsymbol{\theta}_2}(\boldsymbol{x}) \left(\phi(0) + \frac{f_{\boldsymbol{\theta}_1}(\boldsymbol{x})}{f_{\boldsymbol{\theta}_2}(\boldsymbol{x})} \lim_{r\to\infty} \frac{\phi(r)}{r} \right) d\mu(\boldsymbol{x}) \\ &= \phi(0) + \lim_{r\to\infty} \frac{\phi(r)}{r}. \end{aligned}$$

It is clear that $P_{\boldsymbol{\theta}_1} = P_{\boldsymbol{\theta}_2}$ implies $D_\phi(\boldsymbol{\theta}_1, \boldsymbol{\theta}_2) = 0$.

If $S_1 \cap S_2 = \emptyset$, we have

$$\begin{aligned} D_\phi(\boldsymbol{\theta}_1, \boldsymbol{\theta}_2) &= \int_{\mathcal{X}} f_{\boldsymbol{\theta}_2}(\boldsymbol{x}) \phi\left(\tfrac{f_{\boldsymbol{\theta}_1}(\boldsymbol{x})}{f_{\boldsymbol{\theta}_2}(\boldsymbol{x})} \right) d\mu(\boldsymbol{x}) \\ &= \int_{S_1^C \cap S_2} f_{\boldsymbol{\theta}_2}(\boldsymbol{x}) \phi\left(\tfrac{f_{\boldsymbol{\theta}_1}(\boldsymbol{x})}{f_{\boldsymbol{\theta}_2}(\boldsymbol{x})} \right) d\mu(\boldsymbol{x}) + \int_{S_1 \cap S_2^C} f_{\boldsymbol{\theta}_2}(\boldsymbol{x}) \phi\left(\tfrac{f_{\boldsymbol{\theta}_1}(\boldsymbol{x})}{f_{\boldsymbol{\theta}_2}(\boldsymbol{x})} \right) d\mu(\boldsymbol{x}) \\ &= \phi(0) + \lim_{r\to\infty} \frac{\phi(r)}{r}. \end{aligned}$$

Now we are going to establish that if ϕ is strictly convex at $t = 1$, then $D_\phi(\boldsymbol{\theta}_1, \boldsymbol{\theta}_2) = 0$ implies $P_{\boldsymbol{\theta}_1} = P_{\boldsymbol{\theta}_2}$.

In fact, if ϕ is strictly convex at $t = 1$ then

$$\psi\left(\frac{f_{\boldsymbol{\theta}_1}(\boldsymbol{x})}{f_{\boldsymbol{\theta}_2}(\boldsymbol{x})}\right) > 0$$

for $f_{\boldsymbol{\theta}_1}(\boldsymbol{x})/f_{\boldsymbol{\theta}_2}(\boldsymbol{x}) > 1$ and for $f_{\boldsymbol{\theta}_1}(\boldsymbol{x})/f_{\boldsymbol{\theta}_2}(\boldsymbol{x}) < 1$, where ψ is defined in (1.7). If $D_\psi(\boldsymbol{\theta}_1, \boldsymbol{\theta}_2) = 0$ then $f_{\boldsymbol{\theta}_1}(\boldsymbol{x})/f_{\boldsymbol{\theta}_2}(\boldsymbol{x}) \leq 1$ or $f_{\boldsymbol{\theta}_1}(\boldsymbol{x})/f_{\boldsymbol{\theta}_2}(\boldsymbol{x}) \geq 1$. First we suppose that $f_{\boldsymbol{\theta}_1}(\boldsymbol{x})/f_{\boldsymbol{\theta}_2}(\boldsymbol{x}) \leq 1$. We know that

$$D_\phi(\boldsymbol{\theta}_1, \boldsymbol{\theta}_2) = D_\psi(\boldsymbol{\theta}_1, \boldsymbol{\theta}_2) = 0,$$

and

$$\begin{aligned}
0 = D_\psi(\boldsymbol{\theta}_1, \boldsymbol{\theta}_2) &= \int_{\mathcal{X}} f_{\boldsymbol{\theta}_2}(\boldsymbol{x})\psi\left(\frac{f_{\boldsymbol{\theta}_1}(\boldsymbol{x})}{f_{\boldsymbol{\theta}_2}(\boldsymbol{x})}\right) d\mu(\boldsymbol{x}) \\
&= \int_{\mathcal{X}} f_{\boldsymbol{\theta}_2}(\boldsymbol{x})\left(\phi\left(\frac{f_{\boldsymbol{\theta}_1}(\boldsymbol{x})}{f_{\boldsymbol{\theta}_2}(\boldsymbol{x})}\right) - \phi'(1)\left(\frac{f_{\boldsymbol{\theta}_1}(\boldsymbol{x})}{f_{\boldsymbol{\theta}_2}(\boldsymbol{x})} - 1\right)\right) d\mu(\boldsymbol{x}) \\
&= D_\phi(\boldsymbol{\theta}_1, \boldsymbol{\theta}_2) - \phi'(1)\int_{\mathcal{X}} f_{\boldsymbol{\theta}_2}(\boldsymbol{x})\left(\frac{f_{\boldsymbol{\theta}_1}(\boldsymbol{x})}{f_{\boldsymbol{\theta}_2}(\boldsymbol{x})} - 1\right) d\mu(\boldsymbol{x}) \\
&= 0 - \phi'(1)\int_{\mathcal{X}} f_{\boldsymbol{\theta}_2}(\boldsymbol{x})\left(\frac{f_{\boldsymbol{\theta}_1}(\boldsymbol{x})}{f_{\boldsymbol{\theta}_2}(\boldsymbol{x})} - 1\right) d\mu(\boldsymbol{x}) \\
&= -\phi'(1)\int_{\mathcal{X}} \left(\frac{f_{\boldsymbol{\theta}_1}(\boldsymbol{x})}{f_{\boldsymbol{\theta}_2}(\boldsymbol{x})} - 1\right) dP_{\boldsymbol{\theta}_2}.
\end{aligned}$$

Since ϕ is strictly convex at $t = 1$, it must be $P_{\boldsymbol{\theta}_1} = P_{\boldsymbol{\theta}_2}$. For $f_{\boldsymbol{\theta}_1}(\boldsymbol{x})/f_{\boldsymbol{\theta}_2}(\boldsymbol{x}) \geq 1$, the result can be established in the same way.

The strict convexity of ϕ at $t = 1$ implies the strict inequality in (1.11), i.e.,

$$\phi(t) < \phi(0) + t \lim_{r\to\infty} \frac{\phi(r)}{r}, \ \forall t > 0.$$

Then the function

$$l(t) = \phi(0) - \phi(t) + t \lim_{r\to\infty} \frac{\phi(r)}{r}$$

is positive, for any $t > 0$.

If we take $\boldsymbol{x} \in S_1$, i.e., $\boldsymbol{x}$ such that $f_{\boldsymbol{\theta}_1}(\boldsymbol{x}) > 0$, then $t = \frac{f_{\boldsymbol{\theta}_1}(\boldsymbol{x})}{f_{\boldsymbol{\theta}_2}(\boldsymbol{x})} > 0$ and $l\left(\frac{f_{\boldsymbol{\theta}_1}(\boldsymbol{x})}{f_{\boldsymbol{\theta}_2}(\boldsymbol{x})}\right) > 0$.

Therefore,

$$\begin{aligned} D_l(\boldsymbol{\theta}_1, \boldsymbol{\theta}_2) &= \int_{\mathcal{X}} f_{\boldsymbol{\theta}_2}(\boldsymbol{x}) l\left(\tfrac{f_{\boldsymbol{\theta}_1}(\boldsymbol{x})}{f_{\boldsymbol{\theta}_2}(\boldsymbol{x})}\right) d\mu(\boldsymbol{x}) \\ &= \int_{\mathcal{X}} f_{\boldsymbol{\theta}_2}(\boldsymbol{x}) \left(\phi(0) - \phi\left(\tfrac{f_{\boldsymbol{\theta}_1}(\boldsymbol{x})}{f_{\boldsymbol{\theta}_2}(\boldsymbol{x})}\right) + \tfrac{f_{\boldsymbol{\theta}_1}(\boldsymbol{x})}{f_{\boldsymbol{\theta}_2}(\boldsymbol{x})} \lim_{r\to\infty} \tfrac{\phi(r)}{r}\right) d\mu(\boldsymbol{x}) \\ &= -D_\phi(\boldsymbol{\theta}_1, \boldsymbol{\theta}_2) + \phi(0) + \lim_{r\to\infty} \frac{\phi(r)}{r}, \end{aligned}$$

but by (1.10) we have

$$D_\phi(\boldsymbol{\theta}_1, \boldsymbol{\theta}_2) = \phi(0) + \lim_{r\to\infty} \frac{\phi(r)}{r},$$

therefore,

$$D_l(\boldsymbol{\theta}_1, \boldsymbol{\theta}_2) = \int_{\mathcal{X}} f_{\boldsymbol{\theta}_2}(\boldsymbol{x}) l\left(\frac{f_{\boldsymbol{\theta}_1}(\boldsymbol{x})}{f_{\boldsymbol{\theta}_2}(\boldsymbol{x})}\right) d\mu(\boldsymbol{x}) = 0,$$

with

$$l\left(\frac{f_{\boldsymbol{\theta}_1}(\boldsymbol{x})}{f_{\boldsymbol{\theta}_2}(\boldsymbol{x})}\right) > 0.$$

Then, $f_{\boldsymbol{\theta}_2}(\boldsymbol{x}) = 0$, because $D_l(\boldsymbol{\theta}_1, \boldsymbol{\theta}_2) = 0$ and $l\left(\frac{f_{\boldsymbol{\theta}_1}(\boldsymbol{x})}{f_{\boldsymbol{\theta}_2}(\boldsymbol{x})}\right) > 0$, i.e., $\boldsymbol{x} \notin S_2$. This completes the proof.

■

Let $X_1, ..., X_n$ be a sample from $P_{\boldsymbol{\theta}}$, $\boldsymbol{\theta} \in \Theta$. For μ being the Lebesgue measure or a counting measure, let $f_{\boldsymbol{\theta}}(\boldsymbol{x}) = \frac{dP_{\boldsymbol{\theta}}}{d\mu}(\boldsymbol{x})$ where $\boldsymbol{x} = (x_1, ..., x_n)$. Suppose that T is a measurable transformation from $(\mathcal{X}^n, \beta_{\mathcal{X}^n})$ onto a measurable space $(\mathcal{Y}, \beta_{\mathcal{Y}})$. We denote

$$Q_{\boldsymbol{\theta}_i}(A) = P_{\boldsymbol{\theta}_i}\left(T^{-1}(A)\right), \; i = 1, 2, \tag{1.12}$$

with $A \in \beta_{\mathcal{Y}}$ and

$$g_{\boldsymbol{\theta}_i}(\boldsymbol{t}) = \frac{dQ_{\boldsymbol{\theta}_i}}{d\mu}(\boldsymbol{t}), \; f_{\boldsymbol{\theta}_i}(\boldsymbol{x}/\boldsymbol{t}) = \frac{dP_{\boldsymbol{\theta}_i}}{dQ_{\boldsymbol{\theta}_i}}, \; i = 1, 2; \tag{1.13}$$

by $\boldsymbol{t}$ we are denoting the values of T. In this context we have the following property.

Proposition 1.2

Let $\phi \in \Phi^*$ and $Q_{\boldsymbol{\theta}_i}$, $P_{\boldsymbol{\theta}_i}$, $i = 1, 2$, be two probability measures defined in (1.12) and (1.13). Then we have

$$D_\phi(Q_{\boldsymbol{\theta}_1}, Q_{\boldsymbol{\theta}_2}) \leq D_\phi(P_{\boldsymbol{\theta}_1}, P_{\boldsymbol{\theta}_2}).$$

The equality holds if T is sufficient for the probability distributions P_{θ_1} and P_{θ_2}.

Proof. We have

$$\begin{aligned} D_\phi(P_{\theta_1}, P_{\theta_2}) &= \int_{\mathcal{X}} f_{\theta_2}(\boldsymbol{x})\phi\left(\frac{f_{\theta_1}(\boldsymbol{x})}{f_{\theta_2}(\boldsymbol{x})}\right) d\mu(\boldsymbol{x}) \\ &= \int_{\mathcal{X}}\int_{\mathcal{Y}} f_{\theta_2}(\boldsymbol{x}/\boldsymbol{t})\, g_{\theta_2}(\boldsymbol{t})\phi\left(\frac{f_{\theta_1}(\boldsymbol{x})}{f_{\theta_2}(\boldsymbol{x})}\right) d\mu(\boldsymbol{t})d\mu(\boldsymbol{x}) \\ &= \int_{\mathcal{Y}} g_{\theta_2}(\boldsymbol{t})\left(\int_{\mathcal{X}} f_{\theta_2}(\boldsymbol{x}/\boldsymbol{t})\,\phi\left(\frac{f_{\theta_1}(\boldsymbol{x})}{f_{\theta_2}(\boldsymbol{x})}\right) d\mu(\boldsymbol{x})\right) d\mu(\boldsymbol{t}). \end{aligned}$$

Applying Jensen's inequality we obtain

$$D_\phi(P_{\theta_1}, P_{\theta_2}) \geq \int_{\mathcal{Y}} g_{\theta_2}(\boldsymbol{t})\left(\phi\left(\int_{\mathcal{X}} f_{\theta_2}(\boldsymbol{x}/\boldsymbol{t})\,\frac{f_{\theta_1}(\boldsymbol{x})}{f_{\theta_2}(\boldsymbol{x})} d\mu(\boldsymbol{x})\right)\right) d\mu(\boldsymbol{t}).$$

But,

$$\frac{f_{\theta_1}(\boldsymbol{x})}{f_{\theta_2}(\boldsymbol{x})} = \frac{\dfrac{dP_{\theta_1}}{d\mu}}{\dfrac{dP_{\theta_2}}{d\mu}} = \frac{\dfrac{dQ_{\theta_1}}{d\mu}\dfrac{dP_{\theta_1}}{dQ_{\theta_1}}}{\dfrac{dQ_{\theta_2}}{d\mu}\dfrac{dP_{\theta_2}}{dQ_{\theta_2}}} = \frac{g_{\theta_1}(\boldsymbol{t})\, f_{\theta_1}(\boldsymbol{x}/\boldsymbol{t})}{g_{\theta_2}(\boldsymbol{t})\, f_{\theta_2}(\boldsymbol{x}/\boldsymbol{t})}, \tag{1.14}$$

then,

$$D_\phi(P_{\theta_1}, P_{\theta_2}) \geq \int_{\mathcal{Y}} g_{\theta_2}(\boldsymbol{t})\phi\left(\frac{g_{\theta_1}(\boldsymbol{t})}{g_{\theta_2}(\boldsymbol{t})}\right) d\mu(\boldsymbol{t}) = D_\phi(Q_{\theta_1}, Q_{\theta_2}).$$

If ϕ is strictly convex, the equality holds iff

$$\frac{f_{\theta_1}(\boldsymbol{x})}{f_{\theta_2}(\boldsymbol{x})} = \int_{\mathcal{X}} f_{\theta_2}(\boldsymbol{x}/\boldsymbol{t})\,\frac{f_{\theta_1}(\boldsymbol{x})}{f_{\theta_2}(\boldsymbol{x})} d\mu(\boldsymbol{x}), \text{ for all } \boldsymbol{x}.$$

The second term in the previous inequality is equal to $g_{\theta_1}(\boldsymbol{t})/g_{\theta_2}(\boldsymbol{t})$ by (1.14). Then, using the Factorization Theorem, the equality holds if T is sufficient for the probability distributions P_{θ_1} and P_{θ_2}. ■

In the following proposition $\{P_\theta\}_{\theta\in\Theta}$, $\Theta \subset \mathbb{R}$, is a family of probability measures defined on the σ-field of Borel subsets of the real line with monotone likelihood ratio in x, i.e., if for any $\theta_1 < \theta_2$, $f_{\theta_1}(x)$ and $f_{\theta_2}(x)$ are distinct and the ratio $f_{\theta_1}(x) / f_{\theta_2}(x)$ is a nondecreasing function of x. It is also possible to define families of densities with nonincreasing monotone likelihood ratio in x, but such families can be treated by symmetry.

Proposition 1.3

Suppose that the probability distributions $\{P_\theta\}_{\theta \in \Theta}$ *are on the real line,* $\theta \in (a,b) \subset \mathbb{R}$ *and let* P_θ *be absolutely continuous with respect to a* σ*-finite measure* μ *(Lebesgue measure or counting measure). Suppose also that the corresponding density functions or probability mass functions have monotone likelihood ratio in* x*. If* $a < \theta_1 < \theta_2 < \theta_3 < b$ *and the function* ϕ *is continuous, it holds*

$$D_\phi(\theta_1,\theta_2) \le D_\phi(\theta_1,\theta_3), \ \phi \in \Phi^*. \tag{1.15}$$

Proof. We assume that μ is the Lebesgue measure. We define

$$\widetilde{D}_\varphi(\theta_1,\theta_2) = \int_{\mathbb{R}} f_{\theta_1}(x)\,\varphi\left(\frac{f_{\theta_2}(x)}{f_{\theta_1}(x)}\right) dx,$$

and we shall establish

$$\widetilde{D}_\varphi(\theta_1,\theta_2) \le \widetilde{D}_\varphi(\theta_1,\theta_3), \ \varphi \in \Phi^*. \tag{1.16}$$

If (1.16) holds, then (1.15) also holds, because if we consider the function

$$\varphi(t) = t\phi\left(\frac{1}{t}\right) \in \Phi^*,$$

we have

$$\begin{aligned}\widetilde{D}_\varphi(\theta_1,\theta_2) &= \int_{\mathbb{R}} f_{\theta_1}(x)\frac{f_{\theta_2}(x)}{f_{\theta_1}(x)}\phi\left(\frac{f_{\theta_1}(x)}{f_{\theta_2}(x)}\right) dx \\ &= D_\phi(\theta_1,\theta_2).\end{aligned}$$

Since, by hypothesis, the family of distributions $\{P_\theta\}_{\theta\in\Theta\subset\mathbb{R}}$ has monotone nondecreasing likelihood ratio, then

$$h_2(x) = \frac{f_{\theta_2}(x)}{f_{\theta_1}(x)} \quad \text{and} \quad h_3(x) = \frac{f_{\theta_3}(x)}{f_{\theta_1}(x)}$$

are nondecreasing functions of x. The same happens with

$$\frac{h_3(x)}{h_2(x)} = \frac{f_{\theta_3}(x)}{f_{\theta_2}(x)}. \tag{1.17}$$

From (1.17) at the first sight we have three possibilities:

a) $h_3(x) < h_2(x)$, for all x

b) $h_3(x) > h_2(x)$, for all x

c) There exists a number a such that $h_3(x) \leq h_2(x)$ for $x < a$ and $h_3(x) \geq h_2(x)$ for $x > a$.

We know that

$$E_{\theta_1}[h_3(X)] = \int_{\mathbb{R}} f_{\theta_1}(x) \frac{f_{\theta_3}(x)}{f_{\theta_1}(x)} dx = E_{\theta_1}[h_2(X)] = 1.$$

If $E_{\theta_1}[h_3(X)] = E_{\theta_1}[h_2(X)] = 1$, then *a)* and *b)* are not true, hence it should be true *c)*. Using the monotonicity of $h_2(x)$ and $h_3(x)$ we have

$$\{x : h_2(x) \leq b\} \subset \{x : h_3(x) \leq b\}, \quad \text{if} \quad b < h_2(a)$$

and

$$\{x : h_2(x) \leq b\} \supset \{x : h_3(x) \leq b\}, \quad \text{if} \quad b > h_2(a).$$

If we denote

$$F_{h_2(X)}(t) = \Pr_{\theta_1}(h_2(X) \leq t) = \Pr_{\theta_1}(x \in \mathbb{R} : h_2(x) \leq t)$$

$$F_{h_3(X)}(t) = \Pr_{\theta_1}(h_3(X) \leq t) = \Pr_{\theta_1}(x \in \mathbb{R} : h_3(x) \leq t)$$

we have for $t < h_2(a)$

$$F_{h_2(X)}(t) = \Pr_{\theta_1}(x \in \mathbb{R} : h_2(x) \leq t) \leq \Pr_{\theta_1}(x \in \mathbb{R} : h_3(x) \leq t) = F_{h_3(X)}(t),$$

and for $t > h_2(a)$

$$F_{h_2(X)}(t) \geq F_{h_3(X)}(t).$$

Now we shall establish that the statements

a) $E_{\theta_1}[h_3(X)] = E_{\theta_1}[h_2(X)]$

b) $F_{h_2(X)}(t) \leq F_{h_3(X)}(t)$ for $t < h_2(a)$ and $F_{h_2(X)}(t) \geq F_{h_3(X)}(t)$ for $t > h_2(a)$

imply

$$E_{\theta_1}[|h_2(X) - k|] \leq E_{\theta_1}[|h_3(X) - k|] \tag{1.18}$$

for all k.

It is well known that the expectation of a nonnegative random variable X can be written as

$$E[X] = \int_0^\infty (1 - F_X(x))\,dx.$$

In our case,

$$E_{\theta_1}[h_3(X)] = \int_0^\infty \left(1 - F_{h_3(X)}(x)\right) dx = \int_0^\infty \left(1 - F_{h_2(X)}(x)\right) dx = E_{\theta_1}[h_2(X)].$$

Denoting

$$I_1 \equiv \int_0^{h_2(a)} \left(\left(1 - F_{h_3(X)}(x)\right) - \left(1 - F_{h_2(X)}(x)\right)\right) dx$$

and

$$I_2 \equiv \int_{h_2(a)}^\infty \left(\left(1 - F_{h_3(X)}(x)\right) - \left(1 - F_{h_2(X)}(x)\right)\right) dx,$$

we have

$$I_1 = \int_0^{h_2(a)} \left(F_{h_2(X)}(x) - F_{h_3(X)}(x)\right) dx,\ I_2 = \int_{h_2(a)}^\infty \left(F_{h_2(X)}(x) - F_{h_3(X)}(x)\right) dx.$$

Therefore,

$$\begin{aligned} E_{\theta_1}[h_3(X)] - E_{\theta_1}[h_2(X)] = & \int_0^{h_2(a)} \left(F_{h_2(X)}(x) - F_{h_3(X)}(x)\right) dx \\ & + \int_{h_2(a)}^\infty \left(F_{h_2(X)}(x) - F_{h_3(X)}(x)\right) dx \quad = \quad 0. \end{aligned}$$

Finally, we have

$$\int_0^{h_2(a)} \left(F_{h_2(X)}(x) - F_{h_3(X)}(x)\right) dx = \int_{h_2(a)}^\infty \left(F_{h_3(X)}(x) - F_{h_2(X)}(x)\right) dx. \quad (1.19)$$

Now we prove (1.18). It is easy to check that

$$E_{\theta_1}[|h_i(X) - k|] = \int_0^k F_{h_i(X)}(x)\,dx + \int_k^\infty \left(1 - F_{h_i(X)}(x)\right) dx.$$

Assuming that $k \geq h_2(a)$, an analogous proof can be done if $k < h_2(a)$; we have

$$\begin{aligned} E_{\theta_1}[|h_i(X) - k|] \quad = \quad & \int_0^{h_2(a)} F_{h_i(X)}(x)\,dx + \int_{h_2(a)}^k F_{h_i(X)}(x)\,dx \\ & + \int_k^\infty (1 - F_{h_i(X)}(x))dx, \end{aligned}$$

for $i = 2, 3$. Let us define

$$s = E_{\theta_1}\left[\left|h_3(X) - k\right|\right] - E_{\theta_1}\left[\left|h_2(X) - k\right|\right],$$

so that

$$\begin{aligned} s &= \int_0^{h_2(a)} \left(F_{h_3(X)}(x) - F_{h_2(X)}(x)\right) dx - \int_{h_2(a)}^{k} \left(F_{h_2(X)}(x) - F_{h_3(X)}(x)\right) dx \\ &+ \int_k^{\infty} \left(F_{h_2(X)}(x) - F_{h_3(X)}(x)\right) dx. \end{aligned}$$

By (1.19) we have

$$\begin{aligned} \int_0^{h_2(a)} \left(F_{h_3(X)}(x) - F_{h_2(X)}(x)\right) dx &= \int_{h_2(a)}^{\infty} \left(F_{h_2(X)}(x) - F_{h_3(X)}(x)\right) dx \\ &\geq \int_{h_2(a)}^{k} \left(F_{h_2(X)}(x) - F_{h_3(X)}(x)\right) dx. \end{aligned}$$

Then we get that

$$s \geq \int_k^{\infty} \left(F_{h_2(X)}(x) - F_{h_3(X)}(x)\right) dx \geq 0.$$

Thus,

$$E_{\theta_1}\left[\left|h_3(X) - k\right|\right] \geq E_{\theta_1}\left[\left|h_2(X) - k\right|\right]. \tag{1.20}$$

Finally we prove (1.16) or equivalently that

$$E_{\theta_1}\left[\phi\left(h_3(X)\right)\right] \geq E_{\theta_1}\left[\phi\left(h_2(X)\right)\right].$$

Since ϕ is continuous and convex we have

$$\phi(z) - \phi(0) = \int_0^z b(k)\, dk,$$

where b is nondecreasing and bounded in $[0, z]$. Integrating by parts it yields,

$$\phi(z) - \phi(0) = zb(z) - \int_0^z k db(k) = \int_0^z (z - k)\, db(k) + zb(0).$$

Now we consider the function

$$b^*(k) = \begin{cases} b(k) & \text{if } k \in [0, z] \\ c & \text{if } k > z \end{cases}.$$

Then we have

$$\begin{aligned}\phi(z)-\phi(0) &= \int_0^z (z-k)\,db^*(k) + zb^*(0) + \int_z^\infty (z-k)\,db^*(k)\\ &= \int_0^\infty (z-k)\,db^*(k) + zb^*(0),\end{aligned}$$

where we have taken into account that

$$\int_z^\infty (z-k)\,db^*(k) = 0.$$

Therefore

$$\begin{aligned}E[\phi(Z)] &= E\left[\int_0^\infty (Z-k)\,db^*(k) + Zb^*(0) + \phi(0)\right]\\ &= \int_0^\infty\int_0^\infty (z-k)\,db^*(k)\,dF_Z(z) + E[Z]\,b^*(0) + \phi(0)\\ &= \int_0^\infty E[Z-k]\,db^*(k) + E[Z]\,b^*(0) + \phi(0).\end{aligned}$$

But,

$$\begin{aligned}\int_0^\infty E[|Z-k|]\,db^*(k) &= \int_0^\infty \left(\int_0^\infty |z-k|\,dF_Z(z)\right) db^*(k)\\ &= \int_0^\infty \left(\int_0^z (z-k)\,db^*(k)\right.\\ &\quad \left. + \int_z^\infty -(z-k)\,db^*(k)\right) dF_Z(z)\end{aligned}$$

and

$$\int_z^\infty (z-k)\,db^*(k) = 0.$$

Then,

$$\int_0^\infty E[|Z-k|]\,db^*(k) = \int_0^\infty E[Z-k]\,db^*(k),$$

and thus

$$E[\phi(Z)] = \frac{1}{2}\int_0^\infty E[(Z-k)+|Z-k|]\,db^*(k) + E[Z]\,b^*(0) + \phi(0).$$

If we consider $Z \equiv h_2(X)$, we have

$$E_{\theta_1}[\phi(h_2(X))] = \frac{1}{2}\int_0^\infty 1 - k + E[|h_2(X)-k|]\,db^*(k) + b^*(0) + \phi(0)$$

because

$$E_{\theta_1}\left[h_i\left(X\right)\right] = 1, \; i = 2, 3.$$

In the same way

$$E_{\theta_1}\left[\phi\left(h_3(X)\right)\right] = \frac{1}{2}\int_0^{\infty}\left(1 - k + E\left[\left|h_3\left(X\right) - k\right|\right]\right) db^*\left(k\right) + b^*\left(0\right) + \phi\left(0\right).$$

Applying (1.20) we have the desired result. ■

Remark 1.2

It is obvious that if h is a differentiable increasing real mapping, the (h, ϕ)*-divergence measures also satisfy Propositions 1.1, 1.2 and 1.3.*

Remark 1.3

If we consider a function $\phi \in \Phi^*$ *which is strictly convex at* $x = 1$*, the corresponding* ϕ*-divergence is a reflexive distance on the space* $\mathcal{P} = \{P\}_{\theta \in \Theta}$*. It is possible to define a new measure of divergence, based on a given* ϕ*-divergence, in such a way that the new measure of divergence will be not only reflexive but also symmetric. This is possible if we consider the measure of divergence associated with the function* $\varphi\left(t\right) = \phi\left(t\right) + t\phi\left(\frac{1}{t}\right)$*. For more details see Vajda (1995).*

1.3. Other Divergence Measures between Two Probability Distributions

In this Section we present other important divergence measures between two probability distributions that are not, in general, special cases of the ϕ-divergence measures. We consider two groups of measures. The first one corresponding to measures introduced by Burbea and Rao (1982a, 1982b, 1982c) and the second one corresponding to the Bregman distances studied by Bregman (1967). Other important tools in Statistical Information Theory are the Entropy measures. We introduce them because they are necessary for the definition of the R_ϕ-divergence measures introduced and studied by Rao (1982a), Burbea and Rao (1982a, 1982b, 1982c) and Burbea (1983).

In this section we shall assume as in the former section that the integral in the next definitions exists.

1.3.1. Entropy Measures

Let $\boldsymbol{X}$ be a random variable with probability distribution $P_{\boldsymbol{\theta}}$. From a historical perspective the first entropy measure was Shannon's entropy (1948),

$$H(\boldsymbol{X}) \equiv H(P_{\boldsymbol{\theta}}) \equiv H(\boldsymbol{\theta}) = -\int_{\mathcal{X}} f_{\boldsymbol{\theta}}(\boldsymbol{x}) \log f_{\boldsymbol{\theta}}(\boldsymbol{x}) d\mu(\boldsymbol{x}) = E_{\boldsymbol{\theta}} \left[-\log f_{\boldsymbol{\theta}}(\boldsymbol{X}) \right].$$

The Kullback-Leibler divergence is related to Shannon's entropy. If we assume a finite support and the probability distribution $P_{\boldsymbol{\theta}_2}$ is the uniform distribution, we have

$$D_{Kull}(\boldsymbol{\theta}_1, \boldsymbol{\theta}_2) = H(P_{\boldsymbol{\theta}_2}) - H(P_{\boldsymbol{\theta}_1}).$$

The infinite support case may be written in terms of limits.

Rényi (1961) was the first who presented a generalization of Shannon's entropy, given by

$$H_r^1(\boldsymbol{\theta}) = \frac{1}{1-r} \log \int_{\mathcal{X}} f_{\boldsymbol{\theta}}(\boldsymbol{x})^r d\mu(\boldsymbol{x}) = \frac{1}{1-r} \log E_{\boldsymbol{\theta}} \left[f_{\boldsymbol{\theta}}(\boldsymbol{X})^{r-1} \right], \; r > 0, \; r \neq 1.$$

Liese and Vajda (1987) extended Rényi's entropy for all $r \in \mathbb{R} - \{0, 1\}$ by means of the expression

$$H_r^1(\boldsymbol{\theta}) = \frac{1}{r(1-r)} \log \int_{\mathcal{X}} f_{\boldsymbol{\theta}}(\boldsymbol{x})^r d\mu(\boldsymbol{x}) = \frac{1}{r(1-r)} \log E_{\boldsymbol{\theta}} \left[f_{\boldsymbol{\theta}}(\boldsymbol{X})^{r-1} \right], \; r \neq 0, 1. \tag{1.21}$$

In the following, expression in (1.21) will be referred as Rényi's entropy. Rényi's entropy is undefined for $r = -1$ or $r = 0$. However, if we define these cases by the continuous limits of $H_r^1(\boldsymbol{\theta})$ as $r \to 1$ and $r \to 0$, then $H_r^1(\boldsymbol{\theta})$ is continuous in r. It is not difficult to establish that

$$\lim_{r \to 1} H_r^1(\boldsymbol{\theta}) = H(\boldsymbol{\theta}) \quad \text{and} \quad \lim_{r \to 0} H_r^1(\boldsymbol{\theta}) = \int_{\mathcal{X}} \log f_{\boldsymbol{\theta}}(\boldsymbol{x}) d\mu(\boldsymbol{x}).$$

A review about Rényi's entropy for different univariate and k-variate random variables can be seen in Nadarajah and Zografos (2003, 2005) and Zografos and Nadarajah (2005).

Many entropy measures have been introduced; see Mathai and Rathie (1975). In order to present a systematic way of studying the different entropy measures,

Burbea and Rao introduced the so-called ϕ-entropies, by

$$H_{\phi}(\boldsymbol{X}) \equiv H_{\phi}(P_{\boldsymbol{\theta}}) \equiv H_{\phi}(\boldsymbol{\theta}) = \int_{\mathcal{X}} \phi(f_{\boldsymbol{\theta}}(\boldsymbol{x}))\, d\mu(\boldsymbol{x}), \tag{1.22}$$

where $\phi : (0, \infty) \to \mathbb{R}$ is a continuous concave function and

$$\phi(0) = \lim_{t \downarrow 0} \phi(t) \in (-\infty, \infty].$$

Some interesting properties of ϕ-entropies, for univariate discrete random variables with finite support, can be seen in Vajda and Vasek (1985), Vajda and Teboulle (1993), Morales et *al*. (1996) and references therein.

$\phi(x)$	$h(x)$	*Entropy*
$-x\log x$	x	Shannon (1948)
x^r	$[r(1-r)]^{-1}\log x$	Rényi (1961) $(r \neq 0, r \neq 1)$
x^{r-m+1}	$(m-r)^{-1}\log x$	Varma (1966) $(m-1 < r < m, m \geq 1)$
$x^{\frac{r}{m}}$	$(m(m-r))^{-1}\log x$	Varma (1966) $(0 < r < m, m \geq 1)$
$(1-s)^{-1}(x^s - x)$	x	Havrda and Charvat (1967) $(s \neq 1, s > 0)$
$x^{\frac{1}{t}}$	$(t-1)^{-1}(x^t - 1)$	Arimoto (1971) $(t \neq 1, t > 0)$
$x\log x$	$\frac{\exp[(s-1)x] - 1}{(1-s)}$	Sharma and Mittal (1975) $(s \neq 1, s > 0)$
x^r	$\frac{1}{(1-s)}\left(x^{\frac{s-1}{r-1}} - 1\right)$	Sharma and Mittal (1975) $(r \neq 1, s \neq 1, r > 0, s > 0)$
$(1+\lambda x)\log(1+\lambda x)$	$\left(1+\frac{1}{\lambda}\right)\log(1+\lambda) - \frac{x}{\lambda}$	Ferreri (1980) $(\lambda > 0)$
$\frac{x^s + (1-x)^s - 1}{(1-s)}$	x	Kapur (1972) $s \neq 1$
$\frac{x^s - (1+x)^s + 1 + (s-1)^{-1}(2^s - 2)x}{(s-2)}$	x	Burbea (1984)

Table 1.1. ϕ-entropies and (h,ϕ)-entropies

With the ϕ-entropies we encounter the same problem as with the ϕ-divergences: some important entropy measures can not be written as a ϕ-entropy. For this reason, Salicrú et *al*. (1993) defined the (h,ϕ)-entropy as follows,

$$H_\phi^h(\boldsymbol{X}) \equiv H_\phi^h(P_{\boldsymbol{\theta}}) \equiv H_\phi^h(\boldsymbol{\theta}) = h\left(\int_{\mathcal{X}} \phi\left(f_{\boldsymbol{\theta}}(\boldsymbol{x})\right) d\mu(\boldsymbol{x})\right), \tag{1.23}$$

where either $\phi : (0,\infty) \to \mathbb{R}$ is concave and $h : \mathbb{R} \to \mathbb{R}$ is differentiable and increasing, or $\phi : (0,\infty) \to \mathbb{R}$ is convex and $h : \mathbb{R} \to \mathbb{R}$ is differentiable and decreasing. In Table 1.1 we list some important entropy measures based on (1.23).

The following result states some important properties of Shannon's entropy measure.

Proposition 1.4

Let $\boldsymbol{X}\equiv(X_1,...,X_n)$ *and* $\boldsymbol{Y}\equiv(Y_1,...,Y_m)$ *be two continuous random vectors with joint probability density functions* $f_1(\boldsymbol{x})$, $\boldsymbol{x}\in\mathbb{R}^n$ *and* $f_2(\boldsymbol{y})$, $\boldsymbol{y}\in\mathbb{R}^m$, *respectively. We shall assume that* $(\boldsymbol{X},\boldsymbol{Y})$ *is also a continuous random vector with probability density function* $f(\boldsymbol{x},\boldsymbol{y}), (\boldsymbol{x},\boldsymbol{y}) \in \mathbb{R}^{n+m}$. *The conditional probability density of* $\boldsymbol{X}$ *when* $\boldsymbol{Y}=\boldsymbol{y}$ *is given by* $f(\boldsymbol{x},\boldsymbol{y})/f_2(\boldsymbol{y})$. *Then the conditional Shannon entropy of* $\boldsymbol{X}$ *given* $\boldsymbol{Y}=\boldsymbol{y}$ *is defined by*

$$H\left(\boldsymbol{X}/\boldsymbol{Y}=\boldsymbol{y}\right) = -\int_{\mathbb{R}^n} \frac{f(\boldsymbol{x},\boldsymbol{y})}{f_2(\boldsymbol{y})} \log \frac{f(\boldsymbol{x},\boldsymbol{y})}{f_2(\boldsymbol{y})} d\boldsymbol{x}$$

and the conditional Shannon's entropy of $\boldsymbol{X}$ *given* $\boldsymbol{Y}$, *by*

$$H\left(\boldsymbol{X}/\boldsymbol{Y}\right) = -\int_{\mathbb{R}^{n+m}} f(\boldsymbol{x},\boldsymbol{y}) \log \frac{f(\boldsymbol{x},\boldsymbol{y})}{f_2(\boldsymbol{y})} d\boldsymbol{x}d\boldsymbol{y} = \int_{\mathbb{R}^m} f_2(\boldsymbol{y}) H\left(\boldsymbol{X}/\boldsymbol{Y}=\boldsymbol{y}\right) d\boldsymbol{y},$$

assuming the existence of the previous entropy. The following properties are verified by Shannon's entropy:

a) The Shannon's entropy of $\boldsymbol{X}$ *can be negative.*

b) Let $\boldsymbol{\varphi} = (\varphi_1, ..., \varphi_n)$ *be a smooth bijection on* $\mathbb{R}^n$ *and we assume that* $\boldsymbol{Y} = \boldsymbol{\varphi}(\boldsymbol{X})$. *Then,*

$$H\left(\boldsymbol{Y}\right) = H\left(\boldsymbol{X}\right) - \int_{\mathbb{R}^n} f(\boldsymbol{x}) \log \left|J\left(\boldsymbol{\varphi}(\boldsymbol{x})\right)\right| d\boldsymbol{x}$$

where

$$J(\boldsymbol{y}) = \det\left(\frac{\partial \psi_i}{\partial y_j}(\boldsymbol{y})\right)_{i,j=1,...,n}$$

is the determinant of the Jacobian matrix corresponding to the inverse transformation $\boldsymbol{\psi} \equiv (\psi_1, ..., \psi_n)$ *of* $\boldsymbol{\varphi}$.

c) *If* $\boldsymbol{\varphi} = (\varphi_1, ..., \varphi_n)$ *is a linear transformation, with* $\varphi_i(\boldsymbol{x}) = \sum_{j=1}^{n} a_{ij}x_j$, $i = 1, ..., n$, *then* $H(\boldsymbol{Y}) = H(\boldsymbol{X}) + \log|\det(\boldsymbol{A})|$, *where* $\boldsymbol{A} = (a_{ij})_{i,j=1,...,n}$.

d) *It holds*

$$H(\boldsymbol{X}) = -\int_{\mathbb{R}^n} f_1(\boldsymbol{x}) \log f_1(\boldsymbol{x}) d\boldsymbol{x} \leq -\int_{\mathbb{R}^n} f_1(\boldsymbol{x}) \log f_2(\boldsymbol{x}) d\boldsymbol{x}, \qquad (1.24)$$

with equality iff $f_1(\boldsymbol{x}) = f_2(\boldsymbol{x})$ *a.s. This inequality is called Gibbs's lemma for continuous random vectors.*

e) *It holds*

$$H(\boldsymbol{X}, \boldsymbol{Y}) = H(\boldsymbol{Y}) + H(\boldsymbol{X}/\boldsymbol{Y}) = H(\boldsymbol{X}) + H(\boldsymbol{Y}/\boldsymbol{X}).$$

f) *It holds*

$$H(\boldsymbol{X}/(\boldsymbol{Y}_1, \boldsymbol{Y}_2)) \leq H(\boldsymbol{X}/\boldsymbol{Y}_1) \leq H(\boldsymbol{X}).$$

The first inequality turns into equality if and only if the random vector $\boldsymbol{Y}_2$ *is independent of* $\boldsymbol{X}$ *given* $\boldsymbol{Y}_1$ *and the second inequality turns into equality if and only if* $\boldsymbol{Y}_1$ *is independent of* $\boldsymbol{X}$.

g) *We have the chain rule*

$$H(X_1, ..., X_n) = H(X_1) + \sum_{k=2}^{n} H(X_k/X_1, ..., X_{k-1}) \leq \sum_{k=1}^{n} H(X_k).$$

The equality holds if and only if $X_1, ..., X_n$ *are mutually independent.*

h) *We have*

$$\begin{aligned} H(X_1, ..., X_n/Y) &= H(X_1/Y) + \sum_{k=2}^{n} H(X_k/Y, X_1, ..., X_{k-1}) \\ &\leq \sum_{k=1}^{n} H(X_k/Y). \end{aligned}$$

Proof.

a) Let X be a random variable with exponential distribution of parameter $\theta > 0$. Then we have

$$H(\theta) = -\int_0^\infty \theta e^{-\theta x} \log\left(\theta e^{-\theta x}\right) dx = 1 - \log\theta,$$

therefore, if $\theta \in (0, e)$ the entropy is positive and if $\theta \in (e, \infty)$ the entropy is negative.

b) Let $B \subset \mathbb{R}^n$,

$$\begin{aligned} P(\mathbf{Y} \in B) &= P(\boldsymbol{X} \in \psi(B)) \\ &= \int_{\psi(B)} f(\boldsymbol{x}) d\boldsymbol{x} = \int_B f(\psi(\boldsymbol{y})) \left| J(\boldsymbol{y}) \right| d\boldsymbol{y} \end{aligned}$$

being $\psi(B) = \{\psi(\boldsymbol{y}) : \boldsymbol{y} \in B\}$. Therefore the probability density function of the random vector $\mathbf{Y}$ is given by $f(\psi(\boldsymbol{y})) \left| J(\boldsymbol{y}) \right|$ and

$$\begin{aligned} H(\mathbf{Y}) &= -\int_{\mathbb{R}^n} f(\psi(\boldsymbol{y})) \left| J(\boldsymbol{y}) \right| \log \left(f(\psi(\boldsymbol{y})) \left| J(\boldsymbol{y}) \right| \right) d\boldsymbol{y} \\ &= -\int_{\mathbb{R}^n} f(\boldsymbol{x}) \log \left(f(\boldsymbol{x}) \left| J(\varphi(\boldsymbol{x})) \right| \right) d\boldsymbol{x} \\ &= H(\mathbf{X}) - \int_{\mathbb{R}^n} f(\boldsymbol{x}) \log \left| J(\varphi(\boldsymbol{x})) \right| d\boldsymbol{x}. \end{aligned}$$

c) If φ is a linear transformation, we have $J(\boldsymbol{y}) = \det \boldsymbol{A}^{-1}$. Plugging this in *b)*, we get the desired result.

d) Let $A = \{\boldsymbol{x} : f_1(\boldsymbol{x}) > 0\}$. Given $\boldsymbol{x} \in A$, we have

$$\log \frac{f_2(\boldsymbol{x})}{f_1(\boldsymbol{x})} \leq \frac{f_2(\boldsymbol{x})}{f_1(\boldsymbol{x})} - 1,$$

and the equality holds if and only if $f_1(\boldsymbol{x}) = f_2(\boldsymbol{x})$. Therefore,

$$f_1(\boldsymbol{x}) \log \frac{f_2(\boldsymbol{x})}{f_1(\boldsymbol{x})} \leq f_2(\boldsymbol{x}) - f_1(\boldsymbol{x}).$$

If we denote

$$l = -\int_{\mathbb{R}^n} f_1(\boldsymbol{x}) \log f_1(\boldsymbol{x}) d\boldsymbol{x} + \int_{\mathbb{R}^n} f_1(\boldsymbol{x}) \log f_2(\boldsymbol{x}) d\boldsymbol{x},$$

we have that

$$\begin{aligned} l &= -\int_A f_1(\boldsymbol{x}) \log f_1(\boldsymbol{x}) d\boldsymbol{x} + \int_A f_1(\boldsymbol{x}) \log f_2(\boldsymbol{x}) d\boldsymbol{x} \\ &\leq_{(1)} \int_A f_2(\boldsymbol{x}) d\boldsymbol{x} - \int_A f_1(\boldsymbol{x}) d\boldsymbol{x} \\ &\leq_{(2)} \int_{\mathbb{R}^n} f_2(\boldsymbol{x}) d\boldsymbol{x} - \int_{\mathbb{R}^n} f_1(\boldsymbol{x}) d\boldsymbol{x} = 0. \end{aligned}$$

In (1) the equality holds if and only if $f_1(\boldsymbol{x}) = f_2(\boldsymbol{x})$ a.s. in A, and in (2) if and only if $f_2(\boldsymbol{x}) = 0 = f_1(\boldsymbol{x})$ a.s. in A^c. Then the equality holds if and only if $f_1(\boldsymbol{x}) = f_2(\boldsymbol{x})$ a.s.

e) We have

$$\begin{aligned} H(\boldsymbol{X},\boldsymbol{Y}) &= -\int_{\mathbb{R}^m}\int_{\mathbb{R}^m} f(\boldsymbol{x},\boldsymbol{y})\log f(\boldsymbol{x},\boldsymbol{y})d\boldsymbol{x}d\boldsymbol{y} \\ &= -\int_{\mathbb{R}^n}\int_{\mathbb{R}^m} f(\boldsymbol{x},\boldsymbol{y})\log \frac{f(\boldsymbol{x},\boldsymbol{y})}{f_2(\boldsymbol{y})}d\boldsymbol{x}d\boldsymbol{y} \\ &\quad -\int_{\mathbb{R}^n}\int_{\mathbb{R}^m} f(\boldsymbol{x},\boldsymbol{y})\log f_2(\boldsymbol{y})d\boldsymbol{x}d\boldsymbol{y} \\ &= H(\boldsymbol{X}/\boldsymbol{Y})+H(\boldsymbol{Y}). \end{aligned}$$

In a similar way it is possible to establish

$$H(\boldsymbol{X},\boldsymbol{Y}) = H(\boldsymbol{Y}/\boldsymbol{X})+H(\boldsymbol{X}).$$

f) Let

$$g(\boldsymbol{y}_1), g(\boldsymbol{y}_1,\boldsymbol{y}_2), s(\boldsymbol{x},\boldsymbol{y}_1) \text{ and } r(\boldsymbol{x},\boldsymbol{y}_1,\boldsymbol{y}_2)$$

be the probability density functions of the random vectors

$$\boldsymbol{Y}_1, (\boldsymbol{Y}_1,\boldsymbol{Y}_2), (\boldsymbol{X},\boldsymbol{Y}_1) \text{ and } (\boldsymbol{X},\boldsymbol{Y}_1,\boldsymbol{Y}_2)$$

respectively. Thus,

$$\begin{aligned} H(\boldsymbol{X}/(\boldsymbol{Y}_1,\boldsymbol{Y}_2)) &= -\iint g(\boldsymbol{y}_1,\boldsymbol{y}_2)\left(\int \frac{r(\boldsymbol{x},\boldsymbol{y}_1,\boldsymbol{y}_2)}{g(\boldsymbol{y}_1,\boldsymbol{y}_2)}\log\frac{r(\boldsymbol{x},\boldsymbol{y}_1,\boldsymbol{y}_2)}{g(\boldsymbol{y}_1,\boldsymbol{y}_2)}d\boldsymbol{x}\right) \\ &\leq_{(1)} -\iint g(\boldsymbol{y}_1,\boldsymbol{y}_2)\left(\int \frac{r(\boldsymbol{x},\boldsymbol{y}_1,\boldsymbol{y}_2)}{g(\boldsymbol{y}_1,\boldsymbol{y}_2)}\log\frac{s(\boldsymbol{x},\boldsymbol{y}_1)}{g(\boldsymbol{y}_1)}d\boldsymbol{x}\right) \\ &= -\iint s(\boldsymbol{x},\boldsymbol{y}_1)\log\frac{s(\boldsymbol{x},\boldsymbol{y}_1)}{g(\boldsymbol{y}_1)}d\boldsymbol{x}d\boldsymbol{y}_1 \\ &= H(\boldsymbol{X}/\boldsymbol{Y}_1). \end{aligned}$$

The inequality (1) is established applying part *d)* to the functions

$$\frac{r(\boldsymbol{x}/\boldsymbol{y}_1,\boldsymbol{y}_2)}{g(\boldsymbol{y}_1,\boldsymbol{y}_2)} \text{ and } \frac{s(\boldsymbol{x}/\boldsymbol{y}_1)}{g(\boldsymbol{y}_1)}.$$

The equality holds if and only if

$$\frac{r(\boldsymbol{x},\boldsymbol{y}_1,\boldsymbol{y}_2)}{g(\boldsymbol{y}_1,\boldsymbol{y}_2)} = \frac{s(\boldsymbol{x},\boldsymbol{y}_1)}{g(\boldsymbol{y}_1)}.$$

This is equivalent to saying that

$$\frac{r(\boldsymbol{x},\boldsymbol{y}_1,\boldsymbol{y}_2)}{g(\boldsymbol{y}_1)} = \frac{s(\boldsymbol{x},\boldsymbol{y}_1)}{g(\boldsymbol{y}_1)}\frac{g(\boldsymbol{y}_1,\boldsymbol{y}_2)}{g(\boldsymbol{y}_1)}.$$

The first ratio on the right side of the equality is the probability density function of $\boldsymbol{X}$ given $\boldsymbol{Y}_1 = \boldsymbol{y}_1$, while the second one is the probability density function of $\boldsymbol{Y}_2$ given $\boldsymbol{Y}_1 = \boldsymbol{y}_1$. The term on the left side is the probability density function of $(\boldsymbol{X},\boldsymbol{Y}_2)$ given $\boldsymbol{Y}_1 = \boldsymbol{y}_1$. Thus we have proved the first part of *f)*.

The second part is obtained in a similar way.

g) By *e)* we have

$$H\left(X_1, X_2, ..., X_n\right) = H\left(X_1\right) + H\left(X_2, ..., X_n/X_1\right).$$

In the same way as we have established *e)* it is possible to show that

$$H\left(X_2, ..., X_n/X_1\right) = H\left(X_2/X_1\right) + H\left(X_3, ..., X_n/X_1, X_2\right).$$

Repeating the same arguments, we have

$$H\left(X_1, X_2, ..., X_n\right) = H\left(X_1\right) + \sum_{k=2}^{n} H\left(X_k/X_1, ..., X_{k-1}\right).$$

h) The result can be obtained using *g)*. ■

1.3.2. Burbea and Rao Divergence Measures

Based on the concavity property of the (h, ϕ)-entropy, Pardo, L. et *al.* (1993b) introduced the generalized R_ϕ^h-divergence between two probability distributions $P_{\boldsymbol{\theta}_1}$ and $P_{\boldsymbol{\theta}_2}$ by

$$R_\phi^h(P_{\boldsymbol{\theta}_1}, P_{\boldsymbol{\theta}_2}) \equiv R_\phi^h\left(\boldsymbol{\theta}_1, \boldsymbol{\theta}_2\right) = H_\phi^h\left(\frac{P_{\boldsymbol{\theta}_1} + P_{\boldsymbol{\theta}_2}}{2}\right) - \frac{H_\phi^h\left(P_{\boldsymbol{\theta}_1}\right) + H_\phi^h\left(P_{\boldsymbol{\theta}_2}\right)}{2}.$$

For $h\left(x\right) = x$, we have the R_ϕ-divergence of Burbea and Rao (1982a, 1982b, 1982c) and for $\phi\left(x\right) = x \log x$ the information radius of Sibson (1969).

An important family of R_ϕ-divergences is based on the ϕ_α-entropies. This family of entropies (Havrda and Charvat (1967)) is obtained considering the family of functions

$$\phi_\alpha\left(x\right) = \begin{cases} \left(1-\alpha\right)^{-1}\left(x^\alpha - x\right), & \alpha \neq 1 \\ -x \log x, & \alpha = 1. \end{cases}$$

Rao (1982a) used the family of ϕ_α-entropies, associated with random variables with finite support, in genetic diversity between populations. In the particular case of $\alpha = 2$ Gini-Simpson index is obtained. This measure of entropy was introduced by Gini (1912) and by Simpson (1949) in Biometry and its properties have been studied by several authors. Note that if we consider the Gini-Simpson index, the associated R_{ϕ_2}-divergence, for the probability distributions $p_{\boldsymbol{\theta}_1}(x_i)$, $i = 1, ..., M$, and $p_{\boldsymbol{\theta}_2}(x_i)$, $i = 1, ..., M$, is proportional to the square of the Euclidean distance, namely

$$R_{\phi 2}(\boldsymbol{\theta}_1, \boldsymbol{\theta}_2) = \frac{1}{4}\sum_{i=1}^{M} \left(p_{\boldsymbol{\theta}_1}(x_i) - p_{\boldsymbol{\theta}_2}(x_i)\right)^2.$$

Another important family of R_ϕ-divergences is obtained if we consider the Bose-Einstein entropy introduced by Burbea (1984) and given by

$$\phi(x) = \frac{x^s - (1+x)^s + 1 + (s-1)^{-1}(2^s - 2)x}{(s-2)}.$$

The expressions for $s = 2$ or 1 are obtained by continuity. Another interesting family is obtained by considering the Fermi-Dirac entropy introduced by Kapur (1972) and its expression is obtained for

$$\phi(x) = \frac{x^s + (1-x)^s - 1}{(1-s)}.$$

In this case the expression for $s = 1$ is also obtained by continuity.

In this context Pardo, M. C. and Vajda (1997) established that the condition

$$\frac{1}{t}\phi\left(t\frac{u+v}{2}\right) - \frac{\phi(tu) + \phi(tv)}{2t} = \phi\left(\frac{u+v}{2}\right) - \frac{\phi(u) + \phi(v)}{2},$$

valid for all positive t, u, v, implies the identity

$$D_\varphi(P_{\boldsymbol{\theta}_1}, P_{\boldsymbol{\theta}_2}) = R_\phi(P_{\boldsymbol{\theta}_1}, P_{\boldsymbol{\theta}_2}),$$

where $D_\varphi(P_{\boldsymbol{\theta}_1}, P_{\boldsymbol{\theta}_2})$ is the φ-divergence between $P_{\boldsymbol{\theta}_1}$ and $P_{\boldsymbol{\theta}_2}$, for

$$\varphi(x) = \phi\left(\frac{x+1}{2}\right) - \frac{\phi(x) + \phi(1)}{2}.$$

For example, the function $\phi(x) = x\log x - x + 1$ satisfies the above condition.

Burbea and Rao (1982a, 1982b) introduced and studied three new families of divergence measures: L-, K-, and M-divergences, which are defined as

$$K(\boldsymbol{\theta}_1, \boldsymbol{\theta}_2) = \int_{\mathcal{X}} (f_{\boldsymbol{\theta}_1}(\boldsymbol{x}) - f_{\boldsymbol{\theta}_2}(\boldsymbol{x})) \left(\frac{\phi(f_{\boldsymbol{\theta}_1}(\boldsymbol{x}))}{f_{\boldsymbol{\theta}_1}(\boldsymbol{x})} - \frac{\phi(f_{\boldsymbol{\theta}_2}(\boldsymbol{x}))}{f_{\boldsymbol{\theta}_2}(\boldsymbol{x})} \right) d\mu(\boldsymbol{x}),$$

$$L(\boldsymbol{\theta}_1, \boldsymbol{\theta}_2) = \int_{\mathcal{X}} \left[f_{\boldsymbol{\theta}_1}(\boldsymbol{x}) \phi\left(\frac{f_{\boldsymbol{\theta}_2}(\boldsymbol{x})}{f_{\boldsymbol{\theta}_1}(\boldsymbol{x})} \right) + f_{\boldsymbol{\theta}_2}(\boldsymbol{x}) \phi\left(\frac{f_{\boldsymbol{\theta}_1}(\boldsymbol{x})}{f_{\boldsymbol{\theta}_2}(\boldsymbol{x})} \right) \right] d\mu(\boldsymbol{x})$$

and

$$M(\boldsymbol{\theta}_1, \boldsymbol{\theta}_2) = \int_{\mathcal{X}} (\phi(f_{\boldsymbol{\theta}_1}(\boldsymbol{x})) - \phi(f_{\boldsymbol{\theta}_2}(\boldsymbol{x})))^2 \, d\mu(\boldsymbol{x}).$$

We observe that the L-divergence is a special case of Csiszár's φ-divergence with $\varphi(t) = t\phi(t^{-1}) + \phi(t)$, provided $t\phi(t^{-1}) + \phi(t)$ is a convex function. If $\phi(t) = t^{1/2}$, $M(\boldsymbol{\theta}_1, \boldsymbol{\theta}_2)$ is the Matusita's distance (1964). Some applications of K-divergences in statistical problems can be seen in Pérez and Pardo, J. A. (2002, 2003a, 2003b, 2003c, 2004 and 2005).

1.3.3. Bregman's Distances

Bregman (1967) introduced a family of divergences in the following way,

$$B_\varphi(\boldsymbol{\theta}_1, \boldsymbol{\theta}_2) = \int_{\mathcal{X}} (\varphi(f_{\boldsymbol{\theta}_1}(\boldsymbol{x})) - \varphi(f_{\boldsymbol{\theta}_2}(\boldsymbol{x})) - \varphi'(f_{\boldsymbol{\theta}_2}(\boldsymbol{x}))(f_{\boldsymbol{\theta}_1}(\boldsymbol{x}) - f_{\boldsymbol{\theta}_2}(\boldsymbol{x}))) \, d\mu(\boldsymbol{x})$$

for any differentiable convex function $\varphi : (0, \infty) \to \mathbb{R}$ with $\varphi(0) = \lim_{t \to 0} \varphi(t) \in (-\infty, \infty)$. We observe that for $\varphi(t) = t \log t$, $B_\varphi(\boldsymbol{\theta}_1, \boldsymbol{\theta}_2)$ is the Kullback-Leibler divergence and for $\varphi(t) = t^2$ and discrete probability distributions, the Euclidean distance.

Some properties of the R_ϕ-divergences and Bregman's distances as well as their relation with ϕ-divergences have been studied by Pardo, M.C. and Vajda (1997).

1.4. Divergence among k Populations

The measures of divergence previously discussed are designed for two probability distributions. For certain applications such as in the study of taxonomy in biology and genetics, testing if k populations are homogeneous, etc., it might be necessary to measure the overall difference of more than two probability density functions. Matusita (1967, 1973) proposed the first generalization of the Bhattacharyya divergence, in order to express in a qualitative way analogies and differences among k populations and applied it to discriminant analysis techniques. He also derived a lower bound of the Bayes probability of misclassification. An axiomatic foundation in the discrete case was given by Kaufman and Mathai (1973), and some properties of these measures were derived by Toussaint (1974). This author also presented a simple measure of divergence: the J-divergence among k populations.

A general class of divergence measures, called f-dissimilarity among k populations, was defined by Gyorfi and Nemetz (1978) as follows:

$$D_{\mathrm{f}}(\boldsymbol{\theta}_1, ..., \boldsymbol{\theta}_k) = \int_{\mathcal{X}} \mathrm{f}\left(f_{\boldsymbol{\theta}_1}(\boldsymbol{x}), ..., f_{\boldsymbol{\theta}_k}(\boldsymbol{x})\right) d\mu(\boldsymbol{x}), \tag{1.25}$$

where f is a continuous, convex, homogeneous function defined on the set

$$S = \left\{(s_1, ..., s_k) : 0 \leq s_i < \infty,\ i = 1, ..., k\right\}.$$

If $\mathrm{f}(x_1, .., x_k) = -(\prod_{j=1}^k x_j)^{1/k}$ the f-dissimilarity is the negative of Matusita's affinity (1967) of k populations and if $\mathrm{f}(x_1, .., x_k) = -\prod_{j=1}^k x_j^{a_j}, a_j \geq 0$ with $\sum_{j=1}^k a_j = 1$ the f-dissimilarity is the negative affinity introduced by Toussaint, (1974). More examples can be seen in Gyorfi and Nemetz (1978) and Zografos (1998a). The f-dissimilarity leads also to the Csiszar's ϕ-divergence if $\mathrm{f}(x_1, x_2) = x_2\phi(x_1/x_2)$.

Other interesting families of divergence measures among k populations can be seen in Kapur (1988), Sahoo and Wong (1988), Rao (1982a), Toussaint (1978). In Menéndez et *al.* (1992), three different ways of generalizing the information radius for k populations are presented.

Another interesting family is proposed in Burbea and Rao (1982a, 1982b). Given k probability distributions, $P_{\boldsymbol{\theta}_i}$, $i = 1, ..., k$, the ϕ-Jensen difference among the k probability distributions is

$$R_\phi(P_{\boldsymbol{\theta}_1}, ..., P_{\boldsymbol{\theta}_k}) = H_\phi(\lambda_1 P_{\boldsymbol{\theta}_1} + ... + \lambda_k P_{\boldsymbol{\theta}_k}) - \sum_{i=1}^k \lambda_i H_\phi(P_{\boldsymbol{\theta}_i}),$$

where $\sum_{i=1}^{k}\lambda_i = 1$ and H_ϕ is the ϕ-entropy defined in (1.22) . In the particular case of $\phi(t) = -t\log t$, we have the Information Radius for k probability distributions.

1.5. Phi-disparities

The concept of ϕ-disparity appeared first in Lindsay (1994), who found that inference based on statistics of type ϕ-divergence, called ϕ-divergence statistics (obtained by replacing either one or both probability distributions by suitable estimators), requires either bounded differentiability of ϕ or boundedness of ϕ, itself. Since these properties cannot be satisfied on $(0,\infty)$ by functions ϕ figuring in statistically applicable ϕ-divergences (e.g., no such ϕ is bounded on $(0,\infty)$), Menéndez et *al.* (1998a) introduced the ϕ-disparity formally as an extension of the ϕ-divergence. Later, these concepts have been used systematically in many papers.

Definition 1.2

The ϕ-disparity between the probability distributions $P_{\boldsymbol{\theta}_1}$ and $P_{\boldsymbol{\theta}_2}$ is defined by

$$D_\phi(P_{\boldsymbol{\theta}_1}, P_{\boldsymbol{\theta}_2}) = D_\phi(\boldsymbol{\theta}_1, \boldsymbol{\theta}_2) = \int_{\mathcal{X}} f_{\boldsymbol{\theta}_2}(\boldsymbol{x})\phi\left(\frac{f_{\boldsymbol{\theta}_1}(\boldsymbol{x})}{f_{\boldsymbol{\theta}_2}(\boldsymbol{x})}\right) d\mu(\boldsymbol{x}), \tag{1.26}$$

where the function $\phi : (0,\infty) \to [0,\infty)$ is assumed to be continuous, decreasing on $(0,1)$ and increasing on $(1,\infty)$, with $\phi(1) = 0$. The value $\phi(0) \in (0,\infty]$ is defined by the continuous extension.

Remark 1.4

Note that the class of ϕ-disparities contains all ϕ-divergences of Csiszár (see Csiszár (1967), Liese and Vajda (1987) or Vajda (1989)) with $\phi : (0,\infty) \to (0,\infty)$ convex and equal to zero only at 1. Then, the assumed convexity and $\phi(1) = 0$ imply that

$$\frac{\phi(t) - \phi(1)}{t-1} = \frac{\phi(t)}{t-1}$$

is nondecreasing in the domain $t > 0$. Therefore, $\phi(t)$ is increasing in the domain $t > 1$ unless $\phi(t) = 0$ on an interval $(1, t_1)$, and decreasing in the domain $0 < t < 1$ unless $\phi(t) = 0$ in an interval $(t_0, 1)$. But $\phi(t) = 0$ for $t \neq 1$ is excluded by assumptions.

It is easy to verify that the functions

i) $\phi(x) = \min\left((x-1)^2, (1-1/x)^2\right)$

ii) $\phi(x) = \dfrac{x \log x}{(1 + x \log x)}$

iii) $\phi(x) = 1 - \dfrac{2x}{1+x^2}$

iv) $\phi(x) = 1 - \exp\left(-\alpha (x-1)^2\right), \; \alpha > 0,$

are not convex, but they verify the properties of ϕ-disparities.

Some properties about ϕ-disparities can be seen in the cited paper by Menéndez et *al.* (1998a) and also in Menéndez et *al.* (2001a,b,c) and Morales et *al.* (2003, 2004).

1.6. Exercises

1. Show that Shannon's entropy for a random variable X, whose probability density function (p.d.f.), $f(x)$, vanishes outside an interval (a,b) is bounded by the entropy of a uniform distribution in the interval (a,b).

2. Let $(X_1, X_2, ..., X_n, Y)$ be a $(n+1)$-variate random vector with probability density function $f(x_1, x_2, ..., x_n, y)$. State the relation between

$$H(Y/X_1, X_2, \ldots, X_n) \qquad \text{and} \qquad H(Y/X_1, X_2, \ldots, X_k) \qquad (k < n)$$

and find the necessary condition which turns the inequality into equality.

3. Show that Shannon's entropy of a continuous random variable in $\mathbb{R}$ with finite mean μ and variance σ^2 is bounded by Shannon's entropy of a normal distribution with mean μ and variance σ^2.

4. Let X be a random variable with probability density function $f(x)$. Show

$$\int_{\mathbb{R}} x^2 f(x)\, dx \geq \frac{1}{2\pi e} \exp\left(2H(X)\right).$$

5. It is said that the experiment associated with the random variable $\boldsymbol{X}$, with p.d.f. $f_{\boldsymbol{\theta}}(\boldsymbol{x})$, is sufficient for the experiment associated with the random variable $\boldsymbol{Y}$, with p.d.f. $g_{\boldsymbol{\theta}}(\boldsymbol{y})$, if there exists a nonnegative function h on the product space $\mathcal{X} \times \mathcal{Y}$ for which the following relations are satisfied:

$$i)\ g_{\boldsymbol{\theta}}(\boldsymbol{y}) = \int_{\mathcal{X}} h(\boldsymbol{x},\boldsymbol{y}) f_{\boldsymbol{\theta}}(\boldsymbol{x}) d\mu(\boldsymbol{x})$$

$$ii)\ h(\boldsymbol{x},\boldsymbol{y}) \geq 0, \int_{\mathcal{X}} h(\boldsymbol{x},\boldsymbol{y}) d\mu(\boldsymbol{x}) = \int_{\mathcal{Y}} h(\boldsymbol{x},\boldsymbol{y}) d\mu(\boldsymbol{y}) = 1.$$

Show that

$$H(\boldsymbol{Y}) \geq H(\boldsymbol{X}).$$

6. Derive the expression of Shannon's entropy for the following random variables: Beta, Cauchy, Chi-square, Erlang, Exponential, F-Snedecor, Gamma, Laplace, Logistic, Lognormal, Maxwell-Normal, Normal, Normal-generalized, Pareto, Rayleigh and T-Student.

7. Let X be a random variable with probability mass function $P_{\boldsymbol{\theta}}(X = x_i) = p_{\boldsymbol{\theta}}(x_i)$, $i \in \mathbb{N}$. Show that

$$J(\boldsymbol{\theta}_1, \boldsymbol{\theta}_2) \geq 4R(\boldsymbol{\theta}_1, \boldsymbol{\theta}_2),$$

where $R(\boldsymbol{\theta}_1, \boldsymbol{\theta}_2)$ is the R_ϕ-divergence of Burbea and Rao with $\phi(x) = -x \log x$.

8. Suppose that a d-variate random vector $\boldsymbol{X} = (X_1, ..., X_d)$ has a multivariate normal distribution with mean vector $\boldsymbol{\mu} = (\mu_1, ..., \mu_d)^T$ and nonsingular variance-covariance matrix $\boldsymbol{\Sigma}$.

    i) Show that

$$H_r^1(\boldsymbol{\mu}, \boldsymbol{\Sigma}) = \frac{1}{r(1-r)} \log \left(\frac{\det(\boldsymbol{\Sigma})^{-\frac{(r-1)}{2}}}{r^{d/2}} (2\pi)^{-\frac{d(r-1)}{2}} \right)$$

    and

$$H(\boldsymbol{\mu}, \boldsymbol{\Sigma}) = \frac{1}{2} \log \left(\det(\boldsymbol{\Sigma}) (2\pi e)^d \right).$$

    ii) Assuming that X is normal with mean μ and variance σ^2, show using the results obtained in *i)* that

$$H_r^1(\mu, \sigma^2) = \frac{1}{r(1-r)} \log \left(\frac{\left(\sigma^2\right)^{-\frac{(r-1)}{2}}}{r^{1/2}} (2\pi)^{-\frac{(r-1)}{2}} \right)$$

    and

$$H(\mu, \sigma^2) = \frac{1}{2} \log \left(\sigma^2 2\pi e \right).$$

    iii) Show that Shannon's entropy is invariant with respect to orthogonal transformations.

9. Let $\boldsymbol{A} = (a_{ij})_{i,j=1..,d}$ be a symmetric and positive definite matrix. Show that

$$\det(\boldsymbol{A}) \leq \prod_{i=1}^{d} a_{ii} \text{ (Hadamard Theorem).}$$

10. Let $\boldsymbol{X} = (X_1, ..., X_d)$ be a d-variate random vector with p.d.f. given by $f(x_1, ..., x_d)$ and nonsingular variance-covariance matrix $\boldsymbol{\Sigma}$. Show that

$$H(X_1, ..., X_d) \leq \frac{1}{2} \log \left(\det(\boldsymbol{\Sigma}) (2\pi e)^d \right),$$

and equality holds if and only if $\boldsymbol{X} = (X_1, ..., X_d)$ has a multivariate normal distribution with nonsingular variance-covariance matrix $\boldsymbol{\Sigma}$.

11. Show that Rényi's divergence and Kullback-Leibler divergence between two multivariate normal distributions are given respectively by

$$D_r^1\left((\boldsymbol{\mu}_1, \boldsymbol{\Sigma}_1), (\boldsymbol{\mu}_2, \boldsymbol{\Sigma}_2)\right) = \frac{(\boldsymbol{\mu}_1 - \boldsymbol{\mu}_2)^T \left(r\boldsymbol{\Sigma}_2 + (1-r)\boldsymbol{\Sigma}_1\right)^{-1} (\boldsymbol{\mu}_1 - \boldsymbol{\mu}_2)}{2} - \frac{1}{2r(r-1)} \log \frac{\det\left(r\boldsymbol{\Sigma}_2 + (1-r)\boldsymbol{\Sigma}_1\right)}{\det(\boldsymbol{\Sigma}_1)^{1-r} \det(\boldsymbol{\Sigma}_2)^r}$$

and

$$D_{Kull}\left((\boldsymbol{\mu}_1, \boldsymbol{\Sigma}_1), (\boldsymbol{\mu}_2, \boldsymbol{\Sigma}_2)\right) = \tfrac{1}{2}\left((\boldsymbol{\mu}_1 - \boldsymbol{\mu}_2)^T \boldsymbol{\Sigma}_2^{-1} (\boldsymbol{\mu}_1 - \boldsymbol{\mu}_2)\right) + \tfrac{1}{2}\left(trace\left(\boldsymbol{\Sigma}_2^{-1}\boldsymbol{\Sigma}_1 - I\right) + \log \frac{\det(\boldsymbol{\Sigma}_2)}{\det(\boldsymbol{\Sigma}_1)} \right).$$

12. Determine the R_ϕ-Divergence, with $\phi(x) = x - x^2$, between two multivariate normal distributions. Find the expression, as a particular case, for two univariate normal distributions.

13. Determine the Bhattacharyya divergence,

$$B(\boldsymbol{\theta}_1, \boldsymbol{\theta}_2) = -\log \int_{\mathcal{X}} \left(f_{\boldsymbol{\theta}_1}(x) f_{\boldsymbol{\theta}_2}(x)\right)^{1/2} d\mu(\boldsymbol{x}),$$

between two univariate normal distributions.

14. Show that Hellinger's distance

$$D^{He}(\boldsymbol{\theta}_1, \boldsymbol{\theta}_2) = \left(\int_{\mathcal{X}} \left(\sqrt{f_{\boldsymbol{\theta}_1}(\boldsymbol{x})} - \sqrt{f_{\boldsymbol{\theta}_2}(\boldsymbol{x})} \right)^2 d\mu(\boldsymbol{x}) \right)^{1/2}$$

is a metric. Find its expression for two multivariate normal distributions.

15. Evaluate the Rényi's divergence as well as the Kullback-Leibler divergence for two Poisson populations.

16. Let $\boldsymbol{X} = (X_1, ..., X_d)$, $\boldsymbol{Y} = (Y_1, ..., Y_l)$ and $(\boldsymbol{X}, \boldsymbol{Y})$ random vectors with multivariate normal distribution and variance-covariance matrices given by $\boldsymbol{A}, \boldsymbol{B}$ and $\boldsymbol{C}$ respectively. Show that

a) $H(\boldsymbol{X}) - H(\boldsymbol{X}/\boldsymbol{Y}) = \frac{1}{2} \log \dfrac{\det(\boldsymbol{A}) \det(\boldsymbol{B})}{\det(\boldsymbol{C})}.$

b) If $d = l = 1$, then $H(X) - H(X/Y) = -\frac{1}{2}\log\left(1 - \rho(X,Y)^2\right)$, being $\rho(X,Y)$ the correlation coefficient between X and Y.

(It is important to note that the expression $H(\boldsymbol{X}) - H(\boldsymbol{X}/\boldsymbol{Y})$ is called Mutual Information and it is related to the Kullback-Leibler divergence by

$$D_{Kull}\left(P_{\mathbf{XY}}, P_{\mathbf{X}} \times P_{\mathbf{Y}}\right),$$

where by $P_{\mathbf{XY}}$ we are denoting the joint probability distribution of the random variable $(\boldsymbol{X}, \boldsymbol{Y})$.

17. Let $\boldsymbol{X}$ and $\boldsymbol{Y}$ be two d-variate normal distributions with mean vectors $\boldsymbol{\mu}_1$ and $\boldsymbol{\mu}_2$ and variance-covariance matrices $\boldsymbol{\Sigma}_1$ and $\boldsymbol{\Sigma}_2$, respectively. Assume that $\boldsymbol{Z}$ is an arbitrary d-variate continuous random variable with mean vector $\boldsymbol{\mu}_1$ and variance-covariance matrix $\boldsymbol{\Sigma}_1$. Show that

$$D_{Kull}\left(N(\boldsymbol{\mu}_1, \boldsymbol{\Sigma}_1), N(\boldsymbol{\mu}_2, \boldsymbol{\Sigma}_2)\right) \leq D_{Kull}\left(\mathbf{Z}, N(\boldsymbol{\mu}_2, \boldsymbol{\Sigma}_2)\right).$$

1.7. Answers to Exercises

1. Applying (1.24), we have

$$-\int_b^a f(x)\log f(x)\,dx \leq -\int_b^a f(x)\log\frac{1}{b-a}dx = -\log\frac{1}{b-a} = \log(b-a).$$

The result follows because the entropy of a uniform random variable is given by $\log(b-a)$.

2. Applying (1.24) to the probability density functions $f(y/x_1, ..., x_n)$ and $f(y/x_1, ..., x_k)$, we have, denoting by,

$$l = -\int_{\mathbb{R}} f(y/x_1, ..., x_n)\log f(y/x_1, ..., x_n)dy,$$

that

$$l \leq -\int_{\mathbb{R}} f(y/x_1, ..., x_n)\log f(y/x_1, ..., x_k)dy.$$

The equality holds if and only if given $(x_1, ..., x_n)$

$$f(y/x_1, ..., x_n) = f(y/x_1, ..., x_k),\ \forall y.$$

Multiplying by $f(x_1,...,x_n)$, and integrating on $\mathbb{R}^n$, we have

$$\begin{aligned} & -\int_{\mathbb{R}^n} f(x_1,...,x_n)\left(\int_{\mathbb{R}} f(y/x_1,...,x_n)\log f(y/x_1,...,x_n)dy\right)dx_1...dx_n \\ \leq & -\int_{\mathbb{R}^n} f(x_1,...,x_n)\left(\int_{\mathbb{R}} f(y/x_1,...,x_n)\log f(y/x_1,...,x_k)dy\right)dx_1...dx_n, \end{aligned}$$

and

$$H(Y/X_1,...,X_n) \leq H(Y/X_1,...,X_k).$$

The equality holds if and only if

$$f(y/x_1,...,x_n) = f(y/x_1,...,x_k).$$

3. Let X be a random variable with probability density function $f(x)$ with mean μ and variance σ^2. Applying (1.24), we have

$$\begin{aligned} -\int_{\mathbb{R}} f(x)\log f(x)\,dx &\leq -\int_{\mathbb{R}} f(x)\log\left(\frac{1}{\sigma(2\pi)^{1/2}}\exp\left(-\frac{(x-\mu)^2}{2\sigma^2}\right)\right)dx \\ &= -\int_{\mathbb{R}} f(x)\left(\log\frac{1}{\sigma(2\pi)^{1/2}} - \frac{(x-\mu)^2}{2\sigma^2}\right)dx \\ &= \log\left(\sigma(2\pi)^{1/2}\right) + \frac{1}{2} = \log\left(\sigma(2\pi e)^{1/2}\right) \\ &= H(N(\mu,\sigma)). \end{aligned}$$

4. Since $\log x$ is an increasing function it is enough to prove that

$$\log\int_{\mathbb{R}} x^2 f(x)dx \geq \log\frac{1}{2\pi e} + 2H(X).$$

Letting σ^2 be the variance of the random variable X, we know that

$$H(X) \leq \log\left((2\pi e)^{1/2}\sigma\right).$$

Therefore,

$$\log\frac{1}{2\pi e} + 2H(X) \leq \log\frac{1}{2\pi e} + \log\left(2\pi e\sigma^2\right) = \log\sigma^2.$$

On the other hand, as $\sigma^2 \leq \int_{\mathbb{R}} x^2 f(x)dx$, we have

$$\log\sigma^2 \leq \log\int_{\mathbb{R}} x^2 f(x)dx$$

and then

$$\log \frac{1}{2\pi e} + 2H(X) \leq \log \sigma^2 \leq \log \int_{\mathbb{R}} x^2 f(x)dx.$$

5. By hypothesis, $h(\boldsymbol{x}, \boldsymbol{y})$ is a probability density function in $\boldsymbol{y}$. We consider the function $\phi(x) = \log x$ and the random variable taking on the values

$$\frac{g_{\boldsymbol{\theta}}(\boldsymbol{y})}{\int_{\mathcal{Y}} h(\boldsymbol{x}, \boldsymbol{z}) g_{\boldsymbol{\theta}}(\boldsymbol{z}) d\mu(\boldsymbol{z})}$$

with probability density function $h(\boldsymbol{x}, \boldsymbol{y})$. Thus

$$E_{\boldsymbol{\theta}}\left[\frac{g_{\boldsymbol{\theta}}(\boldsymbol{Y})}{\int_{\mathcal{Y}} h(\boldsymbol{x}, \boldsymbol{z}) g_{\boldsymbol{\theta}}(\boldsymbol{z}) d\mu(\boldsymbol{z})}\right] = \int_{\mathcal{Y}} h(\boldsymbol{x}, \boldsymbol{y}) \frac{g_{\boldsymbol{\theta}}(\boldsymbol{y})}{\int_{\mathcal{Y}} h(\boldsymbol{x}, \boldsymbol{z}) g_{\boldsymbol{\theta}}(\boldsymbol{z}) d\mu(\boldsymbol{z})} d\mu(\boldsymbol{y}) = 1$$

and

$$E_{\boldsymbol{\theta}}\left[\phi\left(\frac{g_{\boldsymbol{\theta}}(\boldsymbol{Y})}{\int_{\mathcal{Y}} h(\boldsymbol{x}, \boldsymbol{z}) g_{\boldsymbol{\theta}}(\boldsymbol{z}) d\mu(\boldsymbol{z})}\right)\right]$$

is given by

$$\int_{\mathcal{Y}} h(\boldsymbol{x}, \boldsymbol{y}) \log\left(\frac{g_{\boldsymbol{\theta}}(\boldsymbol{y})}{\int_{\mathcal{Y}} h(\boldsymbol{x}, \boldsymbol{z}) g_{\boldsymbol{\theta}}(\boldsymbol{z}) d\mu(\boldsymbol{z})}\right) d\mu(\boldsymbol{y}).$$

Applying Jensen's inequality, multiplying by $f_{\boldsymbol{\theta}}(\boldsymbol{x})$ and integrating on $\mathcal{X}$

$$\int_{\mathcal{X}} \int_{\mathcal{Y}} f_{\boldsymbol{\theta}}(\boldsymbol{x}) h(\boldsymbol{x}, \boldsymbol{y}) \log\left(\frac{g_{\boldsymbol{\theta}}(\boldsymbol{y})}{\int_{\mathcal{Y}} h(\boldsymbol{x}, \boldsymbol{z}) g_{\boldsymbol{\theta}}(\boldsymbol{z}) d\mu(\boldsymbol{z})}\right) d\mu(\boldsymbol{y}) d\mu(\boldsymbol{x}) \leq 0.$$

Denoting

$$l = \int_{\mathcal{Y}} \int_{\mathcal{X}} f_{\boldsymbol{\theta}}(\boldsymbol{x}) h(\boldsymbol{x}, \boldsymbol{y}) \log g_{\boldsymbol{\theta}}(\boldsymbol{y}) d\mu(\boldsymbol{x}) d\mu(\boldsymbol{y})$$

and

$$m = \int_{\mathcal{Y}} \int_{\mathcal{X}} f_{\boldsymbol{\theta}}(\boldsymbol{x}) h(\boldsymbol{x}, \boldsymbol{y}) \log\left(\int_{\mathcal{Y}} h(\boldsymbol{x}, \boldsymbol{z}) g_{\boldsymbol{\theta}}(\boldsymbol{z}) d\mu(\boldsymbol{z})\right) d\mu(\boldsymbol{x}) d\mu(\boldsymbol{y})$$

we have

$$\begin{aligned} l - m = & \int_{\mathcal{Y}} g_{\boldsymbol{\theta}}(\boldsymbol{y}) \log g_{\boldsymbol{\theta}}(\boldsymbol{y}) d\mu(\boldsymbol{y}) \\ - & \int_{\mathcal{X}} f_{\boldsymbol{\theta}}(\boldsymbol{x}) \left(\log \int_{\mathcal{Y}} h(\boldsymbol{x}, \boldsymbol{z}) g_{\boldsymbol{\theta}}(\boldsymbol{z}) d\mu(\boldsymbol{z}) \right) d\mu(\boldsymbol{x}) \leq 0. \end{aligned}$$

The last equality is obtained integrating in "$\boldsymbol{x}$" the first expression and in "$\boldsymbol{y}$" the second one. Then we have obtained

$$-\int_{\mathcal{Y}} g_{\boldsymbol{\theta}}(\boldsymbol{y}) \log g_{\boldsymbol{\theta}}(\boldsymbol{y}) d\mu(\boldsymbol{y}) \geq -\int_{\mathcal{X}} f_{\boldsymbol{\theta}}(\boldsymbol{x}) \left(\log \int_{\mathcal{Y}} h(\boldsymbol{x}, \boldsymbol{z}) g_{\boldsymbol{\theta}}(\boldsymbol{z}) d\mu(\boldsymbol{z}) \right) d\mu(\boldsymbol{x}). \tag{1.27}$$

The function

$$\gamma(\boldsymbol{x}) = \int_{\mathcal{Y}} h(\boldsymbol{x}, \boldsymbol{z}) g_{\boldsymbol{\theta}}(\boldsymbol{z}) d\mu(\boldsymbol{z})$$

is a probability density function on $\mathcal{X}$ since

$$\begin{aligned} \int_{\mathcal{X}} \gamma(\boldsymbol{x}) d\mu(\boldsymbol{x}) = & \int_{\mathcal{X}} \int_{\mathcal{Y}} h(\boldsymbol{x}, \boldsymbol{z}) g_{\boldsymbol{\theta}}(\boldsymbol{z}) d\mu(\boldsymbol{z}) d\mu(\boldsymbol{x}) \\ = & \int_{\mathcal{Y}} \left(\int_{\mathcal{X}} h(\boldsymbol{x}, \boldsymbol{z}) d\mu(\boldsymbol{x}) \right) g_{\boldsymbol{\theta}}(\boldsymbol{z}) d\mu(\boldsymbol{z}) = 1. \end{aligned}$$

Applying (1.24) to inequality (1.27) we have,

$$-\int_{\mathcal{Y}} g_{\boldsymbol{\theta}}(\boldsymbol{y}) \log g_{\boldsymbol{\theta}}(\boldsymbol{y}) d\mu(\boldsymbol{y}) \geq -\int_{\mathcal{X}} f_{\boldsymbol{\theta}}(\boldsymbol{x}) \log f_{\boldsymbol{\theta}}(\boldsymbol{x}) d\mu(\boldsymbol{x}),$$

i.e.,

$$H(\boldsymbol{Y}) \geq H(\boldsymbol{X}).$$

6. Shannon's entropy for Exponential, Uniform and Normal probability distributions has already been obtained. Now we present the results for Gamma and Beta distributions. The remaining can be obtained in a similar way.

   i) Gamma distribution

   We have

$$\begin{aligned} H(G(a,p)) = & -\int_0^\infty \tfrac{a^p}{\Gamma(p)} e^{-ax} x^{p-1} \left(p \log a - \log \Gamma(p) \right) dx \\ - & \int_0^\infty \tfrac{a^p}{\Gamma(p)} e^{-ax} x^{p-1} \left(-ax + (p-1) \log x \right) dx \\ = & -p \log a + \log \Gamma(p) + p \\ - & (p-1) \int_0^\infty \tfrac{a^p}{\Gamma(p)} e^{-ax} x^{p-1} \log x dx. \end{aligned}$$

To find the desired entropy, we must find the value of

$$\int_0^\infty \frac{a^p}{\Gamma(p)} e^{-ax} x^{p-1} \log x dx.$$

We know that

$$\int_0^\infty a^p e^{-ax} x^{p-1} dx = \Gamma(p).$$

Differentiating with respect to p, we have

$$\int_0^\infty (\log a) a^p e^{-ax} x^{p-1} dx + \int_0^\infty a^p e^{-ax} x^{p-1} \log x dx = \frac{\partial \Gamma(p)}{\partial p},$$

then

$$\int_0^\infty \frac{a^p}{\Gamma(p)} e^{-ax} x^{p-1} \log x dx = \frac{\partial \log \Gamma(p)}{\partial p} - \log a$$

and

$$H(G(a,p)) = \log(\Gamma(p)/a) + (1-p)\Psi(p) + p,$$

where

$$\Psi(p) = \frac{\partial \log \Gamma(p)}{\partial p}, \tag{1.28}$$

is the Digamma function.

ii) Beta distribution

Denoting by

$$l = H(B(a,b))$$

we have

$$\begin{aligned} l = & -\int_0^1 \tfrac{\Gamma(a+b)}{\Gamma(a)\Gamma(b)} x^{a-1} (1-x)^{b-1} \log \left(\tfrac{\Gamma(a+b)}{\Gamma(a)\Gamma(b)} x^{a-1} (1-x)^{b-1} \right) dx \\ = & -\left(\int_0^1 \tfrac{\Gamma(a+b)}{\Gamma(a)\Gamma(b)} x^{a-1} (1-x)^{b-1} \log \tfrac{\Gamma(a+b)}{\Gamma(a)\Gamma(b)} dx \right) \\ - & \ (a-1) \int_0^1 \tfrac{\Gamma(a+b)}{\Gamma(a)\Gamma(b)} x^{a-1} (1-x)^{b-1} \log x dx \\ - & \ (b-1) \int_0^1 \tfrac{\Gamma(a+b)}{\Gamma(a)\Gamma(b)} x^{a-1} (1-x)^{b-1} \log (1-x) dx. \end{aligned}$$

Then

$$\begin{aligned} H(B(a,b)) = & -\log \tfrac{\Gamma(a+b)}{\Gamma(a)\Gamma(b)} \\ & - (a-1) \tfrac{\Gamma(a+b)}{\Gamma(a)\Gamma(b)} \int_0^1 x^{a-1} (1-x)^{b-1} \log x dx \\ & - (b-1) \tfrac{\Gamma(a+b)}{\Gamma(a)\Gamma(b)} \int_0^1 x^{a-1} (1-x)^{b-1} \log (1-x) dx. \end{aligned}$$

Now we have to obtain the following integral $\int_0^1 x^{a-1}(1-x)^{b-1}\log x dx$.

We know

$$\int_0^1 x^{a-1}(1-x)^{b-1}\,dx = \left(\frac{\Gamma(a+b)}{\Gamma(a)\Gamma(b)}\right)^{-1}.$$

Then

$$\log \int_0^1 x^{a-1}(1-x)^{b-1}\,dx = \log\Gamma(a) + \log\Gamma(b) - \log\Gamma(a+b),$$

and differentiating with respect to a, we get the following equality

$$\frac{1}{\int_0^1 x^{a-1}(1-x)^{b-1}dx}\int_0^1 x^{a-1}(1-x)^{b-1}\log x\;dx = \frac{\partial\log\Gamma(a)}{\partial a} - \frac{\partial\log\Gamma(a+b)}{\partial a}.$$

Then

$$\int_0^1 x^{a-1}(1-x)^{b-1}\log x dx = \left(\tfrac{\Gamma(a+b)}{\Gamma(a)\Gamma(b)}\right)^{-1}(\Psi(a) - \Psi(a+b)),$$

with $\Psi(x)$ defined in (1.28).

In a similar way we get

$$\int_0^1 x^{a-1}(1-x)^{b-1}\log(1-x)\,dx = \left(\tfrac{\Gamma(a+b)}{\Gamma(a)\Gamma(b)}\right)^{-1}(\Psi(b) - \Psi(a+b)).$$

Therefore,

$$\begin{aligned} H(B(a,b)) &= -\log\tfrac{\Gamma(a+b)}{\Gamma(a)\Gamma(b)} - (a-1)(\Psi(a) - \Psi(a+b)) \\ &- (b-1)(\Psi(b) - \Psi(a+b)). \end{aligned}$$

In the following table we present without proof the entropy corresponding to the other probability distributions.

Name	*p.d.f.*	*Shannon's Entropy*
Beta	$f(x) = \frac{x^{a-1}(1-x)^{b-1}}{B(a,b)}$ $0 < x < 1,\ a, b > 0$	$\log B(a,b) - (a-1)(\Psi(a)$ $-\Psi(a+b)) - (b-1)$ $\times(\Psi(b) - \Psi(a+b))$
Cauchy	$f(x) = \frac{\lambda}{\pi(\lambda^2+x^2)}$ $-\infty < x < \infty, \lambda > 0$	$\log(4\pi\lambda)$
Chi-square	$f(x) = \frac{x^{\frac{n}{2}-1}e^{-\frac{x}{2}}}{2^{\frac{n}{2}}\Gamma\left(\frac{n}{2}\right)}$ $x > 0, n \in \mathbb{Z}^+$	$\log(2\Gamma\left(\frac{n}{2}\right))$ $+\left(1-\frac{n}{2}\right)\Psi\left(\frac{n}{2}\right)+\frac{n}{2}$
Erlang	$f(x) = \frac{b^n}{(n-1)!}x^{n-1}e^{-bx}$ $x > 0, b > 0, n \in \mathbb{Z}^+$	$(1-n)\Psi(n) + \log\frac{\Gamma(n)}{b} + n$
Exponential	$f(x) = \sigma^{-1}e^{-\frac{x}{\sigma}}; x, \sigma > 0$	$1+\log\sigma$
F-Snedecor	$f(x) = \frac{\left(\frac{mx}{n}\right)^{\frac{m}{2}-1}\left(\frac{m}{n}\right)}{B\left(\frac{m}{2},\frac{n}{2}\right)\left(\frac{mx+n}{n}\right)^{\frac{m+n}{2}}}$ $x > 0, m, n \in \mathbb{Z}^+$	$\log(\frac{m}{n}B\left(\frac{m}{2},\frac{n}{2}\right))$ $+1-\frac{m}{2}\Psi\left(\frac{m}{2}\right)-\left(1+\frac{n}{2}\right)$ $\times\Psi\left(\frac{n}{2}\right)+\frac{m+n}{2}\Psi(\frac{m+n}{2})$
Gamma	$f(x) = \frac{x^{p-1}e^{-ax}}{a^p\Gamma(p)}$ $x, a, b > 0$	$\log(\Gamma(p)/a)$ $+(1-p)\Psi(p)+p$
Laplace	$f(x) = \frac{1}{2}a^{-1}e^{-\frac{\lvert x-\theta\rvert}{a}}$ $-\infty < x < \infty, a > 0$	$1+\log(2a)$
Logistic	$f(x) = e^{-x}(1+e^{-x})^{-2}$ $-\infty < x < \infty$	2
Lognormal	$f(x) = \frac{\exp\left(\frac{-(\log x-\mu)^2}{2\sigma^2}\right)}{\sigma x\sqrt{2\pi}}$ $x > 0$	$\mu + \frac{1}{2}\log\left(2\pi e\sigma^2\right)$
Maxwell-Normal	$f(x) = \left(4\sqrt{\frac{\beta^3}{\pi}}\right)x^2e^{-\beta x^2}$ $x, \beta > 0$	$\log\sqrt{\frac{\pi}{\beta}} + \gamma - \frac{1}{2}$
Normal	$f(x) = \frac{1}{\sigma\sqrt{2\pi}}e^{-\frac{x^2}{2\sigma^2}}$ $-\infty < x < \infty, \sigma > 0$	$\log\left(\sigma\sqrt{2\pi e}\right)$
Generalized Normal	$f(x) = \left(\frac{2\beta^{\frac{\alpha}{2}}}{\Gamma\left(\frac{\alpha}{2}\right)}\right)x^{\alpha-1}e^{-\beta x^2}$ $x, \alpha, \beta > 0$	$\log\left(\frac{\Gamma\left(\frac{\alpha}{2}\right)}{\left(2\beta^{\frac{1}{2}}\right)}\right)$ $-(\frac{(\alpha-1)}{2})\Psi\left(\frac{\alpha}{2}\right)+\frac{\alpha}{2}$
Pareto	$f(x) = \frac{ak^a}{x^{a+1}}$ $x \geq k > 0, a > 0$	$\log\left(\frac{k}{a}\right)+1+\frac{1}{a}$
Rayleigh	$f(x) = \left(\frac{x}{b^2}\right)e^{-\frac{x^2}{2b^2}}$ $x, b > 0$	$1+\log\left(\frac{b}{\sqrt{2}}\right)+\frac{\gamma}{2}$
T-Student	$f(x) = \frac{\left(1+\frac{x^2}{\nu}\right)^{-\frac{(\nu+1)}{2}}}{\nu^{1/2}B\left(\frac{1}{2},\frac{n}{2}\right)}$ $-\infty < x < \infty; \nu \in \mathbb{Z}^+$	$\frac{n+1}{2}\left\{\Psi(\frac{n+1}{2})-\Psi(\frac{n}{2})\right\}$ $+\log(\sqrt{n}B\left(\frac{1}{2},\frac{n}{2}\right))$

Triangular	$f(x) = \begin{cases} \frac{2x}{a} & 0 \leq x \leq a \\ \frac{2(1-x)}{(1-a)} & a \leq x \leq 1 \end{cases}$	$\frac{1}{2} - \log 2$
Uniform	$f(x) = \frac{1}{(b-a)}; a < x < b$	$\log(b-a)$
Weibull	$f(x) = \left(\frac{c}{a}\right) x^{c-1} e^{-x^c/a}; x, c, a > 0$	$\frac{(c-1)\gamma}{c} + \log\left(\frac{a^{1/c}}{c}\right) + 1$

The function Ψ and the value γ appearing in the previous table are given by

$$\begin{aligned} \Psi(x) &= \frac{\partial \log \Gamma(x)}{\partial x} = -\gamma + (x-1) \sum_{k=0}^{\infty} ((k+1)(x+k))^{-1} \\ \gamma &= \text{Euler constant} = 0.5772156649. \end{aligned}$$

More details about the previous table can be seen in Lazo and Rathie (1978).

7. We consider the random variable Z, taking on the values

$$\frac{p_{\boldsymbol{\theta}_1}(x_i) + p_{\boldsymbol{\theta}_2}(x_i)}{2p_{\boldsymbol{\theta}_2}(x_i)}, \; i \in \mathbb{N}$$

with probabilities $p_{\boldsymbol{\theta}_2}(x_i)$, $i \in \mathbb{N}$. We have

$$E_{\theta_2}[Z] = \sum_{i \in \mathbb{N}} \frac{p_{\boldsymbol{\theta}_1}(x_i) + p_{\boldsymbol{\theta}_2}(x_i)}{2p_{\boldsymbol{\theta}_2}(x_i)} p_{\boldsymbol{\theta}_2}(x_i) = 1.$$

If we consider now the convex function $\phi(t) = t \log t$, we have

$$E_{\theta_2}[\phi(Z)] = \sum_{i \in \mathbb{N}} \frac{p_{\boldsymbol{\theta}_1}(x_i) + p_{\boldsymbol{\theta}_2}(x_i)}{2} \log \frac{p_{\boldsymbol{\theta}_1}(x_i) + p_{\boldsymbol{\theta}_2}(x_i)}{2p_{\boldsymbol{\theta}_2}(x_i)}.$$

Applying Jensen's inequality we get

$$0 = \phi(E_{\theta_2}[Z]) \leq \sum_{i \in \mathbb{N}} \frac{p_{\boldsymbol{\theta}_1}(x_i) + p_{\boldsymbol{\theta}_2}(x_i)}{2} \log \frac{p_{\boldsymbol{\theta}_1}(x_i) + p_{\boldsymbol{\theta}_2}(x_i)}{2p_{\boldsymbol{\theta}_2}(x_i)}.$$

Then,

$$-\sum_{i \in \mathbb{N}} \tfrac{p_{\boldsymbol{\theta}_1}(x_i) + p_{\boldsymbol{\theta}_2}(x_i)}{2} \log \tfrac{p_{\boldsymbol{\theta}_1}(x_i) + p_{\boldsymbol{\theta}_2}(x_i)}{2} \leq -\sum_{i \in \mathbb{N}} \tfrac{p_{\boldsymbol{\theta}_1}(x_i) + p_{\boldsymbol{\theta}_2}(x_i)}{2} \log p_{\boldsymbol{\theta}_2}(x_i).$$

On the other hand we can write

$$\begin{aligned} D_{Kull}(\boldsymbol{\theta}_1, \boldsymbol{\theta}_2) &= \sum_{i \in \mathbb{N}} p_{\boldsymbol{\theta}_1}(x_i) \log \frac{p_{\boldsymbol{\theta}_1}(x_i)}{p_{\boldsymbol{\theta}_2}(x_i)} \\ &= \sum_{i \in \mathbb{N}} p_{\boldsymbol{\theta}_1}(x_i) \log p_{\boldsymbol{\theta}_1}(x_i) - \sum_{i \in \mathbb{N}} p_{\boldsymbol{\theta}_1}(x_i) \log p_{\boldsymbol{\theta}_2}(x_i) \\ &= -H(\boldsymbol{\theta}_1) - H(\boldsymbol{\theta}_2) - 2 \sum_{i \in \mathbb{N}} \tfrac{p_{\boldsymbol{\theta}_1}(x_i) + p_{\boldsymbol{\theta}_2}(x_i)}{2} \log p_{\boldsymbol{\theta}_2}(x_i). \end{aligned}$$

Then,

$$\begin{aligned} D_{Kull}(\boldsymbol{\theta}_1,\boldsymbol{\theta}_2) \geq & -H(\boldsymbol{\theta}_1) - H(\boldsymbol{\theta}_2) - 2\sum_{i\in\mathbb{N}} \tfrac{p_{\boldsymbol{\theta}_1}(x_i)+p_{\boldsymbol{\theta}_2}(x_i)}{2} \log \tfrac{p_{\boldsymbol{\theta}_1}(x_i)+p_{\boldsymbol{\theta}_2}(x_i)}{2} \\ = & \ 2H\left(\tfrac{P_{\boldsymbol{\theta}_1}+P_{\boldsymbol{\theta}_2}}{2}\right) - \left(H(P_{\boldsymbol{\theta}_1}) + H(P_{\boldsymbol{\theta}_2})\right). \end{aligned}$$

In a similar way it is possible to prove the statement,

$$D_{Kull}(\boldsymbol{\theta}_2,\boldsymbol{\theta}_1) \geq 2H\left(\frac{P_{\boldsymbol{\theta}_1}+P_{\boldsymbol{\theta}_2}}{2}\right) - \left(H(P_{\boldsymbol{\theta}_1}) + H(P_{\boldsymbol{\theta}_2})\right).$$

Thus $J(\boldsymbol{\theta}_1,\boldsymbol{\theta}_2)$ has the expression

$$D_{Kull}(\boldsymbol{\theta}_1,\boldsymbol{\theta}_2) + D_{Kull}(\boldsymbol{\theta}_2,\boldsymbol{\theta}_1) \geq 4\left(H\left(\frac{P_{\boldsymbol{\theta}_1}+P_{\boldsymbol{\theta}_2}}{2}\right) - \tfrac{H(P_{\boldsymbol{\theta}_1})+H(P_{\boldsymbol{\theta}_2})}{2}\right),$$

and hence

$$J(\boldsymbol{\theta}_1,\boldsymbol{\theta}_2) \geq 4R(\boldsymbol{\theta}_1,\boldsymbol{\theta}_2).$$

8. Let G_d be the family of all d-variate normal distributions, $N(\boldsymbol{\mu},\boldsymbol{\Sigma})$, with mean vector $\boldsymbol{\mu} = (\mu_1,...,\mu_d)^T$ and nonsingular variance-covariance matrix $\boldsymbol{\Sigma}$. A distribution can be specified by an element $(\boldsymbol{\mu},\boldsymbol{\Sigma})$ of the parameter space

$$\Theta = \left\{(\boldsymbol{\mu},\boldsymbol{\Sigma}) : \boldsymbol{\mu} \in \mathbb{R}^d, \boldsymbol{\Sigma} \in P(d,\mathbb{R})\right\},$$

where $P(d,\mathbb{R})$ is the set of all positive definite matrices of order d. The p.d.f.'s of the elements of G_d are given by

$$f_{\boldsymbol{\mu},\boldsymbol{\Sigma}}(x_1,...,x_d) = (2\pi)^{-d/2}\det(\boldsymbol{\Sigma})^{-1/2}\exp\left\{-\frac{1}{2}(\boldsymbol{x}-\boldsymbol{\mu})^T\boldsymbol{\Sigma}^{-1}(\boldsymbol{x}-\boldsymbol{\mu})\right\}$$

with $\boldsymbol{x} = (x_1,...,x_d) \in \mathbb{R}^d$ and $(\boldsymbol{\mu},\boldsymbol{\Sigma}) \in \Theta$.

i) It is clear that

$$H_r^1(\boldsymbol{\mu},\boldsymbol{\Sigma}) = \frac{1}{r(1-r)}\log K_r(\boldsymbol{\mu},\boldsymbol{\Sigma}),\ r \neq 0,1$$

where

$$K_r(\boldsymbol{\mu},\boldsymbol{\Sigma}) = \int_{\mathbb{R}^d}\left((2\pi)^{-d/2}\det(\boldsymbol{\Sigma})^{-1/2}\exp\left\{-\tfrac{(\boldsymbol{x}-\boldsymbol{\mu})^T\boldsymbol{\Sigma}^{-1}(\boldsymbol{x}-\boldsymbol{\mu})}{2}\right\}\right)^r d\boldsymbol{x}.$$

But

$$K_r(\boldsymbol{\mu}, \boldsymbol{\Sigma}) = \frac{\det(\boldsymbol{\Sigma})^{-\frac{(r-1)}{2}}}{r^{d/2}(2\pi)^{\frac{d(r-1)}{2}}} \int_{\mathbb{R}^d} f_{\boldsymbol{\mu},(r^{-1}\boldsymbol{\Sigma})}(x_1,..,x_d)\,d\boldsymbol{x}$$
$$= \frac{\det(\boldsymbol{\Sigma})^{-\frac{(r-1)}{2}}}{r^{d/2}}(2\pi)^{-\frac{d(r-1)}{2}},$$

and we get the first result.

On the other hand,

$$H(\boldsymbol{\mu}, \boldsymbol{\Sigma}) = \lim_{r\to 1} H_r^1(\boldsymbol{\mu}, \boldsymbol{\Sigma}) = \lim_{r\to 1}\left(\tfrac{1}{2r}\log\det(\boldsymbol{\Sigma}) + \tfrac{d}{2r}\log 2\pi - \tfrac{d}{2}\tfrac{\log r}{r(1-r)}\right),$$

but

$$\lim_{r\to 1} -\frac{d}{2}\frac{\log r}{r(1-r)} = \frac{d}{2}.$$

Therefore we have the stated expression for $H(\boldsymbol{\mu}, \boldsymbol{\Sigma})$.

ii) The result follows from *i)* with $d = 1$.

iii) Let $\boldsymbol{Y} = \boldsymbol{L}\boldsymbol{X}$ be a d-variate random vector, where $\boldsymbol{X}$ is a random vector with multivariate normal distribution, with mean vector $\boldsymbol{\mu} = (\mu_1, ..., \mu_d)^T$ and nonsingular variance-covariance matrix $\boldsymbol{\Sigma}$, and $\boldsymbol{L}$ is an orthogonal matrix. We have

$$\det(\boldsymbol{L}^T\boldsymbol{\Sigma}\boldsymbol{L}) = \det(\boldsymbol{L}^T)\det(\boldsymbol{\Sigma})\det(\boldsymbol{L})$$
$$= \det(\boldsymbol{L}^{-1})\det(\boldsymbol{\Sigma})\det(\boldsymbol{L}) = \det(\boldsymbol{\Sigma}).$$

Then,

$$H(\boldsymbol{X}) \equiv H(\boldsymbol{\mu}, \boldsymbol{\Sigma}) = H(\boldsymbol{Y}).$$

9. Let $\boldsymbol{X} = (X_1, ..., X_d)$ be a d-variate random vector with multivariate normal distribution, with mean vector $\boldsymbol{\mu} = (\mu_1, ..., \mu_d)^T$ and nonsingular variance-covariance matrix $\boldsymbol{A} = (a_{ij})_{i,j=1,...,d}$. The variance of the random variable X_i is a_{ii}, for $i = 1, ..., d$.

We know that

$$H(X_1, ..., X_d) \le \sum_{i=1}^{d} H(X_i). \tag{1.29}$$

The marginal distributions of the random variables X_i, $i = 1, ..., d$, are normal with mean μ_i and variance a_{ii}, then

$$H(X_i) = H(\mu_i, a_{ii}) = \log(a_{ii}2\pi e)^{1/2}.$$

On the other hand,

$$H(\boldsymbol{\mu},\boldsymbol{\Sigma}) = H(X_1,...,X_d) = \frac{1}{2}\log\left(\det(\boldsymbol{\Sigma})(2\pi e)\right)^d.$$

Applying (1.29) we have the stated result.

10. Let $\boldsymbol{Y} = (Y_1,...,Y_d)$ be a d-variate random vector with multivariate normal distribution, with mean vector $\boldsymbol{\mu}$ and nonsingular variance-covariance matrix $\boldsymbol{\Sigma}$. By (1.24) we have

$$\begin{aligned} H(\boldsymbol{X}) \leq & -\int_{\mathbb{R}^d} f(x_1,...,x_d)\log\left((2\pi)^{-d/2}\det(\boldsymbol{\Sigma})^{-1/2}\right. \\ & \times \left.\exp\left\{-\tfrac{1}{2}(\boldsymbol{x}-\boldsymbol{\mu})^T\boldsymbol{\Sigma}^{-1}(\boldsymbol{x}-\boldsymbol{\mu})\right\}\right)d\boldsymbol{x} \\ = & \log\left((2\pi)^{d/2}\det(\boldsymbol{\Sigma})^{1/2}\right) \\ & + \tfrac{1}{2}\int_{\mathbb{R}^d} f(x_1,...,x_d)(\boldsymbol{x}-\boldsymbol{\mu})^T\boldsymbol{\Sigma}^{-1}(\boldsymbol{x}-\boldsymbol{\mu})\,d\boldsymbol{x}. \end{aligned}$$

Furthermore, since $\boldsymbol{\Sigma}^{-1}$ is a symmetric nonnegative definite matrix, there exists an orthogonal matrix $\boldsymbol{L}$ such $\boldsymbol{L}^T\boldsymbol{\Sigma}^{-1}\boldsymbol{L} = \boldsymbol{\Lambda}$ for some diagonal matrix $\boldsymbol{\Lambda}$. We shall assume $\boldsymbol{\Lambda} = diag(\lambda_1,...,\lambda_d)$, $\lambda_i > 0$ $\forall i = 1,...,d$. Writing $x_i - \mu_i = u_i$, $i = 1,...,d$, we have

$$H(\boldsymbol{X}) \leq \log\left((2\pi)^{d/2}\det(\boldsymbol{\Sigma})^{1/2}\right) + \frac{1}{2}\int_{\mathbb{R}^d} f_{\boldsymbol{U}}(u_1,...,u_d)\,\boldsymbol{u}^T\boldsymbol{\Sigma}^{-1}\boldsymbol{u}\,d\boldsymbol{u}.$$

Let us make a new change of variables $u_1,...,u_d$ by writing $\boldsymbol{U} = \boldsymbol{L}\boldsymbol{V}$, and note that the Jacobian of this orthogonal transformation is $\det(\boldsymbol{L}) = 1$,

$$\begin{aligned} H(\boldsymbol{X}) \leq & \log\left((2\pi)^{d/2}\det(\boldsymbol{\Sigma})^{1/2}\right) \\ & + \tfrac{1}{2}\int_{\mathbb{R}^d} f_{\boldsymbol{V}}(v_1,...,v_d)\,\boldsymbol{v}^T\boldsymbol{L}^T\boldsymbol{\Sigma}^{-1}\boldsymbol{L}\boldsymbol{v}\,d\boldsymbol{v} \\ = & \log\left((2\pi)^{d/2}\det(\boldsymbol{\Sigma})^{1/2}\right) + \tfrac{1}{2}\int_{\mathbb{R}^d} f_{\boldsymbol{V}}(v_1,...,v_d)\,\boldsymbol{v}^T\boldsymbol{v}\,d\boldsymbol{v} \\ = & \log\left((2\pi)^{d/2}\det(\boldsymbol{\Sigma})^{1/2}\right) + \tfrac{1}{2}\int_{\mathbb{R}^d} f_{\boldsymbol{V}}(v_1,...,v_d)\left(\sum_{i=1}^d \lambda_i v_i^2\right)d\boldsymbol{v} \\ = & \log\left((2\pi)^{d/2}\det(\boldsymbol{\Sigma})^{1/2}\right) + \tfrac{1}{2}\left(\sum_{i=1}^d \lambda_i Var(V_i)\right). \end{aligned}$$

Since $Cov(\boldsymbol{U}) = \boldsymbol{L}Cov(\boldsymbol{V})\boldsymbol{L}^T$ we have $\boldsymbol{L}^{-1}\boldsymbol{\Sigma}(\boldsymbol{L}^T)^{-1} = Cov(\boldsymbol{V})$. Therefore $Cov(\boldsymbol{V}) = \left(\boldsymbol{L}^T\boldsymbol{\Sigma}^{-1}\boldsymbol{L}\right)^{-1} = \boldsymbol{\Lambda}^{-1}$ and

$$H(\boldsymbol{X}) \leq \log\left((2\pi)^{d/2}\det(\boldsymbol{\Sigma})^{1/2}\right) + \frac{1}{2}\sum_{i=1}^d \lambda_i\frac{1}{\lambda_i} = \log\left((2\pi e)^{d/2}\det(\boldsymbol{\Sigma})^{1/2}\right).$$

11. Let us denote

$$K_r^*(\boldsymbol{\theta}_1, \boldsymbol{\theta}_2) = \int_{\mathcal{X}} f_{\boldsymbol{\theta}_1}(\boldsymbol{v})^r f_{\boldsymbol{\theta}_2}(\boldsymbol{v})^{1-r} d\mu(\boldsymbol{v}),$$

where $\boldsymbol{\theta}_1 = (\boldsymbol{\mu}_1, \boldsymbol{\Sigma}_1)$ and $\boldsymbol{\theta}_2 = (\boldsymbol{\mu}_2, \boldsymbol{\Sigma}_2)$. Then

$$\begin{aligned} K_r^*(\boldsymbol{\theta}_1, \boldsymbol{\theta}_2) = & \int_{\mathbb{R}^d} \left((2\pi)^{-d/2} \det(\boldsymbol{\Sigma}_1)^{-1/2} \exp\left\{-\tfrac{(\boldsymbol{x}-\boldsymbol{\mu}_1)^T \boldsymbol{\Sigma}_1^{-1}(\boldsymbol{x}-\boldsymbol{\mu}_1)}{2}\right\}\right)^r \\ \times & \left((2\pi)^{-d/2} \det(\boldsymbol{\Sigma}_2)^{-1/2} \exp\left\{-\tfrac{(\boldsymbol{x}-\boldsymbol{\mu}_2)^T \boldsymbol{\Sigma}_2^{-1}(\boldsymbol{x}-\boldsymbol{\mu}_2)}{2}\right\}\right)^{1-r} d\boldsymbol{x} \\ = & (2\pi)^{-d/2}\left(\det(\boldsymbol{\Sigma}_1)^{-r/2} \det(\boldsymbol{\Sigma}_2)^{-\frac{(1-r)}{2}}\right) \\ \times & \int_{\mathbb{R}^d} \exp\left\{-\tfrac{1}{2}(\boldsymbol{x}-\boldsymbol{\mu}_1)^T \left(r\boldsymbol{\Sigma}_1^{-1}\right)(\boldsymbol{x}-\boldsymbol{\mu}_1)\right\} \\ \times & \exp\left\{-\tfrac{1}{2}(\boldsymbol{x}-\boldsymbol{\mu}_2)^T \left((1-r)\boldsymbol{\Sigma}_2^{-1}\right)(\boldsymbol{x}-\boldsymbol{\mu}_2)\right\} d\boldsymbol{x}. \end{aligned}$$

The expression

$$(\boldsymbol{x}-\boldsymbol{\mu}_1)^T \left(r\boldsymbol{\Sigma}_1^{-1}\right)(\boldsymbol{x}-\boldsymbol{\mu}_1) + (\boldsymbol{x}-\boldsymbol{\mu}_2)^T \left((1-r)\boldsymbol{\Sigma}_2^{-1}\right)(\boldsymbol{x}-\boldsymbol{\mu}_2)$$

can be written as

$$(\boldsymbol{x}-\boldsymbol{\mu}^*)^T \boldsymbol{C}^{-1}(\boldsymbol{x}-\boldsymbol{\mu}^*) + \boldsymbol{B}$$

where

$$\begin{aligned} \boldsymbol{\mu}^* = & \left(r\boldsymbol{\Sigma}_1^{-1} + (1-r)\boldsymbol{\Sigma}_2^{-1}\right)^{-1}\left(r\boldsymbol{\Sigma}_1^{-1}\boldsymbol{\mu}_1 + (1-r)\boldsymbol{\Sigma}_2^{-1}\boldsymbol{\mu}_2\right), \\ \boldsymbol{C}^{-1} = & r\boldsymbol{\Sigma}_1^{-1} + (1-r)\boldsymbol{\Sigma}_2^{-1}, \\ \boldsymbol{B} = & (\boldsymbol{\mu}_1-\boldsymbol{\mu}_2)^T \left(r\boldsymbol{\Sigma}_2 + (1-r)\boldsymbol{\Sigma}_1\right)^{-1} r(1-r)(\boldsymbol{\mu}_1-\boldsymbol{\mu}_2). \end{aligned}$$

Then we have

$$\begin{aligned} K_r^*(\boldsymbol{\theta}_1, \boldsymbol{\theta}_2) = & \det(\boldsymbol{\Sigma}_1)^{-r/2} \det(\boldsymbol{\Sigma}_2)^{-\frac{(1-r)}{2}} \det\left(r\boldsymbol{\Sigma}_1^{-1} + (1-r)\boldsymbol{\Sigma}_2^{-1}\right)^{-1/2} \\ \times & \exp\left\{\tfrac{r(r-1)}{2}(\boldsymbol{\mu}_1-\boldsymbol{\mu}_2)^T \left(r\boldsymbol{\Sigma}_2 + (1-r)\boldsymbol{\Sigma}_1\right)^{-1}(\boldsymbol{\mu}_1-\boldsymbol{\mu}_2)\right\} \end{aligned}$$

or

$$\begin{aligned} K_r^*(\boldsymbol{\theta}_1, \boldsymbol{\theta}_2) = & \frac{\det\left(r\boldsymbol{\Sigma}_1^{-1} + (1-r)\boldsymbol{\Sigma}_2^{-1}\right)^{-1/2}}{\det(\boldsymbol{\Sigma}_1)^{r/2} \det(\boldsymbol{\Sigma}_2)^{\frac{(1-r)}{2}}} \\ \times & \exp\left\{\tfrac{r(r-1)}{2}(\boldsymbol{\mu}_1-\boldsymbol{\mu}_2)^T \left(r\boldsymbol{\Sigma}_2 + (1-r)\boldsymbol{\Sigma}_1\right)^{-1}(\boldsymbol{\mu}_1-\boldsymbol{\mu}_2)\right\}. \end{aligned}$$

Multiplying numerator and denominator in the first term on the right-hand side of the previous expression by

$$\left(\det(\boldsymbol{\Sigma}_1)\det(\boldsymbol{\Sigma}_2)\right)^{-1/2}$$

we have

$$
\begin{aligned}
K_r^*(\boldsymbol{\theta}_1, \boldsymbol{\theta}_2) &= \frac{\det(r\boldsymbol{\Sigma}_2 + (1-r)\boldsymbol{\Sigma}_1)^{-1/2}}{\det(\boldsymbol{\Sigma}_1)^{\frac{r-1}{2}} \det(\boldsymbol{\Sigma}_2)^{-r/2}} \\
&\times \exp\left\{\tfrac{r(r-1)}{2}(\boldsymbol{\mu}_1 - \boldsymbol{\mu}_2)^T (r\boldsymbol{\Sigma}_2 + (1-r)\boldsymbol{\Sigma}_1)^{-1} (\boldsymbol{\mu}_1 - \boldsymbol{\mu}_2)\right\}.
\end{aligned}
$$

From here it is immediate to get the expression of the Rényi's divergence. Regarding the Kullback-Leibler divergence,

$$
\begin{aligned}
D_{Kull}((\boldsymbol{\mu}_1, \boldsymbol{\Sigma}_1), (\boldsymbol{\mu}_2, \boldsymbol{\Sigma}_2)) &= \lim_{r \to 1} D_r^1((\boldsymbol{\mu}_1, \boldsymbol{\Sigma}_1), (\boldsymbol{\mu}_2, \boldsymbol{\Sigma}_2)) \\
&= \tfrac{1}{2}(\boldsymbol{\mu}_1 - \boldsymbol{\mu}_2)^T \boldsymbol{\Sigma}_2^{-1} (\boldsymbol{\mu}_1 - \boldsymbol{\mu}_2) \\
&- \tfrac{1}{2} \lim_{r \to 1} \frac{1}{r(r-1)} \log \frac{\det(r\boldsymbol{\Sigma}_2 + (1-r)\boldsymbol{\Sigma}_1)}{\det(\boldsymbol{\Sigma}_1)^{1-r} \det(\boldsymbol{\Sigma}_2)^r}.
\end{aligned}
$$

But

$$
\frac{\partial \det(\boldsymbol{A})}{\partial \alpha} = \det(\boldsymbol{A}) trace\left(\boldsymbol{A}^{-1} \frac{\partial \boldsymbol{A}}{\partial \alpha}\right).
$$

Denoting

$$
a = \log \frac{\det(r\boldsymbol{\Sigma}_2 + (1-r)\boldsymbol{\Sigma}_1)}{\det(\boldsymbol{\Sigma}_1)^{1-r} \det(\boldsymbol{\Sigma}_2)^r}, \quad b = \frac{\det(\boldsymbol{\Sigma}_1)^{1-r} \det(\boldsymbol{\Sigma}_2)^r}{\det(r\boldsymbol{\Sigma}_2 + (1-r)\boldsymbol{\Sigma}_1)},
$$

and

$$
l = trace\left((r\boldsymbol{\Sigma}_2 + (1-r)\boldsymbol{\Sigma}_1)^{-1} (\boldsymbol{\Sigma}_2 - \boldsymbol{\Sigma}_1)\right),
$$

we have

$$
\begin{aligned}
\frac{\partial a}{\partial r} &= b \frac{\det(r\boldsymbol{\Sigma}_2 + (1-r)\boldsymbol{\Sigma}_1)\ l\ \det(\boldsymbol{\Sigma}_1)^{1-r} \det(\boldsymbol{\Sigma}_2)^r}{\det(\boldsymbol{\Sigma}_1)^{2(1-r)} \det(\boldsymbol{\Sigma}_2)^{2r}} \\
&- b \frac{\det(r\boldsymbol{\Sigma}_2 + (1-r)\boldsymbol{\Sigma}_1)\left(\det(\boldsymbol{\Sigma}_1)^{1-r} \det(\boldsymbol{\Sigma}_2)^r \left(\log \frac{\det(\boldsymbol{\Sigma}_2)}{\det(\boldsymbol{\Sigma}_1)}\right)\right)}{\det(\boldsymbol{\Sigma}_1)^{2(1-r)} \det(\boldsymbol{\Sigma}_2)^{2r}}.
\end{aligned}
$$

From here we have that

$$
\lim_{r \to 1} \frac{\partial a}{\partial r} = trace\left(\boldsymbol{\Sigma}_2^{-1}(\boldsymbol{\Sigma}_2 - \boldsymbol{\Sigma}_1)\right) + (\log\det(\boldsymbol{\Sigma}_2) - \log\det(\boldsymbol{\Sigma}_1)).
$$

Then

$$D_{Kull}\left((\boldsymbol{\mu}_1,\boldsymbol{\Sigma}_1),(\boldsymbol{\mu}_2,\boldsymbol{\Sigma}_2)\right) = \tfrac{1}{2}(\boldsymbol{\mu}_1-\boldsymbol{\mu}_2)^T\boldsymbol{\Sigma}_2^{-1}(\boldsymbol{\mu}_1-\boldsymbol{\mu}_2) - \tfrac{1}{2}trace\left(\boldsymbol{\Sigma}_2^{-1}\left(\boldsymbol{\Sigma}_2-\boldsymbol{\Sigma}_1\right)\right) + \tfrac{1}{2}\log\frac{\det\left(\boldsymbol{\Sigma}_2\right)}{\det\left(\boldsymbol{\Sigma}_1\right)},$$

or

$$D_{Kull}\left((\boldsymbol{\mu}_1,\boldsymbol{\Sigma}_1),(\boldsymbol{\mu}_2,\boldsymbol{\Sigma}_2)\right) = \tfrac{1}{2}(\boldsymbol{\mu}_1-\boldsymbol{\mu}_2)^T\boldsymbol{\Sigma}_2^{-1}(\boldsymbol{\mu}_1-\boldsymbol{\mu}_2) + \tfrac{1}{2}trace\left(\boldsymbol{\Sigma}_2^{-1}\boldsymbol{\Sigma}_1-\boldsymbol{I}\right) + \tfrac{1}{2}\log\frac{\det\left(\boldsymbol{\Sigma}_2\right)}{\det\left(\boldsymbol{\Sigma}_1\right)}.$$

12. We have

$$\begin{aligned} R_\phi\left(\boldsymbol{\theta}_1,\boldsymbol{\theta}_2\right) &= H_\phi\left(\tfrac{P_{\boldsymbol{\theta}_1}+P_{\boldsymbol{\theta}_2}}{2}\right) - \tfrac{H_\phi(P_{\boldsymbol{\theta}_1})+H_\phi(P_{\boldsymbol{\theta}_2})}{2} \\ &= \int_{\mathbb{R}^d}\left(\tfrac{f_{\boldsymbol{\theta}_1}(\boldsymbol{x})+f_{\boldsymbol{\theta}_2}(\boldsymbol{x})}{2} - \left(\tfrac{f_{\boldsymbol{\theta}_1}(\boldsymbol{x})+f_{\boldsymbol{\theta}_2}(\boldsymbol{x})}{2}\right)^2\right)d\mu(\boldsymbol{x}) \\ &- \tfrac{1}{2}\int_{\mathbb{R}^d}\left(f_{\boldsymbol{\theta}_1}(\boldsymbol{x})-f_{\boldsymbol{\theta}_1}(\boldsymbol{x})^2\right)d\mu(\boldsymbol{x}) \\ &- \tfrac{1}{2}\int_{\mathbb{R}^d}\left(f_{\boldsymbol{\theta}_2}(\boldsymbol{x})-f_{\boldsymbol{\theta}_2}(\boldsymbol{x})^2\right)d\mu(\boldsymbol{x}). \end{aligned}$$

Then,

$$\begin{aligned} R_\phi\left(\boldsymbol{\theta}_1,\boldsymbol{\theta}_2\right) &= 1-\int_{\mathbb{R}^d}\left(\tfrac{f_{\boldsymbol{\theta}_1}(\boldsymbol{x})+f_{\boldsymbol{\theta}_2}(\boldsymbol{x})}{2}\right)^2 d\mu(\boldsymbol{x}) - \tfrac{1}{2} - \tfrac{1}{2} \\ &+ \tfrac{1}{2}\left(\int_{\mathbb{R}^d} f_{\boldsymbol{\theta}_1}(\boldsymbol{x})^2 d\mu(\boldsymbol{x}) + \int_{\mathbb{R}^d} f_{\boldsymbol{\theta}_2}(\boldsymbol{x})^2 d\mu(\boldsymbol{x})\right) \end{aligned}$$

and

$$R_\phi\left(\boldsymbol{\theta}_1,\boldsymbol{\theta}_2\right) = \frac{1}{4}\left(K_2(\boldsymbol{\mu}_1,\boldsymbol{\Sigma}_1)+K_2(\boldsymbol{\mu}_2,\boldsymbol{\Sigma}_2)\right) - \frac{1}{2}\int_{\mathbb{R}^d} f_{\boldsymbol{\theta}_1}(\boldsymbol{x})f_{\boldsymbol{\theta}_2}(\boldsymbol{x})d\mu(\boldsymbol{x}).$$

We know

$$K_2(\boldsymbol{\mu},\boldsymbol{\Sigma}) = \int_{\mathbb{R}^d}\left(f_{\boldsymbol{\mu},\boldsymbol{\Sigma}}\left(x_1,...,x_d\right)\right)^2 dx_1...dx_d = \pi^{-d/2}2^{-d}\det\left(\boldsymbol{\Sigma}\right)^{-1/2}.$$

Then it is only necessary to get

$$A = \int_{\mathbb{R}^d} f_{\boldsymbol{\mu}_1,\boldsymbol{\Sigma}_1}\left(x_1,...,x_d\right) f_{\boldsymbol{\mu}_2,\boldsymbol{\Sigma}_2}\left(x_1,...,x_d\right) d\boldsymbol{x},$$

i.e.,

$$\begin{aligned} A = &\int_{\mathbb{R}^d}\left((2\pi)^{-d/2}\det\left(\boldsymbol{\Sigma}_1\right)^{-1/2}\exp\left\{-\tfrac{1}{2}(\boldsymbol{x}-\boldsymbol{\mu}_1)^T\boldsymbol{\Sigma}_1^{-1}(\boldsymbol{x}-\boldsymbol{\mu}_1)\right\}\right) \\ \times &\left((2\pi)^{-d/2}\det\left(\boldsymbol{\Sigma}_2\right)^{-1/2}\exp\left\{-\tfrac{1}{2}(\boldsymbol{x}-\boldsymbol{\mu}_2)^T\boldsymbol{\Sigma}_2^{-1}(\boldsymbol{x}-\boldsymbol{\mu}_2)\right\}\right) d\boldsymbol{x}. \end{aligned}$$

Therefore,

$$A = L \int_{\mathbb{R}^d} \exp\left\{ -\frac{1}{2} \left((\boldsymbol{x}-\boldsymbol{\mu}_1)^T \boldsymbol{\Sigma}_1^{-1} (\boldsymbol{x}-\boldsymbol{\mu}_1) + (\boldsymbol{x}-\boldsymbol{\mu}_1)^T \boldsymbol{\Sigma}_1^{-1} (\boldsymbol{x}-\boldsymbol{\mu}_2) \right) \right\}$$

with

$$L = (2\pi)^{-d} \det(\boldsymbol{\Sigma}_1)^{-1/2} \det(\boldsymbol{\Sigma}_2)^{-1/2}.$$

Now we shall write the expression

$$(\boldsymbol{x}-\boldsymbol{\mu}_1)^T \boldsymbol{\Sigma}_1^{-1} (\boldsymbol{x}-\boldsymbol{\mu}_1) + (\boldsymbol{x}-\boldsymbol{\mu}_2)^T \boldsymbol{\Sigma}_2^{-1} (\boldsymbol{x}-\boldsymbol{\mu}_2)$$

as

$$(\boldsymbol{x}-\boldsymbol{\mu}^*)^T \boldsymbol{C}^{-1} (\boldsymbol{x}-\boldsymbol{\mu}^*) + \boldsymbol{B}.$$

We must get the expression of $\boldsymbol{\mu}^*, \boldsymbol{C}^{-1}$ and $\boldsymbol{B}$. We have

$$\begin{aligned} K_1 &= \boldsymbol{x}^T \boldsymbol{\Sigma}_1^{-1} \boldsymbol{x} - \boldsymbol{x}^T \boldsymbol{\Sigma}_1^{-1} \boldsymbol{\mu}_1 - \boldsymbol{\mu}_1^T \boldsymbol{\Sigma}_1^{-1} \boldsymbol{x} + \boldsymbol{\mu}_1^T \boldsymbol{\Sigma}_1^{-1} \boldsymbol{\mu}_1 \\ &+ \boldsymbol{x}^T \boldsymbol{\Sigma}_2^{-1} \boldsymbol{x} - \boldsymbol{x}^T \boldsymbol{\Sigma}_2^{-1} \boldsymbol{\mu}_2 - \boldsymbol{\mu}_2^T \boldsymbol{\Sigma}_2^{-1} \boldsymbol{x} + \boldsymbol{\mu}_2^T \boldsymbol{\Sigma}_2^{-1} \boldsymbol{\mu}_2 \\ &= \boldsymbol{x}^T (\boldsymbol{\Sigma}_1^{-1} + \boldsymbol{\Sigma}_2^{-1}) \boldsymbol{x} - \boldsymbol{x}^T (\boldsymbol{\Sigma}_1^{-1} \boldsymbol{\mu}_1 + \boldsymbol{\Sigma}_2^{-1} \boldsymbol{\mu}_2) \\ &- (\boldsymbol{\mu}_1^T \boldsymbol{\Sigma}_1^{-1} + \boldsymbol{\mu}_2 \boldsymbol{\Sigma}_2^{-1}) \boldsymbol{x} + \boldsymbol{\mu}_1^T \boldsymbol{\Sigma}_1^{-1} \boldsymbol{\mu}_1 + \boldsymbol{\mu}_2^T \boldsymbol{\Sigma}_2^{-1} \boldsymbol{\mu}_2. \end{aligned}$$

On the other hand

$$\begin{aligned} (\boldsymbol{x}-\boldsymbol{\mu}^*)^T \boldsymbol{C}^{-1} (\boldsymbol{x}-\boldsymbol{\mu}^*) + \boldsymbol{B} &= \boldsymbol{x}^T \boldsymbol{C}^{-1} \boldsymbol{x} - \boldsymbol{x}^T \boldsymbol{C}^{-1} \boldsymbol{\mu}^* \\ &- (\boldsymbol{\mu}^*)^T \boldsymbol{C}^{-1} \boldsymbol{x} + (\boldsymbol{\mu}^*)^T \boldsymbol{C}^{-1} \boldsymbol{\mu}^* + \mathbf{B}. \end{aligned}$$

Therefore

i) $\boldsymbol{C}^{-1} = \boldsymbol{\Sigma}_1^{-1} + \boldsymbol{\Sigma}_2^{-1}$

ii) $\boldsymbol{C}^{-1} \boldsymbol{\mu}^* = \boldsymbol{\Sigma}_1^{-1} \boldsymbol{\mu}_1 + \boldsymbol{\Sigma}_2^{-1} \boldsymbol{\mu}_2$

and

$$\boldsymbol{\mu}^* = (\boldsymbol{\Sigma}_1^{-1} + \boldsymbol{\Sigma}_2^{-1})^{-1} (\boldsymbol{\Sigma}_1^{-1} \boldsymbol{\mu}_1 + \boldsymbol{\Sigma}_2^{-1} \boldsymbol{\mu}_2).$$

Finally we shall get the value of $\boldsymbol{B}$

$$\begin{aligned} \boldsymbol{B} = & -(\boldsymbol{\mu}^*)^T C^{-1} \boldsymbol{\mu}^* + \boldsymbol{\mu}_1^T \boldsymbol{\Sigma}_1^{-1} \boldsymbol{\mu}_1 + \boldsymbol{\mu}^T \boldsymbol{\Sigma}_2^{-1} \boldsymbol{\mu}_2 \\ = & -(\boldsymbol{\Sigma}_1^{-1} \boldsymbol{\mu}_1 + \boldsymbol{\Sigma}_2^{-1} \boldsymbol{\mu}_2)^T (\boldsymbol{\Sigma}_1^{-1} + \boldsymbol{\Sigma}_2^{-1})^{-1} (\boldsymbol{\Sigma}_1^{-1} \boldsymbol{\mu}_1 + \boldsymbol{\Sigma}_2^{-1} \boldsymbol{\mu}_2) \\ + & \ \boldsymbol{\mu}_1^T \boldsymbol{\Sigma}_1^{-1} \boldsymbol{\mu}_1 + \boldsymbol{\mu}_2^T \boldsymbol{\Sigma}_2^{-1} \boldsymbol{\mu}_2. \end{aligned}$$

By elementary calculation, it can be obtained that

$$\begin{aligned}
\boldsymbol{B} &= -\boldsymbol{\mu}^T\boldsymbol{\Sigma}_1^{-1}(\boldsymbol{\Sigma}_1^{-1}+\boldsymbol{\Sigma}_2^{-1})^{-1}\boldsymbol{\Sigma}_1^{-1}\boldsymbol{\mu}_1 - \boldsymbol{\mu}^T\boldsymbol{\Sigma}_1^{-1}(\boldsymbol{\Sigma}_1^{-1}+\boldsymbol{\Sigma}_2^{-1})^{-1}\boldsymbol{\Sigma}_2^{-1}\boldsymbol{\mu}_2 \\
&- \boldsymbol{\mu}^T\boldsymbol{\Sigma}_2^{-1}(\boldsymbol{\Sigma}_1^{-1}+\boldsymbol{\Sigma}_2^{-1})^{-1}\boldsymbol{\Sigma}_1^{-1}\boldsymbol{\mu}_1 \\
&- \boldsymbol{\mu}^T\boldsymbol{\Sigma}_2^{-1}(\boldsymbol{\Sigma}_1^{-1}+\boldsymbol{\Sigma}_2^{-1})^{-1}\boldsymbol{\Sigma}_2^{-1}\boldsymbol{\mu}_2 + \boldsymbol{\mu}^T\boldsymbol{\Sigma}_1^{-1}\boldsymbol{\mu}_1 + \boldsymbol{\mu}^T\boldsymbol{\Sigma}_2^{-1}\boldsymbol{\mu}_2 \\
&= \boldsymbol{\mu}^T\left(\boldsymbol{\Sigma}_1^{-1} - \boldsymbol{\Sigma}_1^{-1}(\boldsymbol{\Sigma}_1^{-1}+\boldsymbol{\Sigma}_2^{-1})^{-1}\boldsymbol{\Sigma}_1^{-1}\right)\boldsymbol{\mu}_1 \\
&- \boldsymbol{\mu}^T\boldsymbol{\Sigma}_1^{-1}(\boldsymbol{\Sigma}_1^{-1}+\boldsymbol{\Sigma}_2^{-1})^{-1}\boldsymbol{\Sigma}_2^{-1}\boldsymbol{\mu}_2 \\
&- \boldsymbol{\mu}^T\boldsymbol{\Sigma}_2^{-1}(\boldsymbol{\Sigma}_1^{-1}+\boldsymbol{\Sigma}_2^{-1})^{-1}\boldsymbol{\Sigma}_1^{-1}\boldsymbol{\mu}_1 \\
&+ \boldsymbol{\mu}^T\left(\boldsymbol{\Sigma}_2^{-1} - \boldsymbol{\Sigma}_2^{-1}(\boldsymbol{\Sigma}_1^{-1}+\boldsymbol{\Sigma}_2^{-1})^{-1}\boldsymbol{\Sigma}_2^{-1}\right)\boldsymbol{\mu}_2.
\end{aligned}$$

We shall study each term of the previous expression.

a) We denote

$$\boldsymbol{S} = \boldsymbol{\Sigma}_1^{-1} - \boldsymbol{\Sigma}_1^{-1}(\boldsymbol{\Sigma}_1^{-1}+\boldsymbol{\Sigma}_2^{-1})^{-1}\boldsymbol{\Sigma}_1^{-1}.$$

We have

$$\begin{aligned}
\boldsymbol{S} &= \boldsymbol{\Sigma}_1^{-1}(\boldsymbol{I} - (\boldsymbol{\Sigma}_1^{-1}+\boldsymbol{\Sigma}_2^{-1})^{-1}\boldsymbol{\Sigma}_1^{-1}) \\
&= \boldsymbol{\Sigma}_1^{-1}\left((\boldsymbol{\Sigma}_1^{-1}+\boldsymbol{\Sigma}_2^{-1})^{-1}(\boldsymbol{\Sigma}_1^{-1}+\boldsymbol{\Sigma}_2^{-1}) - (\boldsymbol{\Sigma}_1^{-1}+\boldsymbol{\Sigma}_2^{-1})^{-1}\boldsymbol{\Sigma}_1^{-1}\right) \\
&= \boldsymbol{\Sigma}_1^{-1}(\boldsymbol{\Sigma}_1^{-1}+\boldsymbol{\Sigma}_2^{-1})^{-1}\left(\boldsymbol{\Sigma}_1^{-1}+\boldsymbol{\Sigma}_2^{-1}-\boldsymbol{\Sigma}_1^{-1}\right) \\
&= \boldsymbol{\Sigma}_1^{-1}(\boldsymbol{\Sigma}_1^{-1}+\boldsymbol{\Sigma}_2^{-1})^{-1}\boldsymbol{\Sigma}_2^{-1} \\
&= \left((\boldsymbol{\Sigma}_1^{-1}+\boldsymbol{\Sigma}_2^{-1})\boldsymbol{\Sigma}_1\right)^{-1}\boldsymbol{\Sigma}_2^{-1} \\
&= (\boldsymbol{I}+\boldsymbol{\Sigma}_2^{-1}\boldsymbol{\Sigma}_1)^{-1}\boldsymbol{\Sigma}_2^{-1} \\
&= (\boldsymbol{\Sigma}_2^{-1}\boldsymbol{\Sigma}_2+\boldsymbol{\Sigma}_2^{-1}\boldsymbol{\Sigma}_1)^{-1}\boldsymbol{\Sigma}_2^{-1} \\
&= (\boldsymbol{\Sigma}_2^{-1}(\boldsymbol{\Sigma}_2+\boldsymbol{\Sigma}_1))^{-1}\boldsymbol{\Sigma}_2^{-1} \\
&= (\boldsymbol{\Sigma}_2+\boldsymbol{\Sigma}_1)^{-1}\boldsymbol{\Sigma}_2\boldsymbol{\Sigma}_2^{-1} \\
&= (\boldsymbol{\Sigma}_2+\boldsymbol{\Sigma}_1)^{-1}.
\end{aligned}$$

Similarly

b) $\boldsymbol{\Sigma}_2^{-1} - \boldsymbol{\Sigma}_2^{-1}(\boldsymbol{\Sigma}_1^{-1}+\boldsymbol{\Sigma}_2^{-1})^{-1}\boldsymbol{\Sigma}_2^{-1} = (\boldsymbol{\Sigma}_1+\boldsymbol{\Sigma}_2)^{-1}$.

c) Also we have

$$\begin{aligned}
\boldsymbol{\Sigma}_1^{-1}(\boldsymbol{\Sigma}_1^{-1}+\boldsymbol{\Sigma}_2^{-1})^{-1}\boldsymbol{\Sigma}_2^{-1} &= ((\boldsymbol{\Sigma}_1^{-1}+\boldsymbol{\Sigma}_2^{-1})\boldsymbol{\Sigma}_1)^{-1}\boldsymbol{\Sigma}_2^{-1} \\
&= (\boldsymbol{\Sigma}_1^{-1}\boldsymbol{\Sigma}_1+\boldsymbol{\Sigma}_2^{-1}\boldsymbol{\Sigma}_1)^{-1}\boldsymbol{\Sigma}_2^{-1} \\
&= (\mathbf{I}+\boldsymbol{\Sigma}_2^{-1}\boldsymbol{\Sigma}_1)^{-1}\boldsymbol{\Sigma}_2^{-1} \\
&= (\boldsymbol{\Sigma}_1+\boldsymbol{\Sigma}_2)^{-1}.
\end{aligned}$$

d) $\boldsymbol{\Sigma}_2^{-1}(\boldsymbol{\Sigma}_1^{-1}+\boldsymbol{\Sigma}_2^{-1})^{-1}\boldsymbol{\Sigma}_1^{-1} = (\boldsymbol{\Sigma}_1+\boldsymbol{\Sigma}_2)^{-1}$.

Therefore

$$\boldsymbol{B} = (\boldsymbol{\mu}_1-\boldsymbol{\mu}_2)^T(\boldsymbol{\Sigma}_2+\boldsymbol{\Sigma}_1)^{-1}(\boldsymbol{\mu}_1-\boldsymbol{\mu}_2)$$

and A is given by

$$\begin{aligned} A = & \frac{\det(\boldsymbol{\Sigma}_1)^{-1/2}\det(\boldsymbol{\Sigma}_2)^{-1/2}}{(2\pi)^d}\exp\left((\boldsymbol{\mu}_1-\boldsymbol{\mu}_2)^T(\boldsymbol{\Sigma}_2+\boldsymbol{\Sigma}_1)^{-1}(\boldsymbol{\mu}_1-\boldsymbol{\mu}_2)\right) \\ \times & \int_{\mathbb{R}^d}\exp\left(-\tfrac{1}{2}(\boldsymbol{x}-\boldsymbol{\mu}^*)^T\left(\boldsymbol{\Sigma}_1^{-1}+\boldsymbol{\Sigma}_2^{-1}\right)(\boldsymbol{x}-\boldsymbol{\mu}^*)\right)d\boldsymbol{x}, \end{aligned}$$

i.e.,

$$\begin{aligned} A = & \frac{\det(\boldsymbol{\Sigma}_1)^{-1/2}\det(\boldsymbol{\Sigma}_2)^{-1/2}}{(2\pi)^{d/2}}\exp\left((\boldsymbol{\mu}_1-\boldsymbol{\mu}_2)^T(\boldsymbol{\Sigma}_2+\boldsymbol{\Sigma}_1)^{-1}(\boldsymbol{\mu}_1-\boldsymbol{\mu}_2)\right) \\ \times & \det\left(\boldsymbol{\Sigma}_{11}^{-1}+\boldsymbol{\Sigma}_2^{-1}\right)^{-1/2}\int_{\mathbb{R}^d} f_{\mu^*,(\boldsymbol{\Sigma}_1^{-1}+\boldsymbol{\Sigma}_2^{-1})}(x_1,...,x_d)\,d\boldsymbol{x} \end{aligned}$$

and then

$$A = \frac{\det(\boldsymbol{\Sigma}_1)^{-1/2}\det(\boldsymbol{\Sigma}_2)^{-1/2}}{(2\pi)^{d/2}\det\left(\boldsymbol{\Sigma}_1^{-1}+\boldsymbol{\Sigma}_2^{-1}\right)^{1/2}}\exp\left((\boldsymbol{\mu}_1-\boldsymbol{\mu}_2)^T(\boldsymbol{\Sigma}_2+\boldsymbol{\Sigma}_1)^{-1}(\boldsymbol{\mu}_1-\boldsymbol{\mu}_2)\right).$$

Finally, we have

$$\begin{aligned} R_\phi(\boldsymbol{\theta}_1,\boldsymbol{\theta}_2) = & \tfrac{1}{4}\left(\pi^{-d/2}2^{-d}\left(\det(\boldsymbol{\Sigma}_1)^{-1/2}+\det(\boldsymbol{\Sigma}_2)^{-1/2}\right)\right) \\ - & \tfrac{1}{2}(2\pi)^{-d/2}\frac{\det(\boldsymbol{\Sigma}_1)^{-1/2}\det(\boldsymbol{\Sigma}_2)^{-1/2}}{\det\left(\boldsymbol{\Sigma}_1^{-1}+\boldsymbol{\Sigma}_2^{-1}\right)^{1/2}} \\ \times & \exp\left((\boldsymbol{\mu}_1-\boldsymbol{\mu}_2)^T(\boldsymbol{\Sigma}_2+\boldsymbol{\Sigma}_1)^{-1}(\boldsymbol{\mu}_1-\boldsymbol{\mu}_2)\right). \end{aligned}$$

In the case of two univariate populations we have

$$\begin{aligned} R_\phi((\mu_1,\sigma_1),(\mu_2,\sigma_2)) = & \tfrac{1}{8\pi^{1/2}}\left(\frac{1}{\sigma_1}+\frac{1}{\sigma_2}\right)-\frac{1}{2\left(2\pi\left(\sigma_1^2+\sigma_2^2\right)\right)^{1/2}} \\ \times & \exp\left(\frac{\mu_1^2+\mu_2^2-2\mu_1\mu_2}{\sigma_1^2+\sigma_2^2}\right). \end{aligned}$$

13. We consider Rényi's divergence for $r = 1/2$.

$$D^1_{r=\frac{1}{2}}(\boldsymbol{\theta}_1,\boldsymbol{\theta}_2) = \frac{1}{\left(-\frac{1}{2}\right)\frac{1}{2}}\log\int_{\mathcal{X}}\sqrt{f_{\boldsymbol{\theta}_1}(\boldsymbol{x})}\sqrt{f_{\boldsymbol{\theta}_2}(\boldsymbol{x})}d\mu(\boldsymbol{x}) = 4B(\boldsymbol{\theta}_1,\boldsymbol{\theta}_2)$$

with $\boldsymbol{\theta}_1 = (\mu_1,\sigma_1)$ and $\boldsymbol{\theta}_2 = (\mu_2,\sigma_2)$.

Then we have

$$\begin{aligned} B((\mu_1,\sigma_1),(\mu_2,\sigma_2)) = & \tfrac{1}{4}D^1_{r=\frac{1}{2}}((\mu_1,\sigma_1),(\mu_2,\sigma_2)) \\ = & \tfrac{1}{4}\frac{(\mu_1-\mu_2)^2}{\sigma_1^2+\sigma_2^2}+\tfrac{1}{2}\log\frac{\sigma_1^2+\sigma_2^2}{2\sigma_1\sigma_2}. \end{aligned}$$

14. First we establish that D^{He} defines a metric.

In fact,

i) $D^{He}(\boldsymbol{\theta}_1, \boldsymbol{\theta}_2) = 0$ if and only if $f_{\boldsymbol{\theta}_1}(\boldsymbol{x}) = f_{\boldsymbol{\theta}_2}(\boldsymbol{x})$ a.s.

ii) The property $D^{He}(\boldsymbol{\theta}_1, \boldsymbol{\theta}_2) = D^{He}(\boldsymbol{\theta}_2, \boldsymbol{\theta}_1)$ is trivial.

iii) Applying the Minkowski's inequality for $p = 2$ to the measurable functions

$$\sqrt{f_{\boldsymbol{\theta}_1}(\boldsymbol{x})} - \sqrt{f_{\boldsymbol{\theta}_2}(\boldsymbol{x})} \text{ and } \sqrt{f_{\boldsymbol{\theta}_2}(\boldsymbol{x})} - \sqrt{f_{\boldsymbol{\theta}_3}(\boldsymbol{x})}$$

we have, denoting

$$l = \left(\int_{\mathcal{X}} \left(\sqrt{f_{\boldsymbol{\theta}_1}(\boldsymbol{x})} - \sqrt{f_{\boldsymbol{\theta}_3}(\boldsymbol{x})} \right)^2 d\mu(\boldsymbol{x}) \right)^{1/2},$$

that

$$\begin{aligned} l \leq & \left(\int_{\mathcal{X}} \left(\sqrt{f_{\boldsymbol{\theta}_1}(\boldsymbol{x})} - \sqrt{f_{\boldsymbol{\theta}_2}(\boldsymbol{x})} \right)^2 d\mu(\boldsymbol{x}) \right)^{1/2} \\ + & \left(\int_{\mathcal{X}} \left(\sqrt{f_{\boldsymbol{\theta}_2}(\boldsymbol{x})} - \sqrt{f_{\boldsymbol{\theta}_3}(\boldsymbol{x})} \right)^2 d\mu(\boldsymbol{x}) \right)^{1/2}, \end{aligned}$$

i.e.,

$$D^{He}(\boldsymbol{\theta}_1, \boldsymbol{\theta}_3) \leq D^{He}(\boldsymbol{\theta}_1, \boldsymbol{\theta}_2) + D^{He}(\boldsymbol{\theta}_2, \boldsymbol{\theta}_3).$$

Now we shall see the relation between $D^{He}(\boldsymbol{\theta}_1, \boldsymbol{\theta}_2)$ and $K_r^*(\boldsymbol{\theta}_1, \boldsymbol{\theta}_2)$. We have

$$\begin{aligned} D^{He}(\boldsymbol{\theta}_1, \boldsymbol{\theta}_2) &= \left(\int_{\mathcal{X}} \left(\sqrt{f_{\boldsymbol{\theta}_1}(\boldsymbol{x})} - \sqrt{f_{\boldsymbol{\theta}_2}(\boldsymbol{x})} \right)^2 d\mu(\boldsymbol{x}) \right)^{1/2} \\ &= \left(2 - 2 \int_{\mathcal{X}} \sqrt{f_{\boldsymbol{\theta}_1}(\boldsymbol{x})} \sqrt{f_{\boldsymbol{\theta}_2}(\boldsymbol{x})} d\mu(\boldsymbol{x}) \right)^{1/2} \\ &= \left(2 \left(1 - K_{\frac{1}{2}}^*(\boldsymbol{\theta}_1, \boldsymbol{\theta}_2) \right) \right)^{1/2}. \end{aligned}$$

Now the expression between two normal populations can be obtained from Exercise 11 in which we have derived the expression of $K_r^*(\boldsymbol{\theta}_1, \boldsymbol{\theta}_2)$.

15. First we calculate

$$\begin{aligned} K_r^*(\theta_1,\theta_2) &= \sum_{x=0}^{\infty}\left(\frac{e^{-\theta_1}\theta_1^x}{x!}\right)^r\left(\frac{e^{-\theta_2}\theta_2^x}{x!}\right)^{1-r} \\ &= \exp\left(-\theta_1 r+(r-1)\theta_2\right)\exp\left(\frac{\theta_1^r}{\theta_2^{r-1}}\right), \end{aligned}$$

then

$$D_r^1(\theta_1,\theta_2)=\frac{1}{r(r-1)}\log K_r^*(\theta_1,\theta_2)=\frac{1}{r(r-1)}\left(\frac{\theta_1^r}{\theta_2^{r-1}}-\theta_1 r+(r-1)\theta_2\right)$$

and

$$D_{Kull}(\theta_1,\theta_2)=\lim_{r\to 1}D_r^1(\theta_1,\theta_2)=\theta_1\log\frac{\theta_1}{\theta_2}+(\theta_2-\theta_1).$$

16. *a)* We know that

$$\begin{aligned} H(\boldsymbol{X}) &= \frac{1}{2}\log\left((2\pi e)^d\det(\boldsymbol{A})\right) \\ H(\boldsymbol{Y}) &= \frac{1}{2}\log\left((2\pi e)^l\det(\boldsymbol{B})\right) \\ H(\boldsymbol{X},\boldsymbol{Y}) &= \frac{1}{2}\log\left((2\pi e)^{d+l}\det(\boldsymbol{C})\right). \end{aligned}$$

Then,

$$\begin{aligned} H(\boldsymbol{X})-H(\boldsymbol{X}/\boldsymbol{Y}) &= H(\boldsymbol{X})+H(\boldsymbol{Y})-H(\boldsymbol{X},\boldsymbol{Y}) \\ &= \frac{1}{2}\log\left((2\pi e)^d\det(\boldsymbol{A})\right)+\log\left((2\pi e)^l\det(\boldsymbol{B})\right) \\ &- \log\left((2\pi e)^{d+l}\det(\boldsymbol{C})\right) \\ &= \frac{1}{2}\log\frac{\det(\boldsymbol{A})\det(\boldsymbol{B})}{\det(\boldsymbol{C})}. \end{aligned}$$

b) Denoting by σ^2 and τ^2 the variances of X and Y respectively and by $\rho=\rho(X,Y)$, we have

$$\boldsymbol{C}=\begin{pmatrix}\sigma^2 & \rho\sigma\tau \\ \rho\sigma\tau & \tau^2\end{pmatrix},$$

then $\det(\boldsymbol{A})=\sigma^2$, $\det(\boldsymbol{B})=\tau^2$, $\det(\boldsymbol{C})=\sigma^2\tau^2\left(1-\rho^2\right)$ and the result is obtained.

17. We know

$$D_{Kull}\left(N(\boldsymbol{\mu}_1, \boldsymbol{\Sigma}_1), N(\boldsymbol{\mu}_2, \boldsymbol{\Sigma}_2)\right) = -H\left(N(\boldsymbol{\mu}_1, \boldsymbol{\Sigma}_1)\right) - \int_{\mathbb{R}^d} f(\boldsymbol{x}) \log g(\boldsymbol{x}) d\boldsymbol{x}$$

$$D_{Kull}\left(\mathbf{Z}, N(\boldsymbol{\mu}_2, \boldsymbol{\Sigma}_2)\right) = -H\left(\mathbf{Z}\right) - \int_{\mathbb{R}^d} t(\boldsymbol{x}) \log g(\boldsymbol{x}) d\boldsymbol{x}$$

where $f(\boldsymbol{x})$, $g(\boldsymbol{x})$ and $t(\boldsymbol{x})$ are the probability density functions associated with $\boldsymbol{X}$, $\boldsymbol{Y}$ and $\boldsymbol{Z}$, respectively.

By Exercise 10

$$-\int_{\mathbb{R}^d} t(\boldsymbol{x}) \log g(\boldsymbol{x}) d\boldsymbol{x} = \frac{1}{2} \log\left((2\pi e)^d \left|\boldsymbol{\Sigma}_2\right|\right) = -\int_{\mathbb{R}^d} f(\boldsymbol{x}) \log g(\boldsymbol{x}) d\boldsymbol{x},$$

and

$$-H\left(N(\boldsymbol{\mu}_1, \boldsymbol{\Sigma}_1)\right) \leq -H\left(\boldsymbol{Z}\right).$$

Therefore

$$D_{Kull}\left(N(\boldsymbol{\mu}_1, \boldsymbol{\Sigma}_1), N(\boldsymbol{\mu}_2, \boldsymbol{\Sigma}_2)\right) \leq D_{Kull}\left(\mathbf{Z}, N(\boldsymbol{\mu}_2, \boldsymbol{\Sigma}_2)\right).$$

2

Entropy as a Measure of Diversity: Sampling Distributions

2.1. Introduction

The term diversity is usually synonymous of "variety" and is simply an indication of the number of different ways a characteristic is present in a group of elements, taking in account the total of elements with each value of the characteristic. For example, we often speak of a "diversity of opinions". While simply accounting for the number of different types of opinions on a topic one can give a rough idea of the "diversity of opinions;" the total of people with the same opinion must be taken into account to get the true sense of the diversity.

The concept of diversity appears in a great number of research areas: ecology, biology, genetics, economics, linguistics, etc. It is in some sense the degree of heterogeneity of the individuals with respect to characteristics under study.

If diversity is defined as "the presence of a great number of different types of industries in a geographical area" (Economics) or "the linguistic differences between the inhabitants of neighboring regions" (Linguistics) or "the number of species in a place as well as the abundance of those species" (Biology), then it would be useful to have a summary statistic to describe the diversity of a characteristic in an area and compare it to that of other areas.

A diversity measure should satisfy certain intuitive conditions which are satisfied by an entropy measure. Later this point will be clarified. Shannon's entropy measure as well as Gini-Simpson index (The expected distance between two individuals drawn at random when the distance is defined as zero if they belong to the same category and unity otherwise, see Exercise 1 of this chapter) have been used as indexes of diversity. We can observe in Exercise 1 that Gini-Simpson index is the ϕ_α-entropy of Havrda and Charvat with $\alpha = 2$ and sometimes is called quadratic entropy. In general, entropy measures can be used as indexes of diversity.

In the rest of the chapter we shall assume the concept of measure of diversity given by Rao (1982a,b). A measure of diversity I is a nonnegative real-valued function defined on the space of probability distributions which reflects the differences between the individuals within a population. Since we are mainly interested in categorical data, we consider the space of the multinomial distributions. We consider a finite population, Π, with N elements that could be classified into M categories or classes $C_1, ..., C_M$ in accordance with a classification process, C. Let $\mathcal{X} = \{C_1, ..., C_M\}$ be the set of the M categories and

$$\Delta_M = \left\{ \boldsymbol{p} = (p_1, ..., p_M)^T : p_i \geq 0, i = 1, ..., M, \ \sum_{i=1}^{M} p_i = 1 \right\}$$

the convex set of probability measures defined on $\mathcal{X}$. A function $I(.)$ mapping Δ_M into the real line is said to be a measure of diversity if it satisfies the following conditions:

i) $I(\boldsymbol{p}) \geq 0, \forall \boldsymbol{p} \in \Delta_M$ and $I(\boldsymbol{p}) = 0$ if and only if $\boldsymbol{p}$ is degenerate.

ii) I is a concave function on Δ_M.

We shall refer to $I(\boldsymbol{p})$ as the diversity within a population Π characterized by the probability distribution $\boldsymbol{p}$.

The condition *i)* is a natural one since a measure of diversity should be nonnegative and should be value zero when all the individuals of a population are identical in accordance with the classification process considered, i.e., when the associated probability measure is concentrated on a particular point of $\mathcal{X}$. The condition *ii)* is motivated by the consideration that the diversity in a mixture of populations should not be smaller than the average of the diversities within individual populations.

From a historical point of view the two most widely used diversity measures are Gini-Simpson index given by

$$H_{GS}(\boldsymbol{p}) = 1 - \sum_{i=1}^{M} p_i^2,$$

and Shannon's entropy given by

$$H(\boldsymbol{p}) = -\sum_{i=1}^{M} p_i \log p_i.$$

Gini-Simpson index was introduced by Gini (1912) and by Simpson (1949) in biological works. Its properties have been studied by various authors, Lieberson (1969), Light and Margolin (1971), Nei (1973), Bhargava and Doyle (1974), Bhargava and Uppuluri (1975), Agresti and Agresti (1978) and Patil and Taille (1982). Rao (1982a,b) gave a characterization of this index. Other references can be obtained from these papers. Regarding Shannon's entropy, some applications of this measure in diversity can be seen in Lewontin (1972) and Pielou (1967, 1975).

Rao (1982a,b), Burbea and Rao (1982a, 1982b) and Nayak (1983, 1985, 1986) investigated the possibility of using other entropy functions as diversity measures. Pardo, J. A. et *al.* (1992), Pardo, L. et *al.* (1992) and Salicrú et *al.* (1993) studied the behavior of the (h,ϕ)-entropies as diversity measures. Disregarding the work context, it is usually very difficult or excessively costly to dispose of a census information (due to population size), so that it is essential to be able to obtain diversity measurement estimates by means of a sample. In this chapter we consider entropy estimates based on samples from unknown populations. So it would be of interest to study stochastic behavior of those estimates. We also consider the natural estimates by replacing p_i's by their maximum likelihood estimators and we derive their asymptotic distributions. We consider this problem for general populations and then we get the corresponding results in multinomial populations as a particular case.

Let $(\mathcal{X}, \beta_{\mathcal{X}}, P_{\boldsymbol{\theta}})_{\boldsymbol{\theta}\in\Theta}$ be the statistical space associated with the random variable $\boldsymbol{X}$, where $\beta_{\mathcal{X}}$ is the σ-field of Borel subsets $A \subset \mathcal{X}$ and $\{P_{\boldsymbol{\theta}}\}_{\boldsymbol{\theta}\in\Theta}$ is a family of probability distributions on the measurable space $(\mathcal{X}, \beta_{\mathcal{X}})$ with Θ an open subset of $\mathbb{R}^{M_0}$, $M_0 \geq 1$. We assume that the probability distributions $P_{\boldsymbol{\theta}}$ are absolutely continuous with respect to a σ-finite measure μ on $(\mathcal{X}, \beta_{\mathcal{X}})$. For simplicity μ is either the Lebesgue measure or a counting measure. We shall obtain the asymptotic distribution of the statistic

$$H^{\phi}(\widehat{\boldsymbol{\theta}}) = \int_{\mathcal{X}} \phi(f_{\widehat{\boldsymbol{\theta}}}(\boldsymbol{x}))d\mu(\boldsymbol{x}),$$

where $\widehat{\boldsymbol{\theta}}$ is the maximum likelihood estimator of $\boldsymbol{\theta}$, being $\phi : [0, \infty) \rightarrow \mathbb{R}$ a concave function, i.e., we work with ϕ-entropies but the results will be extended in a easy way to (h, ϕ)-entropies.

From a historical point of view, the asymptotic behavior of the entropy measures was first studied in multinomial populations and then in general populations. Basharin (1959) gave the asymptotic mean of Shannon's entropy, Lyons and Hutcheson (1979) obtained exact expression for the first four moments of Gini-Simpson index, Bhargava and Uppuluri (1975), for this index, gave the exact distribution for small sample sizes and few classes. Nayak (1985) obtained the asymptotic distribution of the ϕ-entropies in random sampling and Salicrú et *al.* (1993) studied the same problem for the (h, ϕ)-entropies, either in random sampling or in stratified random sampling. Finally, Pardo, L. et *al.* (1997a) studied the problem in general populations.

2.2. Phi-entropies. Asymptotic Distribution

We assume that the statistical space $(\mathcal{X}, \beta_{\mathcal{X}}, P_{\boldsymbol{\theta}})_{\boldsymbol{\theta} \in \Theta}$ satisfies the standard regularity assumptions considered in the parametric asymptotic statistics theory:

i) For all $\boldsymbol{\theta}_1 \neq \boldsymbol{\theta}_2 \in \Theta \subset \mathbb{R}^{M_0}$

$$\mu(\{\boldsymbol{x} \in \mathcal{X} : f_{\boldsymbol{\theta}_1}(\boldsymbol{x}) \neq f_{\boldsymbol{\theta}_2}(\boldsymbol{x})\}) > 0.$$

ii) The set $S_{\mathcal{X}} = \{\boldsymbol{x} \in \mathcal{X}\text{: } f_{\boldsymbol{\theta}}(\boldsymbol{x}) > 0\}$ is independent of $\boldsymbol{\theta}$.

iii) The first, second and third partial derivatives

$$\frac{\partial f_{\boldsymbol{\theta}}(\boldsymbol{x})}{\partial \theta_i}, \frac{\partial^2 f_{\boldsymbol{\theta}}(\boldsymbol{x})}{\partial \theta_i \partial \theta_j}, \frac{\partial^3 f_{\boldsymbol{\theta}}(\boldsymbol{x})}{\partial \theta_i \partial \theta_j \partial \theta_k}, \quad i, j, k = 1, ..., M_0$$

exist everywhere for all $1 \leq i, j, k \leq M_0$.

iv) The first, second and third partial derivatives of $f_{\boldsymbol{\theta}}(\boldsymbol{x})$ with respect to $\boldsymbol{\theta}$ are absolutely bounded by functions α, β and γ with finite integrals

$$\int_{\mathcal{X}} \alpha(\boldsymbol{x})d\mu(\boldsymbol{x}) < \infty, \int_{\mathcal{X}} \beta(\boldsymbol{x})d\mu(\boldsymbol{x}) < \infty \text{ and } \int_{\mathcal{X}} \gamma(\boldsymbol{x})f_{\boldsymbol{\theta}}(\boldsymbol{x})d\mu(\boldsymbol{x}) < \infty.$$

v) For each $\boldsymbol{\theta} \in \Theta$, the Fisher information matrix

$$\mathcal{I}_{\mathcal{F}}(\boldsymbol{\theta}) = \left(\int_{\mathcal{X}} \frac{\partial \log f_{\boldsymbol{\theta}}(\boldsymbol{x})}{\partial \theta_i} \frac{\partial \log f_{\boldsymbol{\theta}}(\boldsymbol{x})}{\partial \theta_j} f_{\boldsymbol{\theta}}(\boldsymbol{x}) d\mu(\boldsymbol{x}) \right)_{i,j=1,\dots,M_0}$$

exists and is positive definite, with elements continuous in the variable $\boldsymbol{\theta}$.

In order to simplify some proofs we shall introduce the notation $o(.)$, $O(.)$, $o_P(.)$ and $O_P(.)$. Sections 14.2-14.4 in Bishop et *al.* (1975) present a detailed study of them.

Given two real number sequences $\{x_n\}_{n \in \mathbb{N}}$ and $\{y_n\}_{n \in \mathbb{N}}$ we say that $x_n = o(y_n)$ (x_n is little o of y_n) as $n \to \infty$ if $x_n/y_n \to 0$ and we say that $x_n = O(y_n)$ (x_n is big O of y_n) as $n \to \infty$ if $|x_n/y_n|$ is bounded.

If we consider vectors $\boldsymbol{x}_n = (x_{n1}, ..., x_{nk})$ the notation $\boldsymbol{x}_n = o(y_n)$ means $\|\boldsymbol{x}_n\| = o(y_n)$ where $\|\boldsymbol{x}_n\|^2 = \boldsymbol{x}_n^T \boldsymbol{x}_n$ and $\boldsymbol{x}_n = O(y_n)$ means $\|\boldsymbol{x}_n\| = O(y_n)$.

Given a sequence of random variables $\{X_n\}_{n \in \mathbb{N}}$ and a sequence of real numbers $\{y_n\}$ we say that $X_n = o_P(y_n)$ as $n \to \infty$ if $X_n/y_n \xrightarrow{P} 0$ and we say that $X_n = O_P(y_n)$ as $n \to \infty$ if X_n/y_n is bounded in probability.

Given a sequence of random vectors $\{\boldsymbol{X}_n\}_{n \in \mathbb{N}}$, where $\boldsymbol{X}_n = (X_{n1}, ..., X_{nk})$, we say that $\boldsymbol{X}_n = o_P(y_n)$ if $\|\boldsymbol{X}_n\| = o_P(y_n)$ and $\boldsymbol{X}_n = O_P(y_n)$ if $\|\boldsymbol{X}_n\| = O_P(y_n)$.

We present without proof some of the most important relations among them:

a) $O(x_n)O(y_n) = O(x_n y_n)$

b) $O(x_n)o(y_n) = o(x_n y_n)$

c) $o(x_n)o(y_n) = o(x_n y_n)$

d) $O(x_n) = O(cx_n)$, $c \neq 0$

e) $o(1) + O(n^{-1/2}) + O(n^{-1}) = o(1)$

f) $O_P(x_n)O_P(y_n) = O_P(x_n y_n)$

g) $O_P(x_n)o_P(y_n) = o_P(x_n y_n)$

h) $o_P(x_n)o_P(y_n) = o_P(x_n y_n)$

i) If $X_n \xrightarrow{L} X \Rightarrow X_n = O_P(1)$

j) $O_P\left(O(\sqrt{n})\right) = O_P(\sqrt{n})$ and $o\left(O_P(x_n)\right) = o_P(x_n)$

k) If $X_n \xrightarrow{L} X \Rightarrow X_n + o_P(1) \xrightarrow{L} X$

The following result was obtained in Pardo, L. et *al.* (1997).

Theorem 2.1

Let $\widehat{\boldsymbol{\theta}}$ *be the maximum likelihood estimator of* $\boldsymbol{\theta}$. *Suppose that i)-v) hold and that, in addition,* $\phi \in C^1([0,\infty))$ *and there exist a measurable and* μ*-integrable function* $F(\boldsymbol{x})$ *such that*

$$\left|\phi'(f_{\boldsymbol{\theta}}(\boldsymbol{x}))\frac{\partial f_{\boldsymbol{\theta}}(\boldsymbol{x})}{\partial \theta_i}\right| < F(\boldsymbol{x}), \quad i = 1, ..., M_0.$$

Then

$$\sqrt{n}\left(H^{\phi}(\widehat{\boldsymbol{\theta}}) - H^{\phi}(\boldsymbol{\theta})\right) \xrightarrow[n\to\infty]{L} N\left(0, \sigma_{\phi}^2(\boldsymbol{\theta})\right),$$

provided $\sigma_{\phi}^2(\boldsymbol{\theta}) > 0$, *where*

$$\sigma_{\phi}^2(\boldsymbol{\theta}) = \boldsymbol{T}^T \mathcal{I}_{\mathcal{F}}(\boldsymbol{\theta})^{-1} \boldsymbol{T}, \tag{2.1}$$

with $\boldsymbol{T}^T = (t_1, ..., t_{M_0})$ *and*

$$t_i = \int_{\mathcal{X}} \phi'(f_{\boldsymbol{\theta}}(\boldsymbol{x}))\frac{\partial f_{\boldsymbol{\theta}}(\boldsymbol{x})}{\partial \theta_i} d\mu(\boldsymbol{x}), \; i = 1, ..., M_0.$$

Proof. The first order Taylor expansion of $H^{\phi}(\widehat{\boldsymbol{\theta}})$ around $\boldsymbol{\theta}$ gives

$$H^{\phi}(\widehat{\boldsymbol{\theta}}) = H^{\phi}(\boldsymbol{\theta}) + \sum_{i=1}^{M_0} t_i(\widehat{\theta}_i - \theta_i) + o\left(\left\|\widehat{\boldsymbol{\theta}} - \boldsymbol{\theta}\right\|\right)$$

being

$$t_i = \frac{\partial H^{\phi}(\boldsymbol{\theta})}{\partial \theta_i} = \int_{\mathcal{X}} \phi'(f_{\boldsymbol{\theta}}(\boldsymbol{x}))\frac{\partial f_{\boldsymbol{\theta}}(\boldsymbol{x})}{\partial \theta_i} d\mu(\boldsymbol{x}), \; i = 1, ..., M_0$$

and $\left\|\widehat{\boldsymbol{\theta}} - \boldsymbol{\theta}\right\|^2 = (\widehat{\boldsymbol{\theta}} - \boldsymbol{\theta})^T(\widehat{\boldsymbol{\theta}} - \boldsymbol{\theta})$.

Since

$$\sqrt{n}(\widehat{\boldsymbol{\theta}} - \boldsymbol{\theta}) \xrightarrow[n\to\infty]{L} N\left(\boldsymbol{0}, \mathcal{I}_{\mathcal{F}}(\boldsymbol{\theta})^{-1}\right), \tag{2.2}$$

then $\sqrt{n}\, o\left(\left\|\widehat{\boldsymbol{\theta}}-\boldsymbol{\theta}\right\|\right)=\sqrt{n}\, o\left(O_P\left(n^{-1/2}\right)\right)=o_P\left(1\right).$

Therefore, the random variables $\sqrt{n}\left(H^{\phi}(\widehat{\boldsymbol{\theta}})-H^{\phi}\left(\boldsymbol{\theta}\right)\right)$ and $\sqrt{n}\boldsymbol{T}^T(\widehat{\boldsymbol{\theta}}-\boldsymbol{\theta})$ have the same asymptotic distribution. By (2.2) we have

$$\sqrt{n}\boldsymbol{T}^T(\widehat{\boldsymbol{\theta}}-\boldsymbol{\theta}) \underset{n\to\infty}{\overset{L}{\longrightarrow}} N\left(0,\boldsymbol{T}^T\mathcal{I}_{\mathcal{F}}\left(\boldsymbol{\theta}\right)^{-1}\boldsymbol{T}\right).$$

This completes the proof. ∎

Remark 2.1

If we consider Shannon's entropy, i.e., $\phi(x)=-x\log x$, *we obtain*

$$t_i=-\int_{\mathcal{X}}\log f_{\boldsymbol{\theta}}(\boldsymbol{x})\frac{\partial f_{\boldsymbol{\theta}}(\boldsymbol{x})}{\partial\theta_i}d\mu(\boldsymbol{x}),\ i=1,2,...,M_0.$$

Corollary 2.1

We consider the (h,ϕ)*-entropies defined in (1.22), then we have*

$$\frac{\sqrt{n}\left(H_h^{\phi}(\widehat{\boldsymbol{\theta}})-H_h^{\phi}\left(\boldsymbol{\theta}\right)\right)}{h'\left(\int_{\mathcal{X}}\phi(f_{\boldsymbol{\theta}}(\boldsymbol{x}))d\mu(\boldsymbol{x})\right)} \underset{n\to\infty}{\overset{L}{\longrightarrow}} N\left(0,\boldsymbol{T}^T\mathcal{I}_{\mathcal{F}}\left(\boldsymbol{\theta}\right)^{-1}\boldsymbol{T}\right),$$

where $\boldsymbol{T}$ *is given in Theorem 2.1.*

Proof. A first order Taylor expansion of $h(y)$ around $y=y_0$ at $y=\widehat{y}$ gives

$$h(\widehat{y})=h(y_0)+h'(y_0)(\widehat{y}-y_0)+o(\widehat{y}-y_0).$$

Now for $y_0=H^{\phi}\left(\boldsymbol{\theta}\right)$ and $\widehat{y}=H^{\phi}(\widehat{\boldsymbol{\theta}})$, we get

$$\begin{aligned} H_h^{\phi}(\widehat{\boldsymbol{\theta}}) &= H_h^{\phi}\left(\boldsymbol{\theta}\right)+h'\left(\int_{\mathcal{X}}\phi(f_{\boldsymbol{\theta}}(\boldsymbol{x}))d\mu(\boldsymbol{x})\right)\left(H^{\phi}(\widehat{\boldsymbol{\theta}})-H^{\phi}\left(\boldsymbol{\theta}\right)\right) \\ &+ o\left(H^{\phi}(\widehat{\boldsymbol{\theta}})-H^{\phi}\left(\boldsymbol{\theta}\right)\right), \end{aligned}$$

and hence

$$\sqrt{n}\left(H_h^{\phi}(\widehat{\boldsymbol{\theta}})-H_h^{\phi}\left(\boldsymbol{\theta}\right)\right)=\sqrt{n}\,h'\left(\int_{\mathcal{X}}\phi(f_{\boldsymbol{\theta}}(\boldsymbol{x}))d\mu(\boldsymbol{x})\right)\left(H^{\phi}(\widehat{\boldsymbol{\theta}})-H^{\phi}\left(\boldsymbol{\theta}\right)\right)+o_P\left(1\right),$$

because $\sqrt{n}\, o\left(H^{\phi}(\widehat{\boldsymbol{\theta}})-H^{\phi}\left(\boldsymbol{\theta}\right)\right)=o_P\left(1\right).$ Therefore,

$$\sqrt{n}\left(H_h^{\phi}(\widehat{\boldsymbol{\theta}})-H_h^{\phi}\left(\boldsymbol{\theta}\right)\right) \underset{n\to\infty}{\overset{L}{\longrightarrow}} N\left(0,h'\left(\int_{\mathcal{X}}\phi(f_{\boldsymbol{\theta}}(\boldsymbol{x}))d\mu(\boldsymbol{x})\right)^2\boldsymbol{T}^T\mathcal{I}_{\mathcal{F}}\left(\boldsymbol{\theta}\right)^{-1}\boldsymbol{T}\right).$$ ∎

Remark 2.2

Under the assumptions of Theorem 2.1, $S_n = n^{1/2}\boldsymbol{T}^T(\widehat{\boldsymbol{\theta}} - \boldsymbol{\theta}) = 0$ *a.s.* $\forall n \in \mathbb{N}$ *iff* $\sigma_\phi^2(\boldsymbol{\theta}) = \boldsymbol{T}^T\mathcal{I}_{\mathcal{F}}(\boldsymbol{\theta})^{-1}\boldsymbol{T} = 0$.

In fact, if $S_n = 0$ *a.s.* $\forall n$, *we have* $\lim_{n\to\infty} Var[S_n] = \sigma_\phi^2(\boldsymbol{\theta}) = 0$. *On the other hand if* $\sigma_\phi^2(\boldsymbol{\theta}) = \boldsymbol{T}^T\mathcal{I}_{\mathcal{F}}(\boldsymbol{\theta})^{-1}\boldsymbol{T} = 0$, *then* $\boldsymbol{T} = \boldsymbol{0}$, *since* $\mathcal{I}_{\mathcal{F}}(\boldsymbol{\theta})$ *is positive definite, and then* $S_n = 0$ *a.s.* $\forall n \in \mathbb{N}$.

If $\boldsymbol{T}^T\mathcal{I}_{\mathcal{F}}(\boldsymbol{\theta})^{-1}\boldsymbol{T} = 0$, we use a second order Taylor expansion to get the asymptotic distribution of $H^\phi(\widehat{\boldsymbol{\theta}})$ in the following theorem:

Theorem 2.2

Assume that i)-v) hold and $\phi \in C^2([0,\infty))$, $\sigma_\phi^2(\boldsymbol{\theta}) = 0$, *and suppose that there exist measurable functions* $F(\boldsymbol{x})$, $G(\boldsymbol{x})$ *and* $H(\boldsymbol{x})$, *such that*

$$\begin{aligned}
\left|\phi'(f_{\boldsymbol{\theta}}(\boldsymbol{x}))\frac{\partial f_{\boldsymbol{\theta}}(\boldsymbol{x})}{\partial\theta_i}\right| &< F(\boldsymbol{x}) \quad i = 1,...,M_0,\\
\left|\phi''(f_{\boldsymbol{\theta}}(\boldsymbol{x}))\frac{\partial f_{\boldsymbol{\theta}}(\boldsymbol{x})}{\partial\theta_i}\frac{\partial f_{\boldsymbol{\theta}}(\boldsymbol{x})}{\partial\theta_j}\right| &< G(\boldsymbol{x}) \quad i,j = 1,...,M_0,\\
\left|\phi'(f_{\boldsymbol{\theta}}(\boldsymbol{x}))\frac{\partial^2 f_{\boldsymbol{\theta}}(\boldsymbol{x})}{\partial\theta_i\partial\theta_j}\right| &< H(\boldsymbol{x}) \quad i,j = 1,...,M_0.
\end{aligned}$$

Then,

$$2n\left(H^\phi(\widehat{\boldsymbol{\theta}}) - H^\phi(\boldsymbol{\theta})\right) \xrightarrow[n\to\infty]{L} \sum_{i=1}^{r}\beta_i Z_i^2,$$

where $Z_1,...,Z_r$ *are independent and identically distributed (iid) normal random variables with mean zero and variance 1,* $r = rank\left(\mathcal{I}_{\mathcal{F}}(\boldsymbol{\theta})^{-1}\boldsymbol{T}\mathcal{I}_{\mathcal{F}}(\boldsymbol{\theta})^{-1}\right)$ *and* β_i'*s are the eigenvalues of the matrix* $\boldsymbol{A}\mathcal{I}_{\mathcal{F}}(\boldsymbol{\theta})^{-1}$, *being,* $\boldsymbol{A} = (a_{ij})_{i,j=1,...,M_0}$ *with*

$$a_{ij} = \int_{\mathcal{X}}\left(\phi''(f_{\boldsymbol{\theta}}(\boldsymbol{x}))\frac{\partial f_{\boldsymbol{\theta}}(\boldsymbol{x})}{\partial\theta_i}\frac{\partial f_{\boldsymbol{\theta}}(\boldsymbol{x})}{\partial\theta_j} + \frac{\partial^2 f_{\boldsymbol{\theta}}(\boldsymbol{x})}{\partial\theta_i\partial\theta_j}\phi'(f_{\boldsymbol{\theta}}(\boldsymbol{x}))\right)d\mu(\boldsymbol{x}).$$

Proof. The second Taylor expansion of $H^\phi(\widehat{\boldsymbol{\theta}})$ around $\boldsymbol{\theta}$ gives

$$\begin{aligned}
H^\phi(\widehat{\boldsymbol{\theta}}) = &\ H^\phi(\boldsymbol{\theta}) + \tfrac{1}{2}\sum_{i=1}^{M_0}\sum_{j=1}^{M_0}\frac{\partial^2 H^\phi(\boldsymbol{\theta})}{\partial\theta_i\partial\theta_j}(\widehat{\theta}_i - \theta_i)(\widehat{\theta}_j - \theta_j) + o\left(\left\|\widehat{\boldsymbol{\theta}} - \boldsymbol{\theta}\right\|^2\right)\\
= &\ H^\phi(\boldsymbol{\theta}) + \tfrac{1}{2}(\widehat{\boldsymbol{\theta}} - \boldsymbol{\theta})^T\boldsymbol{A}(\widehat{\boldsymbol{\theta}} - \boldsymbol{\theta}) + o\left(\left\|\widehat{\boldsymbol{\theta}} - \boldsymbol{\theta}\right\|^2\right),
\end{aligned}$$

where

$$\begin{aligned}\boldsymbol{A} &= (a_{ij})_{i,j=1,\dots,M_0} = \left(\frac{\partial^2 H^\phi(\boldsymbol{\theta})}{\partial\theta_i\partial\theta_j}\right)_{i,j=1,\dots,M_0} \\ &= \left(\int_{\mathcal{X}}\left(\phi''(f_{\boldsymbol{\theta}}(\boldsymbol{x}))\frac{\partial f_{\boldsymbol{\theta}}(\boldsymbol{x})}{\partial\theta_i}\frac{\partial f_{\boldsymbol{\theta}}(\boldsymbol{x})}{\partial\theta_j} + \frac{\partial^2 f_{\boldsymbol{\theta}}(\boldsymbol{x})}{\partial\theta_i\partial\theta_j}\phi'(f_{\boldsymbol{\theta}}(\boldsymbol{x}))\right)d\mu(\boldsymbol{x})\right)_{i,j=1,\dots,M_0}.\end{aligned}$$

But $n\, o(\left\|\widehat{\boldsymbol{\theta}}-\boldsymbol{\theta}\right\|^2) = o_P(1)$, therefore, the asymptotic distribution of the random variables $2n\left(H^\phi(\widehat{\boldsymbol{\theta}}) - H^\phi(\boldsymbol{\theta})\right)$ and $n(\widehat{\boldsymbol{\theta}}-\boldsymbol{\theta})^T\boldsymbol{A}(\widehat{\boldsymbol{\theta}}-\boldsymbol{\theta})$ is the same. We know that

$$\sqrt{n}(\widehat{\boldsymbol{\theta}}-\boldsymbol{\theta}) \xrightarrow[n\to\infty]{L} N(\boldsymbol{0}, \mathcal{I}_{\mathcal{F}}(\boldsymbol{\theta})^{-1});$$

now the result follows by Corollary 2.1 in Dik and Gunst (1985): "Let $\boldsymbol{X}$ a q-variate normal variable with mean vector $\boldsymbol{0}$ and variance-covariance matrix $\boldsymbol{\Sigma}$. Let $\boldsymbol{A}$ be a real symmetric matrix of order q. Let $r = rank(\boldsymbol{\Sigma A \Sigma})$, $r \geq 1$ and let $\beta_1, \dots, \beta_r$, be the nonzero eigenvalues of $\boldsymbol{A\Sigma}$. Then the distribution of the quadratic form $\boldsymbol{X}^T\boldsymbol{A}\boldsymbol{X}$ coincides with the distribution of the random variable $\sum_{i=1}^r \beta_i Z_i^2$, where $Z_1, \dots, Z_r$ are independent, each having a standard normal distribution". ∎

Remark 2.3

In the case of Shannon's entropy the elements a_{ij} are given by

$$a_{ij} = -\int_{\mathcal{X}}\left(\frac{1}{f_{\boldsymbol{\theta}}(\boldsymbol{x})}\frac{\partial f_{\boldsymbol{\theta}}(\boldsymbol{x})}{\partial\theta_i}\frac{\partial f_{\boldsymbol{\theta}}(\boldsymbol{x})}{\partial\theta_j} + \log f_{\boldsymbol{\theta}}(\boldsymbol{x})\frac{\partial^2 f_{\boldsymbol{\theta}}(\boldsymbol{x})}{\partial\theta_i\partial\theta_j}\right)d\mu(\boldsymbol{x})$$

and in the case of the (h,ϕ)-entropies

$$\begin{aligned}a_{ij} &= h'\left(\int_{\mathcal{X}}\phi(f_{\boldsymbol{\theta}}(\boldsymbol{x}))d\mu(\boldsymbol{x})\right) \\ &\times \left(\int_{\mathcal{X}}\left(\phi''(f_{\boldsymbol{\theta}}(\boldsymbol{x}))\frac{\partial f_{\boldsymbol{\theta}}(\boldsymbol{x})}{\partial\theta_i}\frac{\partial f_{\boldsymbol{\theta}}(\boldsymbol{x})}{\partial\theta_j} + \frac{\partial^2 f_{\boldsymbol{\theta}}(\boldsymbol{x})}{\partial\theta_i\partial\theta_j}\phi'(f_{\boldsymbol{\theta}}(\boldsymbol{x}))\right)d\mu(\boldsymbol{x})\right) \\ &+ h''\left(\int_{\mathcal{X}}\phi(f_{\boldsymbol{\theta}}(\boldsymbol{x}))d\mu(\boldsymbol{x})\right) \\ &\times \int_{\mathcal{X}}\phi'(f_{\boldsymbol{\theta}}(\boldsymbol{x}))\frac{\partial f_{\boldsymbol{\theta}}(\boldsymbol{x})}{\partial\theta_i}d\mu(\boldsymbol{x})\int_{\mathcal{X}}\phi'(f_{\boldsymbol{\theta}}(\boldsymbol{x}))\frac{\partial f_{\boldsymbol{\theta}}(\boldsymbol{x})}{\partial\theta_j}d\mu(\boldsymbol{x}).\end{aligned}$$

The derivation of this result step by step can be seen in Pardo, L. et *al.* (1997a).

2.3. Testing and Confidence Intervals for Phi-entropies

The previous results giving the asymptotic distribution of the ϕ-entropy statistics can be used in various settings to construct confidence intervals and to test statistical hypotheses regarding the entropy of a population (diversity). We consider the following tests:

2.3.1. Test for a Predicted Value of the Entropy of a Population (Diversity of a Population)

We are interested in testing

$$H_0 : H^{\phi}(\boldsymbol{\theta}) = D_0,$$

i.e., that the ϕ-entropy is of a certain magnitude D_0, versus one of the three following alternative hypotheses:

$$H_1 : H^{\phi}(\boldsymbol{\theta}) \neq D_0, \; H_1 : H^{\phi}(\boldsymbol{\theta}) > D_0, \; H_1 : H^{\phi}(\boldsymbol{\theta}) < D_0.$$

We can use the test statistic

$$Z_1 = \frac{\sqrt{n}\left(H^{\phi}(\widehat{\boldsymbol{\theta}}) - D_0\right)}{\sigma_{\phi}(\widehat{\boldsymbol{\theta}})},$$

where the expression of $\sigma_{\phi}(\widehat{\boldsymbol{\theta}})$ is obtained from Theorem 2.1 after replacing $\boldsymbol{\theta}$ by the maximum likelihood estimator $\widehat{\boldsymbol{\theta}}$ in $\sigma_{\phi}(\boldsymbol{\theta})$. Using Slutsky's Theorem (see, e.g., Ferguson 1996, Chapter 6) and Theorem 2.1, under H_0, the asymptotic distribution of Z_1 is normal with mean zero and variance one. Therefore in the first case we should reject the null hypothesis if $Z_1 > c_1$ or $Z_1 < c_2$ (where c_1 and c_2 are symmetric, chosen so that the significance level of the test is α, i.e., $c_1 = z_{\alpha/2}$ and $c_2 = -z_{\alpha/2}$). z_a denotes the z-score from the standard normal distribution having right-tailed probability a; this is the $100(1-a)$ percentile of that distribution. In the second case we should reject the null hypothesis if $Z_1 > c$ (then $c = z_\alpha$) and, finally, in the third case if $Z_1 < c$ (then $c = -z_\alpha$).

The power of the two-side test at $t \neq D_0$ is given by the formula

$$\beta_{\phi,n}(t) = 1 - \Phi_n\left(z_{\alpha/2} - \frac{\sqrt{n}\,(t - D_0)}{\sigma_{\phi}(\widehat{\boldsymbol{\theta}})}\right) + \Phi_n\left(-z_{\alpha/2} - \frac{\sqrt{n}\,(t - D_0)}{\sigma_{\phi}(\widehat{\boldsymbol{\theta}})}\right),$$

for a sequence of distributions $\Phi_n(x)$ tending uniformly to the standard normal distribution $\Phi(x)$. We observe that the test is consistent, in the sense of Fraser (1957), i.e., $\beta_{\phi,n}(t)$ tends to one when $n \to \infty$.

2.3.2. Test for the Equality of the Entropies of Two Independent Populations (Equality of Diversities of Two Populations)

We denote $H^{\phi}(\boldsymbol{\theta}_i)$, $i = 1, 2$, the ϕ-entropy associated with the population described by the probability distribution $P_{\boldsymbol{\theta}_i}$, $i = 1, 2$. We are interested in testing

$$H_0 : H^{\phi}(\boldsymbol{\theta}_1) = H^{\phi}(\boldsymbol{\theta}_2)$$

versus one of the three following alternative hypotheses:

$$H_1 : H^{\phi}(\boldsymbol{\theta}_1) \neq H^{\phi}(\boldsymbol{\theta}_2), \; H_1 : H^{\phi}(\boldsymbol{\theta}_1) > H^{\phi}(\boldsymbol{\theta}_2), \; H_1 : H^{\phi}(\boldsymbol{\theta}_1) < H^{\phi}(\boldsymbol{\theta}_2).$$

Using Slutsky's Theorem (see Ferguson 1996, p. 39) and Theorem 2.1, under H_0, the asymptotic distribution of the test statistic

$$Z_2 = \frac{\sqrt{n_1 n_2}\left(H^{\phi}(\widehat{\boldsymbol{\theta}}_1) - H^{\phi}(\widehat{\boldsymbol{\theta}}_2)\right)}{\sqrt{n_2 \sigma_{\phi}^2(\widehat{\boldsymbol{\theta}}_1) + n_1 \sigma_{\phi}^2(\widehat{\boldsymbol{\theta}}_2)}}$$

is normal with mean zero and variance one, where $\sigma_{\phi}^2(\widehat{\boldsymbol{\theta}}_i)$, $i = 1, 2$, are obtained from Theorem 2.1. The critical regions are similar to the ones given in the previous case. We shall assume that the populations are independent.

2.3.3. Test for the Equality of the Entropies of r Independent Populations

We are interested in testing

$$H_0 : H^{\phi}(\boldsymbol{\theta}_1) = H^{\phi}(\boldsymbol{\theta}_2) = ... = H^{\phi}(\boldsymbol{\theta}_r) = D_0$$

(D_0 is a known value) versus the alternative $H_1 : \exists\, i, k \in \{1, ..., r\}$ verifying $H^{\phi}(\boldsymbol{\theta}_i) \neq H^{\phi}(\boldsymbol{\theta}_k)$. We know that

$$\frac{\sqrt{n_i}\left(H^{\phi}(\widehat{\boldsymbol{\theta}}_i) - D_0\right)}{\sigma_{\phi}(\widehat{\boldsymbol{\theta}}_i)} \xrightarrow[n \to \infty]{L} N(0,1) \qquad i = 1, ..., r,$$

where $\sigma_\phi^2(\widehat{\boldsymbol{\theta}}_i)$ is the estimated asymptotic variance for the ith-population given in Theorem 2.1.

Then,

$$\frac{n_i\left(H^\phi(\widehat{\boldsymbol{\theta}}_i)-D_0\right)^2}{\sigma_\phi^2(\widehat{\boldsymbol{\theta}}_i)} \xrightarrow[n\to\infty]{L} \chi_1^2.$$

Therefore,

$$\sum_{i=1}^r \frac{n_i\left(H^\phi(\widehat{\boldsymbol{\theta}}_i)-D_0\right)^2}{\sigma_\phi^2(\widehat{\boldsymbol{\theta}}_i)} \xrightarrow[n\to\infty]{L} \chi_r^2.$$

We reject the null hypothesis if

$$Z_3 \equiv \sum_{i=1}^r \frac{n_i\left(H^\phi(\widehat{\boldsymbol{\theta}}_i)-D_0\right)^2}{\sigma_\phi^2(\widehat{\boldsymbol{\theta}}_i)} > \chi_{r,\alpha}^2,$$

where $\chi_{r,a}^2$ denotes the $100(1-a)$ percentile of the chi-square distribution with r degrees of freedom and it is defined by the equation $\Pr(\chi_r^2 \geq \chi_{r,a}^2) = a$. Now, we assume that D_0 is unknown and we are interested in testing

$$H_0 : H^\phi\left(\boldsymbol{\theta}_1\right) = H^\phi\left(\boldsymbol{\theta}_2\right) = \ldots = H^\phi\left(\boldsymbol{\theta}_r\right).$$

In this case we consider the test statistic

$$Z_4 \equiv \sum_{i=1}^r \frac{n_i\left(H^\phi(\widehat{\boldsymbol{\theta}}_i)-\overline{D}\right)^2}{\sigma_\phi^2(\widehat{\boldsymbol{\theta}}_i)}$$

where

$$\overline{D} = \frac{1}{\sum\limits_{i=1}^r \dfrac{n_i}{\sigma_\phi^2(\widehat{\boldsymbol{\theta}}_i)}} \sum_{i=1}^r \frac{n_i H^\phi(\widehat{\boldsymbol{\theta}}_i)}{\sigma_\phi^2(\widehat{\boldsymbol{\theta}}_i)}. \tag{2.3}$$

To get the asymptotic distribution of Z_4, we do the following decomposition

$$\begin{aligned}\sum_{i=1}^r \frac{n_i\left(H^\phi(\widehat{\boldsymbol{\theta}}_i)-D_0\right)^2}{\sigma_\phi^2(\widehat{\boldsymbol{\theta}}_i)} &= \sum_{i=1}^r \frac{n_i\left(H^\phi(\widehat{\boldsymbol{\theta}}_i)-\overline{D}+\overline{D}-D_0\right)^2}{\sigma_\phi^2(\widehat{\boldsymbol{\theta}}_i)} \\ &= \sum_{i=1}^r \frac{n_i\left(H^\phi(\widehat{\boldsymbol{\theta}}_i)-\overline{D}\right)^2}{\sigma_\phi^2(\widehat{\boldsymbol{\theta}}_i)} + \sum_{i=1}^r \frac{n_i}{\sigma_\phi^2(\widehat{\boldsymbol{\theta}}_i)}\left(\overline{D}-D_0\right)^2.\end{aligned}$$

The asymptotic distribution of the term on the left-hand side of this equality is chi-square with r degrees of freedom. Now we shall establish that the asymptotic distribution of the random variable

$$\sum_{i=1}^{r} \frac{n_i}{\sigma_\phi^2(\widehat{\boldsymbol{\theta}}_i)} \left(\overline{D} - D_0\right)^2$$

is chi-square with one degree of freedom. Using Slutsky's Theorem and Theorem 2.1,

$$\frac{\sqrt{n_i}\left(H^\phi(\widehat{\boldsymbol{\theta}}_i) - D_0\right)}{\sigma_\phi(\widehat{\boldsymbol{\theta}}_i)} \xrightarrow[n\to\infty]{L} N(0,1),$$

then

$$\frac{\sqrt{n_i} H^\phi(\widehat{\boldsymbol{\theta}}_i)}{\sigma_\phi(\widehat{\boldsymbol{\theta}}_i)} \sim N\left(\frac{\sqrt{n_i} D_0}{\sigma_\phi(\widehat{\boldsymbol{\theta}}_i)}, 1\right),$$

where $\sim$ is used to denote "asymptotically distributed as".

Therefore

$$\frac{\sqrt{n_i}}{\sigma_\phi(\widehat{\boldsymbol{\theta}}_i)} \frac{\sqrt{n_i} H^\phi(\widehat{\boldsymbol{\theta}}_i)}{\sigma_\phi(\widehat{\boldsymbol{\theta}}_i)} \sim N\left(\frac{n_i D_0}{\sigma_\phi^2(\widehat{\boldsymbol{\theta}}_i)}, \frac{n_i}{\sigma_\phi^2(\widehat{\boldsymbol{\theta}}_i)}\right),$$

$$\sum_{i=1}^{r} \frac{n_i H^\phi(\widehat{\boldsymbol{\theta}}_i)}{\sigma_\phi^2(\widehat{\boldsymbol{\theta}}_i)} = \sum_{i=1}^{r} \frac{n_i}{\sigma_\phi^2(\widehat{\boldsymbol{\theta}}_i)} \overline{D} \sim N\left(D_0 \sum_{i=1}^{r} \frac{n_i}{\sigma_\phi^2(\widehat{\boldsymbol{\theta}}_i)}, \sum_{i=1}^{r} \frac{n_i}{\sigma_\phi^2(\widehat{\boldsymbol{\theta}}_i)}\right),$$

$$\sqrt{\sum_{i=1}^{r} \frac{n_i}{\sigma_\phi^2(\widehat{\boldsymbol{\theta}}_i)}}\, \overline{D} \sim N\left(D_0 \sqrt{\sum_{i=1}^{r} \frac{n_i}{\sigma_\phi^2(\widehat{\boldsymbol{\theta}}_i)}}, 1\right),$$

finally

$$\sqrt{\sum_{i=1}^{r} \frac{n_i}{\sigma_\phi^2(\widehat{\boldsymbol{\theta}}_i)}} \left(\overline{D} - D_0\right) \xrightarrow[n\to\infty]{L} N(0,1) \quad \text{and} \quad \sum_{i=1}^{r} \frac{n_i}{\sigma_\phi^2(\widehat{\boldsymbol{\theta}}_i)} \left(\overline{D} - D_0\right)^2 \xrightarrow[n\to\infty]{L} \chi_1^2.$$

Until now we have

$$\underbrace{\sum_{i=1}^{r} \frac{n_i\left(H^\phi(\widehat{\boldsymbol{\theta}}_i) - D_0\right)^2}{\sigma_\phi^2(\widehat{\boldsymbol{\theta}}_i)}}_{\chi_r^2} = \sum_{i=1}^{r} \frac{n_i\left(H^\phi(\widehat{\boldsymbol{\theta}}_i) - \overline{D}\right)^2}{\sigma_\phi^2(\widehat{\boldsymbol{\theta}}_i)} + \underbrace{\sum_{i=1}^{r} \frac{n_i}{\sigma_\phi^2(\widehat{\boldsymbol{\theta}}_i)} \left(\overline{D} - D_0\right)^2}_{\chi_1^2}.$$

Applying now the following result (Rao 1973, p. 187. Result *iii*)),

"Let $\boldsymbol{Y}$ be an r-dimensional normal vector with mean vector zero and variance-covariance matrix identity. Let $\boldsymbol{Y}^T\boldsymbol{Y} = \boldsymbol{Y}^T\boldsymbol{A}\boldsymbol{Y} + \boldsymbol{Y}^T\boldsymbol{B}\boldsymbol{Y}$, where $\boldsymbol{Y}^T\boldsymbol{A}\boldsymbol{Y}$ is chi-squared distributed with a degrees of freedom. Then $\boldsymbol{Y}^T\boldsymbol{B}\boldsymbol{Y}$ is chi-squared distributed with $r - a$ degrees of freedom";

the random variable

$$\sum_{i=1}^{r} \frac{n_i \left(H^{\phi}(\widehat{\boldsymbol{\theta}}_i) - \overline{D}\right)^2}{\sigma_{\phi}^2(\widehat{\boldsymbol{\theta}}_i)}$$

follows a chi-square distribution with $r - 1$ degrees of freedom.

In order to verify the hypotheses of the last result, let us denote

$$Y_i = \frac{\sqrt{n_i}\left(H^{\phi}(\widehat{\boldsymbol{\theta}}_i) - D_0\right)}{\sigma_{\phi}(\widehat{\boldsymbol{\theta}}_i)}$$

and $\boldsymbol{Y} = (Y_1, ..., Y_r)^T$. Then we must show that

$$\sum_{i=1}^{r} \frac{n_i}{\sigma_{\phi}^2(\widehat{\boldsymbol{\theta}}_i)} \left(\overline{D} - D_0\right)^2 = \boldsymbol{Y}^T\boldsymbol{A}\boldsymbol{Y}$$

with $rank\,(\boldsymbol{A}) = 1$ and

$$\sum_{i=1}^{r} \frac{n_i \left(H^{\phi}(\widehat{\boldsymbol{\theta}}_i) - \overline{D}\right)^2}{\sigma_{\phi}^2(\widehat{\boldsymbol{\theta}}_i)} = \boldsymbol{Y}^T\boldsymbol{B}\boldsymbol{Y}.$$

Since,

$$\overline{D} = \frac{1}{s}\sum_{i=1}^{r} \frac{n_i H^{\phi}(\widehat{\boldsymbol{\theta}}_i)}{\sigma_{\phi}^2(\widehat{\boldsymbol{\theta}}_i)}$$

with

$$s = \sum_{i=1}^{r} \frac{n_i}{\sigma_{\phi}^2(\widehat{\boldsymbol{\theta}}_i)},$$

we have

$$\begin{aligned}
\sum_{i=1}^{r} \frac{n_i}{\sigma_\phi^2(\widehat{\boldsymbol{\theta}}_i)} \left(\overline{D} - D_0\right)^2 &= s\left(\frac{1}{s}\sum_{i=1}^{r}\frac{n_i H^\phi(\widehat{\boldsymbol{\theta}}_i)}{\sigma_\phi^2(\widehat{\boldsymbol{\theta}}_i)} - D_0\right)^2 \\
&= \frac{1}{s}\left(\sum_{i=1}^{r}\frac{n_i H^\phi(\widehat{\boldsymbol{\theta}}_i)}{\sigma_\phi^2(\widehat{\boldsymbol{\theta}}_i)} - \sum_{i=1}^{r}\frac{n_i}{\sigma_\phi^2(\widehat{\boldsymbol{\theta}}_i)} D_0\right)^2 \\
&= \frac{1}{s}\left(\sum_{i=1}^{r}\frac{n_i\left(H^\phi(\widehat{\boldsymbol{\theta}}_i) - D_0\right)}{\sigma_\phi^2(\widehat{\boldsymbol{\theta}}_i)}\right)^2 \\
&= \frac{1}{s}\left(\sum_{i=1}^{r}\frac{\sqrt{n_i}}{\sigma_\phi(\widehat{\boldsymbol{\theta}}_i)} Y_i\right)^2 \\
&= \frac{1}{s}\sum_{i=1}^{r}\sum_{j=1}^{r}\frac{\sqrt{n_i}}{\sigma_\phi(\widehat{\boldsymbol{\theta}}_i)}\frac{\sqrt{n_j}}{\sigma_\phi(\widehat{\boldsymbol{\theta}}_j)} Y_i Y_j = \boldsymbol{Y}^T \boldsymbol{A} \boldsymbol{Y},
\end{aligned}$$

where $\boldsymbol{A} = (a_{ij})_{i,j=1,\dots,r}$ and

$$a_{ij} = \left(\frac{\sqrt{n_i}}{\sigma_\phi(\widehat{\boldsymbol{\theta}}_i)}\frac{\sqrt{n_j}}{\sigma_\phi(\widehat{\boldsymbol{\theta}}_j)}\right)\left(\sum_{i=1}^{r}\frac{n_i}{\sigma_\phi^2(\widehat{\boldsymbol{\theta}}_i)}\right)^{-1}.$$

We also observe that for all $i, j, m, k = 1, ..., r$, we have

$$\begin{aligned}
\begin{vmatrix} a_{ij} & a_{ik} \\ a_{mj} & a_{mk} \end{vmatrix} &= a_{ij}a_{mk} - a_{ik}a_{mj} \\
&= \left(\frac{\sqrt{n_i}}{\sigma_\phi(\widehat{\boldsymbol{\theta}}_i)}\frac{\sqrt{n_j}}{\sigma_\phi(\widehat{\boldsymbol{\theta}}_j)}\frac{\sqrt{n_m}}{\sigma_\phi(\widehat{\boldsymbol{\theta}}_m)}\frac{\sqrt{n_k}}{\sigma_\phi(\widehat{\boldsymbol{\theta}}_k)}\right. \\
&\quad - \left.\frac{\sqrt{n_i}}{\sigma_\phi(\widehat{\boldsymbol{\theta}}_i)}\frac{\sqrt{n_k}}{\sigma_\phi(\widehat{\boldsymbol{\theta}}_k)}\frac{\sqrt{n_m}}{\sigma_\phi(\widehat{\boldsymbol{\theta}}_m)}\frac{\sqrt{n_j}}{\sigma_\phi(\widehat{\boldsymbol{\theta}}_j)}\right) s^{-2} = 0.
\end{aligned}$$

Then $rank\,(\boldsymbol{A}) = 1$. Finally $\boldsymbol{B} = \boldsymbol{I} - \boldsymbol{A}$ and applying the mentioned result we conclude that the asymptotic distribution of the quadratic form

$$\boldsymbol{Y}^T \boldsymbol{B} \boldsymbol{Y}$$

is chi-square with $r - 1$ degrees of freedom. This completes the proof.

2.3.4. Tests for Parameters

In situations where the ϕ-entropy is one-to-one function of $\boldsymbol{\theta}$ the following tests

$$a)\ H_0 : \boldsymbol{\theta} = \boldsymbol{\theta}_0$$
$$b)\ H_0 : \boldsymbol{\theta}_1 = \boldsymbol{\theta}_2$$
$$c)\ H_0 : \boldsymbol{\theta}_1 = \boldsymbol{\theta}_2 = ... = \boldsymbol{\theta}_r = \boldsymbol{\theta}_0$$
$$d)\ H_0 : \boldsymbol{\theta}_1 = \boldsymbol{\theta}_2 = ... = \boldsymbol{\theta}_r$$

are equivalent to the tests

$$a)\ H_0 : H^{\phi}(\boldsymbol{\theta}) = H^{\phi}(\boldsymbol{\theta}_0)$$
$$b)\ H_0 : H^{\phi}(\boldsymbol{\theta}_1) = H^{\phi}(\boldsymbol{\theta}_2)$$
$$c)\ H_0 : H^{\phi}(\boldsymbol{\theta}_1) = H^{\phi}(\boldsymbol{\theta}_2) = ... = H^{\phi}(\boldsymbol{\theta}_r) = H^{\phi}(\boldsymbol{\theta}_0)$$
$$d)\ H_0 : H^{\phi}(\boldsymbol{\theta}_1) = H^{\phi}(\boldsymbol{\theta}_2) = ... = H^{\phi}(\boldsymbol{\theta}_r).$$

Now we are going to see an important application of the previous results to the problem of equality of variances for independent normal populations.

An Application to Equality of Variances in Normal Populations

Let X_{ij} $(j = 1, 2, ..., n_i;\ i = 1, ..., r)$ be independent normal variables with mean μ_i and variance σ_i^2 for population i, $i = 1, ..., r$. Let

$$\widehat{\mu}_i = \sum_{j=1}^{n_i} \frac{X_{ij}}{n_i} \quad \text{and} \quad \widehat{\sigma}_i^2 = \frac{1}{n_i} \sum_{j=1}^{n_i} (X_{ij} - \widehat{\mu}_i)^2$$

be the maximum likelihood estimators of μ_i and σ_i^2, respectively. We are going to test

$$H_0 : \sigma_1^2 = ... = \sigma_r^2,$$

on the basis of Shannon's entropy.

Shannon's entropy associated with a normal population with mean μ_i and variance σ_i^2 is given by

$$H(\sigma_i^2) \equiv H\left(N(\mu_i, \sigma_i^2)\right) = \log \sqrt{2\pi e \sigma_i^2}.$$

In view of the previous result, testing

$$H_0 : \sigma_1^2 = ... = \sigma_r^2$$

is equivalent to test

$$H_0 : H(\sigma_1^2) = ... = H(\sigma_r^2).$$

On the basis of the result given in Section 2.3.3, the null hypothesis should be rejected if

$$S_{Sha} \equiv Z_4 = \sum_{i=1}^{r} \frac{n_i \left(H(\widehat{\boldsymbol{\theta}}_i) - \overline{D} \right)^2}{\sigma_H^2(\widehat{\boldsymbol{\theta}}_i)} > \chi^2_{r-1,\alpha},$$

where $\sigma_H^2(\widehat{\boldsymbol{\theta}}_i)$ is given in (2.1) for $\phi(x) = -x \log x$, after replacing $\boldsymbol{\theta}_i$ by $\widehat{\boldsymbol{\theta}}_i$ and $\overline{D}$ was defined in (2.3).

Particularly, here

$$\sigma_H^2 (\boldsymbol{\theta}_i) = \boldsymbol{T}_{(i)}^T \mathcal{I}_{\mathcal{F}} \left(\mu_i, \sigma_i^2 \right)^{-1} \boldsymbol{T}_{(i)},$$

where $\boldsymbol{T}_{(i)} = (t_{1i}, t_{2i})^T$ with

$$t_{1i} = \frac{\partial H(\sigma_i^2)}{\partial \mu_i} = 0, \; t_{2i} = \frac{\partial H(\sigma_i^2)}{\partial \sigma_i^2} = \frac{1}{2\sigma_i^2}$$

and

$$\mathcal{I}_{\mathcal{F}} \left(\mu_i, \sigma_i^2 \right)^{-1} = \begin{pmatrix} \sigma_i^2 & 0 \\ 0 & 2\sigma_i^4 \end{pmatrix}.$$

Then $\sigma_H^2 (\boldsymbol{\theta}_i) = \frac{1}{2}$, $i = 1, ..., r$, and asymptotically the null hypothesis

$$H_0 : \sigma_1^2 = ... = \sigma_r^2$$

should be rejected if

$$S_{Sha} \equiv \frac{1}{2} \sum_{j=1}^{r} n_j \left(\log \widehat{\sigma}_j^2 - \log \prod_{j=1}^{r} \left(\widehat{\sigma}_j^2 \right)^{\frac{n_j}{N}} \right)^2 \tag{2.4}$$

is greater than $\chi^2_{r-1,\alpha}$, where $N = \sum_{i=1}^{r} n_i$, i.e., if

$$\frac{1}{2} \sum_{j=1}^{r} n_j \left(v_j - \overline{v} \right)^2 > \chi^2_{r-1,\alpha},$$

where $v_j = \log \widehat{\sigma}_j^2$ and $\overline{v} = \sum_{j=1}^r \frac{n_j}{N} \log \widehat{\sigma}_j^2$ is the weighted average of the $v_j's$.

It is interesting to observe the similarity of the test statistic S_{Sha} with the test statistic, for this problem, proposed by Lehmann (1959, pp. 274-275). This test statistic is given by

$$S_{Leh} \equiv \frac{1}{2} \sum_{j=1}^{r} N_j \left(l_j - \overline{l}\right)^2,$$

where $l_j = \log \frac{1}{N_j} \sum_{j=1}^{n_i} \left(X_{ij} - \widehat{\mu}_i\right)^2, \overline{l} = \sum_{j=1}^r \frac{N_j}{N^*} \log \frac{1}{N_j} \sum_{j=1}^{n_i} \left(X_{ij} - \widehat{\mu}_i\right)^2, N_j = n_j - 1$, with $N^* = N_1 + ... + N_r$. The modification consists of replacing sample size n_j by degrees of freedom $n_j - 1$.

Now we present a study to compare this new test statistic with other well known test statistics. For more details see Pardo et *al.* (1995). Many of the existing parametric and nonparametric tests for homogeneity of variances and some variations of these tests were examined by Conover et *al.* (1981). The purpose of their study is to provide a list of tests with a stable Type I error rate when the normality assumption may not be true, and the sample sizes may be small and/or unequal, and when distributions may be skewed or heavy-tailed. In order to do a comparative study of these tests with the test statistic given in this example, we have done a simulation study similar to that performed by them.

	Normal Distribution			
(n_1, n_2, n_3, n_4)	$(5,5,5,5)$	$(10,10,10,10)$	$(20,20,20,20)$	$(5,5,20,20)$
Shannon	.167 (.488)	.049 (.655)	.074 (.810)	.141 (.770)
Neyman-Pearson	.123 (.460)	.073 (.653)	.064 (.816)	.103 (.761)
Barlett	.041 (.303)	.044 (.591)	.049 (.796)	.044 (.646)
Hartley	.042 (.234)	.044 (.546)	.051 (.781)	.047 (.769)

Table 2.1

Two probability distributions were considered: normal and double exponential. The simulated normal variables were obtained by the Box-Müller method. The double exponential variables were obtained from the inverse cumulative distribution function. Uniform random numbers were generated using a multiplicative congruent generator. Four sets of samples were generated with respective sample sizes $(n_1, n_2, n_3, n_4) = (5,5,5,5)$, $(10,10,10,10)$, $(20,20,20,20)$ and $(5,5,20,20)$. The null hypothesis of equal variances (all equal to 1) was examined along with four alternatives $(\sigma_1^2, \sigma_2^2, \sigma_3^2, \sigma_4^2) = (1,1,1,2), (1,1,1,4), (1,1,1,8)$ and $(1,2,4,8)$. The means were set equal to the standard deviation in each population under the alternative hypothesis. Zero means were used for H_0. Each of these $40\,(2 \times 4 \times 5)$ combinations of distribution types, sample sizes and variances was repeated 10000 times (1000 in Conover et *al.* (1981)), and the Shannon test statistics were computed and compared with their 5 percent nominal critical values 400000 times each. The observed frequency of rejection is reported in Table 2.1 for normal distributions and in Table 2.2 for double exponential distributions. The numbers in the tables are for null hypotheses, while the figures in parentheses represent the averages over the four variance combinations under the alternative hypothesis.

	Double Exponential Distribution			
(n_1, n_2, n_3, n_4)	$(5,5,5,5)$	$(10,10,10,10)$	$(20,20,20,20)$	$(5,5,20,20)$
Shannon	.389 (.602)	.359 (.719)	.356 (.842)	.389 (.818)
Neyman-Pearson	.330 (.564)	.333 (.707)	.341 (.839)	.344 (.801)
Barlett	.179 (.410)	.259 (.653)	.309 (.824)	.237 (.705)
Hartley	.157 (.355)	.237 (.625)	.288 (.811)	.462 (.828)

Table 2.2

The corresponding figures for the asymmetric case were obtained by making the transformation $\sigma X_i^2 + \mu$ where X_i represents the random variable distributed according to the null hypothesis. The two distributions, normal and double exponential, the two sets of sample sizes $(10,10,10,10)$ and $(5,5,20,20)$, and the five variance combinations gave a total of 20 combinations. For each combination, 10000 repetitions were run for each of the four statistics. With the same structure as Tables 2.1 and 2.2, the observed frequencies of rejection are reported in Table 2.3 for normal distributions and double exponential distributions.

	Normal Distribution		Double Exponential Distribution	
(n_1, n_2, n_3, n_4)	$(5,5,5,5)$	$(10,10,10,10)$	$(20,20,20,20)$	$(5,5,20,20)$
Shannon	.701 (.828)	.726 (.839)	.894 (.922)	.906 (.922)
Neyman-Pearson	.679 (.815)	.697 (.877)	.887 (.916)	.892 (.938)
Barlett	.621 (.777)	.595 (.813)	.858 (.896)	.848 (.908)
Hartley	.684 (.764)	.872 (.951)	.849 (.879)	.926 (.971)

Table 2.3

In order to interpret their simulation results, Conover et *al.* (1981) define a test to be robust if the maximum Type I error rate is less than 0.1 for a 5 percent test. In this sense, the four tests considered here are sensitive to departures from normality. If we analyze the numbers in parentheses, the Shannon test appears to have slightly greater power than the remaining three. If we look at the results in Table 5, p. 357, of Conover et *al.* (1981), then we obtain that for the columns 1, 2, 3 and 4 there are just 4, 7, 3 and 0 power values (in parentheses) respectively greater than the corresponding power values of the Shannon statistic. On the other hand, we note that the exact level of the proposed test does not converge very quickly to the asymptotic level 0.05. This point is mentioned in order to highlight the fact that, as in Conover et *al.* (1981), asymptotic critical regions have been used, and therefore the power values have not been calculated for exact $\alpha = 0.05$ level tests. To conclude, Shannon entropy provides a reasonably good test, among the 56 test statistics considered, when the normality assumption holds. Finally, observe that the four considered tests work under the assumptions of normality, so in a strict sense, only the numbers in Table 2.1 are probably of Type I error. An interesting study of this problem, using the quadratic entropy, is presented in Pardo, J. A. et *al.* (1997).

2.3.5. Confidence Intervals

If a sufficiently large sample is available it is possible to construct approximate confidence intervals of any desired level for $H^{\phi}(\boldsymbol{\theta})$. An approximate $(1-\alpha)\,100\%$ confidence interval for $H^{\phi}(\boldsymbol{\theta})$ is

$$\left(H^{\phi}(\widehat{\boldsymbol{\theta}}) - \frac{\sigma_{\phi}(\widehat{\boldsymbol{\theta}})\; z_{\alpha/2}}{\sqrt{n}},\; H^{\phi}(\widehat{\boldsymbol{\theta}}) + \frac{\sigma_{\phi}(\widehat{\boldsymbol{\theta}})\; z_{\alpha/2}}{\sqrt{n}} \right), \tag{2.5}$$

where $\sigma_{\phi}(\widehat{\boldsymbol{\theta}})$ is given in Theorem 2.1 after replacing $\boldsymbol{\theta}$ by its corresponding maximum likelihood estimator $\widehat{\boldsymbol{\theta}}$.

From (2.5) we have

$$-\frac{\sigma_{\phi}(\widehat{\boldsymbol{\theta}})\; z_{\alpha/2}}{\sqrt{n}} \leq H^{\phi}(\widehat{\boldsymbol{\theta}}) - H^{\phi}(\boldsymbol{\theta}) \leq \frac{\sigma_{\phi}(\widehat{\boldsymbol{\theta}})\; z_{\alpha/2}}{\sqrt{n}}. \tag{2.6}$$

If $H^{\phi}(\widehat{\boldsymbol{\theta}})$ is to be used as a point estimate of $H^{\phi}(\boldsymbol{\theta})$, our error, which we shall denote ε, is given by the difference between the value of $H^{\phi}(\widehat{\boldsymbol{\theta}})$ and the unknown value of $H^{\phi}(\boldsymbol{\theta})$. Therefore we can rewrite inequalities (2.6) as

$$-\frac{\sigma_{\phi}(\widehat{\boldsymbol{\theta}})\ z_{\alpha/2}}{\sqrt{n}} \leq \varepsilon \leq \frac{\sigma_{\phi}(\widehat{\boldsymbol{\theta}})\ z_{\alpha/2}}{\sqrt{n}}.$$

This formula can be used to provide the necessary sample size n for a specified confidence coefficient $1-\alpha$ and a desired degree of precision ε. A general formula for determining the desired sample size is

$$n = \left[\frac{\sigma_{\phi}^2(\widehat{\boldsymbol{\theta}}) z_{\alpha/2}^2}{\varepsilon^2}\right] + 1,$$

because the equality $\Pr\left(\left|H^{\phi}(\widehat{\boldsymbol{\theta}}) - H^{\phi}(\boldsymbol{\theta})\right| < \varepsilon\right) = 1-\alpha$ is equivalent to

$$\Pr\left(\left|\frac{\sqrt{n}\left(H^{\phi}(\widehat{\boldsymbol{\theta}}) - H^{\phi}(\boldsymbol{\theta})\right)}{\sigma_{\phi}(\widehat{\boldsymbol{\theta}})}\right| < \frac{\varepsilon\sqrt{n}}{\sigma_{\phi}(\widehat{\boldsymbol{\theta}})}\right) = 1-\alpha,$$

and then $\varepsilon\sqrt{n}/\sigma_{\phi}(\widehat{\boldsymbol{\theta}}) = z_{\alpha/2}$, where by $[.]$ we are representing the integer part function. If two sufficiently large samples are available it is possible to construct approximate confidence intervals of any desired confidence coefficient for the difference $H^{\phi}(\boldsymbol{\theta}_1) - H^{\phi}(\boldsymbol{\theta}_2)$. This interval is given by

$$\left(H^{\phi}(\widehat{\boldsymbol{\theta}}_1) - H^{\phi}(\widehat{\boldsymbol{\theta}}_2)\right) \pm z_{\alpha/2}\sqrt{\frac{\sigma_{\phi}^2(\widehat{\boldsymbol{\theta}}_1)}{n_1} + \frac{\sigma_{\phi}^2(\widehat{\boldsymbol{\theta}}_2)}{n_2}}$$

where $\sigma_{\phi}^2(\widehat{\boldsymbol{\theta}}_1)$ and $\sigma_{\phi}^2(\widehat{\boldsymbol{\theta}}_2)$ are given in Theorem 2.1 after replacing $\boldsymbol{\theta}_1$ and $\boldsymbol{\theta}_2$ by the corresponding maximum likelihood estimators.

If $n = n_1 = n_2$, the sample size, n, necessary for a specified confidence coefficient, $1-\alpha$, and a desired degree of precision, ε, is given by

$$n = \left[\frac{\left(\sigma_{\phi}^2(\widehat{\boldsymbol{\theta}}_1) + \sigma_{\phi}^2(\widehat{\boldsymbol{\theta}}_2)\right) z_{\alpha/2}^2}{\varepsilon^2}\right] + 1.$$

2.4. Multinomial Populations: Asymptotic Distributions

In this section we obtain asymptotic results for multinomial populations as a particular case of the results obtained in the previous sections for general populations.

Let $(\mathcal{X}, \beta_{\mathcal{X}}, P_{\boldsymbol{\theta}})_{\boldsymbol{\theta}\in\Theta}$ be a statistical space, where the sample space $\mathcal{X} = \{x_1, \ldots, x_M\}$ is finite, $\beta_{\mathcal{X}}$ is the family of subsets of $\mathcal{X}$ and

$$\Theta = \left\{ \boldsymbol{\theta} = (p_1, ..., p_{M-1})^T \in \mathbb{R}^{M-1} : \sum_{i=1}^{M-1} p_i < 1, p_i > 0, i = 1, ..., M-1 \right\}.$$

Let $P_{\boldsymbol{\theta}}$ be a probability measure on $(\mathcal{X}, \beta_{\mathcal{X}})$ such that for every $\boldsymbol{\theta} \in \Theta$,

$$P_{\boldsymbol{\theta}}(\{x_i\}) = p_i \text{ if } i = 1, ..., M-1 \qquad P_{\boldsymbol{\theta}}(\{x_M\}) = 1 - \sum_{i=1}^{M-1} p_i = p_M.$$

Let μ be the counting measure on $(\mathcal{X}, \beta_{\mathcal{X}})$ attributing mass one to every $x_i \in \mathcal{X}$. It is clear that $P_{\boldsymbol{\theta}}$ is absolutely continuous with respect to the measure μ for all $\boldsymbol{\theta} \in \Theta$, and

$$f_{\boldsymbol{\theta}}(x_i) = \frac{dP_{\boldsymbol{\theta}}}{d\mu}(x_i) = p_i, \qquad i = 1, ..., M.$$

In this context the ϕ-entropy is given by

$$H^{\phi}(\mathbf{p}) = H^{\phi}(\boldsymbol{\theta}) = \sum_{i=1}^{M} \phi(p_i)$$

being $\mathbf{p} = (p_1, ..., p_{M-1}, p_M)^T \in \Delta_M$.

Let $Y_1 = y_1, ..., Y_n = y_n$ be a random sample from the population $P_{\boldsymbol{\theta}}$. The likelihood associated with this random sample is given by $f_{\boldsymbol{\theta}}(y_1, ..., y_n) = p_1^{n_1} ... p_M^{n_M}$ being n_i the number of times that $Y_j = x_i$, $j = 1, ..., n$. The maximum likelihood estimator of the probabilities p_i is given by $\widehat{p}_i = n_i/n$, $i = 1, ..., M$. Then the maximum likelihood estimators of $\boldsymbol{\theta}$ and $\mathbf{p}$ are given respectively by:

$$\widehat{\boldsymbol{\theta}} = (\widehat{p}_1, ..., \widehat{p}_{M-1})^T \text{ and } \widehat{\mathbf{p}} = (\widehat{p}_1, ..., \widehat{p}_M)^T.$$

In the next result we compute the Fisher information matrix for the considered multinomial model.

Proposition 2.1

The Fisher information matrix in the multinomial model is given by $\mathcal{I}_{\mathcal{F}}(\mathbf{p}) = \mathcal{I}_{\mathcal{F}}(\boldsymbol{\theta}) = (i_{(r,s)})_{r,s=1,...,M-1}$, *where*

$$i_{(r,s)} = \begin{cases} \dfrac{1}{p_r} + \dfrac{1}{p_M} & \text{if } r = s \\[2ex] \dfrac{1}{p_M} & \text{if } r \neq s \end{cases} \qquad r, s = 1, ..., M-1.$$

Proof. The (r,s)th-element $i_{(r,s)}$, of the Fisher information matrix, for $r \neq s$, is

$$\begin{aligned} i_{(r,s)} &= \int_{\mathcal{X}} \frac{\partial \log f_{\boldsymbol{\theta}}(x)}{\partial \theta_r} \frac{\partial \log f_{\boldsymbol{\theta}}(x)}{\partial \theta_s} f_{\boldsymbol{\theta}}(x)\, d\mu(x) \\ &= \sum_{k=1}^{M-1} \frac{\partial \log p_k}{\partial p_r} \frac{\partial \log p_k}{\partial p_s} p_k + \frac{\partial \log \left(1 - \sum_{k=1}^{M-1} p_k\right)}{\partial p_r} \frac{\partial \log \left(1 - \sum_{k=1}^{M-1} p_k\right)}{\partial p_s} p_M \\ &= \left(-\frac{1}{p_M}\right)\left(-\frac{1}{p_M}\right) p_M = \frac{1}{p_M}, \quad r \neq s, \qquad r,s = 1,\dots,M-1, \end{aligned}$$

and the element (r,r) has the expression

$$\begin{aligned} i_{(r,r)} &= \int_{\mathcal{X}} \left(\frac{\partial \log f_{\boldsymbol{\theta}}(x)}{\partial \theta_r}\right)^2 f_{\boldsymbol{\theta}}(x)\, d\mu(x) \\ &= \sum_{k=1}^{M-1} \left(\frac{\partial}{\partial p_r} \log p_k\right)^2 p_k + \left(\frac{\partial}{\partial p_r} \log \left(1 - \sum_{k=1}^{M-1} p_k\right)\right)^2 p_M \\ &= \left(\frac{1}{p_r}\right)^2 p_r + \left(-\frac{1}{p_M}\right)^2 p_M = \frac{1}{p_r} + \frac{1}{p_M}. \end{aligned}$$

■

In the following Proposition the inverse of the Fisher information matrix is obtained.

Proposition 2.2

The inverse of the Fisher information matrix is given by

$$\mathcal{I}_{\mathcal{F}}(\boldsymbol{\theta})^{-1} = diag(\boldsymbol{\theta}) - \boldsymbol{\theta}\boldsymbol{\theta}^T$$

being $\boldsymbol{\theta} = (p_1, \dots, p_{M-1})^T$.

Proof.

By multiplying the matrix $\mathcal{I}_{\mathcal{F}}(\mathbf{p})^{-1} = \mathcal{I}_{\mathcal{F}}(\boldsymbol{\theta})^{-1}$ with elements

$$i_{(r,s)}^{-1} = \begin{cases} p_r(1-p_r) & \text{if } r = s \\ -p_r p_s & \text{if } r \neq s \end{cases} \quad r,s = 1,\dots,M-1,$$

by the matrix $\mathcal{I}_{\mathcal{F}}(\mathbf{p}) = \mathcal{I}_{\mathcal{F}}(\boldsymbol{\theta})$, with elements

$$
i_{(r,s)} = \begin{cases} \dfrac{1}{p_r} + \dfrac{1}{p_M} & \text{if } r = s \\ & \qquad r, s = 1, ..., M-1\,, \\ \dfrac{1}{p_M} & \text{if } r \neq s \end{cases}
$$

we get the identity matrix. To check this equality we note that the (r,s)th-element of the product is given by

$$
\sum_{\substack{j=1 \\ j \neq r}}^{M-1} (-p_r p_j) \frac{1}{p_M} + p_r(1-p_r)\left(\frac{1}{p_s} + \frac{1}{p_M}\right) = 1, \text{ if } r = s,
$$

$$
\sum_{\substack{i=1 \\ i \neq r,s}}^{M-1} (-p_r p_i) \frac{1}{p_M} + \frac{1}{p_M} p_r(1-p_r) - p_r p_s \left(\frac{1}{p_s} + \frac{1}{p_M}\right) = 0, \text{ if } r \neq s.
$$

Then $\mathcal{I}_{\mathcal{F}}(\boldsymbol{\theta})^{-1} \mathcal{I}_{\mathcal{F}}(\boldsymbol{\theta}) = \boldsymbol{I}_{(M-1)\times(M-1)}$ and $\mathcal{I}_{\mathcal{F}}(\boldsymbol{\theta})^{-1}$ is the inverse of $\mathcal{I}_{\mathcal{F}}(\boldsymbol{\theta})$. ∎

Theorem 2.3

The analogical estimator, $H^{\phi}(\widehat{\mathbf{p}})$, obtained by replacing the p_i's by its relative frequencies, $\widehat{p}_i$, in a random sample of size n, verifies

$$
\sqrt{n}\left(H^{\phi}(\widehat{\mathbf{p}}) - H^{\phi}(\mathbf{p})\right) \underset{n\to\infty}{\overset{L}{\longrightarrow}} N\left(0, \sigma_{\phi}^2(\mathbf{p})\right)
$$

whenever $\sigma_{\phi}^2(\mathbf{p}) > 0$, being

$$
\sigma_{\phi}^2(\mathbf{p}) = \boldsymbol{S}^T \boldsymbol{\Sigma}_{\mathbf{p}} \boldsymbol{S} = \sum_{i=1}^{M} s_i^2 p_i - \left(\sum_{i=1}^{M} s_i p_i\right)^2, \tag{2.7}
$$

with $\boldsymbol{S} = (s_1, ..., s_M)^T = (\phi'(p_1), ..., \phi'(p_M))^T$ and $\boldsymbol{\Sigma}_{\mathbf{p}} = diag(\mathbf{p}) - \mathbf{p}\mathbf{p}^T$.

Proof. By Theorem 2.1, we have

$$
\sigma_{\phi}^2(\mathbf{p}) = \boldsymbol{T}^T \mathcal{I}_{\mathcal{F}}(\boldsymbol{\theta})^{-1} \boldsymbol{T},
$$

where

$$
\boldsymbol{T} = (t_1, ..., t_{M-1})^T = (\phi'(p_1) - \phi'(p_M), ..., \phi'(p_{M-1}) - \phi'(p_M))^T.
$$

Now we must show that $\boldsymbol{T}^T \mathcal{I}_{\mathcal{F}}(\boldsymbol{\theta})^{-1} \boldsymbol{T} = \boldsymbol{S}^T \boldsymbol{\Sigma}_{\mathbf{p}} \boldsymbol{S}$. In fact,

$$
\begin{aligned}
\boldsymbol{T}^T \mathcal{I}_{\mathcal{F}}(\boldsymbol{\theta})^{-1} \boldsymbol{T} &= (\phi'(p_1), ..., \phi'(p_{M-1}))\mathcal{I}_{\mathcal{F}}(\boldsymbol{\theta})^{-1} (\phi'(p_1), ..., \phi'(p_{M-1}))^T \\
&- 2(\phi'(p_1), ..., \phi'(p_{M-1}))\mathcal{I}_{\mathcal{F}}(\boldsymbol{\theta})^{-1} (\phi'(p_M), ..., \phi'(p_M))^T \\
&+ (\phi'(p_M), ..., \phi'(p_M))\mathcal{I}_{\mathcal{F}}(\boldsymbol{\theta})^{-1} (\phi'(p_M), ..., \phi'(p_M))^T \\
&= \sum_{i=1}^{M-1} p_i \phi'(p_i)^2 - \sum_{i=1}^{M-1}\sum_{j=1}^{M-1} p_i p_j \phi'(p_i)\phi'(p_j) \\
&- 2\left(\phi'(p_M)\sum_{i=1}^{M-1} p_i\phi'(p_i) - \phi'(p_M)(1-p_M)\sum_{i=1}^{M-1} p_i\phi'(p_i)\right) \\
&+ (1-p_M)\phi'(p_M)^2 - \phi'(p_M)^2 \sum_{i=1}^{M-1}\sum_{j=1}^{M-1} p_i p_j \\
&= \sum_{i=1}^{M-1} p_i \phi'(p_i)^2 - \sum_{i=1}^{M-1}\sum_{j=1}^{M-1} p_i p_j \phi'(p_i)\phi'(p_j) \\
&- 2\phi'(p_M)p_M \sum_{i=1}^{M-1} p_i\phi'(p_i) + p_M\phi'(p_M)^2 - p_M^2\phi'(p_M)^2 \\
&= \sum_{i=1}^{M} p_i\phi'(p_i)^2 - \left(\sum_{i=1}^{M} p_i\phi'(p_i)\right)^2 = \boldsymbol{S}^T\boldsymbol{\Sigma_p}\boldsymbol{S}.
\end{aligned}
$$

Remark 2.4

For the (h, ϕ)*-entropies we have*

$$
\frac{\sqrt{n}}{h'\left(\sum_{i=1}^{M}\phi(p_i)\right)}\left(H_h^{\phi}(\widehat{\mathbf{p}}) - H_h^{\phi}(\mathbf{p})\right) \xrightarrow[n\to\infty]{L} N\left(0, \sigma_\phi^2(\mathbf{p})\right),
$$

where $\sigma_\phi^2(\mathbf{p})$ *was given in (2.7).* ■

A complete study of the statistics based on (h, ϕ)-entropies, in simple random sampling as well as in stratified random sampling, can be seen in Salicrú et *al.* (1993).

Remark 2.5

Consider an experiment whose outcomes belong to one of M $(M \geq 2)$ *mutually exclusive and exhaustive categories,* $E_1, ..., E_M$ *and let* p_i $(0 < p_i < 1)$ *be the probability that the outcome belongs to the ith category* $(i = 1, ..., M)$*. Here* $\sum_{i=1}^M p_i = 1$*. Suppose that the experiment is performed* n *times,* $Y_1, ..., Y_n$*, and that the* n *outcomes are independent. Furthermore, let* N_j *denote the number of these outcomes belonging to category* E_j $(j = 1, ..., M)$*, with* $\sum_{j=1}^M N_j = n$*. The random variable* $(N_1, ..., N_M)$ *has a multinomial distribution with parameters* n *and* $\mathbf{p} = (p_1, ..., p_M)^T$*.*

If we define

$$T_j^{(i} = \begin{cases} 1 & if \quad Y_i \in E_j \\ 0 & otherwise \end{cases}, i = 1, ..., n, \ j = 1, ..., M,$$

the vectors $(T_1^{(i}, ..., T_M^{(i})$, $i = 1, ..., n$, *are functions of* Y_i *and take values on the set*

$$\{(1, 0, ..., 0), ..., (0, ..., 0, 1)\}.$$

We also have:

$$\begin{array}{llll} Y_1 & \to & (T_1^{(1}, ..., T_M^{(1}) & \equiv \ M(1; p_1, ..., p_M) \\ Y_2 & \to & (T_1^{(2}, ..., T_M^{(2}) & \equiv \ M(1; p_1, ..., p_M) \\ ... & ... & \quad ... & \quad\quad ... \\ Y_n & \to & \underbrace{(T_1^{(n}, ..., T_M^{(n})} & \equiv \ M(1; p_1, ..., p_M) \\ & & & \\ & & \underbrace{\left(\sum_{i=1}^{n} T_1^{(i}, ..., \sum_{i=1}^{n} T_M^{(i} \right)}_{(N_1, ..., N_M)} & \equiv \ M(n; p_1, ..., p_M), \end{array}$$

where $\equiv$ *is used to denote "distributed as" and* $M(n; p_1, ..., p_M)$ *denotes the multinomial distribution with parameters* n *and* $p_1, ..., p_M$. *On the other hand, given the sequence of* M*-dimensional random variables* $\{\boldsymbol{U}_n\}_{n \in \mathbb{N}}$ *with*

$$\boldsymbol{U}_n = (U_{1n}, ..., U_{Mn})^T$$

where

$$E[\boldsymbol{U}_n] = \boldsymbol{\mu} \qquad and \qquad Cov[\boldsymbol{U}_n] = \boldsymbol{\Sigma} \qquad \forall n,$$

the Central Limit Theorem states that the random vector

$$\overline{\boldsymbol{U}}_n = \left(\frac{1}{n} \sum_{j=1}^{n} U_{1j}, ..., \frac{1}{n} \sum_{j=1}^{n} U_{Mj} \right)$$

verifies

$$\sqrt{n} \left(\overline{\boldsymbol{U}}_n - \boldsymbol{\mu} \right) \underset{n \to \infty}{\overset{L}{\longrightarrow}} N(\boldsymbol{0}, \boldsymbol{\Sigma}).$$

If we denote

$$\boldsymbol{A} = \left(\frac{N_1 - np_1}{\sqrt{n}}, ..., \frac{N_M - np_M}{\sqrt{n}} \right),$$

we have

$$\begin{aligned} \boldsymbol{A} &= \left(\frac{1}{\sqrt{n}} \left(\sum_{i=1}^{n} T_1^{(i} - np_1 \right), ..., \frac{1}{\sqrt{n}} \left(\sum_{i=1}^{n} T_M^{(i} - np_M \right) \right) \\ &= \left(\sqrt{n} \left(\frac{1}{n} \sum_{i=1}^{n} T_1^{(i} - p_1 \right), ..., \sqrt{n} \left(\frac{1}{n} \sum_{i=1}^{n} T_M^{(i} - p_M \right) \right) \end{aligned}$$

and applying the Central Limit Theorem we have

$$\left(\frac{N_1 - np_1}{\sqrt{n}}, ..., \frac{N_M - np_M}{\sqrt{n}} \right) \xrightarrow[n \to \infty]{L} N\left(\mathbf{0}, \boldsymbol{\Sigma_p}\right),$$

where

$$\boldsymbol{\Sigma_p} = diag\left(\mathbf{p}\right) - \mathbf{p}\mathbf{p}^T.$$

Therefore,

$$\sqrt{n}\left(\widehat{\mathbf{p}} - \mathbf{p}\right) \xrightarrow[n \to \infty]{L} N\left(\mathbf{0}, \boldsymbol{\Sigma_p}\right).$$

Theorem 2.4

If we assume that $\sigma_\phi^2(\mathbf{p}) = 0$, the analogical estimator $H^\phi(\widehat{\mathbf{p}})$, obtained by replacing the p_i's by their relative frequencies, $\widehat{p}_i$, obtained from a random sample of size n, verifies

$$2n\left(H^\phi(\widehat{\mathbf{p}}) - H^\phi(\mathbf{p})\right) \xrightarrow[n \to \infty]{L} \sum_{i=1}^{r} \beta_i Z_i^2$$

where $Z_1, ..., Z_r$ are i.i.d. normal random variables with mean zero and variance 1, $r = rank(\boldsymbol{\Sigma_p B \Sigma_p})$ and β_i's are the eigenvalues of the matrix $\boldsymbol{B\Sigma_p}$ being $\boldsymbol{B}$ the $M \times M$ matrix

$$\boldsymbol{B} = diag(\phi''(p_i)_{i=1,...,M}).$$

Proof. In Theorem 2.2 it was established that the random variables

$$2n\left(H^\phi(\widehat{\boldsymbol{\theta}}) - H^\phi\left(\boldsymbol{\theta}\right)\right) \quad \text{and} \quad \sqrt{n}(\widehat{\boldsymbol{\theta}} - \boldsymbol{\theta})^T \boldsymbol{A} \sqrt{n}(\widehat{\boldsymbol{\theta}} - \boldsymbol{\theta})$$

have the same asymptotic distribution. The elements of the matrix $\boldsymbol{A}$ are given by

$$a_{ij} = \phi''(p_i)\delta_{ij} + \phi''(p_M) \qquad i,j = 1,...,M-1$$

because

$$\begin{aligned}
a_{ij} &= \int_{\mathcal{X}} \left(\phi''\left(f_{\boldsymbol{\theta}}\left(\boldsymbol{x}\right)\right) \frac{\partial f_{\boldsymbol{\theta}}\left(\boldsymbol{x}\right)}{\partial \theta_i} \frac{\partial f_{\boldsymbol{\theta}}\left(\boldsymbol{x}\right)}{\partial \theta_j} + \frac{\partial^2 f_{\boldsymbol{\theta}}\left(\boldsymbol{x}\right)}{\partial \theta_i \partial \theta_j} \phi'\left(f_{\boldsymbol{\theta}}\left(\boldsymbol{x}\right)\right) \right) d\mu\left(\boldsymbol{x}\right) \\
&= \sum_{k=1}^{M-1} \left(\phi''(p_k) \frac{\partial p_k}{\partial p_i} \frac{\partial p_k}{\partial p_j} + \frac{\partial^2 p_k}{\partial p_i \partial p_j} \phi'(p_k) \right) + \phi''(1 - \sum_{k=1}^{M-1} p_k) \\
&\times \frac{\partial \left(1 - \sum_{k=1}^{M-1} p_k\right)}{\partial p_i} \frac{\partial \left(1 - \sum_{k=1}^{M-1} p_k\right)}{\partial p_j} + \frac{\partial^2 \left(1 - \sum_{k=1}^{M-1} p_k\right)}{\partial p_i \partial p_j} \phi' \left(1 - \sum_{k=1}^{M-1} p_k\right).
\end{aligned}$$

Then, for $\boldsymbol{L} = (\widehat{\boldsymbol{\theta}} - \boldsymbol{\theta})^T \boldsymbol{A} (\widehat{\boldsymbol{\theta}} - \boldsymbol{\theta})$ and denoting $\mathbf{1} = (1,..,1)^T$, we have

$$\begin{aligned}
\boldsymbol{L} = & \ (\widehat{\boldsymbol{\theta}} - \boldsymbol{\theta})^T diag(\phi''(p_1), ..., \phi''(p_{M-1}))(\widehat{\boldsymbol{\theta}} - \boldsymbol{\theta}) \\
+ & \ (\widehat{\boldsymbol{\theta}} - \boldsymbol{\theta})^T \phi''(p_M) \mathbf{1}\mathbf{1}^T (\widehat{\boldsymbol{\theta}} - \boldsymbol{\theta}) \\
= & \sum_{i=1}^{M-1} \phi''(p_i)(\widehat{p}_i - p_i)^2 + \phi''(p_M) \sum_{i=1}^{M-1} (\widehat{p}_i - p_i) \sum_{j=1}^{M-1} (\widehat{p}_j - p_j) \\
= & \sum_{i=1}^{M} \phi''(p_i)(\widehat{p}_i - p_i)^2 = (\widehat{\mathbf{p}} - \mathbf{p})^T \boldsymbol{B} (\widehat{\mathbf{p}} - \mathbf{p}),
\end{aligned}$$

being $\boldsymbol{B} = diag(\phi''(p_i)_{i=1,...,M})$.

By applying Remark 2.5 we obtain

$$\sqrt{n}(\widehat{\mathbf{p}} - \mathbf{p}) \xrightarrow[n\to\infty]{L} N(\mathbf{0}, \boldsymbol{\Sigma}_{\mathbf{p}}).$$

Then using the same argument as in Theorem 2.2 we have the desired result.

■

Remark 2.6

In the rest of the book the following result is important (Lemma 3, p. 57, in Ferguson, 1996): Let $\boldsymbol{X}$ *be a k-variate normal random variable with mean vector* $\mathbf{0}$ *and variance-covariance matrix* $\boldsymbol{\Sigma}$. *Then* $\boldsymbol{X}^T\boldsymbol{X}$ *is distributed chi-squared with* r *degrees of freedom if and only if* $\boldsymbol{\Sigma}$ *is a projection of rank* r.

Since $\boldsymbol{\Sigma}$ *is symmetric and squared it is a projection if* $\boldsymbol{\Sigma}^2 = \boldsymbol{\Sigma}$, *and therefore rank*($\boldsymbol{\Sigma}$) =*trace*($\boldsymbol{\Sigma}$).

Corollary 2.2

If we consider $\boldsymbol{u} = (1/M, ..., 1/M)^T$, *we get*

$$\frac{2nM}{\phi''(1/M)}\left(H^{\phi}(\widehat{\mathbf{p}}) - H^{\phi}(\boldsymbol{u})\right) \underset{n\to\infty}{\overset{L}{\longrightarrow}} \chi^2_{M-1}. \tag{2.8}$$

Proof. In this case we have $\boldsymbol{\Sigma}_{\boldsymbol{u}} = diag(\boldsymbol{u}) - \boldsymbol{u}\boldsymbol{u}^T$.

The random variables

$$2nM\left(H^{\phi}(\widehat{\mathbf{p}}) - H^{\phi}(\boldsymbol{u})\right) \text{ and } (\widehat{\mathbf{p}} - \boldsymbol{u})^T Mn\ diag(\phi''(1/M))(\widehat{\mathbf{p}} - \boldsymbol{u})$$

have the same asymptotic distribution, then

$$\frac{2nM}{\phi''(1/M)}\left(H^{\phi}(\widehat{\mathbf{p}}) - H^{\phi}(\boldsymbol{u})\right) \quad \text{and} \quad \sqrt{n}(\widehat{\mathbf{p}} - \boldsymbol{u})^T diag(\boldsymbol{u}^{-1})\sqrt{n}(\widehat{\mathbf{p}} - \boldsymbol{u})$$

have the same asymptotic distribution. But

$$\sqrt{n}(\widehat{\mathbf{p}} - \boldsymbol{u})^T diag(\boldsymbol{u}^{-1})\sqrt{n}(\widehat{\mathbf{p}} - \boldsymbol{u}) = \boldsymbol{X}^T\boldsymbol{X},$$

where

$$\boldsymbol{X} = \sqrt{n}\ diag(\boldsymbol{u}^{-1/2})(\widehat{\mathbf{p}} - \boldsymbol{u})$$

and the asymptotic distribution of $\boldsymbol{X}$ is normal with mean vector $\mathbf{0}$ and variance-covariance matrix

$$\boldsymbol{\Sigma}^* = diag(\boldsymbol{u}^{-1/2})\boldsymbol{\Sigma}_{\boldsymbol{u}} diag(\boldsymbol{u}^{-1/2}).$$

Now we are going to establish that the matrix $\boldsymbol{\Sigma}^*$ is a projection of rank $M-1$.

It is clear that

$$\boldsymbol{\Sigma}^* = diag(\boldsymbol{u}^{-1/2})\boldsymbol{\Sigma}_{\boldsymbol{u}}\ diag(\boldsymbol{u}^{-1/2}) = \boldsymbol{I} - diag(\boldsymbol{u}^{-1/2})\boldsymbol{u}\boldsymbol{u}^T diag\left(\boldsymbol{u}^{-1/2}\right),$$

where $\boldsymbol{I}$ denotes the $M \times M$ identity matrix. Then we have

$$\begin{aligned}\boldsymbol{\Sigma}^*\boldsymbol{\Sigma}^* = &\ \boldsymbol{I} - diag(\boldsymbol{u}^{-1/2})\boldsymbol{u}\boldsymbol{u}^T diag(\boldsymbol{u}^{-1/2}) - diag(\boldsymbol{u}^{-1/2})\boldsymbol{u}\boldsymbol{u}^T diag(\boldsymbol{u}^{-1/2}) \\ \times &\ diag(\boldsymbol{u}^{-1/2})\boldsymbol{u}\boldsymbol{u}^T diag(\boldsymbol{u}^{-1/2})diag(\boldsymbol{u}^{-1/2})\boldsymbol{u}\boldsymbol{u}^T diag(\boldsymbol{u}^{-1/2}).\end{aligned}$$

But

$$\boldsymbol{u}^T diag(\boldsymbol{u}^{-1/2})diag(\boldsymbol{u}^{-1/2})\boldsymbol{u} = 1,$$

then

$$\boldsymbol{\Sigma}^*\boldsymbol{\Sigma}^* = \boldsymbol{I} - diag(\boldsymbol{u}^{-1/2})\boldsymbol{u}\boldsymbol{u}^T diag(\boldsymbol{u}^{-1/2}) = \boldsymbol{\Sigma}^*.$$

Now

$$rank\,(\boldsymbol{\Sigma}^*) = trace(diag(\boldsymbol{u}^{-1/2})\boldsymbol{\Sigma}_{\boldsymbol{u}}\, diag(\boldsymbol{u}^{-1/2})) = trace(diag(\boldsymbol{u}^{-1})\boldsymbol{\Sigma}_{\boldsymbol{u}}),$$

and

$$diag(\boldsymbol{u}^{-1})\boldsymbol{\Sigma}_{\boldsymbol{u}} = \begin{pmatrix} (1-\frac{1}{M}) & -\frac{1}{M} & \cdots & \cdots & -\frac{1}{M} \\ -\frac{1}{M} & (1-\frac{1}{M}) & \cdots & \cdots & -\frac{1}{M} \\ \cdots & \cdots & \cdots & \cdots & \cdots \\ \cdots & \cdots & \cdots & (1-\frac{1}{M}) & -\frac{1}{M} \\ -\frac{1}{M} & -\frac{1}{M} & \cdots & -\frac{1}{M} & (1-\frac{1}{M}) \end{pmatrix}.$$

Therefore,

$$trace(diag(\boldsymbol{u}^{-1})\boldsymbol{\Sigma}_{\boldsymbol{u}}) = M(1-\tfrac{1}{M}) = M-1.$$

Applying Remark 2.6 we get that the asymptotic distribution of the random variable $\boldsymbol{X}^T\boldsymbol{X}$ is chi-square with $M-1$ degrees of freedom.

■

Remark 2.7

In the case of Shannon's entropy we have

$$2n\,(\log M - H(\widehat{\mathbf{p}})) \underset{n\to\infty}{\overset{L}{\longrightarrow}} \chi^2_{M-1},$$

and for the (h,ϕ)*-entropies, Salicrú et al. (1993),*

$$\frac{2nM}{\phi''(1/M)h'(M\phi(1/M))}\left(H_h^\phi(\widehat{\mathbf{p}}) - H_h^\phi(\boldsymbol{u})\right) \underset{n\to\infty}{\overset{L}{\longrightarrow}} \chi^2_{M-1}.$$

Corollary 2.2 as well as Remark 2.7 permit constructing tests of goodness of fit to one distribution.

2.4.1. Test of Discrete Uniformity

If we want to test

$H_0 : p_1 = ... = p_M = \frac{1}{M}$ versus $H_1 : \exists\, i,j \in \{1,...,M\}$ such that $p_i \neq p_j$

we can use the test statistic

$$Z_6 \equiv \frac{2nM}{\phi''(1/M)}\left(H^\phi(\widehat{\mathbf{p}}) - H^\phi(\boldsymbol{u})\right) \tag{2.9}$$

whose asymptotic distribution, by Corollary 2.2, is chi-square with $M-1$ degrees of freedom. We should reject the null hypothesis if $Z_6 > \chi^2_{M-1,\alpha}$.

It is also possible to get the asymptotic power of the test for $\boldsymbol{p}^*=(p_1^*,...,p_M^*)^T \neq (1/M,...,1/M)^T$. This power is given by

$$\begin{aligned}\beta_{\phi,n}(\boldsymbol{p}^*) &= \Pr\left(\tfrac{2nM}{\phi''(1/M)}\left(H^\phi(\widehat{\boldsymbol{p}}) - H^\phi(\boldsymbol{u})\right) > \chi^2_{M-1,\alpha}/H_1 : \boldsymbol{p} = \boldsymbol{p}^*\right) \\ &= 1 - \Phi_n\left(\frac{\sqrt{n}}{\sigma_\phi(\boldsymbol{p}^*)}\left(\frac{\phi''(1/M)\chi^2_{M-1,\alpha}}{2nM} + H^\phi(\boldsymbol{u}) - H^\phi(\boldsymbol{p}^*)\right)\right),\end{aligned}$$

for a sequence of distributions $\Phi_n(x)$ tending uniformly to the standard normal distribution $\Phi(x)$. The expression of $\sigma_\phi(\boldsymbol{p}^*)$ is given in Theorem 2.3.

It is clear that

$$\lim_{n\to\infty} \beta_{\phi,n}(\boldsymbol{p}^*) = 1 - \Phi(-\infty) = 1,$$

i.e., the test is consistent.

The power obtained above can be used to determine approximately the sample size n^* required to achieve a desired power β^* against a given alternative

$$\boldsymbol{p}^*=(p_1^*,...,p_M^*)^T \neq (1/M,...,1/M)^T.$$

The power is approximately

$$1 - \Phi\left(\frac{\sqrt{n}}{\sigma_\phi(\boldsymbol{p}^*)}\left(\frac{\phi''(1/M)\chi^2_{M-1,\alpha}}{2nM} + H^\phi(\boldsymbol{u}) - H^\phi(\boldsymbol{p}^*)\right)\right).$$

If we wish the power to be equal to β^* we must solve the equation

$$\beta^* = 1 - \Phi\left(\frac{\sqrt{n}}{\sigma_\phi(\boldsymbol{p}^*)}\left(\frac{\phi''(1/M)\chi^2_{M-1,\alpha}}{2nM} + H^\phi(\boldsymbol{u}) - H^\phi(\boldsymbol{p}^*)\right)\right).$$

It is not difficult to check that the sample size, n^*, is the solution of the following quadratic equation

$$n^2\left(H^\phi(\boldsymbol{u}) - H^\phi(\boldsymbol{p}^*)\right)^2 - n\sigma_\phi(\boldsymbol{p}^*)(\Phi^{-1}(1-\beta^*))^2 + 2S_\phi(M)\left(H^\phi(\boldsymbol{u}) - H^\phi(\boldsymbol{p}^*)\right) = 0,$$

where

$$S_\phi(M) = \frac{\phi''(1/M)\chi^2_{M-1,\alpha}}{2M}.$$

The solution is given by

$$n^* = \frac{1}{2A}\left(B \pm \sqrt{B^2 - 4A^2 S_\phi(M)}\right)$$

where

$$B = \sigma_\phi(\boldsymbol{p}^*)(\Phi^{-1}(1-\beta^*))^2 \text{ and } A = H^\phi(\boldsymbol{u}) - H^\phi(\boldsymbol{p}^*).$$

The test of discrete uniformity given in this section permits testing the goodness of fit of any distribution, i.e.,

$$H_0 : F(x) = F_0(x) \qquad \forall x \in \mathbb{R},$$

through testing

$$H_0 : p_i = 1/M,\ i = 1, ..., M \qquad \text{versus} \qquad H_1 : \exists\ i, j\ /\ p_i \neq p_j,$$

by simply partitioning the range of the random variable into M intervals with equal probabilities under F_0, and then to test if the observations are from the given discrete uniform distribution. We use the test statistic Z_6 given in (2.9) and we should reject the null hypothesis if $Z_6 > \chi^2_{M-1,\alpha}$.

The particularization of the results obtained in the previous Section for general populations to multinomial populations is vital in order to study the behavior of the diversity among one or various populations. Based on last results we will have the possibility to test: *i)* if the diversity of a population is some specified value, *ii)* if the diversities of two populations are equal, *iii)* if the diversities of several populations are equal, and finally, *iv)* if the population is homogeneous (discrete uniformity) or not.

Remark 2.8

It is interesting to observe that different diversity measures (entropy measures) may give different diversity ordering. For instance, if we consider two populations characterized by the following probability vectors

$$\boldsymbol{p} = (0.3, 0.24, 0.22, 0.12, 0.09, 0.03)^T$$

and

$$\boldsymbol{q} = (0.36, 0.21, 0.16, 0.12, 0.08, 0.07)^T,$$

we have

$$H_{GS}(\boldsymbol{p}) = 0.7806 > H_{GS}(\boldsymbol{q}) = 0.7750$$

while

$$H(\boldsymbol{p}) = 1.61315 < H(\boldsymbol{q}) = 1.631338.$$

Hence it appears the necessity for the definition of ordering of discrete probability distributions. In this sense we have the following results:

Definition 2.1

The probability distribution $\mathbf{p}$ *is majorized by the probability distribution* $\mathbf{q}$*, and we denote,* $\mathbf{p} \prec_m \mathbf{q}$*, iff*

$$\sum_{i=1}^{r} p_{(i)} \leq \sum_{i=1}^{r} q_{(i)}, \; r = 1, ..., M-1$$

where $p_{(1)} \geq p_{(2)} \geq ... \geq p_{(M)}$ *and* $q_{(1)} \geq q_{(2)} \geq ... \geq q_{(M)}$.

Roughly speaking, the sentence "$\boldsymbol{p}$ is majorized by the probability distribution $\boldsymbol{q}$", means that $\boldsymbol{q}$ is less "spread out" than $\boldsymbol{p}$ and the population represented by $\boldsymbol{p}$ has more diversity than the population represented by $\boldsymbol{q}$.

Definition 2.2

A real function f *defined on the simplex* $\triangle_M$ *is said to be schur-concave on* $\triangle_M$ *if*

$$\mathbf{p} \prec_m \mathbf{q} \text{ on } \triangle_M \Longrightarrow f(\mathbf{p}) \geq f(\mathbf{q}),$$

and f *is strictly schur-concave on* $\triangle_M$ *if strict inequality holds whenever* $\mathbf{p} \prec_m \mathbf{q}$ *and* $\mathbf{p}$ *is not a permutation of* $\mathbf{q}$.

Some entropy measures are schur-concave functions on $\triangle_M$. Some examples are: Shannon, Havrda and Charvat, Arimoto, Rényi, etc. These results could be seen in Nayak (1983) as well as in Marshall and Olkin (1979).

2.5. Maximum Entropy Principle and Statistical Inference on Condensed Ordered Data

The Maximum Entropy Principle (MEP) stated for the first time by Jaynes (1957) has had very important applications in Statistical Mechanics, Statistics, Geography, Spatial and Urban Structures, Economics, Marketing, Systems

Analysis, Actuarial Science, Finance, Computer Science, Spectrum Analysis, Pattern Recognition, Search Theory, Operations Research, etc. (e.g., Kapur (1982, 1989)). In Statistics many problems have been studied on the basis of the MEP, but perhaps one of the most important results can be found in the nonparametric density function estimation. This problem was first introduced by Theil and Laitinen (1980) and has been developed by Theil and O'Brien (1980), Theil and Fiebig (1981), Theil and Kidwai (1981a, b), Theil and Lightburn (1981), Rodríguez and Van Ryzin (1985), etc.

In relation with the problem of parametric estimation two interesting papers are: Kapur and Kesaven (1992) and Jiménez and Palacios (1993). These authors present the problem in the following way. Let $Y_1, \ldots, Y_n$ be independent random variables with a common distribution $F_{\boldsymbol{\theta}} \in \{F_{\boldsymbol{\theta}}\}_{\boldsymbol{\theta} \in \Theta}$, being Θ an open subset of $\mathbb{R}^{M_0}$. Let $Y_{(1)} \leq Y_{(2)} \leq \ldots \leq Y_{(n)}$ be the ordered sample, define $Y_{(0)} = -\infty$, $Y_{(n+1)} = \infty$ and consider

$$\begin{aligned} p_1(\boldsymbol{\theta}) &= F_{\boldsymbol{\theta}}(Y_{(1)}), \\ p_i(\boldsymbol{\theta}) &= F_{\boldsymbol{\theta}}(Y_{(i)}) - F_{\boldsymbol{\theta}}(Y_{(i-1)}), \quad i = 2, \ldots, n, \\ p_{n+1}(\boldsymbol{\theta}) &= 1 - F_{\boldsymbol{\theta}}(Y_{(n)}). \end{aligned}$$

In this context they propose to estimate $\boldsymbol{\theta}$ by means of the value $\widetilde{\boldsymbol{\theta}} \in \Theta$, verifying

$$-\sum_{i=1}^{n+1} p_i(\widetilde{\boldsymbol{\theta}}) \log p_i(\widetilde{\boldsymbol{\theta}}) = \max_{\boldsymbol{\theta} \in \Theta} \left(-\sum_{i=1}^{n+1} p_i(\boldsymbol{\theta}) \log p_i(\boldsymbol{\theta}) \right). \tag{2.10}$$

This method can be justified by the fact that the order statistics

$$F_{\boldsymbol{\theta}}(Y_{(1)}), \ldots, F_{\boldsymbol{\theta}}(Y_{(n)})$$

divide the interval $(0,1)$ into $(n+1)$ random subintervals with equal expected length. Therefore, maximizing (2.10) is equivalent to choosing $\boldsymbol{\theta}$ so that the random variables $p_i(\boldsymbol{\theta})$ are close as possible to their expected value $\frac{1}{n+1}$. Also, Ranneby (1984) introduced an interesting method to estimate $\boldsymbol{\theta}$ by using a statistical information theory approach. He proposes to minimize a Kullback–Leibler divergence, or, equivalently, to select the value $\widehat{\boldsymbol{\theta}}$ of $\boldsymbol{\theta}$ which maximizes

$$S_n(\boldsymbol{\theta}) = \frac{1}{n+1} \sum_{i=1}^{n+1} \log(p_i(\boldsymbol{\theta})(n+1)),$$

and it is called the maximum spacing estimate.

The transformed spacings $p_i(\boldsymbol{\theta})$ may be used to get test statistics for testing

$$H_0 : F = F_0, \tag{2.11}$$

see, e.g., Greenwood (1946), Pyke (1965), Cressie (1976) and Hall (1986). Gebert and Kale (1969) indicate that test statistics of this type are useful for detecting departures from hypothesized density functions.

If we assume that the above hypothesis is simple, applying the probability integral transformation F_0 to the sample values $Y_1, \ldots, Y_n$ permits assuming, without loss of generality, that the null hypothesis (2.11) specifies the uniform distribution on the unit interval. Then we wish to test

$$H_0 : G(u) = u, 0 \leq u \leq 1, \tag{2.12}$$

where $G(u)$ is the distribution of $U_i = F(Y_i)$, $i = 1, \ldots, n$. If $U_{(1)} < U_{(2)} < \ldots < U_{(n)}$ are the order statistics from the transformed sample, the spacings are defined by $V_i = U_{(i)} - U_{(i-1)}$ and it is possible to define statistics to test (2.12). See, for instance, Kimball, (1947), Darling (1953), Kale and Godambe (1967), Kirmani and Alam (1974), Pyke (1965), Kale (1969), Sethuraman and Rao (1970) and Rao and Sethuraman (1975). Tests based symmetrically on spacings, namely, of the form $T_n = \frac{1}{n} \sum_{i=1}^{n} h(nV_i)$ are more common among these. Kuo and Rao (1984) demonstrated that among a wide class of such tests, the Greenwood test statistic obtained with $h(x) = x^2$ has maximum efficacy. Jammalamadaka et *al.* (1989) established the asymptotic distribution for $h(x) = x \log x$ under the null hypothesis given in (2.12) and the alternatives given by the densities $f_n(x) = 1 + n^{-1/4} l(x)$, $0 \leq x \leq 1$ where $l(\cdot)$ is assumed to be square integrable and continuously differentiable on $[0, 1]$.

The above tests are known in the literature as test statistics based on first-order spacings. Several authors have proposed generalizations of first-order uniform spacings. Cressie (1976) considers tests statistics of the form

$$S_n^{(m)} = \sum_{i=1}^{n+2-m} g\left((n+1)V_i^{(m)}\right), \qquad m \leq n+1,$$

where g is a "smooth" function, and $V_i^{(m)} = U_{(i-1+m)} - U_{(i-1)}$, $i = 1, \ldots, n+2-m$, are the mth-order spacings or mth-order gaps. See, e.g., Hall (1986), Stephens (1986), Cressie (1979), Del Pino (1979) among others. In this sense, the results for testing normality and uniformity, obtained by Vasicek (1976) and Dudewicz and Van der Meulen (1981) respectively, are important. These authors propose the following test statistics based on Shannon's entropy

$$H_{n,m} = \frac{1}{n}\sum_{i=1}^{n} \log\left(\frac{n}{2m}(Y_{(i+m)} - Y_{(i-m)})\right).$$

Two modified versions of the above test statistic are given by Ebraimi et *al.* (1994).

It is important to remark that previous methods for estimation and testing use all the sample information; however it is very hard to find the asymptotic distribution of the estimators even in the case of specific parametric models. This is obviously not the case for the maximum likelihood principle or the method of moments. Now we propose a method to decrease the amount of sample information in order to be able to have estimators with known asymptotic distribution when the MEP is used.

We consider a population with probability density function $f_{\boldsymbol{\theta}}(x)$. Suppose that we have a random sample $Y_1, \ldots, Y_n$, and we only observe the order statistics

$$Y_{([\frac{n}{M}]+1)}, Y_{([\frac{2n}{M}]+1)}, \ldots, Y_{([\frac{(M-1)n}{M}]+1)}, \tag{2.13}$$

where $[x]$ is the integer part function. We form M cells having boundaries

$$-\infty = c_{0,n} < c_{1,n} < \ldots < c_{M-1,n} < c_{M,n} = \infty,$$

where $c_{i,n} = Y_{([\frac{in}{M}]+1)}$. The observed frequency N_{in} in the ith-cell $(c_{i-1,n}, c_{i,n}]$ is nonrandom, $N_{in} = [\frac{in}{M}] - [\frac{(i-1)n}{M}]$. The vectors of sample and population quantiles of orders $\frac{1}{M}, \frac{2}{M}, \ldots \frac{M-1}{M}$ are $\boldsymbol{c}_n = (c_{1,n}, \ldots, c_{M-1,n})^T$ and $\boldsymbol{c} = (c_1, \ldots, c_{M-1})^T$, respectively. The ith "estimated cell probability" is therefore

$$p_i(\boldsymbol{c}_n; \boldsymbol{\theta}) = F_{\boldsymbol{\theta}}(c_{i,n}) - F_{\boldsymbol{\theta}}(c_{(i-1),n}) \tag{2.14}$$

where $\boldsymbol{\theta} = (\theta_1, \ldots, \theta_{M_0})^T$ and $F_{\boldsymbol{\theta}}$ is the probability distribution function associated with $f_{\boldsymbol{\theta}}(x)$. These are random, unlike the cell frequencies. For $\boldsymbol{\theta} = \boldsymbol{\theta}_0$ (true value of the parameter), we write $p_i = p_i(\boldsymbol{c}; \boldsymbol{\theta}_0) = \frac{i}{M} - \frac{i-1}{M} = \frac{1}{M}$, $i = 1, \ldots, M$. In this context in Menéndez et *al.* (1997a) the following definition was given.

Definition 2.3

Let $Y_1, \ldots, Y_n$ be a random sample from a population described by the random *variable X with probability density function $f_{\boldsymbol{\theta}}(x)$. The MEP estimator based on the order statistics given in (2.13) is defined by*

$$\widehat{\boldsymbol{\theta}}_{Sh} = \arg\max_{\boldsymbol{\theta}\in\Theta} H(\mathbf{p}(\boldsymbol{c}_n;\boldsymbol{\theta})), \tag{2.15}$$

where

$$H(\mathbf{p}(\boldsymbol{c}_n;\boldsymbol{\theta})) = -\sum_{i=1}^{M} p_i(\boldsymbol{c}_n;\boldsymbol{\theta}) \log p_i(\boldsymbol{c}_n;\boldsymbol{\theta}),$$

and $p_i(\boldsymbol{c}_n;\boldsymbol{\theta})$, $i = 1, ..., M$, are defined in (2.14).

We remark that the approach that takes some but not all the sample quantiles has been proposed by Bofinger (1973) for problems of testing goodness of fit with Pearson test statistic. Menéndez et *al.* (1997a) adapted this idea to problems of point estimation. Note also that the approach taken here is applied also to "multiple type II censoring", in which observations between several sets of sample percentiles are unavailable. It is necessary only to take each unobserved interpercentile group as a cell. This is conceptually quite similar to the generality of censoring allowed in the procedures of Turnbull and Weiss (1978). Another interesting approach in which only a relatively small number of order statistics are used in testing for goodness-of-fit is given in Weiss (1974). He establishes that $U_{([n^\delta])}, U_{(2[n^\delta])}, \ldots, U_{(k(n)[n^\delta])}$, with $\delta \in (3/4, 1)$ and $k(n) = [n^{1-\delta}]$, are asymptotically sufficient and can be assumed to have a joint normal distribution for all asymptotic purposes.

Under some assumptions, that can be seen in Menéndez et *al.* (1997a), it can be established that the estimator, $\widehat{\boldsymbol{\theta}}_{Sh}$, verifying the condition (2.15), verifies the following properties,

a) $\widehat{\boldsymbol{\theta}}_{Sh}$ converges in probability to $\boldsymbol{\theta}_0$.

b) $\widehat{\boldsymbol{\theta}}_{Sh}$ is asymptotically efficient, i.e.,

$$\widehat{\boldsymbol{\theta}}_{Sh} = \boldsymbol{\theta}_0 + n^{-1}\boldsymbol{B}_0 \left(\frac{\partial \log L(\boldsymbol{\theta})}{\partial \boldsymbol{\theta}}\right)_{\boldsymbol{\theta}=\boldsymbol{\theta}_0} + o_P(n^{-1/2}),$$

where $\boldsymbol{B}_0$ is a matrix of constants which may depend on $\boldsymbol{\theta}_0$ and $\log L(\boldsymbol{\theta})$ is given by

$$\log L(\boldsymbol{\theta}) = \sum_{i=1}^{M} N_{in} \log p_i(\boldsymbol{c}_n;\boldsymbol{\theta}).$$

c) $n^{1/2}(\widehat{\boldsymbol{\theta}}_{Sh} - \boldsymbol{\theta}_0) \xrightarrow[n\to\infty]{L} N(\mathbf{0}, \mathcal{I}_{\mathcal{F}}(\boldsymbol{\theta}_0)^{-1}).$

Based on sample quantiles, the MEP can be used to test if data come from a given parametric model. In this sense in the cited paper of Menéndez et *al.* (1997a) it was established that

$$T_n = 2n\left(\log M - H(\mathbf{p}(\boldsymbol{c}_n; \widehat{\boldsymbol{\theta}}_{Sh}))\right) \xrightarrow[n\to\infty]{L} \chi^2_{M-M_0-1}.$$

2.6. Exercises

1. Let $\mathcal{X} = \{C_1, ..., C_M\}$ be a set of M categories and let $\boldsymbol{p} = (p_1, ..., p_M)^T$ be a probability measure defined on $\mathcal{X}$. Find the expected distance between two categories drawn at random when the distance is defined as zero if they belong to the same category and unity otherwise.

2. Find the asymptotic variance of the entropies of Rényi, Havrda-Charvat, Arimoto and Sharma-Mittal when the probability vector $\boldsymbol{p} = (p_1, ..., p_M)^T$ is replaced by the estimator $\widehat{\boldsymbol{p}} = (\widehat{p}_1, ..., \widehat{p}_M)^T$ based on a simple random sample of size n.

3. Consider the hypothesis $H_0 : p_1 = ... = p_M = 1/M$ and the test statistics based on the entropies appearing in Exercise 2. Find, in each case, the corresponding expression of the test statistic.

4. The following data correspond to the occupational status by race and year in Walton County, Florida.

Occupational Status	(Observed proportions) White (1880)	White (1885)	Black (1880)	Black (1885)
Professional	0.029	0.093	0.00	0.00
Manager, clerical, proprietor	0.019	0.099	0.007	0.00
Skilled	0.086	0.040	0.007	0.00
Unskilled	0.053	0.073	0.046	0.049
Laborer	0.455	0.517	0.776	0.896
Farmer	0.358	0.178	0.164	0.055
Sample size	209	151	152	144

Source: Nayak, T. K. (1983), Ph. D. Dissertation, University of Pittsburgh.

   a) Compute the diversity indexes based on Shannon's entropy, as well as the asymptotic variances, for the four populations.

   b) Test uniformity in every population with significance level $\alpha = 0.05$.

   c) From data in the previous table one can observe that diversity for the black population has decreased over time. Check this conjecture by using a test statistic with significance level $\alpha = 0.05$ and a 95% confidence interval.

   d) What happens for the white population? Do a similar study to the one given in *c)*.

5. We have obtained from welding dive, four random samples corresponding to bounded areas of 1 squared meter from 5 meters deep. The sampling stations are found in the Spanish coast of Catalonia, from south to the north, from the delta of Ebro river (1) through the area of Maresme (2 and 3) to the Rosas bay (4). The two stations in the area of Maresme have a granite sandy bottom (thick); while the stations situated in delta of Ebro river and in the Rosas bay have a chalky sandy bottom (thin). The descriptions of the communities are listed in Tables 2.4, 2.5, 2.6 and 2.7.

Sampling station 1: Riomar (Delta of Ebro river)
date: 30/06/91 Type of sand: Thin
Table 2.4

Species	**Frequency**
Lentidium mediterraneum	216
Tellina tennis	189
Spisula subtruncata	69
Donax trunculus	51
Mactra corallina	45
Dosina lupinus	40
Carastoderma flancum	37
Otherwise	76
Total	723

Source: Pardo, L., Calvet, C. and Salicrú, M. (1992).

Sampling station 2: Canet de Mar (El Maresme)
Date: 25/03/91 Type of sand: Granite
Table 2.5

Species	**Frequency**
Spisula sultrancata	345
Glycymeris glycymeris	52
Alcanthocardia tuberculada	36
Donax variegatus	36
Donacilla cornea	34
Chamelea gallina	28
Callista chione	18
Otherwise	71
Total	620

Source: Pardo, L., Calvet, C. and Salicrú, M. (1992).

Sampling station 3: Malgrat de Mar (El Maresme)	
Date: 17/04/91 Type of sand: Granite	
Table 2.6	
Species	**Frequency**
Chamelea gallina	440
Spisula subtruncata	377
Callista chione	38
Donax variegatus	35
Glycymeris glycymeris	33
Dosina exoleta	31
Corbula gibba	8
Otherwise	41
Total	1003

Source: Pardo, L., Calvet, C. and Salicrú, M. (1992).

Sampling station 4: Rosas (Rosas bay)	
Date: 09/03/91 Type of sand: Thin	
Table 2.7	
Species	**Frequency**
Ceratoderma glaucum	134
Spisula subtruncata	93
Tapes decussatus	69
Venerupis aurea	59
Loripes lacteus	52
Chamelea gallina	33
Acanthoc. tuberculate	32
Otherwise	135
Total	607

Source: Pardo, L., Calvet, C. and Salicrú, M. (1992).

a) Compute the indexes of diversity based on the Gini-Simpson index for the four stations.

b) Test the uniformity in every station with significance level $\alpha = 0.05$.

c) Give a 95% confidence interval for the diversity in the four populations.

6. Let $X_1, ..., X_n$ a random sample of a normal population, with mean μ and variance σ^2. By using asymptotic properties of Shannon's entropy statistic, prove the following relation $\sqrt{2n} \log \frac{\widehat{\sigma}}{\sigma} \xrightarrow[n \to \infty]{L} N(0,1)$, where $\widehat{\sigma}$ is the maximum likelihood estimator of σ.

7. Two classes of drugs, which are supposed to have some effect in a disease, were tested on 288 individuals. The individuals were classified in four groups according to class of drug they had received. Groups 1 and 2: different drug, group 3: both drugs, group 4: no drug. The records are presented in the following table.

	Group 1	Group 2	Group 3	Group 4
	12	12	13	1
	10	4	14	8
	13	11	14	9
	13	7	17	9
	12	8	11	9
	10	10	11	4
		12	14	0
		5	14	1
$S_i^2 = \frac{1}{n_i-1} \sum_{j=1}^{n_i} (x_{ji} - \overline{x}_i)^2$	1.867	9.696	3.714	16.411

Use the test statistic S_{Sha} given in (2.4) to test the equality of variances in the four groups with significance level $\alpha = 0.05$.

8. The following values correspond to a simple random sample from a population with distribution function F and support the real line: -7.238, -0.804, -0.44, 0.18, -0.02, -1.08, 1.327, 1.98, -0.73, -0.27, -0.32, 0.58, -2.308, -4.37, 0.307, 4.72, 0.32, 0.124, -4.41, 1.98, -0.73, -1.27, -0.32, 0.58, -2.34, -8.19, -12.99, 1.51, 1.09, -4.37. Suppose that F is the Cauchy distribution function, whose probability density function is

$$f(x) = \frac{1}{\pi} \frac{1}{1+x^2} \qquad x \in \mathbb{R}$$

and consider the partition, $(-\infty, -1], (-1, 0], (0, 1], (1, \infty)$.

a) Calculate Shannon's entropy associated with the previous partition.

b) Based on Shannon's entropy test the hypothesis that the previous observations come from a Cauchy population with significance level $\alpha = 0.05$.

c) Calculate the asymptotic power at the point $\boldsymbol{p}^* = (\frac{4}{16}, \frac{4}{16}, \frac{3}{16}, \frac{5}{16})^T$.

9. Let X be a random variable whose density function is $f(x) = |x| e^{-x^2}$, $x \in \mathbb{R}$.

   a) Find an equiprobable partition with four classes in $\mathbb{R}$. Use this partition to give a test statistic, based on Shannon's entropy, for testing if the observations -1.1, 0.42, -0.25, 4.73, -7.2, 12.3, -0.22, 1.8, -0.7, -1.9, 4.77, 2.75, 0.01, -4.2, -0.1, -0.01, 3.23, -0.97, -0.12, -3.2, -0.15, -12.3, 0.75, -0.4, -7.4, 0.27, 1.51, -2.4, -2.67, 0.32 are from the population described by X with significance level $\alpha = 0.05$.

   b) Find the asymptotic power at the point $\boldsymbol{p}^* = (1/2, 1/4, 1/8, 1/8)^T$.

10. We have the following observations from a population with distribution function F and support the real line: 1.491, 2.495, 3.445, 1.108, 3.916, 1.112, 3.422, 2.278, 1.745, 3.131, 0.889, 2.099, 2.693, 2.409, 1.030, 1.437, 0.434, 1.655, 2.130, 1.967, 1.126, 3.113, 2.711, 0.849, 1.904, 1.570, 3.313, 2.599, 2.263, 2.208, 1.6771, 3.173, 1.235, 2.034, 4.007, 2.653, 2.269, 1.774, 4.077, 0.733, 0.061, 1.961, 1.916, 2.607, 2.060, 1.444, -0.357, 0.211, 2.555, 1.157.

    a) Test the hypothesis, with a significance level $\alpha = 0.05$, if the previous observations come from a normal distribution with mean $\mu = 2$ and variance $\sigma^2 = 1.1$, using Shannon's entropy and a partition with 6 equiprobable classes.

    b) Obtain the asymptotic power at the point $\boldsymbol{p}^* = (\frac{1}{15}, \frac{2}{15}, \frac{3}{15}, \frac{3}{15}, \frac{3}{15}, \frac{3}{15})^T$.

11. Let $X_1, ..., X_n$ be a random sample of size n from an exponential distribution of parameter θ. Using Shannon's entropy, give an asymptotic test statistic to test $H_0 : \theta = \theta_0$ versus $H_1 : \theta \neq \theta_0$.

12. Using the entropy of Havrda and Charvat, $H^s(\boldsymbol{\theta})$.

    a) Find the asymptotic distribution of $\sqrt{n}\left(H^s(\widehat{\boldsymbol{\theta}}) - H^s(\boldsymbol{\theta})\right)$, where $\widehat{\boldsymbol{\theta}}$ is the maximum likelihood estimator of $\boldsymbol{\theta}$.

    b) Obtain the asymptotic distribution of $\sqrt{n}\,(H^s(\widehat{\boldsymbol{p}}) - H^s(\boldsymbol{p}))$, as a particular case of the result obtained in *a)*.

13. We consider Laplace's distribution

$$f(x, \theta_1, \theta_2) = \frac{1}{2\theta_2} \exp(-\theta_2^{-1} |x - \theta_1|), \qquad -\infty < x < \infty, \theta_1 \in \mathbb{R}, \theta_2 > 0$$

whose variance is $2\theta_2^2$.

a) Find the expression of the entropy of Havrda and Charvat for this distribution and derive from it the expression of Shannon's entropy.

b) Test, using the entropy of Havrda and Charvat, $H_0 : \theta_2 = \theta^*$ versus $H_1 : \theta_2 \neq \theta^*$.

c) Derive a procedure for testing that the variances of s Laplace independent populations are equal. Give a procedure based on the entropy of Havrda and Charvat.

14. We consider the population given by the following density function

$$f_\theta(x) = \frac{\theta 2^\theta}{x^{\theta+1}} \qquad x \geq 2, \quad \theta > 0.$$

a) Find the test based on Shannon's entropy for testing

$$H_0 : \theta = \theta_0 \qquad \text{against } H_1 : \theta \neq \theta_0.$$

b) Using the test obtained in *a)* and the observations 2.2408, 5.8951, 6.0717, 3.6448, 2.8551, 4.4065, 14.4337, 3.0338, 2.0676, 2.6155, 2.7269, 5.1468, 2.2178, 2.0141, 2.3339, 2.6548, 5.0718, 2.8124, 2.0501, 13.6717, test $H_0 : \theta = 2$ versus $H_1 : \theta \neq 2$ with significance level $\alpha = 0.05$.

15. Find the acceptance region of S_{Sha} given in (2.4) for equality of two variances and compare it with the acceptance region of the test statistics given by Bartlett and Lhemann.

2.7. Answers to Exercises

1. If we consider the distance,

$$d(C_i, C_j) = \begin{cases} 1, & C_i \neq C_j, \quad \Pr(d(C_i, C_j) = 1) = p_i p_j \\ 0 & C_i = C_j \quad \Pr(d(C_i, C_i) = 0) = p_i^2 \end{cases},$$

we have

$$E_{\mathbf{p}}[d] = \sum_{i=1}^{M} \sum_{\substack{j=1 \\ i \neq j}}^{M} p_i p_j = 1 - \sum_{i=1}^{M} p_i^2 = H_{GS}(\mathbf{p}).$$

2. It is a simple exercise to get from Remark 2.4 the following table

Measure	*Asymptotic variance*
Havrda-Charvat	$\left(\frac{s}{s-1}\right)^2 \left(\sum_{i=1}^{M} p_i^{2s-1} - \left(\sum_{i=1}^{M} p_i^s\right)^2\right)$
Rényi	$\left(\frac{1}{r-1}\right)^2 \left(\sum_{i=1}^{M} p_i^r\right)^{-2} \left(\sum_{i=1}^{M} p_i^{2r-1} - \left(\sum_{i=1}^{M} p_i^r\right)^2\right)$
Arimoto	$\frac{1}{(t-1)^2} \left(\left(\sum_{i=1}^{M} p_i^{\frac{2}{t}-1}\right)\left(\sum_{i=1}^{M} p_i^{\frac{1}{t}}\right)^{2(t-1)} - \left(\sum_{i=1}^{M} p_i^{\frac{1}{t}}\right)^{2t}\right)$
Sharma-Mittal $(1,s)$	$\exp\left(2(s-1)\sum_{i=1}^{M} p_i \log p_i\right) \times \left(\sum_{i=1}^{M} p_i (\log p_i)^2 - \left(\sum_{i=1}^{M} p_i \log p_i\right)^2\right)$
Sharma-Mittal (r,s)	$\left(\frac{r}{r-1}\right)^2 \left(\sum_{i=1}^{M} p_i^r\right)^{\frac{2(s-r)}{r-1}} \left(\sum_{i=1}^{M} p_i^{2r-1} - \left(\sum_{i=1}^{M} p_i^r\right)^2\right)$

3. From Corollary 2.2 the expression of the statistic given in (2.8) for the different entropy measures is

Measure	*Statistics*
Havrda-Charvat	$\frac{2n}{(-s)\ M^{1-s}} \left(H^s(\widehat{\mathbf{p}}) - H^s(\boldsymbol{u})\right)$
Rényi	$\frac{2n}{(-r)} \left(H_r^1(\widehat{\mathbf{p}}) - H_r^1(\boldsymbol{u})\right)$
Arimoto	$\frac{2n}{(-t^{-1})\ M^{t-1}} \left({}_tH(\widehat{\mathbf{p}}) -{}_t H(\boldsymbol{u})\right)$
Sharma-Mittal $(1,s)$	$\frac{2n}{-M^{1-s}} \left(H_1^s(\widehat{\mathbf{p}}) - H_1^s(\boldsymbol{u})\right)$
Sharma-Mittal (r,s)	$\frac{2n}{-rM^{1-s}} \left(H_r^s(\widehat{\mathbf{p}}) - H_r^s(\boldsymbol{u})\right)$

4. *a)* First, we obtain the sample entropies, $H(\widehat{\boldsymbol{p}}_i)$, $i = 1, ..., 4$, as well as their asymptotic variances, $\sigma^2(\widehat{\boldsymbol{p}}_i)$, $i = 1, ..., 4$, for the four populations. Based on Theorem 2.3 and taking into account that in the case of Shannon's entropy $\phi(x) = -x \log x$, we have

$$\sigma_H^2(\boldsymbol{p}) = \sum_{i=1}^{M} p_i (\log p_i)^2 - \left(\sum_{i=1}^{M} p_i \log p_i \right)^2 .$$

In the case of the first population we get

$$\begin{aligned} H(\widehat{\boldsymbol{p}}_1) &= -0.029 \log 0.029 - ... - 0.358 \log 0.358 = 1.2707 \\ \sigma_H^2(\widehat{\boldsymbol{p}}_1) &= 0.029 (\log 0.029)^2 + ... + 0.358 (\log 0.358)^2 - H(\widehat{\boldsymbol{p}}_1)^2 \\ &= 0.6822. \end{aligned}$$

In a similar way we obtain the results for the other populations:

	$H(\widehat{\boldsymbol{p}}_i)$	$\sigma_H^2(\widehat{\boldsymbol{p}}_i)$
White (1880)	1.2707	0.6822
White (1885)	1.4179	0.7133
Black (1880)	0.7053	0.8706
Black (1885)	0.4060	0.7546

b) Now we have to test

$$H_0 : \boldsymbol{p}_1 = \boldsymbol{p}_2 = \boldsymbol{p}_3 = \boldsymbol{p}_4 = (1/6, ..., 1/6)^T .$$

We know that $\chi^2_{5,0.05} = 11.07$ and, on the other hand,

	$H(\widehat{\boldsymbol{p}_i})$	$\log M - \frac{\chi^2_{M-1,\alpha}}{2n_i}$
White (1880)	1.2707	1.7653
White (1885)	1.4179	1.7551
Black (1880)	0.7053	1.7553
Black (1885)	0.4060	1.7533

Therefore we should reject the null hypothesis.

c) In this case we must test

$$H_0 : H(\boldsymbol{p}_3) = H(\boldsymbol{p}_4) \qquad \text{versus } H_1 : H(\boldsymbol{p}_3) \neq H(\boldsymbol{p}_4),$$

and we use the test statistic

$$Z_2 = \frac{\sqrt{n_3 n_4}\left(H(\widehat{\mathbf{p}}_3) - H(\widehat{\mathbf{p}}_4)\right)}{\sqrt{n_4\sigma_H^2(\widehat{\mathbf{p}}_3) + n_3\sigma_H^2(\widehat{\mathbf{p}}_4)}}$$

whose value is $Z_2 = 2.857$. On the other hand the critical region is given by $(-\infty, -1.96) \cup (1.96, \infty)$ and we should reject the null hypothesis.

A 95% confidence interval for $H(\mathbf{p}_3) - H(\mathbf{p}_4)$ is given by

$$(H(\widehat{\mathbf{p}}_3) - H(\widehat{\mathbf{p}}_4)) \pm z_{0.025}\left(\frac{\sigma^2(\widehat{\mathbf{p}}_3)}{n_3} + \frac{\sigma^2(\widehat{\mathbf{p}}_4)}{n_4}\right)^{1/2}.$$

After some calculations the following interval is obtained

$$(0.0934,\ 0.5039).$$

This interval does not contain the value zero and then we can conclude that the change is significant.

d) In a similar way we get that the value of the test statistic is $Z_2 = -1.64$ and the confidence interval is

$$(-2.5072,\ 2.128).$$

We can not conclude that the change is significant.

5. *a)* The expression of Gini-Simpson index is

$$H_{GS}(\mathbf{p}) = 1 - \sum_{i=1}^{M} p_i^2.$$

Then

$$\begin{aligned} H_{GS}(\widehat{\mathbf{p}}_1) &= 1 - \left(\tfrac{216}{723}\right)^2 - \left(\tfrac{189}{723}\right)^2 - \left(\tfrac{69}{723}\right)^2 - \left(\tfrac{51}{723}\right)^2 - \left(\tfrac{45}{723}\right)^2 \\ &- \left(\tfrac{40}{723}\right)^2 - \left(\tfrac{37}{723}\right)^2 - \left(\tfrac{76}{723}\right)^2 \\ &= 0.8078. \end{aligned}$$

In a similar way we can get the expression of the Gini-Simpson index for the other populations

$$H_{GS}(\widehat{\mathbf{p}}_2) = 0.6576 \qquad H_{GS}(\widehat{\mathbf{p}}_3) = 0.6598 \qquad H_{GS}(\widehat{\mathbf{p}}_4) = 0.8428.$$

Using Exercise 2 (Havrda-Charvat with $s = 2$) the expression of the asymptotic variance is given by

$$\sigma^2_{GS}(\boldsymbol{p}) = 4\left(\sum_{i=1}^{M} p_i^3 - \left(\sum_{i=1}^{M} p_i^2\right)^2\right);$$

then,

$$\sigma^2_{GS}(\widehat{\boldsymbol{p}}_1) = 0.0419,\ \sigma^2_{GS}(\widehat{\boldsymbol{p}}_2) = 0.2312,\ \sigma^2_{GS}(\widehat{\boldsymbol{p}}_3) = 0.0882,\ \sigma^2_{GS}(\widehat{\boldsymbol{p}}_4) = 0.016.$$

b) From Exercise 3 we have

$$nM\left(\sum_{i=1}^{M} \widehat{p}_i^2 - M^{-1}\right) \underset{n\to\infty}{\overset{L}{\longrightarrow}} \chi^2_{M-1}.$$

The numerical value of the test statistic for each population appears in the following table:

Station	*Statistic*
Riomar	389.15
Canet del Mar	1078.40
Malgrat de Mar	1726.40
Rosas	155.97

But $\chi^2_{7,0.05} = 14.076$ and we should reject the homogeneity in each one of the four populations.

c) A $100\,(1-\alpha)\,\%$ confidence interval for the Gini-Simpson index in each one of the four populations is given by

$$\left(H_{GS}(\widehat{\boldsymbol{p}}) - \frac{\widehat{\sigma}(\widehat{\boldsymbol{p}})}{\sqrt{n}} z_{\alpha/2}, H_{GS}(\widehat{\boldsymbol{p}}) + \frac{\sigma(\widehat{\boldsymbol{p}})}{\sqrt{n}} z_{\alpha/2}\right).$$

In the following table we have the confidence intervals for each one of the four populations

Station	*Confidence Interval*
Riomar	(0.8047, 0.8108)
Canet del Mar	(0.6394, 0.6758)
Malgrat de Mar	(0.6543, 0.6652)
Rosas	(0.8415, 0.8440)

On the basis of the confidence intervals obtained we have the following relations for the diversities: Rosas diversity is greater than Riomar diversity and Riomar diversity is greater than Canet diversity and Canet diversity equal to Malgrat de Mar diversity.

Then in our case the stations with a bottom of thin sand present greater diversity than the stations with a bottom of granitic sand. A study of this problem was made by Pardo, L. et *al.* (1992).

6. For Shannon's entropy we have

$$H(\sigma^2) \equiv H(N(\mu,\sigma^2)) = \log(\sigma^2 2\pi e)^{1/2},$$

then

$$\sqrt{n}\left(H(\widehat{\sigma}^2) - H\left(\sigma^2\right)\right) \xrightarrow[n\to\infty]{L} N\left(0, \boldsymbol{T}^T \mathcal{I}_{\mathcal{F}}\left(\boldsymbol{\theta}\right)^{-1} \boldsymbol{T}\right),$$

where $\boldsymbol{T}^T = (t_1, t_2)$ is given in the example of Section 2.3.4 and $\boldsymbol{\theta} = (\mu, \sigma^2)$,

$$t_1 = \frac{\partial H(\sigma^2)}{\partial \mu} = 0, \quad t_2 = \frac{\partial H(\sigma^2)}{\partial \sigma^2} = \frac{1}{2\sigma^2},$$

and

$$\mathcal{I}_{\mathcal{F}}(\mu,\sigma^2)^{-1} = \begin{pmatrix} \sigma^2 & 0 \\ 0 & 2\sigma^4 \end{pmatrix}.$$

Then $\boldsymbol{T}^T \mathcal{I}_{\mathcal{F}}\left(\boldsymbol{\theta}\right)^{-1} \boldsymbol{T} = 1/2$ and

$$\sqrt{n}\left(H(\widehat{\sigma}^2) - H\left(\sigma^2\right)\right) \xrightarrow[n\to\infty]{L} N\left(0, 1/2\right).$$

Substituting $H(\sigma^2)$ and $H(\hat{\sigma}^2)$ by their expressions we have the statement.

7. Based on (2.4), we reject the null hypothesis

$$H_0 : \sigma_1^2 = \sigma_2^2 = \sigma_3^2 = \sigma_4^2$$

if

$$Z_4 = \frac{1}{2}\sum_{j=1}^{r} n_j \left(\log \widehat{\sigma}_j^2 - \log \prod_{j=1}^{r} \left(\widehat{\sigma}_j^2\right)^{\frac{n_j}{N}}\right)^2 > \chi^2_{r-1,\alpha}.$$

In our case we have

$$\widehat{\sigma}_1^2 = 1.5556,\ \widehat{\sigma}_2^2 = 8.4840,\ \widehat{\sigma}_3^2 = 3.25,\ \widehat{\sigma}_4^2 = 14.359,$$

and $Z_4 = 10.32$. On the other hand we have $\chi^2_{3,0.05} = 7.82$ and then we should reject the null hypothesis.

8. *a)* It is immediate to show that $p_1 = \Pr((-\infty, -1]) = p_2 = \Pr((-1, 0]) = p_3 = \Pr((0, 1]) = p_4 = \Pr((1, \infty)) = 1/4$ and $H(\boldsymbol{u}) = \log 4 = 1.3863$, where $\boldsymbol{u} = (1/4, 1/4, 1/4, 1/4)^T$

   b) For testing

$$H_0 : \text{The observations come from a Cauchy population,}$$

   we could test

$$H_0 : p_1 = p_2 = p_3 = p_4 = 1/4.$$

   The critical region is

$$2n \left(\log 4 - H(\widehat{\mathbf{p}})\right) > \chi^2_{M-1, 0.05}.$$

   In our case,

$$\widehat{p}_1 = 0.333, \widehat{p}_2 = 0.267, \widehat{p}_3 = 0.2, \widehat{p}_4 = 0.2$$

   and then $H(\widehat{\mathbf{p}}) = 1.3625$. Therefore,

$$Z_6 = 2n \left(\log 4 - H(\widehat{\mathbf{p}})\right) = 1.4268.$$

   On the other hand $\chi^2_{3,0.05} = 7.815$. Then the null hypothesis should be not rejected.

   c) If we denote $\boldsymbol{p}^* = \left(\frac{4}{16}, \frac{4}{16}, \frac{3}{16}, \frac{5}{16}\right)^T$, we have

$$H(\boldsymbol{p}^*) = 1.3705 \text{ and } \sigma^2(\boldsymbol{p}^*) = 0.0308.$$

   The power at $\boldsymbol{p}^* = \left(\frac{4}{16}, \frac{4}{16}, \frac{3}{16}, \frac{5}{16}\right)^T$ is given by

$$\begin{aligned}
\beta_{30}(\boldsymbol{p}^*) &= \Pr\left(2n(\log 4 - H(\widehat{\mathbf{p}})) > \chi^2_{3,0.05}\right) \\
&= \Pr\left(-2nH(\widehat{\mathbf{p}}) > \chi^2_{3,0.05} - 2n \log 4\right) \\
&= \Pr\left(\frac{\sqrt{n}}{\sigma(\boldsymbol{p}^*)} \left(H(\widehat{\mathbf{p}}) - H(\boldsymbol{p}^*)\right) < \frac{-\chi^2_{3,0.05} + 2n(\log 4 - H(\boldsymbol{p}^*))}{2\sqrt{n}\sigma(\boldsymbol{p}^*)}\right) \\
&\approx \Pr\left(Z < \frac{-7.815 + 83.1776 - 82.2301}{2 \times 30^{1/2} \times 0.1755}\right) = 0.0002,
\end{aligned}$$

   where Z is a standard normal variable.

9. *a)* First we obtain the equiprobable partition. The first value will be the solution of the equation

$$-\int_{-\infty}^{a_1} x e^{-x^2} dx = \frac{1}{4},$$

and this is given by $a_1 = -0.8305$. The second point is obviously $a_2 = 0$ and the third one, by symmetry, is $a_3 = 0.8325$. Then the partition is given by

$$(-\infty, -0.8305], \ (-0.8305, 0], \ (0, 0.8305], (0.8305, \infty).$$

On the basis of this partition we have

$$\widehat{p}_1 = \frac{1}{3}, \ \widehat{p}_2 = \frac{8}{30}, \ \widehat{p}_3 = \frac{5}{30}, \widehat{p}_4 = \frac{7}{30}.$$

Then

$$H(\widehat{\boldsymbol{p}}) = -\sum_{i=1}^{4} \widehat{p}_i \log \widehat{p}_i = 1.3569$$

and

$$2n \left(\log 4 - H(\widehat{\boldsymbol{p}})\right) = 2 \times 30 \left(\log 4 - 1.3569\right) = 1.7636.$$

But

$$\chi^2_{3,0.05} = 7.8115$$

and there is not statistical evidence to reject the null hypothesis.

b) The power at $\boldsymbol{p}^* = (1/2, 1/4, 1/8, 1/8)^T$ is given by

$$\beta_{30}(\boldsymbol{p}^*) \ \approx \ \Phi\left(\tfrac{\sqrt{30}}{\sigma(\boldsymbol{p}^*)}\left(\log 4 - \tfrac{7.8115}{60} - H(\boldsymbol{p}^*)\right)\right) = 0.6591$$

where

$$H(\boldsymbol{p}^*) = -\frac{1}{2}\log\frac{1}{2} - \frac{1}{4}\log\frac{1}{4} - \frac{1}{8}\log\frac{1}{8} - \frac{1}{8}\log\frac{1}{8} = 1.2130$$

and

$$\sigma^2(\boldsymbol{p}^*) = \sum_{i=1}^{4} p_i^* (\log p_i^*)^2 - \left(\sum_{i=1}^{4} p_i^* \log p_i^*\right)^2 = 0.3303.$$

10. *a)* First we obtain the equiprobable partition. The first value, a_1, must verify

$$\Pr\left(-\infty < N\left(2, 1.1\right) < a_1\right) = \frac{1}{6};$$

this is equivalent to

$$\Pr(-\infty < \frac{X-2}{1.1} < \frac{a_1 - 2}{1.1}) = \frac{1}{6}.$$

Then a_1 verifies

$$\Phi(\frac{a_1 - 2}{1.1}) = 1/6,$$

where Φ denotes the distribution function of the standard normal distribution. Then $a_1 = 2 - 0.97 \times 1.1 = 0.933$ and the first interval is $I_1 = (-\infty, 0.933]$; the interval I_2 is given by $I_2 = (0.933, a_2]$ where a_2 is obtained in such a way that the probability of I_2 is $\frac{1}{6}$. Then $a_2 = 1.527$. In the same way we obtain $I_3 = (1.527, 2]$, $I_4 = (2, 2.473]$, $I_5 = (2.473, 3.067]$ and $I_6 = (3.067, \infty)$.

It is easy to get

$$\widehat{p}_1 = 0.14,\ \widehat{p}_2 = 0.18,\ \widehat{p}_3 = 0.18,\ \widehat{p}_4 = 0.18,\ \widehat{p}_5 = 0.14 \text{ and } \widehat{p}_6 = 0.18.$$

Then

$$H(\widehat{\boldsymbol{p}}) = -\sum_{i=1}^{6} \widehat{p}_i \log \widehat{p}_i = 1.7852,$$

and

$$Z_6 = 2n(\log M - H(\widehat{\boldsymbol{p}})) = 1.1259.$$

On the other hand $\chi^2_{5,0.05} = 11.07$ and we should reject the null hypothesis.

b) It is clear that the power at $\boldsymbol{p}^* = (\frac{1}{15}, \frac{2}{15}, \frac{3}{15}, \frac{3}{15}, \frac{3}{15}, \frac{3}{15})^T$ is

$$\beta_{50}(\boldsymbol{p}^*) \approx \Phi\left(\frac{\sqrt{50}}{\sigma(\boldsymbol{p}^*)}\left(\log 6 - \tfrac{11.07}{100} - H(\boldsymbol{p}^*)\right)\right) = 0.0901$$

where

$$H(\boldsymbol{p}^*) = -\tfrac{1}{15}\log\tfrac{1}{15} - \tfrac{2}{15}\log\tfrac{2}{15} - 4 \times (\tfrac{3}{15}\log\tfrac{3}{15}) = 1.7367,\ n = 50$$

and

$$\sigma(\boldsymbol{p}^*) = \left(\sum_{i=1}^{6} p_i(\log p_i)^2 - \left(\sum_{i=1}^{6} p_i \log p_i\right)^2\right)^{1/2} = 0.2936.$$

11. Shannon's entropy for the exponential distribution is given by

$$H(\theta) = 1 - \log\theta,$$

whereas the maximum likelihood estimator and the Fisher information are

$$\widehat{\theta} = \frac{n}{\sum_{i=1}^{n} X_i} = \overline{X}^{-1} \text{ and } \mathcal{I}_{\mathcal{F}}(\theta) = \theta^{-2}.$$

We know that

$$\sqrt{n}\left(H(\widehat{\theta})-H\left(\theta_{0}\right)\right) \underset{n\to\infty}{\overset{L}{\longrightarrow}} N\left(0,\sigma^{2}\left(\theta_{0}\right)\right),$$

with

$$\sigma^{2}\left(\theta_{0}\right)=\mathcal{I}_{\mathcal{F}}\left(\theta_{0}\right)^{-1}\left(\frac{\partial H\left(\theta_{0}\right)}{\partial\theta}\right)^{2}=\theta_{0}^{2}\theta_{0}^{-2}=1.$$

Then

$$\sqrt{n}(1-\log\widehat{\theta}-1+\log\theta_{0}) \underset{n\to\infty}{\overset{L}{\longrightarrow}} N\left(0,1\right),$$

and

$$\sqrt{n}\log\left(\overline{X}\theta_{0}\right) \underset{n\to\infty}{\overset{L}{\longrightarrow}} N\left(0,1\right),$$

i.e., the null hypothesis should be rejected if

$$\left|\sqrt{n}\log\left(\overline{X}\theta_{0}\right)\right|>z_{\alpha/2}.$$

12. *a)* A first order Taylor expansion gives

$$H^{s}(\widehat{\boldsymbol{\theta}})=H^{s}\left(\boldsymbol{\theta}_{0}\right)+\sum_{i=1}^{M_{0}}\frac{\partial H^{s}\left(\boldsymbol{\theta}_{0}\right)}{\partial\theta_{i}}(\widehat{\theta}_{i}-\theta_{i0})+o\left(\left\|\widehat{\boldsymbol{\theta}}-\boldsymbol{\theta}_{0}\right\|\right).$$

But $\sqrt{n}(\widehat{\boldsymbol{\theta}}-\boldsymbol{\theta}_{0}) \underset{n\to\infty}{\overset{L}{\longrightarrow}} N(\mathbf{0},\mathcal{I}_{\mathcal{F}}\left(\boldsymbol{\theta}_{0}\right)^{-1})$, then $\sqrt{n}\ o\left(\left\|\widehat{\boldsymbol{\theta}}-\boldsymbol{\theta}_{0}\right\|\right)=o_{P}\left(1\right).$
Therefore, the random variables

$$\sqrt{n}\left(H^{s}(\widehat{\boldsymbol{\theta}})-H^{s}\left(\boldsymbol{\theta}_{0}\right)\right) \text{ and } \boldsymbol{T}^{T}\sqrt{n}(\widehat{\boldsymbol{\theta}}-\boldsymbol{\theta}_{0})$$

have the same asymptotic distribution, where $\boldsymbol{T}^{T}=(t_{1},...,t_{M_{0}})$ with

$$t_{i}=\frac{\partial H^{s}\left(\boldsymbol{\theta}_{0}\right)}{\partial\theta_{i}}=\frac{1}{1-s}\int_{\mathcal{X}}sf_{\boldsymbol{\theta}_{0}}(\boldsymbol{x})^{s-1}\frac{\partial f_{\boldsymbol{\theta}_{0}}(\boldsymbol{x})}{\partial\theta_{i}}d\mu(\boldsymbol{x})\qquad s\neq1.$$

Therefore,

$$\sqrt{n}\left(H^{s}(\widehat{\boldsymbol{\theta}})-H^{s}\left(\boldsymbol{\theta}_{0}\right)\right) \underset{n\to\infty}{\overset{L}{\longrightarrow}} N(\mathbf{0},\boldsymbol{T}^{T}\mathcal{I}_{\mathcal{F}}\left(\boldsymbol{\theta}_{0}\right)^{-1}\boldsymbol{T}).$$

b) In this case we have

$$\begin{aligned} H^{s}(\boldsymbol{p}) &= \tfrac{1}{1-s}\left(\sum_{i=1}^{M-1}p_{i}^{s}-1\right)+\tfrac{1}{1-s}(p_{M}^{s}-1),\\ \boldsymbol{T}^{T} &= (t_{1},...,t_{M-1})=\tfrac{s}{1-s}(p_{1}^{s-1}-p_{M}^{s-1},...,p_{M-1}^{s-1}-p_{M}^{s-1}). \end{aligned}$$

Then

$$\begin{aligned}\boldsymbol{T}^T\mathcal{I}_{\mathcal{F}}(\boldsymbol{p})^{-1}\boldsymbol{T} &= \tfrac{s^2}{(1-s)^2}\Big((p_1^{s-1},...,p_{M-1}^{s-1})\mathcal{I}_{\mathcal{F}}(\boldsymbol{p})^{-1}(p_1^{s-1},...,p_{M-1}^{s-1})^T \\ &- 2(p_1^{s-1},...,p_{M-1}^{s-1})\mathcal{I}_{\mathcal{F}}(\boldsymbol{p})^{-1}(p_M^{s-1},...,p_M^{s-1}) \\ &+ (p_M^{s-1},...,p_M^{s-1})\mathcal{I}_{\mathcal{F}}(\boldsymbol{p})^{-1}(p_M^{s-1},...,p_M^{s-1})\Big).\end{aligned}$$

In a similar way to Theorem 2.3 we have

$$\boldsymbol{T}^T\mathcal{I}_{\mathcal{F}}(\boldsymbol{p})^{-1}\boldsymbol{T} = \frac{s^2}{(1-s)^2}\left(\sum_{i=1}^{M} p_i^{2s-1} - \left(\sum_{i=1}^{M} p_i^s\right)^2\right).$$

13. *a)* The entropy of Havrda and Charvat for Laplace distribution is given by

$$\begin{aligned}H^s(\theta_2) &= \frac{1}{1-s}\left(\int_{\mathbb{R}} \frac{1}{(2\theta_2)^s}\exp(-\frac{s}{\theta_2}|x-\theta_1|)dx - 1\right) \\ &= \frac{1}{1-s}(2^{1-s}\theta_2^{1-s}s^{-1}-1), \qquad s \neq 1,\ s>0,\end{aligned}$$

and Shannon's entropy by

$$\lim_{s\to1} H^s(\theta_2) = \lim_{s\to1}\frac{1}{1-s}(2^{1-s}\theta_2^{1-s}s^{-1}-1) = \log 2\theta_2 + 1.$$

b) We want to test

$$H_0 : \theta_2 = \theta^* \text{ versus } H_1 : \theta_2 \neq \theta^*$$

and this is equivalent to test

$$H_0 : \frac{1}{1-s}\left(2^{1-s}\theta_2^{1-s}s^{-1}-1\right) = \frac{1}{1-s}\left(2^{1-s}(\theta^*)^{1-s}s^{-1}-1\right),$$

i.e.,

$$H_0 : H^s(\theta_2) = H^s(\theta^*).$$

For this purpose we use the test statistic

$$Z_1 = \frac{\sqrt{n}\left(H^s(\widehat{\theta}_2) - H^s(\theta^*)\right)}{\sigma(\theta^*)} \xrightarrow[n\to\infty]{L} N(0,1)$$

where

$$\sigma^2(\theta^*) = \boldsymbol{T}^T\mathcal{I}_{\mathcal{F}}(\theta^*)^{-1}\boldsymbol{T}$$

and

$$\boldsymbol{T}^T = \left(\frac{\partial H^s(\theta_2)}{\partial \theta_1}, \frac{\partial H^s(\theta_2)}{\partial \theta_2} \right)_{\theta_2 = \theta^*} = \left(0, \frac{2^{1-s}}{s} (\theta^*)^{-s} \right).$$

Therefore,

$$\begin{aligned} \sigma^2(\theta^*) &= \boldsymbol{T}^T \mathcal{I}_{\mathcal{F}}(\theta^*)^{-1} \boldsymbol{T} \\ &= \left(0, \frac{2^{1-s}}{s} (\theta^*)^{-s} \right) \begin{pmatrix} (\theta^*)^2 & 0 \\ 0 & (\theta^*)^2 \end{pmatrix} \left(0, \frac{2^{1-s}}{s} (\theta^*)^{-s} \right)^T \\ &= \frac{2^{2(1-s)}}{s^2} (\theta^*)^{-2s+2}. \end{aligned}$$

Let us observe that

$$\mathcal{I}_{\mathcal{F}}(\theta_2)^{-1} = \begin{pmatrix} \theta_2^2 & 0 \\ 0 & \theta_2^2 \end{pmatrix},$$

because

$$\begin{aligned} \log f_{\theta_1,\theta_2}(x) &= -\log 2\theta_2 - \frac{1}{\theta_2} |x - \theta_1| \\ \frac{\partial \log f_{\theta_1,\theta_2}(x)}{\partial \theta_2} &= -\frac{1}{\theta_2} + \frac{1}{\theta_2^2} |x - \theta_1| \\ \frac{\partial^2 \log f_{\theta_1,\theta_2}(x)}{\partial \theta_2^2} &= \frac{1}{\theta_2^2} \left(1 - \tfrac{2}{\theta_2} |x - \theta_1| \right). \end{aligned}$$

Then,

$$\mathcal{I}_{\mathcal{F}}(\theta_2) = -E\left[\frac{\partial^2 \log f_{\theta_1,\theta_2}(X)}{\partial \theta_2^2} \right] = -\frac{1}{\theta_2^2} \left(1 - \frac{2}{\theta_2} E\left[|X - \theta_1| \right] \right) = \frac{1}{\theta_2^2}$$

because

$$E\left[|X - \theta_1| \right] = \theta_2.$$

On the other hand we have

$$\frac{\partial \log f_{\theta_1,\theta_2}(x)}{\partial \theta_1} = -\frac{1}{\theta_2} \frac{x - \theta_1}{|x - \theta_1|};$$

therefore,

$$E\left[\left(-\frac{1}{\theta_2} \frac{\theta_1 - X}{|X - \theta_1|} \right)^2 \right] = \frac{1}{\theta_2^2}.$$

It is also clear that

$$E\left[\frac{\partial^2 \log f_{\theta_1,\theta_2}(X)}{\partial \theta_1 \partial \theta_2} \right] = 0,$$

hence

$$\mathcal{I}_{\mathcal{F}}(\theta_2) = \begin{pmatrix} \theta_2^{-2} & 0 \\ 0 & \theta_2^{-2} \end{pmatrix}.$$

It is not difficult to get that

$$Z_1 = \sqrt{n}\frac{1}{1-s}\left(\left(\frac{\widehat{\theta}_2}{\theta^*}\right)^{1-s} - 1\right).$$

The null hypothesis should be rejected if

$$\left|\sqrt{n}\frac{1}{1-s}\left(\left(\frac{\widehat{\theta}_2}{\theta^*}\right)^{1-s} - 1\right)\right| > z_{\alpha/2}$$

where

$$\widehat{\theta}_2 = \frac{1}{n}\left(\left|x_1 - x_{(1/2)}\right| + ... + \left|x_n - x_{(1/2)}\right|\right),$$

with $x_{(1/2)}$ the sample median.

c) In this case we want to test

$$H_0 : \sigma_1^2 = ... = \sigma_s^2,$$

i.e.,

$$H_0 : 2(\theta_2^{(1)})^2 = ... = 2(\theta_2^{(s)})^2, \tag{2.16}$$

where $\left(\theta_1^{(i)}, \theta_2^{(i)}\right)$ represent the parameters in the ith-population.

For testing (2.16) we are going to test

$$H_0 : H^s(\theta_2^{(1)}) = ... = H^s(\theta_2^{(s)}).$$

In this case the test statistic is

$$Z_4 = \sum_{i=1}^{s} \frac{n_i}{\sigma^2(\widehat{\theta_2^{(i)}})}\left(H^s(\widehat{\theta_2^{(i)}}) - \overline{D}\right)^2 \xrightarrow[n\to\infty]{L} \chi_{s-1}^2,$$

being

$$\overline{D} = \frac{1}{\sum\limits_{i=1}^{s} \frac{n_i}{\sigma^2(\widehat{\theta_2^{(i)}})}} \sum_{i=1}^{s} \frac{n_i}{\sigma_i^2(\widehat{\theta_2^{(i)}})} H^s(\widehat{\theta_2^{(i)}}).$$

14. *a)* First, we obtain Shannon's entropy associated with the considered population. We have

$$H(\theta) = -\int_2^\infty \frac{\theta 2^\theta}{x^{\theta+1}} \log \frac{\theta 2^\theta}{x^{\theta+1}} dx = -\log \frac{\theta}{2} + 1 + \frac{1}{\theta}.$$

Then we can test $H_0 : \theta = \theta_0$ by testing $H_0 : H(\theta) = H(\theta_0)$ and in this case we know that

$$Z_1 = \frac{\sqrt{n}}{\sigma(\theta_0)} \left(H(\widehat{\theta}) - H(\theta_0) \right) \underset{n \to \infty}{\overset{L}{\longrightarrow}} N(0,1)$$

with $\sigma^2(\theta_0) = t^2 \mathcal{I}_{\mathcal{F}}(\theta_0)^{-1}$, being

$$t = \left(\frac{\partial H(\theta)}{\partial \theta} \right)_{\theta=\theta_0} = -\frac{(1+\theta_0)}{\theta_0^2}, \quad \mathcal{I}_{\mathcal{F}}(\theta_0) = \theta_0^{-2}.$$

We have

$$\widehat{\theta} = \frac{n}{\sum_{i=1}^n \log \frac{x_i}{2}}.$$

Then we should reject the null hypothesis if

$$\left| \frac{\sqrt{n}}{\sigma(\theta_0)} \left(H(\widehat{\theta}) - H(\theta_0) \right) \right| > z_{\alpha/2}.$$

b) In this case we have $\widehat{\theta} = 1.7$ and

$$\frac{\sqrt{n}}{\sigma(\theta_0)} \left(H(\widehat{\theta}) - H(\theta_0) \right) = 2.9904.$$

As $z_{0.025} = 1.96$ we should reject the null hypothesis.

15. We are going to obtain the acceptance region S_{Sha} given in (2.4) for $r = 2$. In this case, we have

$$\begin{aligned} S_{Sha} = & \ \tfrac{1}{2} \left\{ n_1 \left(\log \widehat{\sigma}_1^2 - \tfrac{n_1}{N} \log \widehat{\sigma}_1^2 - \tfrac{n_2}{N} \log \widehat{\sigma}_2^2 \right)^2 \right. \\ + & \left. \ n_2 \left(\log \widehat{\sigma}_2^2 - \tfrac{n_1}{N} \log \widehat{\sigma}_1^2 - \tfrac{n_2}{N} \log \widehat{\sigma}_2^2 \right)^2 \right\} \\ = & \ \tfrac{1}{2} \left(n_1 \left(\tfrac{n_2}{N} \log \frac{\widehat{\sigma}_1^2}{\widehat{\sigma}_2^2} \right)^2 + n_2 \left(\tfrac{n_1}{N} \log \frac{\widehat{\sigma}_1^2}{\widehat{\sigma}_2^2} \right)^2 \right) \\ = & \ \tfrac{n_1 n_2}{2N} \left(\log \frac{\widehat{\sigma}_1^2}{\widehat{\sigma}_2^2} \right)^2. \end{aligned}$$

Therefore the acceptance region is

$$0 < \frac{n_1 n_2}{2N} \left(\log \frac{\widehat{\sigma}_1^2}{\widehat{\sigma}_2^2} \right)^2 < k$$

and this is equivalent to

$$c_1 < \frac{\widehat{\sigma}_1^2}{\widehat{\sigma}_2^2} < \frac{1}{c_2},$$

where $c_2 < 1$. The acceptance region is given by

$$c < \frac{\widehat{\sigma}_1^2}{\widehat{\sigma}_2^2} < \frac{1}{c} \qquad (0 < c < 1).$$

It is not difficult to establish that

$$S_{Leh} = \frac{(n_1 - 1)(n_1 - 1)}{2(n_1 + n_1 - 2)} \left(\log \frac{\frac{n_1}{n_1 - 1} \widehat{\sigma}_1^2}{\frac{n_2}{n_2 - 1} \widehat{\sigma}_2^2} \right)$$

and the acceptance region is the same.

The Bartlett's test is given by

$$S_{Bart} = n \log \left(\sum_{j=1}^{r} \frac{n_j \widehat{\sigma}_j^2}{n} \right) - \sum_{l=1}^{r} (n_l - 1) \log \left(\frac{n_l \widehat{\sigma}_l^2}{n_l - 1} \right),$$

where $n = \sum_{j=1}^{r} (n_j - 1)$. For $r = 2$, we have

$$S_{Bart} = \frac{1}{2} n \log \left(\frac{1}{4} \left(1 + \left(\frac{n_1 \widehat{\sigma}_1^2}{n_2 \widehat{\sigma}_2^2} \right) \right)^2 \left(\frac{n_2 \widehat{\sigma}_2^2}{n_1 \widehat{\sigma}_1^2} \right) \right)$$

and the acceptance region is equivalent to

$$c < \frac{\widehat{\sigma}_1^2}{\widehat{\sigma}_2^2} < \frac{1}{c}.$$

3

Goodness-of-fit: Simple Null Hypothesis

3.1. Introduction

The problem of goodness-of-fit to a distribution on the real line, $H_0 : F = F_0$, is frequently treated by partitioning the range of data in disjoint intervals and by testing the hypothesis $H_0 : \boldsymbol{p} = \boldsymbol{p}^0$ about the vector of parameters of a multinomial distribution.

Let $\mathcal{P} = \{E_i\}_{i=1,...,M}$ be a partition of the real line $\mathbb{R}$ in M intervals. Let $\boldsymbol{p} = (p_1, \ldots, p_M)^T$ and $\boldsymbol{p}^0 = (p_1^0, \ldots, p_M^0)^T$ be the true and the hypothetical probabilities of the intervals E_i, $i = 1, \ldots, M$, respectively, in such a way that $p_i = \Pr_F(E_i)$, $i = 1, ..., M$, and $p_i^0 = \Pr_{F_0}(E_i) = \int_{E_i} dF_0$, $i = 1, ..., M$.

Let $Y_1, \ldots, Y_n$ be a random sample from F, let $N_i = \sum_{j=1}^n I_{E_i}(Y_j)$, where $I_{E_i}(Y_j) = 1$ if $Y_j \in E_i$ and zero otherwise, and $\widehat{\boldsymbol{p}} = (\widehat{p}_1, ..., \widehat{p}_M)^T$ with $\widehat{p}_i = N_i/n$, $i = 1, \ldots, M$ be the absolute and relative frequencies in the intervals, respectively.

If we wish to test the simple null hypothesis,

$$H_0 : \boldsymbol{p} = \boldsymbol{p}^0, \tag{3.1}$$

the most commonly used test statistics are Pearson's test statistic (or chi-square test statistic), X^2:

$$X^2 \equiv \sum_{i=1}^{M} \frac{(N_i - np_i^0)^2}{np_i^0} \tag{3.2}$$

and the likelihood ratio test statistic, G^2 :

$$G^2 \equiv 2\sum_{i=1}^{M} N_i \log \frac{N_i}{np_i^0}. \tag{3.3}$$

These two test statistics are particular cases of the family of power-divergence test statistics, introduced by Cressie and Read (1984) and given by

$$T_n^{\lambda}(\widehat{\boldsymbol{p}},\boldsymbol{p}^0) = \frac{2n}{\lambda(\lambda+1)}\sum_{i=1}^{M}\widehat{p}_i\left(\left(\frac{\widehat{p}_i}{p_i^0}\right)^{\lambda}-1\right) = \frac{2}{\lambda(\lambda+1)}\sum_{i=1}^{M}N_i\left(\left(\frac{N_i}{np_i^0}\right)^{\lambda}-1\right), \tag{3.4}$$

where $-\infty < \lambda < \infty$. The test statistics $T_n^0(\widehat{\boldsymbol{p}},\boldsymbol{p}^0)$ and $T_n^{-1}(\widehat{\boldsymbol{p}},\boldsymbol{p}^0)$ are defined to be the limits of $T_n^{\lambda}(\widehat{\boldsymbol{p}},\boldsymbol{p}^0)$, as $\lambda \to 0$ and $\lambda \to -1$, respectively. Particular values of λ in (3.4) correspond to well known test statistics: Chi-square test statistic $X^2(\lambda = 1)$, likelihood ratio test statistic G^2 $(\lambda = 0)$, Freeman-Tukey test statistic $(\lambda = -1/2)$, modified likelihood ratio test statistic or minimum discrimination information statistic (Gokhale and Kullback, 1978) $(\lambda = -1)$, Neyman-modified test statistic or modified chi-square test statistic $(\lambda = -2)$ and Cressie-Read test statistic $(\lambda = 2/3)$. The expressions of the test statistics X^2 and G^2 are given in (3.2) and (3.3) respectively. The expressions of the other test statistics are given below:

i) $\lambda = -2$ (Modified chi-square test statistic)

$$T_n^{-2}(\widehat{\boldsymbol{p}},\boldsymbol{p}^0) = n\sum_{i=1}^{M}\frac{(p_i^0-\widehat{p}_i)^2}{\widehat{p}_i} = \sum_{i=1}^{M}\frac{(np_i^0-N_i)^2}{N_i}.$$

ii) $\lambda = -1$ $(\lambda \to -1)$ (Modified likelihood ratio test statistic)

$$T_n^{-1}(\widehat{\boldsymbol{p}},\boldsymbol{p}^0) = 2n\sum_{i=1}^{M}p_i^0\log\left(\frac{p_i^0}{\widehat{p}_i}\right) = 2\sum_{i=1}^{M}N_i\log\left(\frac{np_i^0}{N_i}\right).$$

iii) $\lambda = -1/2$ (Freeman-Tukey test statistic)

$$T_n^{-1/2}(\widehat{\boldsymbol{p}},\boldsymbol{p}^0) = 8n\left(1-\sum_{i=1}^{M}\sqrt{p_i^0\widehat{p}_i}\right) = 8n\left(1-\sum_{i=1}^{M}\sqrt{\frac{p_i^0N_i}{n}}\right).$$

iv) $\lambda = 2/3$ (Cressie-Read test statistic)

$$T_n^{2/3}(\widehat{\boldsymbol{p}}, \boldsymbol{p}^0) = \tfrac{9}{5}n\left(\sum_{i=1}^{M}\widehat{p}_i\left(\frac{\widehat{p}_i}{p_i^0}\right)^{2/3} - 1\right).$$

Although the power-divergence test statistics yield an important flexible family, it is possible to consider a more general family of test statistics for testing (3.1) and containing (3.4) as a particular case: ϕ-divergence test statistics, which are defined by

$$T_n^{\phi}(\widehat{\boldsymbol{p}}, \boldsymbol{p}^0) = \frac{2n}{\phi''(1)}\sum_{i=1}^{M} p_i^0 \phi\left(\frac{\widehat{p}_i}{p_i^0}\right), \quad \phi \in \Phi^*. \tag{3.5}$$

In all the chapter we shall assume that $\phi(x)$ is twice continuously differentiable for $x > 0$ with the second derivative $\phi''(1) \neq 0$.

Cressie and Read (1984) obtained the asymptotic distribution of the power-divergence test statistic $T_n^{\lambda}(\widehat{\boldsymbol{p}}, \boldsymbol{p}^0)$ under $H_0 : \boldsymbol{p} = \boldsymbol{p}^0$ for any $\lambda \in \mathbb{R}$ and Zografos et *al.* (1990) extended the result to the family $T_n^{\phi}(\widehat{\boldsymbol{p}}, \boldsymbol{p}^0)$ under $H_0 : \boldsymbol{p} = \boldsymbol{p}^0$ for any $\phi \in \Phi^*$. This result will be proved in Section 3.2., but not only under the null hypothesis but also under contiguous alternative hypotheses. A review about ϕ-divergence test statistics can be seen in Cressie and Pardo (2002a). A usual practice is to increase the number of intervals M as the sample size n increases. The large-sample theory of the usual chi-square test statistic for increasing M is available in the case of a simple null hypothesis (Holst 1972, Morris 1975, Cressie and Read 1984, Menéndez et *al.* 1998b). In this situation the behavior of the ϕ-divergence test statistic $T_n^{\phi}(\widehat{\boldsymbol{p}}, \boldsymbol{p}^0)$ is studied in Section 3.3. Finally, in Section 3.4., we study some nonstandard problems on the basis of ϕ-divergence test statistics. More concretely we consider the following problems: *a)* Goodness-of-fit with quantile characterization, *b)* Goodness-of-fit with dependent observations and *c)* Goodness-of-fit with misclassified data.

Cox (2002) provided some perspective on the importance, historically and contemporarily, of the chi-square test statistic and Rao (2002) reviewed the early work on the chi-square test statistic, its use in practice and recent contributions to alternative tests. Some interesting books in relation to the techniques of goodness-of-fit are: Agresti (2002), D'Agostino and Stephens (1986), Bishop et *al.* (1975) and Greenwood and Nikulin (1996).

3.2. Phi-divergences and Goodness-of-fit with Fixed Number of Classes

It is well known that Pearson (1900) proved that $X^2 \underset{n\to\infty}{\overset{L}{\longrightarrow}} \chi^2_{M-1}$, with X^2 given in (3.2). Note that the power-divergence test statistic, $T_n^{\lambda}(\widehat{\mathbf{p}},\mathbf{p}^0)$, coincides with the test statistic X^2 for $\lambda = 1$. This result was later extended to the likelihood ratio test statistic and to the modified chi-square test statistic by Neyman and Pearson (1928) and Neyman (1949). Later Cressie and Read (1984) established that $T_n^{\lambda}(\widehat{\mathbf{p}},\mathbf{p}^0) \underset{n\to\infty}{\overset{L}{\longrightarrow}} \chi^2_{M-1}$ under $H_0 : \mathbf{p} = \mathbf{p}^0$ for any $\lambda \in \mathbb{R}$. Zografos et *al*. (1990) proved that $T_n^{\phi}(\widehat{\mathbf{p}},\mathbf{p}^0) \underset{n\to\infty}{\overset{L}{\longrightarrow}} \chi^2_{M-1}$ under $H_0 : \mathbf{p} = \mathbf{p}^0$ for any $\phi \in \Phi^*$. We obtain, in this Section, the asymptotic distribution of the ϕ-divergence test statistic $T_n^{\phi}(\widehat{\mathbf{p}},\mathbf{p}^0)$ under the null hypothesis $H_0 : \mathbf{p} = \mathbf{p}^0$, under the alternative hypothesis $H_1 : \mathbf{p} = \mathbf{p}^* \neq \mathbf{p}^0$ and under contiguous alternative hypotheses that will be formulated later.

Theorem 3.1

Under the null hypothesis $H_0 : \mathbf{p} = \mathbf{p}^0 = (p_1^0, \ldots, p_M^0)^T$, the asymptotic distribution of the ϕ-divergence test statistic, $T_n^{\phi}(\widehat{\mathbf{p}},\mathbf{p}^0)$, is chi-square with $M - 1$ degrees of freedom.

Proof. Let $g : \mathbb{R}^M \longrightarrow \mathbb{R}^+$ be a function defined by

$$g(y_1, ..., y_M) = \sum_{i=1}^{M} p_i^0 \phi\left(\frac{y_i}{p_i^0}\right). \tag{3.6}$$

A second order Taylor expansion of g around $\mathbf{p}^0$ at $\widehat{\mathbf{p}} = (\widehat{p}_1, \ldots, \widehat{p}_M)^T$ gives

$$\begin{aligned} g(\widehat{p}_1, \ldots, \widehat{p}_M) = \ & g(p_1^0, \ldots, p_M^0) + \sum_{i=1}^{M} \left(\frac{\partial g(y_1, ..., y_M)}{\partial y_i}\right)_{\mathbf{p}=\mathbf{p}^0} (\widehat{p}_i - p_i^0) \\ + \ & \frac{1}{2} \sum_{i=1}^{M} \sum_{j=1}^{M} \left(\frac{\partial^2 g(y_1, ..., y_M)}{\partial y_i \partial y_j}\right)_{\mathbf{p}=\mathbf{p}^0} (\widehat{p}_i - p_i^0)(\widehat{p}_j - p_j^0) \\ + \ & o\left(\left\|\widehat{\mathbf{p}} - \mathbf{p}^0\right\|^2\right). \end{aligned}$$

But,

$$g(\widehat{p}_1, \ldots, \widehat{p}_M) = D_\phi(\widehat{\mathbf{p}},\mathbf{p}^0),\ g\left(p_1^0, \ldots, p_M^0\right) = D_\phi(\mathbf{p}^0,\mathbf{p}^0) = \phi(1) = 0$$

and

$$\left(\frac{\partial g(\boldsymbol{y})}{\partial y_i}\right)_{\boldsymbol{p}=\boldsymbol{p}^0}=\phi'(1),\ \left(\frac{\partial^2 g(\boldsymbol{y})}{\partial y_i \partial y_j}\right)_{\boldsymbol{p}=\boldsymbol{p}^0}=\begin{cases}\phi''(1)\frac{1}{p_i^0} & j=i\\ 0 & j\neq i\end{cases},$$

where $\boldsymbol{y}=(y_1,...,y_M)$. Therefore we have

$$D_\phi(\widehat{\boldsymbol{p}},\boldsymbol{p}^0)=\frac{1}{2}\phi''(1)\sum_{i=1}^{M}\frac{1}{p_i^0}\left(\widehat{p}_i-p_i^0\right)^2+o\left(\left\|\widehat{\boldsymbol{p}}-\boldsymbol{p}^0\right\|^2\right).$$

But

$$n\,o\left(\left\|\widehat{\boldsymbol{p}}-\boldsymbol{p}^0\right\|^2\right)=o_P(1),$$

since by Remark 2.5 $\sqrt{n}\left(\widehat{\boldsymbol{p}}-\boldsymbol{p}^0\right)\underset{n\to\infty}{\overset{L}{\longrightarrow}}N\left(\boldsymbol{0},\boldsymbol{\Sigma}_{\boldsymbol{p}^0}\right)$, where

$$\boldsymbol{\Sigma}_{\boldsymbol{p}^0}=diag\left(\boldsymbol{p}^0\right)-\boldsymbol{p}^0\left(\boldsymbol{p}^0\right)^T.$$

Then the random variables

$$T_n^{\phi}(\widehat{\boldsymbol{p}},\boldsymbol{p}^0)=\frac{2n}{\phi''(1)}D_\phi(\widehat{\boldsymbol{p}},\boldsymbol{p}^0) \tag{3.7}$$

and

$$n\sum_{i=1}^{M}\frac{1}{p_i^0}(\widehat{p}_i-p_i^0)^2$$

have the same asymptotic distribution. But

$$n\sum_{i=1}^{M}\frac{1}{p_i^0}\left(\widehat{p}_i-p_i^0\right)^2=\sqrt{n}\left(\widehat{\boldsymbol{p}}-\boldsymbol{p}^0\right)^T\boldsymbol{C}\sqrt{n}\left(\widehat{\boldsymbol{p}}-\boldsymbol{p}^0\right), \tag{3.8}$$

where $\boldsymbol{C}$ is a $M\times M$ matrix given by $\boldsymbol{C}=diag\left(\left(\boldsymbol{p}^0\right)^{-1}\right)$.

Then, we have

$$\sqrt{n}\left(\widehat{\boldsymbol{p}}-\boldsymbol{p}^0\right)^T\boldsymbol{C}\sqrt{n}\left(\widehat{\boldsymbol{p}}-\boldsymbol{p}^0\right)=\boldsymbol{X}^T\boldsymbol{X},$$

where $\boldsymbol{X}=\sqrt{n}\,diag\left(\left(\boldsymbol{p}^0\right)^{-1/2}\right)\left(\widehat{\boldsymbol{p}}-\boldsymbol{p}^0\right)$. The asymptotic distribution of the random variable $\boldsymbol{X}$ is normal with mean vector $\boldsymbol{0}$ and variance-covariance matrix given by

$$\boldsymbol{L}=diag\left(\left(\boldsymbol{p}^0\right)^{-1/2}\right)\boldsymbol{\Sigma}_{\boldsymbol{p}^0}diag\left(\left(\boldsymbol{p}^0\right)^{-1/2}\right).$$

We are going to prove that $\boldsymbol{L}$ is a projection of rank $M-1$.

It is clear that

$$\boldsymbol{L} = \boldsymbol{I} - diag\left(\left(\mathbf{p}^0\right)^{-1/2}\right)\mathbf{p}^0\left(\mathbf{p}^0\right)^T diag\left(\left(\mathbf{p}^0\right)^{-1/2}\right),$$

and

$$\begin{aligned}
\boldsymbol{L}\times\boldsymbol{L} = \ & \boldsymbol{I} - diag\left(\left(\mathbf{p}^0\right)^{-1/2}\right)\mathbf{p}^0\left(\mathbf{p}^0\right)^T diag\left(\left(\mathbf{p}^0\right)^{-1/2}\right) \\
- \ & diag\left(\left(\mathbf{p}^0\right)^{-1/2}\right)\mathbf{p}^0\left(\mathbf{p}^0\right)^T diag\left(\left(\mathbf{p}^0\right)^{-1/2}\right) \\
+ \ & diag\left(\left(\mathbf{p}^0\right)^{-1/2}\right)\mathbf{p}^0\left(\mathbf{p}^0\right)^T diag\left(\left(\mathbf{p}^0\right)^{-1/2}\right) diag\left(\left(\mathbf{p}^0\right)^{-1/2}\right) \\
\times \ & \mathbf{p}^0\left(\mathbf{p}^0\right)^T diag\left(\left(\mathbf{p}^0\right)^{-1/2}\right) \\
= \ & \boldsymbol{I} - diag\left(\left(\mathbf{p}^0\right)^{-1/2}\right)\mathbf{p}^0\left(\mathbf{p}^0\right)^T diag\left(\left(\mathbf{p}^0\right)^{-1/2}\right) = \boldsymbol{L},
\end{aligned}$$

because,

$$\left(\mathbf{p}^0\right)^T diag\left(\left(\mathbf{p}^0\right)^{-1/2}\right) diag\left(\left(\mathbf{p}^0\right)^{-1/2}\right)\mathbf{p}^0 = 1.$$

On the other hand

$$rank\left(\boldsymbol{L}\right) = rank\left(diag\left(\left(\mathbf{p}^0\right)^{-1}\right)\boldsymbol{\Sigma}_{\mathbf{p}^0}\right) = rank\left(\boldsymbol{C}\boldsymbol{\Sigma}_{\mathbf{p}^0}\right) = trace\left(\boldsymbol{C}\boldsymbol{\Sigma}_{\mathbf{p}^0}\right),$$

but

$$\boldsymbol{C}\boldsymbol{\Sigma}_{\mathbf{p}^0} = \left(\delta_{ij} - p_j^0\right)_{i,j=1,\ldots,M},$$

then

$$trace\left(\boldsymbol{C}\boldsymbol{\Sigma}_{\mathbf{p}^0}\right) = \sum_{j=1}^{M}\left(1 - p_j^0\right) = M - 1.$$

By Remark 2.6, we have

$$T_n^{\phi}(\widehat{\mathbf{p}}, \mathbf{p}^0) = \frac{2n}{\phi''(1)} D_{\phi}(\widehat{\mathbf{p}}, \mathbf{p}^0) \xrightarrow[n\to\infty]{L} \chi^2_{M-1}.$$

∎

Corollary 3.1

Under the null hypothesis $H_0 : \mathbf{p} = \mathbf{p}^0$, *the asymptotic distribution of the* ϕ-*divergence test statistic,* $T_n^{\phi}(\mathbf{p}^0, \widehat{\mathbf{p}})$, *is chi-square with* $M-1$ *degrees of freedom.*

Proof. We consider the function $\varphi(x) = x\phi\left(x^{-1}\right)$. If $\phi \in \Phi^*$ then $\varphi \in \Phi^*$ and from Theorem 3.1 we have

$$T_n^{\varphi}(\widehat{\mathbf{p}}, \mathbf{p}^0) \xrightarrow[n\to\infty]{L} \chi^2_{M-1}.$$

Taking into account that $\varphi''(1) = \phi''(1)$, we have

$$\begin{aligned} T_n^{\varphi}(\widehat{\mathbf{p}}, \mathbf{p}^0) = & \frac{2n}{\varphi''(1)} D_{\varphi}(\widehat{\mathbf{p}}, \mathbf{p}^0) = \frac{2n}{\varphi''(1)} \sum_{j=1}^{M} p_i^0 \varphi\left(\frac{\widehat{p}_i}{p_i^0}\right) \\ = & \frac{2n}{\phi''(1)} \sum_{j=1}^{M} p_i^0 \frac{\widehat{p}_i}{p_i^0} \phi\left(\frac{p_i^0}{\widehat{p}_i}\right) = T_n^{\phi}(\mathbf{p}^0, \widehat{\mathbf{p}}), \end{aligned}$$

and this completes the proof. ∎

Remark 3.1

a) In the case of Kullback-Leibler divergence, we have

$$T_n^0(\widehat{\mathbf{p}}, \mathbf{p}^0) = 2nD_{Kull}(\widehat{\mathbf{p}}, \mathbf{p}^0) \underset{n\to\infty}{\overset{L}{\longrightarrow}} \chi^2_{M-1}$$

and

$$T_n^0(\mathbf{p}^0, \widehat{\mathbf{p}}) = 2nD_{Kull}(\mathbf{p}^0, \widehat{\mathbf{p}}) \underset{n\to\infty}{\overset{L}{\longrightarrow}} \chi^2_{M-1}.$$

The first test statistic is the likelihood ratio test and the second one is the modified likelihood ratio test.

b) In the case of (h, ϕ)-divergences the asymptotic distribution of the test statistics

$$T_n^{\phi,h}(\widehat{\mathbf{p}}, \mathbf{p}^0) = \frac{2n}{h'(0)\,\phi''(1)} D_{\phi}^h(\widehat{\mathbf{p}}, \mathbf{p}^0)$$

and

$$T_n^{\phi,h}(\mathbf{p}^0, \widehat{\mathbf{p}}) = \frac{2n}{h'(0)\,\phi''(1)} D_{\phi}^h(\mathbf{p}^0, \widehat{\mathbf{p}})$$

is chi-square with $M-1$ degrees of freedom.

Based on Theorem 3.1, if the sample size is large enough, one can use the $100(1-\alpha)$ percentile, $\chi^2_{M-1,\alpha}$, of the chi-square with $M-1$ degrees of freedom, defined by the equation $\Pr(\chi^2_{M-1} \geq \chi^2_{M-1,\alpha}) = \alpha$, to propose the decision rule:

$$\text{“Reject } H_0, \text{ with a significance level } \alpha, \text{ if } T_n^{\phi}(\widehat{\mathbf{p}}, \mathbf{p}^0) > \chi^2_{M-1,\alpha}\text{”} \tag{3.9}$$

(or $T_n^{\phi}(\mathbf{p}^0, \widehat{\mathbf{p}}) > \chi^2_{M-1,\alpha}$).

This is the goodness-of-fit test based on the ϕ-divergence test statistic.

In the following theorem we present an approximation of the power function for the testing procedure given in (3.9).

Theorem 3.2

Let $\mathbf{p}^* = (p_1^*, \ldots, p_M^*)^T$ *be a probability distribution with* $\mathbf{p}^* \neq \mathbf{p}^0$*. The power of the test with decision rule given in (3.9), at* $\mathbf{p}^* = (p_1^*, \ldots, p_M^*)^T$*, is*

$$\beta_{n,\phi}\left(p_1^*, ..., p_M^*\right) = 1 - \Phi_n\left(\frac{1}{\sigma_1(\mathbf{p}^*)}\left(\frac{\phi''(1)}{2\sqrt{n}}\chi^2_{M-1,\alpha} - \sqrt{n}D_\phi(\mathbf{p}^*, \mathbf{p}^0)\right)\right),$$

where Φ_n *tends uniformly to the standard normal distribution function* $\Phi(x)$ *and*

$$\sigma_1^2(\mathbf{p}^*) \quad = \quad \sum_{i=1}^M p_i^*\left(\phi'\left(\frac{p_i^*}{p_i^0}\right)\right)^2 - \left(\sum_{i=1}^M p_i^*\phi'\left(\frac{p_i^*}{p_i^0}\right)\right)^2. \qquad (3.10)$$

Proof. First we establish that under the hypothesis $H_1 : \mathbf{p} = \mathbf{p}^* \neq \mathbf{p}^0$ we have

$$\sqrt{n}\left(D_\phi(\widehat{\mathbf{p}}, \mathbf{p}^0) - D_\phi(\mathbf{p}^*, \mathbf{p}^0)\right) \underset{n\to\infty}{\overset{L}{\longrightarrow}} N\left(0, \sigma_1^2(\mathbf{p}^*)\right),$$

whenever $\sigma_1^2(\mathbf{p}^*) > 0$ and with $\sigma_1^2(\mathbf{p}^*)$ given in (3.10).

A first order Taylor expansion of the function g, given in (3.6), around $\mathbf{p}^* = (p_1^*, \ldots, p_M^*)^T$ at $\widehat{\mathbf{p}} = (\widehat{p}_1, \ldots, \widehat{p}_M)^T$ gives

$$D_\phi(\widehat{\mathbf{p}}, \mathbf{p}^0) = D_\phi\left(\mathbf{p}^*, \mathbf{p}^0\right) + \sum_{i=1}^M \left(\frac{\partial D_\phi(\mathbf{p}, \mathbf{p}^0)}{\partial p_i}\right)_{\mathbf{p}=\mathbf{p}^*} (\widehat{p}_i - p_i^*) + o\left(\|\widehat{\mathbf{p}} - \mathbf{p}^*\|\right)$$

where

$$\left(\frac{\partial D_\phi(\mathbf{p}, \mathbf{p}^0)}{\partial p_i}\right)_{\mathbf{p}=\mathbf{p}^*} = \phi'\left(\frac{p_i^*}{p_i^0}\right), \; i = 1, ..., M.$$

Under the hypothesis $H_1 : \mathbf{p} = \mathbf{p}^*$, we have that

$$\sqrt{n}\left(\widehat{\mathbf{p}} - \mathbf{p}^*\right) \underset{n\to\infty}{\overset{L}{\longrightarrow}} N\left(\mathbf{0}, \mathbf{\Sigma}_{\mathbf{p}^*}\right),$$

with $\mathbf{\Sigma}_{\mathbf{p}^*} = diag\left(\mathbf{p}^*\right) - \mathbf{p}^*\left(\mathbf{p}^*\right)^T$, then $\sqrt{n}\, o\left(\|\widehat{\mathbf{p}} - \mathbf{p}^*\|\right) = o_P\left(1\right)$. Therefore the asymptotic distribution of the random variables

$$\sqrt{n}\left(D_\phi(\widehat{\mathbf{p}}, \mathbf{p}^0) - D_\phi(\mathbf{p}^*, \mathbf{p}^0)\right) \quad \text{and} \quad \sqrt{n}\sum_{i=1}^M t_i\left(\widehat{p}_i - p_i^*\right),$$

with

$$t_i = \phi'\left(\frac{p_i^*}{p_i^0}\right), \qquad i = 1, ..., M,$$

is the same.

But

$$\sqrt{n}\sum_{i=1}^{M} t_i\left(\widehat{p}_i - p_i^*\right) = \sqrt{n}\boldsymbol{T}^T\left(\widehat{\mathbf{p}} - \mathbf{p}^*\right)$$

converges in law to a normal distribution with mean zero and variance $\boldsymbol{T}^T\boldsymbol{\Sigma}_{\mathbf{p}^*}\boldsymbol{T}$, where $\boldsymbol{T} = (t_1, ..., t_M)^T$. It is not difficult to establish that

$$\boldsymbol{T}^T\boldsymbol{\Sigma}_{\mathbf{p}^*}\boldsymbol{T} = \sigma_1^2(\mathbf{p}^*).$$

Then

$$\begin{aligned}\beta_{n,\phi}\left(p_1^*, ..., p_M^*\right) &= \Pr\left(T_n^{\phi}(\widehat{\mathbf{p}}, \mathbf{p}^0) > \chi^2_{M-1,\alpha} / \ H_1 : \mathbf{p} = \mathbf{p}^*\right)\\ &= 1 - \Phi_n\left(\frac{1}{\sigma_1(\mathbf{p}^*)}\left(\frac{\phi''(1)}{2\sqrt{n}}\chi^2_{M-1,\alpha} - \sqrt{n}D_\phi(\mathbf{p}^*, \mathbf{p}^0)\right)\right),\end{aligned}$$

where $\Phi_n(x)$ tends uniformly to the standard normal distribution function $\Phi(x)$ and $\sigma_1^2(\mathbf{p}^*)$ is given in (3.10). This completes the proof. ∎

Based on this result an approximation of the power function of the test, with decision rule given in (3.9), at $\mathbf{p}^* = (p_1^*, ..., p_M^*)^T$, is

$$\beta_{n,\phi}\left(p_1^*, ..., p_M^*\right) \simeq 1 - \Phi\left(\frac{1}{\sigma_1(\mathbf{p}^*)}\left(\frac{\phi''(1)}{2\sqrt{n}}\chi^2_{M-1,\alpha} - \sqrt{n}D_\phi(\mathbf{p}^*, \mathbf{p}^0)\right)\right),$$

where Φ is the standard normal distribution function.

It is clear that $\lim_{n\to\infty}\beta_{n,\phi}\left(p_1^*, ..., p_M^*\right) = 1$, i.e., the test is consistent.

Corollary 3.2

In a similar way to the previous theorem it is possible to establish that

$$\sqrt{n}\left(D_\phi(\mathbf{p}^0, \widehat{\mathbf{p}}) - D_\phi(\mathbf{p}^0, \mathbf{p}^*)\right) \xrightarrow[n\to\infty]{L} N\left(0, \sigma_2(\mathbf{p}^*)\right)$$

where $\sigma_2(\mathbf{p}^*)$ *is given by*

$$\sigma_2^2(\mathbf{p}^*) = \sum_{i=1}^{M} p_i^* s_i^2 - \left(\sum_{i=1}^{M} p_i^* s_i\right)^2,$$

with

$$s_i = \phi\left(\frac{p_i^0}{p_i^*}\right) - \frac{p_i^0}{p_i^*}\phi'\left(\frac{p_i^0}{p_i^*}\right), \; i = 1, ..., M.$$

Corollary 3.3

a) In the case of the Kullback-Leibler divergence measure, we have

$$\begin{aligned}\sigma_1^2(\boldsymbol{p}^*) &= \sum_{i=1}^M p_i^*\left(\log\frac{p_i^*}{p_i^0}\right)^2 - \left(\sum_{i=1}^M p_i^* \log\frac{p_i^*}{p_i^0}\right)^2, \\ \sigma_2^2(\boldsymbol{p}^*) &= \sum_{i=1}^M \frac{\left(p_i^0\right)^2}{p_i^*} - 1.\end{aligned}$$

b) In the case of (h, ϕ)-divergences we have

$$\begin{aligned}\sigma_1^2(\boldsymbol{p}^*) &= \sum_{i=1}^M p_i^*\left(h'\left(D_\phi(\boldsymbol{p}^*, \boldsymbol{p}^0)\right)\phi'\left(\frac{p_i^*}{p_i^0}\right)\right)^2 \\ &- \left(\sum_{i=1}^M p_i^* h'\left(D_\phi(\boldsymbol{p}^*, \boldsymbol{p}^0)\right)\phi'\left(\frac{p_i^*}{p_i^0}\right)\right)^2\end{aligned}$$

and

$$\begin{aligned}\sigma_2^2(\boldsymbol{p}^*) &= \sum_{i=1}^M p_i^*\left(\left(h'\left(D_\phi(\boldsymbol{p}^0, \boldsymbol{p}^*)\right)\right)\left(\phi\left(\frac{p_i^0}{p_i^*}\right) - \frac{p_i^0}{p_i^*}\phi'\left(\frac{p_i^0}{p_i^*}\right)\right)\right)^2 \\ &- \left(\sum_{i=1}^M p_i\left(h'\left(D_\phi(\boldsymbol{p}^0, \boldsymbol{p}^*)\right)\right)\left(\phi\left(\frac{p_i^0}{p_i^*}\right) - \phi'\left(\frac{p_i^0}{p_i^*}\right)\right)\right)^2.\end{aligned}$$

Proof. Part *a)* is a simple exercise. We prove part *b)*. If we consider the function

$$g\left(y_1, ..., y_M\right) = h\left(\sum_{i=1}^M p_i^0 \phi\left(\frac{y_i}{p_i^0}\right)\right)$$

and its first Taylor's expansion around $\boldsymbol{p}^* = \left(p_1^*, ..., p_M^*\right)^T$ at $\widehat{\boldsymbol{p}}$, we get

$$g\left(\widehat{p}_1, ..., \widehat{p}_M\right) = g\left(p_1^*, ..., p_M^*\right) + \sum_{i=1}^M \left(\frac{\partial g}{\partial y_i}\right)_{\boldsymbol{y}=\boldsymbol{p}^*}\left(\widehat{p}_i - p_i^*\right) + o\left(\|\widehat{\boldsymbol{p}} - \boldsymbol{p}^*\|\right).$$

Therefore the random variables

$$\sqrt{n}\left(g\left(\widehat{p}_1, ..., \widehat{p}_M\right) - g\left(p_1^*, ..., p_M^*\right)\right) \text{ and } \sqrt{n}\sum_{i=1}^M\left(\left(\frac{\partial g}{\partial y_i}\right)_{\boldsymbol{y}=\boldsymbol{p}^*}\left(\widehat{p}_i - p_i^*\right)\right)$$

have the same asymptotic distribution, i.e.,

$$\sqrt{n}\left(D_{\phi}^{h}(\widehat{\boldsymbol{p}},\boldsymbol{p}^{0})-D_{\phi}^{h}(\boldsymbol{p}^{*},\boldsymbol{p}^{0})\right) \text{ and } \sqrt{n}\sum_{i=1}^{M}t_{i}^{*}\left(\widehat{p}_{i}-p_{i}^{*}\right)$$

have the same asymptotic distribution, where

$$t_{i}^{*}=h'\left(D_{\phi}(\boldsymbol{p}^{*},\boldsymbol{p}^{0})\right)\phi'\left(\frac{p_{i}^{*}}{p_{i}^{0}}\right),\ i=1,...,M.$$

In a similar way to Theorem 3.1 one can get $\sigma_2^2(\boldsymbol{p}^*)$. ■

In order to produce a nontrivial asymptotic power, Cochran (1952) suggested using a set of local alternatives contiguous to the null hypothesis as n increases. Consider the multinomial probability vector

$$\boldsymbol{p}_n \equiv \boldsymbol{p}^0+\boldsymbol{d}/\sqrt{n},$$

where $\boldsymbol{d}=(d_1,...,d_M)^T$ is a fixed $M\times 1$ vector such that $\sum_{j=1}^{M}d_j=0$, and recall that n is the total-count parameter of the multinomial distribution. As $n\to\infty$ the sequence of probability vectors $\{\boldsymbol{p}_n\}_{n\in N}$ converge to the probability vector $\boldsymbol{p}^0$ in the null hypothesis at the rate $O\left(n^{-1/2}\right)$. We say that

$$H_{1,n}:\boldsymbol{p}=\boldsymbol{p}_n\equiv\boldsymbol{p}^0+\boldsymbol{d}/\sqrt{n} \tag{3.11}$$

is a sequence of contiguous alternative hypotheses, here contiguous to the null hypothesis $\boldsymbol{p}^0$. Our interest is to study the asymptotic behavior of the test power under contiguous alternative hypotheses, i.e.,

$$\beta_{n,\phi}(\boldsymbol{p}_n)=\Pr\left(T_n^{\phi}(\widehat{\boldsymbol{p}},\boldsymbol{p}^0)>\chi^2_{M-1,\alpha}/H_{1,n}:\ \boldsymbol{p}=\boldsymbol{p}_n\right). \tag{3.12}$$

In what follows we show that under the alternative hypotheses $H_{1,n}$, as $n\to\infty$, $T_n^{\phi}(\widehat{\boldsymbol{p}},\boldsymbol{p}^0)$ converges in distribution to a noncentral chi-square random variable with noncentrality parameter δ, where δ is given in Theorem 3.3, and $M-1$ degrees of freedom $\left(\chi^2_{M-1}(\delta)\right)$. Lehmann (1959) argues that contiguous alternative hypotheses are the only alternative hypotheses of interest.

It is interesting to observe that if we consider a point $\boldsymbol{p}^*\neq\boldsymbol{p}^0$, we can write $\boldsymbol{p}^*=\boldsymbol{p}^0+n^{-1/2}\left(\sqrt{n}\left(\boldsymbol{p}^*-\boldsymbol{p}^0\right)\right)$, and if we define $\boldsymbol{p}_n\equiv\boldsymbol{p}^0+\boldsymbol{d}/\sqrt{n}$ with $\boldsymbol{d}=\sqrt{n}\left(\boldsymbol{p}^*-\boldsymbol{p}^0\right)$ we can use the expression given in (3.12) to get an approximation of the power function at $\boldsymbol{p}^*$.

Theorem 3.3

The asymptotic distribution of the ϕ-divergence test statistic $T_n^{\phi}(\widehat{\mathbf{p}}, \mathbf{p}^0)$, under the contiguous alternative hypotheses (3.11), is noncentral chi-square with $M-1$ degrees of freedom and noncentrality parameter $\boldsymbol{\delta}$ given by

$$\boldsymbol{\delta} = \boldsymbol{d}^T diag\left(\left(\mathbf{p}^0\right)^{-1}\right)\boldsymbol{d}.$$

Proof. We can write

$$\sqrt{n}\left(\widehat{\mathbf{p}} - \mathbf{p}^0\right) = \sqrt{n}\left(\widehat{\mathbf{p}} - \mathbf{p}_n\right) + \sqrt{n}\left(\mathbf{p}_n - \mathbf{p}^0\right) = \sqrt{n}\left(\widehat{\mathbf{p}} - \mathbf{p}_n\right) + \boldsymbol{d},$$

and under the hypothesis $H_{1,n} : \mathbf{p} = \mathbf{p}_n \equiv \mathbf{p}^0 + \frac{\boldsymbol{d}}{\sqrt{n}}$, we have

$$\sqrt{n}\left(\widehat{\mathbf{p}} - \mathbf{p}_n\right) \underset{n\to\infty}{\overset{L}{\longrightarrow}} N\left(\mathbf{0}, \boldsymbol{\Sigma}_{\mathbf{p}^0}\right)$$

and

$$\sqrt{n}\left(\widehat{\mathbf{p}} - \mathbf{p}^0\right) \underset{n\to\infty}{\overset{L}{\longrightarrow}} N\left(\boldsymbol{d}, \boldsymbol{\Sigma}_{\mathbf{p}^0}\right).$$

By (3.8) in Theorem 3.1 we have

$$T_n^{\phi}(\widehat{\mathbf{p}}, \mathbf{p}^0) = \frac{2n}{\phi''(1)} D_{\phi}(\widehat{\mathbf{p}}, \mathbf{p}^0) = \sqrt{n}(\widehat{\mathbf{p}} - \mathbf{p}^0)^T \boldsymbol{C} \sqrt{n}(\widehat{\mathbf{p}} - \mathbf{p}^0) + o\left(\left\|\widehat{\mathbf{p}} - \mathbf{p}^0\right\|^2\right).$$

Then

$$\begin{aligned} T_n^{\phi}(\widehat{\mathbf{p}}, \mathbf{p}^0) &= \sqrt{n}\left(diag\left(\left(\mathbf{p}^0\right)^{-1/2}\right)\left(\widehat{\mathbf{p}} - \mathbf{p}^0\right)\right)^T \sqrt{n}\left(diag\left(\left(\mathbf{p}^0\right)^{-1/2}\right)\left(\widehat{\mathbf{p}} - \mathbf{p}^0\right)\right) \\ &+ o\left(\left\|\widehat{\mathbf{p}} - \mathbf{p}^0\right\|\right)^2 = \boldsymbol{X}^T\boldsymbol{X} + o\left(\left\|\widehat{\mathbf{p}} - \mathbf{p}^0\right\|\right)^2 \end{aligned}$$

being the asymptotic distribution of $\boldsymbol{X}$ multivariate normal with mean vector $diag\left(\left(\mathbf{p}^0\right)^{-1/2}\right)\boldsymbol{d}$ and variance-covariance matrix

$$diag\left(\left(\mathbf{p}^0\right)^{-1/2}\right)\boldsymbol{\Sigma}_{\mathbf{p}^0} diag\left(\left(\mathbf{p}^0\right)^{-1/2}\right).$$

Applying Lemma (Ferguson 1996 p. 63) "Suppose that $\boldsymbol{X}$ is $N(\boldsymbol{\mu}, \boldsymbol{\Sigma})$. If $\boldsymbol{\Sigma}$ is idempotent of rank M and $\boldsymbol{\Sigma}\boldsymbol{\mu} = \boldsymbol{\mu}$, the distribution of $\boldsymbol{X}^T\boldsymbol{X}$ is noncentral chi-square with $M-1$ degrees of freedom and noncentrality parameter $\boldsymbol{\delta} = \boldsymbol{\mu}^T\boldsymbol{\mu}$", the result follows if we establish that

$$diag\left(\left(\mathbf{p}^0\right)^{-1/2}\right)\boldsymbol{\Sigma}_{\mathbf{p}^0} diag\left(\left(\mathbf{p}^0\right)^{-1/2}\right) diag\left(\left(\mathbf{p}^0\right)^{-1/2}\right)\boldsymbol{d} = diag\left(\left(\mathbf{p}^0\right)^{-1/2}\right)\boldsymbol{d},$$

because in Theorem 3.1 it was proved that

$$\begin{aligned} \boldsymbol{L} = & \; diag\left(\left(\boldsymbol{p}^0\right)^{-1/2}\right)\boldsymbol{\Sigma}_{\boldsymbol{p}^0} diag\left(\left(\boldsymbol{p}^0\right)^{-1/2}\right) \\ = & \; \boldsymbol{I} - diag\left(\left(\boldsymbol{p}^0\right)^{-1/2}\right)\mathbf{p}^0\left(\boldsymbol{p}^0\right)^T diag\left(\left(\boldsymbol{p}^0\right)^{-1/2}\right), \end{aligned}$$

is a projection of rank $M-1$.

We denote by $\boldsymbol{U}$

$$diag\left(\left(\boldsymbol{p}^0\right)^{-1/2}\right)\boldsymbol{\Sigma}_{\boldsymbol{p}^0} diag\left(\left(\boldsymbol{p}^0\right)^{-1/2}\right) diag\left(\left(\boldsymbol{p}^0\right)^{-1/2}\right)\boldsymbol{d} = \boldsymbol{L} diag\left(\left(\boldsymbol{p}^0\right)^{-1/2}\right)$$

and we have

$$\begin{aligned} \boldsymbol{U} = & \left(\boldsymbol{I} - diag\left(\left(\boldsymbol{p}^0\right)^{-1/2}\right)\mathbf{p}^0\left(\boldsymbol{p}^0\right)^T diag\left(\left(\boldsymbol{p}^0\right)^{-1/2}\right)\right) diag\left(\left(\boldsymbol{p}^0\right)^{-1/2}\right)\boldsymbol{d} \\ = & \; diag\left(\left(\boldsymbol{p}^0\right)^{-1/2}\right)\boldsymbol{d} - diag\left(\left(\boldsymbol{p}^0\right)^{-1/2}\right)\mathbf{p}^0\left(\boldsymbol{p}^0\right)^T diag\left(\left(\boldsymbol{p}^0\right)^{-1}\right)\boldsymbol{d} \\ = & \; diag\left(\left(\boldsymbol{p}^0\right)^{-1/2}\right)\boldsymbol{d}, \end{aligned}$$

since $\left(\boldsymbol{p}^0\right)^T diag\left(\left(\boldsymbol{p}^0\right)^{-1}\right)\boldsymbol{d} = \mathbf{0}$. ■

The asymptotic distribution of the ϕ-divergence test statistic in a stratified random sampling was obtained in Morales et *al.* (1994) .

3.3. Phi-divergence Test Statistics under Sparseness Assumptions

In the previous Section we have established the asymptotic distribution of the ϕ-divergence test statistic assuming M fixed and letting the sample size n tend to infinity. A different approach lets both n and M tend to infinity, but at the same rate so that n/M remains constant. In Bishop et *al.* (1975, p. 410) the following can be seen "...One reason for looking at this special type of asymptotic comes from practical considerations. Typically, multinomial data arrive in the form of a cross-classification of discrete variables. In many situations there are a large number of variables which can be used to cross-classify each observation, and if all of these variables are used the data would be spread too thinly over the cells in the resulting multidimensional contingency table. Thus if the investigator uses a subset of the variables to keep the average number of observations from becoming too small, he is in effect choosing M so that n/M

is moderate." Holst (1972) pointed out the following "it is rather unnatural to keep M fixed when $n \to \infty$; instead we should have that $\lim_{n\to\infty} M = \infty$." For equiprobable cells Hoeffding (1965, p. 372) showed that Pearson test statistic is much more powerful than likelihood ratio test statistic against near alternative $\boldsymbol{p}$ satisfying $\max_i \left|p_i - M^{-1}\right| = O\left(M^{-1}\right)$, for moderate significance levels, with n/M moderate and M large.

In this Section we assume that the partition size M depends on the sample size n, i.e., $M = M_n$ with $\lim_{n\to\infty} M_n = \infty$. The intervals or classes depend, in general, on n. For this reason we denote the partition by $\mathcal{P}_n$ and its elements by E_{nj}, $j = 1, ..., M_n$. Then we have $\mathcal{P}_n = \{E_{n1}, ..., E_{nM_n}\}$, with $1 < M_n \leq \infty$. We denote $\gamma_n = n/M_n$ and we assume that

$$\lim_{n\to\infty} \gamma_n = \gamma \in (0, \infty). \tag{3.13}$$

Let $\boldsymbol{p}_n = (p_{n1}, ..., p_{nM_n})^T$ be the vector of probabilities verifying $p_{ni} = \Pr(E_{ni})$, $i = 1, ..., M_n$, and let $\widehat{\boldsymbol{p}}_n = (\widehat{p}_{n1}, ..., \widehat{p}_{nM_n})^T$ be the relative frequency vector based on a random sample of size n, $Y_1, ..., Y_n$, i.e., $\widehat{p}_{ni} = N_{ni}/n$ being N_{ni} the number of elements in the class E_{ni}, $i = 1, ..., M_n$. We write $\boldsymbol{N}_n = (N_{n1}, ..., N_{nM_n})$ to denote the vector of absolute frequencies.

Assumption (3.13) is realistic in goodness-of-fit testing, where partitions $\mathcal{P}_n$ are usually specified so that all observed frequencies $N_{ni} = n\widehat{p}_{ni}$ were approximately the same and relatively large. If we denote the desired level of cell frequencies by γ then we obtain from the condition $N_{ni} = \gamma + O_P(n)$, and from the law of large numbers condition $\widehat{p}_{ni} = p_{ni} + o_P(1)$, that the ratio n/M_n, the expected number of observations in each cell, must be close to γ. On the other hand, (3.13) with small γ means that many cells are sparsely frequented. Thus (3.13) is also known as the sparseness assumption; see Section 4.3 in Read and Cressie (1988) devoted to testing under this assumption.

The first results of this kind were published by Holst (1972). He developed a Poissonization technique leading under (3.13) to asymptotic distribution of Pearson test statistic and likelihood ratio test statistic. The so called Poissonization technique is originating from the fact that the vector $\boldsymbol{Z} = (Z_1, ..., Z_M)$ of independent Poisson random variables with $E[\boldsymbol{Z}] = (\lambda_1, ..., \lambda_M)^T$ is under the condition $Z_1 + ... + Z_M = n$ multinomially distributed with parameters n and $\boldsymbol{p} = (\lambda_1/n, ..., \lambda_M/n)^T$. Thus if $M = M_n$ and the expectation of $\boldsymbol{Z} = \boldsymbol{Z}_n$ is $n\boldsymbol{p}_n$

then the conditional distribution of $\boldsymbol{Z}_n$ given $Z_{n1}+...+Z_{nM_n}=n$ coincides with the unconditional distribution of above defined $\boldsymbol{N}_n$.

Morris (1975) derived a central limit law for a sequence of X^2 and G^2 test statistics, each measuring the fit of a known multinomial to a count data set with M cells. He gave conditions under which X^2 and G^2, suitably standardized, approach normality as M tends to infinity. The closeness of these distributions to the normal, for selected sparse multinomial, was examined in a simulation study by Koehler and Larntz (1980). Dale (1986) obtained the asymptotic distribution of X^2 and G^2 test statistics for product-multinomial model and later Morales et *al*. (2003) obtained, for this model, the asymptotic distribution of the ϕ-divergence test statistic.

Holst (1972) considered the test statistic

$$S_n = \sum_{i=1}^{M_n} \Phi_n(N_{ni}, i/M_n)$$

where $\Phi_n : [0,\infty) \times [0,1] \rightarrow \mathbb{R}$ is a measurable function satisfying the condition

$$|\Phi_n(u,v)| \leq c_1\, e^{c_2 u}$$

for some $c_1, c_2 \in \mathbb{R}$ not depending on n. He proved that the conditions:

i) $\limsup_n \max_{1\leq i\leq M_n} n\, p_{in} < \infty$

ii) $0 < \liminf_n \frac{\sigma_n^2}{n} \leq \limsup_n \frac{\sigma_n^2}{n} < \infty$

imply

$$\frac{S_n - \mu_n}{\sigma_n} \underset{n\rightarrow\infty}{\overset{L}{\longrightarrow}} N(0,1),$$

being

$$\mu_n = \sum_{i=1}^{M_n} E\left[\Phi_n(Z_{ni}, i/M_n)\right]$$

and

$$\sigma_n^2 = \sum_{i=1}^{M_n} Var\left[\Phi_n(Z_{ni}, i/M_n)\right] - n^{-1}\left(\sum_{i=1}^{M_n} Cov\left[Z_{ni}, \Phi_n(Z_{ni}, i/M_n)\right]\right)^2,$$

where Z_{ni}, $i = 1, ..., M_n$ are Poisson random variables with parameters np_{ni} $(E[Z_{ni}] = np_{ni})$.

If we consider $\boldsymbol{q}_n = (q_{n1}, ..., q_{nM_n})^T$ with $q_{ni} > 0$, $1 \leq i \leq M_n$ and the continuous piecewise linear function $g_n : [0,1] \to \mathbb{R}$ defined by $g_n(0) = 1$ and

$$g_n(i/M_n) = n\, q_{ni} \quad \text{if } 1 \leq i \leq M_n$$

and we define for $u \in [0, \infty)$ and $v \in [0,1]$

$$\Phi_n(u,v) = g_n(v)\, \phi\left(\frac{u}{g_n(v)}\right),$$

then

$$nD_\phi(\widehat{\boldsymbol{p}}_n, \boldsymbol{q}_n) = \sum_{i=1}^{M_n} n\, q_{ni}\, \phi\left(\frac{N_{ni}}{n\, q_{ni}}\right) = \sum_{i=1}^{M_n} g_n(i/M_n)\, \phi\left(\frac{N_{ni}}{g_n(i/M_n)}\right) = S_n.$$

To use the limit theorem of Holst it is necessary to establish that there exist $c_1, c_2 \in \mathbb{R}$ not depending on n and verifying

$$|\Phi_n(u,v)| = \left| g_n(v)\, \phi\left(\frac{u}{g_n(v)}\right)\right| \leq c_1 e^{c_2 u}.$$

We know that $\phi(t) \geq 0$ and let us suppose

$$\phi(0) = \lim_{t \downarrow 0} \phi(t) < \infty \quad \text{and} \quad \lim_{t \to \infty} \frac{\log \phi(t)}{t} < \infty. \tag{3.14}$$

Since $\phi(t)$ is convex in the domain $0 \leq t < \infty$, the function $f(t) = (\phi(t) - \phi(0))/t$ is nondecreasing. Hence if $t_0 > 0$ then

$$0 \leq \phi(t) \leq \phi(0) + t\, \frac{\phi(t_0) - \phi(0)}{t_0} \qquad 0 \leq t \leq t_0.$$

Further, by (3.14) there exist a positive value t_0 and $c \in \mathbb{R}$ such that

$$\phi(t) \leq e^{c\,t} \qquad t > t_0.$$

Then there exist c_1^*, $c_2^* \in \mathbb{R}$ verifying $0 \leq \phi(t) \leq c_1^*\, e^{c_2^* t} \quad \forall\, t \geq 0$.

Therefore for each $0 < \tau < \infty$ we have the relation

$$\tau \phi\left(\frac{u}{\tau}\right) \leq \tau\, c_1^*\, e^{c_2^* u/\tau}.$$

Since $\phi_u(\tau) = \tau\,\phi(u/\tau)$, $0 \le u < \infty$, are convex functions of τ, we have for each $0 < \tau_1 < \tau_2 < \infty$ and $0 \le u < \infty$

$$\sup_{\tau_1 \le \tau \le \tau_2} \left| \tau\,\phi\left(\frac{u}{\tau}\right) \right| \le \max\left\{ \tau_1\,\phi\left(\frac{u}{\tau_1}\right), \tau_2\,\phi\left(\frac{u}{\tau_2}\right) \right\}.$$

The assumption

$$0 < \liminf_n \min_{1 \le i \le M_n} n\,q_{ni} \le \liminf_n \max_{1 \le i \le M_n} n\,q_{ni} < \infty \tag{3.15}$$

implies the existence of $0 < \tau_1 < \tau_2 < \infty$ and n_0 with the property $\tau_1 \le n\,q_{ni} \le \tau_2$ for all $1 \le i \le M_n$ and $n > n_0$.

It follows from here that under (3.15) the function $\Phi_n(u, v)$ satisfies for all $n > n_0$ and $c_1 = \max\{\tau_1 c_1^*, \tau_2 c_1^*\}$, $c_2 = \max\{c_2^*/\tau_1, c_2^*/\tau_2\}$ that

$$|\Phi_n(u, v)| \le c_1 e^{c_2 u}.$$

It is easy to check that in this case $\mu_n = n\,\mu_{\phi,n}$ and $\sigma_n^2 = n\,\sigma_{\phi,n}^2$ for

$$\mu_{\phi,n} = \sum_{i=1}^{M_n} E\left[q_{ni}\,\phi\left(\frac{Z_{ni}}{n\,q_{ni}}\right)\right]$$

and

$$\sigma_{\phi,n}^2 = n \sum_{i=1}^{M_n} Var\left[q_{ni}\,\phi\left(\frac{Z_{ni}}{n\,q_{ni}}\right)\right] - \left(\sum_{i=1}^{M_n} Cov\left[Z_{ni}, q_{ni}\,\phi\left(\frac{Z_{ni}}{n\,q_{ni}}\right)\right]\right)^2,$$

where Z_{ni} are Poisson random variables with $E\left[Z_{ni}\right] = n\,p_{ni}$.

Now the condition *ii)* given previously is equivalent to

$$0 < \liminf_n \sigma_{\phi,n}^2 \le \limsup_n \sigma_{\phi,n}^2 < \infty.$$

Then we have established the following result for $D_\phi(\widehat{\boldsymbol{p}}_n, \boldsymbol{q}_n)$.

Theorem 3.4
If the null hypothesis $H_0 : \boldsymbol{p} = \boldsymbol{p}_n$ *satisfies*

$$\limsup_n \max_{1 \le i \le M_n} n\,p_{in} < \infty,$$

the conditions:

i) $\phi(0) = \lim_{t \downarrow 0} \phi(t) < \infty$ *and* $\lim_{t \to \infty} \frac{\log \phi(t)}{t} < \infty,$

ii) Given $\boldsymbol{q}_n = (q_{n_1}, ..., q_{nM_n})^T$ *with* $q_{ni} > 0, \forall\, 1 \leq i \leq M_n$

$$0 < \liminf_n \min_{1 \leq i \leq M_n} n\, q_{ni} \leq \limsup \max_{1 \leq i \leq M_n} n\, q_{ni} < \infty$$

and

iii)

$$0 < \liminf_n \sigma^2_{\phi,n} \leq \limsup_n \sigma^2_{\phi,n} < \infty$$

imply that

$$\sqrt{n}(D_\phi(\widehat{\boldsymbol{p}}_n, \boldsymbol{q}_n) - \mu_{\phi,n})/\sigma_{\phi,n} \xrightarrow[n\to\infty]{L} N(0,1).$$

This result was established by Menéndez et *al.* (1998b). Now we present some interesting results in relation to the assumptions in the previous theorem.

Proposition 3.1

If we assume that

$$q_{ni} = \frac{1}{M_n}, \qquad 1 \leq i \leq M_n, \tag{3.16}$$

then it holds

$$0 < \liminf_n \min_{1 \leq i \leq M_n} n\, q_{ni} \leq \limsup_n \max_{1 \leq i \leq M_n} n\, q_{ni} < \infty.$$

Proof. This result is immediate since we have assumed that $\lim_{n\to\infty} n/M_n = \gamma \in (0, \infty)$. ∎

Proposition 3.2

If $\boldsymbol{q}_n = \boldsymbol{p}_n$, *the condition*

$$0 < \liminf_n \min_{1 \leq i \leq M_n} n\, q_{ni} \leq \limsup_n \max_{1 \leq i \leq M_n} n\, q_{ni} < \infty$$

implies the condition

$$\limsup_n \max_{1 \leq i \leq M_n} n\, p_{in} < \infty.$$

If also $\boldsymbol{p}_n = \left(\frac{1}{M_n}, ..., \frac{1}{M_n}\right)^T$, *the conditions* $\lim_n \sup \max_{1 \leq i \leq M_n} n\, p_{in} < \infty$ *and*

$$0 < \liminf_n \min_{1 \leq i \leq M_n} n\, q_{ni} \leq \limsup_n \max_{1 \leq i \leq M_n} n\, q_{ni} < \infty$$

hold and in this case

$$Z_{n_i} = Poisson\,(nq_{ni}) = Poisson\left(\frac{n}{M_n}\right) = Poisson\,(\gamma_n)\,,$$

where $\gamma_n = n/M_n$ *and then*

$$\mu_{\phi,n} = E\left[\phi\left(\frac{Z_n}{\gamma_n}\right)\right]$$

and

$$\sigma^2_{\phi,n} = \gamma_n Var\left[\phi\left(\frac{Z_n}{\gamma_n}\right)\right] - Cov^2\left[Z_n, \phi\left(\frac{Z_n}{\gamma_n}\right)\right].$$

Proof. The result follows by previous theorem. ■

In the mentioned Section 4.3 of Read and Cressie (1988) they considered the power-divergence test statistic. We point out that this family of test statistics is obtained as a particular case of $D_\phi(\widehat{\boldsymbol{p}}_n, \boldsymbol{q}_n)$ by taking $\phi(x) \equiv \phi_{(\lambda)}(x) = \frac{1}{\lambda(\lambda+1)}\left(x^{\lambda+1} - x - \lambda(x-1)\right)$, $\lambda \neq 0, -1$,where $\phi_{(0)}(x) = \lim_{\lambda \to 0} \phi_{(\lambda)}(x)$ and $\phi_{(-1)}(x) = \lim_{\lambda \to -1} \phi_{(\lambda)}(x)$. Using Holst's theorem, Read and Cressie (1988) obtained under the uniform hypothesis (3.16) asymptotic distributions of the test statistics $D_{\phi_{(\lambda)}}(\widehat{\boldsymbol{p}}_n, \boldsymbol{q}_n)$ for every $\lambda > -1$. For nonuniform $\boldsymbol{p}_n$ they established a similar result, but only for $\lambda > -1$ and integer-valued. For the proof they referred to Corollary 4.1 and Theorem 5.2 of Morris (1975), who employed a Poissonization idea alternative to Holst's, leading to a similar result under slightly weaker assumptions about $\boldsymbol{p}_n$, but only for integer orders $\lambda > -1$. They formulated a conjecture that their asymptotic result can be extended to every $\lambda > -1$. The theorem presented in this Section, established in Menéndez et *al.* (1998b), provides asymptotic distributions also for test statistics not admitted in the theory of Morris, i.e., for the power-divergence of all orders $\lambda > -1$, since the corresponding functions $\phi_{(\lambda)}(x)$ satisfy the condition $\phi_{(\lambda)}(0) < \infty$ if and only if $\lambda > -1$. This in particular means that the previous theorem confirms the conjecture of Read and Cressie.

3.4. Nonstandard Problems: Tests Statistics based on Phi-divergences

3.4.1. Goodness-of-fit with Quantile Characterization

The quantile characterization is an alternative method for testing goodness-of-fit and, perhaps, has some advantage over that using $T_n^{\phi}(\widehat{\mathbf{p}}, \mathbf{p}^0)$ as discussed previously in Bofinger (1973), Durbin (1978), Menéndez et *al.* (2001b, 2001c) and others. The hypothetical and empirical quantile functions are defined as

$$F_0^{-1}(\pi) = \inf\{x : F_0(x) > \pi\} \text{ and } F_n^{-1}(\pi) = \inf\{x : F_n(x) > \pi\},$$

respectively, for every $\pi \in (0,1)$. $F_n(x)$ is the empirical distribution function defined in Section 1 of Chapter 1.

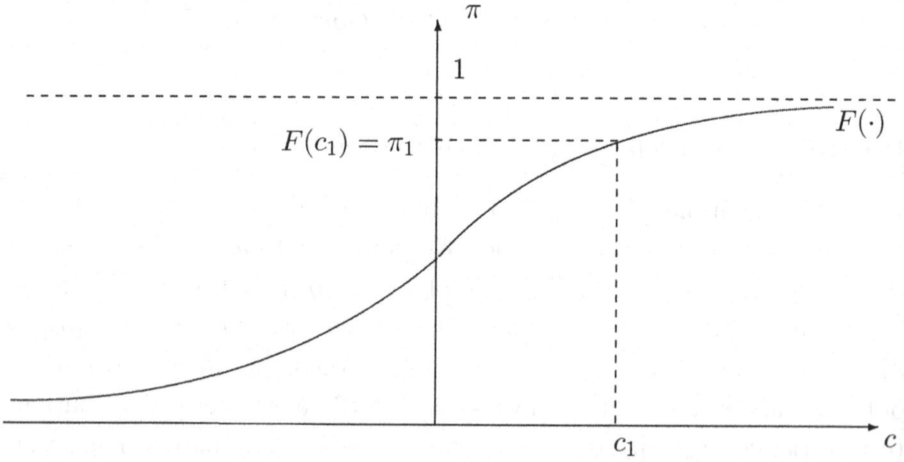

Figure 3.1. Distribution function.

Data are reduced by considering a partition $\boldsymbol{\pi} = (\pi_1, ..., \pi_{M-1}) \in (0,1)^{M-1}$ with

$$\pi_0 = 0 < \pi_1 < ... < \pi_{M-1} < 1 = \pi_M, \tag{3.17}$$

and by applying the functions F_0^{-1} and F_n^{-1} to $\boldsymbol{\pi}$. (See Figures 3.1 and 3.2.) Let $Y_1, ..., Y_n$ be a random sample of size n. Hypothetical and empirical quantile vectors are calculated, respectively, as follows

$$\begin{aligned} \boldsymbol{c} = & \ (c_1, ..., c_{M-1}) = \left(F_0^{-1}(\pi_1), ..., F_0^{-1}(\pi_{M-1})\right), \\ \boldsymbol{Y}_n = & \ \left(Y_{n_1}, ..., Y_{n_{M-1}}\right) = \left(F_n^{-1}(\pi_1), ..., F_n^{-1}(\pi_{M-1})\right), \end{aligned}$$

where $Y_{n_i} = Y_{(n_i)}$ $(n_i = [n\pi_i] + 1,\ i = 1, ..., M-1)$ is the n_ith-order statistic.

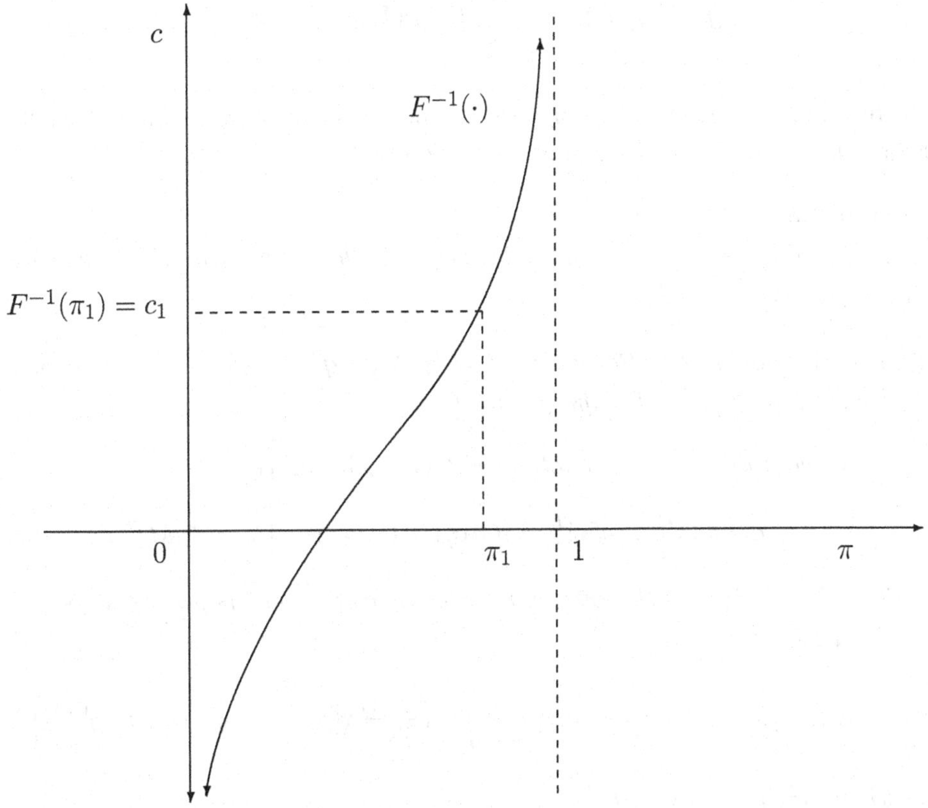

Figure 3.2. Quantile function.

Hypothetical and empirical probability vectors, $\boldsymbol{q}^0$ and $p(\boldsymbol{Y}_n)$, are calculated by

$$\boldsymbol{q}^0 = (q_1^0, ..., q_M^0)^T = (F_0(c_j) - F_0(c_{j-1}) : 1 \leq j \leq M)^T = (\pi_j - \pi_{j-1} : 1 \leq j \leq M)^T,$$

and

$$\boldsymbol{p}(\boldsymbol{Y}_n) = (p_1(\boldsymbol{Y}_n), ..., p_M(\boldsymbol{Y}_n))^T = \left(F_0(Y_{n_j}) - F_0(Y_{n_{j-1}}) : 1 \leq j \leq M\right)^T,$$

where $n_0 = 0$, $n_M = +\infty$, $Y_{n_0} = -\infty$ and $Y_{n_M} = +\infty$.

Once we have calculated the probability vectors $\boldsymbol{q}^0$ and $p(\boldsymbol{Y}_n)$, different test statistics can be used to test

$$H_0 : F = F_0. \tag{3.18}$$

The ϕ-divergence test statistic is given by

$$T_n^{\phi}(\boldsymbol{p}(\boldsymbol{Y}_n), \boldsymbol{q}^0) = \frac{2n}{\phi''(1)} D_\phi(\boldsymbol{p}(\boldsymbol{Y}_n), \boldsymbol{q}^0), \ \phi \in \Phi^*.$$

In relation with this family of ϕ-divergence test statistics in Menéndez et *al.* (1998a, 2001a) the following result was obtained:

Theorem 3.5

Let $F_0(x)$ be continuous and increasing in the neighborhood of each $c_j = F_0^{-1}(\pi_j)$, $j = 1, ..., M-1$.

i) The decision rule " Reject H_0 if $T_n^{\phi}(\boldsymbol{p}(\boldsymbol{Y}_n), \boldsymbol{q}^0) > \chi^2_{M-1,\alpha}$" defines a test for testing (3.18) with significance level α.

ii) Let G be a distribution function with $G \neq F_0$ and consider

$$\boldsymbol{q}^* = (q_1^*, ..., q_M^*)^T = (G(c_j) - G(c_{j-1}) : 1 \leq j \leq M)^T.$$

The power $\beta_{n,\phi}$ of the test with decision rule given in part i) satisfies, at the alternative $\boldsymbol{q}^*$,

$$\beta_{n,\phi}(q_1^*, ..., q_M^*) = 1 - \Phi_n\left(\frac{1}{\sigma_\phi(\boldsymbol{q}^*)}\left(\frac{\phi''(1)}{2\sqrt{n}}\chi^2_{M-1,\alpha} - \sqrt{n} D_\phi(\boldsymbol{q}^*, \boldsymbol{q}^0)\right)\right),$$

where $\sigma_\phi^2(\boldsymbol{q}^*)$ is given by

$$\sigma_\phi^2(\boldsymbol{q}^*) = \sum_{i=1}^{M} q_i^* \left(\phi'\left(\frac{q_i^*}{q_i^0}\right)\right)^2 - \left(\sum_{i=1}^{M} q_i^* \phi'\left(\frac{q_i^*}{q_i^0}\right)\right)^2,$$

and $\Phi_n(x)$ is a sequence of distribution functions tending uniformly to the standard normal distribution function $\Phi(x)$.

iii) The test given in part i) is consistent in the sense of Fraser (1957), that is, for every alternative $\boldsymbol{q}^* \neq \boldsymbol{q}^0$,

$$\lim_{n\to\infty} \beta_{n,\phi}(q_1^*, ..., q_M^*) = 1 \text{ for all } \alpha \in (0,1).$$

iv) If we consider contiguous alternative hypotheses

$$H_{1,n} : \boldsymbol{q}_n = \boldsymbol{q}^0 + \frac{1}{\sqrt{n}}(\boldsymbol{q}^* - \boldsymbol{q}^0),$$

the asymptotic distribution of the ϕ-divergence test statistic, $T_n^{\phi}(\boldsymbol{p}(\boldsymbol{Y}_n), \boldsymbol{q}^0)$, given in part i), is noncentral chi-square with $M-1$ degrees of freedom and noncentrality parameter δ given by

$$\delta = \sum_{j=1}^{M} \frac{\left(q_j^* - q_j^0\right)^2}{q_j^0}.$$

The proof of this Theorem is given in Menéndez et *al.* (2001a). An example for comparing the procedure given in this Section to the procedure given in Section 3.2 can be seen in this paper.

3.4.2. Goodness-of-fit with Dependent Observations

In this Section we study the ϕ-divergence test statistic for testing the stationary distribution as well as the matrix of transition probabilities in a Markov chain.

Stationary Distribution

Methods of statistical inference established for stationary independent data are often applied to dependent data. Investigations into the effects of Markov dependence seem to have been initiated by Bartlett (1951), who showed that such tests need no longer have the "usual" asymptotic distribution. Later this problem has been considered in many papers, see for instance Moore (1982), Glesser and Moore (1983a, 1983b), Molina et *al.* (2002) and references therein. Tavaré and Altham (1983) presented for irreducible aperiodic Markov chains a goodness-of-fit test for the stationary distribution, under simple null hypothesis, based on Pearson test statistic. In this Section we present a methodology, studied in Menéndez et *al.* (1997b, 1999a), for specification of critical values and powers of the ϕ-divergence test statistic in the framework of general statistical models with stationary data. The general methodology is illustrated in the model considered by Tavaré and Altham (1983).

Let $\boldsymbol{Y} = \{Y_k,\ k \geq 0\}$ be a stationary sequence of random variables taking on values in the sample space $\mathcal{X} \subset \mathbb{R}$, and F the distribution function of components Y_k on $\mathcal{X}$. We consider the problem of testing the hypothesis $H_0 : F = F_0$

based on a realization of length n from $\boldsymbol{Y}$. We consider the fixed partition $\mathcal{P} = \{E_i\}_{i=1,\dots,M}$ of $\mathcal{X}$. Let $N_i = \sum_{j=1}^{n} I_{E_i}(Y_j)$ be the number of observations in E_i, $i = 1, ..., M$. In other words, we consider the classical goodness-of-fit tests for vectors $\widehat{\mathbf{p}} = (\widehat{p}_1, \dots, \widehat{p}_M)^T$, $\widehat{p}_i = N_i/n$, of the observed cell frequencies and vectors of the theoretical cell frequencies $\boldsymbol{p} = (p_1, \dots, p_M)^T$, where $p_i = \Pr_F(E_i)$, $i = 1, ..., M$. The hypothesis F_0 is indicated by writing $\boldsymbol{p}^0 = (p_1^0, \dots, p_M^0)^T$, with $p_i^0 = \Pr_{F_0}(E_i)$, $i = 1, ..., M$, and it is assumed that all the components of $\boldsymbol{p}^0$ are nonzero.

In this context Menéndez et *al*. (1997b) established the following result.

Theorem 3.6

If the model satisfies the regularity assumptions:

i) Under the null hypothesis $H_0 : \mathbf{p} = \mathbf{p}^0$, *for* $n \to \infty$, $\widehat{p}_i = p_i^0 + o_P(1)$, *for all* $1 \le i \le M$.

ii) The autocorrelation structure of the model verifies

$$\sqrt{n}\left((p_1^0)^{-1/2} \left(\widehat{p}_1 - p_1^0\right), ..., (p_M^0)^{-1/2} \left(\widehat{p}_M - p_M^0\right) \right) \underset{n\to\infty}{\overset{L}{\longrightarrow}} N(\mathbf{0}, \mathbf{V}),$$

where $\boldsymbol{V}$ *is a given matrix.*

Then, the ϕ*-divergence test statistic*

$$T_n^{\phi}(\widehat{\mathbf{p}}, \mathbf{p}^0) = \frac{2n}{\phi''(1)} D_\phi(\widehat{\mathbf{p}}, \mathbf{p}^0)$$

converges in law to a random variable $\boldsymbol{X}$, *defined by*

$$\boldsymbol{X} \equiv \sum_{i=1}^{M} \rho_i Z_i^2, \tag{3.19}$$

where ρ_i *are the eigenvalues of the matrix* $\mathbf{V}$ *and* Z_i, $i = 1, ..., M$, *are independent standard normal variables.*

Remark 3.2

Based on the previous theorem we should reject the null hypothesis $F = F_0$ *if*

$$T_n^{\phi}(\widehat{\mathbf{p}}, \mathbf{p}^0) > Q_\alpha,$$

where Q_α *is the* $100\,(1-\alpha)$*-percentile of the random variable* $\boldsymbol{X}$ *given in (3.19). The matrix* $\mathbf{V}$ *and, consequently, the eigenvalues* ρ_i *may not be specified uniquely* *by the null hypothesis F_0 (uniquely the marginal distribution of components Y_i is specified). If $\boldsymbol{V}$ depends continuously on the model parameters which remain free under F_0, and there exist consistent estimates of these parameters leading to the estimate $\boldsymbol{V}_n$ of the matrix $\boldsymbol{V}$, then we can use the tests*

$$T_n^{\phi}(\widehat{\mathbf{p}}, \boldsymbol{p}^0) > Q_{n\alpha},$$

where $Q_{n\alpha}$ is the $100\,(1-\alpha)$-percentile of

$$\boldsymbol{X}_n \equiv \sum_{i=1}^{M} \rho_{ni} Z_i^2,$$

and ρ_{ni} are the eigenvalues of the matrix $\boldsymbol{V}_n$. The continuity argument leads to the conclusion that ρ_{ni} estimates consistently ρ_i, i.e., $Q_{n\alpha}$ estimates consistently Q_α.

Theorem 3.6 and Remark 3.2 assert that all members of the family of ϕ-divergence test statistics $T_n^{\phi}(\widehat{\mathbf{p}}, \boldsymbol{p}^0)$ are asymptotically equivalent from the point of view of the test size

$$\alpha = \Pr\left(T_n^{\phi}(\widehat{\mathbf{p}}, \boldsymbol{p}^0) > Q_{n\alpha}/\boldsymbol{p} = \boldsymbol{p}^0\right)$$

and preferences between them are based on the test powers. The power for $\boldsymbol{p}^* \neq \boldsymbol{p}^0$ is

$$\beta_{n,\phi}(\boldsymbol{p}^*) = \Pr\left(T_n^{\phi}(\widehat{\mathbf{p}}, \boldsymbol{p}^0) > Q_{n\alpha}/\boldsymbol{p} = \boldsymbol{p}^*\right).$$

The previous ϕ-divergence test statistic, $T_n^{\phi}(\widehat{\mathbf{p}}, \boldsymbol{p}^0)$, can be applied to stationary irreducible aperiodic Markov chains. We consider a random sample of size n from a stationary irreducible aperiodic Markov chain $\boldsymbol{Y} = \{Y_k,\ k \geq 0\}$ with state space $\{1, ..., M\}$. By $\boldsymbol{P} = (p(i,j))_{i,j=1,...,M}$ we denote the matrix of transition probabilities of this chain and by $\boldsymbol{p}^0$ the stationary distribution, i.e., solution of the equation $\boldsymbol{p}^0 = \boldsymbol{P}(\boldsymbol{p}^0)^T$. We assume that $\boldsymbol{P}$ is from the class of irreducible aperiodic stochastic matrices with one ergodic class so that the solution is unique. The irreducibility means that there are no transient states, i.e., $p_i \neq 0$ for all $1 \leq i \leq M$. The decomposition, considered in the previous theorem, may be defined by $E_i = \{i\}$, $i = 1, ..., M$ and the assumptions *i)* and *ii)* hold (cf., e.g., Billingsley (1961a)). Since the probability distributions associated with the distribution functions F and F_0, in this case, coincide with the vectors $\boldsymbol{p}$ and $\boldsymbol{p}^0$, we consider the hypothesis $H_0 : \boldsymbol{p} = \boldsymbol{p}^0$ about the stationary distribution of the chain matrix $\boldsymbol{P}$. Since no states are transient, the hypothesis satisfies the condition that all the components of $\boldsymbol{p}^0$ are nonzero.

In the model under consideration the goodness-of-fit test of the null hypothesis $H_0 : \boldsymbol{p} = \boldsymbol{p}^0$ based on the Pearson's test statistic has been considered by Tavaré and Altham (1983). This result was extended by Menéndez et *al.* (1997b, 1999a) to the ϕ-divergence test statistic. These authors established that in this case the ϕ-divergence test statistic $T_n^{\phi}(\widehat{\boldsymbol{p}}, \boldsymbol{p}^0)$ converges in law to a random variable $\boldsymbol{X}$, defined by

$$\boldsymbol{X} \equiv \sum_{i=1}^{M-1} \frac{1+\lambda_i}{1-\lambda_i} Z_i^2, \tag{3.20}$$

where $\lambda_1, ..., \lambda_{M-1}$ are the nonunit eigenvalues of the chain matrix $\boldsymbol{P}$.

If $\boldsymbol{P} = (p(i,j))_{i,j=1,...,M}$ has identical rows then it is reversible and all its nonunit eigenvalues are zero. Thus Theorem 3.6 implies that if data $Y_1, ..., Y_n$ are independent then all ϕ-divergence test statistics $T_n^{\phi}(\widehat{\boldsymbol{p}}, \boldsymbol{p}^0)$ are asymptotically distributed chi-squared with $M-1$ degrees of freedom. More generally, if $\boldsymbol{P} = (1-\pi)\boldsymbol{I}_{M\times M} + \pi\boldsymbol{1}(\boldsymbol{p}^0)^T$, where $0 < \pi \leq 1$ and $\boldsymbol{1} = (1, ..., 1)^T$, then the nonunit eigenvalues of $\boldsymbol{P}$ are all equal $1-\pi$. Therefore all ϕ-divergence test statistics $\frac{\pi}{2-\pi} T_n^{\phi}(\widehat{\boldsymbol{p}}, \boldsymbol{p}^0)$ tend in law to a chi-square distribution with $M-1$ degrees of freedom.

If the matrix $\boldsymbol{P}$ is not known we can use the relative frequencies $\widehat{p}(i,j) = v_{ij}/v_{i*}$, where

$$v_{ij} = \sum_{k=2}^{n} I_{\{(i,j)\}}(Y_{k-1}, Y_k) \text{ and } v_{i*} = \sum_{j=1}^{M} v_{ij},$$

to estimate the transition probabilities $p(i,j)$ of the matrix $\boldsymbol{P}$ consistently (cf., Billingsley (1961a)). By $I_{\{(i,j)\}}(Y_{k-1}, Y_k)$ we are denoting the function defined as

$$I_{\{(i,j)\}}(Y_{k-1}, Y_k) = \begin{cases} 1 & \text{if } \; Y_{k-1} = i \text{ and } Y_k = j \\ 0 & \quad \text{otherwise} \end{cases}.$$

Since the eigenvalues λ_i considered in (3.20) are continuous functions of the elements $p(i,j)$ of $\boldsymbol{P}$, the substitution $p(i,j) = \widehat{p}(i,j)$ in these functions leads to consistent estimates $\widehat{\lambda}_i$ of λ_i.

The binary Markov model, considered in Exercise 13, was studied in Menéndez et *al.* (1997b) and they found that for the power-divergence test statistics the value $\lambda = -2$ (Modified chi-square test statistic) is optimal in the sense of the power. It is interesting to observe that in the case of independent data $T_n^{\phi(-2)}(\widehat{\boldsymbol{p}}, \boldsymbol{p}^0)$ is rarely optimal in this sense.

Chain Markov and Order

We consider a random sample of size n from an irreduccible homogeneous Markov chain, $\boldsymbol{Y} = \{Y_k,\ k \geq 0\}$, with state space $\{1, ..., M\}$ and matrix of transition probabilities given by $\boldsymbol{P} = (p(i,j))_{i,j=1,...,M}$. Billingsley (1961a, 1961b) considered the problem of testing the hypothesis

$$H_0 : \boldsymbol{P}{=}\boldsymbol{P}^0 = \left(p^0(i,j)\right)_{i,j=1,...,M}$$

on the basis of the likelihood ratio test statistic and chi-square test statistic. These results were extended in Menéndez et *al.* (1999b) by considering the ϕ-divergence test statistic. They considered the family of ϕ-divergence test statistics given by

$$T_n^{\phi}(\widehat{\boldsymbol{P}},\boldsymbol{P}^0) = \frac{2n}{\phi''(1)} \sum_{i=1}^{M} \frac{v_{i*}}{n} \sum_{j=1}^{M} p^0(i,j)\phi\left(\frac{\widehat{p}(i,j)}{p^0(i,j)}\right), \tag{3.21}$$

and established that its asymptotic distribution is chi-square with $c - M$ degrees of freedom, where c is the number of elements in $C_0 = \left\{(i,j) : p_{ij}^0 > 0\right\}$. The test statistics given by Billingsley appear as a particular case of the ϕ-divergence test statistic given in (3.21). Interesting papers applicable to stationary finite-state irreducible Markov chains have been written by Azalarov and Narkhuzhaev (1987, 1992), Mirvalev and Narkhuzhaev (1990), Ivchenko and Medvedev (1990), Basawa and Prakasa Rao (1980) and Rousas (1979).

In Markov chains, the future evolution of the chain is conditionally independent of the past given the present state. It is, however, possible that the dependence relation is more complicated. In a second order Markov chain, the future evolution of the chain over times $n+1, n+2, ...$ is independent of the past given the states at times n and $n-1$. In a rth order Markov chain, r consecutive states must be conditioned upon for the future to be independent of the past. In a rth order Markov chain, the transition probabilities are $r+1$ dimensional, r for the present and one for the future. The case $r = 0$ corresponds to a sequence of independent trials, while $r = 1$ corresponds to the usual Markov chain. It is important to be able to determine the order of the chain and for simplicity to find the lowest acceptable value of r. This problem has been studied using the likelihood ratio test statistic and the chi-square test statistic by Billingsley (1961a, 1961b) and using ϕ-divergence test statistics by Menéndez et *al.* (2001d).

3.4.3. Misclassified Data

The theory of goodness-of-fit tests, in the analysis of categorical data, has been developed extensively. One of the difficulties, often encountered in practice, is the possibility of a false classification of one or more individuals into the respective categories or classes. This problem was first discussed by Bross (1954) for the case of two categories. Bross established that the sample proportion is a biased estimate of the proportion and the bias is a function of the amount of misclassification of the data. Mote and Anderson (1965) studied the effect of misclassification on Pearson's test statistic. If errors of misclassification are ignored, the size of the test will increase and the asymptotic power will be reduced. If we consider the family of ϕ-divergence test statistics $T_n^{\phi}(\widehat{\boldsymbol{p}}, \boldsymbol{p}^0)$, a similar study to the one realized by Mote and Anderson (1965), gives analogous results. Then for the goodness-of-fit tests, the usual test requires modification when there are misclassification errors.

In order to solve the difficulties involved in inference from a sample of categorical data, obtained by using a fallible classifying mechanism, Tenenbein (1970, 1971, 1972) presented double sampling methods for estimating the multinomial proportions in the presence of misclassification errors. Hochberg (1977) extended the use of Tenenbein's double sample schemes for modeling and testing hypotheses on the true parameters from general multidimensional contingency tables with misclassification errors. Cheng et *al.* (1998) used also the Tenenbein's double sampling scheme for introducing an adjusted chi-square test and the likelihood ratio test.

In Pardo, L. and Zografos (2000), the family of ϕ-divergence test statistics for testing goodness-of-fit when the categorical data are subject to misclassification was considered.

The double sample scheme is used in the context of the following experimental situation. Suppose that we have two methods of collecting the data: one which is error-free but expensive and the second which is fallible but inexpensive. An obvious dilemma results for the researcher, specially when funds are limited. Should they sacrifice accuracy for quantity? Diamond and Lilienfield (1962) discuss an experimental situation in public health research where the true classification device is the physician's examination whereas the fallible classifier is a questionnaire completed by the patient. The Tenenbein's double sample scheme gives to the researcher another alternative which incorporates a balance between both measurement methods and their respective cost. The scheme suggests that, at the first stage, a sample of n units is drawn and the true and fallible classifications are obtained for each unit, and at the second stage a sample of $N - n$ units is drawn and the fallible classification is obtained for each unit. Then, there are a total of n units in the sample which have been classified by both the true and fallible devices. The multinomial proportions can be estimated from the available data without going to the extreme of obtaining the true classification for all N units in the sample.

We denote by Y the random variable associated with the true measurement, taking on the value "i" if the sampling unit belongs in fact to category E_i, $i = 1, ..., M$, and by Y_0 the random variable associated with the fallible measurement, taking on the value "j" if the sampling unit is classified by the fallible device as being in category E_j, $j = 1, ..., M$. Let us denote the marginal probabilities of Y and Y^0, by

$$p_i = \Pr(Y = i), \quad \pi_j = \Pr(Y^0 = j), \ i, j = 1, ..., M,$$

respectively, with $\sum_{i=1}^{M} p_i = \sum_{j=1}^{M} \pi_j = 1$. To describe misclassification we define θ_{ij} to be the probability that a unit, which, in fact, belongs to the category E_i, is classified in the category E_j. Thus

$$\theta_{ij} = \Pr(Y^0 = j | Y = i), \ i, j = 1, ..., M,$$

and it is clear that

$$\sum_{j=1}^{M} \theta_{ij} = 1 \text{ and } \pi_j = \sum_{i=1}^{M} p_i \theta_{ij}, \ i, j = 1, ..., M.$$

In this situation a double sampling scheme can be described as follows:

i) A sample of n units is drawn and the true and fallible classifications, denoted by $Y_1, ..., Y_n$ and $Y_1^0, ..., Y_n^0$, respectively, are obtained for each unit. We denote by n_{ij} the number of units in the sample whose true category is E_i, $i = 1, ..., M$, and whose fallible category is E_j, $j = 1, ..., M$, and by $n_{i*} = \sum_{j=1}^{M} n_{ij}$, $n_{*j} = \sum_{i=1}^{M} n_{ij}$.

ii) A further sample of $N-n$ units is drawn and the fallible classifications $Y_{n+1}^0, ..., Y_N^0$, are obtained for each unit. We denote by

$$m_j = \sum_{k=n+1}^{N} I_{\{j\}}(Y_k^0),\ j = 1, ..., M,$$

the number of units whose fallible category is E_j, $j = 1, ..., M$, and by $(m_1, ..., m_M)^T$ the vector of frequencies associated with the random sample $Y_{n+1}^0, ..., Y_N^0$. By $I_{\{j\}}(Y_k^0)$ we are denoting

$$I_{\{j\}}(Y_k^0) = \begin{cases} 1 & \text{if } Y_k^0 = j \\ 0 & \text{otherwise} \end{cases}.$$

The joint likelihood function associated with the observed data

$$(Y_1, Y_1^0), ..., (Y_n, Y_n^0), Y_{n+1}^0, ..., Y_N^0,$$

is given by

$$L(\boldsymbol{p}, \boldsymbol{\Theta}) = \prod_{i=1}^{M}\prod_{j=1}^{M} (p_i\theta_{ij})^{n_{ij}} (\pi_j - p_i\theta_{ij})^{n_{*j}-n_{ij}} \prod_{k=1}^{M} \pi_k^{m_k},$$

with $\boldsymbol{p} = (p_1, ..., p_M)^T$, $\boldsymbol{\Theta} = (\theta_{ij})_{i,j=1,..,M}$ (c.f., Cheng et *al.* (1998)). Then the maximum likelihood estimators (cf., Tenenbein (1972)) are

$$\widehat{p}_i = \sum_{j=1}^{M} \frac{(m_j + n_{*j})n_{ij}}{Nn_{*j}} \text{ and } \widehat{\theta}_{ij} = \frac{(m_j + n_{*j})n_{ij}}{Nn_{*j}\widehat{p}_i},\ i, j = 1, ..., M.$$

Using the above expression for $\widehat{p}_i$, and assuming that $n/N \to f > 0$, as $N \to \infty$, the asymptotic distribution of $(\widehat{p}_1, ..., \widehat{p}_{M-1})$ is

$$\sqrt{N}(\widehat{p}_1 - p_1, ..., \widehat{p}_{M-1} - p_{M-1}) \underset{N\to\infty}{\overset{L}{\to}} N(\boldsymbol{0}, \boldsymbol{\Sigma}),$$

where the asymptotic variance-covariance matrix is defined (cf., Tenenbein (1972), Cheng et *al.* (1998)) by

$$\boldsymbol{\Sigma} = (\sigma_{ij})_{i,j=1,..,M},\ \sigma_{ij} = \begin{cases} \frac{p_i q_i}{f}[1 - (1-f)K_i], & i = j \\ \left(1 - \frac{1}{f}\right) \sum\limits_{k=1}^{M} \lambda_{ik}\lambda_{jk}\pi_k - p_i q_j, & i \neq j \end{cases}, \quad (3.22)$$

with $q_i = 1 - p_i$, $K_i = \left(Corr\left[I_{\{i\}}(Y), E[I_{\{i\}}(Y)\,|Y^0]\right]\right)^2 = \frac{p_i}{q_i}\left(\sum_{k=1}^{M}\frac{\theta_{ik}^2}{\pi_k} - 1\right)$ and $\lambda_{ij} = E[I_{\{i\}}(Y)\,|Y^0 = j] = \frac{p_i\theta_{ij}}{\pi_j}$, for $i, j = 1, ..., M-1$.

Under the simple null hypothesis

$$H_0 : \mathbf{p}{=}\mathbf{p}^0 = (p_1^0, ..., p_M^0)^T,$$

Cheng et *al.* (1998) established that

$$N\sum_{i=1}^{M-1}\sum_{j=1}^{M-1}(\widehat{p}_i - p_i^0)\widehat{\tau}_{ij}(\widehat{p}_j - p_j^0) \underset{N\to\infty}{\overset{L}{\longrightarrow}} \chi^2_{M-1},$$

with $\mathbf{\Sigma}^{-1} = (\tau_{ij})_{i,j=1,..,M}$, and $\widehat{\tau}_{ij}$ the maximum likelihood estimator of τ_{ij}, $i, j = 1, ..., M-1$.

Next result regarding the ϕ-divergence test statistic for misclassified data was established in the cited paper of Pardo, L. and Zografos (2000).

Theorem 3.7

Based on Tenenbein's (1972) double sampling scheme, let $\widehat{\mathbf{p}}{=}(\widehat{p}_1, ..., \widehat{p}_M)^T$ *be with* $\widehat{p}_i = \sum_{j=1}^{M}\frac{(m_j + n_{*j})n_{ij}}{Nn_{*j}}$, *and assume that* $n/N \to f > 0$, *as* $N \to \infty$. *Then we have*

i) Under the hypothesis $H_0 :\ \mathbf{p}{=}\mathbf{p}^0 = (p_1^0, ..., p_M^0)^T$

$$T_N^{\phi}(\widehat{\mathbf{p}},\mathbf{p}^0) = \frac{2N}{\phi''(1)}D_{\phi}(\widehat{\mathbf{p}},\mathbf{p}^0) \underset{N\to\infty}{\overset{L}{\longrightarrow}} \sum_{i=1}^{r}\lambda_i Z_i^2,$$

where $Z_1, ..., Z_r$ *are independent and identically distributed normal random variables with mean zero and variance* 1 *and* λ_i, $i = 1, ..., r$ *are the eigenvalues of the matrix* $\mathbf{A\Sigma}$, *being* $\mathbf{A}$ *the diagonal matrix with elements* $\left(p_i^0\right)^{-1}$, $i = 1, ..., M-1$, $\mathbf{\Sigma}$ *the asymptotic variance-covariance matrix of the random vector* $\sqrt{N}(\widehat{\mathbf{p}}{-}\mathbf{p}^0)$ *given in (3.22) and* $r = rank(\mathbf{A\Sigma A})$.

ii) Let $\mathbf{p}^* = (p_1^*, \ldots, p_M^*)^T$ *be a probability distribution with* $\mathbf{p}^* \neq \mathbf{p}^0$, *then*

$$\beta_{N,\phi}(\mathbf{p}^*) = 1 - \Phi_N\left(\frac{\sqrt{N}}{\sigma_{\phi}(\mathbf{p}^*)}\left(\frac{k_\alpha}{2N}\phi''(1) - D_{\phi}(\mathbf{p}^*,\mathbf{p}^0)\right)\right),$$

where $\sigma_\phi^2(\mathbf{p}^*) = \boldsymbol{T}^T\Sigma\boldsymbol{T}$, $\boldsymbol{T} = (t_1, ..., t_M)^T$ *and*

$$t_j = \left(\frac{\partial D_\phi(\mathbf{p}, \mathbf{p}^0)}{\partial p_j}\right)_{\mathbf{p}=\mathbf{p}^*}.$$

The value k_α *verifies*

$$\Pr\left(T_N^\phi(\widehat{\mathbf{p}},\mathbf{p}^0) > k_\alpha \;/\; H_0 : \mathbf{p} = \mathbf{p}^0\right) = \alpha,$$

and $\Phi_N(x)$ *is a sequence of distribution functions that tends uniformly to the standard normal distribution* $\Phi(x)$.

3.4.4. Goodness-of-fit for and against Order Restrictions

In some situations the probability vector $\boldsymbol{p} = (p_1, ..., p_M)^T$ exhibits a trend. If, for example, the classes $\{E_i\}_{i=1,...,M}$ have the same length in $\mathbb{R}$ and the original probability density function is unimodal, there is a positive integer k such that

$$p_1 \leq ... \leq p_{k-1} \leq p_k \geq ... \geq p_{M-1}.$$

Some interesting examples in which the probability vector $\boldsymbol{p}$ exhibits a trend can be seen in Robertson (1978), Lee (1987), Robertson et *al.* (1988) and many others cited there.

Statistical inference concerning a set of multinomial parameters under order restrictions has been studied since Chacko (1966) considered the maximum likelihood estimation of multinomial parameters subject to a simple order restriction. He also obtained the asymptotic null distribution of a chi-square type test statistic for testing homogeneity of a set of multinomial parameters against the simple order. The asymptotic null distribution of this test statistic is a mixture of chi-square distributions, which is called a chi-bar-square distribution. The mixing coefficients, which are called level probabilities, depend upon the multinomial parameter set as well as the order restriction.

Robertson (1966) found maximum likelihood estimates of multinomial parameters subject to a partial order restriction and also Robertson (1978) generalized Chacko's result to the one-sample likelihood ratio test of the equality of two multinomial parameters (one is known) against a partial order restriction. He showed that the asymptotic null distribution of this test statistic is chi-bar-square and the level probabilities depend upon the known multinomial parameter only through the sets on which the known parameter is constant. He also considered the likelihood ratio test of an order restriction as a null hypothesis. He showed that the asymptotic null distribution of this test statistic is also chi-bar-square and that the level probabilities depend upon the true parameter only through the sets on which the true parameter is constant.

Chi-square type tests have been studied by several researchers. Lee (1987) considered chi-square type tests for and against an order restriction on a set of multinomial parameters. He compared three test procedures, namely: *i)* the likelihood ratio test statistic, *ii)* the Pearson chi-square test statistic and *iii)* the modified chi-square test statistic. He showed that all three test statistics have the same asymptotic null distribution which is of chi-bar-square type. Menéndez et *al.* (2002) considered the ϕ-divergence test statistic for testing the equality of two multinomial parameters, one is known, against a partial order restriction and also for testing an order restriction as a null hypothesis. They established that the ϕ-divergence test statistics, under the null hypothesis, are asymptotically chi-bar-squared distributed. Other interesting results in this area, using ϕ-divergence test statistics, can be seen in Menéndez et *al.* (2003a, 2003c).

3.5. Exercises

1. The following data represent the number of injured players in a random sample of 200 soccer matches:

Injured players	0	1	2	3	≥ 4
Soccer matches	82	90	20	7	1

   a) Test the hypothesis that the distribution is Poisson with parameter $\lambda = 0.775$. Take significance level $\alpha = 0.05$ and Freeman-Tukey test statistic.

   b) Find the power of the test at $\boldsymbol{p}^* = (0.45, 0.35, 0.1, 0.05, 0.05)^T$.

2. Find the expression of the test statistic for goodness-of-fit based on divergence measures of Pearson, Matusita ($a = 1/2$), Balakrishnan-Sanghvi, Rathie-Kanappan, Power-divergence, Rukhin and Rényi.

3. Consider the divergence measure $D_s^r(\boldsymbol{p}, \boldsymbol{q})$. Find the asymptotic variance for the statistic $D_s^r(\widehat{\mathbf{p}}, \mathbf{p}^0)$ and as a special case the asymptotic variance of the statistic associated with Rényi divergence.

4. Find the asymptotic variance of the estimated entropy of order r and degree s as a special case of the estimated divergence of order r and degree s.

5. Let $\boldsymbol{u} = (1/M, ..., 1/M)^T$. Find the asymptotic distribution of the estimated entropy of order r and degree s as a special case of the result obtained in Remark 3.1.

6. Let ϕ be a concave function with continuous second derivative.

   a) Find the asymptotic distribution of the test statistic $8nR_\phi(\widehat{\mathbf{p}}, \mathbf{p}^0)$ being R_ϕ the divergence measure introduced in Chapter 1 and $\mathbf{p}^0 = (p_1^0, ..., p_M^0)^T$.

   b) Let $\mathbf{p}^0 = (1/M, ..., 1/M)^T$. Show that

$$S_\phi(\widehat{\mathbf{p}}, \mathbf{p}^0) = -\frac{8nM}{\phi''(1/M)} R_\phi(\widehat{\mathbf{p}}, \mathbf{p}^0) \underset{n\to\infty}{\overset{L}{\longrightarrow}} \chi^2_{M-1},$$

   provided $\phi''(1/M) < 0$.

   c) Let $\phi(x) = -x \log x$. Find the asymptotic distribution for $8nR_\phi(\widehat{\mathbf{p}}, \mathbf{p}^0)$.

7. For the study of an urban system in a given region we suppose that a certain function of service f is more implanted relatively in the cities of major dimensions than in the smaller ones. For this purpose a spatial index is used. This index is given by

$$I_f = \frac{A_{fi}A_{fr}}{A_{ti}A_{tr}},$$

where:

A_{fi} represents the number of addresses of the function f in the city i.

A_{fr} represents the number of addresses of the function f in the region r.

A_{ti} represents the total number of addresses of the function f in the city i.

A_{tr} represents the total number of addresses of the function f in the region r.

The cities are ordered in six levels taking into account I_f the largest (6) to the smallest (1). From previous studies the expected frequencies of each type of city according to its dimension are known.

The theoretical and observed frequencies into the six categories are as follow:

	Levels						
	1	2	3	4	5	6	*Total*
Theoretical	1	2	5	7	15	20	50
Observed	1	3	6	10	17	13	50

Using the test statistic given in *c)* of the previous exercise, analyze if the presence of the function f is associated with the city dimension, taking as significance level $\alpha = 0.01$.

8. We want to find a model to predict the probability of winning at a greyhound race track in Australia. Data collected on 595 races give the starting numbers of the eight dogs included in each race ordered according to the race finishing positions (the starting numbers are always the digits 1,...,8; 1 denotes the dog started on the fence, 2 denotes the second from the fence, etc.). We assume throughout that the initial positions are assigned at random to each of the eight dogs. We group the results into eight cells according to which starting number comes in first. Now we want to test

the hypothesis that all starting numbers have an equal chance of coming in first regardless of the positions of the other seven dogs, that is,

$$H_0 : p_i = 1/8;\ i = 1, ..., 8.$$

For this purpose use the test statistic given in Exercise 6 part *b)* with the function

$$\phi_a(x) = \begin{cases} \frac{1}{1-a}(x^a - x) & \text{si } a \neq 1 \\ -x \log x & \text{si } a = 1 \end{cases}$$

for $a = 1$, 13/7 and 2 and significance level $\alpha = 0.05$.

The theoretical and observed frequencies into the eight categories are as follows:

Dog i	*Observed*	*Expected*
1	0.175	0.125
2	0.16	0.125
3	0.111	0.125
4	0.106	0.125
5	0.104	0.125
6	0.097	0.125
7	0.101	0.125
8	0.146	0.125

Source: Haberman (1978, p. 2).

9. Let $\boldsymbol{p}_n = (1/M_n, ..., 1/M_n)^T$ and we consider the notation established in Section 3.3. Find the asymptotic distribution of the test statistic

$$\chi^2(\widehat{\boldsymbol{p}}_n, \boldsymbol{p}_n) = \sum_{i=1}^{M_n} p_{ni}\left(1 - \frac{N_{ni}}{np_{ni}}\right)^2.$$

10. Let $\boldsymbol{p}_n = \boldsymbol{q}_n$. Suppose that the following inequality holds

$$0 < \liminf_n \min_{1 \leq i \leq M_n} n\, q_{ni} \leq \limsup_n \max_{1 \leq i \leq M_n} n\, q_{ni} < \infty.$$

Show that for $\phi(x) = (1 - x)^2$,

$$\mu_{\phi,n} = \frac{M_n}{n}$$

and

$$\sigma^2_{\phi,n} = \frac{M_n}{n}\left(\frac{1}{M_n\, n}\sum_{i=1}^{M_n}\frac{1}{p_{ni}} - \gamma_n^{-1} + 2\right).$$

11. Consider the power divergence family. Find the expression of the mean and asymptotic variance corresponding to the family of test statistics based on it for $\boldsymbol{q}_n = \boldsymbol{p}_n = (1/M_n, ..., 1/M_n)^T$ and $\lambda > -1$.

12. We consider the Markov chain with states 1 and 2 and stochastic matrix $\mathbf{P}$ of transition probabilities given by

$$\mathbf{P} = \begin{pmatrix} 1-\beta & \beta \\ \gamma & 1-\gamma \end{pmatrix}; \; 0 < \beta, \gamma \leq 1, \; \beta + \gamma < 2.$$

Find the expression of the ϕ-divergence test statistic for testing

$$H_0 : \mathbf{p} = \mathbf{p}^0 = (1/2, 1/2)^T.$$

13. Find the expression of the ϕ-divergence test statistic given in Theorem 3.7 if $\boldsymbol{p}^0 = (p_0, q_0)^T$, with $q_0 = 1 - p_0$, $0 < p_0 < 1$.

14. Obtain the expression of the ϕ-divergence test statistic given in Theorem 3.7 for

$$\phi(x) = \frac{1}{2}(x-1)^2.$$

15. A sample of $n = 100$ units is doubly classified by true and fallible methods and a second random sample of 400 measurements is taken and classified by the fallible method. The sample sizes n and N are respectively 100 and 500. The following table shows the obtained data.

		Fallible Device			
		0	1		
True Device	0	61	7	68	
	1	1	31	32	First Sample
		62	38	100	
		218	182	400	Second Sample

Test $H_0 : \boldsymbol{p} = \boldsymbol{p}^0 = (1/2, 1/2)^T$ by using the test statistic based on the power-divergence family with $\lambda = 1$, -1 and 0 and significance level $\alpha = 0.05$.

16. With the notation of Section 3.3 establish that if $N_{ni} = \gamma + O_P(n)$ and $\widehat{p}_{ni} = p_{ni} + o_P(1)$, then

$$\gamma = np_{ni} + O_P(1). \tag{3.23}$$

17. We consider the population given by the distribution function

$$F_0(x) = \begin{cases} 0 & \text{if } x \leq 2 \\ 1 - \frac{4}{x^2} & \text{if } x > 2 \end{cases}.$$

Use the procedure given in Section 3.4.1 to study if the observations

2.2408	5.8951	6.0717	3.6448	2.8551	4.4065	14.4337
3.0338	2.0676	2.6155	2.7269	5.1468	2.2178	2.0141
2.3339	2.6548	5.0718	2.8124	2.0501	13.6717	

are from $F_0(x)$ on the basis of the power-divergence test statistic for $\lambda = -2, -1, -1/2, 0, 2/3$ and 1 and significance level $\alpha = 0.05$.

18. Find the expression of the asymptotic variance of

$$\sqrt{n}\left(D_\phi(\widehat{\boldsymbol{p}}, \boldsymbol{p}^0) - D_\phi(\boldsymbol{p}^*, \boldsymbol{p}^0)\right),$$

with $\boldsymbol{p}^0 \neq \boldsymbol{p}^*$, using the divergence measures of Pearson, Matusita ($a = 1/2$), Balakrishnan-Sanghvi, Rathie-Kanappan, Power-divergence and Rukhin.

3.6. Answers to Exercises

1. If we consider the partition of the sample space given by

$$E_1 = \{0\},\ E_2 = \{1\},\ E_3 = \{2\},\ E_4 = \{3\} \text{ and } E_5 = \{x \in \mathbb{N} : x \geq 4\},$$

we have $\boldsymbol{p}^0 = (0.4607, 0.3570, 0.1383, 0.0357, 0.0083)^T$. Now our problem consists of testing $H_0 : \boldsymbol{p} = \boldsymbol{p}^0$, using the test statistic obtained from the power-divergence test statistics with $\lambda = -1/2$. The test statistic obtained in this case is the Freeman-Tukey's test statistic and we should reject the null hypothesis if

$$T_n^{-1/2}(\widehat{\boldsymbol{p}}, \boldsymbol{p}^0) = 8n\left(1 - \sum_{i=1}^{5}\sqrt{p_i^0 \widehat{p}_i}\right) > \chi^2_{M-1,\alpha}.$$

From the data we get $\widehat{\boldsymbol{p}} = (82/200, 90/200, 20/200, 7/200, 1/200)^T$, and then

$$T_n^{-1/2}(\widehat{\boldsymbol{p}}, \boldsymbol{p}^0) = 8.2971.$$

On the other hand $\chi^2_{4,\,0.05} = 9.4877$. Therefore we should not reject the null hypothesis.

The asymptotic power at the point $\boldsymbol{p}^* = (0.45, 0.35, 0.1, 0.05, 0.05)^T$ is given by

$$\beta_{200,\phi_{(-1/2)}}(\boldsymbol{p}^*) \simeq 1 - \Phi\left(\frac{\sqrt{200}}{\sigma_1(\boldsymbol{p}^*)}\left(\frac{\phi''_{(-1/2)}(1)\,\chi^2_{M-1,\alpha}}{400} - D_{\phi_{(-1/2)}}(\boldsymbol{p}^*, \boldsymbol{p}^0)\right)\right).$$

It is not difficult to establish that

$$\sigma_1(\boldsymbol{p}^*) = 4\left(1 - \left(\sum_{i=1}^{5}\sqrt{p_i^0}\sqrt{p_i^*}\right)^2\right),$$

then

$$\sigma_1(\boldsymbol{p}^*) = 0.2955,\ \ D_{\phi_{(-1/2)}}(\boldsymbol{p}^*, \boldsymbol{p}^0) = 0.0\,4391 \text{ and } \phi''_{(-1/2)}(1) = 1.$$

With these values we get $\beta_{200,\phi_{(-1/2)}}(\boldsymbol{p}^*) \simeq 0.8330$.

2. In Theorem 3.1 it was established that

$$T_n^{\phi}(\widehat{\boldsymbol{p}}, \boldsymbol{p}^0) = \frac{2n}{\phi''(1)}\sum_{i=1}^{M} p_i^0 \phi\left(\frac{\widehat{p}_i}{p_i^0}\right) \xrightarrow[n\to\infty]{L} \chi^2_{M-1}.$$

For Pearson divergence, $\phi(x) = \frac{1}{2}(x-1)^2$, then

$$X^2 \equiv T_n^{\phi}(\widehat{\boldsymbol{p}}, \boldsymbol{p}^0) = \sum_{i=1}^{M}\frac{(N_i - p_i^0)^2}{np_i^0}.$$

This is the classical chi-square test statistic.

For Matusita $(a = 1/2)$, $\phi(x) = (1-\sqrt{x})^2$, then

$$T_n^{\phi}(\widehat{\boldsymbol{p}}, \boldsymbol{p}^0) = 4n\sum_{i=1}^{M}\left(\sqrt{p_i^0} - \sqrt{\widehat{p}_i}\right)^2.$$

It is not difficult to establish that this test statistic coincides with the Freeman-Tukey test statistic.

For Balakrishnan-Sanghvi, $\phi(x) = \frac{(x-1)^2}{(x+1)^2}$, then

$$T_n^{\phi}(\widehat{\boldsymbol{p}}, \boldsymbol{p}^0) = 4n\sum_{i=1}^{M}\left(\frac{\widehat{p}_i - p_i^0}{\widehat{p}_i + p_i^0}\right)^2.$$

For Rathie-Kanappan, $\phi(x) = \frac{x^s - x}{s-1}$, then

$$T_n^s(\widehat{\boldsymbol{p}}, \boldsymbol{p}^0) = \frac{2n}{s(s-1)} \left(\sum_{i=1}^{M} \frac{(\widehat{p}_i)^s}{(p_i^0)^{s-1}} - 1 \right).$$

For $\phi(x) = \phi_{(\lambda)}(x)$, where $\phi_{(\lambda)}(x)$ was given in the first chapter, then

$$T_n^{\lambda}(\widehat{\boldsymbol{p}}, \boldsymbol{p}^0) = \frac{2n}{\lambda(\lambda+1)} \left(\sum_{i=1}^{M} \frac{(\widehat{p}_i)^{\lambda+1}}{(p_i^0)^{\lambda}} - 1 \right), \; \lambda \neq 0, -1.$$

For Rukhin,

$$\phi_a(x) = \frac{(1-x)^2}{2(a + (1-a)x)}, \qquad a \in [0,1],$$

and

$$T_n^{\phi_a}(\widehat{\boldsymbol{p}}, \boldsymbol{p}^0) = n \sum_{j=1}^{M} \frac{(p_j^0 - \widehat{p}_j)^2}{ap_j^0 + (1-a)\widehat{p}_j} \qquad 0 \leq a \leq 1. \tag{3.24}$$

Finally, for Rényi divergence we have

$$h(x) = \frac{\log(r(r-1)x + 1)}{r(r-1)} \text{ and } \phi(x) = \frac{(x^r - r(x-1) - 1)}{r(r-1)}$$

and the family of Rényi test statistics is given by

$$T_n^r(\widehat{\boldsymbol{p}}, \boldsymbol{p}^0) = \frac{2n}{r(r-1)} \log \left(\sum_{j=1}^{M} (p_j^0)^{1-r} (\widehat{p}_j)^r - 1 \right).$$

3. The divergence measure $D_r^s(\boldsymbol{p}, \boldsymbol{q})$ is a (h, ϕ)-divergence, with

$$h(x) = \frac{1}{s-1} \left((1 + r(r-1)x)^{\frac{s-1}{r-1}} - 1 \right); \; s, \; r \neq 1,$$

and

$$\phi(x) = \frac{x^r - r(x-1) - 1}{r(r-1)}; \quad r \neq 0, 1.$$

By Corollary 3.3, the asymptotic variance of $D_r^s(\widehat{\boldsymbol{p}}, \boldsymbol{p}^0)$ is given by

$$\frac{r^2 \left(\sum_{i=1}^{M} (p_i^*)^r (p_i^0)^{1-r} \right)^{2\frac{s-r}{r-1}}}{(r-1)^2} \left(\sum_{i=1}^{M} (p_i^*)^{2r-1} (p_i^0)^{2(1-r)} - \left(\sum_{i=1}^{M} (p_i^*)^r (p_i^0)^{1-r} \right)^2 \right), \tag{3.25}$$

because

$$h'(x) = r\left(1 + r\left(r-1\right)x\right)^{\frac{s-r}{r-1}},$$

$$D_\phi(\boldsymbol{p}^*, \boldsymbol{p}^0) = \frac{1}{r\left(r-1\right)}\left(\sum_{i=1}^{M}\left(p_i^*\right)^r\left(p_i^0\right)^{1-r} - 1\right)$$

and

$$h'\left(D_\phi(\boldsymbol{p}^*, \boldsymbol{p}^0)\right) = r\left(\sum_{i=1}^{M}\left(p_i^*\right)^r\left(p_i^0\right)^{1-r}\right)^{\frac{s-r}{r-1}}.$$

We know that

$$\lim_{s\to 1} D_r^s(\widehat{\boldsymbol{p}}, \boldsymbol{p}^0) = rD_r^1(\widehat{\boldsymbol{p}}, \boldsymbol{p}^0),$$

therefore for Rényi test statistic the asymptotic variance is

$$\frac{\left(\sum\limits_{i=1}^{M}\left(p_i^*\right)^r\left(p_i^0\right)^{1-r}\right)^{-2}}{\left(r-1\right)^2}\left(\sum_{i=1}^{M}\left(p_i^*\right)^{2r-1}\left(p_i^0\right)^{2(1-r)} - \left(\sum_{i=1}^{M}\left(p_i^*\right)^r\left(p_i^0\right)^{1-r}\right)^2\right).$$

4. Denoting the uniform distribution by $\boldsymbol{u} = (1/M, ..., 1/M)^T$, it is immediate to show that

$$D_r^s\left(\boldsymbol{p}, \boldsymbol{u}\right) = M^{1-s}\left(H_r^s\left(\boldsymbol{u}\right) - H_r^s\left(\boldsymbol{p}\right)\right),$$

where

$$H_r^s\left(\boldsymbol{p}\right) = \frac{1}{1-s}\left(\left(\sum_{i=1}^{M} p_i^r\right)^{\frac{s-1}{r-1}} - 1\right)$$

is the entropy of r-order and s-degree (Entropy of Sharma and Mittal). From Exercise 3, we have

$$\sqrt{n}\left(D_r^s(\widehat{\boldsymbol{p}}, \boldsymbol{u}) - D_r^s(\boldsymbol{p}^*, \boldsymbol{u})\right) \xrightarrow[n\to\infty]{L} N\left(0, \sigma_1^2(\boldsymbol{p}^*)\right)$$

where $\sigma_1^2(\boldsymbol{p}^*)$ is obtained from (3.25) replacing $\boldsymbol{p}^0$ by $\boldsymbol{u}$. Therefore,

$$\sqrt{n}\left(H_r^s\left(\boldsymbol{p}^*\right) - H_r^s\left(\widehat{\boldsymbol{p}}\right)\right) \xrightarrow[n\to\infty]{L} N\left(0, \sigma_{r,s}^2(\boldsymbol{p}^*)\right),$$

where $\sigma_{r,s}^2(\boldsymbol{p}^*) = \sigma_1^2(\boldsymbol{p}^*)/M^{2(1-s)}$ and

$$\sigma_1^2(\boldsymbol{p}^*) = \frac{r^2}{\left(r-1\right)^2}\left(\sum_{i=1}^{M}\left(p_i^*\right)^r\right)^{2\frac{s-r}{r-1}}\left(\sum_{i=1}^{M}\left(p_i^*\right)^{2r-1} - \left(\sum_{i=1}^{M}\left(p_i^*\right)^r\right)^2\right).$$

5. By Remark 3.1(b), we have

$$\frac{2n}{h'(0)\,\phi''(1)} D_{\phi}^{h}(\widehat{\boldsymbol{p}}, \boldsymbol{u}) \underset{n\to\infty}{\overset{L}{\longrightarrow}} \chi_{M-1}^{2}.$$

But in our case we have $h'(0) = r$ and $\phi''(1) = 1$. Then,

$$\frac{2n}{r} D_{r}^{s}(\widehat{\boldsymbol{p}}, \boldsymbol{u}) \underset{n\to\infty}{\overset{L}{\longrightarrow}} \chi_{M-1}^{2}.$$

On the other hand,

$$D_{r}^{s}(\boldsymbol{p}, \boldsymbol{u}) = M^{1-s}\left(H_{r}^{s}(\boldsymbol{u}) - H_{r}^{s}(\boldsymbol{p})\right),$$

therefore,

$$\frac{2nM^{s-1}}{r}\left(H_{r}^{s}(\boldsymbol{u}) - H_{r}^{s}(\widehat{\boldsymbol{p}})\right) \underset{n\to\infty}{\overset{L}{\longrightarrow}} \chi_{M-1}^{2}.$$

6. *a*) A second order Taylor expansion gives that the random variables

$$S_{n}^{\phi}(\widehat{\boldsymbol{p}}, \boldsymbol{p}^{0}) \equiv 8nR_{\phi}(\widehat{\boldsymbol{p}}, \boldsymbol{p}^{0})$$

and

$$\sqrt{n}\left(\widehat{\boldsymbol{p}} - \boldsymbol{p}^{0}\right)^{T} \boldsymbol{A} \sqrt{n}\left(\widehat{\boldsymbol{p}} - \boldsymbol{p}^{0}\right),$$

with $\boldsymbol{A} = diag\left(-\phi''(p_1), ..., -\phi''(p_M)\right)$, have the same asymptotic distribution. Then

$$8nR_{\phi}(\widehat{\boldsymbol{p}}, \boldsymbol{p}^{0}) \underset{n\to\infty}{\overset{L}{\longrightarrow}} \sum_{i=1}^{r} \lambda_i Z_i^2$$

where $Z_1, ..., Z_r$ are independent, identically distributed normal random variables with mean zero and variance 1, $r = rank\left(\boldsymbol{\Sigma}_{\boldsymbol{p}^0} \boldsymbol{A} \boldsymbol{\Sigma}_{\boldsymbol{p}^0}\right)$ and $\beta_i's$ are the eigenvalues of the matrix $\boldsymbol{A}\boldsymbol{\Sigma}_{\boldsymbol{p}^0}$.

b) In this case we have

$$\boldsymbol{A} = diag\left(-\phi''(1/M), ..., -\phi''(1/M)\right).$$

Then $S_{n}^{\phi}(\widehat{\boldsymbol{p}}, \boldsymbol{p}^{0})$ and

$$\sqrt{n}\left(\widehat{\boldsymbol{p}} - \boldsymbol{p}^{0}\right)^{T} diag\left(\boldsymbol{u}^{-1}\right) \sqrt{n}\left(\widehat{\boldsymbol{p}} - \boldsymbol{p}^{0}\right),$$

with $\boldsymbol{u} = (1/M, ..., 1/M)^{T}$, have the same asymptotic distribution. This asymptotic distribution, see Theorem 3.1, is chi-square with $M-1$ degrees of freedom.

c) In this case we have $\boldsymbol{A} = diag\left(\left(\boldsymbol{p}^0\right)^{-1}\right)$, because $\phi''(p_i^0) = 1/p_i^0$. Therefore

$$8nR_\phi(\widehat{\boldsymbol{p}}, \boldsymbol{p}^0) \xrightarrow[n\to\infty]{L} \chi^2_{M-1}.$$

7. We have

$$\begin{aligned}
400R_\phi(\widehat{\boldsymbol{p}}, \boldsymbol{p}^0) &= 400\left(\sum_{i=1}^{6} \tfrac{\widehat{p}_i \log \widehat{p}_i + p_i^0 \log p_i^0}{2} - \sum_{i=1}^{6} \tfrac{\widehat{p}_i + p_i^0}{2} \log \tfrac{\widehat{p}_i + p_i^0}{2}\right) \\
&= 400\left(\tfrac{0.02 \log 0.02 + 0.02 \log 0.02}{2} - \left(\tfrac{0.02+0.02}{2} \log \tfrac{0.02+0.02}{2}\right)\right. \\
&+ \;\cdots\cdots\cdots \\
&+ \left.\tfrac{0.26 \log 0.26 + 0.4 \log 0.4}{2} - \left(\tfrac{0.26+0.4}{2} \log \tfrac{0.26+0.4}{2}\right)\right) \\
&= 4.89182
\end{aligned}$$

and $\chi^2_{5,\ 0.01} = 15.086$. Then the null hypothesis H_0 should not be rejected.

8. In the following table we have computed the test statistics for the different values of a,

a	1	13/7	2
$S_{\phi_a}(\widehat{\boldsymbol{p}}, \boldsymbol{p}^0)$	29.1768	30.5175	30.788

But $\chi^2_{7,\ 0.05} = 14.07$, and the null hypothesis should be rejected.

9. The expression of $\chi^2(\widehat{\boldsymbol{p}}_n, \boldsymbol{p}_n)$ is obtained from $D_\phi(\widehat{\boldsymbol{p}}_n, \boldsymbol{p}_n)$ with $\phi(x) = (x-1)^2$. Applying Proposition 3.2 we get

$$\frac{\sqrt{n}\left(\chi^2(\widehat{\boldsymbol{p}}_n, \boldsymbol{p}_n) - E\left[\phi\left(\frac{Z_n}{\gamma_n}\right)\right]\right)}{\sqrt{\gamma_n Var\left[\phi\left(\frac{Z_n}{\gamma_n}\right)\right] - Cov^2\left[Z_n, \phi\left(\frac{Z_n}{\gamma_n}\right)\right]}} \xrightarrow[n\to\infty]{L} N(0,1),$$

where Z_n is a Poisson random variable with parameter γ_n. Then,

$$\mu_{\phi,n} = E\left[\phi\left(\frac{Z_n}{\gamma_n}\right)\right] = E\left[\frac{Z_n}{\gamma_n} - 1\right]^2 = \frac{1}{\gamma_n} = \frac{M_n}{n}$$

and

$$\begin{aligned}
\sigma^2_{\phi,n} &= \gamma_n Var\left[\phi\left(\frac{Z_n}{\gamma_n}\right)\right] - Cov^2\left[Z_n, \phi\left(\frac{Z_n}{\gamma_n}\right)\right] \\
&= \gamma_n \left(E\left[\frac{Z_n}{\gamma_n} - 1\right]^4 - \left(E\left[\frac{Z_n}{\gamma_n} - 1\right]^2\right)^2\right) \\
&\quad - \left(E\left[(Z_n - \gamma_n)\left(\frac{Z_n}{\gamma_n} - 1\right)^2\right]\right)^2 \\
&= \gamma_n \left(\frac{\gamma_n + 3\gamma_n^2}{\gamma_n^4} - \left(\frac{1}{\gamma_n}\right)^2\right) - \left[\frac{1}{\gamma_n}\right]^2 \\
&= \frac{1}{\gamma_n^2} + \frac{3}{\gamma_n} - \frac{1}{\gamma_n} - \frac{1}{\gamma_n^2} = \frac{2}{\gamma_n} = \frac{2M_n}{n},
\end{aligned}$$

because if Z is a Poisson variable with $E[Z] = \lambda$ then for every $a > 0$,

$$\begin{aligned}
E\left[\tfrac{Z}{a} - 1\right] &= \tfrac{\lambda - a}{a} \\
E\left[\left(\tfrac{Z}{a} - 1\right)^2\right] &= \tfrac{\lambda + (\lambda - a)^2}{a^2} \\
E\left[(Z - \lambda)\left(\tfrac{Z}{a} - 1\right)^2\right] &= \tfrac{\lambda + 2\lambda(\lambda - a)}{a^2} \\
E\left[\left(\tfrac{Z}{a} - 1\right)^4\right] &= \tfrac{\lambda + 3\lambda^2 + 4\lambda(\lambda - a) + 6\lambda(\lambda - a)^2 + (\lambda - a)^4}{a^4}.
\end{aligned}$$

Therefore,

$$\frac{n}{\sqrt{2M_n}}\left(\chi^2(\widehat{\boldsymbol{p}}_n, \boldsymbol{p}_n) - \frac{M_n}{n}\right) \xrightarrow[n\to\infty]{L} N(0,1).$$

10. By Theorem 3.4 and using the previous formulae, we have

$$\mu_{\phi,n} = \sum_{i=1}^{M_n} E\left[p_{ni}\phi\left(\frac{Z_{ni}}{np_{ni}}\right)\right] = \sum_{i=1}^{M_n} p_{ni}\frac{1}{n\,p_{ni}} = \frac{M_n}{n}$$

and

$$\begin{aligned}
\sigma^2_{\phi,n} &= n\sum_{i=1}^{M_n} p_{ni}^2\left(\frac{n\,p_{ni}+3(n\,p_{ni})^2}{(n\,p_{ni})^4}-\left(\frac{1}{n\,p_{ni}}\right)^2\right)-\left(\sum_{i=1}^{M_n} p_{n_i}\frac{1}{n\,p_{ni}}\right)^2\\
&= n\sum_{i=1}^{M_n}\left(\frac{1}{n^3\,p_{ni}}+\frac{3n^2}{n^4}-\frac{1}{n^2}\right)-\left(\sum_{i=1}^{M_n}\frac{1}{n}\right)^2\\
&= \tfrac{1}{n^2}\sum_{i=1}^{M_n}\frac{1}{p_{ni}}+\frac{2M_n}{n}-\left(\frac{M_n}{n}\right)^2\\
&= \frac{M_n}{n}\left(\frac{1}{M_n\,n}\sum_{i=1}^{M_n}\frac{1}{p_{ni}}-\gamma_n^{-1}+2\right).
\end{aligned}$$

11. In this case we have for $\lambda > -1$ and $\lambda \neq 0$

$$\begin{aligned}
\mu_{\phi_{(\lambda)},n} &= E\left[\phi_{(\lambda)}\left(\frac{Z_n}{\gamma_n}\right)\right]=E\left[\tfrac{1}{\lambda(\lambda+1)}\left(\left(\frac{Z_n}{\gamma_n}\right)^{\lambda+1}-\frac{Z_n}{\gamma_n}\right)\right]\\
&= \tfrac{1}{\lambda(\lambda+1)}\left(E\left[\left(\frac{Z_n}{\gamma_n}\right)^{\lambda+1}\right]-1\right)
\end{aligned}$$

and

$$\begin{aligned}
\sigma^2_{\phi_{(\lambda)},n} &= \gamma_n Var\left[\tfrac{1}{\lambda(\lambda+1)}\left(\left(\frac{Z_n}{\gamma_n}\right)^{\lambda+1}-\frac{Z_n}{\gamma_n}\right)\right]\\
&- Cov^2\left[Z_n,\tfrac{1}{\lambda(\lambda+1)}\left(\left(\frac{Z_n}{\gamma_n}\right)^{\lambda+1}-\frac{Z_n}{\gamma_n}\right)\right]\\
&= \gamma_n\tfrac{1}{\lambda^2(\lambda+1)^2}\left(Var\left[\left(\frac{Z_n}{\gamma_n}\right)^{\lambda+1}\right]-\gamma_n Cov^2\left[\frac{Z_n}{\gamma_n},\left(\frac{Z_n}{\gamma_n}\right)^{\lambda+1}\right]\right).
\end{aligned}$$

For $\lambda = 0$, we get

$$\mu_{\phi_{(0)},n} = E\left[\frac{Z_n}{\gamma_n}\log\frac{Z_n}{\gamma_n}\right]$$

and

$$\sigma^2_{\phi_{(0)},n} = \gamma_n\left(Var\left[\frac{Z_n}{\gamma_n}\log\frac{Z_n}{\gamma_n}\right]-\gamma_n Cov^2\left[\frac{Z_n}{\gamma_n},\frac{Z_n}{\gamma_n}\log\frac{Z_n}{\gamma_n}\right]\right),$$

where Z_n is a Poisson random variable with mean γ_n. In the case of a zero observed cell frequency, $D_{\phi_{(\lambda)}}(\widehat{\boldsymbol{p}}_n,\boldsymbol{p}_n)$ is undefined for $\lambda \leq -1$, since it requires taking positive powers of $p_{ni}/\widehat{p}_{ni}$ where $N_{ni}=0$.

12. The class of possible Markov matrices $\boldsymbol{P}$ is given by

$$\begin{pmatrix} 1-\beta & \beta \\ \gamma & 1-\gamma \end{pmatrix}, \; 0<\beta,\gamma<1, \; \alpha+\gamma<2$$

and the subclass of possible Markov matrices satisfying the condition $\boldsymbol{p}^T = \boldsymbol{p}^T\boldsymbol{P}$ is

$$\begin{pmatrix} 1-\beta & \beta \\ p_1\beta/p_2 & 1-p_1\beta/p_2 \end{pmatrix}, \; 0\leq\beta\leq\min\left(1,\frac{p_2}{p_1}\right), \; \beta<1,$$

where $\boldsymbol{p}=(p_1,p_2)^T$ and the subclass satisfying $(\boldsymbol{p}^0)^T = (\boldsymbol{p}^0)^T\boldsymbol{P}$ is given by

$$\begin{pmatrix} 1-\beta & \beta \\ \beta & 1-\beta \end{pmatrix}, \; 0<\beta<1.$$

The nonunit eigenvalue of this matrix is $\lambda = 1-2\beta$. Then we have

$$T_n^{\phi}(\widehat{\boldsymbol{p}},\boldsymbol{p}^0) \underset{n\to\infty}{\overset{L}{\longrightarrow}} \frac{1+\lambda}{1-\lambda}Z^2,$$

where Z is a standard normal variable.

Therefore,

$$T_n^{\phi}(\widehat{\boldsymbol{p}},\boldsymbol{p}^0) \underset{n\to\infty}{\overset{L}{\longrightarrow}} \frac{1-\beta}{\beta}Z^2.$$

But if we consider the matrix

$$\begin{pmatrix} \widehat{p}(1,1) & 1-\widehat{p}(1,1) \\ 1-\widehat{p}(2,2) & \widehat{p}(2,2) \end{pmatrix},$$

the nonunit eigenvalue is $\widehat{\lambda} = -1+\widehat{p}(2,2)+\widehat{p}(1,1)$. Then we have

$$\frac{1+\widehat{\lambda}}{1-\widehat{\lambda}} = \frac{\widehat{p}(2,2)+\widehat{p}(1,1)}{2-\widehat{p}(2,2)+\widehat{p}(1,1)}$$

and

$$T_n^{\phi}(\widehat{\boldsymbol{p}},\boldsymbol{p}^0)\frac{2-\widehat{p}(2,2)+\widehat{p}(1,1)}{\widehat{p}(2,2)+\widehat{p}(1,1)} \underset{n\to\infty}{\overset{L}{\longrightarrow}} \chi_1^2.$$

Therefore we should reject the null hypothesis if

$$T_n^{\phi}(\widehat{\boldsymbol{p}},\boldsymbol{p}^0)\frac{2-\widehat{p}(2,2)+\widehat{p}(1,1)}{\widehat{p}(2,2)+\widehat{p}(1,1)} > \chi_{1,\alpha}^2.$$

A simulation study to choose the best value of the parameter λ in the power-divergence test statistics can be seen in Menéndez et *al.* (1997b).

13. In this case the null hypothesis is

$$H_0 : \mathbf{p} = \mathbf{p}^0,$$

where p is the probability of having outcome 1 for the binary observation and $q = 1 - p$ is the probability of having outcome 0. The misclassification probabilities are $\theta = \Pr(Y^0 = 0|Y = 1)$ and $\psi = \Pr(Y^0 = 1|Y = 0)$.

It is immediate that $\pi = \Pr(Y^0 = 1) = p(1 - \theta) + q\psi$. Denote by n_{ij} the number of units in the validation subsample whose true category is i and fallible category is j; $i, j = 0, 1$, and

$$m_k = \sum_{j=n+1}^{N} I_{\{k\}}\left(Y_j^0\right),\ k = 0, 1.$$

In this context, the maximum likelihood estimators of the probabilities p, θ and ψ are respectively $\widehat{p} = \frac{n_{11}}{n_{*1}} \frac{n_{*1}+m_1}{N} + \frac{n_{10}}{n_{*0}} \frac{n_{*0}+m_0}{N}$, $\widehat{\theta} = \frac{n_{10}}{n_{*0}} \frac{n_{*0}+m_0}{N\,\widehat{p}}$ and $\widehat{\psi} = \frac{n_{01}}{n_{*1}} \frac{n_{*1}+m_1}{N(1-\widehat{p})}$.

The matrix $\boldsymbol{A\Sigma}$, appearing in Theorem 3.7

$$\begin{aligned} \boldsymbol{A\Sigma} &= \begin{pmatrix} p_0^{-1} Var(\widehat{p}) & p_0^{-1} Cov(\widehat{p}, 1 - \widehat{p}) \\ q_0^{-1} Cov(1 - \widehat{p},\, \widehat{p}) & q_0^{-1} Var(1 - \widehat{p}) \end{pmatrix} \\ &= \begin{pmatrix} p_0^{-1} Var(\widehat{p}) & -p_0^{-1} Var(\widehat{p}) \\ -q_0^{-1} Var(\widehat{p}) & q_0^{-1} Var(\widehat{p}) \end{pmatrix}, \end{aligned}$$

with

$$Var(\widehat{p}) = \frac{p_0 q_0}{f}\left(1 - (1 - f)\frac{p_0 q_0}{\pi(1 - \pi)}(1 - \theta - \psi)^2\right),\ q_0 = 1 - p_0,$$

and f is given by limit as $N \to \infty$ of n/N with $n = n_{*0} + n_{*1}$. It can be easily seen that the unique nonzero eigenvalue of $\boldsymbol{A\Sigma}$ is

$$\mu = \frac{1}{f}\left(1 - (1 - f)\frac{p_0 q_0}{\pi(1 - \pi)}(1 - \theta - \psi)^2\right).$$

Then we have

$$\frac{2N}{\mu \phi''(1)} D_\phi(\widehat{\mathbf{p}}, \mathbf{p}^0) \underset{N \to \infty}{\overset{L}{\longrightarrow}} \chi_1^2.$$

If we denote $\widehat{\mu} = \frac{1}{f}\left(1 - (1 - f)\frac{p_0 q_0}{\pi(1-\pi)}(1 - \widehat{\theta} - \widehat{\psi})^2\right)$, then

$$\frac{2N}{\widehat{\mu}\,\phi''(1)}\left[p_0 \phi\left(\frac{\widehat{p}}{p_0}\right) + q_0 \phi\left(\frac{\widehat{q}}{q_0}\right)\right] \underset{N \to \infty}{\overset{L}{\longrightarrow}} \chi_1^2, \tag{3.26}$$

with $q_0 = 1 - p_0$ and $\widehat{q} = 1 - \widehat{p}$.

14. For $\phi(x) = \frac{1}{2}(x-1)^2$, we get

$$D_\phi(\widehat{\boldsymbol{p}}, \boldsymbol{p}^0) = \frac{1}{2}(\widehat{p} - p_0)^2 / p_0 q_0,$$

and

$$\frac{2N}{\widehat{\mu}\,\phi''(1)} D_\phi(\widehat{\boldsymbol{p}}, \boldsymbol{p}^0) = \frac{N(\widehat{p} - p_0)^2}{p_0 q_0 \frac{1}{f}\left(1 - (1-f)\frac{p_0 q_0}{\pi(1-\pi)}(1 - \widehat{\theta} - \widehat{\psi})^2\right)}.$$

This test statistic has been studied in Cheng et *al*. (1998).

15. For testing the null hypothesis $H_0 : \boldsymbol{p} = (1/2, 1/2)^T$ on the basis of a random sample of size $N-n$, and ignoring the possibility of misclassification, it holds

$$\frac{2\,(N-n)}{\phi''_{(\lambda)}(1)} D_{\phi_{(\lambda)}}(\widehat{\boldsymbol{p}}, \boldsymbol{p}^0) \underset{N\to\infty}{\overset{L}{\longrightarrow}} \chi^2_{M-1}.$$

The expression of the family of test statistics given in Theorem 3.7, for the power-divergence test statistics, can be written as

$$\frac{2N}{\widehat{\mu}} D_{\phi_{(\lambda)}}(\widehat{\boldsymbol{p}}, \boldsymbol{p}^0) = \begin{cases} \frac{2N}{\widehat{\mu}\lambda(\lambda+1)}\left(\frac{\widehat{p}^{\lambda+1}}{p_0^\lambda} + \frac{\widehat{q}^{\lambda+1}}{q_0^\lambda} - 1\right), & \lambda \neq -1, 0 \\ \frac{2N}{\widehat{\mu}} \log\left(\left(\frac{\widehat{p}}{p_0}\right)^{\widehat{p}}\left(\frac{\widehat{q}}{q_0}\right)^{\widehat{q}}\right), & \lambda = 0 \\ -\frac{2N}{\widehat{\mu}} \log\left(\left(\frac{\widehat{p}}{p_0}\right)^{p_0}\left(\frac{\widehat{q}}{q_0}\right)^{q_0}\right), & \lambda = -1 \end{cases}, \quad (3.27)$$

with $q_0 = 1 - p_0$, $\widehat{q} = 1 - \widehat{p}$ and $\widehat{\mu} = \frac{1}{f}(1 - (1-f)\frac{p_0 q_0}{\pi(1-\pi)}(1 - \widehat{\theta} - \widehat{\psi})^2)$.

From the data we have

$$\widehat{p} = 0.368,\ \widehat{\theta} = 0.0245,\ \widehat{\psi} = 0.129 \text{ and } \widehat{\pi} = 0.44.$$

If we do not consider the misclassification and we use only the second sample (218, 182), we have that the sampling proportions are 0.545 and 0.455 respectively, denoted by $\widehat{\boldsymbol{p}}_1 = (\widehat{p}_1, \widehat{q}_1 = 1 - \widehat{p}_1)^T = (0.545, 0.455)^T$.

For testing the hypothesis $H_0 : \boldsymbol{p} = (1/2, 1/2)^T$, we have

$$2(N-n) D_{\phi_{(\lambda)}}(\widehat{\boldsymbol{p}}, \boldsymbol{p}^0) = \begin{cases} \frac{800}{2}\left(\frac{\widehat{p}_1^2}{1/2} + \frac{\widehat{q}_1^2}{1/2} - 1\right) = 3.240, & \lambda = 1 \\ 800 \log\left(\left(\frac{\widehat{p}_1}{1/2}\right)^{\widehat{p}_1}\left(\frac{\widehat{q}_1}{1/2}\right)^{\widehat{q}_1}\right) = 3.244, & \lambda = 0 \\ -800 \log\left(\left(\frac{\widehat{p}_1}{1/2}\right)^{\frac{1}{2}}\left(\frac{\widehat{q}_1}{1/2}\right)^{\frac{1}{2}}\right) = 3.253, & \lambda = -1 \end{cases},$$

and $\chi^2_{1,0.05} = 3.841$. Then for $\lambda = 1$, $\lambda = 0$ and $\lambda = -1$, the null hypothesis H_0 should not be rejected if the size of the test is $\alpha = 0.05$.

Now, we are going to get the expression of the family of test statistics, given in Section 3.4.3, when we have a double sample data. First we obtain the expression for $\widehat{\mu}$,

$$\widehat{\mu} = \frac{1}{f}\left(1 - (1-f)\frac{p_0 q_0}{\widehat{\pi}(1-\widehat{\pi})}(1 - \widehat{\theta} - \widehat{\psi})^2\right) = 2.0919.$$

Then, based on (3.27), we have

$$\frac{2N}{\widehat{\mu}} D_{\phi(\lambda)}\left(\widehat{\boldsymbol{p}}, \boldsymbol{p}^0\right) = \begin{cases} \frac{2\times 500}{2\times 2.0919}\left(\frac{\widehat{p}^2}{1/2} + \frac{\widehat{q}^2}{1/2} - 1\right) = 16.659, & \lambda = 1 \\ \frac{2\times 500}{2.0919}\log\left(\left(\frac{\widehat{p}}{1/2}\right)^{\widehat{p}}\left(\frac{\widehat{q}}{1/2}\right)^{\widehat{q}}\right) = 16.858, & \lambda = 0 \\ -\frac{2\times 500}{2.0919}\log\left(\left(\frac{\widehat{p}}{1/2}\right)^{\frac{1}{2}}\left(\frac{\widehat{q}}{1/2}\right)^{\frac{1}{2}}\right) = 17.267, & \lambda = -1 \end{cases},$$

because in this case $\widehat{p} = 0.368$ and $\widehat{q} = 0.632$. Thus if we consider double sample data, H_0 should be rejected. Therefore we can see that if one uses only the fallible data then one may be led to the wrong conclusion to accept H_0.

16. To establish (3.23) we have to prove if for $\eta > 0$ there exists a constant $k(\eta)$ such as if $n \geq n(\eta)$, then

$$\Pr\left(|\gamma - np_{ni}| > k(\eta)\right) < \eta.$$

Denoting $q = \Pr\left(|\gamma - np_{ni}| > k(\eta)\right)$, we have

$$\begin{aligned} q &= \Pr\left(|\gamma - N_{ni} + N_{ni} - np_{ni}| > k(\eta)\right) \\ &\leq \Pr\left(|\gamma - N_{ni}| > \tfrac{k(\eta)}{2}\right) + \Pr\left(|N_{ni} - np_{ni}| > \tfrac{k(\eta)}{2}\right) \\ &= \Pr\left(\left|\tfrac{\gamma}{n} - \tfrac{N_{ni}}{n}\right| > \tfrac{k(\eta)}{2n}\right) + \Pr\left(\left|\tfrac{N_{ni}}{n} - \tfrac{np_{ni}}{n}\right| > \tfrac{k(\eta)}{2n}\right) \\ &\leq \Pr\left(\left|\tfrac{\gamma}{n} - \tfrac{N_{ni}}{n}\right| > \tfrac{k(\eta)}{2}\right) + \Pr\left(\left|\tfrac{N_{ni}}{n} - p_{ni}\right| > \tfrac{k(\eta)}{2}\right). \end{aligned}$$

Using the assumption $N_{ni} = \gamma + O_P(n)$, we have that given $\eta/2$ there exist a constant $k_1\left(\frac{\eta}{2}\right)$ and $n^* = n(\eta)$ such that for all $n \geq n^*$,

$$\Pr\left(\left|\frac{\gamma}{n} - \frac{N_{ni}}{n}\right| > k_1\left(\frac{\eta}{2}\right)\right) \leq \frac{\eta}{2}.$$

Using the assumption $\widehat{p}_{ni} = p_{ni} + o_P(1)$ we have that given $k_1\left(\frac{\eta}{2}\right)$ and $\frac{\eta}{2} > 0$ there exists a number $n^{**} = n\left(k_1\left(\frac{\eta}{2}\right), \eta\right)$ such that for $n \geq n^{**}$,

$$\Pr\left(\left|\frac{N_{ni}}{n} - p_{ni}\right| > k_1\left(\frac{\eta}{2}\right)\right) \leq \frac{\eta}{2}.$$

If we consider $k(\eta) = k_1\left(\frac{\eta}{2}\right)$ and $n = \max(n^*, n^{**})$, we have that for all $\eta > 0$,

$$\Pr\left(|\gamma - np_{ni}| > k(\eta)\right) \leq \frac{\eta}{2} + \frac{\eta}{2}.$$

17. We wish to test $H_0 : F = F_0$ using the family of power-divergence test statistics

$$T_n^{\lambda}(\boldsymbol{p}(\boldsymbol{Y}_n), \boldsymbol{q}^0) = \frac{2n}{\lambda(\lambda+1)} \sum_{i=1}^{M} p_i(\boldsymbol{Y}_n)\left(\left(\frac{p_i(\boldsymbol{Y}_n)}{q_i^0}\right)^{\lambda} - 1\right),$$

where $p_i(\boldsymbol{Y}_n)$ and q_i^0, $i = 1, ..., M$ must be calculated. We assume $M = 4$ and consider the partition of the unit interval defined by

$$\pi_0 = 0,\ \pi_1 = 1/4,\ \pi_2 = 2/4,\ \pi_3 = 3/4,\ \pi_4 = 1.$$

We have

$$\boldsymbol{q}^0 = \left(q_1^0, ..., q_4^0\right)^T = (F_0(c_j) - F_0(c_{j-1}) : 1 \leq j \leq 4)^T = \left(\frac{1}{4}, \frac{1}{4}, \frac{1}{4}, \frac{1}{4}\right)^T,$$

and

$$\boldsymbol{p}(\boldsymbol{Y}_n) = (F_0(Y_{(6)}), F_0(Y_{(11)}) - F_0(Y_{(6)}), F_0(Y_{(16)}) - F_0(Y_{(11)}), 1 - F_0(Y_{(16)}))^T,$$

because $Y_{n_i} = Y_{(n_i)}$ and $n_i = [20 \times \pi_i] + 1$, $i = 1, 2, 3$.

We have

$$\boldsymbol{p}(\boldsymbol{Y}_n) = (0.26566, 0.24364, 0.3397, 0.151)^T,$$

because $Y_{(6)} = 2.3339$, $Y_{(11)} = 2.8551$ and $Y_{(16)} = 5.1468$.

In the following table we report the values of the power-divergence test statistics $T_n^{\lambda}(\boldsymbol{p}(\boldsymbol{Y}_n), \boldsymbol{q}^0)$ for different values of λ.

	$\lambda = -2$	$\lambda = -1$	$\lambda = -1/2$	$\lambda = 0$	$\lambda = 2/3$	$\lambda = 1$
T_n^{λ}	1.7936	1.6259	1.5643	1.5153	1.4676	1.4506

On the other hand $\chi^2_{3,\ 0.05} = 7.815$. Therefore we should not reject the null hypothesis.

18. The expression of the asymptotic variance is given by

$$\sigma_1^2(\boldsymbol{p}^*) = \sum_{i=1}^M p_i^* \left(\phi'\left(\frac{p_i^*}{p_i^0}\right)\right)^2 - \left(\sum_{i=1}^M p_i^* \phi'\left(\frac{p_i^*}{p_i^0}\right)\right)^2.$$

Then we have

Divergence	**Asymptotic variance**
Pearson	$\sum_{i=1}^M \frac{(p_i^*)^3}{(p_i^0)^2} - \left(\sum_{i=1}^M \frac{(p_i^*)^2}{p_i^0}\right)^2$
Matusita $(a=1/2)$	$1 - \left(\sum_{i=1}^M \sqrt{p_i^*}\sqrt{p_i^0}\right)^2$
Balakrishnan	$\sum_{i=1}^M p_i^* \left(\frac{4(p_i^* - p_i^0)(p_i^0)^2}{(p_i^* + p_i^0)^3}\right)^2 - \left(\sum_{i=1}^M p_i^* \frac{4(p_i^* - p_i^0)(p_i^0)^2}{(p_i^* + p_i^0)^3}\right)^2$
Rathie-Kanappan	$\frac{s^2}{(1-s)^2}\left(\sum_{i=1}^M (p_i^*)^{2s-1}(p_i^0)^{2(1-s)} - \left(\sum_{i=1}^M (p_i^*)^s (p_i^0)^{1-s}\right)^2\right)$
Power-divergence	$\frac{1}{\lambda^2}\left(\sum_{i=1}^M (p_i^*)^{2\lambda+1}(p_i^0)^{-2\lambda} - \left(\sum_{i=1}^M (p_i^*)^{\lambda+1}(p_i^0)^{-\lambda}\right)^2\right)$
Rukhin	$\sum_{i=1}^M p_i^* \frac{(p_i^0 - p_i^*)^2 \left(p_i^0(a+1) + (a-1)p_i^*\right)^2}{(ap_i^0 + (1-a)p_i^*)^4} - \left(\sum_{i=1}^M p_i^* \frac{(p_i^0 - p_i^*)\left(p_i^0(a+1) + (a-1)p_i^*\right)}{(ap_i^0 + (1-a)p_i^*)^2}\right)^2$

4

Optimality of Phi-divergence Test Statistics in Goodness-of-fit

4.1. Introduction

In the previous chapter we have studied the family of ϕ-divergence test statistics, $T_n^{\phi}(\widehat{\boldsymbol{p}}, \boldsymbol{p}^0)$, for the problem of goodness-of-fit. If we denote by $F_{T_n^{\phi}(\widehat{\boldsymbol{p}},\boldsymbol{p}^0)}(t)$ the exact distribution of $T_n^{\phi}(\widehat{\boldsymbol{p}}, \boldsymbol{p}^0)$, for fixed ϕ, we established that

$$F_{T_n^{\phi}(\widehat{\boldsymbol{p}},\boldsymbol{p}^0)}(t) = F_{\chi^2_{M-1}}(t) + o(1) \qquad \text{as } n \to \infty, \tag{4.1}$$

under the null hypothesis

$$H_0 : \boldsymbol{p} = \boldsymbol{p}^0. \tag{4.2}$$

Based on (4.1) we considered for the problem of goodness-of-fit given in (4.2) the decision rule

$$\text{“Reject, with significance level } \alpha, \ H_0 \text{ if } T_n^{\phi}(\widehat{\boldsymbol{p}}, \boldsymbol{p}^0) > \chi^2_{M-1,\alpha}\text{”}. \tag{4.3}$$

Now in this chapter we shall present some criteria to choose the best function ϕ in some sense. In Section 4.2, Pitman asymptotic efficiency (contiguous alternative hypotheses), Bahadur efficiency and some asymptotic approximations of the power function for the ϕ-divergence test statistic are studied.

Result (4.1) is asymptotic and it is only valid for large sample sizes but in finite samples is frequently assumed to hold approximately in order to calculate critical regions for the ϕ-divergence test statistic in goodness-of-fit tests. Then it will be important to give some methods to improve the accuracy of the ϕ-divergence test statistic in those situations where the sample size cannot be assumed large. We shall study with the ϕ-divergence test statistic the same procedures studied previously by Read and Cressie (1988) with the power-divergence family of test statistics. In Section 4.3 we investigate the criterion based on the speed of convergence of the exact moments of the ϕ-divergence test statistic, $T_n^{\phi}(\widehat{\boldsymbol{p}},\boldsymbol{p}^0)$, to its asymptotic moments. The exact distribution of every member of the family of ϕ-divergence test statistics, $T_n^{\phi}(\widehat{\boldsymbol{p}},\boldsymbol{p}^0)$ (see (4.1)) differs from chi-square by $o(1)$. In Section 4.4 a closer approximation to the exact distribution is obtained by extracting the ϕ-dependent second order component from the $o(1)$ term. In Section 4.5 comparisons of exact power based on exact critical regions are presented and finally in Section 4.6 comparisons between the exact distribution of the ϕ-divergence test statistics, $T_n^{\phi}(\widehat{\boldsymbol{p}},\boldsymbol{p}^0)$, and the different asymptotic approximations are studied.

Throughout the chapter we shall assume that $\phi \in \Phi^*$ is 4 times continuously differentiable in the neighborhood of 1 and $\phi''(1) \neq 0$.

4.2. Asymptotic Efficiency

Power functions are usually difficult to evaluate and we mostly have to be content with approximations based on limit results. In this Section we consider three approaches in order to choose an optimal ϕ-divergence test statistic: Pitman efficiency, Bahadur efficiency and comparisons based on some approximations to the power function.

4.2.1. Pitman Asymptotic Relative Efficiency

For a probability vector $\boldsymbol{p}^* \neq \boldsymbol{p}^0$, we established that

$$\lim_{n\to\infty} \beta_{n,\phi}(\boldsymbol{p}^*) = \Pr\left(T_n^{\phi}(\widehat{\boldsymbol{p}},\boldsymbol{p}^0) > \chi^2_{M-1,\alpha} \;/\; H_1 : \boldsymbol{p} = \boldsymbol{p}^*\right) = 1.$$

Hence to get a limit value less than 1, we must consider a sequence of contiguous alternative hypotheses

$$H_{1,n} : \boldsymbol{p}_n = \mathbf{p}^0 + \mathbf{d}/\sqrt{n},$$

where $\boldsymbol{d} = (d_1, ..., d_M)^T$ is a fixed $M \times 1$ vector such that $\sum_{j=1}^{M} d_j = 0$, because in this case

$$\lim_{n\to\infty} \beta_{n,\phi}(\boldsymbol{p}_n) = 1 - G_{\chi^2_{M-1}(\delta)}\left(\chi^2_{M-1,\alpha}\right), \tag{4.4}$$

where $G_{\chi^2_{M-1}(\delta)}$ is the distribution function of a noncentral chi-square random variable with $M-1$ degrees of freedom and noncentrality parameter

$$\delta = \mathbf{d}^T diag\left(\left(\mathbf{p}^0\right)^{-1}\right)\mathbf{d}. \tag{4.5}$$

In this context we can consider the Pitman asymptotic relative efficiency to compare the behavior of two test statistics. Let us consider two ϕ-divergence test statistics $T_n^{\phi_1}(\widehat{\boldsymbol{p}}, \boldsymbol{p}^0)$ and $T_n^{\phi_2}(\widehat{\boldsymbol{p}}, \boldsymbol{p}^0)$ and let us suppose that for a given n and a significance level α there exists a number N_n such that

$$\beta_{n,\phi_1}(\boldsymbol{p}_n) = \beta_{N_n,\phi_2}(\boldsymbol{p}_{N_n}) \underset{n\to\infty}{\longrightarrow} \beta \text{ (any asigned value)} < 1,$$

that is, the powers are equal and that $N_n \to \infty$ as $n \to \infty$. By $\boldsymbol{p}_n$ and $\boldsymbol{p}_{N_n}$ we are denoting contiguous alternative hypotheses,

$$\boldsymbol{p}_n = \mathbf{p}^0 + \mathbf{d}/\sqrt{n} \text{ and } \boldsymbol{p}_{N_n} = \mathbf{p}^0 + \mathbf{d}/\sqrt{N_n}.$$

Definition 4.1

The Pitman asymptotic relative efficiency of $T_n^{\phi_2}(\widehat{\boldsymbol{p}}, \boldsymbol{p}^0)$ *with respect* $T_n^{\phi_1}(\widehat{\boldsymbol{p}}, \boldsymbol{p}^0)$ *is given by*

$$\lim_{n\to\infty} \frac{n}{N_n}.$$

For more details about the Pitman asymptotic efficiency see Rao (1973, pp. 467-470), Read and Cressie (1988, pp. 54-55) or Pitman (1979, pp. 56-62).

Based on (4.4) the asymptotic power functions $\beta_{n,\phi_1}(\boldsymbol{p}_n)$ and $\beta_{N_n,\phi_2}(\boldsymbol{p}_N)$ will be equal because both of them are based in $G_{\chi^2_{M-1}(\delta)}\left(\chi^2_{M-1,\alpha}\right)$, with δ given on (4.5). For this reason the Pitman asymptotic efficiency is 1 for any ϕ-divergence test statistics $T_n^{\phi_1}(\widehat{\boldsymbol{p}}, \boldsymbol{p}^0)$ and $T_n^{\phi_2}(\widehat{\boldsymbol{p}}, \boldsymbol{p}^0)$. We can conclude that using Pitman asymptotic relative efficiency is not possible to discriminate between the ϕ-divergence test statistics.

4.2.2. Bahadur Efficiency

Let $\{Y_k\}_{k\in N}$ be a sequence of independent and identically distributed random variables from F and we denote by $s = (y_1, y_2, ...)$ a possible outcome of the previous sequence. We consider the problem of goodness-of-fit

$$H_0 : F = F_0$$

by partitioning the range of data in disjoint intervals, $\{E_i\}_{i=1,..,M}$, by testing the hypothesis $H_0 : \mathbf{p} = \mathbf{p}^0$, i.e., we consider the notation introduced in Section 3.1. We denote $N_i = \sum_{j=1}^n I_{E_i}(Y_j)$. We can see that N_i is based on the first n components of the sequence $\{Y_k\}_{k\in N}$. We denote by $n_i(s) = \sum_{j=1}^n I_{E_i}(y_j)$ and by $\widehat{\mathbf{p}}(s)$ the associated probability vector with the first n components of the outcome "s". We consider a test statistic $H_n(\widehat{\mathbf{p}}, \mathbf{p}^0)$ for testing the null hypothesis $H_0 : \mathbf{p} = \mathbf{p}^0$, and its outcome based on $(y_1, y_2, ..., y_n)$ given by $H_n(\widehat{\mathbf{p}}(s), \mathbf{p}^0)$. Here $H_n(\widehat{\mathbf{p}}, \mathbf{p}^0)$ denotes a general family of test statistics, not necessarily the same as the family of the ϕ-divergence test statistics $T_n^{\phi}(\widehat{\mathbf{p}}, \mathbf{p}^0)$ considered in (4.1).

We denote by $F_{H_n(\widehat{\mathbf{p}},\mathbf{p}^0)}(t)$ the distribution function of $H_n(\widehat{\mathbf{p}}, \mathbf{p}^0)$ under the null hypothesis $H_0 : \mathbf{p} = \mathbf{p}^0$. The level attained by $H_n(\widehat{\mathbf{p}}, \mathbf{p}^0)$ is defined by

$$L_n(s) = 1 - F_{H_n(\widehat{\mathbf{p}},\mathbf{p}^0)}\left(H_n(\widehat{\mathbf{p}}(s), \mathbf{p}^0)\right).$$

Bahadur (1971) pointed out that in typical cases L_n is asymptotically uniform distributed over $(0, 1)$, under the null hypothesis $H_0 : \mathbf{p} = \mathbf{p}^0$ and $L_n \to 0$ exponentially fast (with probability one) under $\mathbf{p} \neq \mathbf{p}^0$. Now we define the exact Bahadur slope.

Definition 4.2

We consider $\mathbf{p} \neq \mathbf{p}^0$ *and* $\mathbf{p} \in \Delta_M^+$, *where*

$$\Delta_M^+ = \left\{ \mathbf{p} = (p_1, ..., p_M)^T : p_i > 0, i = 1, ..., M, \ \sum_{i=1}^M p_i = 1 \right\}.$$

We shall say that the sequence $\left\{H_n(\widehat{\mathbf{p}}, \mathbf{p}^0)\right\}_{n\in N}$ *has exact Bahadur slope* $c(\mathbf{p})$, $0 < c(\mathbf{p}) < \infty$, *if*

$$\lim_{n\to\infty} \log L_n(s) = -\frac{1}{2} c(\mathbf{p}), \tag{4.6}$$

with probability one when $n \to \infty$.

This definition is motivated by Bahadur (1971) in the following terms "Consider the Fisherian transformation $V_n(s) = -2\log L_n(s)$. Then, in typical cases, $V_n \to \chi_2^2$ in distribution in the null case. Suppose now that a non-null $\boldsymbol{p}$ is obtained and that (4.6) holds, with $0 < c(\boldsymbol{p}) < \infty$. Suppose we plot, for a given s, the sequence of points $\{(n, V_n(s)) : n = 1, 2, ...\}$ in the uv-plane. It follows from (4.6) that, for almost all s, this sequence of points moves out to infinity in the direction of a ray from the origin, the angle between the ray and the u-axis, on which axis the sample size n is being plotted, being $\tan^{-1} c(\boldsymbol{p})$".

Given $\varepsilon > 0$, $0 < \varepsilon < 1$, and s, let $N = N(\varepsilon, s)$ be the smallest integer m such that $L_n(s) < \varepsilon$ for all $n \geq m$ and let $N = \infty$ if no such m exists. Then N is the sample size required for the sequence $\{H_n(\widehat{\mathbf{p}}, \mathbf{p}^0)\}_{n\in N}$ in order to become significant (and remains significant) at the level ε. But the most important fact pointed out by Bahadur (1971) is that, for small ε, N is approximately inversely proportional to the exact Bahadur slope, i.e., if (4.6) holds and $0 < c(\boldsymbol{p}) < \infty$, then

$$\lim_{\varepsilon\to 0} N(\varepsilon, s)/2\log(1/\varepsilon) = c(\boldsymbol{p})^{-1}, \tag{4.7}$$

with probability one when $\varepsilon \to 0$ (see Theorem 7.1 in Bahadur (1971)).

Suppose that $\{H_n^1(\widehat{\mathbf{p}}, \mathbf{p}^0)\}_{n\in N}$ and $\{H_n^2(\widehat{\mathbf{p}}, \mathbf{p}^0)\}_{n\in N}$ are two sequences of test statistics such that $H_n^i(\widehat{\mathbf{p}}, \mathbf{p}^0)$ has exact Bahadur slope $c_i(\mathbf{p})$, $0 < c_i(\mathbf{p}) < \infty$, $i = 1, 2$. From (4.7) if $N^{(i)}(\varepsilon, s)$ is the sample size required to make $H_n^i(\widehat{\mathbf{p}}, \mathbf{p}^0)$ significant at level ε, we have

$$\frac{N^{(2)}(\varepsilon, s)}{N^{(1)}(\varepsilon, s)} \to \frac{c_1(\mathbf{p})}{c_2(\mathbf{p})},$$

with probability one when $\varepsilon \to 0$. Consequently $c_1(\mathbf{p})/c_2(\mathbf{p})$ is a measure of the asymptotic efficiency of $H_n^1(\widehat{\mathbf{p}}, \mathbf{p}^0)$ relative to $H_n^2(\widehat{\mathbf{p}}, \mathbf{p}^0)$.

Based on this relation we can give the following definition:

Definition 4.3

Let $\mathbf{p} \in \triangle_M^+$. A measure of the asymptotic efficiency of $H_n^1(\widehat{\mathbf{p}}, \mathbf{p}^0)$ relative to $H_n^2(\widehat{\mathbf{p}}, \mathbf{p}^0)$ is given by $c_1(\mathbf{p})/c_2(\mathbf{p})$.

The following result given in Bahadur (1971) (Theorem 7.2) in general populations and adapted to multinomial populations in Lemma 3.1 in Cressie and Read (1984) will be important to establish the main theorem in this Section.

Lemma 4.1

If $H_n(\widehat{\mathbf{p}}, \mathbf{p}^0)$ is a test statistic for the simple null hypothesis $H_0 : \mathbf{p} = \mathbf{p}^0$, based on the first n observations of the sequence $\{Y_k\}_{k \in N}$, verifying:

i) There exists a function $b : \Delta_M^+ \to \mathbb{R}$ with $-\infty < b(\mathbf{p}) < \infty$, such that

$$n^{-1/2} H_n(\widehat{\mathbf{p}}, \mathbf{p}^0) \to b(\mathbf{p})$$

with probability one as $n \to \infty$ for each $\mathbf{p} \in \Delta_M^+$ and $\mathbf{p} \neq \mathbf{p}^0$.

ii) For each t in an open interval I, there exists a continuous function f, verifying

$$n^{-1} \log \Pr\left\{H_n(\widehat{\mathbf{p}}, \mathbf{p}^0) \geq \sqrt{n}t \;/\; \mathbf{p} = \mathbf{p}^0\right\} \to -f(t) \text{ as } n \to \infty$$

and $\left\{b(\mathbf{p}) / \mathbf{p} \neq \mathbf{p}^0\right\} \subset I$.

Then, the exact Bahadur slope for $H_n(\widehat{\mathbf{p}}, \mathbf{p}^0)$ is given by $c(\mathbf{p}) = 2f(b(\mathbf{p}))$, for each $\mathbf{p} \in \Delta_M^+$ and $\mathbf{p} \neq \mathbf{p}^0$.

Theorem 4.1 describes a useful method of finding the exact Bahadur slope for the ϕ-divergence test statistic, $T_n^{\phi}(\widehat{\mathbf{p}}, \mathbf{p}^0)$. It is necessary to give some concepts before establishing its proof.

Definition 4.4

Let $A \subset \Delta_M^+$ and $A_n = \Delta_{M,n} \cap A$, where

$$\Delta_{M,n} = \left\{ \boldsymbol{v} = (v_1, ..., v_M)^T : v_j = i_j/n,\ i_j \in \mathbb{N} - \{0\},\ j = 1, ..., M,\ \sum_{j=1}^{M} i_j = n \right\}.$$

Let us say that A is $\mathbf{p}^0$-regular, $\mathbf{p}^0 \in \Delta_M^+$, if

$$\lim_{n\to\infty} \inf_{\boldsymbol{v} \in A_n} D_{Kull}(\boldsymbol{v}, \mathbf{p}^0) = \inf_{\boldsymbol{v} \in A} D_{Kull}(\boldsymbol{v}, \mathbf{p}^0).$$

Some properties of this definition are given in the following proposition; its proof is in Bahadur (1971).

Lemma 4.2

Let $A \subset \Delta_M^+$ and $\overline{A^0}$ the closure of A^0 (interior of A). We have:

i) If $A \subset \overline{A^0}$ (e.g., if A is open) A is $\mathbf{p}^0$-regular

ii) If A is $\mathbf{p}^0$-regular and $\widehat{\mathbf{p}} \in \Delta_M^+$, there exists a positive constant, $\delta(M)$, depending only on M, such that

$$\delta(M)\, n^{-\frac{M-1}{2}} \exp\left(-n \inf_{\boldsymbol{v}\in A_n} D_{Kull}(\boldsymbol{v},\mathbf{p}^0)\right) \leq \Pr_{\boldsymbol{p}^0}\{\widehat{\mathbf{p}} \in A\}$$
$$\leq (n+1)^M \exp\left(-n \inf_{\boldsymbol{v}\in A_n} D_{Kull}(\boldsymbol{v},\mathbf{p}^0)\right).$$

Theorem 4.1 describes a practical way to get the exact Bahadur slope for the ϕ-divergence test statistic and it is based on Lemmas 4.1 and 4.2.

Theorem 4.1

Let ϕ continuously differentiable and verifying

$$\phi(0) + \lim_{r\to\infty} \frac{\phi(r)}{r} < \infty.$$

Then, the sequence of test statistics

$$\{H_n(\widehat{\mathbf{p}},\mathbf{p}^0)\}_{n\in\mathbb{N}} = \left\{\sqrt{T_n^{\phi}(\widehat{\mathbf{p}},\mathbf{p}^0)}\right\}_{n\in\mathbb{N}}$$

verifies

i) Under $\mathbf{p} \in \Delta_M^+$ and $\mathbf{p} \neq \mathbf{p}^0$,

$$n^{-1/2}\sqrt{T_n^{\phi}(\widehat{\mathbf{p}},\mathbf{p}^0)} \to \sqrt{\frac{2}{\phi''(1)} D_{\phi}(\mathbf{p},\mathbf{p}^0)}$$

with probability 1 as $n\to\infty$.

ii) Under the null hypothesis $H_0 : \mathbf{p} = \mathbf{p}^0$,

$$n^{-1}\log \Pr\left\{\sqrt{T_n^{\phi}(\widehat{\mathbf{p}},\mathbf{p}^0)} \geq n^{1/2}t \;/\; \mathbf{p} = \mathbf{p}^0\right\} \to -\inf_{\boldsymbol{v}\in A_{\phi,t}} D_{Kull}(\boldsymbol{v},\mathbf{p}^0)$$

as $n\to\infty$ for each t in an open interval with

$$A_{\phi,t} = \left\{\boldsymbol{v} : \boldsymbol{v} \in \Delta_M^+ \text{ and } \sqrt{\frac{2}{\phi''(1)} D_{Kull}(\boldsymbol{v},\mathbf{p}^0)} \geq t\right\}.$$

iii) The exact Bahadur slope of the ϕ-divergence test statistic $T_n^{\phi}(\widehat{\mathbf{p}},\mathbf{p}^0)$ is given by

$$c_{\phi}(\mathbf{p}) = \inf_{\boldsymbol{v}\in B_{\phi}} 2D_{Kull}(\boldsymbol{v},\mathbf{p}^0),$$

where

$$B_\phi = \left\{ \boldsymbol{v} : \boldsymbol{v} \in \Delta_M^+ \text{ and } D_\phi(\boldsymbol{v}, \boldsymbol{p}^0) \geq D_\phi(\boldsymbol{p}, \boldsymbol{p}^0) \right\},$$

$\boldsymbol{p} \in \Delta_M^+$ *and* $\boldsymbol{p} \neq \boldsymbol{p}^0$.

Proof. We shall establish the proof step by step.

i) We know that under $\boldsymbol{p} \neq \boldsymbol{p}^0$, $\widehat{\boldsymbol{p}} \xrightarrow[n\to\infty]{c.s.} \boldsymbol{p}$. Now the result follows by the continuity of $D_\phi(\boldsymbol{v}, \boldsymbol{p}^0)$ with respect to the first argument. Condition *i)* of Lemma 4.1 is verified with

$$b(\boldsymbol{p}) = \sqrt{\frac{2}{\phi''(1)} D_\phi(\ \boldsymbol{p}, \boldsymbol{p}^0)}.$$

ii) It is well known that

$$0 < D_\phi(\boldsymbol{v}, \boldsymbol{p}^0) \leq \phi(0) + \lim_{r\to\infty} \frac{\phi(r)}{r};$$

therefore if we denote by $\gamma_1 = \sqrt{\frac{2}{\phi''(1)} \left(\phi(0) + \lim_{r\to\infty} \frac{\phi(r)}{r} \right)}$ we have that the range of b, as $\boldsymbol{p}$ varies over $\Delta_M^+ - \left\{ \boldsymbol{p}^0 \right\}$, is $I = (0, \gamma_1)$. For $t \in I$, the event $\sqrt{T_n^\phi(\widehat{\boldsymbol{p}}, \boldsymbol{p}^0)} \geq n^{1/2}t$ is equivalent to the event

$$\widehat{\boldsymbol{p}} \in A_{\phi,t} = \left\{ \boldsymbol{v} : \boldsymbol{v} \in \Delta_M^+ \text{ and } \sqrt{\frac{2}{\phi''(1)} D_\phi(\boldsymbol{v}, \boldsymbol{p}^0)} \geq t \right\}.$$

The probability vector $\widehat{\boldsymbol{p}}$ takes its values on the lattice $\Delta_{M,n}$. We consider the set

$$A_n = A_{\phi,t} \cap \Delta_{M,n}.$$

The continuity of $D_\phi(\boldsymbol{v}, \boldsymbol{p}^0)$ in $\boldsymbol{v}$ implies that $A_{\phi,t}$ is the closure of its interior and by Lemma 4.2, $\boldsymbol{p}^0$-regular for any $\boldsymbol{v} \in \Delta_{M,n}$. By *ii)* in Lemma 4.2, we have under $\boldsymbol{p} \in \Delta_M^+$

$$n^{-1} \log \Pr \left\{ \widehat{\boldsymbol{p}} \in A_{\phi,t} / \boldsymbol{p} = \boldsymbol{p}^0 \right\} = - \inf_{\boldsymbol{v} \in A_{\phi,t}} D_{Kull}(\boldsymbol{v}, \boldsymbol{p}^0),$$

i.e.,

$$n^{-1} \log \Pr \left(\sqrt{T_n^\phi(\boldsymbol{v}, \boldsymbol{p}^0)} \geq \sqrt{n}t / \boldsymbol{p} = \boldsymbol{p}^0 \right) = - \inf_{\boldsymbol{v} \in A_{\phi,t}} D_{Kull}(\boldsymbol{v}, \boldsymbol{p}^0).$$

Therefore we have the condition *ii)* of Lemma 4.1 with

$$f(t) = \inf_{\boldsymbol{v} \in A_{\phi,t}} D_{Kull}(\boldsymbol{v}, \boldsymbol{p}^0),\ t \in I = (0, \gamma_1)$$

and f is a continuous function, by the continuity of $D_{Kull}(\boldsymbol{v}, \boldsymbol{p}^0)$, in I and of course $I \supset \{b(\boldsymbol{p}) / \boldsymbol{p} \in \Delta_M^+ \text{ and } \boldsymbol{p} \neq \boldsymbol{p}^0\}$.

iii) Applying Lemma 4.1 we have that the exact Bahadur slope for $\sqrt{T_n^{\phi}(\boldsymbol{v}, \boldsymbol{p}^0)}$, with ϕ fixed, is $2f(\boldsymbol{p})$; i.e., $c_{\phi}(\boldsymbol{p}) \equiv 2f(\boldsymbol{p}) = 2\inf_{\boldsymbol{v} \in B_{\phi}} D_{Kull}(\boldsymbol{v}, \boldsymbol{p}^0)$, where $B_{\phi} = A_{\phi, f(\boldsymbol{p})}$.

Since the function $g(x) = x^2$ is a strictly monotonic increasing function of $x > 0$, then the level obtained by the ϕ-divergence test statistic $T_n^{\phi}(\widehat{\boldsymbol{p}}, \boldsymbol{p}^0)$ will be the same as the one attained by $H_n(\widehat{\boldsymbol{p}}, \boldsymbol{p}^0)$ for every n and s. Hence the exact Bahadur slopes of both sequences $\{H_n(\widehat{\boldsymbol{p}}, \boldsymbol{p}^0)\}_{n \in N}$ and $\left\{T_n^{\phi}(\widehat{\boldsymbol{p}}, \boldsymbol{p}^0)\right\}_{n \in N}$ will be the same as Bahadur (1971, see Remark 2, p. 27) pointed out.

■

Remark 4.1

It is necessary to consider H_n instead of T_n^{ϕ} due to condition i) in Lemma 4.1.

Remark 4.2

If we consider $\phi(x) = x \log x - x + 1$ we have

$$B_{Kull} = \{\boldsymbol{v} : \boldsymbol{v} \in \Delta_M^+ \text{ and } D_{Kull}(\boldsymbol{v}, \boldsymbol{p}^0) \geq D_{Kull}(\boldsymbol{p}, \boldsymbol{p}^0)\}$$

and then

$$c_{Kull}(\boldsymbol{p}) = 2D_{Kull}(\boldsymbol{p}, \boldsymbol{p}^0), \ \boldsymbol{p} \in \Delta_M^+ \text{ and } \boldsymbol{p} \neq \boldsymbol{p}^0.$$

Therefore

$$c_{\phi}(\boldsymbol{p}) = \inf_{\boldsymbol{v} \in B_{\phi}} 2D_{Kull}(\boldsymbol{v}, \boldsymbol{p}^0) \leq 2D_{Kull}(\boldsymbol{p}, \boldsymbol{p}^0) = c_{Kull}(\boldsymbol{p})$$

because $\boldsymbol{p} \in B_{\phi}$. Then we have

$$c_{\phi}(\boldsymbol{p}) / c_{Kull}(\boldsymbol{p}) \leq 1, \text{ for all } \boldsymbol{v} \in \Delta_M^+ \ \boldsymbol{p} \neq \boldsymbol{p}^0.$$

The likelihood ratio test obtained for $\phi(x) = x \log x - x + 1$ has maximal Bahadur efficiency among all the ϕ-divergence test statistics. However, in the same way as Cressie and Read (1984) pointed out with the power-divergence test statistic, other family members can be equally efficient if there does not exist a probability vector $\boldsymbol{v}$ satisfying both $D_{Kull}(\boldsymbol{v}, \boldsymbol{p}^0) < D_{Kull}(\boldsymbol{p}, \boldsymbol{p}^0)$ and $D_{\phi}(\boldsymbol{v}, \boldsymbol{p}^0) \geq D_{\phi}(\boldsymbol{p}, \boldsymbol{p}^0)$.

Computing measures of large sample performance often leads to different conclusions. We have seen that, when M is fixed, the likelihood ratio test statistic is superior to any ϕ-divergence test statistic in terms of exact Bahadur efficiency, but all ϕ-divergence test statistics are equivalent in terms of Pitman efficiency. There are interesting results in relation to the chi-square test statistic and likelihood ratio test statistic when the number of cells is not fixed. Quine and Robinson (1985) studied the problem when $M_n \to \infty$ and they found that chi-square test statistic has Pitman efficiency 1 and Bahadur efficiency 0 relative to likelihood ratio test statistic. But if $n/M_n \to \gamma \in (0, \infty)$, i.e., sparseness assumption, chi-square test statistic is strictly superior to likelihood ratio test statistic in the Pitman sense but still has Bahadur efficiency 0. Similar results were found by Kallenberg et *al.* (1985). An interesting overview of this problem can be seen in Drost et *al.* (1989). The results presented by Quine and Robinson (1985) can be extended to the family of ϕ-divergence test statistics. This is an open problem.

4.2.3. Approximations to the Power Function: Comparisons

In Chapter 3 we obtained two approximations to the power function, associated with the decision rule given in (4.3), via approximations to the limiting alternative distribution of the ϕ-divergence test statistic, one for contiguous alternative hypotheses and another for fixed alternatives. Given the contiguous alternative hypothesis $\boldsymbol{p}_n$ we established that

$$\lim_{n\to\infty} \beta_{n,\phi}(\boldsymbol{p}_n) = 1 - G_{\chi^2_{M-1}(\delta)}\left(\chi^2_{M-1,\alpha}\right),$$

where δ was given in (4.5).

Then every ϕ-divergence test statistic has the same asymptotic distribution. The distribution $G_{\chi^2_{M-1}(\delta)}$ gives a fair approximation to the power function of the chi-square test statistic when $n \geq 100$. The approximation is not as good for the likelihood ratio test statistic; see Kallenberg et *al.* (1985) and Broffitt and Randles (1977). Drost et *al.* (1989) pointed out that in the case of the power-divergence family the approximation for values of $\lambda \neq 1$ is very poor. The approach based on $G_{\chi^2_{M-1}(\delta)}$ was considered for the first time by Patnaik (1949). Slakter (1968) simulated the power of the chi-square test statistic in many cases and compared it to the approach given by $G_{\chi^2_{M-1}(\delta)}$. He was very pessimistic. An interesting study is also given in Haber (1980).

The other asymptotic approximation has the expression

$$\beta_{n,\phi}(\boldsymbol{p}^*) \approx 1 - \Phi\left(\frac{1}{\sigma_1(\boldsymbol{p}^*)}\left(\frac{\phi''(1)}{2\sqrt{n}}\chi^2_{M-1,\alpha} - \sqrt{n}D_\phi(\boldsymbol{p}^*, \boldsymbol{p}^0)\right)\right),$$

where $\sigma_1(\boldsymbol{p}^*)$ is given in (3.10).

Broffit and Randles (1977) compared the two previous approximations, in the case of the chi-square test statistic, and they conclude that for large values of the true power the normal approximation is better, but for moderate values of the power, the approximation $G_{\chi^2_{M-1}(\delta)}$ is better. Menéndez et *al.* (2001a) considered a model in which the normal approximation, $\beta_{n,\phi}(\boldsymbol{p}^*)$, of the power function is very poor for the family of the power-divergence test statistics. Drost et *al.* (1989) proposed two new approximations to the power function for the power-divergence test statistic. Both the computation and results on asymptotic error rates suggest that the new approximations are greatly superior to the traditional power approximations. Sekiya et *al.* (1999) proposed a normal approximation based on the normalizing transformation of the power-divergence test statistic. Sekiya and Taneichi (2004) using multivariate Edgeworth expansion for a continuous distribution showed how the normal approximation can be improved. Obtaining these approximations for the ϕ-divergence test statistic is an open problem.

4.3. Exact and Asymptotic Moments: Comparison

The speed of convergence of the exact moments to the asymptotic moments, in the family of ϕ-divergence test statistics, gives us information about the speed of convergence of the exact distribution to its asymptotic distribution.

We shall consider a second order Taylor expansion of the first three exact moments of the ϕ-divergence test statistic, $T_n^\phi(\widehat{\boldsymbol{p}}, \boldsymbol{p}^0)$, and we compare them to the corresponding moments of a chi-square distribution with $M-1$ degrees of freedom. The sizes of the correction terms will give information about the errors that we are doing when we use the asymptotic distribution instead of the exact distribution.

The method was used by Cressie and Read (1984). We therefore omit its justification or motivation. We consider the problem under the null hypothesis as well as under contiguous alternative hypotheses.

4.3.1. Under the Null Hypothesis

We denote

$$\mu_\beta(T_n^\phi(\widehat{\mathbf{p}},\mathbf{p}^0)) = E\left[\left(T_n^\phi(\widehat{\mathbf{p}},\mathbf{p}^0)\right)^\beta\right],\ \beta = 1,2,3,$$

and we shall establish, in Propositions 4.1, 4.2 and 4.3 that

$$\mu_\beta\left(T_n^\phi(\widehat{\mathbf{p}},\mathbf{p}^0)\right) = \mu_\beta\left(\chi^2_{M-1}\right) + \frac{1}{n}\, f_\phi^\beta + O(n^{-3/2}),\ \beta = 1,2,3,$$

where $\mu_\beta(\chi^2_{M-1}) = E\left[\left(\chi^2_{M-1}\right)^\beta\right]$. We point out that $E\left[\left(\chi^2_{M-1}\right)^\beta\right] = M-1$, M^2-1 or M^3+3M^2-M-3, if $\beta = 1,2$ or 3, respectively.

Then f_ϕ^β controls the speed at which the first exact three moments, about the origin, of the ϕ-divergence test statistic $T_n^\phi(\widehat{\mathbf{p}},\mathbf{p}^0)$, converge to the first three moments of a chi-square distribution with $M-1$ degrees of freedom. The function ϕ for which $f_\phi^\beta = 0$, $\beta = 1,2,3$, will be the best.

In the next propositions we shall obtain the second order Taylor expansion of the first three moments, with respect to the origin, of the ϕ-divergence test statistic.

Proposition 4.1
It holds

$$E\left[T_n^\phi(\widehat{\mathbf{p}},\mathbf{p}^0)\right] = M - 1 + \frac{1}{n} f_\phi^1 + O(n^{-3/2}),$$

where

$$f_\phi^1 = \frac{\phi'''(1)}{3\phi''(1)}\,(2-3M+S) + \frac{\phi^{IV}(1)}{4\phi''(1)}\,(1-2M+S) \tag{4.8}$$

and $S = \sum_{j=1}^M (p_j^0)^{-1}$.

Proof. Let us denote

$$W_j = \frac{1}{\sqrt{n}}\left(N_j - np_j^0\right),\ j = 1, ..., M.$$

A fourth order Taylor expansion of $D_\phi(\widehat{\boldsymbol{p}},\boldsymbol{p}^0)$ around $\boldsymbol{p}^0$ gives

$$\begin{aligned}
D_\phi(\widehat{\boldsymbol{p}},\boldsymbol{p}^0) &= \sum_{j=1}^{M}\left(\frac{\partial D_\phi(\boldsymbol{p},\boldsymbol{p}^0)}{\partial p_j}\right)_{\boldsymbol{p}=\boldsymbol{p}^0}\frac{W_j}{\sqrt{n}}+\frac{1}{2!}\sum_{j=1}^{M}\left(\frac{\partial^2 D_\phi(\boldsymbol{p},\boldsymbol{p}^0)}{\partial p_j^2}\right)_{\boldsymbol{p}=\boldsymbol{p}^0}\frac{W_j^2}{n}\\
&+\frac{1}{3!}\sum_{j=1}^{M}\left(\frac{\partial^3 D_\phi(\boldsymbol{p},\boldsymbol{p}^0)}{\partial p_j^3}\right)_{\boldsymbol{p}=\boldsymbol{p}^0}\frac{W_j^3}{n\sqrt{n}}\\
&+\frac{1}{4!}\sum_{j=1}^{M}\left(\frac{\partial^4 D_\phi(\boldsymbol{p},\boldsymbol{p}^0)}{\partial p_j^4}\right)_{\boldsymbol{p}=\boldsymbol{p}^0}\frac{W_j^4}{n^2}+O_P(n^{-5/2}).
\end{aligned}$$

But

$$\begin{aligned}
\left(\frac{\partial D_\phi(\boldsymbol{p},\boldsymbol{p}^0)}{\partial p_j}\right)_{\boldsymbol{p}=\boldsymbol{p}^0} &= \left(\phi'\left(\frac{p_j}{p_j^0}\right)\right)_{\boldsymbol{p}=\boldsymbol{p}^0} = \phi'(1),\\
\left(\frac{\partial^2 D_\phi(\boldsymbol{p},\boldsymbol{p}^0)}{\partial p_j^2}\right)_{\boldsymbol{p}=\boldsymbol{p}^0} &= \left(\frac{1}{p_j^0}\phi''\left(\frac{p_j}{p_j^0}\right)\right)_{\boldsymbol{p}=\boldsymbol{p}^0} = \frac{1}{p_j^0}\phi''(1),\\
\left(\frac{\partial^3 D_\phi(\boldsymbol{p},\boldsymbol{p}^0)}{\partial p_j^3}\right)_{\boldsymbol{p}=\boldsymbol{p}^0} &= \left(\frac{1}{\left(p_j^0\right)^2}\phi'''\left(\frac{p_j}{p_j^0}\right)\right)_{\boldsymbol{p}=\boldsymbol{p}^0} = \frac{1}{\left(p_j^0\right)^2}\phi'''(1),\\
\left(\frac{\partial^4 D_\phi(\boldsymbol{p},\boldsymbol{p}^0)}{\partial p_j^4}\right)_{\boldsymbol{p}=\boldsymbol{p}^0} &= \left(\frac{1}{\left(p_j^0\right)^3}\phi^{IV}\left(\frac{p_j}{p_j^0}\right)\right)_{\boldsymbol{p}=\boldsymbol{p}^0} = \frac{1}{\left(p_j^0\right)^3}\phi^{IV}(1).
\end{aligned}$$

Then

$$\begin{aligned}
T_n^\phi(\widehat{\boldsymbol{p}},\boldsymbol{p}^0) &= \frac{2n}{\phi''(1)}D_\phi(\widehat{\boldsymbol{p}},\boldsymbol{p}^0)=\sum_{j=1}^{M}\frac{W_j^2}{p_j^0}+\frac{\phi'''(1)}{3\sqrt{n}\phi''(1)}\sum_{j=1}^{M}\frac{W_j^3}{\left(p_j^0\right)^2}\\
&+\frac{\phi^{IV}(1)}{12n\phi''(1)}\sum_{j=1}^{M}\frac{W_j^4}{\left(p_j^0\right)^3}+O_P(n^{-3/2}).
\end{aligned} \tag{4.9}$$

By (4.9) we can write

$$\begin{aligned}
E\left[T_n^\phi(\widehat{\boldsymbol{p}},\boldsymbol{p}^0)\right] &= \sum_{j=1}^{M}\frac{E\left[W_j^2\right]}{p_j^0}+\frac{\phi'''(1)}{3\sqrt{n}\phi''(1)}\sum_{j=1}^{M}\frac{E\left[W_j^3\right]}{\left(p_j^0\right)^2}\\
&+\frac{\phi^{IV}(1)}{12n\phi''(1)}\sum_{j=1}^{M}\frac{E\left[W_j^4\right]}{\left(p_j^0\right)^3}+O(n^{-3/2}),
\end{aligned} \tag{4.10}$$

since

$$E\left[O_P(n^{-3/2})\right]=O(n^{-3/2}).$$

The moment-generating function of a multinomial random variable,

$$\boldsymbol{N}=(N_1,...,N_M)^T,$$

with parameters n and $\boldsymbol{p}^0$ is

$$M_{\mathbf{N}}(\boldsymbol{t}) = E[\exp(\boldsymbol{t}^T \boldsymbol{N})] = \left(p_1^0 \exp(t_1) + ... + p_M^0 \exp(t_M)\right)^n$$

with $\boldsymbol{t} = (t_1, ..., t_M)^T$. The moment generating function of the M-dimensional random variable $\boldsymbol{W} = \frac{1}{\sqrt{n}}\left(\boldsymbol{N} - n\boldsymbol{p}^0\right)$ is thus given by

$$\begin{aligned} M_{\mathbf{W}}(\boldsymbol{t}) &= E[\exp(\boldsymbol{t}^T \boldsymbol{W})] = E[\exp(\boldsymbol{t}^T \left(\mathbf{N}/\sqrt{n} - \sqrt{n}\mathbf{p}^0\right))] \\ &= \exp(-\sqrt{n}\boldsymbol{t}^T \mathbf{p}^0) E[\exp(\boldsymbol{t}^T \mathbf{N}/\sqrt{n})] \\ &= \exp(-\sqrt{n}\boldsymbol{t}^T \mathbf{p}^0) M(\boldsymbol{t}/\sqrt{n}), \end{aligned} \tag{4.11}$$

and the ath-moment of W_j about the origin by

$$E[W_j^a] = \left(\frac{\partial^a M_{\mathbf{W}}(\boldsymbol{t})}{\partial t_j^a}\right)_{\boldsymbol{t}=\mathbf{0}} \tag{4.12}$$

for $j = 1, ..., M, \quad a = 1, 2, ...$

From (4.11) and (4.12) we have

$$\begin{aligned} E[W_j^2] &= -(p_j^0)^2 + p_j^0 \\ E[W_j^3] &= n^{-1/2}\left(2(p_j^0)^3 - 3(p_j^0)^2 + p_j^0\right) \\ E[W_j^4] &= 3(p_j^0)^4 - 6(p_j^0)^3 + 3(p_j^0)^2 \\ &+ n^{-1}\left(-6(p_j^0)^4 + 12(p_j^0)^3 - 7(p_j^0)^2 + p_j^0\right), \end{aligned}$$

and substituting these expressions in (4.10) the proof is complete. ■

Proposition 4.2

It holds

$$E\left[T_n^{\phi}(\widehat{\boldsymbol{p}}, \boldsymbol{p}^0)^2\right] = M^2 - 1 + \frac{1}{n} f_{\phi}^2 + O(n^{-3/2}),$$

where

$$\begin{aligned} f_{\phi}^2 &= (2 - 2M - M^2 + S) + \tfrac{2\phi'''(1)}{3\phi''(1)}\left(10 - 13M - 6M^2(M+8)S\right) \\ &+ \tfrac{1}{3}\left(\tfrac{\phi'''(1)}{\phi''(1)}\right)^2 \left(4 - 6M - 3M^2 + 5S\right) \\ &+ \tfrac{\phi^{IV}(1)}{2\phi''(1)}\left(3 - 5M - 2M^2 + (M+3)S\right) \end{aligned} \tag{4.13}$$

and $S = \sum_{j=1}^M (p_j^0)^{-1}$.

Proof.

Squaring and taking expectations in (4.9) we get

$$
\begin{aligned}
E\left[\left(T_n^{\phi}(\widehat{\boldsymbol{p}},\boldsymbol{p}^0)\right)^2\right] &= \sum_{j=1}^{M}\frac{E[W_j^4]}{\left(p_j^0\right)^2}+\sum_{j\neq i}^{M}\frac{E[W_j^2W_i^2]}{p_j^0p_i^0}\\
&+ \frac{2\phi'''(1)}{3\sqrt{n}\phi''(1)}\left(\sum_{j=1}^{M}\frac{E[W_j^5]}{\left(p_j^0\right)^3}+\sum_{j\neq i}^{M}\frac{E[W_j^2W_i^3]}{p_j^0\left(p_i^0\right)^2}\right)\\
&+ \frac{1}{n}\left(\frac{\phi^{IV}(1)}{6\phi''(1)}\left(\sum_{j=1}^{M}\frac{E[W_j^6]}{(p_j^0)^4}+\sum_{j\neq i}^{M}\frac{E[W_j^2W_i^4]}{p_j^0(p_i^0)^3}\right)\right.\\
&+ \left.\left(\frac{\phi'''(1)}{3\phi''(1)}\right)^2\left(\sum_{j=1}^{M}\frac{E[W_j^6]}{(p_j^0)^4}+\sum_{j\neq i}^{M}\frac{E[W_j^3W_i^3]}{(p_j^0)^2\,(p_i^0)^2}\right)\right)+O(n^{-3/2}).
\end{aligned}
\tag{4.14}
$$

By (4.11) we have

$$
E[W_j^aW_i^b]=\left(\frac{\partial^{a+b}M_{\mathbf{W}}(\boldsymbol{t})}{\partial t_j^a\partial t_i^b}\right)_{\boldsymbol{t}=\mathbf{0}} \tag{4.15}
$$

for $j,\ i=1,...,M$ and $a,b=1,2,\ldots$. Then

$$
\begin{aligned}
E[W_j^5] &= n^{-1/2}\left(-20(p_j^0)^5+50(p_j^0)^4-40(p_j^0)^3+10(p_j^0)^2\right)\\
&+ n^{-3/2}\left(24(p_j^0)^5-60(p_j^0)^4+50(p_j^0)^3-15(p_j^0)^2+p_j^0\right),
\end{aligned}
$$

and

$$
\begin{aligned}
E[W_j^6] &= -15(p_j^0)^6+45(p_j^0)^5-45(p_j^0)^4+15(p_j^0)^3\\
&+ n^{-1}\left(130(p_j^0)^6-390(p_j^0)^5+415(p_j^0)^4-180(p_j^0)^3+25(p_j^0)^2\right)\\
&+ O(n^{-2}).
\end{aligned}
$$

For $j\neq i$ both fixed, we define $p_{ab}=(p_j^0)^a(p_i^0)^b$; then

$$
\begin{aligned}
E[W_j^2W_i^2] &= 3p_{22}-p_{21}-p_{12}+p_{11}+n^{-1}\left(-6p_{22}+2p_{21}+2p_{12}-p_{11}\right),\\
E[W_j^2W_i^3] &= n^{-1/2}\left(-20p_{23}+5p_{13}+15p_{22}-6p_{12}-p_{21}+p_{11}\right),\\
E[W_j^2W_i^4] &= -15p_{24}+18p_{23}+3p_{14}-3p_{22}-6p_{13}+3p_{12}\\
&+ \tfrac{1}{n}(130p_{24}-156p_{23}-26p_{14}+41p_{22}+42p_{13}-p_{21}-17p_{12}+p_{11})\\
&+ O(n^{-2}),\\
E[W_j^3W_i^3] &= -15p_{33}+9p_{32}+9p_{23}-9p_{22}+n^{-1}\left(130p_{33}-78p_{32}-78p_{23}\right.\\
&+ \left.63p_{22}+5p_{31}+5p_{13}-6p_{21}-6p_{12}+p_{11}\right)+O(n^{-2}),
\end{aligned}
$$

and substituting these expressions in (4.14) and simplifying we have

$$E\left[\left(T_n^{\phi}(\widehat{\boldsymbol{p}},\boldsymbol{p}^0)\right)^2\right] = M^2 - 1 + \frac{1}{n} f_{\phi}^2 + O(n^{-3/2}),$$

where

$$\begin{aligned} f_{\phi}^2 &= (2 - 2M - M^2 + S) \\ &+ \tfrac{2\phi'''(1)}{3\phi''(1)} \left(10 - 13M - 6M^2 + (M+8)S\right) \\ &+ \tfrac{1}{3}\left(\tfrac{\phi'''(1)}{\phi''(1)}\right)^2 (4 - 6M - 3M^2 + 5S) \\ &+ \tfrac{\phi^{IV}(1)}{2\phi''(1)} \left(3 - 5M - 2M^2 + (M+3)\, S\right). \end{aligned}$$

To obtain the previous expression we have used the following relations,

$$\sum_{j=1}^{M} \left(p_j^0\right)^2 + \sum_{j\neq i}^{M} p_j^0 p_i^0 = 1; \quad \sum_{j\neq i}^{M} p_j^0 = M - 1$$

and

$$\sum_{j\neq i}^{M} \frac{p_j^0}{p_i^0} = S - M.$$

■

Proposition 4.3
It holds

$$E\left[\left(T_n^{\phi}(\widehat{\boldsymbol{p}},\boldsymbol{p}^0)\right)^3\right] = M^3 + 3M^2 - M - 3 + \tfrac{1}{n} f_{\phi}^3 + O(n^{-3/2})$$

where

$$\begin{aligned} f_{\phi}^3 &= \left(26 - 24M - 21M^2 - 3M^3 + (19+3M)\, S\right) \\ &+ \tfrac{\phi'''(1)}{\phi''(1)} \left(70 - 81M - 64M^2 - 9M^3 + \left(65 + 18M + M^2\right) S\right) \\ &+ \left(\tfrac{\phi'''(1)}{\phi''(1)}\right)^2 \left(20 - 26M - 21M^2 - 3M^3 + (25+5M)\, S\right) \\ &+ \tfrac{3\phi^{IV}(1)}{4\phi''(1)} \left(15 - 22M - 15M^2 - 2M^3 + \left(15 + 8M + M^2\right) S\right) \end{aligned} \tag{4.16}$$

and $S = \sum_{j=1}^{M} (p_j^0)^{-1}$.

Proof.

Cubing and taking expectations in (4.9) we have

$$\begin{aligned}
E\left[T_n^{\phi}(\widehat{\boldsymbol{p}},\boldsymbol{p}^0)^3\right] &= \sum_{j=1}^{M}\frac{E[W_j^6]}{p_{0j}^3}+3\sum_{j\neq i}^{M}\frac{E[W_j^4W_i^2]}{(p_j^0)^2p_i^0}+\sum_{j\neq i\neq k}^{M}\frac{E[W_j^2W_i^2W_k^2]}{p_j^0p_i^0p_k^0}\\
&+\frac{\phi'''(1)}{\sqrt{n}\phi''(1)}\left(\sum_{j=1}^{M}\frac{E[W_j^7]}{(p_j^0)^4}+\sum_{j\neq i}^{M}\frac{E[W_j^4W_i^3]}{(p_j^0)^2(p_i^0)^2}+2\sum_{j\neq i}^{M}\frac{E[W_j^2W_i^5]}{p_j^0(p_i^0)^3}\right.\\
&+\left.\frac{E[W_j^2W_i^2W_k^3]}{p_j^0p_i^0(p_k^0)^2}\right)+\frac{1}{n}\left(\frac{1}{3}\left(\frac{\phi'''(1)}{\phi''(1)}\right)^2\left(\sum_{j=1}^{M}\frac{E[W_j^8]}{(p_j^0)^5}\right.\right.\\
&+\left.\sum_{j\neq i}^{M}\frac{E[W_j^6W_i^2]}{(p_j^0)^4p_i^0}+2\sum_{j\neq i}^{M}\frac{E[W_j^3W_i^5]}{(p_j^0)^2(p_i^0)^3}+\sum_{j\neq i\neq k}^{M}\frac{E[W_j^3W_i^3W_k^2]}{(p_j^0)^2(p_i^0)^2p_k^0}\right)\\
&+\frac{\phi^{IV}(1)}{4\phi''(1)}\left(\sum_{j=1}^{M}\frac{E[W_j^8]}{(p_j^0)^5}+\sum_{j\neq i}^{M}\frac{E[W_j^4W_i^4]}{(p_j^0)^2(p_i^0)^3}+2\sum_{j\neq i}^{M}\frac{E[W_j^2W_i^6]}{p_j^0(p_i^0)^4}\right.\\
&+\left.\left.\sum_{j\neq i\neq k}^{M}\frac{E[W_j^2W_i^2W_k^4]}{p_j^0p_i^0(p_k^0)^3}\right)\right)+O(n^{-3/2}).
\end{aligned}\tag{4.17}$$

Using again (4.11) we get

$$\begin{aligned}
E[W_j^7] &= n^{-1/2}\left(210(p_j^0)^7-735(p_j^0)^6+945(p_j^0)^5-525(p_j^0)^4+105(p_j^0)^3\right)\\
&+ O(n^{-3/2}),\\
E[W_j^8] &= 105(p_j^0)^8-420(p_j^0)^7+630(p_j^0)^6-420(p_j^0)^5+105(p_j^0)^4+O(n^{-1}).
\end{aligned}$$

For $j\neq i\neq k$ all fixed, we define

$$p_{ab}=(p_j^0)^a(p_i^0)^b \text{ and } p_{abc}=(p_j^0)^a(p_i^0)^b(p_k^0)^c.$$

Then

$$\begin{aligned}
E[W_j^4W_i^3] &= n^{-1/2}\,(210p_{43}-105p_{42}-210p_{33}+3p_{41}+144p_{32}+36p_{23}\\
&- 6p_{31}-39p_{22}+3p_{21})+O(n^{-3/2}),\\
E[W_j^4W_i^4] &= 105p_{44}-90\,(p_{43}+p_{34})+9\,(p_{24}+3p_{42})+108p_{33}\\
&- 18(p_{32}+p_{23})+9p_{22}+O(n^{-1}),\\
E[W_j^5W_i^2] &= n^{-1/2}\,(210p_{52}-35p_{51}-350p_{42}+80p_{41}+150p_{32}-55p_{31}\\
&- 10p_{22}+10p_{21})+O(n^{-3/2}),\\
E[W_j^5W_i^3] &= 105p_{53}-45p_{52}-150p_{43}+90p_{42}+45p_{33}-45p_{32}+O(n^{-1}),\\
E[W_j^6W_i^2] &= 105p_{62}-15p_{61}-225p_{52}+45p_{51}+135p_{42}-45p_{41}\\
&- 15p_{32}+15p_{31}+O(n^{-1}),
\end{aligned}$$

$$\begin{aligned}E[W_j^2 W_i^2 W_k^2] &= -15p_{222} + 3\,(p_{122} + p_{212} + p_{221}) - (p_{112} + p_{121} + p_{211}) + p_{111}\\ &+ n^{-1}\,(130p_{222} - 26\,(p_{122} + p_{212} + p_{221}) + 7\,(p_{112} + p_{121} + p_{211})\\ &- 3p_{111}) + O(n^{-2}),\end{aligned}$$

$$\begin{aligned}E[W_j^2 W_i^2 W_k^3] &= n^{-1/2}\,(210p_{223} - 105p_{222} - 35(p_{123} + p_{213}) + 8p_{113} + 24(p_{112}\\ &+ p_{212}) + 3p_{221} - 9p_{112} - (p_{211} + p_{121}) + p_{111}) + O(n^{-3/2}),\end{aligned}$$

$$\begin{aligned}E[W_j^2 W_i^3 W_k^3] &= 105p_{233} - 15p_{133} - 45(p_{223} + p_{232}) + 9\,(p_{132} + p_{123})\\ &+ 27p_{222} - 9p_{122} + O(n^{-1}),\end{aligned}$$

$$\begin{aligned}E[W_j^2 W_i^2 W_k^4] &= 105p_{224} - 15\,(p_{124} + p_{214}) - 90p_{223} + 3p_{114} + 18\,(p_{123} + p_{213})\\ &+ 9p_{222} - 6p_{113} - 3\,(p_{122} + p_{212}) + 3p_{112} + O(n^{-1}),\end{aligned}$$

and by (4.17) we have

$$E\left[\left(T_n^{\phi}(\widehat{\mathbf{p}}, \mathbf{p}^0)\right)^3\right] = M^3 + 3M^2 - M - 3 + \frac{1}{n} f_{\phi}^3 + O(n^{-3/2}),$$

where

$$\begin{aligned}f_{\phi}^3 &= (26 - 24M - 21M^2 - 3M^3 + (19 + 3M)\,S)\\ &+ \tfrac{\phi'''(1)}{\phi''(1)}\,(70 - 81M - 64M^2 - 9M^3 + (65 + 18M + M^2)\,S)\\ &+ \left(\tfrac{\phi'''(1)}{\phi''(1)}\right)^2 (20 - 26M - 21M^2 - 3M^3 + (25 + 5M)\,S)\\ &+ \tfrac{3\phi^{IV}(1)}{4\phi''(1)}\,(15 - 22M - 15M^2 - 2M^3 + (15 + 8M + M^2)\,S)\,.\end{aligned}$$

To obtain the previous expression we have used the following relations:

$$\sum_{j\neq i}^{M} \frac{(p_i^0)^2}{p_i^0} + \sum_{j\neq i\neq k}^{M} \frac{p_j^0 p_i^0}{p_k^0} = S - 2M + 1$$

and

$$\sum_{j=1}^{M} (p_j^0)^3 + 3\sum_{j\neq i}^{M} (p_j^0)^2 p_i^0 + \sum_{j\neq i\neq k}^{M} p_j^0 p_i^0 p_k^0 = 1.$$

■

Remark 4.3

It is clear that f_{ϕ}^i, $i = 1, 2, 3$, *control the speed at which the first three exact moments, about the origin, of the* ϕ*-divergence test statistic,* $T_n^{\phi}(\widehat{\mathbf{p}}, \mathbf{p}^0)$, *converge to the three first moments, about the origin, of a chi-square random variable with* $M - 1$ *degrees of freedom.*

Let us consider a function $\phi \in \Phi^*$ *depending on a parameter "a". In the following we shall denote it by* $\phi \equiv \phi_a$. *In this context as optimal in the sense of the moments of order* β *we consider the values of "a" from the set* R_β *of roots of the equations* $f^i_{\phi_a} = 0$, $i = 1, 2, 3$. *Strictly speaking these expansions are valid only as* $n \to \infty$ *for* $M < \infty$ *fixed.*

If now we consider the null hypothesis $H_0 : \mathbf{p} = \mathbf{p}^0 = (1/M, ..., 1/M)^T$, *we have* $\sum_{j=1}^M (p_j^0)^{-1} = M^2$. *However for* M *increasing the roots of the equation* $f^1_{\phi_a} = 0$ *converge to the roots of the equation*

$$4\phi_a'''(1) + 3\phi_a^{IV}(1) = 0, \qquad \text{for } \phi_a''(1) \neq 0, \tag{4.18}$$

since the equation (4.8) can be written as

$$4\phi_a'''(1) \left(\frac{2 - 3M + M^2}{1 - 2M + M^2} \right) + 3\phi_a^{IV}(1) = 0.$$

Then the roots of this last equation converge to the roots of the equation (4.18) as $M \to \infty$. *In relation with the equations* $f^i_{\phi_a} = 0$, $i = 2, 3$, *it is possible to apply similar arguments.*

If we consider the family of the power-divergence test statistics, $\phi \equiv \phi_{(\lambda)}$, *we have that the roots of the equation (4.18) are* $\lambda = 1$ *and* $\lambda = 2/3$. *These values were found directly by Read and Cressie (1988).*

Example 4.1

We consider the family of ϕ*-divergence measures given by*

$$\phi_a(x) = \frac{(1-x)^2}{2\,(a + (1-a)\,x)}, \qquad a \in [0, 1], \tag{4.19}$$

i.e., the family of Rukhin's divergence measures. This family was introduced by Rukhin (1994). The associated family of test statistics has the expression

$$T_n^{\phi_a}(\widehat{\mathbf{p}}, \mathbf{p}^0) = n \sum_{j=1}^M \frac{(p_j^0 - \widehat{p}_j)^2}{a p_j^0 + (1-a)\widehat{p}_j} \qquad 0 \leq a \leq 1. \tag{4.20}$$

Some properties of this test statistic can be seen in Pardo, M. C. and J. A. Pardo (1999). It is observed immediately that $T_n^{\phi_0}(\widehat{\mathbf{p}}, \mathbf{p}^0)$ *is the modified chi-square test statistic and* $T_n^{\phi_1}(\widehat{\mathbf{p}}, \mathbf{p}^0)$ *is the chi-square test statistic.*

For $a \in [0, 1]$ *we have*

$$\phi_a'''(x) = -\frac{3(1-a)^3}{(a+(1-a)x)^4} \quad \text{and} \quad \phi_a^{IV}(x) = \frac{12(1-a)^4}{(a+(1-a)x)^5}$$

and the equation (4.18) becomes $36a^2 - 60a + 24 = 0$. *Then the roots of the equation (4.18) are* $a = 1$ *and* $a = 2/3$.

It seems interesting to know how large M *has to be for using the roots* $a = 1$ *and* $a = 2/3$. *If* M *is not large we must use the roots of the equations* $f^i_{\phi_a} = 0, i = 1, 2, 3$ *given in (4.8), (4.13) and (4.16). These solutions are given in Tables 4.1, 4.2, 4.3 and 4.4 as the number of classes* M *increases and* $\sum_{j=1}^M (p_j^0)^{-1}$ *changes. In particular we have considered, for* $\sum_{j=1}^M (p_j^0)^{-1}$, *the values* M^2, M^3, M^4 *and* M^5.

	M	2	3	4	5	10	20	40	50	100	200...∞
$f^1_{\phi_a}$	a_1	1.00	1.00	1.00	1.00	1.00	1.00	1.00	1.00	1.00	1.00...1.00
	a_2	1.00	0.83	0.78	0.75	0.70	0.68	0.67	0.67	0.67	0.67... 2/3
$f^2_{\phi_a}$	a_1	1.45	1.20	1.13	1.10	1.05	1.02	1.01	1.01	1.00	1.00...1.00
	a_2	0.55	0.59	0.60	0.61	0.63	0.64	0.65	0.66	0.66	0.66... 2/3
$f^3_{\phi_a}$	a_1	1.31	1.09	1.05	1.03	1.01	1.00	1.00	1.00	1.00	1.00...1.00
	a_2	0.69	0.67	0.66	0.65	0.65	0.66	0.66	0.66	0.66	0.66... 2/3

Table 4.1. Roots ($a_1 > a_2$) for $f^i_{\phi_a} = 0$, $i = 1, 2, 3$, for $\sum_{j=1}^M (p_j^0)^{-1} = M^2$.

	M	2	3	4	5	10	20	40	50	100	200...∞
$f^1_{\phi_a}$	a_1	1.00	1.00	1.00	1.00	1.00	1.00	1.00	1.00	1.00	1.00...1.00
	a_2	0.73	0.70	0.68	0.68	0.67	0.67	0.67	0.67	0.67	0.67...2/3
$f^2_{\phi_a}$	a_1	1.06	1.03	1.02	1.01	1.00	1.00	1.00	1.00	1.00	1.00...1.00
	a_2	0.55	0.57	0.58	0.59	0.61	0.63	0.65	0.65	0.65	0.66...2/3
$f^3_{\phi_a}$	a_1	0.93	0.93	0.93	0.93	0.95	0.97	0.98	0.98	0.99	0.99...1.00
	a_2	0.62	0.62	0.62	0.62	0.63	0.64	0.65	0.65	0.66	0.66...2/3

Table 4.2. Roots ($a_1 > a_2$) for $f^i_{\phi_a} = 0, i = 1, 2, 3$, for $\sum_{j=1}^M (p_j^0)^{-1} = M^3$.

	M	2	3	4	5	10	20	40	50	100	200...∞
$f^1_{\phi_a}$	a_1	1.00	1.00	1.00	1.00	1.00	1.00	1.00	1.00	1.00	1.00...1.00
	a_2	0.69	0.67	0.67	0.67	0.67	0.67	0.67	0.67	0.67	0.67...2/3
$f^2_{\phi_a}$	a_1	1.02	1.01	1.00	1.00	1.00	1.00	1.00	1.00	1.00	1.00...1.00
	a_2	0.55	0.57	0.58	0.58	0.61	0.63	0.65	0.65	0.66	0.66...2/3
$f^3_{\phi_a}$	a_1	0.91	0.91	0.92	0.93	0.95	0.96	0.98	0.98	0.99	0.99...1.00
	a_2	0.61	0.61	0.61	0.61	0.63	0.64	0.65	0.65	0.66	0.66...2/3

Table 4.3. Roots ($a_1 > a_2$) for $f^i_{\phi_a} = 0$, $i = 1, 2, 3$, for $\sum_{j=1}^M (p_j^0)^{-1} = M^4$.

	M	2	3	4	5	10	20	40	50	100	200...∞
$f_{\phi_a}^1$	a_1	1.00	1.00	1.00	1.00	1.00	1.00	1.00	1.00	1.00	1.00...1.00
	a_2	0.68	0.67	0.67	0.67	0.67	0.67	0.67	0.67	0.67	0.67...2/3
$f_{\phi_a}^2$	a_1	1.01	1.00	1.00	1.00	1.00	1.00	1.00	1.00	1.00	1.00...1.00
	a_2	0.55	0.57	0.57	0.59	0.61	0.63	0.65	0.65	0.66	0.66...2/3
$f_{\phi_a}^3$	a_1	0.91	0.91	0.92	0.92	0.95	0.96	0.98	0.98	0.99	0.99...1.00
	a_2	0.60	0.60	0.61	0.61	0.63	0.64	0.65	0.65	0.66	0.66...2/3

Table 4.4. Roots ($a_1 > a_2$) for $f_{\phi_a}^i = 0$, $i = 1, 2, 3$, for $\sum_{j=1}^M (p_j^0)^{-1} = M^5$.

For $M \geq 20$ all roots are within ± 0.05 of the limiting roots $a = 1$ and $a = 2/3$. Therefore for $M > 20$, choosing $a = 1$ or $a = 2/3$ the convergence of the first three moments to those of a random variable chi-square with $M - 1$ degrees of freedom is faster. For $M \geq 4$, most of the roots are within ± 0.1 of the limiting roots $a = 1$ and $a = 2/3$. This suggests that the "a"-range $[0.6, 1]$ is optimal for all M not too small. For $M < 4$ we should use the previous table.

4.3.2. Under Contiguous Alternative Hypotheses

Under the contiguous alternative hypotheses, $H_{1,n}$ given in Section 3.2, it was established that the asymptotic distribution of the ϕ-divergence test statistic, $T_n^\phi(\widehat{\boldsymbol{p}}, \boldsymbol{p}^0)$, is noncentral chi-square with $M - 1$ degrees of freedom and noncentrality parameter $\delta = \sum_{j=1}^M d_j^2/p_j^0$. In the same way as under the null hypothesis we shall establish in this case

$$E\left[T_n^\phi(\widehat{\boldsymbol{p}}, \boldsymbol{p}^0)\right] = M - 1 + \delta + \frac{1}{\sqrt{n}} g_\phi^1 + O\left(n^{-1}\right) \tag{4.21}$$

$$E\left[T_n^\phi(\widehat{\boldsymbol{p}}, \boldsymbol{p}^0)^2\right] = M^2 - 1 + 2(M-1)\delta + \delta^2 + \frac{1}{\sqrt{n}} g_\phi^2 + O\left(n^{-1}\right). \tag{4.22}$$

Based on (4.21) and (4.22) it is possible to give conditions to the function ϕ for improving the approximations of the exact moments to the asymptotic moments for the different values of M, $\boldsymbol{p}^0$ and $\boldsymbol{d}$. The convergence speed of the exact moments to the asymptotic moments gives information about the convergence speed of the exact distribution to the asymptotic distribution.

We are going to get the expressions of g_ϕ^i, $i = 1, 2$. We shall omit the third moment and the expansion is only considered until the order $O\left(n^{-1/2}\right)$. Note that

$$T_n^{\phi}(\widehat{\boldsymbol{p}},\boldsymbol{p}^0) = \frac{2n}{\phi''(1)} D_n^{\phi}(\widehat{\boldsymbol{p}},\boldsymbol{p}^0) = \sum_{j=1}^{M} \frac{W_j^2}{p_j^0} + \frac{\phi'''(1)}{3\sqrt{n}\phi''(1)} \sum_{j=1}^{M} \frac{W_j^3}{(p_j^0)^2} + O_P(n^{-1}),$$

where $W_j = n^{-1/2}(N_j - np_j^0)$, $j = 1, ..., M$.

Consider the random variable

$$V_j = n^{1/2}(\widehat{p}_j - p_j), j = 1, ..., M,$$

where $\widehat{p}_j = N_j/n$ and $\boldsymbol{N} = (N_1, \ldots, N_M)$ is a multinomial random variable with parameters n and $\boldsymbol{p} = (p_1, \ldots, p_M)^T$, with $p_j = p_j^0 + n^{-1/2} d_j$, $j = 1, ..., M$. Then we have that $W_j = V_j + d_j$, $j = 1, ..., M$, and the moments of W_j can be obtained from the moments of V_j, $j = 1, ..., M$.

We know that

$$E\left[T_n^{\phi}(\widehat{\boldsymbol{p}},\boldsymbol{p}^0)\right] = \sum_{j=1}^{M} \frac{E\left[W_j^2\right]}{p_j^0} + \frac{\phi'''(1)}{3\sqrt{n}\phi''(1)} \sum_{j=1}^{M} \frac{E\left[W_j^3\right]}{(p_j^0)^2} + O(n^{-1}) \tag{4.23}$$

and

$$E[W_j^2] = E[(V_j + d_j)^2] = -(p_j^0)^2 + p_j^0 + d_j^2 + n^{-1/2}(d_j - 2p_j^0 d_j) + O(n^{-1}),$$

$$E[W_j^3] = E[(V_j + d_j)^3] = d_j^3 - 3d_j(p_j^0)^2 + 3d_j p_j^0 + O(n^{-1/2}).$$

Substituting the expressions of $E[W_j^2]$ and $E[W_j^3]$ in (4.23) we have

$$\begin{aligned} E\left[T_n^{\phi}(\widehat{\boldsymbol{p}},\boldsymbol{p}^0)\right] &= M - 1 + \sum_{j=1}^{M} \frac{d_j^2}{p_j^0} \\ &+ \frac{1}{\sqrt{n}} \left(\sum_{j=1}^{M} \frac{d_j}{p_j^0} + \frac{\phi'''(1)}{3\phi''(1)} \left(\sum_{j=1}^{M} \frac{d_j^3}{(p_j^0)^2} + 3 \sum_{j=1}^{M} \frac{d_j}{p_j^0} \right) \right) + O(n^{-1}) \\ &= M - 1 + \delta + \tfrac{1}{\sqrt{n}} g_\phi^1 + O\left(n^{-1}\right), \end{aligned} \tag{4.24}$$

where

$$g_\phi^1 = \sum_{j=1}^{M} \frac{d_j}{p_j^0} + \frac{\phi'''(1)}{3\phi''(1)} \left(\sum_{j=1}^{M} \frac{d_j^3}{(p_j^0)^2} + 3 \sum_{j=1}^{M} \frac{d_j}{p_j^0} \right).$$

On the other hand,

$$
\begin{aligned}
E\left[\left(T_n^{\phi}(\widehat{\boldsymbol{p}},\boldsymbol{p}^0)\right)^2\right] &= \sum_{j=1}^{M}\frac{E[W_j^4]}{\left(p_j^0\right)^2}+\sum_{j\neq i}^{M}\frac{E[W_j^2W_i^2]}{p_j^0p_i^0}\\
&+ \frac{2\phi'''(1)}{3\sqrt{n}\phi''(1)}\left(\sum_{j=1}^{M}\frac{E[W_j^5]}{\left(p_j^0\right)^3}+\sum_{j\neq i}^{M}\frac{E[W_j^2W_i^3]}{p_j^0\left(p_i^0\right)^3}\right)+O(n^{-1});
\end{aligned} \tag{4.25}
$$

and by using the previous procedure we have

$$
\begin{aligned}
E[W_j^4] &= 3(p_j^0)^4-6(p_j^0)^3+3(p_j^0)^2+6p_j^0d_j^2-6(p_j^0)^2d_j^2+d_j^4\\
&+ n^{-1/2}\left(20(p_j^0)^3d_j-30(p_j^0)^2d_j+10p_j^0d_j+6d_j^3-12p_j^0d_j^3\right),
\end{aligned}
$$

$$
\begin{aligned}
E[W_j^2W_i^2] &= 3p_{22}-p_{21}-p_{12}+p_{11}+\left(p_{01}-p_{02}\right)d_j^2-4p_{11}d_jd_i\\
&+ \left(p_{10}-p_{20}\right)d_j^2+d_j^2d_i^2+n^{-1/2}\left(6p_{21}d_i+6p_{12}d_j-p_{20}d_i\right.\\
&- 2p_{11}d_j-p_{02}d_j-2p_{11}d_i+p_{01}d_j+p_{10}d_i+2(2p_{12}-p_{11})d_j\\
&+ (d_i-2p_{01}d_i)d_j^2+2(2p_{21}-p_{11})d_i-4(p_{10}d_j+p_{01}d_j)d_jd_i\\
&+ \left.(d_j-2p_{10}d_j)d_i^2\right),
\end{aligned}
$$

$$
E[W_j^5] = 5(3p_{0j}^4-6p_{0j}^3+3p_{0j}^2)d_j+10(p_{0j}-p_{0j}^2)d_j^3+d_j^5+O(n^{-1/2}),
$$

$$
\begin{aligned}
E[W_j^2W_i^3] &= 6(p_{13}-p_{12})d_j+3(3p_{22}-p_{21}-p_{12}+p_{11})d_j+3(p_{01}-p_{02})d_j^2d_i\\
&- 6p_{11}d_jd_i^2+(p_{10}-p_{20})d_i^3+d_j^2d_i^3+O(n^{-1/2}),
\end{aligned}
$$

where $p_{ab}=(p_j^0)^a(p_i^0)^b$.

Substituting the previous expressions in (4.25) we get

$$
\begin{aligned}
E\left[\left(T_n^{\phi}(\widehat{\boldsymbol{p}},\boldsymbol{p}^0)\right)^2\right] &= M^2-1+2(M-1)\sum_{j=1}^{M}\frac{d_j^2}{p_j^0}+\left(\sum_{j=1}^{M}\frac{d_j^2}{p_j^0}\right)^2\\
&+ \tfrac{1}{\sqrt{n}}\left(\left(2(M+3)\sum_{j=1}^{M}\frac{d_j}{p_j^0}+2\sum_{j=1}^{M}\frac{d_j}{p_j^0}\sum_{j=1}^{M}\frac{d_j^2}{p_j^0}\right.\right.\\
&+ \left.4\sum_{j=1}^{M}\frac{d_j^3}{\left(p_j^0\right)^2}\right)+\frac{\phi'''(1)}{\phi''(1)}\left(2(M+3)\sum_{j=1}^{M}\frac{d_j}{p_j^0}\right.\\
&+ 2\sum_{j=1}^{M}\frac{d_j}{p_j^0}\sum_{j=1}^{M}\frac{d_j^2}{p_j^0}+\frac{2}{3}\sum_{j=1}^{M}\frac{d_j^2}{p_j^0}\sum_{j=1}^{M}\frac{d_j^3}{\left(p_j^0\right)^2}\\
&+ \left.\left.\tfrac{2}{3}(M+5)\sum_{j=1}^{M}\frac{d_j^3}{\left(p_j^0\right)^2}\right)\right)+O(n^{-1})\\
&= M^2-1+2(M-1)\delta+\delta^2+\tfrac{1}{\sqrt{n}}g_{\phi}^2+O\left(n^{-1}\right),
\end{aligned} \tag{4.26}
$$

where g_ϕ^2 is given by

$$\begin{aligned} g_\phi^2 &= \left(\left(2(M+3)\sum_{j=1}^{M}\frac{d_j}{p_j^0}+2\left(\sum_{j=1}^{M}\frac{d_j}{p_j^0}\right)\left(\sum_{j=1}^{M}\frac{d_j^2}{p_j^0}\right)+4\sum_{j=1}^{M}\frac{d_j^3}{\left(p_j^0\right)^2}\right)\right. \\ &+ \frac{\phi'''(1)}{\phi''(1)}\left(2(M+3)\sum_{j=1}^{M}\frac{d_j}{p_j^0}+2\left(\sum_{j=1}^{M}\frac{d_j}{p_j^0}\right)\left(\sum_{j=1}^{M}\frac{d_j^2}{p_j^0}\right)\right. \\ &+ \left.\left.\frac{2}{3}\left(\sum_{j=1}^{M}\frac{d_j^2}{p_j^0}\right)\left(\sum_{j=1}^{M}\frac{d_j^3}{\left(p_j^0\right)^2}\right)+\frac{2}{3}(M+5)\left(\sum_{j=1}^{M}\frac{d_j^3}{\left(p_j^0\right)^2}\right)\right)\right). \end{aligned}$$

From (4.24) and (4.26) we observe that the first two moments under the contiguous alternative hypotheses coincide with the moments of a noncentral chi-square distribution with $M-1$ degrees of freedom and noncentrality parameter $\delta=\sum_{j=1}^{M}\frac{d_j^2}{p_{0j}}$ plus terms g_ϕ^1 and g_ϕ^2 of order $O(n^{-1/2})$ that depend on

$$\phi,\ M,\ \sum_{j=1}^{M}\frac{d_j}{p_{0j}},\ \sum_{j=1}^{M}\frac{d_j^2}{p_{0j}} \text{ and } \sum_{j=1}^{M}\frac{d_j^3}{p_{0j}^2},$$

plus a term of order $O(n^{-1})$.

The term of order $O(n^{-1/2})$ is cancelled, in the first moment, if we choose ϕ verifying

$$\frac{\phi'''(1)}{\phi''(1)}=\left(-3\sum_{j=1}^{M}\frac{d_j}{p_j^0}\right)\left(\sum_{j=1}^{M}\frac{d_j^3}{(p_j^0)^2}+3\sum_{j=1}^{M}\frac{d_j}{p_j^0}\right)^{-1}$$

and for the second moment if

$$\begin{aligned} \frac{\phi'''(1)}{\phi''(1)} &= -\left(2(M+3)\sum_{j=1}^{M}\frac{d_j}{p_{0j}}+2\left(\sum_{j=1}^{M}\frac{d_j}{p_j^0}\right)\left(\sum_{j=1}^{M}\frac{d_j^2}{p_j^0}\right)+4\sum_{j=1}^{M}\frac{d_j^3}{\left(p_j^0\right)^2}\right) \\ &\times \left(2(M+3)\sum_{j=1}^{M}\frac{d_j}{p_j^0}+2\left(\sum_{j=1}^{M}\frac{d_j}{p_j^0}\right)\left(\sum_{j=1}^{M}\frac{d_j^2}{p_j^0}\right)\right. \\ &+ \left.\frac{2}{3}\left(\sum_{j=1}^{M}\frac{d_j^3}{\left(p_j^0\right)^2}\right)\left(\sum_{j=1}^{M}\frac{d_j^2}{p_j^0}+(M+5)\right)\right)^{-1}. \end{aligned}$$

Example 4.2

If we consider the family of divergence measures given by (4.19) and $p_j^0=1/M$, $j=1,...,M$, *the* $O\left(n^{-1/2}\right)$ *correction factors in the first and second moment are* $a=1$ *and*

$$a=1-\frac{2}{M\sum_{j=1}^{M}d_j^2+M+5},$$

respectively. Since $M \geq 1$ *and* $\sum_{j=1}^{M} d_j^2 \geq 0$ *this value belongs to the interval* $(0,1)$ *and, when* M *increases, tends to* 1. *This result hints at the Pearson's test statistic* $(a = 1)$ *having closest distribution to the approximate noncentral chi-square under contiguous alternative hypotheses. The same happens with the power-divergence family. See Cressie and Read (1984, p. 454).*

4.3.3. Corrected Phi-divergence Test Statistic

We know that

$$\mu = E\left[\chi^2_{M-1}\right] = M - 1 \text{ and } \sigma^2 = Var\left[\chi^2_{M-1}\right] = 2(M-1).$$

We can modify $T_n^{\phi}(\widehat{\mathbf{p}}, \mathbf{p}^0)$ (we shall denote the corrected ϕ-divergence test statistic by ${}^{\mathbf{c}}T_n^{\phi}(\widehat{\mathbf{p}}, \mathbf{p}^0)$), in such a way that

$$E[{}^{\mathbf{c}}T_n^{\phi}(\widehat{\mathbf{p}}, \mathbf{p}^0)] = \mu + o(n^{-1})$$

and

$$Var[{}^{\mathbf{c}}T_n^{\phi}(\widehat{\mathbf{p}}, \mathbf{p}^0] = \sigma^2 + o(n^{-1}).$$

We know that

$$E[T_n^{\phi}(\widehat{\mathbf{p}}, \mathbf{p}^0)] = \mu + f_{\phi}^1/n + o(n^{-1})$$

and

$$Var[T_n^{\phi}(\widehat{\mathbf{p}}, \mathbf{p}^0)] = \sigma^2 + b_{\phi}/n + o(n^{-1})$$

where f_{ϕ}^1 was defined in (4.8) and b_{ϕ} is given by

$$\begin{aligned} b_{\phi} &= (2 - 2M - M^2 + S) + \tfrac{2\phi'''(1)}{\phi''(1)}\left(4 - 6M - M^2 + 3S\right) \\ &+ \tfrac{1}{3}\left(\tfrac{\phi'''(1)}{\phi''(1)}\right)^2 \left(4 - 6M - 3M^2 + 5S\right) \\ &+ \tfrac{2\phi^{IV}(1)}{\phi''(1)}\left(1 - 2M + S\right), \end{aligned}$$

where $S = \sum_{j=1}^{M}(p_j^0)^{-1}$.

We define

$${}^{\mathbf{c}}T_n^{\phi}(\widehat{\mathbf{p}}, \mathbf{p}^0) = \frac{T_n^{\phi}(\widehat{\mathbf{p}}, \mathbf{p}^0) - \gamma_{\phi}}{\sqrt{\delta_{\phi}}}$$

in such a way that

$$E[{}^{\mathbf{c}}T_n^{\phi}(\widehat{\mathbf{p}}, \mathbf{p}^0)] = \mu + o(n^{-1})$$

and

$$Var[{}^{\mathbf{c}}T_n^{\phi}(\widehat{\boldsymbol{p}},\boldsymbol{p}^0)] = \sigma^2 + o(n^{-1}).$$

To do this it is necessary to consider

$$\gamma_\phi = \mu\left(1-\sqrt{\delta_\phi}\right) + f_\phi^1/n \quad \text{and} \quad \delta_\phi = 1 + b_\phi/n\sigma^2,$$

i.e.,

$$\begin{aligned}
\delta_\phi &= 1 + \tfrac{1}{2(M-1)n}\left((2-2M-M^2+S)\right.\\
&+ \tfrac{2\phi'''(1)}{\phi''(1)}\left(4-6M-M^2+3S\right)\\
&+ \tfrac{1}{3}\left(\tfrac{\phi'''(1)}{\phi''(1)}\right)^2\left(4-6M-3M^2+5S\right)\\
&+ \left.\tfrac{2\phi^{IV}(1)}{\phi''(1)}\left(1-2M+S\right)\right),
\end{aligned}$$

and

$$\begin{aligned}
\gamma_\phi &= (M-1)\left(1-\sqrt{\delta_\phi}\right) + \tfrac{1}{n}\left(\tfrac{\phi'''(1)}{3\phi''(1)}\left(2-3M+S\right)\right.\\
&+ \left.\tfrac{\phi^{IV}(1)}{4\phi''(1)}\left(1-2M+S\right)\right).
\end{aligned}$$

Example 4.3

We consider the family of divergence measures given in (4.19) and $p_j^0 = 1/M, j = 1, ..., M$. *It is immediate to get*

$$\gamma_a = (M-1)\left(1-\delta_a^{1/2}\right) + \frac{1}{n}(1-a)\left(-\left(2-3M+M^2\right)+3(1-a)\left(M-1\right)^2\right)$$

and

$$\delta_a = 1 + \frac{1}{n}\left(-1-6(1-a)\left(M-2\right)+3(1-a)^2\left(5M-6\right)\right).$$

4.4. A Second Order Approximation to the Exact Distribution

In this Section we shall present an approximation of the exact distribution of $T_n^{\phi}(\widehat{\boldsymbol{p}},\boldsymbol{p}^0)$ extracting the ϕ-dependent second order component from the $o(1)$ term in (4.1). This second order component was obtained by Yarnold (1972) for the chi- square test statistic under the null hypothesis (the approximation consists of a term of multivariate Edgeworth expansion for a continuous distribution and a discontinuous term), by Siatoni and Fujikoshi (1984) for the likelihood ratio test statistic and Freeman-Tukey test statistic, by Read (1984a) for the power-divergence test statistic and by Menéndez et *al.* (1997c) for the ϕ-divergence test statistic $T_n^{\phi}(\widehat{\boldsymbol{p}},\boldsymbol{p}^0)$.

Let $(N_1,...,N_M)$ be a M-dimensional random vector multinomially distributed with parameters n and $\boldsymbol{p}^0$. We consider the $(M-1)$-dimensional random vector $\widetilde{\boldsymbol{w}}=(W_1,...,W_{M-1})$, defined by

$$W_j=\sqrt{n}\left(N_j/n-p_j^0\right),\ j=1,...,M-1.$$

The random vector $\widetilde{\boldsymbol{w}}$ takes values on the lattice

$$L=\left\{\widetilde{\boldsymbol{w}}=(w_1,...,w_{M-1})^T:\widetilde{\boldsymbol{w}}=\sqrt{n}\left(\boldsymbol{m}/n-\widetilde{\boldsymbol{p}}^0\right),\boldsymbol{m}\in K\right\},$$

where $\widetilde{\boldsymbol{p}}^0=(p_1^0,...,p_{M-1}^0)^T$ and K is given by

$$K=\left\{\widetilde{\boldsymbol{m}}=(n_1,...,n_{M-1})^T: n_j \text{ is a nonnegative integer},\ j=1,...,M-1 \text{ with } \textstyle\sum_{j=1}^{M-1}n_j\leq n\right\}.$$

Siatoni and Fujikoshi (1984) established, under the null hypothesis, the following asymptotic expansion for the probability mass function of the random vector $\widetilde{\boldsymbol{w}}$,

$$\Pr\left(\widetilde{\boldsymbol{W}}=\widetilde{\boldsymbol{w}}\right)=n^{-(M-1)/2}g\left(\widetilde{\boldsymbol{w}}\right)\left(1+\frac{1}{\sqrt{n}}h_1\left(\widetilde{\boldsymbol{w}}\right)+\frac{1}{n}h_2\left(\widetilde{\boldsymbol{w}}\right)+O\left(n^{-3/2}\right)\right),$$

where

- $g\left(\widetilde{\boldsymbol{w}}\right)=(2\pi)^{-(M-1)/2}\left|\boldsymbol{\Sigma}_{\widetilde{\boldsymbol{p}}^0}\right|^{-1/2}\exp\left(-\frac{1}{2}\widetilde{\boldsymbol{w}}^T\boldsymbol{\Sigma}_{\widetilde{\boldsymbol{p}}^0}^{-1}\widetilde{\boldsymbol{w}}\right)$
- $h_1\left(\widetilde{\boldsymbol{w}}\right)=-\frac{1}{2}\sum_{j=1}^{M}\frac{w_j}{p_j^0}+\frac{1}{6}\sum_{j=1}^{M}\frac{w_j^3}{(p_j^0)^2},$
- $h_2\left(\widetilde{\boldsymbol{w}}\right)=\frac{1}{2}\left(h_1\left(\widetilde{\boldsymbol{w}}\right)\right)^2+\frac{1}{12}\left(1-S\right)+\frac{1}{4}\sum_{j=1}^{M}\frac{w_j^2}{(p_j^0)^2}-\frac{1}{12}\sum_{j=1}^{M}\frac{w_j^4}{(p_j^0)^3}$

being

$$\boldsymbol{\Sigma}_{\widetilde{\boldsymbol{p}}^0}=diag\left(\widetilde{\boldsymbol{p}}^0\right)-\widetilde{\boldsymbol{p}}^0\left(\widetilde{\boldsymbol{p}}^0\right)^T \text{ and } w_M=-\sum_{j=1}^{M-1}w_j.$$

This expansion provides a local Edgeworth approximation for the probability of $\widetilde{\boldsymbol{w}}$ at each point $\widetilde{\boldsymbol{w}}$ in L. If $\widetilde{\boldsymbol{w}}$ had a continuous distribution function it would be possible to use the standard Edgeworth expansion to calculate the probability of any set B, by

$$\Pr\left(\widetilde{\boldsymbol{w}} \in B\right) = \int_B \cdots \int g\left(\widetilde{\boldsymbol{w}}\right)\left(1 + \frac{1}{\sqrt{n}} h_1\left(\widetilde{\boldsymbol{w}}\right) + \frac{1}{n} h_2\left(\widetilde{\boldsymbol{w}}\right)\right) d\,\widetilde{\boldsymbol{w}} + O\left(n^{-3/2}\right). \tag{4.27}$$

But $\widetilde{\boldsymbol{w}}$ has a lattice distribution and Yarnold (1972) indicated that in this case that expression is not valid and established for "extended convex sets" (convex set whose sections parallel to each coordinate axis are all intervals) B the following result

$$\Pr\left(\widetilde{\boldsymbol{W}} \in B\right) = J_1 + J_2 + J_3 + O\left(n^{-3/2}\right),$$

where J_1 is the Edgeworth's expansion for a continuous distribution given in (4.27), while J_2 is a term to account for the discontinuity of $\widetilde{\boldsymbol{w}}$. This term is $O\left(n^{-1/2}\right)$ and J_3 term is $O\left(n^{-1}\right)$ and it has a very complicated expression (see Siatoni and Fujikoshi (1984)).

We denote $\boldsymbol{p}^* = \left(\frac{\widetilde{\boldsymbol{w}}}{\sqrt{n}} + \widetilde{\boldsymbol{p}}^0, \widetilde{w}_M = \frac{w_M}{\sqrt{n}} + p_M^0\right)^T$, $w_M = -\sum_{i=1}^{M-1} w_i$, and we consider the extended convex set

$$B^{\phi}(b) = \left\{\widetilde{\boldsymbol{w}}\colon\ \widetilde{\boldsymbol{w}} \in L \text{ and } T_n^{\phi}(\boldsymbol{p}^*, \boldsymbol{p}^0) < b\right\}. \tag{4.28}$$

We have (see Menéndez et *al*. (1997c)), under the null hypothesis, that the distribution function of $T_n^{\phi}(\widehat{\boldsymbol{p}}, \boldsymbol{p}^0)$ can be expressed by

$$\Pr(T_n^{\phi}(\widehat{\boldsymbol{p}}, \boldsymbol{p}^0) < b) = \Pr\left(\widetilde{\boldsymbol{W}} \in B^{\phi}(b)\right) = J_1^{\phi} + J_2^{\phi} + J_3^{\phi} + O\left(n^{-3/2}\right)$$

being

$$J_1^{\phi} = \Pr(\chi^2_{M-1} < b) + \frac{1}{24n}\sum_{j=0}^{3} r_j^{\phi} \Pr\left(\chi^2_{M-1+2j} < b\right) + O\left(n^{-3/2}\right),$$

where

- $r_0^{\phi} = 2(1-S)$
- $r_1^{\phi} = 6\frac{\phi'''(1)}{\phi''(1)}(S - M^2) - 3\left(\frac{\phi^{IV}(1)}{\phi''(1)}\right)(S - 2M + 1)$
 $+ \left(\frac{\phi'''(1)}{\phi''(1)}\right)^2 (5S - 3M^2 - 6M + 4) + 3(3S - M^2 - 2M)$

$$\cdot \; r_2^{\phi} = -2\left(\frac{\phi'''(1)}{\phi''(1)}\right)(8S-6M^2-6M+4)+3\left(\frac{\phi^{IV}(1)}{\phi''(1)}\right)(S-2M+1)$$
$$- \; 2\left(\frac{\phi'''(1)}{\phi''(1)}\right)^2(5S-3M^2-6M+4)-6\left(-M^2+2S-2M+1\right)$$

and

$$\cdot \; r_3^{\phi} = \left(\frac{\phi'''(1)}{\phi''(1)}+1\right)^2(5S-3M^2-6M+4).$$

J_2^{ϕ} is a discontinuous $O\left(n^{-1}\right)$ term to account for the discontinuity in $\widetilde{\boldsymbol{W}}$ and can be approximated to first order by

$$\widehat{J}_2^{\phi} = \left(N^{\phi}(b)-n^{(M-1)/2}V^{\phi}(b)\right)e^{-b/2}\Bigg/\left((2\pi n)^{(M-1)}\prod_{j=1}^{M}p_j^0\right)^{1/2},$$

where $N^{\phi}(b)$ is the number of lattice points in $B^{\phi}(b)$, and

$$V^{\phi}(b) = \frac{(\pi b)^{(M-1)/2}}{\Gamma((M+1)/2)}\left(\prod_{i=1}^{M}p_j^0\right)^{1/2}\left(1+\frac{b}{24(M+1)n}(l_1-l_2)\right),$$

where

$$l_1 = \left(\frac{\phi'''(1)}{\phi''(1)}\right)^2(5S-3M^2-6M+4) \text{ and } l_2 = 3\left(\frac{\phi^{IV}(1)}{\phi''(1)}\right)(S-2M+1),$$

being $S=\sum_{j=1}^{M}\frac{1}{p_j^0}$. This term $\widehat{J}_2^{\phi}$ does not look so complicated but it is very hard to obtain it when n and M are not small, because of the difficulty for getting $V^{\phi}(b)$.

Finally, $J_3^{\phi}=O\left(n^{-1}\right)$ and its expression is too complicated. Neglecting the J_3^{ϕ} term Menéndez et *al.* (1997c) proposed, in the same way as Read (1984a), the approximation

$$\Pr(T_n^{\phi}(\widehat{\boldsymbol{p}},\boldsymbol{p}^0)<b)\simeq J_1^{\phi}+\widehat{J}_2^{\phi}.$$

The proof of this result can be seen in Menéndez et *al.* (1997c). A similar result has been established in Taneichi et *al.* (2001a, b) under contiguous alternative hypotheses.

Read (1984a) studied the usefulness of this approximation to the exact distribution for the power-divergence test statistics and established that it is externally close to the exact distribution for n so small as 10, and furthermore it provides a substantial improvement over the chi-square first order approximation.

Example 4.4

For the family of divergence measures given in (4.19), we have for $p_j^0 = 1/M$, $j = 1, ..., M$

$$F_{Ed}^{\phi_a}(b) \equiv \Pr(T_n^{\phi}(\widehat{\boldsymbol{p}}, \boldsymbol{p}^0) < b) \simeq J_1^a + \widehat{J}_2^a$$

being

$$J_1^a = \Pr(\chi_{M-1}^2 < b) + \frac{1}{24n} \sum_{j=0}^{3} r_j^a \Pr\left(\chi_{M-1+2j}^2 < b\right) + O\left(n^{-3/2}\right),$$

where

- $r_0^a = \ 2(1 - M^2)$
- $r_1^a = \ 6M(M-1)(1-3(1-a)^2$
- $r_2^a = \ 12(1-a)(M^2-3M+2) + 36(1-a)^2(M-1) - 6(M-1)^2$

and

- $r_3^a = \ 2(3a-2)^2(M^2-3M+2).$

The term $\widehat{J}_2^a$ *has the expression*

$$\widehat{J}_2^a = \left(N^a(b) - n^{(M-1)/2} V^a(b)\right) e^{-b/2} \Big/ \left((2\pi n)^{(M-1)} (1/M)^M\right)^{1/2}$$

where $N^a(b)$ *is the number of lattice points in* $B^{\phi_a}(b)$, $B^{\phi_a}(b)$ *is defined in the same way as* $B^{\phi}(b)$ *in (4.28) by replacing* ϕ *by* ϕ_a, *and*

$$\begin{aligned} V^a(b) = & \ \tfrac{(\pi b)^{(M-1)/2}}{\Gamma((M+1)/2)} \left(\tfrac{1}{M}\right)^{M/2} \left(1 + \tfrac{3b}{4(M+1)n}(1-a)^2(M-M^2)\right) \\ + & \ O(n^{-3/2}). \end{aligned}$$

4.5. Exact Powers Based on Exact Critical Regions

We consider a function ϕ, some specified alternative hypotheses and a significance level α and we are going to get the exact power function, using the exact critical value (without any reference to asymptotic) for the ϕ-divergence test statistic $T_n^{\phi}(\widehat{\boldsymbol{p}}, \boldsymbol{p}^0)$. We shall use this exact power function to compare the members of the family of ϕ-divergence test statistics $T_n^{\phi}(\widehat{\boldsymbol{p}}, \boldsymbol{p}^0)$. This approach is similar to one used by West and Kempthorne (1972) and Haber (1984) to compare the power of the chi-square test statistic and likelihood ratio test statistic, by Read (1984b) to compare the power-divergence test statistic and by Pardo, M. C. and Pardo, J. A. (1999) to compare the family of Rukhin test statistics given in (4.20). We restrict our attention to the equiprobable null model $H_0 : \boldsymbol{p}^0 = (1/M, ..., 1/M)^T$ and we consider alternative models where one of the M probabilities is perturbed, and the rest are adjusted so that they still sum to 1,

$$H_1 : p_i = \begin{cases} \dfrac{M-1-\delta}{M(M-1)} & \text{if } i = 1, ..., M-1 \\ \dfrac{1+\delta}{M} & \text{if } i = M \end{cases} \tag{4.29}$$

where $-1 \leq \delta \leq M - 1$.

We are going to justify a little bit the equiprobable null model. Sturges (1926) initiated the study of the choice of cell and recommended that the cell would be chosen to have equal probabilities with $M = 1 + 2.303 \log_{10} n$. Mann and Wald (1942) for a sample size n recommended $M = 4\left(2n^2 z_\alpha^{-1}\right)^{1/5}$ where z_α denotes the $100(1-\alpha)$ percentile of the standard normal distribution. Schorr (1974) confirmed that the "optimum" M is smaller than the value given by Mann and Wald and he suggested using $M = 2n^{2/5}$. Greenwood and Nikulin (1996) suggested using $M \leq \min(1/\alpha, \log n)$. Cohen and Sackrowitz (1975) proved that the tests which lead to reject the above hypothesis if $\sum_{i=1}^{M} h_i(x_i) > c$, where c is positive, h_i, $i = 1, ..., M$, are convex functions and $x_i \geq 0$, $i = 1, ..., M$, are unbiased for equal cell probabilities. Bednarski and Ledwina (1978) stated that if tests of fit are based on continuous functions, then in general they are biased for testing an arbitrary simple hypothesis.

In order to get the exact power for each ϕ-divergence test statistic $T_n^\phi(\widehat{\boldsymbol{p}}, \boldsymbol{p}^0)$, it is necessary to choose a significance level α and calculate the associated critical region. To do this, we will see first the way to get exact $100(1-\alpha)$ percentiles, $t_{n,M,\alpha}^\phi$, corresponding to the exact distribution of $T_n^\phi(\widehat{\boldsymbol{p}}, \boldsymbol{p}^0)$.

The distribution function of $T_n^\phi(\widehat{\boldsymbol{p}}, \boldsymbol{p}^0)$ under the null hypothesis $H_0 : \boldsymbol{p} = \boldsymbol{p}^0$ is

$$F_{T_n^\phi(\widehat{\boldsymbol{p}}, \boldsymbol{p}^0)}(t) = \Pr_{\boldsymbol{p}^0}\left(T_n^\phi(\widehat{\boldsymbol{p}}, \boldsymbol{p}^0) \leq t\right) = 1 - \Pr_{\boldsymbol{p}^0}\left(T_n^\phi(\widehat{\boldsymbol{p}}, \boldsymbol{p}^0) > t\right),$$

where

$$\Pr_{\boldsymbol{p}^0}\left(T_n^{\phi}(\widehat{\boldsymbol{p}},\boldsymbol{p}^0) > t\right) = \sum_{(n_1,\ldots,n_M)\in A_{M,t}^n} \Pr_{\boldsymbol{p}^0}\left(N_1 = n_1,\ldots,N_M = n_M\right),$$

$$A_{M,t}^n = \left\{(n_1,\ldots,n_M) \in (N \cup \{0\})^M / n_1 + \ldots + n_M = n, T_n^{\phi}(\widehat{\boldsymbol{p}},\boldsymbol{p}^0) > t\right\}$$

and

$$\Pr_{\boldsymbol{p}^0}\left(N_1 = n_1,\ldots,N_M = n_M\right) = \frac{n!}{n_1!\ldots n_M!}(p_1^0)^{n_1}\ldots(p_M^0)^{n_M}.$$

The set of upper tail probabilities of $T_n^{\phi}(\widehat{\boldsymbol{p}},\boldsymbol{p}^0)$ is

$$\mathcal{U}_{n,M}^{\phi} = \left\{\alpha \in (0,1) : \exists t > 0 \text{ with } \Pr_{\boldsymbol{p}^0}\left(T_n^{\phi}(\widehat{\boldsymbol{p}},\boldsymbol{p}^0) > t\right) = \alpha\right\}.$$

$100(1-\alpha)$ percentiles $t_{n,M,\alpha}^{\phi}$ of $T_n^{\phi}(\widehat{\boldsymbol{p}},\boldsymbol{p}^0)$ are obtained for any $\alpha \in \mathcal{U}_{n,m}^{\phi}$ through the equation $\alpha = \Pr_{\boldsymbol{p}^0}\left(T_n^{\phi}(\widehat{\boldsymbol{p}},\boldsymbol{p}^0) > t_{n,M,\alpha}^{\phi}\right)$. In general it will not be possible to get an exact percentile and therefore we shall consider an approximate $100(1-\alpha)$ percentile that we are going to define. If $\alpha \in (0,1) - \mathcal{U}_{n,M}^{\phi}$, we consider

$$\alpha_1 = \max\left\{\alpha_0 \in (0,\alpha] : \exists t > 0 \text{ with } \Pr_{\boldsymbol{p}^0}\left(T_n^{\phi}(\widehat{\boldsymbol{p}},\boldsymbol{p}^0) > t\right) = \alpha_0\right\},$$

so that t_{n,M,α_1}^{ϕ} is defined as the approximate $100(1-\alpha)$ percentile. We calculate the approximate percentiles for α, M, n and ϕ all fixed. This process can be divided into four steps:

1. Generate all the elements $\boldsymbol{x}_M = (n_1,\ldots,n_M)$ of

   $$A_M^n = \left\{(n_1,\ldots,n_M) \in (N \cup \{0\})^M / n_1 + \ldots + n_M = n\right\}$$

   and calculate the corresponding probabilities $\Pr_{\boldsymbol{p}^0}(N_1 = n_1,\ldots,N_M = n_M)$.

2. For each $\boldsymbol{x}_M \in A_M^n$, calculate the test statistics $T_n^{\phi}(\widehat{\boldsymbol{p}},\boldsymbol{p}^0)$.

3. Put $T_n^{\phi}(\widehat{\boldsymbol{p}},\boldsymbol{p}^0)$ and $\Pr_{\boldsymbol{p}^0}(N_1 = n_1,\ldots,N_M = n_M)$ in increasing order with respect to the values of $T_n^{\phi}(\widehat{\boldsymbol{p}},\boldsymbol{p}^0)$.

4. Calculate the approximate $100(1-\alpha)$ percentile t_{n,M,α_1}^{ϕ}.

We use randomized tests in order to decide with probability $\gamma^{\phi}_{n,M,\alpha}$ the rejection of the hypothesis $H_0 : \boldsymbol{p}^0 = (1/M, ..., 1/M)^T$ when the test statistic takes on the value t^{ϕ}_{n,M,α_1}. Let $\varphi\left(T^{\phi}_n(\widehat{\boldsymbol{p}}, \boldsymbol{p}^0)\right)$ be a function giving the probability of rejecting H_0 when the ϕ-divergence test statistic $T^{\phi}_n(\widehat{\boldsymbol{p}}, \boldsymbol{p}^0)$ is observed. This function is defined by the formula

$$\varphi\left(T^{\phi}_n(\widehat{\boldsymbol{p}}, \boldsymbol{p}^0)\right) = \begin{cases} 1 & \text{if } T^{\phi}_n(\widehat{\boldsymbol{p}}, \boldsymbol{p}^0) > t^{\phi}_{n,M,\alpha_1} \\ \gamma^{\phi}_{n,M,\alpha} & \text{if } T^{\phi}_n(\widehat{\boldsymbol{p}}, \boldsymbol{p}^0) = t^{\phi}_{n,M,\alpha_1} \\ 0 & \text{if } T^{\phi}_n(\widehat{\boldsymbol{p}}, \boldsymbol{p}^0) < t^{\phi}_{n,M,\alpha_1} \end{cases}. \tag{4.30}$$

We have

$$\begin{aligned} \alpha = & \; E_{\boldsymbol{p}^0}\left[\varphi\left(T^{\phi}_n(\widehat{\boldsymbol{p}}, \boldsymbol{p}^0)\right)\right] \\ = & \; 1 \times \Pr_{\boldsymbol{p}^0}\left(T^{\phi}_n(\widehat{\boldsymbol{p}}, \boldsymbol{p}^0) > t^{\phi}_{n,M,\alpha_1}\right) + \gamma^{\phi}_{n,M,\alpha} \Pr_{\boldsymbol{p}^0}\left(T^{\phi}_n(\widehat{\boldsymbol{p}}, \boldsymbol{p}^0) = t^{\phi}_{n,M,\alpha_1}\right), \end{aligned}$$

and

$$\gamma^{\phi}_{n,M,\alpha} = \frac{\alpha - \Pr\left(T^{\phi}_n(\widehat{\boldsymbol{p}}, \boldsymbol{p}^0) > t^{\phi}_{n,M,\alpha_1}\right)}{\Pr\left(T^{\phi}_n(\widehat{\boldsymbol{p}}, \boldsymbol{p}^0) = t^{\phi}_{n,M,\alpha_1}\right)}.$$

Let us consider $\boldsymbol{p} = (p_1, ..., p_M)^T$ with p_i given in (4.29). The exact power function of the test $\varphi\left(T^{\phi}_n(\widehat{\boldsymbol{p}}, \boldsymbol{p}^0)\right)$, defined in (4.30), at $\boldsymbol{p}$ is

$$\begin{aligned} \beta_{\phi,n}(\boldsymbol{p}) = & \; E_{\boldsymbol{p}}\left[\varphi\left(T^{\phi}_n(\widehat{\boldsymbol{p}}, \boldsymbol{p}^0)\right)\right] \\ = & \; \Pr_{\boldsymbol{p}}\left(T^{\phi}_n(\widehat{\boldsymbol{p}}, \boldsymbol{p}^0) > t^{\phi}_{n,M,\alpha_1}\right) + \gamma^{\phi}_{n,M,\alpha} \Pr_{\boldsymbol{p}}\left(T^{\phi}_n(\widehat{\boldsymbol{p}}, \boldsymbol{p}^0) = t^{\phi}_{n,M,\alpha_1}\right). \end{aligned}$$

A nice and extensive study about this problem can be seen in Marhuenda (2003). Now we present a practical example using the family of divergence measures given in (4.19). The study corresponding to the power-divergence test statistic can be seen in Cressie and Read (1988).

Example 4.5

We consider the family of Rukhin test statistics, $T^{\phi_a}_n(\widehat{\boldsymbol{p}}, \boldsymbol{p}^0)$, given in (4.20). Table 4.5 presents exact powers for the randomized test (4.30) based on $\alpha = 0.05$ for different values of δ and "a", $n = 10$, 20 and $M = 5$. For $\delta < 0$ the power decreases as "a" increases. For $\delta > 0$ the reverse occurs. For alternatives with $\delta > 0$ we should choose "a" as large as other restrictions will allow in order to obtain the best power. For alternatives with $\delta < 0$ we should choose "a" as *small as possible. If we wish to choose a test with reasonable power against these alternatives, for every δ value, we should choose $a \in [0.6, 0.7]$. The reason is that there is a marked reduction in power as "a" moves from 0.7 to 1 and a marked increase in power as "a" moves from* 0 *to* 0.6.

	$(n = 10, M = 5)$						$(n = 20, M = 5)$				
	δ						δ				
a	-0.9	-0.5	0.5	1	1.5	a	-.9	-.5	.5	1	1.5
0	.1670	.0819	.0707	.1392	.2600	0	.5903	.1281	.0784	.1470	.2573
.1	.1637	.0805	.0719	.1412	.2617	.1	.5853	.1272	.0787	.1473	.2574
.2	.1620	.0797	.0728	.1426	.2630	.2	.5658	.1239	.0794	.1521	.2836
.3	.1670	.0818	.0711	.1400	.2607	.3	.5688	.1269	.0840	.1864	.3853
.4	.1517	.0766	.0755	.1570	.3053	.4	.5264	.1265	.0942	.2453	.5159
.5	.1517	.0766	.0755	.1570	.3053	.5	.4485	.1224	.1023	.2770	.5627
.6	.1516	.0774	.0797	.1749	.3417	.6	.3842	.1182	.1105	.3165	.6245
2/3	.1429	.0772	.0753	.1677	.3353	2/3	.3785	.1177	.1121	.3241	.6366
.7	.1367	.0758	.0833	.1977	.4056	.7	.3738	.1170	.1121	.3242	.6367
.8	.1313	.0746	.0844	.2028	.4056	.8	.3218	.1121	.1186	.3551	.6805
.9	.1372	.0767	.0833	.2009	.4032	.9	.2853	.1087	.1205	.3627	.6891
1	.1247	.0737	.0858	.2099	.4212	1	.2731	.1068	.1229	.3722	.7004

Table 4.5. Exact power functions for the randomized size $\alpha = 0.05$ test of the symmetric hypothesis.

4.6. Small Sample Comparisons for the Phi-divergence Test Statistics

The literature contains many simulation studies concerning the accuracy of using the chi-square distribution tail function $F_{\chi^2_{M-1}}$ as an approximation to $F_{T_n^{\phi}(\widehat{\boldsymbol{p}},\boldsymbol{p}^0)}$ for the chi-square test statistic and the likelihood ratio test statistic G^2 (e.g., Good et *al.* 1970; Roscoe and Byars, 1971; Tate and Hyer, 1973; Margolin and Light, 1974; Radlow and Alf, 1975; Chapman, 1976; Larntz, 1978; Kotze and Gokhale, 1980; Lawal, 1984; Kallenberg et *al.* 1985; Hosmane, 1986; Koehler, 1986; Rudas, 1986). Much of what follows in this section generalizes these studies. Two criteria used by Read (1984b) for comparing the family of power-divergence test statistics are proposed here for small n. A study in relation to Lin test statistic can be seen in Menéndez et *al.* (1997d) and in relation with Rukhin test statistic in Pardo, M. C. and Pardo, J. A. (1999).

We have obtained four different asymptotic approximations for the exact distribution of the test statistic $T_n^{\phi}(\widehat{\boldsymbol{p}}, \boldsymbol{p}^0)$. We shall assume that $\boldsymbol{p}^0 = (\frac{1}{M}, ..., \frac{1}{M})^T$. The first one and the second one were obtained in Chapter 3. The first one is derived under the assumption that the number of classes is finite, $F_{\chi^2_{M-1}}(b) = \Pr\left(\chi^2_{M-1} \leq b\right)$ and the second one, $F_N(b) = \Pr\left(N(0,1) \leq (b - \mu_{\phi,n})/\sigma_{\phi,n}\right)$, when the number of classes increases to infinity. We must remark that in this case

$$\frac{\sqrt{n}}{\sigma_{\phi,n}}\left(D_n^{\phi}(\widehat{\boldsymbol{p}}, \boldsymbol{p}_0) - \mu_{\phi,n}\right) \xrightarrow[n\to\infty]{L} N(0,1)$$

where $\mu_{\phi,n}$ and $\sigma_{\phi,n}$ are defined in Proposition 3.2. In this Chapter we have presented two other approaches. In Section 2 we have considered

$$F_C(b) = \Pr\left(\chi^2_{M-1} \leq (b - \gamma_\phi)/\delta_\phi^{1/2}\right)$$

and finally in Section 3 we have considered $F_{Ed}(b) \simeq J_1^{\phi} + \widehat{J}_2^{\phi}$.

Two criteria, considered by Cressie and Read (1988), are presented to compare these four asymptotic approximations for small n. Criterion 1 consists of recording the maximum approximation error incurred by each of the four approximations to the exact distribution, $F_{T_n^{\phi}(\widehat{\boldsymbol{p}},\boldsymbol{p}^0)}(b)$, of the test statistic $T_n^{\phi}(\widehat{\boldsymbol{p}}, \boldsymbol{p}^0)$. We calculate

$$\max_{\widehat{\boldsymbol{p}}}\left|F_{T_n^{\phi}(\widehat{\boldsymbol{p}},\boldsymbol{p}^0)}\left(T_n^{\phi}(\widehat{\boldsymbol{p}}, \boldsymbol{p}^0)\right) - F_i\left(T_n^{\phi}(\widehat{\boldsymbol{p}}, \boldsymbol{p}^0)\right)\right|,$$

for ϕ fixed and $i = \chi^2_{M-1}$, N, C, Ed. The sign associated with the maximum difference is also recorded. So, we know if the maximum error is an overestimate or an underestimate.

Criterion 2 consists of assessing the accuracy for the approximation in calculating the size of a test. We use the standard approximation $F_{\chi^2_{M-1}}$ to give a test with approximate significance level α. We choose c_α such that $1 - F_{\chi^2_{M-1}}(c_\alpha) = \alpha$, i.e., $c_\alpha = \chi^2_{M-1,\alpha}$. We calculate $1 - F_i(\chi^2_{M-1,\alpha})$, $i = T_n^{\phi}(\widehat{\boldsymbol{p}}, \boldsymbol{p}^0)$, N, C, Ed, and assess how they vary for different functions ϕ. There are two reasons to take the critical c_α to be the $(1-\alpha)$ percentile of a chi-square distribution with $M-1$ degrees of freedom. On the one hand, this is the most commonly used for the tests based on Pearson and likelihood ratio test. On the other hand, the critical region obtained from this approximation is independent of ϕ.

In the next example we present a practical study based on the divergence measures given in (4.19).

Example 4.6

We consider the family of Rukhin test statistics, $T_n^{\phi_a}(\widehat{\mathbf{p}},\mathbf{p}^0)$, *with* $\mathbf{p}^0 = (1/M, ..., 1/M)^T$, *given in (4.20). Figures 4.1 and 4.2 illustrate the maximum approximation error resulting from using the approximations* $F_{\chi^2_{M-1}}$, F_C, F_{Ed} *and* F_N *for the exact distribution of* $T_n^{\phi}(\widehat{\mathbf{p}},\mathbf{p}^0)$ *which are labelled Apr1, Apr2, Apr3 and Apr4 respectively on the graphs. The results are illustrated for specific values of "a" in the range [0,1], number of classes* $M = 5$ *and sample sizes* $n = 10$ *and* $n = 20$. *The approach* F_N *is obtained in Exercise 1. At first glance, it is clear that the optimal parameter values are between 0.6 and 1 for the cases considered here. As n increases from 10 to 20, the error curves flatten over this range; furthermore, the size of the maximum errors decreases overall except for Apr4. But this behavior of the approximation* F_N *is not surprising when one recalls that it relies on M increasing with n. In general the maximum error associated with the approximations* $F_{\chi^2_{M-1}}$, F_C *and* F_{Ed} *can be seen to be negative and of a similar order. However the normal approximation* F_N *has larger maximum error than the others and of the opposite sign. Note that* F_N *is not defined for* $a = 0$ *for the reason given in Exercise 1. Secondly, we assess the accuracy of the approximation in calculating the size of a test. We use the standard approximation* $F_{\chi^2_{M-1}}$ *to give a test with approximate significance level* α, *i.e., choose* c_α *such that* $1 - F_{\chi^2_{M-1}}(c_\alpha) = \alpha$, *i.e.,* $c_\alpha = \chi^2_{M-1,\alpha}$. *We calculate* $1 - F_i(\chi^2_{M-1,\alpha})$, $i = T_n^{\phi_a}(\widehat{\mathbf{p}},\mathbf{p}^0)$, *N, C, Ed, and assess how they vary for different values of "a". The results are illustrated in Figures 4.3 and 4.4 again for the specific values of "a", sample sizes and number of classes used for criterion 1, and* $\alpha = 0.1$. *There is a close agreement between the exact and nominal levels obtained for* F_C *and* F_{Ed} *than for* $F_{\chi^2_{M-1}}$. *In general, the normal approximation is clearly poor in comparison to the other two approximations and tends to overestimate the true level for all "a" values considered. But this result must be expected because this approximation is an asymptotic result in n and M. Due to the value of reference the chi-square approximation, in the pictures, is represented by a line in* 0.1. *The exact and nominal levels for the approximation* $F_{\chi^2_{M-1}}$ *are quite similar for all* $a \in [0.6, 1]$ *and it is not too similar outside this interval. If we want to use the approximation given by* $F_{\chi^2_{M-1}}$ *we should use* $a \in [0.6, 1]$ *for small M. From the result obtained in relation with the criterion based on the speed of the convergence of the exact moment, when M increases, the optimal values were* $a = 2/3$ *and* $a = 1$. *This suggests that the range* $a \in [0.6, 1]$ *is also optimum for big values of M. If we wish to consider values outside the interval* $[0.6, 1]$ *the approximation* *based on F_C is a good alternative to the approximation F_{Ed} since the approximation F_C is easier to get than the approximation F_{Ed}.*

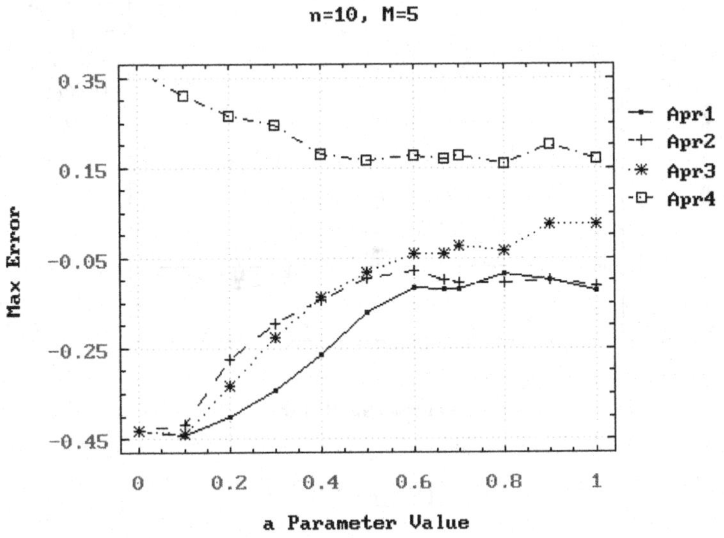

Figure 4.1

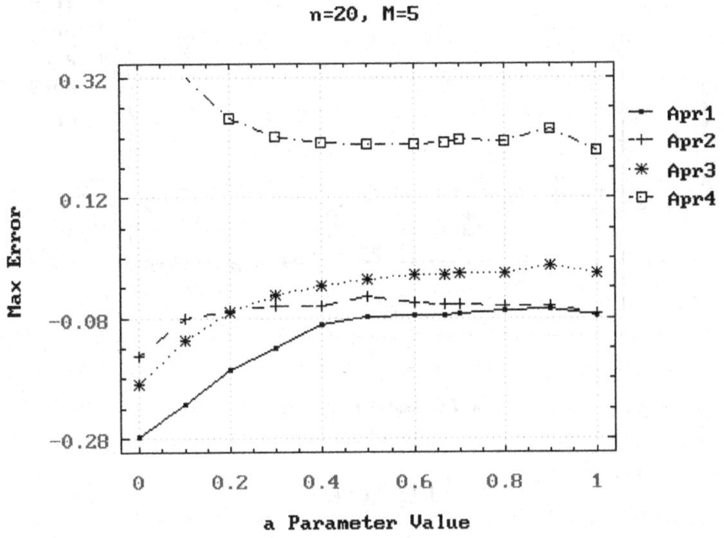

Figure 4.2

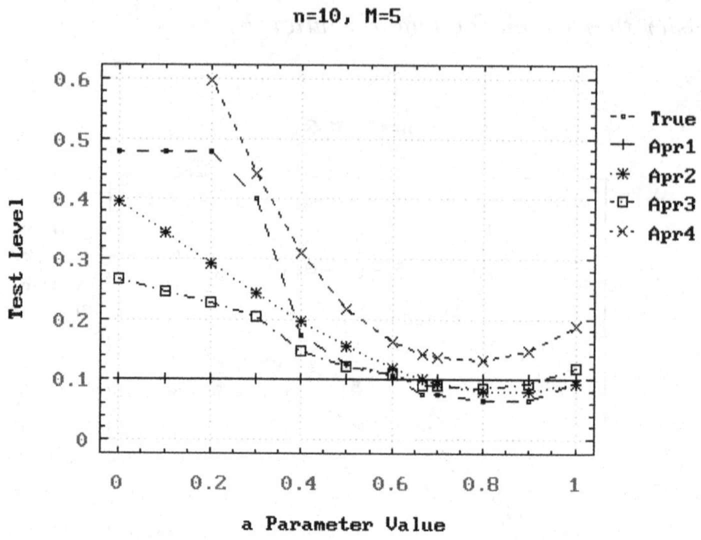

Figure 4.3

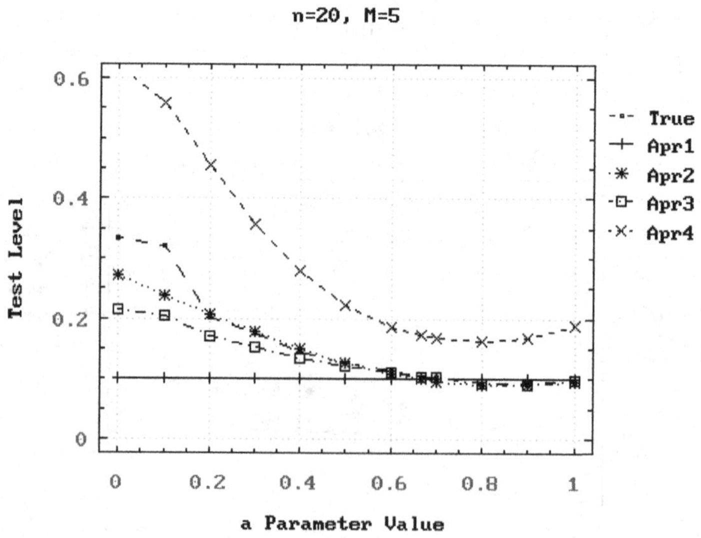

Figure 4.4

4.7. Exercises

1. Consider the family of divergence measures, $D_n^{\phi_a}(\widehat{\boldsymbol{p}}, \boldsymbol{p}^0)$, where ϕ_a is given in (4.19) and $\boldsymbol{p}^0 = (1/M, ..., 1/M)^T$. Find the asymptotic distribution obtained in Proposition 3.2 when the number of classes M_n verifies

$$\lim_{n\to\infty} \frac{n}{M_n} = \gamma \in (0, \infty).$$

2. Consider the family of test statistics associated with Lin divergence measure (Lin, 1991) defined by

$$\phi_a(x) = \frac{1}{a(1-a)} (ax \log x - (ax + (1-a)) \log (ax + 1 - a)), \ a \in (0,1),$$

and

$$\phi_0(x) = \lim_{a\to 0} \phi_a(x) = x \log x - x + 1, \ \phi_1(x) = \lim_{a\to 1} \phi_a(x) = -\log x - 1 + x.$$

Is there some optimum value of "a" in goodness-of-fit according to the moment criterion given in (4.18)?

3. Consider the family of test statistics associated with the "harmonic mean divergence" given by

$$\phi_r(x) = (1-x)/2 - 2^{1/r}\left(1 + x^{-r}\right)^{-1/r}, \ r > 0.$$

Find the optimum values of "r" in goodness-of-fit according to the moment criterion given in (4.18).

4. Consider the family of divergence measures given in Exercise 2 and the following alternatives

$$H_1 : \pi_i = \begin{cases} \dfrac{M-1-\delta}{M(M-1)} & \text{if } i = 1, ..., M-1 \\ \dfrac{1+\delta}{M} & \text{if } i = M \end{cases}.$$

Find the exact power for $\alpha = 0.05$, $n = 20$ and $M = 5$.

5. Let $\phi : (0, \infty) \to \mathbb{R}$ be a concave function with continuous fourth derivative verifying $\phi''(1/M) < 0$. Find what conditions the function ϕ must verify to be optimum the statistic $S_n^{\phi}(\widehat{\boldsymbol{p}}, \boldsymbol{p}^0)$, given in Exercise 6 of Chapter 3, according to the moment criterion with $\boldsymbol{p}^0 = (1/M, ..., 1/M)^T$.

6. Consider the family of R_ϕ-divergences given by

$$\phi_a(x) = \begin{cases} \frac{1}{1-a}\left(x^a - x\right) & if \quad a \neq 1 \\ -x \log x & if \quad a = 1 \end{cases}.$$

Find the values of "a" for which the test statistic $S_n^{\phi_a}(\widehat{\boldsymbol{p}}, \boldsymbol{p}^0)$, $\boldsymbol{p}^0 = (1/M, ..., 1/M)^T$ is optimum according to the moment criterion as $M \to \infty$.

7. Consider the power-divergence family of test statistics. Find the expression of the corrected test statistic given in Subsection 4.3.3 of this chapter.

8. Consider the family of test statistics given in Exercise 2. Find the expression of the corrected test statistic given in Subsection 4.3.3 of this chapter, with $\boldsymbol{p}^0 = (1/M, ..., 1/M)^T$.

9. Consider the test statistic $S_n^{\phi}(\widehat{\boldsymbol{p}}, \boldsymbol{p}^0)$, with $\boldsymbol{p}^0 = (1/M, ..., 1/M)^T$. Using the results given in Exercise 5, find the expression of the corrected test statistic given in Subsection 4.3.3 of this chapter.

10. Find the expression of the approximation based on Edgeworth's expansion associated with Lin test statistic in the case of equiprobable hypothesis.

11. Find the expression of the approximation based on Edgeworth's expansion associated with Pearson-test statistic.

12. Find the expression of the approximation based on Edgeworth's expansion associated with the power-divergence test statistics for $\lambda \neq -1$ and $\lambda \neq 0$.

4.8. Answers to Exercises

1. We denote by $D_n^{\phi_a}(\widehat{\boldsymbol{p}}, \boldsymbol{p}^0)$ the family of divergence measures obtained with the functions defined in (4.19). By Proposition 3.2 we have

$$\frac{\sqrt{n}}{\sigma_{n,a}}\left(D_n^{\phi_a}(\widehat{\boldsymbol{p}}, \boldsymbol{p}^0) - \mu_{n,a}\right) \xrightarrow[n\to\infty]{L} N(0,1)$$

where

$$\mu_{n,a} = E\left[\phi_a\left(\frac{Z_n}{\gamma_n}\right)\right] = E\left[\frac{(\gamma_n - Z_n)^2}{2\gamma_n(\gamma_n a + (1-a)Z_n)}\right], \quad a \in [0,1]$$

and

$$\sigma_{n,a}^2 = \gamma_n Var\left[\frac{(\gamma_n - Z_n)^2}{2\gamma_n(\gamma_n a + (1-a)Z_n)}\right] - Cov^2\left[Z_n, \frac{(\gamma_n - Z_n)^2}{2\gamma_n(\gamma_n a + (1-a)Z_n)}\right],$$

$a \in [0,1]$. Therefore,

$$\sqrt{n}\left(D_n^{\phi_a}(\widehat{\mathbf{p}}, \mathbf{p}^0) - \mu_a\right) \underset{n\to\infty}{\overset{L}{\longrightarrow}} N\left(0, \sigma_a^2\right)$$

where

$$\mu_a = E\left[\frac{(\gamma - Z)^2}{2\gamma\left(\gamma a + (1-a)Z\right)}\right]$$

and

$$\sigma_a^2 = \gamma Var\left[\frac{(\gamma - Z)^2}{2\gamma\left(\gamma a + (1-a)Z\right)}\right] - Cov^2\left[Z, \frac{(\gamma - Z)^2}{2\gamma\left(\gamma a + (1-a)Z\right)}\right], \ a \in [0,1]$$

where Z is a Poisson random variable with parameter γ.

Finally, under the null hypothesis of equiprobability we have

$$F_N(b) = \Pr(N(0,1) \leq (b - \mu_a)/\sigma_a).$$

We can observe that μ_0 and σ_0^2 do not exist because $E\left[Z^{-1}\right]$ does not exist. For this reason this approximation is valid only for $a \in (0,1]$.

2. We can observe that for $a \to 0$ we have the likelihood ratio test statistic and for $a \to 1$ the modified likelihood ratio test statistic. It is clear that for $a = 0$ and $a = 1$, the equation (4.18) does not hold. Now we consider $a \in (0,1)$. We have $\phi_a'''(1) = -(1+a)$ and $\phi_a^{IV}(1) = 2(a^2+a+1)$ and the equation (4.18) becomes

$$-4(1+a) + 6\left(a^2 + a + 1\right) = 0.$$

This equation does not have any real-valued solution for $a \in (0,1)$, so there is not an optimum value according to this criterion.

3. It is immediate to check that $\phi_r'''(1) = -3/8 - 3/8r$ and $\phi_r^{IV}(1) = 15/16 + 7/8r - 3/16r^2 - 1/8r^3$, then the equation (4.18) becomes

$$4\left(-\frac{3}{8} - \frac{3}{8}r\right) + 3\left(\frac{15}{16} + \frac{7}{8}r - \frac{3}{16}r^2 - \frac{1}{8}r^3\right) = 0.$$

The positive root of the equation (4.18) is $r = -\frac{1}{4} + \frac{1}{4}\sqrt{57}$.

4. Using the procedure given in Section 4.4, we have computed the exact powers for the randomized test (4.30) with $\alpha = 0.05$ and for different values of δ and a. Taking $M = 5$ and $n = 10$, we have obtained the powers

	$(n = 20, M = 5)$				
	δ				
a	$-.9$	$-.5$	$.5$	1	1.5
0	.4463	.1213	.1076	.3056	.6103
.1	.4841	.1244	.1012	.2765	.5678
.2	.4948	.1252	.1001	.2718	.5588
.3	.5260	.1267	.0953	.2521	.5279
.4	.5336	.1268	.0933	.2424	.5128
.5	.5627	.1281	.0886	.2205	.4705
.6	.5689	.1272	.0849	.1971	.4222
.7	.5670	.1251	.0821	.1759	.3646
.8	.5844	.1282	.0800	.1640	.3361
.9	.5849	.1270	.0790	.1503	.2792
1	.5803	.1262	.0789	.1474	.2574

For alternatives with $\delta > 0$, we must choose "a" as larger as other restrictions permit getting the best power. On the contrary, for alternatives $\delta < 0$ we must choose "a" as smaller as possible. Given a fixed "a" the power rises as $|\delta|$ increases. If we want to choose a test with a reasonable power against the given alternatives for any value of δ, we should choose $a \in [0.3, 0.6]$.

5. Using the procedure given for the family of ϕ-divergences we have

$$\begin{aligned} E\left[S_n^{\phi}(\widehat{\mathbf{p}},\mathbf{p}^0)\right] &= M-1+\frac{1}{n}f_\phi^1+O\left(n^{-3/2}\right) \\ E\left[\left(S_n^{\phi}(\widehat{\mathbf{p}},\mathbf{p}^0)\right)^2\right] &= M^2-1+\frac{1}{n}f_\phi^2+O\left(n^{-3/2}\right) \\ E\left[\left(S_n^{\phi}(\widehat{\mathbf{p}},\mathbf{p}^0)\right)^3\right] &= M^3+3M^2-M-3+\frac{1}{n}f_\phi^3+O\left(n^{-3/2}\right) \end{aligned}$$

where

$$f_\phi^1 = \tfrac{\phi'''(1/M)}{2\phi''(1/M)}\left(\tfrac{2}{M}-3+M\right)+\tfrac{7\phi^{IV}(1/M)}{16\phi''(1/M)}\left(\tfrac{1}{M^2}-\tfrac{2}{M}+1\right)$$

$$\begin{aligned} f_\phi^2 &= -2M+2+\left(\tfrac{10}{M}-13+2M+M^2\right)\tfrac{\phi'''(1/M)}{\phi''(1/M)} \\ &+ \tfrac{1}{4}\left(\tfrac{\phi'''(1/M)}{\phi''(1/M)}\right)^2\left(\tfrac{12}{M^2}-\tfrac{18}{M}+6\right) \\ &+ \tfrac{7\phi^{IV}(1/M)}{8\phi''(1/M)}\left(\tfrac{3}{M^2}-\tfrac{5}{M}+1+M\right) \end{aligned}$$

and

$$\begin{aligned} f_\phi^3 &= 26 - 24M - 2M^2 + \tfrac{\phi'''(1/M)}{2\phi''(1/M)}\left(\tfrac{210}{M} - 243 + 3M + 27M^2 + 3M^3\right) \\ &+ \tfrac{7\phi^{IV}(1/M)}{16\phi''(1/M)}\left(\tfrac{45}{M^2} - \tfrac{66}{M} + 18M + 3M^2\right) \\ &+ \tfrac{1}{4}\left(\tfrac{\phi'''(1/M)}{\phi''(1/M)}\right)^2\left(\tfrac{180}{M^2} - \tfrac{234}{M} + 36 + 18M\right). \end{aligned}$$

Therefore, the functions ϕ under which the asymptotic moments are closer to the exact moments, for fixed M, are those for which it holds

$$f_\phi^i = 0, \ i = 1, 2, 3.$$

This happens because the second order expansions of the first three moments of S_n^ϕ are the same as the first three moments of a chi-square distribution χ^2_{M-1} plus the correction factor of order $O\left(n^{-1}\right)$, f_ϕ^i, $i = 1, 2, 3$, respectively.

6. We have to solve the equations $f_\phi^i = 0$, $i = 1, 2, 3$ given in the previous exercise for the given function ϕ_a as $M \to \infty$.

   Solving the first equation,

$$\begin{aligned} f_{\phi_a}^1 &= \tfrac{7}{48}\left(3 - 6M + 3M^2\right)a^2 + \left(-\tfrac{81}{48}M^2 + \tfrac{138}{48}M - \tfrac{57}{48}\right)a \\ &+ \left(\tfrac{13}{8}M^2 - \tfrac{18}{8}M + \tfrac{5}{8}\right) = 0 \end{aligned}$$

   and making $M \to \infty$, we have

$$a = \frac{81 \pm \sqrt{81^2 - 6552}}{42};$$

   therefore, the solutions of the equation are $a = 2$ and $a = 13/7$. On the other hand, solving the second equation $f_{\phi_a}^2 = 0$, we have

$$\begin{aligned} &\left(45 - 71M + 19M^2 + 7M^3\right)a^2 + \left(-27M^3 - 67M^2 + 215M - 121\right)a \\ + &\left(26M^3 + 58M^2 - 162M + 78\right) = 0 \end{aligned}$$

   and making $M \to \infty$, we get

$$a = \frac{27 \pm \sqrt{27^2 - 728}}{14},$$

   i.e., $a = 2$ and $a = 13/7$.

Finally to approximate the third asymptotic moment to the exact one, we have

$$\begin{aligned} f^3_{\phi_a} &= \left(1035 - 1398M + 144M^2 + 198M^3 + 21M^4\right) a^2 \\ &+ \left(-81M^4 - 702M^3 - 552M^2 + 4110M - 2775\right) a \\ &+ \left(78M^4 + 612M^3 + 496M^2 - 3012M + 1826\right) = 0 \end{aligned}$$

and as $M \to \infty$, we obtain

$$a = \frac{81 \pm \sqrt{81^2 - 6552}}{42};$$

therefore the solutions of the equation are $a = 2$ and $a = 13/7$.

This result is valid for large M. If M is small we should use the next table, which has the roots "a" of the equations $f^i_{\phi_a} = 0, i = 1, 2, 3,$ for fixed values of M that increase to ∞.

	M	2	3	4	5	10	20	40	50	100	200	500
$f^1_{\phi_a}$	a_1	3.0	2.42	2.23	2.14	2.0	2.0	2.0	2.0	2.0	2.0	2.0
	a_2	2.0	2.0	2.0	2.0	2.98	1.91	1.88	1.88	1.86	1.86	1.85
$f^2_{\phi_a}$	a_1	3.34	2.52	2.31	2.21	2.07	2.02	2.0	2.0	2.0	2.0	2.0
	a_2	1.65	1.68	1.7	1.71	1.76	1.8	1.83	1.83	1.84	1.85	1.85
$f^3_{\phi_a}$	a_1	3.69	2.62	2.37	2.27	2.10	2.04	2.01	2.01	2.0	2.0	2.0
	a_2	1.3	1.41	1.47	1.51	1.62	1.72	1.78	1.79	1.82	1.84	1.85

Values of the roots ($a_1 > a_2$) of $f^i_{\phi_a} = 0, \ i = 1, 2, 3.$

In this table we observe that for $M > 20$ we can use the previous result since the first order factors of the three first moments are closer to 0 for $a = 2$ and $a = 13/7$. For $M \leq 20$, it would be reasonable to choose one test statistic $S_n^{\phi_a}$ with $a \in [1.5, 2]$. For more details see Pardo, M. C. (1999).

7. In this case, the corrected test statistic is given by

$$^{c}T_n^{\phi_{(\lambda)}}(\widehat{\boldsymbol{p}}, \boldsymbol{p}^0) \equiv \frac{T_n^{\phi_{(\lambda)}}(\widehat{\boldsymbol{p}}, \boldsymbol{p}^0) - \gamma_{\phi_{(\lambda)}}}{\sqrt{\delta_{\phi_{(\lambda)}}}}$$

where $\gamma_{\phi_{(\lambda)}}$ and $\delta_{\phi_{(\lambda)}}$ are obtained from the theoretical results given in Subsection 4.3.3.

Since $\phi''_{(\lambda)}(1) = 1$, $\phi'''_{(\lambda)}(1) = \lambda - 1$, $\phi^{IV}_{(\lambda)}(1) = (\lambda - 1)(\lambda - 2)$ we have

$$\begin{aligned} \delta_{\phi_{(\lambda)}} &= 1 + \tfrac{1}{2(M-1)n}\left((2 - 2M - M^2 + S)\right. \\ &+ (\lambda - 1)(8 - 12M - 2M^2 + 6S) \\ &+ \tfrac{1}{3}(\lambda - 1)^2\left(4 - 6M - 3M^2 + 5S\right) \\ &+ \left.(\lambda - 1)(\lambda - 2)(2 - 4M + 2S)\right) \end{aligned}$$

and

$$\begin{aligned} \gamma_{\phi(\lambda)} &= (M-1)\left(1-\sqrt{\delta_{\phi(\lambda)}}\right)+\tfrac{1}{n}\left(\tfrac{1}{3}(\lambda-1)(2-3M+S)\right) \\ &+ \frac{(\lambda-1)(\lambda-2)}{4}(1-2M+S). \end{aligned}$$

8. Similarly to the previous exercise and taking into account that

$$\phi_a''(1)=1,\ \phi_a'''(1)=-(1+a),\ \phi_a^{IV}(1)=2(a^2+a+1)$$

and $\sum_{j=1}^{M} p_j^0 = M^2$, for $a\in(0,1)$, we have

$$\begin{aligned} \gamma_a &= (M-1)\left(1-\delta_a^{1/2}\right)-\tfrac{1}{3n}(a+1)\left(2-3M+M^2\right) \\ &+ \tfrac{a^2+a+1}{2n}\left(1-2M+M^2\right) \end{aligned}$$

and

$$\begin{aligned} \delta_a &= 1-\frac{1}{n}-\tfrac{2(a+1)}{(M-1)n}\left(2-3M+M^2\right)+\tfrac{(a+1)^2}{3n(M-1)}\left(2-3M+M^2\right) \\ &+ 2\tfrac{a^2+a+1}{n(M-1)}\left(1-2M+M^2\right). \end{aligned}$$

For $a=0$, we have the likelihood ratio test statistic and for $a=1$ the modified likelihood ratio test statistic.

9. We know that

$$E\left[S_n^{\phi}(\widehat{\boldsymbol{p}},\boldsymbol{p}^0)\right]=M-1+a_{\phi}/n+o\left(n^{-1}\right)$$

and

$$Var\left[S_n^{\phi}(\widehat{\boldsymbol{p}},\boldsymbol{p}^0)\right]=2(M-1)+b_{\phi}/n+o\left(n^{-1}\right)$$

where

$$a_{\phi}=\frac{\phi'''(1/M)}{2\phi''(1/M)}\left(\frac{2}{M}-3+M\right)+\frac{7\phi^{IV}(1/M)}{16\phi''(1/M)}\left(\frac{1}{M^2}-\frac{2}{M}+1\right)$$

and

$$\begin{aligned} b_{\phi} &= -2M+2+\tfrac{\phi'''(1/M)}{\phi''(1/M)}\left(\tfrac{12}{M}-18+6M\right)+\tfrac{1}{2}\left(\tfrac{\phi'''(1/M)}{\phi''(1/M)}\right)^2 \\ &\times \left(\tfrac{6}{M^2}-\tfrac{9}{M}+3\right)+\tfrac{7\phi^{IV}(1/M)}{2\phi''(1/M)}\left(\tfrac{1}{M^2}-\tfrac{2}{M}+1\right). \end{aligned}$$

We define

$$^{c}S_n^{\phi}(\widehat{\boldsymbol{p}},\boldsymbol{p}^0)=\frac{S_n^{\phi}(\widehat{\boldsymbol{p}},\boldsymbol{p}^0)-\gamma_{\phi}}{\sqrt{\delta_{\phi}}}$$

in such a way that

$$E\left[{}^{c}S_n^{\phi}(\widehat{\boldsymbol{p}},\boldsymbol{p}^0)\right] = M-1+o\left(n^{-1}\right)$$

and

$$Var\left[{}^{c}S_n^{\phi}(\widehat{\boldsymbol{p}},\boldsymbol{p}^0)\right] = 2\left(M-1\right)+o\left(n^{-1}\right).$$

For this purpose it is necessary to consider

$$\gamma_\phi = (M-1)\left(1-\sqrt{\delta_\phi}\right)+a_\phi/n \quad \text{and} \quad \delta_\phi = b_\phi/n2\left(M-1\right),$$

i.e.,

$$\begin{aligned}\delta_\phi &= 1-\tfrac{1}{n}+\tfrac{1}{n(M-1)}\left(\tfrac{\phi'''(1/M)}{\phi''(1/M)}\left(\tfrac{6}{M}-9+3M\right)\right.\\ &+ \left.\tfrac{1}{4}\left(\tfrac{\phi'''(1/M)}{\phi''(1/M)}\right)^2\left(\tfrac{6}{M^2}-\tfrac{9}{M}+3\right)+\tfrac{7\phi^{IV}(1/M)}{4\phi''(1/M)}\left(\tfrac{1}{M^2}-\tfrac{2}{M}+1\right)\right)\end{aligned}$$

and

$$\begin{aligned}\gamma_\phi &= (M-1)\left(1-\sqrt{\delta_\phi}\right)\\ &+ \tfrac{1}{2n}\left(\tfrac{\phi'''(1/M)}{\phi''(1/M)}\left(\tfrac{2}{M}-3+M\right)+\tfrac{7\phi^{IV}(1/M)}{8\phi''(1/M)}\left(\tfrac{1}{M^2}-\tfrac{2}{M}+1\right)\right).\end{aligned}$$

10. It is immediate to show that Lin test statistic is

$$\begin{aligned}T_n^{\phi_a}(\widehat{\boldsymbol{p}},\boldsymbol{p}^0) &= 2n\left(\tfrac{1}{1-a}\sum_{i=1}^{M}\widehat{p}_i\log\tfrac{\widehat{p}_i}{a\widehat{p}_i+(1-a)p_i^0}\right.\\ &+ \left.\tfrac{1}{a}\sum_{i=1}^{M}p_i^0\log\tfrac{p_i^0}{a\widehat{p}_i+(1-a)p_i^0}\right), \quad a\in(0,1).\end{aligned}$$

The expression of the approximation based on Edgeworth's expansion associated with this test statistic for $\boldsymbol{p}^0=(1/M,...,1/M)^T$ is

$$\Pr(T_n^{\phi_a}(\widehat{\boldsymbol{p}},\boldsymbol{p}^0)<b)\simeq J_1^a+\widehat{J}_2^a$$

being

$$J_1^a = \Pr(\chi^2_{M-1}<b)+\frac{1}{24n}\sum_{j=0}^{3}r_j^a\Pr\left(\chi^2_{M-1+2j}<b\right)+O\left(n^{-3/2}\right),$$

where

- $r_0^a = 2(1-M^2)$
- $r_1^a =$ -6$(a^2+a+1)(M-1)^2$+2$(1+a)^2(M^2-3M+2)$+6(M^2-M)

· $r_2^a = $ -4$a(a+1)\left(M^2-3M+2\right)$+6$\left(a^2+a+1\right)(M-1)^2-6(M^2-M)$

and

· $r_3^a = \; 2\,(a+2)^2\,(M^2-3M+2)$.

The term $\widehat{J}_2^a$ has the expression

$$\widehat{J}_2^a = \left(N^a(b) - n^{(M-1)/2}V^a(b)\right)e^{-c/2} \Big/ \left((2\pi n)^{(M-1)}\,(1/M)^M\right)^{1/2},$$

where $N^a(b)$ is the number of points $(w_1,\ldots,w_{M-1})$ satisfying

$$w_i = \sqrt{n}\left(\frac{N_i}{n}-\frac{1}{M}\right),\; N_i = 0,1,2,\ldots$$

such that $\sum_{i=1}^M N_i = n$ and $T_n^{\phi_a}(\widehat{\boldsymbol{p}},\boldsymbol{p}^0) < b$, and

$$\begin{aligned} V^a(b) &= \tfrac{(\pi c)^{(M-1)/2}}{\Gamma((M+1)/2)}\,(1/M)^{M/2}\left(1+\tfrac{c}{12(M+1)n}\left((1+a)^2(M^2-3M+2)\right.\right.\\ &- \left.\left.3(a^2+a+1)(M-1)^2\right)\right)+O(n^{-3/2}). \end{aligned}$$

11. In this case $\phi(x) = (x-1)^2/2$, then for the test statistic

$$X^2 = n\sum_{i=1}^{M}\frac{(\widehat{p}_i-p_i^0)^2}{p_i^0}$$

we have $\Pr(X^2<b) \simeq J_1^\phi + \widehat{J}_2^\phi$, being

$$\begin{aligned} J_1^\phi &= \Pr(\chi^2_{M-1}<b) + \tfrac{1}{24n}\left(\Pr(\chi^2_{M-1}<b)2(1-S)\right.\\ &+ \left.\Pr(\chi^2_{M+5}<b)(5S-3M^2-6M+4)\right) \end{aligned}$$

and

$$\widehat{J}_2^\phi = \left(N^\phi(b) - n^{(M-1)/2}V^\phi(b)\right)e^{-b/2} \Big/ \left((2\pi n)^{(M-1)}\prod_{i=1}^{M}p_j^0\right)^{1/2},$$

where $N^\phi(b)$ is the number of points $(w_1,\ldots,w_{M-1})$ satisfying

$$w_i = \sqrt{n}\left(\frac{N_i}{n}-\frac{1}{M}\right),\; N_i = 0,1,2,\ldots$$

such that $\sum_{i=1}^M N_i = n$ and $X^2<b$ and

$$V^\phi(b) = \frac{(\pi b)^{(M-1)/2}}{\Gamma((M+1)/2)}\left(\prod_{i=1}^{M}p_j^0\right)^{1/2}$$

being $S=\sum_{j=1}^M (p_j^0)^{-1}$.

12. Using similar arguments to that used in the previous exercises and taking into account that $\phi''_{(\lambda)}(1) = 1$, $\phi'''_{(\lambda)}(1) = \lambda - 1$, $\phi^{IV}_{(\lambda)}(1) = (\lambda-1)(\lambda-2)$ we have $\Pr\left(T_n^{\phi(\lambda)}(\widehat{\boldsymbol{p}}, \boldsymbol{p}^0) < b\right) \simeq J_1^\lambda + \widehat{J_2^\lambda}$ being

$$J_1^\lambda = \Pr(\chi^2_{M-1} < b) + \frac{1}{24n}\sum_{j=0}^{3} r_j^\lambda \Pr\left(\chi^2_{M-1+2j} < b\right) + O\left(n^{-3/2}\right),$$

where

- $r_0^\lambda = 2(1-S)$
- $r_1^\lambda = 6(\lambda-1)\left(S-M^2\right) - 3(\lambda-1)(\lambda-2)(S-2M+1) + (\lambda-1)^2\left(5S-3M^2-6M+4\right) + 3(3S-M^2-2M)$
- $r_2^\lambda = (\lambda-1)\left(-2(8S-6M^2-6M+4) + 3(\lambda-2)(S-2M+1)\right) - 2(\lambda-1)^2\left(5S-3M^2-6M+4\right) - 6\left(-M^2+2S-2M+1\right)$

and

- $r_3^\lambda = \lambda^2(5S-3M^2-6M+4).$

The term $\widehat{J_2^\lambda}$ has the expression

$$\widehat{J_2^\lambda} = \left(N^\lambda(b) - n^{(M-1)/2}V^\lambda(b)\right)e^{-b/2}\left((2\pi n)^{(M-1)}\prod_{i=1}^{M} p_j^0\right)^{1/2}$$

where $N^\lambda(b)$ is the number of X-values such that $T_n^{\phi(\lambda)}(\widehat{\boldsymbol{p}}, \boldsymbol{p}^0) < b$ and

$$V^\phi(b) = \frac{(\pi b)^{(M-1)/2}}{\Gamma((M+1)/2)}\left(\prod_{i=1}^{M} p_j^0\right)^{1/2}\left(1 + \frac{b}{24(M+1)n}(l_1 - l_2)\right),$$

where

$$l_1 = (\lambda-1)^2\left(5S-3M^2-6M+4\right)$$

and

$$l_2 = 3(\lambda-1)(\lambda-2)(S-2M+1),$$

being $S = \sum_{j=1}^{M}(p_j^0)^{-1}$. For more details see Cressie and Read (1988).

5

Minimum Phi-divergence Estimators

5.1. Introduction

In this chapter we consider a wide class of estimators which can be used when the data are discrete, either the underlying distribution is discrete or is continuous but the observations are classified into groups. The latter situation can occur either by experimental reasons or because the estimation problem without grouped data is not easy to solve; see Fryer and Robertson (1972). Some examples in which it is not possible to find the maximum likelihood estimator based on the original data can be seen in Le Cam (1990). For example, when we consider distributions resulting from the mixture of two normal populations, whose probability density function is given by

$$f_{\boldsymbol{\theta}}(x) = w\frac{1}{\sqrt{2\pi}\sigma_1}\exp\left(-\frac{1}{2}\left(\frac{x-\mu_1}{\sigma_1}\right)^2\right)+(1-w)\frac{1}{\sqrt{2\pi}\sigma_2}\exp\left(-\frac{1}{2}\left(\frac{x-\mu_2}{\sigma_2}\right)^2\right),$$

the likelihood function is not a bounded function. In this situation

$$\boldsymbol{\theta} = (\mu_1, \mu_2, \sigma_1, \sigma_2, w), \mu_1, \mu_2 \in \mathbb{R}, \sigma_1, \sigma_2 > 0 \text{ and } w \in (0,1),$$

and the likelihood function for a random sample of size n, $x_1, ..., x_n$ is given by

$$L(\boldsymbol{\theta}; x_1, \ldots, x_n) = \prod_{j=1}^{n} f_{\boldsymbol{\theta}}(x_j).$$

If we consider $\mu_2 = \mu_1 = x_i$ for some i $(i = 1, \ldots, n)$, then $f_{\boldsymbol{\theta}}(x_i) > w(\sqrt{2\pi}\sigma_1)^{-1}$ and

$$f_{\boldsymbol{\theta}}(x_j) > (1-w)(\sqrt{2\pi}\sigma_2)^{-1} \exp\left(-\frac{1}{2}\left(\frac{x_j - x_i}{\sigma_2}\right)^2\right) \qquad \text{for } j \neq i.$$

Therefore

$$L(\boldsymbol{\theta}; x_1, \ldots, x_n) > (2\pi)^{-n/2} w(1-w)^{n-1} \sigma_1^{-1} \sigma_2^{-(n-1)} \exp\left(-\frac{1}{2} \sum_{j=1, j\neq i}^{n} \left(\frac{x_j - x_i}{\sigma_2}\right)^2\right)$$

and choosing σ_1 sufficiently small it is possible to do L as big as it is desired. Then there are not values $w, \sigma_1, \sigma_2, \mu_1$ and μ_2 that maximize L.

Several authors have paid attention to estimation of the unknown parameters of a mixture of two unspecified normal densities. There are mainly three different approaches: moments, maximum likelihood and minimum distance.

The moment solution to the problem of estimating the five parameters of an arbitrary mixture of two unspecified normal densities was studied by Karl Pearson (1894). Although many random phenomena have subsequently been shown to follow this distribution until 1966 this estimation problem was not considered. Important applications of mixture modeling occur in satellite remote-sensing of agricultural characteristics: specifically, the use of spectral measurements of light intensity to determine crop types, distributions of wind velocities and distributions of physical dimensions of various mass produced items. Hassenblad (1966) seems to have been the first to reopen the question. Since then the problem has also attracted the attention of Cohen (1967) who showed how the computation of Pearson's moment method can be lightened to some extent.

Since the likelihood function is not a bounded function, the objective in the maximum likelihood approach is to find an appropriate local maximum. Since closed-form solutions of the likelihood equations do not exist, they must be solved by using iterative techniques. Day (1969) and Behboodian (1970) find an appropriate local maximum of the likelihood function by using iterative techniques.

Minimum distance estimation (MDE), in general, was presented for the first time by Wolfowitz (1957) and it provides a convenient method of consistently estimating unknown parameters. An extensive bibliography for minimum distance estimators is in Parr (1981). In the mixture model setting, Choi and Bulgren

(1968) and MacDonald (1971) estimated the mixture proportions (assuming the component distributions were known) by minimizing the sum-of-squares distance between the empirical and theoretical distribution functions. Quandt and Ramsey (1978) estimated the parameters in the mixture model by minimizing the sum-of-squares distance between the empirical and theoretical moment generating functions. Kumar et *al.* (1979), however, showed that this technique is highly sensitive to starting values. Bryant and Paulson (1983) examined the empirical characteristic function in this setting. Fryer and Robertson (1972) considered the MDE for grouped data and finally by using the families of R_ϕ-divergences and ϕ-divergences in Pardo, M. C. (1997b, 1999b) the problem of estimating the parameters of a mixture of normal distributions was considered.

In this chapter we study the minimum ϕ-divergence estimator, considered in Morales et *al.* (1995), for grouped data.

5.2. Maximum Likelihood and Minimum Phi-divergence Estimators

Let $(\mathcal{X}, \beta_{\mathcal{X}}, P_{\boldsymbol{\theta}})_{\boldsymbol{\theta}\in\Theta}$ be the statistical space associated with the random variable $\boldsymbol{X}$, where $\beta_{\mathcal{X}}$ is the σ-field of Borel subsets $A \subset \mathcal{X}$ and $\{P_{\boldsymbol{\theta}}\}_{\boldsymbol{\theta}\in\Theta}$ is a family of probability distributions defined on the measurable space $(\mathcal{X}, \beta_{\mathcal{X}})$ with Θ an open subset of $\mathbb{R}^{M_0}$, $M_0 \geq 1$. Let $\mathcal{P} = \{E_i\}_{i=1,\ldots,M}$ be a partition of $\mathcal{X}$. The formula $\Pr_{\boldsymbol{\theta}}(E_i) = p_i(\boldsymbol{\theta})$, $i = 1, \ldots, M$, defines a discrete statistical model. Let $Y_1, \ldots, Y_n$ be a random sample from the population described by the random variable $\boldsymbol{X}$, let $N_i = \sum_{j=1}^n I_{E_i}(Y_j)$ and $\widehat{p}_i = N_i/n$, $i = 1, \ldots, M$. Estimating $\boldsymbol{\theta}$ by maximum likelihood method, under the discrete statistical model, consists of maximizing for fixed $n_1, \ldots, n_M$,

$$\Pr{}_{\boldsymbol{\theta}}(N_1 = n_1, \ldots, N_M = n_M) = \frac{n!}{n_1! \ldots n_M!} p_1(\boldsymbol{\theta})^{n_1} \times \ldots \times p_M(\boldsymbol{\theta})^{n_M} \tag{5.1}$$

or, equivalently,

$$\log \Pr{}_{\boldsymbol{\theta}}(N_1 = n_1, \ldots, N_M = n_M) = -nD_{Kull}(\widehat{\boldsymbol{p}}, \boldsymbol{p}(\boldsymbol{\theta})) + k \tag{5.2}$$

where $\widehat{\boldsymbol{p}} = (\hat{p}_1, \ldots, \hat{p}_M)^T$, $\boldsymbol{p}(\boldsymbol{\theta}) = (p_1(\boldsymbol{\theta}), \ldots, p_M(\boldsymbol{\theta}))^T$ and k is independent of $\boldsymbol{\theta}$. The equality (5.2) is easy to check because, if we denote $l(\boldsymbol{\theta}) = \log \Pr_{\boldsymbol{\theta}}(N_1 =$

$n_1, \ldots, N_M = n_M)$, we have

$$
\begin{aligned}
l(\boldsymbol{\theta}) &= \log \frac{n!}{n_1! \ldots n_M!} + n \sum_{i=1}^{M} \widehat{p}_i \log p_i(\boldsymbol{\theta}) \\
&= \log \frac{n!}{n_1! \ldots n_M!} - n \sum_{i=1}^{M} \widehat{p}_i \log \frac{1}{p_i(\boldsymbol{\theta})} + n \sum_{i=1}^{M} \widehat{p}_i \log \widehat{p}_i - n \sum_{i=1}^{M} \widehat{p}_i \log \widehat{p}_i \\
&= \log \frac{n!}{n_1! \ldots n_M!} - n \sum_{i=1}^{M} \widehat{p}_i \log \frac{\widehat{p}_i}{p_i(\boldsymbol{\theta})} + n \sum_{i=1}^{M} \widehat{p}_i \log \widehat{p}_i \\
&= -n \sum_{i=1}^{M} \widehat{p}_i \log \frac{\widehat{p}_i}{p_i(\boldsymbol{\theta})} + k = -n D_{Kull}(\widehat{\mathbf{p}}, \mathbf{p}(\boldsymbol{\theta})) + k.
\end{aligned}
$$

Then, estimating $\boldsymbol{\theta}$ with the maximum likelihood estimator of the discrete model is equivalent to minimizing the Kullback-Leibler divergence on $\boldsymbol{\theta} \in \Theta \subseteq \mathbb{R}^{M_0}$. Since Kullback-Leibler divergence is not the unique divergence measure, we can choose as estimator of $\boldsymbol{\theta}$ the value $\widetilde{\boldsymbol{\theta}}$ verifying

$$D(\widehat{\mathbf{p}}, \mathbf{p}(\widetilde{\boldsymbol{\theta}})) = \inf_{\boldsymbol{\theta} \in \Theta \subseteq R^{M_0}} D(\widehat{\mathbf{p}}, \mathbf{p}(\boldsymbol{\theta})), \tag{5.3}$$

where D is a divergence measure.

In the following we assume that there exists a function

$$\mathbf{p}(\boldsymbol{\theta}) = (p_1(\boldsymbol{\theta}), \ldots, p_M(\boldsymbol{\theta}))^T$$

that maps each $\boldsymbol{\theta} = (\theta_1, \ldots, \theta_{M_0})^T$ into a point in Δ_M, where Δ_M was defined in Chapter 2, Section 2.1. As $\boldsymbol{\theta}$ ranges over the values of Θ, $\mathbf{p}(\boldsymbol{\theta})$ ranges over a subset T of Δ_M. When we assume that a given model is "correct", we are just assuming that there exists a value $\boldsymbol{\theta}_0 \in \Theta$ such that $\mathbf{p}(\boldsymbol{\theta}_0) = \boldsymbol{\pi}$, where $\boldsymbol{\pi}$ is the true value of the multinomial probability, i.e., $\boldsymbol{\pi} \in T$.

Definition 5.1

Let $Y_1, \ldots, Y_n$ be a random sample from a population described by the random variable $\boldsymbol{X}$ with associated statistical space $(\mathcal{X}, \beta_{\mathcal{X}}, P_{\boldsymbol{\theta}})_{\boldsymbol{\theta} \in \Theta}$. The minimum ϕ-divergence estimator of $\boldsymbol{\theta}_0$ is any $\widehat{\boldsymbol{\theta}}_\phi \in \Theta$ verifying

$$D_\phi(\widehat{\mathbf{p}}, \mathbf{p}(\widehat{\boldsymbol{\theta}}_\phi)) = \inf_{\boldsymbol{\theta} \in \Theta \subseteq \mathbb{R}^{M_0}} D_\phi(\widehat{\mathbf{p}}, \mathbf{p}(\boldsymbol{\theta})).$$

In other words, the minimum ϕ-divergence estimator satisfies the condition

$$\widehat{\boldsymbol{\theta}}_\phi = \arg \inf_{\boldsymbol{\theta} \in \Theta \subseteq \mathbb{R}^{M_0}} D_\phi(\widehat{\mathbf{p}}, \mathbf{p}(\boldsymbol{\theta})). \tag{5.4}$$

This method chooses the point of T closest to $\widehat{\mathbf{p}}$ in the sense of the ϕ-divergence chosen.

Remark 5.1

If we consider the family of the power-divergence measures we obtain the minimum power-divergence estimator studied by Cressie and Read (1984). This is given by the condition

$$\widehat{\boldsymbol{\theta}}_{(\lambda)} = \arg \inf_{\boldsymbol{\theta} \in \Theta \subseteq \mathbb{R}^{M_0}} D_{\phi_{(\lambda)}}(\widehat{\mathbf{p}}, \mathbf{p}(\boldsymbol{\theta})), \tag{5.5}$$

where

$$D_{\phi_{(\lambda)}}(\widehat{\mathbf{p}}, \mathbf{p}(\boldsymbol{\theta})) = \frac{1}{\lambda(\lambda+1)} \sum_{i=1}^{M} \widehat{p}_i \left(\left(\frac{\widehat{p}_i}{p_i(\boldsymbol{\theta})} \right)^{\lambda} - 1 \right).$$

For $\lambda \to 0$ we obtain the maximum likelihood estimator, for $\lambda = 1$ the minimum chi-square estimator, for $\lambda = -2$ the minimum modified chi-square estimator (or minimum Neyman modified estimator), for $\lambda \to -1$ the minimum modified likelihood estimator (or minimum discrimination information estimator), for $\lambda = -0.5$ Freeman-Tukey estimator and for $\lambda = 2/3$ Cressie-Read estimator.

We know that Kullback-Leibler divergence measure can be obtained from the power-divergence measure with $\lambda = 0$. For this reason in the rest of the chapter in order to distinguish between the MLE based on original data and the MLE associated with the discrete model we shall denote by $\widehat{\boldsymbol{\theta}}_{(0)}$ the maximum likelihood estimator in the discrete model and by $\widehat{\boldsymbol{\theta}}$ the maximum likelihood estimator based on the original data.

We present an example to clarify all the notation and concepts introduced until now.

Example 5.1

Suppose that n independent and identical distributed Poisson variables with mean θ ($\theta > 0$) are observed, and let the observations be truncated at $x = 2$. Let N_1, N_2 and N_3 be the number of observations taking on the values 0, 1 and 2 or more, respectively. Then $\boldsymbol{N} = (N_1, N_2, N_3)$ has the trinomial distribution $(n; p_1(\theta), p_2(\theta), p_3(\theta))$, where $p_1(\theta) = \Pr_\theta(X = 0) = e^{-\theta}$, $p_2(\theta) = \Pr_\theta(X = 1) = \theta e^{-\theta}$ and $p_3(\theta) = \Pr_\theta(X \geq 2) = 1 - (1 + \theta)e^{-\theta}$.

If we consider Cressie-Read estimator, we have to get the minimum in θ of *the function*

$$D_{\phi_{(2/3)}}(\widehat{\boldsymbol{p}},\boldsymbol{p}(\theta)) = \tfrac{9}{10}\left[\widehat{p}_1\left(\left(\tfrac{\widehat{p}_1}{\exp(-\theta)}\right)^{2/3}-1\right)+\widehat{p}_2\left(\left(\tfrac{\widehat{p}_2}{\theta\exp(-\theta)}\right)^{2/3}-1\right)\right.$$
$$\left. + \ \widehat{p}_3\left(\left(\tfrac{\widehat{p}_3}{1-(1+\theta)\exp(-\theta)}\right)^{2/3}-1\right)\right].$$

Now if we assume, for instance, that $\widehat{\mathbf{p}} = (0.2, 0.3, 0.5)^T$ *we obtain* $\widehat{\theta}_{(2/3)} =$ 1.6595.

Geometrically, Δ_3 is the triangle side ABC depicted in Figure 5.1, that we represent in the plane through the triangle of Figure 5.2.

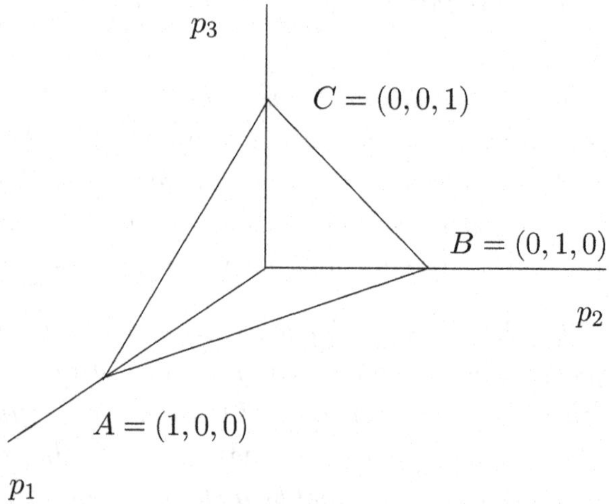

Figure 5.1. Set of probability distributions for Example 5.1.

As θ varies over $\mathbb{R}^+ = [0, \infty)$,

$$\boldsymbol{p}(\theta) = (e^{-\theta}, \theta e^{-\theta}, 1-(1+\theta)e^{-\theta})^T$$

traces out a curve in Δ_3. This curve is the subset T. When $\theta \to 0$, $\boldsymbol{p}(\theta) \to (1, 0, 0)^T$, and when $\theta \to \infty$, $\boldsymbol{p}(\theta) \to (0, 0, 1)^T$. Thus, the boundary points of θ in this example correspond to the boundary points of Δ_3. Figure 5.2 shows the relationships between Δ_3, T, $\boldsymbol{\pi}$ and $\widehat{\boldsymbol{p}}$ in this example. If the Poisson model is incorrect, then the true value of $\boldsymbol{\pi}$ does not generally lie on the curve, although in principle it can. Because of the discreteness of the multinomial distribution, it often happens that $\widehat{\boldsymbol{p}}$ does not lie on T (as is the case in the figure). The estimation method based on the minimum distance leads to a point in T closest to $\widehat{\boldsymbol{p}}$ in the sense of the distance chosen.

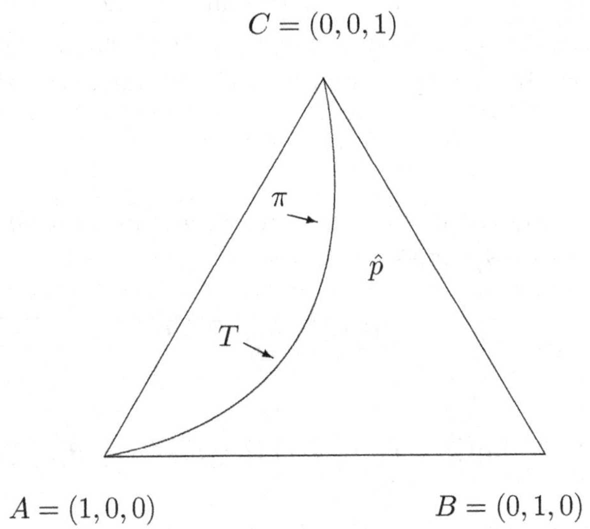

Figure 5.2. Relation between Δ_3, T, $\boldsymbol{\pi}$ and $\widehat{\boldsymbol{p}}$ for Example 5.1.

From a historical point of view the maximum likelihood estimator in grouped data or the minimum Kullback-Leibler divergence estimator was considered for the first time by Fisher (1925), the minimum chi-square estimator and the minimum modified chi-square estimator by Neyman (1949). Matusita (1955), using the measure of divergence that has his name, studied the corresponding estimator for $a = 1/2$ (Freeman-Tukey estimator). Read and Cressie (1988) considered the minimum power-divergence estimators and finally Morales et *al.* (1995) studied the minimum ϕ-divergence estimators. The results obtained by Read and Cressie (1988) as well as the results obtained in Morales et *al.* (1995) are generalizations of the results given by Birch (1964) for the maximum likelihood estimator. The results we present in this chapter follow the line of that given by Pardo, M. C. (1997a) for the R_ϕ-divergences, which are a generalization of the procedure given by Cox (1984) in relation to the maximum likelihood estimator.

5.2.1. Minimum Power-divergence Estimators in Normal and Weibull Populations

In this Section we consider the minimum power-divergence estimator, given in (5.5), of the Normal and Weibull parameters for different values of λ. Moreover, we compare the results obtained with the maximum likelihood estimator based on the original data and the estimator based on minimizing the Kolmogorov distance (Kolmogorov estimator). A similar study was carried out for Weibull distribution by Pardo, M.C. (1997a) using the family of R_ϕ-divergences to define a minimum distance estimator.

Definition 5.2

Kolmogorov estimator for the parameter $\boldsymbol{\theta}$, *of a distribution family* $\{F_{\boldsymbol{\theta}}\}_{\boldsymbol{\theta}\in\Theta}$, *is defined as the value* $\widehat{\boldsymbol{\theta}}_{Ko} \in \Theta$ *verifying*

$$D_n(\widehat{\boldsymbol{\theta}}_{Ko}) = \min_{\boldsymbol{\theta}\in\Theta} D_n(\boldsymbol{\theta}),$$

for

$$D_n(\boldsymbol{\theta}) = \sup_{x\in\mathbb{R}}\{\mid F_n(x) - F_{\boldsymbol{\theta}}(x) \mid\} = \max\{D_n^+(\boldsymbol{\theta}), D_n^-(\boldsymbol{\theta})\},$$

where

$$D_n^+(\boldsymbol{\theta}) = \sup_{x\in\mathbb{R}}\{F_n(x) - F_{\boldsymbol{\theta}}(x)\} = \max\{0, \max_{i=1,\dots,n}\{\tfrac{i}{n} - F_{\boldsymbol{\theta}}(x_{(i)})\}\},$$

$$D_n^-(\boldsymbol{\theta}) = \sup_{x\in\mathbb{R}}\{F_{\boldsymbol{\theta}}(x) - F_n(x)\} = \max\{0, \max_{i=1,\dots,n}\{F_{\boldsymbol{\theta}}(x_{(i)}) - \tfrac{i-1}{n}\}\},$$

F_n *is the empirical distribution function of the sample* $x_1, \dots, x_n$ *and* $x_{(1)} \leq x_{(2)} \leq \dots \leq x_{(n)}$ *are the order statistics.*

It is well known that a random variable X has a Weibull distribution, $We\,(b,c)$, with parameters $\boldsymbol{\theta} = (b,c)$, b, c, $b > 0$, $c > 0$, if the distribution function is given by

$$F_{\boldsymbol{\theta}}(x) = 1 - \exp\left(-\left(x/b\right)^c\right), \qquad x \geq 0.$$

The general scheme for calculating the minimum power-divergence estimator is as follows:

Step 1: We fix

a) sample size (n),

b) number of classes in the partition (M),

c) number of simulated samples (N):

We consider the partition of the sample space, $\{E_i\}_{i=1,...,M}$, $E_i = (a_{i-1}, a_i]$, $i = 1, \ldots, M$, where the values $a_0, ..., a_M$ are obtained from

$$\int_{a_{i-1}}^{a_i} f_{\boldsymbol{\theta}}(x)dx = 1/M, \quad i = 1, ..., M.$$

Step 2: Given λ fixed, do for $i = 1$ to N

a) Generate a random sample of size n,

b) Calculate the relative frequencies, $\widehat{p}_l$, of $E_l = (a_{l-1}, a_l]$, $l = 1, ..., M$,

c) Minimize on $\boldsymbol{\theta}$ the function $D_{\phi_{(\lambda)}}(\widehat{\mathbf{p}}, \boldsymbol{p}(\boldsymbol{\theta}))$. Estimator $\widehat{\boldsymbol{\theta}}_{(\lambda),i}=(\widehat{\theta}^1_{(\lambda),i}, ..., \widehat{\theta}^{M_0}_{(\lambda),i})$ is obtained.

The minimum of the function $D_{\phi_{(\lambda)}}(\widehat{\mathbf{p}}, \boldsymbol{p}(\boldsymbol{\theta}))$ has been obtained using the subroutine ZXMIN of the package IMSL. In other parts of the book we use the Newton-Raphson method to get minimum ϕ-divergence estimators.

Step 3: Let $\hat{\boldsymbol{\theta}}_{(\lambda)} = (\hat{\theta}^1_{(\lambda)}, ..., \hat{\theta}^{M_0}_{(\lambda)})$ be the mean of the values obtained by minimizing the function $D_{\phi_{(\lambda)}}$ in step 2(c) for all the samples and *mse* the mean squared error of the estimated parameters, i.e.,

$$\hat{\theta}^j_{(\lambda)} = \frac{1}{N}\sum_{i=1}^{N} \hat{\theta}^j_{\phi_{(\lambda)},i}, \; mse(\theta_j) = \frac{1}{N}\sum_{i=1}^{N}\left(\hat{\theta}^j_{\phi_{(\lambda)},i} - \theta_j\right)^2 \text{ and } mse = \frac{1}{M_0}\sum_{j=1}^{M_0} mse\,(\theta_j)\,.$$

$We(1,1)$		$n=20$	$n=40$	$n=60$
$\hat{\boldsymbol{\theta}}$	$\widehat{b}$	.998783	.994317	.994150
	$\widehat{c}$	1.063596	1.029655	1.019258
	mse	.055893	.025969	.014047
$\widehat{\boldsymbol{\theta}}_{Ko}$	$\widehat{b}$	.984651	1.006958	.978376
	$\widehat{c}$	1.565195	1.137134	1.185521
	mse	1.023289	.121828	.108134
$\hat{\boldsymbol{\theta}}_{(-2)}$	$\widehat{b}$	1.008785	1.010294	.985119
	$\widehat{c}$	1.423182	1.084708	1.101808
	mse	.776172	.106446	.080867
$\hat{\boldsymbol{\theta}}_{(-1)}$	$\widehat{b}$	1.013693	1.006479	.980422
	$\widehat{c}$	1.406272	1.067542	1.087102
	mse	.756604	.097466	.078229
$\hat{\boldsymbol{\theta}}_{(-0.5)}$	$\widehat{b}$	.992916	1.000374	.977708
	$\widehat{c}$	1.380089	1.069138	1.081392
	mse	.739701	.103279	.07735
$\hat{\boldsymbol{\theta}}_{(0)}$	$\widehat{b}$	.976527	.995429	.9735
	$\widehat{c}$	1.369282	1.052883	1.085483
	mse	.739195	.093768	.077515
$\hat{\boldsymbol{\theta}}_{(1)}$	$\widehat{b}$	.974568	.982324	.967026
	$\widehat{c}$	1.405006	1.06289	1.083276
	mse	.765451	.098793	.080175

Table 5.1. Estimators for parameters of a Weibull (1,1).

Tables 5.1 and 5.2, show the maximum likelihood estimator with original data, $\hat{\boldsymbol{\theta}}$, Kolmogorov estimator, $\widehat{\boldsymbol{\theta}}_{Ko}$, and the minimum power-divergence estimator $\hat{\boldsymbol{\theta}}_{(\lambda)}$ for Normal and Weibull populations with parameters $b=1$, $c=1$, for Weibull populations and $\mu=0$, $\sigma=1$, for Normal populations. These values have been calculated by computer simulation for 1000 samples, number of classes $M=6$ and sample sizes $n=20,40$ and 60. The sums of the mean squared errors of the two parameters also appear in both tables.

The programs which calculate the $\widehat{\boldsymbol{\theta}}_{Ko}$ and the $\hat{\boldsymbol{\theta}}_{(\lambda)}$ need initial estimates. For Weibull populations these estimates have been calculated by the Dannenbring (1997) method, i.e.,

$$\widehat{b}=x_{([.6321n]+1)} \text{ and } \widehat{c}=\frac{\log(\log 2)}{\log(x_{(1/2)}/\widehat{b})}$$

where $x_{(1/2)}$ is the sample median.

In the case of normal populations we have taken as initial values the maximum likelihood estimators based on original data.

As we expected, the mean squared error (*mse*) associated with the maximum likelihood estimator based on the original Weibull and Normal values, $\hat{\boldsymbol{\theta}}$, is smaller than that associated with the minimum-power divergence estimators, $\hat{\boldsymbol{\theta}}_{(\lambda)}$. On the other hand the *mse* associated with the Kolmogorov estimator $\widehat{\boldsymbol{\theta}}_{Ko}$ is greater than the *mse* associated with all minimum power-divergence estimators, $\hat{\boldsymbol{\theta}}_{(\lambda)}$, for Weibull populations, although the Kolmogorov estimator, $\widehat{\boldsymbol{\theta}}_{Ko}$, is based on the original data and the minimum power-divergence estimators, $\hat{\boldsymbol{\theta}}_{(\lambda)}$, classify the original data into classes.

$N(0,1)$		$n=20$	$n=40$	$n=60$
$\hat{\boldsymbol{\theta}}$	$\widehat{\mu}$	-.011351	-.004091	-.002958
	$\widehat{\sigma}$	.960474	.978162	.984269
	mse	.038515	.018526	.012468
$\widehat{\boldsymbol{\theta}}_{Ko}$	$\widehat{\mu}$	-.014687	-.004947	-.003634
	$\widehat{\sigma}$	.972221	.985606	.987717
	mse	.043739	.021340	.014099
$\hat{\boldsymbol{\theta}}_{(-2)}$	$\widehat{\mu}$	-.012443	-.005091	-.005469
	$\widehat{\sigma}$	1.020231	1.007078	1.001058
	mse	.059777	.02851	.017253
$\hat{\boldsymbol{\theta}}_{(-1)}$	$\widehat{\mu}$	-.010154	-.005765	-.003588
	$\widehat{\sigma}$	1.002308	.995969	.993877
	mse	.048961	.024424	.015981
$\hat{\boldsymbol{\theta}}_{(-0.5)}$	$\widehat{\mu}$	-.010375	-.005316	-.002832
	$\widehat{\sigma}$	.991389	.989804	.992492
	mse	.044468	.023099	.015205
$\hat{\boldsymbol{\theta}}_{(0)}$	$\widehat{\mu}$	-.009577	-.006597	-.003736
	$\widehat{\sigma}$	.984915	.986656	.989282
	mse	.044277	.022511	.01498
$\hat{\boldsymbol{\theta}}_{(1)}$	$\widehat{\mu}$	-.009477	-.00711	-.002969
	$\widehat{\sigma}$	.970271	.979003	.981616
	mse	.039804	.021464	.014725

Table 5.2. Estimators for parameters of a Normal (0,1).

In Normal populations the *mse* associated with $\widehat{\boldsymbol{\theta}}_{Ko}$ coincides, more or less, with the *mse* associated with $\hat{\boldsymbol{\theta}}_{(1)}$. For each sample size, $\hat{\boldsymbol{\theta}}_{(1)}$ is the best of the minimum power-divergence estimators $\hat{\boldsymbol{\theta}}_{(\lambda)}$ for all sample sizes. It does not happen the same with Weibull populations because it depends on the sample size. For $n = 20$ and 40 it seems that $\hat{\boldsymbol{\theta}}_{(0)}$ is the best estimator and for $n = 60$, the best is $\hat{\boldsymbol{\theta}}_{(-0.5)}$.

From the comparison study we carried out, the new family of estimators we have introduced is a good alternative when it is necessary to classify the data. In Weibull populations, every member of the family of the minimum power-divergence estimator is even better than Kolmogorov estimator which is based on original data. In relation to Normal populations the results are still better, since $\hat{\boldsymbol{\theta}}_{(1)}$ is as good as both estimators based on original data $\hat{\boldsymbol{\theta}}$ and $\hat{\boldsymbol{\theta}}_{K_0}$.

5.3. Properties of the Minimum Phi-divergence Estimator

Throughout the Section, we assume that the model is correct, so that $\boldsymbol{\pi} = \boldsymbol{p}(\boldsymbol{\theta}_0)$, and $M_0 < M - 1$. Furthermore, we restrict ourselves to unknown parameters $\boldsymbol{\theta}_0$ satisfying the regularity conditions 1-6 introduced by Birch (1964):

1. $\boldsymbol{\theta}_0$ is an interior point of Θ.

2. $\pi_i = p_i(\boldsymbol{\theta}_0) > 0$ for $i = 1, \ldots, M$. Thus $\boldsymbol{\pi} = (\pi_1, \ldots, \pi_M)^T$ is an interior point of the set Δ_M.

3. The mapping $\boldsymbol{p} : \Theta \to \Delta_M$ is totally differentiable at $\boldsymbol{\theta}_0$ so that the partial derivatives of $p_i(\boldsymbol{\theta}_0)$ with respect to each θ_j exist at $\boldsymbol{\theta}_0$ and $p_i(\boldsymbol{\theta})$ has a linear approximation at $\boldsymbol{\theta}_0$ given by:

$$p_i(\boldsymbol{\theta}) = p_i(\boldsymbol{\theta}_0) + \sum_{j=1}^{M} (\theta_j - \theta_{0j}) \frac{\partial p_i(\boldsymbol{\theta}_0)}{\partial \theta_j} + o(\|\boldsymbol{\theta} - \boldsymbol{\theta}_0\|)$$

   where $o(\|\boldsymbol{\theta} - \boldsymbol{\theta}_0\|)$ denotes a function verifying

$$\lim_{\boldsymbol{\theta} \to \boldsymbol{\theta}_0} \frac{o(\|\boldsymbol{\theta} - \boldsymbol{\theta}_0\|)}{\|\boldsymbol{\theta} - \boldsymbol{\theta}_0\|} = 0.$$

4. The Jacobian matrix

$$\boldsymbol{J}(\boldsymbol{\theta}_0) = \left(\frac{\partial \boldsymbol{p}(\boldsymbol{\theta})}{\partial \boldsymbol{\theta}} \right)_{\boldsymbol{\theta} = \boldsymbol{\theta}_0} = \left(\frac{\partial p_i(\boldsymbol{\theta}_0)}{\partial \theta_j} \right)_{\substack{i=1,\ldots,M \\ j=1,\ldots,M_0}} \tag{5.6}$$

is of full rank (i.e., of rank M_0).

5. The inverse mapping $\boldsymbol{p}^{-1} : T \rightarrow \Theta$ is continuous at $\boldsymbol{p}(\boldsymbol{\theta}_0) = \boldsymbol{\pi}$.

6. The mapping $\boldsymbol{p} : \Theta \rightarrow \Delta_M$ is continuous at every point $\boldsymbol{\theta} \in \Theta$.

We first derive the Fisher information matrix for the multinomial model. In some parts of the book it will be necessary to distinguish between the Fisher information matrix corresponding to the original model and the Fisher information matrix corresponding to an associated discretized model, i.e., the Fisher information matrix corresponding to a multinomial model. For this reason we denote $\mathcal{I}_{\mathcal{F}}(\boldsymbol{\theta})$ the Fisher information matrix associated with the original model and $\boldsymbol{I}_F(\boldsymbol{\theta})$ the Fisher information matrix associated with the multinomial model.

Proposition 5.1

We consider the multinomial model defined in (5.1) based on a random sample of size n, with $n = \sum_{i=1}^{M} n_i$. Then the $M_0 \times M_0$ Fisher information matrix associated with the random sample of size n is given by

$$\begin{aligned}
\boldsymbol{I}_F^{(n)}(\boldsymbol{\theta}) &= \left(i_{(r,s)}(\boldsymbol{\theta})\right)_{r,s=1,\ldots,M_0} \\
&= n\left(\sum_{j=1}^{M} \frac{1}{p_j(\boldsymbol{\theta})} \frac{\partial p_j(\boldsymbol{\theta})}{\partial \theta_r} \frac{\partial p_j(\boldsymbol{\theta})}{\partial \theta_s}\right)_{r,s=1,\ldots,M_0} \\
&= n\boldsymbol{A}(\boldsymbol{\theta})^T \boldsymbol{A}(\boldsymbol{\theta})
\end{aligned}$$

where $\boldsymbol{A}(\boldsymbol{\theta})$ is a $M \times M_0$ matrix given by

$$\boldsymbol{A}(\boldsymbol{\theta}) = diag\left(\boldsymbol{p}(\boldsymbol{\theta})^{-1/2}\right) \boldsymbol{J}(\boldsymbol{\theta}). \tag{5.7}$$

For $n = 1$, $\boldsymbol{I}_F^{(1)}(\boldsymbol{\theta}) \equiv \boldsymbol{I}_F(\boldsymbol{\theta}) = \boldsymbol{A}(\boldsymbol{\theta})^T \boldsymbol{A}(\boldsymbol{\theta})$.

Proof.

We have

$$\log \Pr_{\boldsymbol{\theta}}(N_1 = n_1, \ldots, N_M = n_M) = k + \sum_{j=1}^{M} n_j \log p_j(\boldsymbol{\theta}),$$

where k is independent of $\boldsymbol{\theta}$.

Therefore,

$$\frac{\partial \log \Pr_{\boldsymbol{\theta}}(N_1 = n_1, \ldots, N_M = n_M)}{\partial \theta_r} = \sum_{j=1}^{M} \frac{n_j}{p_j(\boldsymbol{\theta})} \frac{\partial p_j(\boldsymbol{\theta})}{\partial \theta_r},$$

and

$$
\begin{aligned}
i_{(r,s)}(\boldsymbol{\theta}) &= E\left[\frac{\partial \log \Pr_{\boldsymbol{\theta}}(.)}{\partial \theta_r}\frac{\partial \log \Pr_{\boldsymbol{\theta}}(.)}{\partial \theta_s}\right] \\
&= E\left[\sum_{i,j=1}^{M} \frac{N_i}{p_i(\boldsymbol{\theta})}\frac{\partial p_i(\boldsymbol{\theta})}{\partial \theta_r}\frac{N_j}{p_j(\theta)}\frac{\partial p_j(\boldsymbol{\theta})}{\partial \theta_s}\right] \\
&= E\left[\sum_{i=1}^{M} \frac{N_i^2}{p_i^2(\boldsymbol{\theta})}\frac{\partial p_i(\boldsymbol{\theta})}{\partial \theta_r}\frac{\partial p_i(\boldsymbol{\theta})}{\partial \theta_s}\right] + E\left[\sum_{\substack{i,j=1\\ i\neq j}}^{M} \frac{N_i}{p_i(\boldsymbol{\theta})}\frac{\partial p_i(\boldsymbol{\theta})}{\partial \theta_r}\frac{N_j}{p_j(\boldsymbol{\theta})}\frac{\partial p_j(\boldsymbol{\theta})}{\partial \theta_s}\right].
\end{aligned}
$$

But

$$
E\left[N_i N_j\right] = \begin{cases} np_i\left(\boldsymbol{\theta}\right)\left(1-p_i\left(\boldsymbol{\theta}\right)\right)+n^2 p_i^2\left(\boldsymbol{\theta}\right) & \text{if } i=j \\ -np_i\left(\boldsymbol{\theta}\right)p_j\left(\boldsymbol{\theta}\right)+n^2 p_i\left(\boldsymbol{\theta}\right)p_j\left(\boldsymbol{\theta}\right) & \text{if } i\neq j, \end{cases}
$$

then we have

$$
\begin{aligned}
i_{(r,s)}(\boldsymbol{\theta}) =& \sum_{i=1}^{M} \frac{np_i\left(\boldsymbol{\theta}\right)\left(1-p_i\left(\boldsymbol{\theta}\right)\right)+n^2 p_i\left(\boldsymbol{\theta}\right)^2}{p_i^2(\boldsymbol{\theta})}\frac{\partial p_i(\boldsymbol{\theta})}{\partial \theta_r}\frac{\partial p_i(\boldsymbol{\theta})}{\partial \theta_s} \\
&+ \sum_{\substack{i,j=1\\ i\neq j}}^{M} \frac{-np_i\left(\boldsymbol{\theta}\right)p_j\left(\boldsymbol{\theta}\right)+n^2 p_i\left(\boldsymbol{\theta}\right)p_j\left(\boldsymbol{\theta}\right)}{p_i(\boldsymbol{\theta})p_j\left(\boldsymbol{\theta}\right)}\frac{\partial p_i(\boldsymbol{\theta})}{\partial \theta_r}\frac{\partial p_j(\boldsymbol{\theta})}{\partial \theta_s} \\
=& \; n\sum_{i=1}^{M} \frac{1}{p_i\left(\boldsymbol{\theta}\right)}\frac{\partial p_i(\boldsymbol{\theta})}{\partial \theta_r}\frac{\partial p_i(\boldsymbol{\theta})}{\partial \theta_s} - n\left(1-n\right)\sum_{i=1}^{M}\frac{\partial p_i(\boldsymbol{\theta})}{\partial \theta_r}\frac{\partial p_i(\boldsymbol{\theta})}{\partial \theta_s} \\
&- n\sum_{\substack{i,j=1\\ i\neq j}}^{M}\frac{\partial p_i(\boldsymbol{\theta})}{\partial \theta_r}\frac{\partial p_j(\boldsymbol{\theta})}{\partial \theta_s} + n^2\sum_{\substack{i,j=1\\ i\neq j}}^{M}\frac{\partial p_i(\boldsymbol{\theta})}{\partial \theta_r}\frac{\partial p_j(\boldsymbol{\theta})}{\partial \theta_s} \\
=& \; n\sum_{i=1}^{M} \frac{1}{p_i\left(\boldsymbol{\theta}\right)}\frac{\partial p_i(\boldsymbol{\theta})}{\partial \theta_r}\frac{\partial p_i(\boldsymbol{\theta})}{\partial \theta_s} + \left(n^2-n\right)\sum_{i,j=1}^{M}\frac{\partial p_i(\boldsymbol{\theta})}{\partial \theta_r}\frac{\partial p_j(\boldsymbol{\theta})}{\partial \theta_s} \\
=& \; n\sum_{i=1}^{M} \frac{1}{p_i\left(\boldsymbol{\theta}\right)}\frac{\partial p_i(\boldsymbol{\theta})}{\partial \theta_r}\frac{\partial p_i(\boldsymbol{\theta})}{\partial \theta_s}.
\end{aligned}
$$

By (5.7)

$$\boldsymbol{A}(\boldsymbol{\theta}) = diag\left(\boldsymbol{p}(\boldsymbol{\theta})^{-1/2}\right)\boldsymbol{J}(\boldsymbol{\theta}),$$

therefore

$$\boldsymbol{A}(\boldsymbol{\theta})^T\boldsymbol{A}(\boldsymbol{\theta}) = \left(\sum_{j=1}^{M}\frac{1}{p_j(\boldsymbol{\theta})}\frac{\partial p_j(\boldsymbol{\theta})}{\partial\theta_r}\frac{\partial p_j(\boldsymbol{\theta})}{\partial\theta_s}\right)_{\substack{r=1,\dots,M_0\\ s=1,\dots,M_0}}.$$

■

5.3.1. Asymptotic Properties

In order to prove the following theorem we use the Implicit Function Theorem: Let

$$\boldsymbol{F} = (F_1, \dots, F_{M_0}) : \mathbb{R}^{M+M_0} \to \mathbb{R}^{M_0}$$

be continuously differentiable in an open set $D \subset \mathbb{R}^{M+M_0}$, containing the point

$$\left(\boldsymbol{x}_0 = \left(x_1^0, \dots, x_M^0\right), \boldsymbol{y}_0 = \left(y_1^0, \dots, y_{M_0}^0\right)\right)$$

for which $\boldsymbol{F}(\boldsymbol{x}_0, \boldsymbol{y}_0) = \boldsymbol{0}$. Further, suppose that the matrix

$$\boldsymbol{J}_F = \left(\frac{\partial F_i}{\partial y_j}\right)_{\substack{i=1,\dots,M_0\\ j=1,\dots,M_0}}$$

is nonsingular at $(\boldsymbol{x}_0, \boldsymbol{y}_0)$. Then there exists a neighborhood U of $(\boldsymbol{x}_0, \boldsymbol{y}_0)$ such that the matrix $\boldsymbol{J}_F$ is nonsingular for all $(\boldsymbol{x}, \boldsymbol{y}) \in U$, an open set $A \subset \mathbb{R}^M$ containing $\boldsymbol{x}_0$ and a continuously differentiable function $\boldsymbol{g} = (g_1, \dots, g_{M_0}) : A \to \mathbb{R}^{M_0}$ such that

$$\{(\boldsymbol{x}, \boldsymbol{y}) \in U : \boldsymbol{F}(\boldsymbol{x}, \boldsymbol{y}) = 0\} = \{(\boldsymbol{x}, g(\boldsymbol{x})) : \boldsymbol{x} \in A\}$$

(see, e.g., p. 148 in Fleming (1977)).

Theorem 5.1

Let $\phi \in \Phi^*$ be a twice continuously differentiable function in $x > 0$ with $\phi''(1) > 0$ and $\boldsymbol{\pi} = \boldsymbol{p}(\boldsymbol{\theta}_0)$. Under the Birch regularity conditions and assuming that the function $\boldsymbol{p}$: $\Theta \longrightarrow \Delta_M$ has continuous second partial derivatives in a neighborhood of $\boldsymbol{\theta}_0$, it holds

$$\widehat{\boldsymbol{\theta}}_\phi = \boldsymbol{\theta}_0 + \boldsymbol{I}_F(\boldsymbol{\theta}_0)^{-1} A(\boldsymbol{\theta}_0)^T diag\left(\boldsymbol{\pi}^{-1/2}\right)(\widehat{\boldsymbol{p}} - \boldsymbol{\pi}) + o(\| \widehat{\boldsymbol{p}} - \boldsymbol{\pi} \|),$$

where $\widehat{\boldsymbol{\theta}}_{\phi}$ is unique in a neighborhood of $\boldsymbol{\theta}_0$.

Proof.

Let l^M be the interior of the unit M-dimensional cube with $\Delta_M \subset l^M$. Let V be a neighborhood of $\boldsymbol{\theta}_0$ on which $\boldsymbol{p} : \Theta \to \Delta_M$ has continuous second partial derivatives.

Let

$$\boldsymbol{F} = (F_1, ..., F_{M_0}) : l^M \times V \to \mathbb{R}^{M_0}$$

be defined by

$$F_j\left(\widetilde{p}_1, ..., \widetilde{p}_M; \theta_1, ..., \theta_{M_0}\right) = \frac{\partial D_{\phi}(\widetilde{\mathbf{p}}, \boldsymbol{p}(\boldsymbol{\theta}))}{\partial \theta_j}, \ \forall j = 1, ..., M_0.$$

It holds

$$F_j\left(\pi_1, ..., \pi_M; \theta_{01}, ..., \theta_{0M_0}\right) = 0, \ \forall j = 1, ..., M_0$$

due to

$$\frac{\partial D_{\phi}(\widetilde{\mathbf{p}}, \boldsymbol{p}(\boldsymbol{\theta}))}{\partial \theta_j} = \sum_{l=1}^{M} \left(\phi\left(\frac{\widetilde{p}_l}{p_l(\boldsymbol{\theta})}\right) - \phi'\left(\frac{\widetilde{p}_l}{p_l(\boldsymbol{\theta})}\right) \frac{\widetilde{p}_l}{p_l(\boldsymbol{\theta})} \right) \frac{\partial p_l(\boldsymbol{\theta})}{\partial \theta_j} \qquad \forall j = 1, ..., M_0.$$

Since

$$\begin{aligned} \frac{\partial}{\partial \theta_r}\left(\frac{\partial D_{\phi}(\widetilde{\mathbf{p}}, \boldsymbol{p}(\boldsymbol{\theta}))}{\partial \theta_j}\right) &= -\sum_{l=1}^{M} \phi'\left(\frac{\widetilde{p}_l}{p_l(\boldsymbol{\theta})}\right) \frac{\widetilde{p}_l}{p_l(\boldsymbol{\theta})^2} \frac{\partial p_l(\boldsymbol{\theta})}{\partial \theta_r} \frac{\partial p_l(\boldsymbol{\theta})}{\partial \theta_j} \\ &+ \sum_{l=1}^{M} \phi''\left(\frac{\widetilde{p}_l}{p_l(\boldsymbol{\theta})}\right) \frac{\widetilde{p}_l}{p_l(\boldsymbol{\theta})^2} \frac{\partial p_l(\boldsymbol{\theta})}{\partial \theta_r} \frac{\partial p_l(\boldsymbol{\theta})}{\partial \theta_j} \frac{\widetilde{p}_l}{p_l(\boldsymbol{\theta})} \\ &+ \sum_{l=1}^{M} \phi'\left(\frac{\widetilde{p}_l}{p_l(\boldsymbol{\theta})}\right) \frac{\widetilde{p}_l}{p_l(\boldsymbol{\theta})^2} \frac{\partial p_l(\boldsymbol{\theta})}{\partial \theta_r} \frac{\partial p_l(\boldsymbol{\theta})}{\partial \theta_j} \\ &+ \sum_{l=1}^{M} \frac{\partial^2 p_l(\boldsymbol{\theta})}{\partial \theta_j \partial \theta_r} \left(\phi\left(\frac{\widetilde{p}_l}{p_l(\boldsymbol{\theta})}\right) - \phi'\left(\frac{\widetilde{p}_l}{p_l(\boldsymbol{\theta})}\right) \frac{\widetilde{p}_l}{p_l(\boldsymbol{\theta})} \right), \end{aligned}$$

the $M_0 \times M_0$ matrix $\boldsymbol{J_F}$ associated with the function $\boldsymbol{F}$ at the point $(\boldsymbol{p}(\boldsymbol{\theta}_0), \boldsymbol{\theta}_0)$

is given by

$$\begin{aligned}
\frac{\partial \boldsymbol{F}}{\partial \boldsymbol{\theta}_0} &= \left(\frac{\partial \boldsymbol{F}}{\partial \boldsymbol{\theta}}\right)_{(\widetilde{\mathbf{p}},\boldsymbol{\theta})=(\pi_1,...,\pi_M;\theta_{01},...,\theta_{0M_0})} \\
&= \left(\left(\frac{\partial}{\partial \theta_r}\left(\frac{\partial D_\phi(\widetilde{\mathbf{p}},\boldsymbol{p}(\boldsymbol{\theta}))}{\partial \theta_j}\right)\right)_{\substack{j=1,...,M_0\\ r=1,...,M_0}}\right)_{(\widetilde{\mathbf{p}},\boldsymbol{\theta})=(\pi_1,...,\pi_M;\theta_{01},...,\theta_{0M_0})} \\
&= \phi''(1)\left(\sum_{l=1}^{M}\frac{1}{p_l(\boldsymbol{\theta}_0)}\frac{\partial p_l(\boldsymbol{\theta}_0)}{\partial \theta_r}\frac{\partial p_l(\boldsymbol{\theta}_0)}{\partial \theta_j}\right)_{\substack{r=1,...,M_0\\ j=1,...,M_0}} \\
&= \phi''(1)\,\boldsymbol{A}(\boldsymbol{\theta}_0)^T\boldsymbol{A}(\boldsymbol{\theta}_0).
\end{aligned}$$

We recall that if $\boldsymbol{B}$ is a $p\times q$ matrix with rank p and $\boldsymbol{C}$ is a $q\times s$ matrix with $rank(\boldsymbol{C}) = q$ then $rank(\boldsymbol{BC}) = p$. By taking

$$\boldsymbol{B} = \boldsymbol{J}(\boldsymbol{\theta}_0)^T_{M_0\times M} \text{ and } \boldsymbol{C} = diag\left(\boldsymbol{p}(\boldsymbol{\theta}_0)^{-1/2}\right)_{M\times M}$$

it follows that $\boldsymbol{A}(\boldsymbol{\theta}_0)^T = \boldsymbol{BC}$ has rank M_0 by condition 4 of Birch. Also,

$$rank\left(\boldsymbol{A}(\boldsymbol{\theta}_0)^T\boldsymbol{A}(\boldsymbol{\theta}_0)\right) = rank\left(\boldsymbol{A}(\boldsymbol{\theta}_0)\boldsymbol{A}(\boldsymbol{\theta}_0)^T\right) = rank\left(\boldsymbol{A}(\boldsymbol{\theta}_0)\right) = M_0.$$

Therefore, the $M_0\times M_0$ matrix $\dfrac{\partial \boldsymbol{F}}{\partial \boldsymbol{\theta}_0}$ is nonsingular at $(\pi_1,...,\pi_M;\theta_{01},...,\theta_{0M_0})$.

Applying the Implicit Function Theorem there exists a neighborhood U of $(\boldsymbol{p}(\boldsymbol{\theta}_0),\boldsymbol{\theta}_0)$ such that the matrix $\boldsymbol{J}_F$ is nonsingular (in our case $\boldsymbol{J}_F$ in $(\boldsymbol{p}(\boldsymbol{\theta}_0),\boldsymbol{\theta}_0)$ is positive definite and then $\boldsymbol{J}_F$ is positive definite for all $(\widetilde{\mathbf{p}},\boldsymbol{\theta})\in U$, because $\boldsymbol{F}$ is continuously differentiable). Also, there exists a continuously differentiable function $\widetilde{\boldsymbol{\theta}}:A\subset l^M\to\mathbb{R}^{M_0}$ such that $\boldsymbol{p}(\boldsymbol{\theta}_0)\in A$ and

$$\{(\widetilde{\mathbf{p}},\boldsymbol{\theta})\in U : \boldsymbol{F}(\widetilde{\mathbf{p}},\boldsymbol{\theta}) = \mathbf{0}\} = \left\{(\widetilde{\mathbf{p}},\widetilde{\boldsymbol{\theta}}(\widetilde{\mathbf{p}})) : \widetilde{\mathbf{p}}\in A\right\}. \tag{5.8}$$

We can observe that $\widetilde{\boldsymbol{\theta}}(\boldsymbol{p}(\boldsymbol{\theta}_0))$ is an arg inf of

$$\Psi(\boldsymbol{\theta}) = D_\phi(\boldsymbol{p}(\boldsymbol{\theta}_0),\boldsymbol{p}(\boldsymbol{\theta}))$$

because $\boldsymbol{p}(\boldsymbol{\theta}_0)\in A$ and then

$$\boldsymbol{F}\left(\boldsymbol{p}(\boldsymbol{\theta}_0),\widetilde{\boldsymbol{\theta}}(\boldsymbol{p}(\boldsymbol{\theta}_0))\right) = \frac{\partial D_\phi\left(\boldsymbol{p}(\boldsymbol{\theta}_0),\boldsymbol{p}\left(\widetilde{\boldsymbol{\theta}}(\boldsymbol{p}(\boldsymbol{\theta}_0))\right)\right)}{\partial \boldsymbol{\theta}} = 0$$

and also by (5.8), $(\boldsymbol{p}(\boldsymbol{\theta}_0), \widetilde{\boldsymbol{\theta}}(\boldsymbol{p}(\boldsymbol{\theta}_0))) \in U$ and then $\boldsymbol{J}_F$ is positive definite at $\left(\boldsymbol{p}(\boldsymbol{\theta}_0), \widetilde{\boldsymbol{\theta}}(\boldsymbol{p}(\boldsymbol{\theta}_0))\right)$. Therefore,

$$D_\phi\left(\boldsymbol{p}(\boldsymbol{\theta}_0), \boldsymbol{p}\left(\widetilde{\boldsymbol{\theta}}(\boldsymbol{p}(\boldsymbol{\theta}_0))\right)\right) = \inf_{\boldsymbol{\theta} \in \Theta} D_\phi(\boldsymbol{p}(\boldsymbol{\theta}_0), \boldsymbol{p}(\boldsymbol{\theta}))$$

and by the ϕ-divergence properties $\widetilde{\boldsymbol{\theta}}(\boldsymbol{p}(\boldsymbol{\theta}_0)) = \boldsymbol{\theta}_0$.

Applying the chain rule, we get

$$\frac{\partial \boldsymbol{F}(\widetilde{\boldsymbol{p}}, \widetilde{\boldsymbol{\theta}}(\widetilde{\boldsymbol{p}}))}{\partial \widetilde{\boldsymbol{p}}} + \frac{\partial \boldsymbol{F}(\widetilde{\boldsymbol{p}}, \widetilde{\boldsymbol{\theta}}(\widetilde{\boldsymbol{p}}))}{\partial \widetilde{\boldsymbol{\theta}}(\widetilde{\boldsymbol{p}})} \frac{\partial \widetilde{\boldsymbol{\theta}}(\widetilde{\boldsymbol{p}})}{\partial \widetilde{\boldsymbol{p}}} = \boldsymbol{0}$$

and, for $\widetilde{\boldsymbol{p}}=\boldsymbol{\pi}$,

$$\frac{\partial \boldsymbol{F}}{\partial \boldsymbol{\pi}} + \frac{\partial \boldsymbol{F}}{\partial \boldsymbol{\theta}_0} \frac{\partial \boldsymbol{\theta}_0}{\partial \boldsymbol{\pi}} = \boldsymbol{0}.$$

Further we know that

$$\frac{\partial \boldsymbol{F}}{\partial \boldsymbol{\theta}_0} = \phi''(1) \boldsymbol{A}(\boldsymbol{\theta}_0)^T \boldsymbol{A}(\boldsymbol{\theta}_0)$$

and we shall establish later that the $M_0 \times M$ matrix $\dfrac{\partial \boldsymbol{F}}{\partial \boldsymbol{\pi}}$ is

$$\frac{\partial \boldsymbol{F}}{\partial \boldsymbol{\pi}} = -\phi''(1) \boldsymbol{A}(\boldsymbol{\theta}_0)^T diag\left(\boldsymbol{p}(\boldsymbol{\theta}_0)^{-1/2}\right). \tag{5.9}$$

Therefore the $M_0 \times M$ matrix $\dfrac{\partial \boldsymbol{\theta}_0}{\partial \boldsymbol{\pi}}$ is

$$\frac{\partial \boldsymbol{\theta}_0}{\partial \boldsymbol{\pi}} = (\boldsymbol{A}(\boldsymbol{\theta}_0)^T \boldsymbol{A}(\boldsymbol{\theta}_0))^{-1} \boldsymbol{A}(\boldsymbol{\theta}_0)^T diag\left(\boldsymbol{p}(\boldsymbol{\theta}_0)^{-1/2}\right).$$

The Taylor expansion of the function $\widetilde{\boldsymbol{\theta}}$ around $\boldsymbol{\pi}$ yields

$$\widetilde{\boldsymbol{\theta}}(\widetilde{\boldsymbol{p}}) = \widetilde{\boldsymbol{\theta}}(\boldsymbol{\pi}) + \left(\frac{\partial \widetilde{\boldsymbol{\theta}}(\widetilde{\boldsymbol{p}})}{\partial \widetilde{\boldsymbol{p}}}\right)_{\widetilde{\boldsymbol{p}}=\boldsymbol{\pi}} (\widetilde{\boldsymbol{p}} - \boldsymbol{\pi}) + o(\| \widetilde{\boldsymbol{p}} - \boldsymbol{\pi} \|).$$

As $\widetilde{\boldsymbol{\theta}}(\boldsymbol{\pi}) = \boldsymbol{\theta}_0$ we obtain from here

$$\widetilde{\boldsymbol{\theta}}(\widetilde{\boldsymbol{p}}) = \boldsymbol{\theta}_0 + \left(A(\boldsymbol{\theta}_0)^T A(\boldsymbol{\theta}_0)\right)^{-1} A(\boldsymbol{\theta}_0)^T diag\left(\boldsymbol{p}(\boldsymbol{\theta}_0)^{-1/2}\right)(\widetilde{\boldsymbol{p}} - \boldsymbol{\pi}) + o(\| \widetilde{\boldsymbol{p}} - \boldsymbol{\pi} \|).$$

We know that $\widehat{\boldsymbol{p}} \underset{n \to \infty}{\overset{c.s.}{\longrightarrow}} \boldsymbol{\pi} = \boldsymbol{p}(\boldsymbol{\theta}_0)$, so that $\widehat{\boldsymbol{p}} \in A$ and, consequently, $\widetilde{\boldsymbol{\theta}}(\widehat{\boldsymbol{p}})$ is the unique solution of the system of equations

$$\frac{\partial D_\phi\left(\widehat{\boldsymbol{p}},\boldsymbol{p}(\widetilde{\boldsymbol{\theta}}(\widehat{\boldsymbol{p}}))\right)}{\partial\theta_j}=0,\qquad j=1,...,M_0$$

and also $\left(\widehat{\boldsymbol{p}},\widetilde{\boldsymbol{\theta}}(\widehat{\boldsymbol{p}})\right)\in U$. Therefore, $\widetilde{\boldsymbol{\theta}}(\widehat{\boldsymbol{p}})$ is the minimum ϕ-divergence estimator, $\widehat{\boldsymbol{\theta}}_\phi$, satisfying the relation

$$\begin{aligned}\widehat{\boldsymbol{\theta}}_\phi &= \boldsymbol{\theta}_0+\left(\boldsymbol{A}(\boldsymbol{\theta}_0)^T\boldsymbol{A}(\boldsymbol{\theta}_0)\right)^{-1}\boldsymbol{A}(\boldsymbol{\theta}_0)^T diag\left(\boldsymbol{p}\left(\boldsymbol{\theta}_0\right)^{-1/2}\right)\left(\widehat{\boldsymbol{p}}-\boldsymbol{p}(\boldsymbol{\theta}_0)\right)\\ &\quad +o(\|\widehat{\boldsymbol{p}}-\boldsymbol{p}(\boldsymbol{\theta}_0)\|).\end{aligned}$$

Finally, we establish the formula (5.9). We calculate the (i,j)th-element of the $M_0\times M$ matrix $\dfrac{\partial \boldsymbol{F}}{\partial\boldsymbol{\pi}}$,

$$\begin{aligned}\frac{\partial}{\partial p_i}\left(\frac{\partial D_\phi\left(\boldsymbol{p},\boldsymbol{p}\left(\boldsymbol{\theta}\right)\right)}{\partial\theta_j}\right) &= \frac{\partial}{\partial p_i}\left(\sum_{l=1}^{M}\left(\phi\left(\frac{p_l}{p_l(\boldsymbol{\theta})}\right)-\phi'\left(\frac{p_l}{p_l(\boldsymbol{\theta})}\right)\frac{p_l}{p_l(\boldsymbol{\theta})}\right)\frac{\partial p_l\left(\boldsymbol{\theta}\right)}{\partial\theta_j}\right)\\ &= \frac{1}{p_i(\boldsymbol{\theta})}\left(-\frac{p_i}{p_i(\boldsymbol{\theta})}\phi''\left(\frac{p_i}{p_i(\boldsymbol{\theta})}\right)\right)\frac{\partial p_i\left(\boldsymbol{\theta}\right)}{\partial\theta_j}\end{aligned}$$

and for $(\pi_1,...,\pi_M;\theta_{01},...,\theta_{0M_0})$ we have

$$\left(\frac{\partial}{\partial p_i}\left(\frac{\partial D_\phi(\boldsymbol{p},\boldsymbol{p}\left(\boldsymbol{\theta}_0\right))}{\partial\theta_j}\right)\right)=-\frac{1}{p_i\left(\boldsymbol{\theta}_0\right)}\phi''\left(1\right)\frac{\partial p_i\left(\boldsymbol{\theta}_0\right)}{\partial\theta_j}.$$

Since $\boldsymbol{A}(\boldsymbol{\theta}_0)=diag\left(\boldsymbol{p}\left(\boldsymbol{\theta}_0\right)^{-1/2}\right)\boldsymbol{J}\left(\boldsymbol{\theta}_0\right)$ then (5.9) holds. ∎

Theorem 5.2

Under the assumptions of Theorem 5.1, it holds

$$\sqrt{n}(\widehat{\boldsymbol{\theta}}_\phi-\boldsymbol{\theta}_0)\xrightarrow[n\to\infty]{L}N(\boldsymbol{0},\boldsymbol{I}_F\left(\boldsymbol{\theta}_0\right)^{-1}).$$

Proof.

Applying the Central Limit Theorem, we get

$$\sqrt{n}(\widehat{\boldsymbol{p}}-\boldsymbol{p}(\boldsymbol{\theta}_0))\xrightarrow[n\to\infty]{L}N(\boldsymbol{0},\boldsymbol{\Sigma}_{\boldsymbol{p}(\boldsymbol{\theta}_0)})$$

being $\boldsymbol{\Sigma}_{\boldsymbol{p}(\boldsymbol{\theta}_0)}$ the $M\times M$ matrix

$$\boldsymbol{\Sigma}_{\boldsymbol{p}(\boldsymbol{\theta}_0)}=diag\left(\boldsymbol{p}\left(\boldsymbol{\theta}_0\right)\right)-\boldsymbol{p}\left(\boldsymbol{\theta}_0\right)\boldsymbol{p}\left(\boldsymbol{\theta}_0\right)^T.$$

Consequently

$$\sqrt{n}(\widehat{\boldsymbol{\theta}}_{\phi}-\boldsymbol{\theta}_0) \xrightarrow[n\to\infty]{L} N\left(\mathbf{0},\boldsymbol{\Sigma}^*\right),$$

where

$$\begin{aligned}\boldsymbol{\Sigma}^* = & \left(\boldsymbol{A}(\boldsymbol{\theta}_0)^T\boldsymbol{A}(\boldsymbol{\theta}_0)\right)^{-1}\boldsymbol{A}(\boldsymbol{\theta}_0)^T diag\left(\mathbf{p}\left(\boldsymbol{\theta}_0\right)^{-1/2}\right)\boldsymbol{\Sigma}_{\mathbf{p}(\boldsymbol{\theta}_0)} \\ \times & \ diag\left(\mathbf{p}\left(\boldsymbol{\theta}_0\right)^{-1/2}\right)\boldsymbol{A}(\boldsymbol{\theta}_0)\left(\boldsymbol{A}(\boldsymbol{\theta}_0)^T\boldsymbol{A}(\boldsymbol{\theta}_0)\right)^{-1},\end{aligned}$$

because $\widehat{\boldsymbol{\theta}}_{\phi}$ has the expression

$$\boldsymbol{\theta}_0+\left(\boldsymbol{A}(\boldsymbol{\theta}_0)^T\boldsymbol{A}(\boldsymbol{\theta}_0)\right)^{-1}\boldsymbol{A}(\boldsymbol{\theta}_0)^T diag\left(\mathbf{p}\left(\boldsymbol{\theta}_0\right)^{-1/2}\right)\left(\widehat{\mathbf{p}}-\mathbf{p}(\boldsymbol{\theta}_0)\right)+o(\|\,\widehat{\mathbf{p}}-\mathbf{p}(\boldsymbol{\theta}_0)\,\|).$$

For the $M\times M$ matrix

$$\boldsymbol{B}=diag\left(\mathbf{p}\left(\boldsymbol{\theta}_0\right)^{-1/2}\right)\boldsymbol{\Sigma}_{\mathbf{p}(\boldsymbol{\theta}_0)}diag\left(\mathbf{p}\left(\boldsymbol{\theta}_0\right)^{-1/2}\right),$$

we have

$$\begin{aligned}\boldsymbol{B}= & \ diag\left(\mathbf{p}\left(\boldsymbol{\theta}_0\right)^{-1/2}\right)\left(diag\left(\mathbf{p}\left(\boldsymbol{\theta}_0\right)\right)-\mathbf{p}\left(\boldsymbol{\theta}_0\right)\mathbf{p}\left(\boldsymbol{\theta}_0\right)^T\right)diag\left(\mathbf{p}\left(\boldsymbol{\theta}_0\right)^{-1/2}\right) \\ = & \ diag\left(\mathbf{p}\left(\boldsymbol{\theta}_0\right)^{-1/2}\right)diag\left(\mathbf{p}\left(\boldsymbol{\theta}_0\right)\right)diag\left(\mathbf{p}\left(\boldsymbol{\theta}_0\right)^{-1/2}\right) \\ - & \ diag\left(\mathbf{p}\left(\boldsymbol{\theta}_0\right)^{-1/2}\right)\mathbf{p}\left(\boldsymbol{\theta}_0\right)\mathbf{p}\left(\boldsymbol{\theta}_0\right)^T diag\left(\mathbf{p}\left(\boldsymbol{\theta}_0\right)^{-1/2}\right) \\ = & \ \boldsymbol{I}_{M\times M}-\sqrt{\mathbf{p}\left(\boldsymbol{\theta}_0\right)}\sqrt{\mathbf{p}\left(\boldsymbol{\theta}_0\right)}^T.\end{aligned}$$

Therefore,

$$\begin{aligned}\boldsymbol{\Sigma}^*= & \left(\boldsymbol{A}(\boldsymbol{\theta}_0)^T\boldsymbol{A}(\boldsymbol{\theta}_0)\right)^{-1} \\ - & \left(\boldsymbol{A}(\boldsymbol{\theta}_0)^T\boldsymbol{A}(\boldsymbol{\theta}_0)\right)^{-1}\boldsymbol{A}(\boldsymbol{\theta}_0)^T\sqrt{\mathbf{p}\left(\boldsymbol{\theta}_0\right)}\sqrt{\mathbf{p}\left(\boldsymbol{\theta}_0\right)}^T\boldsymbol{A}(\boldsymbol{\theta}_0)\left(\boldsymbol{A}(\boldsymbol{\theta}_0)^T\boldsymbol{A}(\boldsymbol{\theta}_0)\right)^{-1}.\end{aligned}$$

Finally we establish that $\sqrt{\mathbf{p}\left(\boldsymbol{\theta}_0\right)}^T\boldsymbol{A}(\boldsymbol{\theta}_0)=\mathbf{0}_{1\times M_0}$.

We have

$$\boldsymbol{A}(\boldsymbol{\theta}_0) = \begin{pmatrix} p_1(\boldsymbol{\theta}_0)^{-1/2} \dfrac{\partial p_1(\boldsymbol{\theta}_0)}{\partial \theta_1} & \cdots & p_1(\boldsymbol{\theta}_0)^{-1/2} \dfrac{\partial p_1(\boldsymbol{\theta}_0)}{\partial \theta_{M_0}} \\ . & \cdots & . \\ . & \cdots & . \\ . & \cdots & . \\ p_M(\boldsymbol{\theta}_0)^{-1/2} \dfrac{\partial p_M(\boldsymbol{\theta}_0)}{\partial \theta_1} & \cdots & p_M(\boldsymbol{\theta}_0)^{-1/2} \dfrac{\partial p_M(\boldsymbol{\theta}_0)}{\partial \theta_{M_0}} \end{pmatrix}$$

then

$$\begin{aligned} \sqrt{\boldsymbol{p}(\boldsymbol{\theta}_0)^T}\boldsymbol{A}(\boldsymbol{\theta}_0) &= \left(p_1(\boldsymbol{\theta}_0)^{1/2}, ..., p_M(\boldsymbol{\theta}_0)^{1/2}\right)\boldsymbol{A}(\boldsymbol{\theta}_0) \\ &= \left(\sum_{i=1}^M \frac{\partial p_i(\boldsymbol{\theta}_0)}{\partial \theta_1}, ..., \sum_{i=1}^M \frac{\partial p_i(\boldsymbol{\theta}_0)}{\partial \theta_{M_0}}\right) = \mathbf{0}_{1\times M_0}. \end{aligned}$$

Finally

$$\boldsymbol{\Sigma}^* = \left(\boldsymbol{A}(\boldsymbol{\theta}_0)^T\boldsymbol{A}(\boldsymbol{\theta}_0)\right)^{-1} = \boldsymbol{I}_F(\boldsymbol{\theta}_0)^{-1}.$$

■

Based on the last two theorems we have that the minimum ϕ-divergence estimator is a BAN (Best Asymptotically Normal) estimator.

5.3.2. Minimum Phi-divergence Functional Robustness

In this section we consider deviations from the discrete model,

$$\boldsymbol{p}(\boldsymbol{\theta}) = (p_1(\boldsymbol{\theta}), \ldots, p_M(\boldsymbol{\theta}))^T,$$

given by $\boldsymbol{p}_\varepsilon(\boldsymbol{\theta}) = (1-\varepsilon)\boldsymbol{p}(\boldsymbol{\theta}) + \varepsilon\boldsymbol{p}$ for $\varepsilon > 0$, $\boldsymbol{\theta} \in \Theta$ and $\boldsymbol{p} \in \Delta_M$. Let $\boldsymbol{\theta}_\phi^\varepsilon(\boldsymbol{p})$ be the vector that minimizes the function

$$g_\varepsilon(\boldsymbol{p}, \boldsymbol{\theta}) = \sum_{i=1}^M p_i(\boldsymbol{\theta}, \varepsilon)\phi\left(\frac{p_i}{p_i(\boldsymbol{\theta}, \varepsilon)}\right),$$

where $\boldsymbol{p}_\varepsilon(\boldsymbol{\theta}) = (p_1(\boldsymbol{\theta}, \varepsilon), \ldots, p_M(\boldsymbol{\theta}, \varepsilon))^T$. In order to guarantee the robustness of $\boldsymbol{\theta}_\phi(\boldsymbol{p})$, we have to verify that slight deviations of $\boldsymbol{p}(\boldsymbol{\theta})$ correspond to slight deviations of $\boldsymbol{\theta}_\phi^\varepsilon(\boldsymbol{p})$ or, analytically, that $\lim_{\varepsilon\to 0}\boldsymbol{\theta}_\phi^\varepsilon(\boldsymbol{p}) = \boldsymbol{\theta}_\phi(\boldsymbol{p})$. The following theorem gives conditions that guarantee the functional robustness.

Theorem 5.3

Let the assumptions of Theorem 5.1 be fulfilled and in addition Θ *is a compact set. Then*

$$\lim_{\varepsilon \to 0} \boldsymbol{\theta}_{\phi}^{\varepsilon}(\boldsymbol{p}) = \boldsymbol{\theta}_{\phi}(\boldsymbol{p}).$$

Proof. Let $\{\varepsilon_n\}$ be an arbitrary sequence of positive numbers verifying $\lim_{n\to\infty} \varepsilon_n = 0$. Since ϕ is continuous and $p_i(\boldsymbol{\theta}, \epsilon_n) \underset{\varepsilon_n \to 0}{\longrightarrow} p_i(\boldsymbol{\theta}), i = 1, \ldots, M$, we get that $g_{\varepsilon_n}(\boldsymbol{p}, \boldsymbol{\theta}) \underset{\varepsilon_n \to 0}{\longrightarrow} g_0(\boldsymbol{p}, \boldsymbol{\theta})$, $\forall \boldsymbol{\theta} \in \Theta$. Since Θ is compact the pointwise convergence implies the uniform convergence. Consequently $\lim_{\varepsilon_n \to 0} \sup_{\boldsymbol{\theta} \in \Theta} \mid g_{\varepsilon_n}(\boldsymbol{p}, \boldsymbol{\theta}) - g_0(\boldsymbol{p}, \boldsymbol{\theta}) \mid = 0$ which implies that

$$\lim_{\varepsilon_n \to 0} \mid \inf_{\boldsymbol{\theta} \in \Theta} g_{\varepsilon_n}(\boldsymbol{p}, \boldsymbol{\theta}) - \inf_{\boldsymbol{\theta} \in \Theta} g_0(\boldsymbol{p}, \boldsymbol{\theta}) \mid = 0,$$

or equivalently

$$\lim_{\varepsilon_n \to 0} \mid g_{\varepsilon_n}(\boldsymbol{p}, \boldsymbol{\theta}_{\phi}^{\varepsilon_n}(\boldsymbol{p})) - g_0(\boldsymbol{p}, \boldsymbol{\theta}_{\phi}(\boldsymbol{p})) \mid = 0.$$

So, we have proved that

$$\lim_{\varepsilon_n \to 0} g_{\varepsilon_n}(\boldsymbol{p}, \boldsymbol{\theta}_{\phi}^{\varepsilon_n}(\boldsymbol{p})) = g_0(\boldsymbol{p}, \boldsymbol{\theta}_{\phi}(\boldsymbol{p})). \tag{5.10}$$

If $\lim_{\varepsilon_n \to 0} \boldsymbol{\theta}_{\phi}^{\varepsilon_n}(\boldsymbol{p}) \neq \boldsymbol{\theta}_{\phi}(\boldsymbol{p})$, the compactness of Θ guarantees the existence of a subsequence $\{\boldsymbol{\theta}_{\phi}^{\delta_n}(\boldsymbol{p})\} \subset \{\boldsymbol{\theta}_{\phi}^{\varepsilon_n}(\boldsymbol{p})\}$ such that $\lim_{\delta_n \to 0} \boldsymbol{\theta}_{\phi}^{\delta_n}(\boldsymbol{p}) = \boldsymbol{\theta}_* \neq \boldsymbol{\theta}_{\phi}(\boldsymbol{p})$. By (5.10), $g_0(\boldsymbol{p}, \boldsymbol{\theta}_*) = g_0(\boldsymbol{p}, \boldsymbol{\theta}_{\phi}(\boldsymbol{p}))$ for $\boldsymbol{\theta}_* \neq \boldsymbol{\theta}_{\phi}(\boldsymbol{p})$ which contradicts the assumed uniqueness of $\boldsymbol{\theta}_{\phi}(\boldsymbol{p})$. The statement of theorem follows from here since the sequence $\{\varepsilon_n\}$ can be chosen arbitrarily. ■

A more general way of studying the robustness is to assume that the true distribution $\boldsymbol{\pi} \in \Delta_M$ verifies $\|\boldsymbol{\pi} - \boldsymbol{p}(\boldsymbol{\theta})\| < \varepsilon$ for some $\boldsymbol{\theta} \in \Theta$ and to prove that if ε is small, the $\boldsymbol{\theta}_{\phi}(\boldsymbol{\pi})$ value is near to $\boldsymbol{\theta}_{\phi}(\boldsymbol{p}(\boldsymbol{\theta})) = \boldsymbol{\theta}$.

Theorem 5.4

Let the assumptions of Theorem 5.1 hold and let $\boldsymbol{\pi} \in \Delta_M$. *Then*

$$\lim_{\|\boldsymbol{\pi} - \boldsymbol{p}(\boldsymbol{\theta})\| \to 0} \boldsymbol{\theta}_{\phi}(\boldsymbol{\pi}) = \boldsymbol{\theta}_{\phi}(\boldsymbol{p}(\boldsymbol{\theta})) = \boldsymbol{\theta}.$$

Proof. It is immediate because $\boldsymbol{\theta}_{\phi}$ is a continuous function. ■

5.4. Normal Mixtures: Minimum Phi-divergence Estimator

In this section we report the results of a simulation experiment designed to compare empirically some minimum power-divergence estimators $\widehat{\boldsymbol{\theta}}_{(\lambda)}$ that can be seen in Pardo, M.C. (1999b). For the parameters of a mixture of two normal populations we analyze the efficiency as well as the robustness of the chosen estimators. We consider estimators corresponding to $\lambda = -2, -1, -0.5, 0, 2/3$ and 1. Simulations reported in this section are based on mixing proportions 0.25, 0.5 and 0.75. For each of these mixing proportions, firstly, we consider mixtures of the densities $f_1(x)$ and $f_2(x)$ where $f_1(x)$ is the density for the random variable $X = aY$ and $f_2(x)$ is the density associated with $X = Y + b$ where $a > 0$, $b > 0$ and Y is standard normal. Secondly, we assume that Y is a Student's t distribution with two or four degrees of freedom, or double exponential, to study the robustness under symmetric departures from normality. Thus, "a" is the ratio of scale parameters which we take to be 1 and $\sqrt{2}$ while "b" is selected to provide the overlap desired between the two distributions. We consider "overlap" as the probability of misclassification, using the rule: Classify an observation x as being from population 1 if $x < x_c$ and from population 2 if $x \geq x_c$, where x_c is the unique point between μ_1 and μ_2 such that $wf_1(x_c) = (1-w)f_2(x_c)$ with $\mu_i = \int_{\mathbb{R}} x_i f_i(x)\, dx$, $i = 1, 2$. The overlaps examined in the current study are 0.03 and 0.1.

For each set of considered configurations, 5000 samples of size $n = 100$ were generated from the corresponding mixture distribution, and for each considered sample the minimum power-divergence estimators $\widehat{\boldsymbol{\theta}}_{(-2)}, \widehat{\boldsymbol{\theta}}_{(-1)}, \widehat{\boldsymbol{\theta}}_{(-1/2)}, \widehat{\boldsymbol{\theta}}_{(0)}, \widehat{\boldsymbol{\theta}}_{(2/3)}$ and $\widehat{\boldsymbol{\theta}}_{(1)}$ were obtained. Our implementation of the minimum power-divergence estimator, $\widehat{\boldsymbol{\theta}}_{(\lambda)}$, employed the IMSL subroutine ZXMIN which minimizes a function of various variables. Although all these estimation procedures provide estimators of all five parameters, Tables 5.3, 5.5, 5.7 and 5.9 present only estimations for w; however, in Table 5.11 we present estimations for all the parameters.

For either of the estimators proposed in the previous section to be used in practice, one must have starting values for the iterative procedures. We choose an ad hoc quasi-clustering technique used by Woodward et *al.* (1984) that is easy to implement. They allow as possible values for the initial estimate of w only the values 0.1,...,0.9. For each of these values of w, the sample is divided into two subsamples, $Y_1, Y_2, ..., Y_{n_1}$ and $Y_{n_1+1}, Y_{n_1+2}, ..., Y_n$, where Y_i is the ith-order statistic and n_1 is "nw" rounded to the nearest integer. So, $\widehat{w}$ is that value at which $w(1-w)(x^1_{(1/2)} - x^2_{(1/2)})^2$ is maximized,

$$\widehat{\mu}_1 = x^1_{(1/2)}, \widehat{\mu}_2 = x^2_{(1/2)}, \ \widehat{\sigma}_1^2 = ((x^1_{(1/2)} - x^1_{(0.25)})/.6745)^2$$

and

$$\widehat{\sigma}_2^2 = ((x^2_{(0.75)} - x^2_{(1/2)})/.6745)^2,$$

where $x^j_{(q)}$ is the qth-percentile from the jth-subsample. In Table 5.3 we present summary results of the simulation carried out to compare the performance of the estimators for mixtures of normal components. Estimators of the bias and mean squared error (mse) based on the simulations are given by

$$\widehat{Bias} = \frac{1}{n_s}\sum_{i=1}^{n_s}(\widehat{w}_i - w) \text{ and } \widehat{mse} = \frac{1}{n_s}\sum_{i=1}^{n_s}(\widehat{w}_i - w)^2,$$

where n_s is the number of samples and $\widehat{w}_i$ denotes an estimate of w for the ith-sample. It should be noted that $n\widehat{mse}$ is the quantity actually given in the tables. We also table empirical measures of the relative efficiencies of each estimator, $\widehat{\boldsymbol{\theta}}_{(\lambda)}$, with respect to the MLE for grouped data $\widehat{\boldsymbol{\theta}}_{(0)}$. The relative efficiency is given by

$$\widehat{E} = \frac{\widehat{mse}(\neq \widehat{\boldsymbol{\theta}}_{(0)})}{\widehat{mse}(\widehat{\boldsymbol{\theta}}_{(0)})}.$$

We use a minimax criterion to find the best estimator. Let

$$S = \{-2, \ -1, \ -.5, \ 0, \ 2/3, \ 1\}$$

be the values of λ considered for $\widehat{\boldsymbol{\theta}}_{(\lambda)}$ studied. Let $\beta_{MIN}(P_j) = \inf_{\lambda \in S} n\widehat{mse}_\lambda(P_j)$, $j = 1, ..., 10$ where P_j, $j = 1, ..., 10$ is each one of the mixture of normal under consideration. If we define the inefficiency function as $i_\lambda(P_j) = n\widehat{mse}_\lambda(P_j) - \beta_{MIN}(P_j)$, $j = 1, ..., 10$, then, $\eta(\lambda) = \max_j(i_\lambda(P_j))$ can be considered as a inefficiency measure of the minimum power-divergence estimator, $\widehat{\boldsymbol{\theta}}_{(\lambda)}$, $\lambda \in S$. We consider the two best estimators under this criterion. Analyzing the results of Table 5.3, we can see in Table 5.4 that the minimum chi-square estimator, $\widehat{\boldsymbol{\theta}}_{(1)}$, and minimum modified chi-square estimator, $\widehat{\boldsymbol{\theta}}_{(-2)}$, are the best.

			.1 Overlap			.03 Overlap		
w	a	Estimator	$\widehat{Bias}$	$n\ \widehat{mse}$	$\widehat{E}$	$\widehat{Bias}$	$n\ \widehat{mse}$	$\widehat{E}$
.25	1	$\widehat{\boldsymbol{\theta}}_{(-2)}$	.062	1.997	1.02	.037	.767	1.02
		$\widehat{\boldsymbol{\theta}}_{(-1)}$	.059	1.934	.99	.038	.797	1.06
		$\widehat{\boldsymbol{\theta}}_{(-.5)}$	.063	1.937	.99	.037	.767	1.02
		$\widehat{\boldsymbol{\theta}}_{(0)}$	.062	1.960		.036	.752	
		$\widehat{\boldsymbol{\theta}}_{(2/3)}$	.062	2.053	1.05	.035	.771	1.03
		$\widehat{\boldsymbol{\theta}}_{(1)}$	.061	1.918	.98	.035	.760	1.01
.5	1	$\widehat{\boldsymbol{\theta}}_{(-2)}$	-.015	1.579	1.05	-.012	.707	.86
		$\widehat{\boldsymbol{\theta}}_{(-1)}$	-.015	1.565	1.04	-.014	.757	.92
		$\widehat{\boldsymbol{\theta}}_{(-.5)}$	-.015	1.455	.97	-.012	.719	.87
		$\widehat{\boldsymbol{\theta}}_{(0)}$	-.014	1.504		.011	.826	
		$\widehat{\boldsymbol{\theta}}_{(2/3)}$	-.018	1.594	1.06	-.010	.800	.97
		$\widehat{\boldsymbol{\theta}}_{(1)}$	-.016	1.529	1.02	-.012	.781	.95
.25	$\sqrt{2}$	$\widehat{\boldsymbol{\theta}}_{(-2)}$	-.023	.947	1.01	-.001	.588	1.05
		$\widehat{\boldsymbol{\theta}}_{(-1)}$	-.025	1.021	1.09	-.002	.573	1.03
		$\widehat{\boldsymbol{\theta}}_{(-.5)}$	-.025	.949	1.01	-.002	.567	1.02
		$\widehat{\boldsymbol{\theta}}_{(0)}$	-.026	.939		-.001	.557	
		$\widehat{\boldsymbol{\theta}}_{(2/3)}$	-.026	.920	.98	-.002	.567	1.02
		$\widehat{\boldsymbol{\theta}}_{(1)}$	-.024	.951	1.01	-.001	.564	1.01
.5	$\sqrt{2}$	$\widehat{\boldsymbol{\theta}}_{(-2)}$	-.102	2.135	1.01	-.057	1.003	.95
		$\widehat{\boldsymbol{\theta}}_{(-1)}$	-.100	2.176	1.02	-.062	1.057	1.00
		$\widehat{\boldsymbol{\theta}}_{(-.5)}$	-.101	2.152	1.01	-.061	1.079	1.02
		$\widehat{\boldsymbol{\theta}}_{(0)}$	-.101	2.123		-.060	1.054	
		$\widehat{\boldsymbol{\theta}}_{(2/3)}$	-.101	2.120	1.00	-.063	1.124	1.07
		$\widehat{\boldsymbol{\theta}}_{(1)}$	-.098	2.043	.96	-.061	1.079	1.02
.75	$\sqrt{2}$	$\widehat{\boldsymbol{\theta}}_{(-2)}$	-.168	4.563	.98	-.080	1.475	.95
		$\widehat{\boldsymbol{\theta}}_{(-1)}$	-.167	4.700	1.01	-.078	1.658	1.07
		$\widehat{\boldsymbol{\theta}}_{(-.5)}$	-.167	4.628	.99	-.080	1.620	1.04
		$\widehat{\boldsymbol{\theta}}_{(0)}$	-.166	4.667		-.076	1.553	
		$\widehat{\boldsymbol{\theta}}_{(2/3)}$	-.167	4.704	1.01	-.073	1.379	.89
		$\widehat{\boldsymbol{\theta}}_{(1)}$	-.165	4.653	1.00	-.075	1.478	.95

Table 5.3. Simulation results for mixtures of normal components.

	$\widehat{\boldsymbol{\theta}}_{(-2)}$	$\widehat{\boldsymbol{\theta}}_{(-1)}$	$\widehat{\boldsymbol{\theta}}_{(-.5)}$	$\widehat{\boldsymbol{\theta}}_{(0)}$	$\widehat{\boldsymbol{\theta}}_{(2/3)}$	$\widehat{\boldsymbol{\theta}}_{(1)}$
$\eta(.)$	**0.124**	0.279	0.241	0.174	0.141	**0.099**

Table 5.4. Inefficiencies.

In Table 5.5 we display the results for the nonnormal components (double exponential components). In this case (see Table 5.6), in accordance with our criterion, the best are $\widehat{\boldsymbol{\theta}}_{(0)}$ and $\widehat{\boldsymbol{\theta}}_{(2/3)}$.

			.1 Overlap			.03 Overlap		
w	a	Estimator	$\widehat{Bias}$	$n\ \widehat{mse}$	$\widehat{E}$	$\widehat{Bias}$	$n\ \widehat{mse}$	$\widehat{E}$
.25	1	$\widehat{\boldsymbol{\theta}}_{(-2)}$	.056	1.000	1.00	.057	.780	1.05
		$\widehat{\boldsymbol{\theta}}_{(-1)}$	.055	.975	.98	.056	.780	1.05
		$\widehat{\boldsymbol{\theta}}_{(-.5)}$	.057	1.052	1.06	.055	.771	1.04
		$\widehat{\boldsymbol{\theta}}_{(0)}$	.054	.995		.055	.740	
		$\widehat{\boldsymbol{\theta}}_{(2/3)}$	.054	.961	.97	.055	.776	1.05
		$\widehat{\boldsymbol{\theta}}_{(1)}$	.053	.996	1.00	.053	.754	1.02
.5	1	$\widehat{\boldsymbol{\theta}}_{(-2)}$	-.007	.682	.96	-.002	.425	1.07
		$\widehat{\boldsymbol{\theta}}_{(-1)}$	-.007	.716	1.00	-.002	.381	.96
		$\widehat{\boldsymbol{\theta}}_{(-.5)}$	-.002	.703	.99	-.002	.391	.99
		$\widehat{\boldsymbol{\theta}}_{(0)}$	-.004	.713		.000	.397	
		$\widehat{\boldsymbol{\theta}}_{(2/3)}$	-.003	.738	1.04	-.001	.390	.98
		$\widehat{\boldsymbol{\theta}}_{(1)}$	-.002	.712	1.00	.001	.405	1.02
.25	$\sqrt{2}$	$\widehat{\boldsymbol{\theta}}_{(-2)}$	.013	.669	1.00	.036	.535	.97
		$\widehat{\boldsymbol{\theta}}_{(-1)}$	.012	.638	.96	.034	.517	.93
		$\widehat{\boldsymbol{\theta}}_{(-0.5)}$	.013	.659	.99	.036	.552	1.00
		$\widehat{\boldsymbol{\theta}}_{(0)}$	.014	.666		.036	.554	
		$\widehat{\boldsymbol{\theta}}_{(2/3)}$	.012	.656	.99	.036	.562	1.02
		$\widehat{\boldsymbol{\theta}}_{(1)}$	.015	.662	.99	.036	.555	1.00
.5	$\sqrt{2}$	$\widehat{\boldsymbol{\theta}}_{(-2)}$	.052	1.157	1.07	-.029	.547	1.01
		$\widehat{\boldsymbol{\theta}}_{(-1)}$	-.049	1.070	.99	-.027	.526	.97
		$\widehat{\boldsymbol{\theta}}_{(-.5)}$	-.051	1.111	1.03	-.029	.541	1.00
		$\widehat{\boldsymbol{\theta}}_{(0)}$	-.048	1.079		-.027	.541	
		$\widehat{\boldsymbol{\theta}}_{(2/3)}$	-.050	1.039	.96	-.026	.558	1.03
		$\widehat{\boldsymbol{\theta}}_{(1)}$	-.050	1.124	1.04	-.028	.591	1.09
.75	$\sqrt{2}$	$\widehat{\boldsymbol{\theta}}_{(-2)}$	-.106	2.191	1.14	-.078	1.167	1.03
		$\widehat{\boldsymbol{\theta}}_{(-1)}$	-.104	2.135	1.11	-.078	1.185	1.04
		$\widehat{\boldsymbol{\theta}}_{(-.5)}$	-.096	1.841	.96	-.078	1.181	1.04
		$\widehat{\boldsymbol{\theta}}_{(0)}$	-.098	1.921		-.075	1.136	
		$\widehat{\boldsymbol{\theta}}_{(2/3)}$	-.095	1.908	.99	-.075	1.163	1.02
		$\widehat{\boldsymbol{\theta}}_{(1)}$	-.094	1.952	1.02	-.072	1.071	.94

Table 5.5. Simulation results for mixtures of double exponential components.

	$\widehat{\boldsymbol{\theta}}_{(-2)}$	$\widehat{\boldsymbol{\theta}}_{(-1)}$	$\widehat{\boldsymbol{\theta}}_{(-.5)}$	$\widehat{\boldsymbol{\theta}}_{(0)}$	$\widehat{\boldsymbol{\theta}}_{(2/3)}$	$\widehat{\boldsymbol{\theta}}_{(1)}$
$\eta(.)$	0.35	0.294	0.091	**0.08**	**0.067**	0.111

Table 5.6. Inefficiencies.

In Table 5.7 we present the results for the nonnormal components (t-Student with 4 degrees of freedom, t(4)).

			.1 Overlap			.03 Overlap		
w	a	Estimator	$\widehat{Bias}$	$n\ \widehat{mse}$	$\widehat{E}$	$\widehat{Bias}$	$n\ \widehat{mse}$	$\widehat{E}$
.25	1	$\widehat{\boldsymbol{\theta}}_{(-2)}$	.084	2.699	1.00	.049	.947	1.04
		$\widehat{\boldsymbol{\theta}}_{(-1)}$	.083	2.663	.99	.047	.901	.99
		$\widehat{\boldsymbol{\theta}}_{(-.5)}$	.081	2.642	.98	.047	.913	1.00
		$\widehat{\boldsymbol{\theta}}_{(0)}$	.080	2.695		.049	.911	
		$\widehat{\boldsymbol{\theta}}_{(2/3)}$	.077	2.576	.96	.046	.890	.98
		$\widehat{\boldsymbol{\theta}}_{(1)}$	.081	2.604	.97	.048	.931	1.02
.5	1	$\widehat{\boldsymbol{\theta}}_{(-2)}$	-.008	1.697	1.02	-.008	.827	1.04
		$\widehat{\boldsymbol{\theta}}_{(-1)}$	-.008	1.657	.99	-.010	.793	1.00
		$\widehat{\boldsymbol{\theta}}_{(-0.5)}$	-.008	1.598	.96	-.012	.838	1.06
		$\widehat{\boldsymbol{\theta}}_{(0)}$	-.010	1.668		.006	.794	
		$\widehat{\boldsymbol{\theta}}_{(2/3)}$	-.011	1.745	1.05	-.008	.837	1.06
		$\widehat{\boldsymbol{\theta}}_{(1)}$	-.005	1.735	1.04	-.006	.809	1.02
.25	$\sqrt{2}$	$\widehat{\boldsymbol{\theta}}_{(-2)}$	-.008	1.007	1.01	.008	.582	1.03
		$\widehat{\boldsymbol{\theta}}_{(-1)}$	-.007	1.055	1.06	.008	.554	.98
		$\widehat{\boldsymbol{\theta}}_{(-.5)}$	-.007	1.033	1.03	.008	.564	1.00
		$\widehat{\boldsymbol{\theta}}_{(0)}$	-.007	.999		.006	.565	
		$\widehat{\boldsymbol{\theta}}_{(2/3)}$	-.010	.979	.98	.007	.553	.98
		$\widehat{\boldsymbol{\theta}}_{(1)}$	-.007	.967	.97	.007	.574	1.01
.5	$\sqrt{2}$	$\widehat{\boldsymbol{\theta}}_{(-2)}$	-.093	2.296	.92	-.060	1.191	1.01
		$\widehat{\boldsymbol{\theta}}_{(-1)}$	-.098	2.539	1.01	-.061	1.18	.999
		$\widehat{\boldsymbol{\theta}}_{(-.5)}$	-.099	2.594	1.04	-.060	1.151	.98
		$\widehat{\boldsymbol{\theta}}_{(0)}$	-.099	2.504		-.058	1.177	
		$\widehat{\boldsymbol{\theta}}_{(2/3)}$	-.099	2.48	.99	-.061	1.232	1.05
		$\widehat{\boldsymbol{\theta}}_{(1)}$	-.097	2.531	1.01	-.060	1.188	1.01
.75	$\sqrt{2}$	$\widehat{\boldsymbol{\theta}}_{(-2)}$	-.189	5.818	1.06	-.100	1.989	1.06
		$\widehat{\boldsymbol{\theta}}_{(-1)}$	-.186	5.748	1.05	-.102	2.105	1.13
		$\widehat{\boldsymbol{\theta}}_{(-0.5)}$	-.187	5.778	1.053	-.098	1.876	1.00
		$\widehat{\boldsymbol{\theta}}_{(0)}$	-.181	5.487		-.096	1.870	
		$\widehat{\boldsymbol{\theta}}_{(2/3)}$	-.183	5.699	1.04	-.098	1.86	.99
		$\widehat{\boldsymbol{\theta}}_{(1)}$	-.182	5.541	1.01	-.099	1.893	1.01

Table 5.7. Simulation results for mixtures of t-Student (4) components.

	$\widehat{\boldsymbol{\theta}}_{(-2)}$	$\widehat{\boldsymbol{\theta}}_{(-1)}$	$\widehat{\boldsymbol{\theta}}_{(-0.5)}$	$\widehat{\boldsymbol{\theta}}_{(0)}$	$\widehat{\boldsymbol{\theta}}_{(2/3)}$	$\widehat{\boldsymbol{\theta}}_{(1)}$
$\eta(.)$	0.331	0.261	0.298	**0.208**	**0.212**	0.235

Table 5.8. Inefficiencies.

For t-Student(4) we have that $\widehat{\boldsymbol{\theta}}_{(0)}$ and $\widehat{\boldsymbol{\theta}}_{(2/3)}$ are again the best. Now we present the results for a t-Student with 2 degrees of freedom, t(2).

			.1 Overlap			.03 Overlap		
w	a	Estimator	$\widehat{Bias}$	$n\ \widehat{mse}$	$\widehat{E}$	$\widehat{Bias}$	$n\ \widehat{mse}$	$\widehat{E}$
.25	1	$\widehat{\boldsymbol{\theta}}_{(-2)}$	.111	6.742	1.00	.059	1.468	1.00
		$\widehat{\boldsymbol{\theta}}_{(-1)}$	.113	6.963	1.03	.060	1.509	1.02
		$\widehat{\boldsymbol{\theta}}_{(-.5)}$	.110	6.755	1.00	.057	1.493	1.01
		$\widehat{\boldsymbol{\theta}}_{(0)}$	.112	6.732		.057	1.473	
		$\widehat{\boldsymbol{\theta}}_{(2/3)}$	.107	6.746	1.00	.058	1.491	1.01
		$\widehat{\boldsymbol{\theta}}_{(1)}$	.109	6.839	1.02	.058	1.495	1.01
.5	1	$\widehat{\boldsymbol{\theta}}_{(-2)}$	-.012	3.900	.98	-.011	1.193	1.07
		$\widehat{\boldsymbol{\theta}}_{(-1)}$	-.017	3.876	.98	-.009	1.136	1.02
		$\widehat{\boldsymbol{\theta}}_{(-.5)}$	-.016	3.903	.99	-.011	1.136	1.02
		$\widehat{\boldsymbol{\theta}}_{(0)}$	-.017	3.960		-.006	1.112	
		$\widehat{\boldsymbol{\theta}}_{(2/3)}$	-.016	3.874	.98	-.007	1.006	.995
		$\widehat{\boldsymbol{\theta}}_{(1)}$	-.017	3.995	1.01	-.005	1.138	1.02
.25	$\sqrt{2}$	$\widehat{\boldsymbol{\theta}}_{(-2)}$	.038	4.736	.99	.011	.832	1.03
		$\widehat{\boldsymbol{\theta}}_{(-1)}$	.037	4.588	.96	.011	.869	1.07
		$\widehat{\boldsymbol{\theta}}_{(-.5)}$	.037	4.715	.98	.009	.836	1.03
		$\widehat{\boldsymbol{\theta}}_{(0)}$	.037	4.797		.013	.809	
		$\widehat{\boldsymbol{\theta}}_{(2/3)}$	.032	4.740	.99	.016	.815	1.01
		$\widehat{\boldsymbol{\theta}}_{(1)}$	.028	4.520	.94	.011	.802	.99
.5	$\sqrt{2}$	$\widehat{\boldsymbol{\theta}}_{(-2)}$	-.119	4.497	.93	-.072	1.874	1.01
		$\widehat{\boldsymbol{\theta}}_{(-1)}$	-.126	4.829	1.00	-.070	1.794	.96
		$\widehat{\boldsymbol{\theta}}_{(-.5)}$	-.124	4.551	.94	-.071	1.796	.96
		$\widehat{\boldsymbol{\theta}}_{(0)}$	-.127	4.839		-.071	1.862	
		$\widehat{\boldsymbol{\theta}}_{(2/3)}$	-.126	4.827	1.00	-.069	1.799	.97
		$\widehat{\boldsymbol{\theta}}_{(1)}$	-.126	4.795	.99	-.069	1.702	.91
.75	$\sqrt{2}$	$\widehat{\boldsymbol{\theta}}_{(-2)}$	-.229	11.064	.99	-.121	3.353	1.06
		$\widehat{\boldsymbol{\theta}}_{(-1)}$	-.234	11.112	1.00	-.117	3.177	1.00
		$\widehat{\boldsymbol{\theta}}_{(-.5)}$	-.234	11.374	1.02	-.117	3.096	.98
		$\widehat{\boldsymbol{\theta}}_{(0)}$	-.231	11.145		-.113	3.171	
		$\widehat{\boldsymbol{\theta}}_{(2/3)}$	-.236	11.463	1.03	-.111	3.075	.97
		$\widehat{\boldsymbol{\theta}}_{(1)}$	-.230	11.102	1.00	-.114	3.184	1.00

Table.5.9. Simulation results for mixtures of t-Student (2) components.

For t-Student(2), the best are $\widehat{\boldsymbol{\theta}}_{(-2)}$ and $\widehat{\boldsymbol{\theta}}_{(1)}$, although $\widehat{\boldsymbol{\theta}}_{(1)}$ has less bias.

	$\widehat{\boldsymbol{\theta}}_{(-2)}$	$\widehat{\boldsymbol{\theta}}_{(-1)}$	$\widehat{\boldsymbol{\theta}}_{(-.5)}$	$\widehat{\boldsymbol{\theta}}_{(0)}$	$\widehat{\boldsymbol{\theta}}_{(2/3)}$	$\widehat{\boldsymbol{\theta}}_{(1)}$
$\eta(.)$	**0.278**	0.332	0.31	0.342	0.399	**0.298**

Table 5.10. Inefficiencies.

			.1 Overlap				
w	a	Estimator	μ_1	σ_1	μ_2	σ_2	w
			Normal				
.25	1	$\widehat{\boldsymbol{\theta}}_{(-2)}$	1.02	1.04	1.03	.99	1.02
		$\widehat{\boldsymbol{\theta}}_{(-1)}$	1.31	1.05	1.02	1.00	.99
		$\widehat{\boldsymbol{\theta}}_{(-.5)}$	.95	1.01	1.02	.99	.99
		$\widehat{\boldsymbol{\theta}}_{(2/3)}$	1.08	1.05	1.02	.98	1.05
		$\widehat{\boldsymbol{\theta}}_{(1)}$	1.00	1.07	.99	.99	.98
.5	1	$\widehat{\boldsymbol{\theta}}_{(-2)}$	1.01	1.09	1.03	.97	1.05
		$\widehat{\boldsymbol{\theta}}_{(-1)}$	1.83	1.04	1.04	.99	1.04
		$\widehat{\boldsymbol{\theta}}_{(-.5)}$	1.00	1.05	.99	.96	.97
		$\widehat{\boldsymbol{\theta}}_{(2/3)}$	1.04	1.03	1.04	1.01	1.06
		$\widehat{\boldsymbol{\theta}}_{(1)}$	1.00	1.01	.98	.97	1.02
.25	$\sqrt{2}$	$\widehat{\boldsymbol{\theta}}_{(-2)}$	1.00	.95	1.03	1.06	1.01
		$\widehat{\boldsymbol{\theta}}_{(-1)}$	.98	1.01	1.10	1.03	1.09
		$\widehat{\boldsymbol{\theta}}_{(-.5)}$	.97	1.02	1.07	1.06	1.01
		$\widehat{\boldsymbol{\theta}}_{(2/3)}$	.98	1.11	1.00	.99	.98
		$\widehat{\boldsymbol{\theta}}_{(1)}$	.97	1.00	1.01	1.04	1.01
.5	$\sqrt{2}$	$\widehat{\boldsymbol{\theta}}_{(-2)}$	1.01	1.08	1.02	1.02	1.01
		$\widehat{\boldsymbol{\theta}}_{(-1)}$	.99	1.06	1.03	1.05	1.02
		$\widehat{\boldsymbol{\theta}}_{(-.5)}$	.96	1.01	1.02	.97	1.01
		$\widehat{\boldsymbol{\theta}}_{(2/3)}$	.99	.99	.97	1.02	1.00
		$\widehat{\boldsymbol{\theta}}_{(1)}$	1.00	1.02	.96	.91	.96
.75	$\sqrt{2}$	$\widehat{\boldsymbol{\theta}}_{(-2)}$	1.00	1.00	.96	.97	.98
		$\widehat{\boldsymbol{\theta}}_{(-1)}$	1.11	1.10	.96	1.00	1.00
		$\widehat{\boldsymbol{\theta}}_{(-.5)}$	1.06	1.09	.93	.98	.99
		$\widehat{\boldsymbol{\theta}}_{(2/3)}$	1.00	1.02	.93	1.05	1.01
		$\widehat{\boldsymbol{\theta}}_{(1)}$	1.02	1.03	.95	.98	1.00

Table 5.11. Estimated relative efficiencies of the estimators relative to the MLE for mixture.

w	a	Estimator	.03 Overlap				
			μ_1	σ_1	μ_2	σ_2	w
			Normal				
.25	1	$\widehat{\boldsymbol{\theta}}_{(-2)}$	.93	1.03	.98	1.00	1.02
		$\widehat{\boldsymbol{\theta}}_{(-1)}$	.92	.89	1.04	1.01	1.06
		$\widehat{\boldsymbol{\theta}}_{(-0.5)}$	.94	.94	.98	.98	1.02
		$\widehat{\boldsymbol{\theta}}_{(2/3)}$	1.03	1.02	.99	.97	1.03
		$\widehat{\boldsymbol{\theta}}_{(1)}$	.95	1.11	1.01	.96	1.01
.5	1	$\widehat{\boldsymbol{\theta}}_{(-2)}$	.98	1.04	.99	1.04	.86
		$\widehat{\boldsymbol{\theta}}_{(-1)}$	.97	1.01	.93	1.16	.92
		$\widehat{\boldsymbol{\theta}}_{(-.5)}$	.96	.99	.96	.98	1.02
		$\widehat{\boldsymbol{\theta}}_{(2/3)}$	.99	.99	.99	.99	.97
		$\widehat{\boldsymbol{\theta}}_{(1)}$	1.01	1.01	.99	.94	.95
.25	$\sqrt{2}$	$\widehat{\boldsymbol{\theta}}_{(-2)}$	1.01	1.04	1.04	1.02	1.05
		$\widehat{\boldsymbol{\theta}}_{(-1)}$	1.15	1.21	1.05	.98	1.03
		$\widehat{\boldsymbol{\theta}}_{(-5)}$	.98	.93	1.00	.98	1.02
		$\widehat{\boldsymbol{\theta}}_{(2/3)}$	1.06	1.04	1.01	1.03	1.02
		$\widehat{\boldsymbol{\theta}}_{(1)}$	1.10	1.03	1.00	1.00	1.01
.5	$\sqrt{2}$	$\widehat{\boldsymbol{\theta}}_{(-2)}$	.96	1.07	1.03	1.12	.95
		$\widehat{\boldsymbol{\theta}}_{(-1)}$	.99	1.04	1.20	1.07	1.00
		$\widehat{\boldsymbol{\theta}}_{(-.5)}$	1.04	1.00	.97	.95	1.02
		$\widehat{\boldsymbol{\theta}}_{(2/3)}$	1.04	.99	1.03	1.00	1.07
		$\widehat{\boldsymbol{\theta}}_{(1)}$	1.05	1.02	1.09	1.01	1.02
.75	$\sqrt{2}$	$\widehat{\boldsymbol{\theta}}_{(-2)}$	.95	.93	.90	.92	.95
		$\widehat{\boldsymbol{\theta}}_{(-1)}$	.97	.93	1.03	1.05	1.07
		$\widehat{\boldsymbol{\theta}}_{(-.5)}$	1.04	.97	.99	.99	1.04
		$\widehat{\boldsymbol{\theta}}_{(2/3)}$	.99	.96	.93	.98	.89
		$\widehat{\boldsymbol{\theta}}_{(1)}$	.95	.96	.99	1.01	.95

Table 5.11 (Continuation).

Although our emphasis has been put on the estimation of the mixing proportion, the estimation routines used here obtain estimations for all five parameters. So, it seems obvious to question about whether the results shown for w are similar for the rest of the parameters μ_1, σ_1, μ_2 and σ_2. In Table 5.11 we display empirical relative efficiencies for all the parameters for normal and t(4) mixtures. From the table we see that the results for the other parameters with t(4) do not exhibit in general patterns similar to those shown for w. The Freeman-Tukey estimator, $\widehat{\boldsymbol{\theta}}_{(-.5)}$, seems to be the most efficient for μ_1, σ_1, μ_2 and σ_2 in Normal and t(4) mixtures.

			.1 Overlap				
w	a	Estimator	μ_1	σ_1	μ_2	σ_2	w
		t-Student(4)					
.25	1	$\widehat{\boldsymbol{\theta}}_{(-2)}$	1.01	1.01	1.03	1.00	1.00
		$\widehat{\boldsymbol{\theta}}_{(-1)}$	1.03	1.07	1.03	1.00	.99
		$\widehat{\boldsymbol{\theta}}_{(-.5)}$	1.07	1.07	1.03	1.01	.98
		$\widehat{\boldsymbol{\theta}}_{(2/3)}$	1.02	1.33	1.00	1.03	.96
		$\widehat{\boldsymbol{\theta}}_{(1)}$	1.05	1.05	1.01	1.01	.97
.5	1	$\widehat{\boldsymbol{\theta}}_{(-2)}$	.96	1.03	.99	1.07	1.02
		$\widehat{\boldsymbol{\theta}}_{(-1)}$	.96	.97	.99	1.01	.99
		$\widehat{\boldsymbol{\theta}}_{(0)}$	.93	1.00	.96	1.03	.96
		$\widehat{\boldsymbol{\theta}}_{(2/3)}$	.95	1.00	1.05	1.05	1.05
		$\widehat{\boldsymbol{\theta}}_{(1)}$	1.05	1.14	1.01	.99	1.04
.25	$\sqrt{2}$	$\widehat{\boldsymbol{\theta}}_{(-2)}$	.99	1.18	1.03	1.02	1.01
		$\widehat{\boldsymbol{\theta}}_{(-1)}$	1.00	.91	1.02	1.05	1.06
		$\widehat{\boldsymbol{\theta}}_{(-.5)}$	.97	1.02	1.03	1.02	1.03
		$\widehat{\boldsymbol{\theta}}_{(2/3)}$	1.02	.91	.98	.99	.98
		$\widehat{\boldsymbol{\theta}}_{(1)}$	1.02	.96	1.00	.95	.97
.5	$\sqrt{2}$	$\widehat{\boldsymbol{\theta}}_{(-2)}$	.97	.92	.90	.89	.92
		$\widehat{\boldsymbol{\theta}}_{(-1)}$	.99	1.27	.97	1.03	1.01
		$\widehat{\boldsymbol{\theta}}_{(-.5)}$	1.03	.84	.90	.94	1.04
		$\widehat{\boldsymbol{\theta}}_{(2/3)}$	1.07	.88	.93	.93	.99
		$\widehat{\boldsymbol{\theta}}_{(1)}$	1.03	.85	.93	.90	1.01
.75	$\sqrt{2}$	$\widehat{\boldsymbol{\theta}}_{(-2)}$	1.05	.88	1.09	1.07	1.06
		$\widehat{\boldsymbol{\theta}}_{(-1)}$	1.05	.84	1.03	1.00	1.05
		$\widehat{\boldsymbol{\theta}}_{(-.5)}$	1.05	.85	1.05	1.03	1.05
		$\widehat{\boldsymbol{\theta}}_{(2/3)}$	1.07	1.04	1.04	1.03	1.04
		$\widehat{\boldsymbol{\theta}}_{(1)}$	1.04	.98	1.01	.97	1.01

Table 5.11. (Continuation).

To sum up, the Freeman-Tukey estimator seems to be, in general, the most efficient for all the parameters except the proportion parameter. For the proportion parameter we recommend to use $\widehat{\boldsymbol{\theta}}_{(2/3)}$, Cressie-Read estimator. It is the best alternative for the component distributions double exponential and t(4), almost as good as the best at the true model. It performs only worse for t(2) when the departure from normality is more extreme. Finally, the minimum chi-square estimator is preferable for extreme departures from normality.

			.03 Overlap				
w	a	Estimator	μ_1	σ_1	μ_2	σ_2	w
			t-Student(4)				
.25	1	$\widehat{\boldsymbol{\theta}}_{(-2)}$	1.01	.96	.99	1.02	1.04
		$\widehat{\boldsymbol{\theta}}_{(-1)}$	.98	.97	.99	1.01	.99
		$\widehat{\boldsymbol{\theta}}_{(-.5)}$	1.02	.97	1.02	1.08	1.00
		$\widehat{\boldsymbol{\theta}}_{(2/3)}$	1.03	1.19	.96	.96	.98
		$\widehat{\boldsymbol{\theta}}_{(1)}$	1.07	1.16	.98	.99	1.02
.5	1	$\widehat{\boldsymbol{\theta}}_{(-2)}$	1.03	1.04	1.14	1.11	1.04
		$\widehat{\boldsymbol{\theta}}_{(-1)}$	1.04	1.36	1.01	1.08	1.00
		$\widehat{\boldsymbol{\theta}}_{(-.5)}$	1.02	1.01	.96	1.04	1.06
		$\widehat{\boldsymbol{\theta}}_{(2/3)}$	1.03	1.06	.94	.99	1.06
		$\widehat{\boldsymbol{\theta}}_{(1)}$	1.06	1.13	.91	.94	1.02
.25	$\sqrt{2}$	$\widehat{\boldsymbol{\theta}}_{(-2)}$	.86	.92	1.02	1.06	1.03
		$\widehat{\boldsymbol{\theta}}_{(-1)}$	.99	.97	.98	1.04	.98
		$\widehat{\boldsymbol{\theta}}_{(-.5)}$	.93	.96	1.03	1.00	1.00
		$\widehat{\boldsymbol{\theta}}_{(2/3)}$	.92	1.38	1.01	1.01	.98
		$\widehat{\boldsymbol{\theta}}_{(1)}$	.91	.91	1.01	.98	1.01
.5	$\sqrt{2}$	$\widehat{\boldsymbol{\theta}}_{(-2)}$	.95	.86	1.15	1.16	1.01
		$\widehat{\boldsymbol{\theta}}_{(-1)}$	.99	.79	1.10	1.15	1.00
		$\widehat{\boldsymbol{\theta}}_{(-.5)}$	1.04	.80	.99	1.01	.98
		$\widehat{\boldsymbol{\theta}}_{(2/3)}$	1.08	.77	1.05	1.01	1.05
		$\widehat{\boldsymbol{\theta}}_{(1)}$	1.05	.77	1.13	1.06	1.01
.75	$\sqrt{2}$	$\widehat{\boldsymbol{\theta}}_{(-2)}$	1.03	1.16	1.09	.99	1.06
		$\widehat{\boldsymbol{\theta}}_{(-1)}$	2.95	1.13	1.22	1.18	1.13
		$\widehat{\boldsymbol{\theta}}_{(-.5)}$	.98	1.12	.99	1.19	1.00
		$\widehat{\boldsymbol{\theta}}_{(2/3)}$	1.00	1.03	1.01	.98	.99
		$\widehat{\boldsymbol{\theta}}_{(1)}$	1.02	1.02	1.26	.95	1.01

Table 5.11. (Continuation).

5.5. Minimum Phi-divergence Estimator with Constraints: Properties

A new problem appears if we suppose that we have ν real-valued functions $f_1(\boldsymbol{\theta}), ..., f_\nu(\boldsymbol{\theta})$ which constrain the parameter $\boldsymbol{\theta}$:

$$f_m(\boldsymbol{\theta}) = 0,\ m = 1, 2, ..., \nu.$$

The problem to obtain the maximum likelihood estimator subject to constraints was considered, for the first time, by Aitchison and Silvey (1958) in the context of a random variable whose distribution function F depends on M_0 parameters $\theta_1, ..., \theta_{M_0}$ which are not mathematically independent but satisfy ν functional relationships. The Lagrangian multiplier method is used to find the maximum likelihood estimator of the parameters, which are found numerically by a process of iteration. The estimator is shown to have an asymptotic normal distribution. Silvey (1959) further discussed the Lagrangian method and the mathematical conditions for the existence of the maximum likelihood estimators. Diamond et *al.* (1960) considered the restricted maximum likelihood estimator, but in multinomial populations. Haber and Brown (1986) proposed a two-step algorithm for obtaining maximum likelihood estimators of the expected frequencies in log-linear models with expected frequencies subject to linear constraints. Problems of this type in multinomial models with the log-linear parameterization have been described by Bhapkar (1979), Bonett (1989), Gokhale (1973), Haber (1985), Haberman (1974) and Wedderburn (1974).

In this Section we consider the minimum ϕ-divergence estimator subject to some constraints: The restricted minimum ϕ-divergence estimator, $\widehat{\theta}_{\phi}^{(r)}$, which is seen to be a generalization of the maximum likelihood estimator, subject to constraints, was studied for the first time in Pardo, J. A. et *al.* (2002).

We suppose that we have ν $(\nu < M_0)$ real-valued functions $f_1(\boldsymbol{\theta}), ..., f_\nu(\boldsymbol{\theta})$ that constrain the parameter $\boldsymbol{\theta} \in \Theta \subset \mathbb{R}^{M_0}$, $f_m(\boldsymbol{\theta}) = 0$, $m = 1, ..., \nu$, and such that

i) Every $f_m(\boldsymbol{\theta})$ has continuous second partial derivatives,

ii) The $\nu \times M_0$ matrix

$$\boldsymbol{B}(\boldsymbol{\theta}) = \left(\frac{\partial f_m(\boldsymbol{\theta})}{\partial \theta_k}\right)_{\substack{m=1,...,\nu \\ k=1,...,M_0}}$$

is of rank ν.

Under these assumptions the restricted minimum ϕ-divergence estimator of $\boldsymbol{\theta}$ is $\widehat{\boldsymbol{\theta}}_{\phi}^{(r)} \in \Theta$ satisfying the condition

$$D_\phi(\widehat{\boldsymbol{p}}, \boldsymbol{p}(\widehat{\boldsymbol{\theta}}_{\phi}^{(r)})) = \inf_{\left\{\boldsymbol{\theta} \in \Theta \subseteq \mathbb{R}^{M_0} : f_m(\boldsymbol{\theta}) = 0, m=1,...,\nu\right\}} D_\phi(\widehat{\boldsymbol{p}}, \boldsymbol{p}(\boldsymbol{\theta})).$$

In the cited paper of Pardo, J. A. et *al.* (2002) it was established, under the conditions of Theorem 5.1 and assuming the previous conditions *i)* and *ii)* about the functions

$$f_m(\boldsymbol{\theta}) = 0, \; m = 1, ..., \nu,$$

that $\widehat{\boldsymbol{\theta}}_{\phi}^{(r)}$ has the following expansion

$$\boldsymbol{\theta}_0 + \boldsymbol{H}(\boldsymbol{\theta}_0)\boldsymbol{I}_F(\boldsymbol{\theta}_0)^{-1}\boldsymbol{A}(\boldsymbol{\theta}_0)^T diag\left(\mathbf{p}(\boldsymbol{\theta}_0)^{-1/2}\right)(\widehat{\mathbf{p}} - \mathbf{p}(\boldsymbol{\theta}_0)) + o(\|\widehat{\mathbf{p}} - \mathbf{p}(\boldsymbol{\theta}_0)\|)$$

where $\widehat{\boldsymbol{\theta}}_{\phi}^{(r)}$ is unique in a neighborhood of $\boldsymbol{\theta}_0$ and the $M_0 \times M_0$ matrix $\boldsymbol{H}(\boldsymbol{\theta}_0)$ is defined by

$$\boldsymbol{H}(\boldsymbol{\theta}_0) = \boldsymbol{I}_{M_0\times M_0} - \boldsymbol{I}_F(\boldsymbol{\theta}_0)^{-1}\boldsymbol{B}(\boldsymbol{\theta}_0)^T\left(\boldsymbol{B}(\boldsymbol{\theta}_0)\boldsymbol{I}_F(\boldsymbol{\theta}_0)^{-1}\boldsymbol{B}(\boldsymbol{\theta}_0)^T\right)^{-1}\boldsymbol{B}(\boldsymbol{\theta}_0).$$

In the cited paper of Pardo, J. A. et *al.* (2002) it is also established that

$$\sqrt{n}(\widehat{\boldsymbol{\theta}}_{\phi}^{(r)} - \boldsymbol{\theta}_0) \underset{n\to\infty}{\overset{L}{\to}} N(\mathbf{0}, \boldsymbol{\Sigma}_*), \tag{5.11}$$

where the $M_0 \times M_0$ matrix $\boldsymbol{\Sigma}_*$ is given by

$$\boldsymbol{I}_F(\boldsymbol{\theta}_0)^{-1}\left(\boldsymbol{I}_{M_0\times M_0} - \boldsymbol{B}(\boldsymbol{\theta}_0)^T\left(\boldsymbol{B}(\boldsymbol{\theta}_0)\boldsymbol{I}_F(\boldsymbol{\theta}_0)^{-1}\boldsymbol{B}(\boldsymbol{\theta}_0)^T\right)^{-1}\boldsymbol{B}(\boldsymbol{\theta}_0)\right)\boldsymbol{I}_F(\boldsymbol{\theta}_0)^{-1}.$$

In Chapters 6 and 8 we describe some important applications of this result.

5.6. Exercises

1. Let X, Y and Z be Lognormal, Gamma and Weibull random variables. Let $g_{\boldsymbol{\theta}}(x)$ be the p.d.f. obtained from the mixture of the three random variables,
$$g_{\boldsymbol{\theta}}(x) = \pi_1 f_{\mu,\sigma}(x) + \pi_2 f_{a,p}(x) + \pi_3 f_{c,d}(x),$$
with $0 < \pi_i < 1$, $i = 1, 2, 3$, $\sum_{i=1}^{3} \pi_i = 1$, and $\boldsymbol{\theta} = (\pi_1, \pi_2, \mu, \sigma, a, p, c, d)$. We consider a random sample of size n from the mixture $g_{\boldsymbol{\theta}}(x)$. Prove that the corresponding likelihood function is not a bounded function.

2. Let X be a random variable with probability mass function
$$\Pr{}_{\boldsymbol{\theta}}(X = x_1) = \frac{1}{3} - \theta, \; \Pr{}_{\boldsymbol{\theta}}(X = x_2) = \frac{2}{3} - \theta \text{ and } \Pr{}_{\boldsymbol{\theta}}(X = x_3) = 2\theta,$$
with $0 < \theta < \frac{1}{3}$. Find the minimum ϕ-divergence estimator of θ, with
$$\phi(x) = \frac{1}{2}\left(\frac{1}{x} + x - 2\right).$$

3. Let $Y_1, ..., Y_n$ be a sample from the population X with probability mass function
$$\begin{array}{lll} p_1(\theta) = & \Pr(X = 1) & = \frac{1}{4}(2 + \theta) \\ p_2(\theta) = & \Pr(X = 2) & = \frac{1}{4}(1 - \theta) \\ p_3(\theta) = & \Pr(X = 3) & = \frac{1}{4}(1 - \theta) \\ p_4(\theta) = & \Pr(X = 4) & = \frac{1}{4}\theta, \end{array}$$
$\theta \in (0, 1)$.

   a) Find the minimum ϕ-divergence estimator with $\phi(x) = x \log x - x + 1$.

   b) Find its asymptotic distribution.

4. Let $Y_1, ..., Y_n$ be a sample from a Bernoulli random variable of parameter θ. Find the minimum ϕ-divergence estimator of θ with $\phi(x) = \frac{1}{2}(x - 1)^2$ as well as its asymptotic distribution.

5. Let $\boldsymbol{X}$ be a random variable with associated statistical space given by
$$(\mathcal{X}, \beta_{\mathcal{X}}, P_{\boldsymbol{\theta}})_{\boldsymbol{\theta} \in \Theta \subset \mathbb{R}^{M_0}}$$
with $\mathcal{X} = \mathbb{R}^d$. We consider the discretized model given by the partition $\mathcal{P} = \{E_i\}_{i=1,...,M}$ and $p_i(\boldsymbol{\theta}) = \Pr_{\boldsymbol{\theta}}(E_i), i = 1, .., M$. Find the asymptotic

distribution of the probability vector $\boldsymbol{p}(\widehat{\boldsymbol{\theta}}_\phi) = \left(p_1(\widehat{\boldsymbol{\theta}}_\phi), ..., p_M(\widehat{\boldsymbol{\theta}}_\phi)\right)^T$, being $\widehat{\boldsymbol{\theta}}_\phi$ the minimum ϕ-divergence estimator.

6. Find the asymptotic distribution of Shannon's entropy, $H(\boldsymbol{p}(\widehat{\boldsymbol{\theta}}_\phi))$.

7. Let $Y_1, ..., Y_n$ be a sample from a Bernoulli random variable of parameter θ. Using the result given in Exercise 6 find the asymptotic distribution of Shannon's entropy associated with the Bernoulli model when the parameter θ is estimated by the minimum ϕ-divergence estimator.

8. Solve Exercise 7 without using Exercise 6.

9. Let $Y_1, ..., Y_n$ be a sample of a random variable, X, with probability mass function,
$$\begin{array}{rcl} p_1(\theta) = & \Pr(X=-1) & = \theta^2 \\ p_2(\theta) = & \Pr(X=0) & = (1-\theta)^2 \\ p_3(\theta) = & \Pr(X=1) & = 2\theta(1-\theta) \end{array}$$
where $\theta \in (0,1)$. Find the minimum ϕ-divergence estimator of θ with $\phi(x) = x\log x - x + 1$ as well as its asymptotic distribution.

10. We consider a $2\times 3\times 2$ contingency table and let $\widehat{\boldsymbol{p}} = (\widehat{p}_{111}, ..., \widehat{p}_{232})^T$ be the nonparametric estimator of the unknown probability vector $\boldsymbol{p} = (p_{111}, ..., p_{232})^T$ where $p_{ijk} = \Pr(X=i, Y=j, K=k)$. We consider the parameter space
$$\Theta = \left\{\boldsymbol{\theta} : \boldsymbol{\theta} = (p_{ijk};\ i=1,2, j=1,2,3, k=1,2;\ (i,j,k) \neq (2,3,2))^T\right\}$$
and we denote by $\boldsymbol{p}(\boldsymbol{\theta}) = (p_{111}, ..., p_{232})^T = \boldsymbol{p}$ the probability vector characterizing the model, with
$$p_{232} = 1 - \sum_{\substack{i=1,2;\ j=1,2,3;\ k=1,2 \\ (i,j,k)\neq(2,3,2)}} p_{ijk}.$$
We assume the following constraints about the parameter $\boldsymbol{\theta}$
$$\begin{array}{rcl} f_1(p_{111}, ..., p_{231}) = & p_{111}p_{212} - p_{112}p_{211} = & 0 \\ f_2(p_{111}, ..., p_{231}) = & p_{121}p_{222} - p_{122}p_{221} = & 0\ . \\ f_3(p_{111}, ..., p_{231}) = & p_{131}p_{232} - p_{132}p_{231} = & 0 \end{array}$$
Find the asymptotic distribution of the minimum ϕ-divergence estimator with the previous constraints.

5.7. Answers to Exercises

1. We have

$$L(\boldsymbol{\theta}; x_1, \ldots, x_n) = \prod_{j=1}^{n} g_{\boldsymbol{\theta}}(x_j) = \prod_{j=1}^{n} \left(\pi_1 f_{\mu,\sigma}(x_j) + \pi_2 f_{a,p}(x_j) + \pi_3 f_{c,d}(x_j)\right),$$

with $\boldsymbol{\theta} = (\pi_1, \pi_2, \mu, \sigma, a, p, c, d)$. If we consider the particular point of $\boldsymbol{\theta}$ given by $(1/3, 1/3, \log x_1, \sigma, a, p, c, d)$, where x_1 is the first sample value, we have

$$\begin{aligned} L(\boldsymbol{\theta}; x_1, \ldots, x_n) = {} & \tfrac{1}{3}\left(\tfrac{1}{\sqrt{2\pi}\sigma x_1} \exp\left(\tfrac{-(\log x_1 - \log x_1)^2}{2\sigma^2}\right) + f_{a,p}(x_1) + f_{c,d}(x_1)\right) \\ & \times \prod_{j=2}^{n} \left(\pi_1 f_{\mu,\sigma}(x_j) + \pi_2 f_{a,p}(x_j) + \pi_3 f_{c,d}(x_j)\right) \end{aligned}$$

and choosing σ sufficiently small it is possible to do L as large as it is desired.

2. We consider a random sample of size n, $y_1, ..., y_n$, and we denote by n_i, $i = 1, 2, 3$, the number of values in the sample that coincide with x_i. We have to obtain the minimum, at θ, of the function

$$g(\theta) = \tfrac{1}{2} \sum_{j=1}^{3} p_j(\theta) \left(\frac{p_j(\theta)}{\widehat{p}_j} + \frac{\widehat{p}_j}{p_j(\theta)} - 2\right) = \tfrac{1}{2} \sum_{j=1}^{3} \frac{n}{n_j} \left(p_j(\theta) - \widehat{p}_j\right)^2,$$

where $p_j(\theta) = \Pr_{\boldsymbol{\theta}}(X = x_j)$, $j = 1, 2, 3$. Differentiating with respect to θ and setting the result equal to zero, we have

$$g'(\theta) = \frac{n}{n_1} \left(\frac{1}{3} - \theta - \frac{n_1}{n}\right)(-1) + \frac{n}{n_2} \left(\frac{2}{3} - \theta - \frac{n_2}{n}\right)(-1) + \frac{2n}{n_3} \left(2\theta - \frac{n_3}{n}\right),$$

and from $g'(\theta) = 0$ we obtain

$$\widehat{\theta}_\phi(y_1, ..., y_n) = \frac{\dfrac{1}{3n_1} + \dfrac{2}{3n_2}}{\dfrac{1}{n_1} + \dfrac{1}{n_2} + \dfrac{4}{n_3}}.$$

It is easy to prove that this point corresponds to a minimum.

3. *a)* It will be necessary to obtain the minimum, at θ, of the function

$$\begin{aligned} g(\theta) & = D(\widehat{\boldsymbol{p}}, \boldsymbol{p}(\theta)) \\ & = \widehat{p}_1 \log \frac{\widehat{p}_1}{p_1(\theta)} + \widehat{p}_2 \log \frac{\widehat{p}_2}{p_2(\theta)} + \widehat{p}_3 \log \frac{\widehat{p}_3}{p_3(\theta)} + \widehat{p}_4 \log \frac{\widehat{p}_4}{p_4(\theta)} \\ & = -\widehat{p}_1 \log(2 + \theta) - \widehat{p}_2 \log(1 - \theta) - \widehat{p}_3 \log(1 - \theta) - \widehat{p}_4 \log \theta + c. \end{aligned}$$

Differentiating with respect to θ and setting the result equal to zero, we have

$$g'(\theta) = -\frac{\widehat{p}_1}{2+\theta} + \frac{\widehat{p}_2}{1-\theta} + \frac{\widehat{p}_3}{1-\theta} - \frac{\widehat{p}_4}{\theta} = 0.$$

Then

$$\theta^2 \left(\widehat{p}_1 + \widehat{p}_2 + \widehat{p}_3 + \widehat{p}_4\right) + \theta \left(2\widehat{p}_2 + 2\widehat{p}_3 - \widehat{p}_1 + \widehat{p}_4\right) - 2\widehat{p}_4 = 0,$$

i.e.,

$$\theta^2 + \theta \left(1 - 2\widehat{p}_1 + \widehat{p}_2 + \widehat{p}_3\right) - 2\widehat{p}_4 = 0.$$

Therefore,

$$\widehat{\theta}_\phi \left(y_1, ..., y_n\right) = \frac{\left(-1 + 2\widehat{p}_1 - \widehat{p}_2 - \widehat{p}_3\right) + \left(\left(-1 + 2\widehat{p}_1 - \widehat{p}_2 - \widehat{p}_3\right)^2 + 8\widehat{p}_4\right)^{1/2}}{2}.$$

b) In this case the matrix *diag* $\left(\boldsymbol{p}\left(\theta\right)^{-1/2}\right)$ is given by

$$\begin{pmatrix} 2\left(2+\theta\right)^{-1/2} & 0 & 0 & 0 \\ 0 & 2\left(1-\theta\right)^{-1/2} & 0 & 0 \\ 0 & 0 & 2\left(1-\theta\right)^{-1/2} & 0 \\ 0 & 0 & 0 & 2\theta^{-1/2} \end{pmatrix},$$

and

$$\frac{\partial p_1\left(\theta\right)}{\partial\theta} = \frac{1}{4}, \frac{\partial p_2\left(\theta\right)}{\partial\theta} = -\frac{1}{4}, \frac{\partial p_3\left(\theta\right)}{\partial\theta} = -\frac{1}{4} \text{ and } \frac{\partial p_4\left(\theta\right)}{\partial\theta} = \frac{1}{4}.$$

Therefore $\boldsymbol{A}\left(\theta\right)$ is given by

$$\left(2^{-1}\left(2+\theta\right)^{-1/2}, -2^{-1}\left(1-\theta\right)^{-1/2}, -2^{-1}\left(1-\theta\right)^{-1/2}, 2^{-1}\theta^{-1/2}\right)^T.$$

Then

$$\begin{aligned} \boldsymbol{A}\left(\theta\right)^T \boldsymbol{A}\left(\theta\right) &= \tfrac{1}{4}\left(2+\theta\right)^{-1} + \tfrac{1}{4}\left(1-\theta\right)^{-1} + \tfrac{1}{4}\left(1-\theta\right)^{-1} + \tfrac{1}{4}\theta^{-1} \\ &= \tfrac{1}{4}\left(\frac{1}{2+\theta} + \frac{2}{1-\theta} + \frac{1}{\theta}\right) = \frac{1}{2}\frac{2\theta+1}{\left(2+\theta\right)\left(1-\theta\right)\theta}. \end{aligned}$$

Finally,

$$\sqrt{n}\left(\widehat{\theta}_\phi\left(Y_1, ..., Y_n\right) - \theta_0\right) \underset{n\to\infty}{\overset{L}{\longrightarrow}} N\left(0, \frac{2\left(2+\theta_0\right)\left(1-\theta_0\right)\theta_0}{2\theta_0+1}\right).$$

4. We have to minimize, at θ, the function

$$
\begin{aligned}
g(\theta) &= \sum_{i=1}^{2} p_i(\theta)\left(1-\frac{n_i}{np_i(\theta)}\right)^2 \\
&= \sum_{i=1}^{2} \frac{(np_i(\theta)-n_i)^2}{n^2 p_i(\theta)} \\
&= \frac{(np_1(\theta)-n_1)^2}{n^2 p_1(\theta)} + \frac{(np_2(\theta)-n_2)^2}{n^2 p_2(\theta)} \\
&= \frac{(np_1(\theta)-n_1)^2}{n^2 p_1(\theta)} + \frac{(np_1(\theta)-n_1)^2}{n^2 p_2(\theta)} \\
&= \frac{(np_1(\theta)-n_1)^2}{n^2 p_1(\theta)\, p_2(\theta)} \\
&= (n\theta-n_1)^2 \frac{1}{n^2\theta(1-\theta)}.
\end{aligned}
$$

It is clear that

$$\widehat{\theta}_\phi(y_1, ..., y_n) = n_1/n.$$

Now the asymptotic distribution is given by

$$\sqrt{n}\left(\widehat{\theta}_\phi(Y_1, ..., Y_n) - \theta_0\right) \xrightarrow[n\to\infty]{L} N\left(0, \left(\boldsymbol{A}(\theta_0)^T \boldsymbol{A}(\theta_0)\right)^{-1}\right).$$

But

$$diag\left(\boldsymbol{p}(\theta)^{-1/2}\right) = \begin{pmatrix} \theta^{-1/2} & 0 \\ 0 & (1-\theta)^{-1/2} \end{pmatrix}$$

and

$$\frac{\partial p_1(\theta)}{\partial\theta} = 1, \quad \frac{\partial p_2(\theta)}{\partial\theta} = -1,$$

then

$$\boldsymbol{A}(\theta) = \begin{pmatrix} \theta^{-1/2} & 0 \\ 0 & (1-\theta)^{-1/2} \end{pmatrix}\begin{pmatrix} 1 \\ -1 \end{pmatrix} = \begin{pmatrix} \theta^{-1/2} \\ -(1-\theta)^{-1/2} \end{pmatrix}$$

and

$$\boldsymbol{A}(\theta)^T \boldsymbol{A}(\theta) = \begin{pmatrix} \theta^{-1/2}, & -(1-\theta)^{-1/2} \end{pmatrix}\begin{pmatrix} \theta^{-1/2} \\ -(1-\theta)^{-1/2} \end{pmatrix} = \frac{1}{\theta(1-\theta)}.$$

Therefore

$$\sqrt{n}(\widehat{\theta}_\phi(Y_1, ..., Y_n) - \theta_0) \xrightarrow[n\to\infty]{L} N(0, \theta_0(1-\theta_0)).$$

This result could be obtained using directly the Central Limit Theorem

$$\frac{\sqrt{n}\left(\frac{N_1}{n}-\theta_0\right)}{\theta_0\left(1-\theta_0\right)} \underset{n\to\infty}{\overset{L}{\longrightarrow}} N\left(0,1\right).$$

5. We have

$$\boldsymbol{p}(\widehat{\boldsymbol{\theta}}_\phi)=\boldsymbol{p}\left(\boldsymbol{\theta}_0\right)+\left(\frac{\partial p_i\left(\boldsymbol{\theta}_0\right)}{\partial\theta_j}\right)_{\substack{i=1,\ldots,M\\ j=1,\ldots,M_0}}\left(\widehat{\boldsymbol{\theta}}_\phi-\boldsymbol{\theta}_0\right)+o\left(\left\|\widehat{\boldsymbol{\theta}}_\phi-\boldsymbol{\theta}_0\right\|\right).$$

But $\sqrt{n}\,o\left(\left\|\widehat{\boldsymbol{\theta}}_\phi-\boldsymbol{\theta}_0\right\|\right)=o_P(1)$, therefore

$$\sqrt{n}\left(\boldsymbol{p}(\widehat{\boldsymbol{\theta}}_\phi)-\boldsymbol{p}\left(\boldsymbol{\theta}_0\right)\right)\underset{n\to\infty}{\overset{L}{\longrightarrow}} N\left(\boldsymbol{0},\boldsymbol{\Sigma}_*\right),$$

with $\boldsymbol{\Sigma}_*=\boldsymbol{J}(\boldsymbol{\theta}_0)\left(\boldsymbol{A}(\boldsymbol{\theta}_0)^T\boldsymbol{A}(\boldsymbol{\theta}_0)\right)^{-1}\boldsymbol{J}(\boldsymbol{\theta}_0)^T$, but

$$\boldsymbol{A}\left(\boldsymbol{\theta}_0\right)=diag\left(\boldsymbol{p}\left(\boldsymbol{\theta}_0\right)^{-1/2}\right)\boldsymbol{J}(\boldsymbol{\theta}_0);$$

therefore we have that $\sqrt{n}\left(\boldsymbol{p}(\widehat{\boldsymbol{\theta}}_\phi)-\boldsymbol{p}\left(\boldsymbol{\theta}_0\right)\right)$ converges to a normal distribution with mean vector zero and variance-covariance matrix given by

$$diag\left(\boldsymbol{p}\left(\boldsymbol{\theta}_0\right)^{1/2}\right)\boldsymbol{A}\left(\boldsymbol{\theta}_0\right)\left(\boldsymbol{A}(\boldsymbol{\theta}_0)^T\boldsymbol{A}(\boldsymbol{\theta}_0)\right)^{-1}\boldsymbol{A}(\boldsymbol{\theta}_0)^T diag\left(\boldsymbol{p}\left(\boldsymbol{\theta}_0\right)^{1/2}\right).$$

6. We have

$$\begin{aligned}H(\boldsymbol{p}(\widehat{\boldsymbol{\theta}}_\phi))= &\ H\left(\boldsymbol{p}\left(\boldsymbol{\theta}_0\right)\right)+\left(\frac{\partial H\left(\boldsymbol{p}\left(\boldsymbol{\theta}_0\right)\right)}{\partial p_i\left(\boldsymbol{\theta}\right)}\right)_{i=1,\ldots,M}\left(\boldsymbol{p}(\widehat{\boldsymbol{\theta}}_\phi)-\boldsymbol{p}\left(\boldsymbol{\theta}_0\right)\right)\\ &+\ o\left(\left\|\boldsymbol{p}\left(\widehat{\boldsymbol{\theta}}_\phi\right)-\boldsymbol{p}\left(\boldsymbol{\theta}_0\right)\right\|\right).\end{aligned}$$

On the other hand

$$\boldsymbol{p}(\widehat{\boldsymbol{\theta}}_\phi)=\boldsymbol{p}\left(\boldsymbol{\theta}_0\right)+\boldsymbol{J}\left(\boldsymbol{\theta}_0\right)\left(\widehat{\boldsymbol{\theta}}_\phi-\boldsymbol{\theta}_0\right)+o\left(\left\|\widehat{\boldsymbol{\theta}}_\phi-\boldsymbol{\theta}_0\right\|\right)$$

with

$$\sqrt{n}\left(\boldsymbol{p}(\widehat{\boldsymbol{\theta}}_\phi)-\boldsymbol{p}\left(\boldsymbol{\theta}_0\right)\right)\underset{n\to\infty}{\overset{L}{\longrightarrow}} N\left(\boldsymbol{0},\boldsymbol{\Sigma}_*\right),$$

being

$$\boldsymbol{\Sigma}_*=\boldsymbol{J}(\boldsymbol{\theta}_0)\left(\boldsymbol{A}(\boldsymbol{\theta}_0)^T\boldsymbol{A}(\boldsymbol{\theta}_0)\right)^{-1}\boldsymbol{J}(\boldsymbol{\theta}_0)^T.$$

Therefore

$$\sqrt{n}\left(H(\boldsymbol{p}(\widehat{\boldsymbol{\theta}}_{\phi})) - H(\boldsymbol{p}(\boldsymbol{\theta}_0))\right)$$

converges in law to a normal distribution with mean vector zero and variance-covariance matrix

$$\left(\frac{\partial H(\boldsymbol{p}(\boldsymbol{\theta}_0))}{\partial p_i(\boldsymbol{\theta})}\right)_{i=1,\ldots,M} \boldsymbol{\Sigma}_* \left(\left(\frac{\partial H(\boldsymbol{p}(\boldsymbol{\theta}_0))}{\partial p_i(\boldsymbol{\theta})}\right)_{i=1,\ldots,M}\right)^T.$$

7. We know

$$H(\boldsymbol{p}(\theta)) = -p_1(\theta)\log p_1(\theta) - p_2(\theta)\log p_2(\theta).$$

On the other hand

$$\begin{aligned}\frac{\partial H(\boldsymbol{p}(\theta))}{\partial p_2(\theta)} &= -1 - \log p_2(\theta) = -1 - \log(1-\theta)\\ \frac{\partial H(\boldsymbol{p}(\theta))}{\partial p_1(\theta)} &= -1 - \log p_1(\theta) = -1 - \log\theta\end{aligned}$$

and

$$\frac{\partial p_1(\theta)}{\partial \theta} = 1, \frac{\partial p_2(\theta)}{\partial \theta} = -1,$$

then we have

$$\begin{aligned}\boldsymbol{\Sigma}_* &= \boldsymbol{J}(\theta)(\boldsymbol{A}(\theta)^T\boldsymbol{A}(\theta))^{-1}\boldsymbol{J}(\theta)^T = \begin{pmatrix} 1 \\ -1 \end{pmatrix}\theta(1-\theta)\begin{pmatrix} 1 & -1 \end{pmatrix}\\ &= \theta(1-\theta)\begin{pmatrix} 1 & -1 \\ -1 & 1 \end{pmatrix}\end{aligned}$$

because

$$\left(\boldsymbol{A}(\theta)^T\boldsymbol{A}(\theta)\right)^{-1} = \theta(1-\theta).$$

Therefore, denoting

$$\boldsymbol{L} = \left(\frac{\partial H(\boldsymbol{p}(\theta))}{\partial p_i(\theta)}\right)_{i=1,2} \boldsymbol{\Sigma}_* \left(\left(\frac{\partial H(\boldsymbol{p}(\theta))}{\partial p_i(\theta)}\right)_{i=1,2}\right)^T,$$

we have

$$\begin{aligned}\boldsymbol{L} &= \begin{pmatrix} -1-\log\theta, & -1-\log(1-\theta) \end{pmatrix}\theta(1-\theta)\begin{pmatrix} 1 & -1 \\ -1 & 1 \end{pmatrix}\\ &\times \begin{pmatrix} -1-\log\theta \\ -1-\log(1-\theta) \end{pmatrix}\\ &= \theta(1-\theta)\left(\log\frac{\theta}{1-\theta}\right)^2,\end{aligned}$$

i.e.,

$$\sqrt{n}\left(H(\boldsymbol{p}(\widehat{\boldsymbol{\theta}}_{\phi}))-H\left(\boldsymbol{p}\left(\boldsymbol{\theta}_0\right)\right)\right) \underset{n\to\infty}{\overset{L}{\longrightarrow}} N\left(0,\theta_0\left(1-\theta_0\right)\left(\log\frac{\theta_0}{1-\theta_0}\right)^2\right).$$

8. We have

$$H\left(\boldsymbol{p}\left(\theta\right)\right)=-\left(1-\theta\right)\log\left(1-\theta\right)-\theta\log\theta.$$

Then

$$H(\boldsymbol{p}(\widehat{\theta}_{\phi}))-H\left(\boldsymbol{p}\left(\theta\right)\right)=\left(-\log\frac{\theta}{\left(1-\theta\right)}\right)(\widehat{\theta}_{\phi}-\theta)+o\left(\left|\widehat{\theta}_{\phi}-\theta\right|\right)$$

but

$$\sqrt{n}(\widehat{\theta}_{\phi}\left(Y_1,...,Y_n\right)-\theta_0) \underset{n\to\infty}{\overset{L}{\longrightarrow}} N\left(0,\theta_0\left(1-\theta_0\right)\right)$$

and now the result is immediate.

9. We have to minimize, at θ, the function

$$\begin{aligned} g\left(\theta\right) &= \widehat{p}_1\log\frac{\widehat{p}_1}{p_1\left(\theta\right)}+\widehat{p}_2\log\frac{\widehat{p}_2}{p_2\left(\theta\right)}+\widehat{p}_3\log\frac{\widehat{p}_3}{p_3\left(\theta\right)} \\ &= \widehat{p}_1\log\widehat{p}_1-\widehat{p}_1\log p_1\left(\theta\right)+\widehat{p}_2\log\widehat{p}_2-\widehat{p}_2\log p_2\left(\theta\right)+\widehat{p}_3\log\widehat{p}_3 \\ &- \widehat{p}_3\log p_3\left(\theta\right) \\ &= c-\widehat{p}_1\log p_1\left(\theta\right)-\widehat{p}_2\log p_2\left(\theta\right)-\widehat{p}_3\log p_3\left(\theta\right) \\ &= c-2\widehat{p}_1\log\theta-\widehat{p}_2 2\log\left(1-\theta\right)-\widehat{p}_3\log\left(2\theta\left(1-\theta\right)\right). \end{aligned}$$

But

$$g'\left(\theta\right)=\frac{-\left(1-\theta\right)2\widehat{p}_1+2\widehat{p}_2\theta-\widehat{p}_3\left(1-2\theta\right)}{\theta\left(1-\theta\right)},$$

then we have

$$\widehat{\theta}_{\phi}\left(y_1,...,y_n\right)=\frac{2\widehat{p}_1+\widehat{p}_3}{2}.$$

We know that

$$\sqrt{n}\left(\widehat{\theta}_{\phi}\left(Y_1,...,Y_n\right)-\theta_0\right) \underset{n\to\infty}{\overset{L}{\longrightarrow}} N\left(0,\left(\boldsymbol{A}\left(\theta_0\right)^T\boldsymbol{A}\left(\theta_0\right)\right)^{-1}\right),$$

but

$$
\begin{aligned}
\boldsymbol{A}(\theta) &= diag\left(\boldsymbol{p}(\theta)^{-1/2}\right)\boldsymbol{J}(\theta) \\
&= \begin{pmatrix} \theta^{-1} & 0 & 0 \\ 0 & (1-\theta)^{-1} & 0 \\ 0 & 0 & (2\theta(1-\theta))^{-\frac{1}{2}} \end{pmatrix} \begin{pmatrix} 2\theta \\ -2(1-\theta) \\ 2(1-2\theta) \end{pmatrix} \\
&= \begin{pmatrix} 2 \\ -2 \\ \frac{2(1-2\theta)}{\sqrt{2\theta(1-\theta)}} \end{pmatrix},
\end{aligned}
$$

then we have

$$\boldsymbol{A}(\theta)^T \boldsymbol{A}(\theta) = \frac{2}{\theta(1-\theta)}$$

and therefore

$$\sqrt{n}\left(\widehat{\theta}_\phi(Y_1, ..., Y_n) - \theta_0\right) \underset{n\to\infty}{\overset{L}{\longrightarrow}} N\left(0, 2^{-1}\theta_0(1-\theta_0)\right).$$

10. We can observe that the constraints given can be written in the way

$$\frac{p_{111}p_{212}}{p_{112}p_{211}} = 1 \quad \frac{p_{121}p_{222}}{p_{122}p_{221}} = 1 \quad \text{and} \quad \frac{p_{131}p_{232}}{p_{132}p_{231}} = 1,$$

i.e., using the odds ratios. These odds ratios characterize the model of conditional independence of the first and third variables given the second variable. In this case we know that the maximum likelihood estimator without constraints is given by

$$\widehat{p}_{ijk} = \frac{n_{ijk}}{n}$$

where n_{ijk} is the number of elements in the random sample of size n in the cell (i, j, k). We denote by

$$\widehat{\boldsymbol{\theta}} = \left(\widehat{p}_{111}, \widehat{p}_{112}, \widehat{p}_{121}, \widehat{p}_{122}, \widehat{p}_{131}, \widehat{p}_{132}, \widehat{p}_{211}, \widehat{p}_{212}, \widehat{p}_{221}, \widehat{p}_{222}, \widehat{p}_{231}\right)^T,$$

and

$$\boldsymbol{\Sigma}_{\boldsymbol{\theta}} = diag(\boldsymbol{\theta}) - \boldsymbol{\theta}\boldsymbol{\theta}^T.$$

It is well known that

$$\sqrt{n}(\widehat{\boldsymbol{\theta}} - \boldsymbol{\theta}_0) \underset{n\to\infty}{\overset{L}{\longrightarrow}} N(\boldsymbol{0}, \boldsymbol{\Sigma}_{\boldsymbol{\theta}_0}),$$

and the restricted maximum likelihood estimator is given by

$$\widehat{p}_{ijk}^{*} = \frac{\widehat{p}_{i*k}\widehat{p}_{*jk}}{\widehat{p}_{**k}},$$

where

$$\widehat{p}_{i*k} = \sum_{j=1}^{3} \widehat{p}_{ijk}^{*},\ \widehat{p}_{*jk} = \sum_{i=1}^{2} \widehat{p}_{ijk}^{*} \text{ and } \widehat{p}_{**k} = \sum_{i=1}^{2}\sum_{j=1}^{3} \widehat{p}_{ijk}^{*}.$$

The vector $\widehat{\boldsymbol{\theta}}_{\phi}^{(r)}$ with components

$$(\widehat{\theta}_{\phi,111}^{(r)}, \widehat{\theta}_{\phi,112}^{(r)}, \widehat{\theta}_{\phi,121}^{(r)}, \widehat{\theta}_{\phi,122}^{(r)}, \widehat{\theta}_{\phi,131}^{(r)}, \widehat{\theta}_{\phi,132}^{(r)}, \widehat{\theta}_{\phi,211}^{(r)}, \widehat{\theta}_{\phi,212}^{(r)}, \widehat{\theta}_{\phi,221}^{(r)}, \widehat{\theta}_{\phi,222}^{(r)}, \widehat{\theta}_{\phi,231}^{(r)})^T$$

denotes the restricted minimum ϕ-divergence estimator. Using (5.11), we have

$$\sqrt{n}(\widehat{\boldsymbol{\theta}}_{\phi}^{(r)} - \boldsymbol{\theta}_0) \underset{n\to\infty}{\overset{L}{\longrightarrow}} N(\mathbf{0}, \boldsymbol{\Sigma}_*),$$

where $\boldsymbol{\Sigma}_*$ is given by

$$\boldsymbol{I}_F(\boldsymbol{\theta}_0)^{-1}\left(\boldsymbol{I} - \boldsymbol{B}(\boldsymbol{\theta}_0)^T(\boldsymbol{B}(\boldsymbol{\theta}_0)\boldsymbol{I}_F(\boldsymbol{\theta}_0)^{-1}\boldsymbol{B}(\boldsymbol{\theta}_0)^T)^{-1}\boldsymbol{B}(\boldsymbol{\theta}_0)\right)\boldsymbol{I}_F(\boldsymbol{\theta}_0)^{-1},$$

because $\boldsymbol{\Sigma}_{\boldsymbol{\theta}_0} = \boldsymbol{I}_F(\boldsymbol{\theta}_0)^{-1}$. The matrix $\boldsymbol{B}(\boldsymbol{\theta})$ is given by $\boldsymbol{B}(\boldsymbol{\theta}) = (\boldsymbol{A}_1, \boldsymbol{A}_2)$, where

$$\boldsymbol{A}_1 = \begin{pmatrix} p_{212} & -p_{211} & 0 & 0 & 0 & 0 \\ 0 & 0 & p_{222} & -p_{221} & 0 & 0 \\ -p_{131} & -p_{131} & -p_{131} & -p_{131} & 1-2p_{131} & -p_{131}-p_{231} \end{pmatrix}$$

and

$$\boldsymbol{A}_2 = \begin{pmatrix} -p_{112} & p_{111} & 0 & 0 & 0 \\ 0 & 0 & -p_{122} & p_{121} & 0 \\ -p_{131} & -p_{131} & -p_{131} & -p_{131} & -p_{131}-p_{132} \end{pmatrix}.$$

6

Goodness-of-fit: Composite Null Hypothesis

6.1. Introduction

In Chapter 3 we considered a random sample $Y_1, \ldots, Y_n$ from F and we studied the problem of goodness-of-fit when F is completely known, i.e., $H_0 : F = F_0$. It is common to wish to test the composite hypothesis that the distribution function F is a member of a parametric family $\{F_{\boldsymbol{\theta}}\}_{\boldsymbol{\theta} \in \Theta}$, where Θ is an open subset in $\mathbb{R}^{M_0}$, i.e., we are interested in testing

$$H_0 : F = F_{\boldsymbol{\theta}}. \tag{6.1}$$

An approach to this problem is to consider a discrete statistical model associated with the original model. In order to do this we consider a partition $\mathcal{P} = \{E_i\}_{i=1,\ldots,M}$ of the original sample space. Now the probabilities of the elements of the partition, E_i, $i = 1, ..., M$, depend on the unknown parameter $\boldsymbol{\theta}$, i.e.,

$$p_i(\boldsymbol{\theta}) = \Pr_{\boldsymbol{\theta}}(E_i) = \int_{E_i} dF_{\boldsymbol{\theta}}, \; i = 1, ..., M.$$

The hypothesis given in (6.1) can be tested by the hypotheses

$$H_0 : \boldsymbol{p} = \boldsymbol{p}(\boldsymbol{\theta}_0) \in T \text{ for some unknown } \boldsymbol{\theta}_0 \in \Theta \tag{6.2}$$

versus

$$H_1 : \boldsymbol{p} \in \Delta_M - T,$$

with

$$T = \left\{ \boldsymbol{p}(\boldsymbol{\theta}) = (p_1(\boldsymbol{\theta}), ..., p_M(\boldsymbol{\theta}))^T \in \Delta_M : \boldsymbol{\theta} \in \Theta \right\},$$

$\Theta \subset \mathbb{R}^{M_0}$ open and $M_0 < M - 1$.

Pearson recommended (see D'Agostino and Stephens (1986, p. 65)) estimating $\boldsymbol{\theta}$ by an estimator $\widetilde{\boldsymbol{\theta}}$ based on $Y_1, \ldots, Y_n$ and testing (6.2), by using the chi-square test statistic,

$$X^2\left(\widetilde{\boldsymbol{\theta}}\right) \equiv \sum_{i=1}^{M} \frac{(N_i - np_i(\widetilde{\boldsymbol{\theta}}))^2}{np_i(\widetilde{\boldsymbol{\theta}})}. \qquad (6.3)$$

Pearson thought that if $\widetilde{\boldsymbol{\theta}}$ was a consistent estimator of $\boldsymbol{\theta}$, the asymptotic distribution of the test statistic (6.3) would coincide with the distribution of the test statistic, X^2, given in Chapter 3 for testing the null hypothesis (3.1). He was wrong. Fisher (1924) established that the asymptotic distribution of the test statistic (6.3) was not chi-square with $M-1$ degrees of freedom and also that depends on the method of estimation. Fisher pointed out that the right method of estimation was the method of the maximum likelihood for the discretized model or equivalently the maximum likelihood estimator based on grouped data. In the previous chapter, it was established that this estimator, $\widehat{\boldsymbol{\theta}}_{(0)}$, coincides with the minimum Kullback-Leibler divergence estimator. It is interesting to note that Fisher also established that the likelihood ratio test statistic,

$$G^2\left(\widehat{\boldsymbol{\theta}}_{(0)}\right) \equiv 2\sum_{i=1}^{M} N_i \log \frac{N_i}{np_i(\widehat{\boldsymbol{\theta}}_{(0)})}, \qquad (6.4)$$

is asymptotically equivalent to $X^2\left(\widehat{\boldsymbol{\theta}}_{(0)}\right)$ and that the minimum chi-square estimator, $\widehat{\boldsymbol{\theta}}_{(1)}$, is asymptotically equivalent to $\widehat{\boldsymbol{\theta}}_{(0)}$. Finally, he established that the asymptotic distribution of the test statistics $X^2\left(\widehat{\boldsymbol{\theta}}_{(0)}\right)$, $X^2\left(\widehat{\boldsymbol{\theta}}_{(1)}\right)$, $G^2\left(\widehat{\boldsymbol{\theta}}_{(0)}\right)$, $G^2\left(\widehat{\boldsymbol{\theta}}_{(1)}\right)$ is the same and is chi-square with $M - M_0 - 1$ degrees of freedom. It is interesting to note that Neyman (1949) established that the minimum modified chi-square estimator, $\widehat{\boldsymbol{\theta}}_{(-2)}$, is also asymptotically equivalent to $\widehat{\boldsymbol{\theta}}_{(0)}$.

In this chapter we consider, for testing (6.2), the family of ϕ-divergence test statistics

$$T_n^{\phi_1}\left(\widehat{\boldsymbol{\theta}}_{\phi_2}\right) \equiv \frac{2n}{\phi_1''(1)} D_{\phi_1}\left(\widehat{\boldsymbol{p}}, \boldsymbol{p}(\widehat{\boldsymbol{\theta}}_{\phi_2})\right),$$

and we study its asymptotic distribution under different situations. We observe that for $\phi_1(x) = \frac{1}{2}(x-1)^2$ and $\phi_2(x) = x\log x - x + 1$ we get the test statistic $X^2\left(\widehat{\boldsymbol{\theta}}_{(0)}\right)$; for $\phi_1(x) = \phi_2(x) = x\log x - x + 1$ we have the likelihood ratio test statistic, $G^2\left(\widehat{\boldsymbol{\theta}}_{(0)}\right)$. The test statistic $X^2\left(\widehat{\boldsymbol{\theta}}_{(1)}\right)$ is obtained for $\phi_1(x) = \phi_2(x) = \frac{1}{2}(x-1)^2$ and the test statistic $G^2\left(\widehat{\boldsymbol{\theta}}_{(1)}\right)$ for $\phi_1(x) = x\log x - x + 1$ and $\phi_2(x) = \frac{1}{2}(x-1)^2$.

We shall assume that $\phi_1, \phi_2 \in \Phi^*$ are twice continuously differentiable in $x > 0$ with the second derivatives $\phi_1''(1) \neq 0$ and $\phi_2''(1) \neq 0$.

6.2. Asymptotic Distribution with Fixed Number of Classes

In the same way as in the case of the simple null hypothesis, we state three theorems. The first one gives the asymptotic distribution under the null hypothesis given in (6.2), the second one under an alternative hypothesis different from the hypothesis given in (6.2) and finally the third one under contiguous alternative hypotheses.

Theorem 6.1

Under the null hypothesis given in (6.2), and assuming the regularity conditions given in Theorem 5.1, we have

$$T_n^{\phi_1}\left(\widehat{\boldsymbol{\theta}}_{\phi_2}\right) = \frac{2n}{\phi_1''(1)} D_{\phi_1}\left(\widehat{\boldsymbol{p}}, \boldsymbol{p}(\widehat{\boldsymbol{\theta}}_{\phi_2})\right) \underset{n\to\infty}{\overset{L}{\longrightarrow}} \chi^2_{M-M_0-1},$$

where $\phi_1, \phi_2 \in \Phi^*$.

Proof.

First, we obtain the asymptotic distribution of the random vector

$$\sqrt{n}\left(\widehat{\boldsymbol{p}} - \boldsymbol{p}(\widehat{\boldsymbol{\theta}}_{\phi_2})\right).$$

We know that

$$\begin{aligned} \widehat{\boldsymbol{\theta}}_{\phi_2} - \boldsymbol{\theta}_0 = & \left(\boldsymbol{A}(\boldsymbol{\theta}_0)^T \boldsymbol{A}(\boldsymbol{\theta}_0)\right)^{-1} \boldsymbol{A}(\boldsymbol{\theta}_0)^T diag\left(\boldsymbol{p}(\boldsymbol{\theta}_0)^{-1/2}\right)\left(\widehat{\boldsymbol{p}} - \boldsymbol{p}(\boldsymbol{\theta}_0)\right) \\ + & \; o\left(\|\widehat{\boldsymbol{p}} - \boldsymbol{p}(\boldsymbol{\theta}_0)\|\right) \end{aligned}$$

and

$$\boldsymbol{p}(\widehat{\boldsymbol{\theta}}_{\phi_2}) - \boldsymbol{p}(\boldsymbol{\theta}_0) = \boldsymbol{J}(\boldsymbol{\theta}_0)(\widehat{\boldsymbol{\theta}}_{\phi_2} - \boldsymbol{\theta}_0) + o\left(\left\|\widehat{\boldsymbol{\theta}}_{\phi_2} - \boldsymbol{\theta}_0\right\|\right),$$

where $\boldsymbol{J}(\boldsymbol{\theta}_0)$ and $\boldsymbol{A}(\boldsymbol{\theta}_0)$ were defined in (5.6) and (5.7), respectively. Therefore

$$\begin{aligned}\boldsymbol{p}(\widehat{\boldsymbol{\theta}}_{\phi_2}) - \boldsymbol{p}(\boldsymbol{\theta}_0) = {} & \boldsymbol{J}(\boldsymbol{\theta}_0)\left(\boldsymbol{A}(\boldsymbol{\theta}_0)^T\boldsymbol{A}(\boldsymbol{\theta}_0)\right)^{-1}\boldsymbol{A}(\boldsymbol{\theta}_0)^T diag\left(\boldsymbol{p}(\boldsymbol{\theta}_0)^{-1/2}\right)\\ \times {} & (\widehat{\boldsymbol{p}} - \boldsymbol{p}(\boldsymbol{\theta}_0)) + o\left(\left\|\widehat{\boldsymbol{\theta}}_{\phi_2} - \boldsymbol{\theta}_0\right\|\right) + o\left(\|\widehat{\boldsymbol{p}} - \boldsymbol{p}(\boldsymbol{\theta}_0))\|\right).\end{aligned}$$

In the following we denote by $\boldsymbol{L}$ the $M \times M$ matrix

$$\boldsymbol{L} = \boldsymbol{J}(\boldsymbol{\theta}_0)\left(\boldsymbol{A}(\boldsymbol{\theta}_0)^T\boldsymbol{A}(\boldsymbol{\theta}_0)\right)^{-1}\boldsymbol{A}(\boldsymbol{\theta}_0)^T diag\left(\boldsymbol{p}(\boldsymbol{\theta}_0)^{-1/2}\right)$$

and by $\boldsymbol{I}$ the $M \times M$ identity matrix. We have

$$\begin{aligned}\begin{pmatrix}\widehat{\boldsymbol{p}} - \boldsymbol{p}(\boldsymbol{\theta}_0)\\ \boldsymbol{p}(\widehat{\boldsymbol{\theta}}_{\phi_2}) - \boldsymbol{p}(\boldsymbol{\theta}_0)\end{pmatrix}_{2M\times 1} = {} & \begin{pmatrix}\boldsymbol{I}\\ \boldsymbol{L}\end{pmatrix}_{2M\times M}(\widehat{\boldsymbol{p}} - \boldsymbol{p}(\boldsymbol{\theta}_0))_{M\times 1}\\ + {} & \begin{pmatrix}\boldsymbol{0}_{M\times 1}\\ o\left(\left\|\left(\widehat{\boldsymbol{\theta}}_{\phi_2} - \boldsymbol{\theta}_0\right)\right\|\right) + o\left(\|\widehat{\boldsymbol{p}} - \boldsymbol{p}(\boldsymbol{\theta}_0))\|\right)\end{pmatrix}.\end{aligned}$$

Then

$$\sqrt{n}\begin{pmatrix}\widehat{\boldsymbol{p}} - \boldsymbol{p}(\boldsymbol{\theta}_0)\\ \boldsymbol{p}(\widehat{\boldsymbol{\theta}}_{\phi_2}) - \boldsymbol{p}(\boldsymbol{\theta}_0)\end{pmatrix}_{2M\times 1} \xrightarrow[n\to\infty]{L} N\left(\boldsymbol{0}, \begin{pmatrix}\boldsymbol{I}\\ \boldsymbol{L}\end{pmatrix}\boldsymbol{\Sigma}_{\boldsymbol{p}(\boldsymbol{\theta}_0)}\left(\boldsymbol{I}, \boldsymbol{L}^T\right)\right),$$

because

$$\sqrt{n}\,(\widehat{\boldsymbol{p}} - \boldsymbol{p}(\boldsymbol{\theta}_0)) \xrightarrow[n\to\infty]{L} N\left(\boldsymbol{0}, \boldsymbol{\Sigma}_{\boldsymbol{p}(\boldsymbol{\theta}_0)}\right),$$

i.e.,

$$\sqrt{n}\begin{pmatrix}\widehat{\boldsymbol{p}} - \boldsymbol{p}(\boldsymbol{\theta}_0)\\ \boldsymbol{p}(\widehat{\boldsymbol{\theta}}_{\phi_2}) - \boldsymbol{p}(\boldsymbol{\theta}_0)\end{pmatrix}_{2M\times 1} \xrightarrow[n\to\infty]{L} N\left(\boldsymbol{0}, \begin{pmatrix}\boldsymbol{\Sigma}_{\boldsymbol{p}(\boldsymbol{\theta}_0)} & \boldsymbol{\Sigma}_{\boldsymbol{p}(\boldsymbol{\theta}_0)}\boldsymbol{L}^T\\ \boldsymbol{L}\boldsymbol{\Sigma}_{\boldsymbol{p}(\boldsymbol{\theta}_0)} & \boldsymbol{L}\boldsymbol{\Sigma}_{\boldsymbol{p}(\boldsymbol{\theta}_0)}\boldsymbol{L}^T\end{pmatrix}\right).$$

Therefore

$$\sqrt{n}\left(\widehat{\boldsymbol{p}} - \boldsymbol{p}(\widehat{\boldsymbol{\theta}}_{\phi_2})\right) \xrightarrow[n\to\infty]{L} N\left(\boldsymbol{0}, \boldsymbol{\Sigma}_{\boldsymbol{p}(\boldsymbol{\theta}_0)} - \boldsymbol{\Sigma}_{\boldsymbol{p}(\boldsymbol{\theta}_0)}\boldsymbol{L}^T - \boldsymbol{L}\boldsymbol{\Sigma}_{\boldsymbol{p}(\boldsymbol{\theta}_0)} + \boldsymbol{L}\boldsymbol{\Sigma}_{\boldsymbol{p}(\boldsymbol{\theta}_0)}\boldsymbol{L}^T\right).$$

A second order Taylor expansion of $D_{\phi_1}(\boldsymbol{p},\boldsymbol{q})$ around $(\boldsymbol{p}(\boldsymbol{\theta}_0),\boldsymbol{p}(\boldsymbol{\theta}_0))$ at $(\widehat{\boldsymbol{p}},\boldsymbol{p}(\widehat{\boldsymbol{\theta}}_{\phi_2}))$ is given, denoting $l = D_{\phi_1}\left(\widehat{\boldsymbol{p}},\boldsymbol{p}(\widehat{\boldsymbol{\theta}}_{\phi_2})\right)$, by

$$\begin{aligned} l = {} & \tfrac{1}{2}\sum_{i,j=1}^{M}\left(\frac{\partial^2 D_{\phi_1}(\boldsymbol{p},\boldsymbol{q})}{\partial p_i \partial p_j}\right)_{(\boldsymbol{p}(\boldsymbol{\theta}_0),\boldsymbol{p}(\boldsymbol{\theta}_0))}(\widehat{p}_i - p_i(\boldsymbol{\theta}_0))(\widehat{p}_j - p_j(\boldsymbol{\theta}_0)) \\ + {} & \tfrac{1}{2}2\sum_{i,j=1}^{M}\left(\frac{\partial^2 D_{\phi_1}(\boldsymbol{p},\boldsymbol{q})}{\partial p_i \partial q_j}\right)_{(\boldsymbol{p}(\boldsymbol{\theta}_0),\boldsymbol{p}(\boldsymbol{\theta}_0))}(\widehat{p}_i - p_i(\boldsymbol{\theta}_0))(p_j(\widehat{\boldsymbol{\theta}}_{\phi_2}) - p_j(\boldsymbol{\theta}_0)) \\ + {} & \tfrac{1}{2}\sum_{i,j=1}^{M}\left(\frac{\partial^2 D_{\phi_1}(\boldsymbol{p},\boldsymbol{q})}{\partial q_i \partial q_j}\right)_{(\boldsymbol{p}(\boldsymbol{\theta}_0),\boldsymbol{p}(\boldsymbol{\theta}_0))}(p_i(\widehat{\boldsymbol{\theta}}_{\phi_2}) - p_i(\boldsymbol{\theta}_0))(p_j(\widehat{\boldsymbol{\theta}}_{\phi_2}) - p_j(\boldsymbol{\theta}_0)) \\ + {} & o\left(\|\widehat{\boldsymbol{p}} - \boldsymbol{p}(\boldsymbol{\theta}_0)\|^2\right) + o\left(\left\|\boldsymbol{p}(\widehat{\boldsymbol{\theta}}_{\phi_2}) - \boldsymbol{p}(\boldsymbol{\theta}_0)\right\|^2\right). \end{aligned}$$

But

$$\frac{\partial D_{\phi_1}(\boldsymbol{p},\boldsymbol{q})}{\partial p_i} = \phi_1'\left(\frac{p_i}{q_i}\right), \quad \frac{\partial D_{\phi_1}(\boldsymbol{p},\boldsymbol{q})}{\partial q_i} = \phi_1\left(\frac{p_i}{q_i}\right) - \frac{p_i}{q_i}\phi_1'\left(\frac{p_i}{q_i}\right)$$

and

$$\left(\frac{\partial^2 D_{\phi_1}(\boldsymbol{p},\boldsymbol{q})}{\partial p_i \partial p_j}\right)_{(\boldsymbol{p}(\boldsymbol{\theta}_0),\boldsymbol{p}(\boldsymbol{\theta}_0))} = \begin{cases} \phi_1''(1)\dfrac{1}{p_i(\boldsymbol{\theta}_0)} & i=j \\ 0 & i \neq j \end{cases}.$$

In the same way

$$\left(\frac{\partial^2 D_{\phi_1}(\boldsymbol{p},\boldsymbol{q})}{\partial p_i \partial q_j}\right)_{(\boldsymbol{p}(\boldsymbol{\theta}_0),\boldsymbol{p}(\boldsymbol{\theta}_0))} = \begin{cases} -\phi_1''(1)\dfrac{1}{p_i(\boldsymbol{\theta}_0)} & i=j \\ 0 & i \neq j \end{cases}.$$

Finally,

$$\frac{\partial^2 D_{\phi_1}(\boldsymbol{p},\boldsymbol{q})}{\partial q_i \partial q_j} = \begin{cases} \phi_1''\left(\dfrac{p_i}{q_i}\right)\dfrac{p_i}{q_i^2} & i=j \\ 0 & i \neq j \end{cases}$$

and

$$\left(\frac{\partial^2 D_{\phi_1}(\boldsymbol{p},\boldsymbol{q})}{\partial q_i \partial q_j}\right)_{(\boldsymbol{p}(\boldsymbol{\theta}_0),\boldsymbol{p}(\boldsymbol{\theta}_0))} = \begin{cases} \phi_1''(1)\dfrac{1}{p_i(\boldsymbol{\theta}_0)} & i=j \\ 0 & i \neq j \end{cases}.$$

Therefore

$$
\begin{aligned}
l = & \ \tfrac{1}{2}\phi_1''(1)\sum_{i=1}^{M}\frac{1}{p_i(\boldsymbol{\theta}_0)}(\widehat{p}_i - p_i(\boldsymbol{\theta}_0))^2 \\
- & \ \tfrac{1}{2}\phi_1''(1)\,2\sum_{i=1}^{M}\frac{1}{p_i(\boldsymbol{\theta}_0)}(\widehat{p}_i - p_i(\boldsymbol{\theta}_0))(p_i(\widehat{\boldsymbol{\theta}}_{\phi_2}) - p_i(\boldsymbol{\theta}_0)) \\
+ & \ \tfrac{1}{2}\phi_1''(1)\sum_{i=1}^{M}\frac{1}{p_i(\boldsymbol{\theta}_0)}(p_i(\widehat{\boldsymbol{\theta}}_{\phi_2}) - p_i(\boldsymbol{\theta}_0))^2 + o\left(\|\widehat{\boldsymbol{p}} - \boldsymbol{p}(\boldsymbol{\theta}_0)\|^2\right) \\
+ & \ o\left(\left\|\boldsymbol{p}(\widehat{\boldsymbol{\theta}}_{\phi_2}) - \boldsymbol{p}(\boldsymbol{\theta}_0)\right\|^2\right) \\
= & \ \tfrac{1}{2}\phi_1''(1)\sum_{i=1}^{M}\frac{1}{p_i(\boldsymbol{\theta}_0)}\left\{(\widehat{p}_i - p_i(\boldsymbol{\theta}_0))^2\right. \\
- & \ \left.2(\widehat{p}_i - p_i(\boldsymbol{\theta}_0))(p_i(\widehat{\boldsymbol{\theta}}_{\phi_2}) - p_i(\boldsymbol{\theta}_0)) + (p_i(\widehat{\boldsymbol{\theta}}_{\phi_2}) - p_i(\boldsymbol{\theta}_0))^2\right\} + o_P\left(n^{-1}\right) \\
= & \ \tfrac{1}{2}\phi_1''(1)\sum_{i=1}^{M}\frac{1}{p_i(\boldsymbol{\theta}_0)}(\widehat{p}_i - p_i(\widehat{\boldsymbol{\theta}}_{\phi_2}))^2 + o_P\left(n^{-1}\right)
\end{aligned}
$$

and

$$
\begin{aligned}
\frac{2D_{\phi_1}\left(\widehat{\boldsymbol{p}}, \boldsymbol{p}(\widehat{\boldsymbol{\theta}}_{\phi_2})\right)}{\phi_1''(1)} &= \sum_{i=1}^{M}\frac{1}{p_i(\boldsymbol{\theta}_0)}(\widehat{p}_i - p_i(\widehat{\boldsymbol{\theta}}_{\phi_2}))^2 + o_P\left(n^{-1}\right) \\
&= (\widehat{\boldsymbol{p}} - \boldsymbol{p}(\widehat{\boldsymbol{\theta}}_{\phi_2}))^T\boldsymbol{C}(\widehat{\boldsymbol{p}} - \boldsymbol{p}(\widehat{\boldsymbol{\theta}}_{\phi_2})) + o_P\left(n^{-1}\right)
\end{aligned}
$$

being $\boldsymbol{C} = diag\left(\boldsymbol{p}(\boldsymbol{\theta}_0)^{-1}\right)$.

On the one hand

$$
\sqrt{n}(\widehat{\boldsymbol{p}} - \boldsymbol{p}(\widehat{\boldsymbol{\theta}}_{\phi_2})) \xrightarrow[n\to\infty]{L} N\left(\mathbf{0}, \boldsymbol{\Sigma}_{\boldsymbol{p}(\boldsymbol{\theta}_0)} - \boldsymbol{\Sigma}_{\boldsymbol{p}(\boldsymbol{\theta}_0)}\boldsymbol{L}^T - \boldsymbol{L}\boldsymbol{\Sigma}_{\boldsymbol{p}(\boldsymbol{\theta}_0)} + \boldsymbol{L}\boldsymbol{\Sigma}_{\boldsymbol{p}(\boldsymbol{\theta}_0)}\boldsymbol{L}^T\right)
$$

and on the other hand

$$
\begin{aligned}
\frac{2n}{\phi_1''(1)}D_{\phi_1}\left(\widehat{\boldsymbol{p}}, \boldsymbol{p}(\widehat{\boldsymbol{\theta}}_{\phi_2})\right) &= n(\widehat{\boldsymbol{p}} - \boldsymbol{p}(\widehat{\boldsymbol{\theta}}_{\phi_2}))^T\boldsymbol{C}(\widehat{\boldsymbol{p}} - \boldsymbol{p}(\widehat{\boldsymbol{\theta}}_{\phi_2})) + o_P(1) \\
&= \boldsymbol{X}^T\boldsymbol{X} + o_P(1),
\end{aligned}
$$

where

$$
\boldsymbol{X} = \sqrt{n}\, diag\left(\boldsymbol{p}(\boldsymbol{\theta}_0)^{-1/2}\right)(\widehat{\boldsymbol{p}} - \boldsymbol{p}(\widehat{\boldsymbol{\theta}}_{\phi_2}))
$$

is a M-variate normal random variable with mean vector $\mathbf{0}$ and variance-covariance matrix $\boldsymbol{T}^*$ given by

$$
diag\left(\boldsymbol{p}(\boldsymbol{\theta}_0)^{-1/2}\right)\boldsymbol{W}(\boldsymbol{\theta}_0)\, diag\left(\boldsymbol{p}(\boldsymbol{\theta}_0)^{-1/2}\right), \qquad (6.5)
$$

where $\boldsymbol{W}(\boldsymbol{\theta}_0) = \boldsymbol{\Sigma}_{\boldsymbol{p}(\boldsymbol{\theta}_0)} - \boldsymbol{\Sigma}_{\boldsymbol{p}(\boldsymbol{\theta}_0)}\boldsymbol{L}^T - \boldsymbol{L}\boldsymbol{\Sigma}_{\boldsymbol{p}(\boldsymbol{\theta}_0)} + \boldsymbol{L}\boldsymbol{\Sigma}_{\boldsymbol{p}(\boldsymbol{\theta}_0)}\boldsymbol{L}^T$.

Therefore $\boldsymbol{X}^T\boldsymbol{X}$ (see Remark 2.6) is chi-squared distributed with r degrees of freedom if $\boldsymbol{T}^*$ is a projection of rank r.

The matrix $\boldsymbol{T}^*$ can be written as

$$\begin{aligned}
\boldsymbol{T}^* &= diag\left(\boldsymbol{p}(\boldsymbol{\theta}_0)^{-1/2}\right)\left(\boldsymbol{\Sigma}_{\boldsymbol{p}(\boldsymbol{\theta}_0)} - \boldsymbol{\Sigma}_{\boldsymbol{p}(\boldsymbol{\theta}_0)}\boldsymbol{L}^T - \boldsymbol{L}\boldsymbol{\Sigma}_{\boldsymbol{p}(\boldsymbol{\theta}_0)} + \boldsymbol{L}\boldsymbol{\Sigma}_{\boldsymbol{p}(\boldsymbol{\theta}_0)}\boldsymbol{L}^T\right) \\
&\times\ diag\left(\boldsymbol{p}(\boldsymbol{\theta}_0)^{-1/2}\right) \\
&= diag\left(\boldsymbol{p}(\boldsymbol{\theta}_0)^{-1/2}\right)\boldsymbol{\Sigma}_{\boldsymbol{p}(\boldsymbol{\theta}_0)} diag\left(\boldsymbol{p}(\boldsymbol{\theta}_0)^{-1/2}\right) \\
&-\ diag\left(\boldsymbol{p}(\boldsymbol{\theta}_0)^{-1/2}\right)\boldsymbol{\Sigma}_{\boldsymbol{p}(\boldsymbol{\theta}_0)}\boldsymbol{L}^T diag\left(\boldsymbol{p}(\boldsymbol{\theta}_0)^{-1/2}\right) \\
&-\ diag\left(\boldsymbol{p}(\boldsymbol{\theta}_0)^{-1/2}\right)\boldsymbol{L}\boldsymbol{\Sigma}_{\boldsymbol{p}(\boldsymbol{\theta}_0)} diag\left(\boldsymbol{p}(\boldsymbol{\theta}_0)^{-1/2}\right) \\
&+\ diag\left(\boldsymbol{p}(\boldsymbol{\theta}_0)^{-1/2}\right)\boldsymbol{L}\boldsymbol{\Sigma}_{\boldsymbol{p}(\boldsymbol{\theta}_0)}\boldsymbol{L}^T diag\left(\boldsymbol{p}(\boldsymbol{\theta}_0)^{-1/2}\right).
\end{aligned}$$

Let $\boldsymbol{S}$ denote the $M \times M$ matrix

$$\boldsymbol{S} = diag\left(\boldsymbol{p}(\boldsymbol{\theta}_0)^{-1/2}\right)\boldsymbol{\Sigma}_{\boldsymbol{p}(\boldsymbol{\theta}_0)} diag\left(\boldsymbol{p}(\boldsymbol{\theta}_0)^{-1/2}\right).$$

We have

$$\begin{aligned}
\boldsymbol{S} =&\ diag\left(\boldsymbol{p}(\boldsymbol{\theta}_0)^{-1/2}\right)(diag(\boldsymbol{p}(\boldsymbol{\theta}_0)) - \boldsymbol{p}(\boldsymbol{\theta}_0)\boldsymbol{p}(\boldsymbol{\theta}_0)^T) diag\left(\boldsymbol{p}(\boldsymbol{\theta}_0)^{-1/2}\right) \\
=&\ \boldsymbol{I} - diag\left(\boldsymbol{p}(\boldsymbol{\theta}_0)^{-1/2}\right)\boldsymbol{p}(\boldsymbol{\theta}_0)\boldsymbol{p}(\boldsymbol{\theta}_0)^T diag\left(\boldsymbol{p}(\boldsymbol{\theta}_0)^{-1/2}\right) \\
=&\ \boldsymbol{I} - \boldsymbol{p}(\boldsymbol{\theta}_0)^{1/2}\left(\boldsymbol{p}(\boldsymbol{\theta}_0)^{1/2}\right)^T.
\end{aligned}$$

We define the matrix $\boldsymbol{B}$

$$\boldsymbol{B} = diag\left(\boldsymbol{p}(\boldsymbol{\theta}_0)^{-1/2}\right)\boldsymbol{L}\boldsymbol{\Sigma}_{\boldsymbol{p}(\boldsymbol{\theta}_0)} diag\left(\boldsymbol{p}(\boldsymbol{\theta}_0)^{-1/2}\right)$$

and we are going to express it in a simpler way,

$$\begin{aligned}
\boldsymbol{B} =&\ diag\left(\boldsymbol{p}(\boldsymbol{\theta}_0)^{-1/2}\right)\boldsymbol{L}\ \left(diag(\boldsymbol{p}(\boldsymbol{\theta}_0)) - \boldsymbol{p}(\boldsymbol{\theta}_0)\boldsymbol{p}(\boldsymbol{\theta}_0)^T\right) diag\left(\boldsymbol{p}(\boldsymbol{\theta}_0)^{-1/2}\right) \\
=&\ diag\left(\boldsymbol{p}(\boldsymbol{\theta}_0)^{-1/2}\right)\boldsymbol{L}\ diag\left(\boldsymbol{p}(\boldsymbol{\theta}_0)^{1/2}\right) \\
-&\ diag\left(\boldsymbol{p}(\boldsymbol{\theta}_0)^{-1/2}\right)\boldsymbol{L}\boldsymbol{p}(\boldsymbol{\theta}_0)\boldsymbol{p}(\boldsymbol{\theta}_0)^T diag\left(\boldsymbol{p}(\boldsymbol{\theta}_0)^{-1/2}\right)
\end{aligned}$$

and

$$\boldsymbol{p}(\boldsymbol{\theta}_0)^T = \sqrt{\boldsymbol{p}(\boldsymbol{\theta}_0)^T} diag\left(\boldsymbol{p}(\boldsymbol{\theta}_0)^{1/2}\right).$$

Then we get

$$\begin{aligned}
\boldsymbol{B} = & \; diag\left(\boldsymbol{p}(\boldsymbol{\theta}_0)^{-1/2}\right) \boldsymbol{L} \; diag\left(\boldsymbol{p}(\boldsymbol{\theta}_0)^{1/2}\right) \\
- & \; diag\left(\boldsymbol{p}(\boldsymbol{\theta}_0)^{-1/2}\right) \boldsymbol{L} \; diag\left(\boldsymbol{p}(\boldsymbol{\theta}_0)^{1/2}\right) \sqrt{\boldsymbol{p}(\boldsymbol{\theta}_0)} \sqrt{\boldsymbol{p}(\boldsymbol{\theta}_0)}^T \\
= & \; diag\left(\boldsymbol{p}(\boldsymbol{\theta}_0)^{-1/2}\right) \boldsymbol{L} \; diag\left(\boldsymbol{p}(\boldsymbol{\theta}_0)^{1/2}\right) (\boldsymbol{I} - \sqrt{\boldsymbol{p}(\boldsymbol{\theta}_0)} \sqrt{\boldsymbol{p}(\boldsymbol{\theta}_0)}^T) \\
= & \; \boldsymbol{K} \, \boldsymbol{S},
\end{aligned}$$

being $\boldsymbol{K}$ the $M \times M$ matrix

$$\boldsymbol{K} = diag\left(\boldsymbol{p}(\boldsymbol{\theta}_0)^{-1/2}\right) \boldsymbol{L} \; diag\left(\boldsymbol{p}(\boldsymbol{\theta}_0)^{1/2}\right).$$

Therefore

$$\boldsymbol{T}^* = \boldsymbol{S} - \boldsymbol{KS} - \boldsymbol{SK}^T + \boldsymbol{KSK}^T$$

where

$$\begin{aligned}
\boldsymbol{K} = & \; diag\left(\boldsymbol{p}(\boldsymbol{\theta}_0)^{-1/2}\right) \; \boldsymbol{L} \; diag\left(\boldsymbol{p}(\boldsymbol{\theta}_0)^{1/2}\right) \\
= & \; diag\left(\boldsymbol{p}(\boldsymbol{\theta}_0)^{-1/2}\right) \boldsymbol{J}(\boldsymbol{\theta}_0) \; (\boldsymbol{A}(\boldsymbol{\theta}_0)^T \boldsymbol{A}(\boldsymbol{\theta}_0))^{-1} \boldsymbol{A}(\boldsymbol{\theta}_0)^T diag\left(\boldsymbol{p}(\boldsymbol{\theta}_0)^{-1/2}\right) \\
\times & \; diag\left(\boldsymbol{p}(\boldsymbol{\theta}_0)^{1/2}\right).
\end{aligned}$$

But

$$\boldsymbol{A}(\boldsymbol{\theta}_0) = diag\left(\boldsymbol{p}(\boldsymbol{\theta}_0)^{-1/2}\right) \boldsymbol{J}(\boldsymbol{\theta}_0),$$

then

$$\boldsymbol{K} = \boldsymbol{A}(\boldsymbol{\theta}_0) \left(\boldsymbol{A}(\boldsymbol{\theta}_0)^T \boldsymbol{A}(\boldsymbol{\theta}_0)\right)^{-1} \boldsymbol{A}(\boldsymbol{\theta}_0)^T$$

and

$$\begin{aligned}
\boldsymbol{T}^* = & \; \boldsymbol{S} - \boldsymbol{A}(\boldsymbol{\theta}_0) \left(\boldsymbol{A}(\boldsymbol{\theta}_0)^T \boldsymbol{A}(\boldsymbol{\theta}_0)\right)^{-1} \boldsymbol{A}(\boldsymbol{\theta}_0)^T \boldsymbol{S} - \boldsymbol{SK} \\
+ & \; \boldsymbol{A}(\boldsymbol{\theta}_0) \left(\boldsymbol{A}(\boldsymbol{\theta}_0)^T \boldsymbol{A}(\boldsymbol{\theta}_0)\right)^{-1} \boldsymbol{A}(\boldsymbol{\theta}_0)^T \; \boldsymbol{S} \, \boldsymbol{A}(\boldsymbol{\theta}_0) \left(\boldsymbol{A}(\boldsymbol{\theta}_0)^T \boldsymbol{A}(\boldsymbol{\theta}_0)\right)^{-1} \boldsymbol{A}(\boldsymbol{\theta}_0)^T \\
= & \; \boldsymbol{I} - \sqrt{\boldsymbol{p}(\boldsymbol{\theta}_0)} \sqrt{\boldsymbol{p}(\boldsymbol{\theta}_0)}^T \\
- & \; \boldsymbol{A}(\boldsymbol{\theta}_0) \left(\boldsymbol{A}(\boldsymbol{\theta}_0)^T \boldsymbol{A}(\boldsymbol{\theta}_0)\right)^{-1} \boldsymbol{A}(\boldsymbol{\theta}_0)^T \left(\boldsymbol{I} - \sqrt{\boldsymbol{p}(\boldsymbol{\theta}_0)} \sqrt{\boldsymbol{p}(\boldsymbol{\theta}_0)}^T\right) \\
- & \; \left(\boldsymbol{I} - \sqrt{\boldsymbol{p}(\boldsymbol{\theta}_0)} \sqrt{\boldsymbol{p}(\boldsymbol{\theta}_0)}^T\right) \boldsymbol{A}(\boldsymbol{\theta}_0) \left(\boldsymbol{A}(\boldsymbol{\theta}_0)^T \boldsymbol{A}(\boldsymbol{\theta}_0)\right)^{-1} \boldsymbol{A}(\boldsymbol{\theta}_0)^T \\
+ & \; \boldsymbol{A}(\boldsymbol{\theta}_0) \left(\boldsymbol{A}(\boldsymbol{\theta}_0)^T \boldsymbol{A}(\boldsymbol{\theta}_0)\right)^{-1} \boldsymbol{A}(\boldsymbol{\theta}_0)^T \left(\boldsymbol{I} - \sqrt{\boldsymbol{p}(\boldsymbol{\theta}_0)} \sqrt{\boldsymbol{p}(\boldsymbol{\theta}_0)}^T\right) \\
\times & \; \boldsymbol{A}(\boldsymbol{\theta}_0) \left(\boldsymbol{A}(\boldsymbol{\theta}_0)^T \boldsymbol{A}(\boldsymbol{\theta}_0)\right)^{-1} \boldsymbol{A}(\boldsymbol{\theta}_0)^T.
\end{aligned}$$

We know that

$$\sqrt{\boldsymbol{p}(\boldsymbol{\theta}_0)}^T \boldsymbol{A}(\boldsymbol{\theta}_0) = \boldsymbol{0}_{1\times M_0}$$

and hence

$$\boldsymbol{T}^* = \boldsymbol{I} - \sqrt{\boldsymbol{p}(\boldsymbol{\theta}_0)}\sqrt{\boldsymbol{p}(\boldsymbol{\theta}_0)}^T - \boldsymbol{A}(\boldsymbol{\theta}_0)\left(\boldsymbol{A}(\boldsymbol{\theta}_0)^T\boldsymbol{A}(\boldsymbol{\theta}_0)\right)^{-1}\boldsymbol{A}(\boldsymbol{\theta}_0)^T.$$

Now we are going to establish that the matrix $\boldsymbol{T}^*$ is idempotent,

$$\begin{aligned}(\boldsymbol{T}^*)^2 = \; & \boldsymbol{I} - \sqrt{\boldsymbol{p}(\boldsymbol{\theta}_0)}\sqrt{\boldsymbol{p}(\boldsymbol{\theta}_0)}^T - \boldsymbol{A}(\boldsymbol{\theta}_0)\left(\boldsymbol{A}(\boldsymbol{\theta}_0)^T\boldsymbol{A}(\boldsymbol{\theta}_0)\right)^{-1}\boldsymbol{A}(\boldsymbol{\theta}_0)^T \\ - \; & \sqrt{\boldsymbol{p}(\boldsymbol{\theta}_0)}\sqrt{\boldsymbol{p}(\boldsymbol{\theta}_0)}^T + \sqrt{\boldsymbol{p}(\boldsymbol{\theta}_0)}\sqrt{\boldsymbol{p}(\boldsymbol{\theta}_0)}^T\sqrt{\boldsymbol{p}(\boldsymbol{\theta}_0)}\sqrt{\boldsymbol{p}(\boldsymbol{\theta}_0)}^T \\ + \; & \sqrt{\boldsymbol{p}(\boldsymbol{\theta}_0)}\sqrt{\boldsymbol{p}(\boldsymbol{\theta}_0)}^T\boldsymbol{A}(\boldsymbol{\theta}_0)(\boldsymbol{A}(\boldsymbol{\theta}_0)^T\boldsymbol{A}(\boldsymbol{\theta}_0))^{-1}\boldsymbol{A}(\boldsymbol{\theta}_0)^T \\ - \; & \boldsymbol{A}(\boldsymbol{\theta}_0)\left(\boldsymbol{A}(\boldsymbol{\theta}_0)^T\boldsymbol{A}(\boldsymbol{\theta}_0)\right)^{-1}\boldsymbol{A}(\boldsymbol{\theta}_0)^T \\ + \; & \boldsymbol{A}(\boldsymbol{\theta}_0)\left(\boldsymbol{A}(\boldsymbol{\theta}_0)^T\boldsymbol{A}(\boldsymbol{\theta}_0)\right)^{-1}\boldsymbol{A}(\boldsymbol{\theta}_0)^T\sqrt{\boldsymbol{p}(\boldsymbol{\theta}_0)}\sqrt{\boldsymbol{p}(\boldsymbol{\theta}_0)}^T \\ + \; & \boldsymbol{A}(\boldsymbol{\theta}_0)\left(\boldsymbol{A}(\boldsymbol{\theta}_0)^T\boldsymbol{A}(\boldsymbol{\theta}_0)\right)^{-1}\boldsymbol{A}(\boldsymbol{\theta}_0)^T\boldsymbol{A}(\boldsymbol{\theta}_0)\left(\boldsymbol{A}(\boldsymbol{\theta}_0)^T\boldsymbol{A}(\boldsymbol{\theta}_0)\right)^{-1}\boldsymbol{A}(\boldsymbol{\theta}_0)^T,\end{aligned}$$

but

$$\begin{aligned}\boldsymbol{A}(\boldsymbol{\theta}_0)\left(\boldsymbol{A}(\boldsymbol{\theta}_0)^T\boldsymbol{A}(\boldsymbol{\theta}_0)\right)^{-1}\boldsymbol{A}(\boldsymbol{\theta}_0)^T\sqrt{\boldsymbol{p}(\boldsymbol{\theta}_0)}\sqrt{\boldsymbol{p}(\boldsymbol{\theta}_0)}^T &= \boldsymbol{0}_{M\times 1} \\ \sqrt{\boldsymbol{p}(\boldsymbol{\theta}_0)}^T\sqrt{\boldsymbol{p}(\boldsymbol{\theta}_0)} &= 1,\end{aligned}$$

then

$$\begin{aligned}(\boldsymbol{T}^*)^2 = \; & \boldsymbol{I} - \sqrt{\boldsymbol{p}(\boldsymbol{\theta}_0)}\sqrt{\boldsymbol{p}(\boldsymbol{\theta}_0)}^T - \boldsymbol{A}(\boldsymbol{\theta}_0)\left(\boldsymbol{A}(\boldsymbol{\theta}_0)^T\boldsymbol{A}(\boldsymbol{\theta}_0)\right)^{-1}\boldsymbol{A}(\boldsymbol{\theta}_0)^T \\ = \; & \boldsymbol{I} - \sqrt{\boldsymbol{p}(\boldsymbol{\theta}_0)}\sqrt{\boldsymbol{p}(\boldsymbol{\theta}_0)}^T - \boldsymbol{A}(\boldsymbol{\theta}_0)\left(\boldsymbol{A}(\boldsymbol{\theta}_0)^T\boldsymbol{A}(\boldsymbol{\theta}_0)\right)^{-1}\boldsymbol{A}(\boldsymbol{\theta}_0)^T \\ = \; & \boldsymbol{T}^*.\end{aligned}$$

In relation to rank of $\boldsymbol{T}^*$ we have

$$\begin{aligned}rank(\boldsymbol{T}^*) = trace(\boldsymbol{T}^*) = \; & trace(\boldsymbol{I}) - trace\left(\sqrt{\boldsymbol{p}(\boldsymbol{\theta}_0)}\sqrt{\boldsymbol{p}(\boldsymbol{\theta}_0)}^T\right) \\ - \; & trace\left(\boldsymbol{A}(\boldsymbol{\theta}_0)\left(\boldsymbol{A}(\boldsymbol{\theta}_0)^T\boldsymbol{A}(\boldsymbol{\theta}_0)\right)^{-1}\boldsymbol{A}(\boldsymbol{\theta}_0)^T\right) \\ = \; & M - 1 - trace\left(\left(\boldsymbol{A}(\boldsymbol{\theta}_0)^T\boldsymbol{A}(\boldsymbol{\theta}_0)\right)^{-1}\left(\boldsymbol{A}(\boldsymbol{\theta}_0)^T\boldsymbol{A}(\boldsymbol{\theta}_0)\right)\right) \\ = \; & M - 1 - M_0,\end{aligned}$$

because

$$\left(\boldsymbol{A}(\boldsymbol{\theta}_0)^T\boldsymbol{A}(\boldsymbol{\theta}_0)\right)^{-1}\left(\boldsymbol{A}(\boldsymbol{\theta}_0)^T\boldsymbol{A}(\boldsymbol{\theta}_0)\right) = \boldsymbol{I}_{M_0\times M_0}.$$

Then the ϕ-divergence test statistic $T_n^{\phi_1}\left(\widehat{\boldsymbol{\theta}}_{\phi_2}\right)$ is asymptotically distributed chi-squared with $M-1-M_0$ degrees of freedom. ■

Based on the previous result we should reject the null hypothesis given in (6.1), with a significance level α, if

$$T_n^{\phi_1}\left(\widehat{\boldsymbol{\theta}}_{\phi_2}\right) > \chi^2_{M-M_0-1,\alpha}. \tag{6.6}$$

We present, in the sequel, an approximation for the power function of the previous test statistic. Let $\boldsymbol{q} = (q_1, ..., q_M)^T$ be a point at the alternative hypothesis. If the alternative hypothesis $\boldsymbol{q}$ is true, we have that $\widehat{\boldsymbol{p}}$ tends, in probability, to $\boldsymbol{q}$. We denote by $\boldsymbol{\theta}_a$ the point on Θ verifying

$$\boldsymbol{\theta}_a = \arg\min_{\boldsymbol{\theta}\in\Theta} D_{\phi_2}(\boldsymbol{q}, \boldsymbol{p}(\boldsymbol{\theta})),$$

and we have that $\boldsymbol{p}(\widehat{\boldsymbol{\theta}}_{\phi_2})$ tends, in probability, to $\boldsymbol{p}(\boldsymbol{\theta}_a)$. In the next theorem, we use the following assumption,

$$\sqrt{n}\left(\left(\widehat{\boldsymbol{p}}, \boldsymbol{p}(\widehat{\boldsymbol{\theta}}_{\phi_2})\right) - (\boldsymbol{q}, \boldsymbol{p}(\boldsymbol{\theta}_a))\right) \underset{n\to\infty}{\overset{L}{\longrightarrow}} N(\mathbf{0}, \boldsymbol{\Sigma}), \tag{6.7}$$

under the alternative hypothesis $\boldsymbol{q}$, where

$$\boldsymbol{\Sigma} = \begin{pmatrix} \boldsymbol{\Sigma}_{11} & \boldsymbol{\Sigma}_{12} \\ \boldsymbol{\Sigma}_{21} & \boldsymbol{\Sigma}_{22} \end{pmatrix}, \quad \boldsymbol{\Sigma}_{11} = diag(\boldsymbol{q}) - \boldsymbol{q}\boldsymbol{q}^T \text{ and } \boldsymbol{\Sigma}_{12} = \boldsymbol{\Sigma}_{21}.$$

Theorem 6.2

The asymptotic power for the test given by (6.6) at the alternative hypothesis $\boldsymbol{q}$, assuming condition (6.7), is given by

$$\beta_{n,\phi_1}(\boldsymbol{q}) = 1 - \Phi_n\left(\frac{1}{\sigma_{\phi_1}(\boldsymbol{q})}\left(\frac{\phi_1''(1)}{2\sqrt{n}}\chi^2_{M-M_0-1,\alpha} - \sqrt{n}D_{\phi_1}(\boldsymbol{q}, \boldsymbol{p}(\boldsymbol{\theta}_a))\right)\right),$$

where

$$\sigma^2_{\phi_1}(\boldsymbol{q}) = \mathbf{Z}^T\boldsymbol{\Sigma}_{11}\mathbf{Z} + 2\mathbf{Z}^T\boldsymbol{\Sigma}_{12}\mathbf{S} + \mathbf{S}^T\boldsymbol{\Sigma}_{22}\mathbf{S}, \tag{6.8}$$

being

$$\mathbf{Z}^T = \left(\frac{\partial D_{\phi_1}(\boldsymbol{v}, \boldsymbol{p}(\boldsymbol{\theta}_a))}{\partial \boldsymbol{v}}\right)_{\boldsymbol{v}=\boldsymbol{q}},$$

$$\mathbf{S}^T = \left(\frac{\partial D_{\phi_1}(\boldsymbol{q}, \boldsymbol{w})}{\partial \boldsymbol{w}}\right)_{\boldsymbol{w}=\boldsymbol{p}(\boldsymbol{\theta}_a)},$$

and $\Phi_n(x)$ is a sequence of distribution functions tending uniformly to the standard normal distribution function $\Phi(x)$.

Proof. A first-order Taylor expansion of the divergence measure gives

$$D_{\phi_1}\left(\widehat{\boldsymbol{p}},\boldsymbol{p}(\widehat{\boldsymbol{\theta}}_{\phi_2})\right)=D_{\phi_1}(\boldsymbol{q},\boldsymbol{p}\left(\boldsymbol{\theta}_a\right))+\mathbf{Z}^T(\widehat{\boldsymbol{p}}-\boldsymbol{q})+\mathbf{S}^T(\boldsymbol{p}(\widehat{\boldsymbol{\theta}}_{\phi_2})-\boldsymbol{p}\left(\boldsymbol{\theta}_a\right))+o_P(n^{-1/2}).$$

Then

$$\sqrt{n}\left(D_{\phi_1}\left(\widehat{\boldsymbol{p}},\boldsymbol{p}(\widehat{\boldsymbol{\theta}}_{\phi_2})\right)-D_{\phi_1}(\boldsymbol{q},\boldsymbol{p}\left(\boldsymbol{\theta}_a\right))\right)\underset{n\to\infty}{\overset{L}{\longrightarrow}}N\left(0,\sigma^2_{\phi_1}(\boldsymbol{q})\right),$$

for $\sigma^2_{\phi_1}(\boldsymbol{q})$ given in (6.8). ■

The previous result is not easy to apply. For this reason we are going to consider a sequence of contiguous alternatives that approach the null hypothesis at rate of $O\left(n^{-1/2}\right)$. Consider the contiguous alternative hypotheses,

$$H_{1,n}:\boldsymbol{p}_n=\boldsymbol{p}\left(\boldsymbol{\theta}_0\right)+\frac{1}{\sqrt{n}}\boldsymbol{d}, \tag{6.9}$$

where $\boldsymbol{d}\equiv(d_1,...,d_M)^T$ is a fixed vector such that $\sum_{j=1}^M d_j=0$, with $\boldsymbol{p}_n\neq\boldsymbol{p}\left(\boldsymbol{\theta}_0\right)$, $\boldsymbol{\theta}_0$ unknown and $\boldsymbol{\theta}_0\in\Theta$. Then we have the following result:

Theorem 6.3

The asymptotic distribution of the ϕ_1-divergence test statistic $T_n^{\phi_1}\left(\widehat{\boldsymbol{\theta}}_{\phi_2}\right)$, under the contiguous alternative hypotheses given in (6.9), is noncentral chi-square with $M-M_0-1$ degrees of freedom and noncentrality parameter δ given by

$$\begin{aligned}\delta=\;&\boldsymbol{d}^T diag\left(\boldsymbol{p}\left(\boldsymbol{\theta}_0\right)^{-1/2}\right)\left(\boldsymbol{I}-\boldsymbol{A}\left(\boldsymbol{\theta}_0\right)\left(\boldsymbol{A}\left(\boldsymbol{\theta}_0\right)^T\boldsymbol{A}\left(\boldsymbol{\theta}_0\right)\right)^{-1}\boldsymbol{A}\left(\boldsymbol{\theta}_0\right)^T\right)\\ \times\;& diag\left(\boldsymbol{p}\left(\boldsymbol{\theta}_0\right)^{-1/2}\right)\boldsymbol{d}.\end{aligned}$$

Proof. We have

$$\begin{aligned}\sqrt{n}\left(\widehat{\boldsymbol{p}}-\boldsymbol{p}\left(\boldsymbol{\theta}_0\right)\right)&=\sqrt{n}\left(\widehat{\boldsymbol{p}}-\boldsymbol{p}_n+\boldsymbol{p}_n-\boldsymbol{p}\left(\boldsymbol{\theta}_0\right)\right)\\&=\sqrt{n}\left(\widehat{\boldsymbol{p}}-\boldsymbol{p}_n\right)+\sqrt{n}\left(\boldsymbol{p}_n-\boldsymbol{p}\left(\boldsymbol{\theta}_0\right)\right)\\&=\sqrt{n}\left(\widehat{\boldsymbol{p}}-\boldsymbol{p}_n\right)+\boldsymbol{d},\end{aligned}$$

then under the alternatives (6.9),

$$\sqrt{n}(\widehat{\boldsymbol{p}}-\boldsymbol{p}\left(\boldsymbol{\theta}_0\right))\underset{n\to\infty}{\overset{L}{\longrightarrow}}N\left(\boldsymbol{d},\boldsymbol{\Sigma}_{\boldsymbol{p}(\boldsymbol{\theta}_0)}\right)$$

because under the alternative hypotheses given in (6.9),

$$\sqrt{n}(\widehat{\boldsymbol{p}}-\boldsymbol{p}_n)\underset{n\to\infty}{\overset{L}{\longrightarrow}}\left(\boldsymbol{0},\boldsymbol{\Sigma}_{\boldsymbol{p}(\boldsymbol{\theta}_0)}\right).$$

But

$$\begin{aligned}\sqrt{n}(\widehat{\boldsymbol{\theta}}_{\phi_2}-\boldsymbol{\theta}_0) &= \sqrt{n}\left(\boldsymbol{A}(\boldsymbol{\theta}_0)^T\boldsymbol{A}(\boldsymbol{\theta}_0)\right)^{-1}\boldsymbol{A}(\boldsymbol{\theta}_0)^T \, diag\left(\boldsymbol{p}(\boldsymbol{\theta}_0)^{-1/2}\right)\\ &\times (\widehat{\boldsymbol{p}}-\boldsymbol{p}(\boldsymbol{\theta}_0))+o_P(1)\end{aligned}$$

and

$$\sqrt{n}(\widehat{\boldsymbol{p}}-\boldsymbol{p}(\boldsymbol{\theta}_0))=\sqrt{n}(\widehat{\boldsymbol{p}}-\boldsymbol{p}_n)+\boldsymbol{d},$$

then

$$\begin{aligned}\sqrt{n}(\widehat{\boldsymbol{\theta}}_{\phi_2}-\boldsymbol{\theta}_0) = &\ \sqrt{n}\left(\boldsymbol{A}(\boldsymbol{\theta}_0)^T\boldsymbol{A}(\boldsymbol{\theta}_0)\right)^{-1}\boldsymbol{A}(\boldsymbol{\theta}_0)^T \, diag\left(\boldsymbol{p}(\boldsymbol{\theta}_0)^{-1/2}\right)\\ \times &\ \left((\widehat{\boldsymbol{p}}-\boldsymbol{p}_n)+n^{-1/2}\boldsymbol{d}\right)+o_P(1)\\ = &\ \sqrt{n}\left(\boldsymbol{A}(\boldsymbol{\theta}_0)^T\boldsymbol{A}(\boldsymbol{\theta}_0)\right)^{-1}\boldsymbol{A}(\boldsymbol{\theta}_0)^T \, diag\left(\boldsymbol{p}(\boldsymbol{\theta}_0)^{-1/2}\right)(\widehat{\boldsymbol{p}}-\boldsymbol{p}_n)\\ + &\ \left(\boldsymbol{A}(\boldsymbol{\theta})^T\boldsymbol{A}(\boldsymbol{\theta}_0)\right)^{-1}\boldsymbol{A}(\boldsymbol{\theta})^T \, diag\left(\boldsymbol{p}(\boldsymbol{\theta}_0)^{-1/2}\right)\boldsymbol{d}+o_P(1).\end{aligned}$$

Therefore the random vector $l(\widehat{\boldsymbol{\theta}}_{\phi_2})\equiv\sqrt{n}\left(\boldsymbol{p}(\widehat{\boldsymbol{\theta}}_{\phi_2})-\boldsymbol{p}(\boldsymbol{\theta}_0)\right)$ can be written as

$$\begin{aligned}l(\widehat{\boldsymbol{\theta}}_{\phi_2}) &= \boldsymbol{J}(\boldsymbol{\theta}_0)\sqrt{n}(\widehat{\boldsymbol{\theta}}_{\phi_2}-\boldsymbol{\theta}_0)+o_P(1)\\ &= \sqrt{n}\boldsymbol{J}(\boldsymbol{\theta}_0)\left(\boldsymbol{A}(\boldsymbol{\theta}_0)^T\boldsymbol{A}(\boldsymbol{\theta}_0)\right)^{-1}\boldsymbol{A}(\boldsymbol{\theta}_0)^T \, diag\left(\boldsymbol{p}(\boldsymbol{\theta}_0)^{-1/2}\right)(\widehat{\boldsymbol{p}}-\boldsymbol{p}_n)\\ &\quad +\boldsymbol{J}(\boldsymbol{\theta}_0)\left(\boldsymbol{A}(\boldsymbol{\theta}_0)^T\boldsymbol{A}(\boldsymbol{\theta}_0)\right)^{-1}\boldsymbol{A}(\boldsymbol{\theta}_0)^T \, diag\left(\boldsymbol{p}(\boldsymbol{\theta}_0)^{-1/2}\right)\boldsymbol{d}+o_P(1).\end{aligned}$$

If we denote

$$\boldsymbol{L}=\boldsymbol{J}(\boldsymbol{\theta}_0)\left(\boldsymbol{A}(\boldsymbol{\theta}_0)^T\boldsymbol{A}(\boldsymbol{\theta}_0)\right)^{-1}\boldsymbol{A}(\boldsymbol{\theta}_0)^T \, diag\left(\boldsymbol{p}(\boldsymbol{\theta}_0)^{-1/2}\right),$$

and by $\boldsymbol{I}$ the $M\times M$ identity matrix, we have

$$\sqrt{n}\left(\boldsymbol{p}(\widehat{\boldsymbol{\theta}}_{\phi_2})-\boldsymbol{p}(\boldsymbol{\theta}_0)\right)=\sqrt{n}\boldsymbol{L}(\widehat{\boldsymbol{p}}-\boldsymbol{p}_n)+\boldsymbol{L}\boldsymbol{d}+o_P(1)$$

and

$$\sqrt{n}\begin{pmatrix}\widehat{\boldsymbol{p}}-\boldsymbol{p}(\boldsymbol{\theta}_0)\\ \boldsymbol{p}(\widehat{\boldsymbol{\theta}}_{\phi_2})-\boldsymbol{p}(\boldsymbol{\theta}_0)\end{pmatrix}=\sqrt{n}\begin{pmatrix}\boldsymbol{I}\\ \boldsymbol{L}\end{pmatrix}(\widehat{\boldsymbol{p}}-\boldsymbol{p}_n)+\begin{pmatrix}\boldsymbol{d}\\ \boldsymbol{L}\boldsymbol{d}\end{pmatrix}+\begin{pmatrix}0\\ o_P(1)\end{pmatrix}.$$

Hence

$$\sqrt{n}\begin{pmatrix}\widehat{\boldsymbol{p}}-\boldsymbol{p}(\boldsymbol{\theta}_0)\\ \boldsymbol{p}(\widehat{\boldsymbol{\theta}}_{\phi_2})-\boldsymbol{p}(\boldsymbol{\theta}_0)\end{pmatrix}\xrightarrow[n\to\infty]{L}N\left(\begin{pmatrix}\boldsymbol{d}\\ \boldsymbol{L}\boldsymbol{d}\end{pmatrix},\begin{pmatrix}\boldsymbol{\Sigma}_{\boldsymbol{p}(\boldsymbol{\theta}_0)} & \boldsymbol{\Sigma}_{\boldsymbol{p}(\boldsymbol{\theta}_0)}\boldsymbol{L}^T\\ \boldsymbol{L}\boldsymbol{\Sigma}_{\boldsymbol{p}(\boldsymbol{\theta}_0)} & \boldsymbol{L}\boldsymbol{\Sigma}_{\boldsymbol{p}(\boldsymbol{\theta}_0)}\boldsymbol{L}^T\end{pmatrix}\right)$$

and

$$\sqrt{n}(\widehat{\mathbf{p}}-\mathbf{p}(\widehat{\boldsymbol{\theta}}_{\phi_2})) \underset{n\to\infty}{\overset{L}{\longrightarrow}} N\left((\boldsymbol{I}-\boldsymbol{L})\boldsymbol{d}, \boldsymbol{\Sigma}_{\mathbf{p}(\boldsymbol{\theta}_0)} - \boldsymbol{\Sigma}_{\mathbf{p}(\boldsymbol{\theta}_0)}\boldsymbol{L}^T - \boldsymbol{L}\boldsymbol{\Sigma}_{\mathbf{p}(\boldsymbol{\theta}_0)} + \boldsymbol{L}\boldsymbol{\Sigma}_{\mathbf{p}(\boldsymbol{\theta}_0)}\boldsymbol{L}^T\right).$$

On the other hand

$$\begin{aligned}\frac{2n}{\phi_1''(1)} D_{\phi_1}\left(\widehat{\mathbf{p}}, \mathbf{p}(\widehat{\boldsymbol{\theta}}_{\phi_2})\right) &= n(\widehat{\mathbf{p}}-\mathbf{p}(\widehat{\boldsymbol{\theta}}_{\phi_2}))^T diag\left(\mathbf{p}(\boldsymbol{\theta}_0)^{-1}\right)(\widehat{\mathbf{p}}-\mathbf{p}(\widehat{\boldsymbol{\theta}}_{\phi_2})) + o_P(1) \\ &= \boldsymbol{X}^T\boldsymbol{X} + o_P(1)\end{aligned}$$

with

$$\boldsymbol{X} = \sqrt{n} diag\left(\mathbf{p}(\boldsymbol{\theta}_0)^{-1/2}\right)(\widehat{\mathbf{p}}-\mathbf{p}(\widehat{\boldsymbol{\theta}}_{\phi_2}))$$

and

$$\boldsymbol{X} \underset{n\to\infty}{\overset{L}{\longrightarrow}} N\left(diag\left(\mathbf{p}(\boldsymbol{\theta}_0)^{-1/2}\right)(\boldsymbol{I}-\boldsymbol{L})\boldsymbol{d}, \boldsymbol{T}^*\right)$$

where $\boldsymbol{T}^*$ is defined in (6.5).

By Theorem 6.1 the rank of the matrix $\boldsymbol{T}^*$ is $M - M_0 - 1$. If we prove that $\boldsymbol{T}^*\boldsymbol{\mu} = \boldsymbol{\mu}$, where

$$\boldsymbol{\mu} = diag\left(\mathbf{p}(\boldsymbol{\theta}_0)^{-1/2}\right)(\boldsymbol{I}_{M\times M} - \boldsymbol{L})\boldsymbol{d},$$

we will have established that the test statistic $T_n^{\phi_1}\left(\widehat{\boldsymbol{\theta}}_{\phi_2}\right)$ is asymptotically distributed as a noncentral chi-square distribution with $M - M_0 - 1$ degrees of freedom and noncentrality parameter $\delta = \boldsymbol{\mu}^T\boldsymbol{\mu}$.

We know $diag\left(\mathbf{p}(\boldsymbol{\theta}_0)^{1/2}\right)\boldsymbol{A}(\boldsymbol{\theta}_0) = \boldsymbol{J}(\boldsymbol{\theta}_0)$, then

$$\begin{aligned}\boldsymbol{\mu} &= \left(diag\left(\mathbf{p}(\boldsymbol{\theta}_0)^{-1/2}\right)\boldsymbol{d} - diag\left(\mathbf{p}(\boldsymbol{\theta}_0)^{-1/2}\right) diag\left(\mathbf{p}(\boldsymbol{\theta}_0)^{1/2}\right)\boldsymbol{A}(\boldsymbol{\theta}_0)\right. \\ &\times \left.\left(\boldsymbol{A}(\boldsymbol{\theta}_0)^T\boldsymbol{A}(\boldsymbol{\theta}_0)\right)^{-1}\boldsymbol{A}(\boldsymbol{\theta}_0)^T diag\left(\mathbf{p}(\boldsymbol{\theta}_0)^{-1/2}\right)\right)\boldsymbol{d} \\ &= \left(diag\left(\mathbf{p}(\boldsymbol{\theta}_0)^{-1/2}\right) - \boldsymbol{A}(\boldsymbol{\theta}_0)\left(\boldsymbol{A}(\boldsymbol{\theta}_0)^T\boldsymbol{A}(\boldsymbol{\theta}_0)\right)^{-1}\right. \\ &- \left.\boldsymbol{A}(\boldsymbol{\theta}_0)^T diag\left(\mathbf{p}(\boldsymbol{\theta}_0)^{-1/2}\right)\right)\boldsymbol{d}\end{aligned}$$

By Theorem 6.1 the matrix $\boldsymbol{T}^*$ has the expression

$$\boldsymbol{T}^* = \boldsymbol{I} - \sqrt{\mathbf{p}(\boldsymbol{\theta}_0)}\sqrt{\mathbf{p}(\boldsymbol{\theta}_0)^T} - \boldsymbol{A}(\boldsymbol{\theta}_0)\left(\boldsymbol{A}(\boldsymbol{\theta}_0)^T\boldsymbol{A}(\boldsymbol{\theta}_0)\right)^{-1}\boldsymbol{A}(\boldsymbol{\theta}_0)^T.$$

Then we have

$$\boldsymbol{T}^{*}\boldsymbol{\mu}= \left(\boldsymbol{I}-\sqrt{\boldsymbol{p}(\boldsymbol{\theta}_0)}\sqrt{\boldsymbol{p}(\boldsymbol{\theta}_0)}^T-\boldsymbol{A}(\boldsymbol{\theta}_0)\left(\boldsymbol{A}(\boldsymbol{\theta}_0)^T\boldsymbol{A}(\boldsymbol{\theta})\right)^{-1}\boldsymbol{A}(\boldsymbol{\theta}_0)^T\right)\boldsymbol{\mu}$$

and

$$\begin{aligned}\boldsymbol{T}^{*}\boldsymbol{\mu}= & \ \boldsymbol{\mu}-\sqrt{\boldsymbol{p}(\boldsymbol{\theta}_0)}\sqrt{\boldsymbol{p}(\boldsymbol{\theta}_0)}^T\left(diag\left(\boldsymbol{p}(\boldsymbol{\theta}_0)^{-1/2}\right)\right.\\ - & \left.\boldsymbol{A}(\boldsymbol{\theta}_0)\left(\boldsymbol{A}(\boldsymbol{\theta}_0)^T\boldsymbol{A}(\boldsymbol{\theta}_0)\right)^{-1}\boldsymbol{A}(\boldsymbol{\theta}_0)^T diag\left(\boldsymbol{p}(\boldsymbol{\theta}_0)^{-1/2}\right)\right)\boldsymbol{d}\\ - & \ \boldsymbol{A}(\boldsymbol{\theta}_0)\left(\boldsymbol{A}(\boldsymbol{\theta}_0)^T\boldsymbol{A}(\boldsymbol{\theta}_0)\right)^{-1}\boldsymbol{A}(\boldsymbol{\theta}_0)^T diag\left(\boldsymbol{p}(\boldsymbol{\theta}_0)^{-1/2}\right)\boldsymbol{d}\\ + & \ \boldsymbol{A}(\boldsymbol{\theta}_0)\left(\boldsymbol{A}(\boldsymbol{\theta}_0)^T\boldsymbol{A}(\boldsymbol{\theta}_0)\right)^{-1}\boldsymbol{A}(\boldsymbol{\theta}_0)^T\boldsymbol{A}(\boldsymbol{\theta}_0)\left(\boldsymbol{A}(\boldsymbol{\theta}_0)^T\boldsymbol{A}(\boldsymbol{\theta}_0)\right)^{-1}\\ \times & \ \boldsymbol{A}(\boldsymbol{\theta}_0)^T diag\left(\boldsymbol{p}(\boldsymbol{\theta}_0)^{-1/2}\right)\boldsymbol{d}.\end{aligned}$$

Therefore

$$\begin{aligned}\boldsymbol{T}^{*}\boldsymbol{\mu}= & \left(\boldsymbol{I}-\boldsymbol{A}(\boldsymbol{\theta}_0)\left(\boldsymbol{A}(\boldsymbol{\theta}_0)^T\boldsymbol{A}(\boldsymbol{\theta}_0)\right)^{-1}\boldsymbol{A}(\boldsymbol{\theta}_0)^T\right)diag\left(\boldsymbol{p}(\boldsymbol{\theta}_0)^{-1/2}\right)\boldsymbol{d}\\ + & \sqrt{\boldsymbol{p}(\boldsymbol{\theta}_0)}\sqrt{\boldsymbol{p}(\boldsymbol{\theta}_0)}^T\boldsymbol{A}(\boldsymbol{\theta}_0)\left(\boldsymbol{A}(\boldsymbol{\theta}_0)^T\boldsymbol{A}(\boldsymbol{\theta}_0)\right)^{-1}\\ \times & \ \boldsymbol{A}(\boldsymbol{\theta}_0)^T diag\left(\boldsymbol{p}(\boldsymbol{\theta}_0)^{-1/2}\right)\boldsymbol{d}\\ - & \ \boldsymbol{A}(\boldsymbol{\theta}_0)\left(\boldsymbol{A}(\boldsymbol{\theta}_0)^T\boldsymbol{A}(\boldsymbol{\theta}_0)\right)^{-1}\boldsymbol{A}(\boldsymbol{\theta}_0)^T diag\left(\boldsymbol{p}(\boldsymbol{\theta}_0)^{-1/2}\right)\boldsymbol{d}\\ + & \ \boldsymbol{A}(\boldsymbol{\theta}_0)\left(\boldsymbol{A}(\boldsymbol{\theta}_0)^T\boldsymbol{A}(\boldsymbol{\theta}_0)\right)^{-1}\boldsymbol{A}(\boldsymbol{\theta}_0)^T diag\left(\boldsymbol{p}(\boldsymbol{\theta}_0)^{-1/2}\right)\boldsymbol{d}\\ = & \left(\boldsymbol{I}-\boldsymbol{A}(\boldsymbol{\theta}_0)\left(\boldsymbol{A}(\boldsymbol{\theta}_0)^T\boldsymbol{A}(\boldsymbol{\theta}_0)\right)^{-1}\boldsymbol{A}(\boldsymbol{\theta}_0)^T\right)diag\left(\boldsymbol{p}(\boldsymbol{\theta}_0)^{-1/2}\right)\boldsymbol{d}=\boldsymbol{\mu},\end{aligned}$$

and finally

$$\frac{2n}{\phi_1''(1)}D_{\phi_1}\left(\widehat{\boldsymbol{p}},\boldsymbol{p}(\widehat{\boldsymbol{\theta}}_{\phi_2})\right)\underset{n\to\infty}{\overset{L}{\longrightarrow}}\chi^2_{M-M_0-1}(\delta)$$

being $\delta=\boldsymbol{\mu}^T\boldsymbol{\mu}$ and

$$\boldsymbol{\mu}=\left(\boldsymbol{I}-\boldsymbol{A}(\boldsymbol{\theta}_0)\left(\boldsymbol{A}(\boldsymbol{\theta}_0)^T\boldsymbol{A}(\boldsymbol{\theta}_0)\right)^{-1}\boldsymbol{A}(\boldsymbol{\theta})^T\right)diag\left(\boldsymbol{p}(\boldsymbol{\theta}_0)^{-1/2}\right)\boldsymbol{d}.$$

If we denote $\boldsymbol{U}=\boldsymbol{I}-\boldsymbol{A}(\boldsymbol{\theta}_0)\left(\boldsymbol{A}(\boldsymbol{\theta}_0)^T\boldsymbol{A}(\boldsymbol{\theta}_0)\right)^{-1}\boldsymbol{A}(\boldsymbol{\theta}_0)^T$, we have

$$\begin{aligned}\boldsymbol{U}^T\boldsymbol{U} = & \ \boldsymbol{I} - \boldsymbol{A}(\boldsymbol{\theta}_0)\left(\boldsymbol{A}(\boldsymbol{\theta}_0)^T\boldsymbol{A}(\boldsymbol{\theta}_0)\right)^{-1}\boldsymbol{A}(\boldsymbol{\theta}_0)^T \\ - & \ \boldsymbol{A}(\boldsymbol{\theta}_0)\left(\boldsymbol{A}(\boldsymbol{\theta}_0)^T\boldsymbol{A}(\boldsymbol{\theta}_0)\right)^{-1}\boldsymbol{A}(\boldsymbol{\theta}_0)^T \\ + & \ \boldsymbol{A}(\boldsymbol{\theta}_0)\left(\boldsymbol{A}(\boldsymbol{\theta}_0)^T\boldsymbol{A}(\boldsymbol{\theta}_0)\right)^{-1}\boldsymbol{A}(\boldsymbol{\theta}_0)^T\boldsymbol{A}(\boldsymbol{\theta}_0) \\ \times & \left(\boldsymbol{A}(\boldsymbol{\theta}_0)^T\boldsymbol{A}(\boldsymbol{\theta}_0)\right)^{-1}\boldsymbol{A}(\boldsymbol{\theta}_0)^T \\ = & \ \boldsymbol{I} - \boldsymbol{A}(\boldsymbol{\theta}_0)\left(\boldsymbol{A}(\boldsymbol{\theta}_0)^T\boldsymbol{A}(\boldsymbol{\theta}_0)\right)^{-1}\boldsymbol{A}(\boldsymbol{\theta}_0)^T.\end{aligned}$$

Therefore

$$\begin{aligned}\delta = & \ \boldsymbol{\mu}^T\boldsymbol{\mu} = \boldsymbol{d}^T diag\left(\boldsymbol{p}(\boldsymbol{\theta}_0)^{-1/2}\right)\left(\boldsymbol{I} - \boldsymbol{A}(\boldsymbol{\theta}_0)\left(\boldsymbol{A}(\boldsymbol{\theta}_0)^T\boldsymbol{A}(\boldsymbol{\theta}_0)\right)^{-1}\boldsymbol{A}(\boldsymbol{\theta}_0)^T\right) \\ & \times diag\left(\boldsymbol{p}(\boldsymbol{\theta}_0)^{-1/2}\right)\boldsymbol{d}.\end{aligned}$$

■

The result presented in this theorem was obtained for $\phi(x) = \frac{1}{2}(x-1)^2$, chi-square test statistic, by Diamond et *al.* (1960). For the ϕ-divergence test statistic the result was obtained by Menéndez et *al.* (2001a).

6.3. Nonstandard Problems: Test Statistics Based on Phi-divergences

6.3.1. Maximum Likelihood Estimator Based on Original Data and Test Statistics Based on Phi-divergences

If we only have the number N_i, $i = 1, ..., M$, of observations falling into the ith of the M cells, it is clear that we have to estimate the unknown parameters using a minimum ϕ-divergence estimator. If the original observations $Y_1, ..., Y_n$ are available, one is tempted to use more efficient estimators, such as the maximum likelihood estimator, $\widehat{\boldsymbol{\theta}}$, from the likelihood function $\prod_{i=1}^n f_{\boldsymbol{\theta}}(y_i)$, i.e., the maximum likelihood estimator based on the original data. One may reasonably expect this procedure to provide more powerful tests than those based on the grouped data; at the same time the estimates usually are simpler and easier to obtain.

Before proceeding to the main result we shall consider an example.

Example 6.1

Let $Y_1, ..., Y_n$ be a random sample from a normal population with mean θ and variance 1. If we consider the partition of $\mathbb{R}$ given by

$$E_1 = (-\infty, 0] \text{ and } E_2 = (0, \infty),$$

we have

$$\begin{aligned} \Pr_\theta(E_1) &= p_1(\theta) = \int_{-\infty}^{0} \frac{1}{\sqrt{2\pi}} \exp\left(-\frac{(x-\theta)^2}{2}\right) dx \\ \Pr_\theta(E_2) &= p_2(\theta) = 1 - p_1(\theta) = \int_{0}^{\infty} \frac{1}{\sqrt{2\pi}} \exp\left(-\frac{(x-\theta)^2}{2}\right) dx. \end{aligned}$$

It is well known that the maximum likelihood estimator of θ, based on the original observations, is given by

$$\widehat{\theta} = \overline{Y}, \tag{6.10}$$

and the maximum likelihood estimator based on grouped data is obtained maximizing

$$\Pr_\theta(N_1 = n_1, N_2 = n_2) = \frac{n_1! n_2!}{n!} p_1(\theta)^{n_1} p_2(\theta)^{n_2}.$$

We consider

$$\phi(x) = \frac{1}{2}(1-x)^2,$$

and we are going to study the asymptotic distribution of the ϕ-divergence test statistic $T_n^\phi\left(\widehat{\theta}\right)$, where $\widehat{\theta}$ is the maximum likelihood estimator given in (6.10), i.e., the maximum likelihood estimator based on the original data. We have

$$\begin{aligned} T_n^\phi\left(\widehat{\theta}\right) &= n \sum_{i=1}^{2} \frac{\left(\widehat{p}_i - p_i\left(\overline{Y}\right)\right)^2}{p_i\left(\overline{Y}\right)} = n \left(\frac{\left(\widehat{p}_1 - p_1\left(\overline{Y}\right)\right)^2}{p_1\left(\overline{Y}\right)} + \frac{\left(\widehat{p}_2 - p_2\left(\overline{Y}\right)\right)^2}{p_2\left(\overline{Y}\right)} \right) \\ &= n \left(\frac{\left(\widehat{p}_1 - p_1\left(\overline{Y}\right)\right)^2}{p_1\left(\overline{Y}\right)} + \frac{\left(\widehat{p}_1 - p_1\left(\overline{Y}\right)\right)^2}{p_2\left(\overline{Y}\right)} \right) = n \frac{\left(\widehat{p}_1 - p_1\left(\overline{Y}\right)\right)^2}{p_1\left(\overline{Y}\right) p_2\left(\overline{Y}\right)}. \end{aligned}$$

But

$$p_1\left(\overline{Y}\right) p_2\left(\overline{Y}\right) \xrightarrow[n\to\infty]{P} p_1(\theta) p_2(\theta),$$

and therefore (see Ferguson 1996, p. 39) the asymptotic distribution of $T_n^\phi\left(\widehat{\theta}\right)$ coincides with the asymptotic distribution of

$$R_n^\phi\left(\widehat{\theta}\right) = n \frac{\left(\widehat{p}_1 - p_1\left(\overline{Y}\right)\right)^2}{p_1(\theta) p_2(\theta)}.$$

We have

$$p_1\left(\overline{Y}\right) = p_1(\theta) + \frac{\partial p_1(\theta)}{\partial\theta}\left(\overline{Y}-\theta\right) + o_P\left(n^{-1/2}\right),$$

but

$$\begin{aligned}\frac{\partial p_1(\theta)}{\partial\theta} &= \int_{-\infty}^{0}\frac{1}{\sqrt{2\pi}}(x-\theta)\exp\left(-\frac{1}{2}(x-\theta)^2\right)dx\\ &= \left(-\frac{1}{\sqrt{2\pi}}\exp\left(-\frac{1}{2}(x-\theta)^2\right)\right)_{-\infty}^{0} = -\frac{1}{\sqrt{2\pi}}\exp\left(-\frac{\theta^2}{2}\right),\end{aligned}$$

then we get

$$\sqrt{n}\left(p_1\left(\overline{Y}\right)-p_1(\theta)\right) = -\frac{\sqrt{n}}{\sqrt{2\pi}}\exp\left(-\frac{\theta^2}{2}\right)\left(\overline{Y}-\theta\right)+o_P(1).$$

Now we are going to establish the asymptotic distribution of the random variable

$$\sqrt{n}\left(\widehat{p}_1-p_1(\theta),\overline{Y}-\theta\right)^T$$

that is a bivariate normal random variable with mean vector $(0,0)^T$ *and variance-covariance matrix with elements*

$$\begin{aligned}Var\left[\sqrt{n}\left(\widehat{p}_1-p_1(\theta)\right)\right] &= nVar\left[\widehat{p}_1\right] = n\frac{1}{n^2}Var\left[N_1\right]\\ &= n\frac{1}{n^2}np_1(\theta)\,p_2(\theta) = p_1(\theta)\,p_2(\theta)\end{aligned}$$

$$\begin{aligned}Var\left[\sqrt{n}\left(\overline{Y}-\theta\right)\right] &= nVar\left[\overline{Y}-\theta\right] = nVar\left[\overline{Y}\right]\\ &= n\frac{1}{n}Var\left[Y_i\right] = 1.\end{aligned}$$

Denoting by $T = \frac{1}{n}\sum_{i=1}^{n}\left(I_{(-\infty,0]}(Y_i)-p_1(\theta)\right)$, *we have*

$$\begin{aligned}Cov\left[\sqrt{n}\left(\widehat{p}_1-p_1(\theta),\overline{Y}-\theta\right)\right] &= nCov\left[T,\tfrac{1}{n}\sum_{i=1}^{n}(Y_i-\theta)\right]\\ &= Cov\left[I_{(-\infty,0]}(Y_i)-p_1(\theta),Y_i-\theta\right]\\ &= E\left[\left(I_{(-\infty,0]}(Y_i)-p_1(\theta)\right)(Y_i-\theta)\right]\\ &- E\left[I_{(-\infty,0]}(Y_i)-p_1(\theta)\right]E\left[Y_i-\theta\right]\\ &= E\left[\left(I_{(-\infty,0]}(Y_i)\right)(Y_i-\theta)\right]\\ &= \int_{-\infty}^{0}\tfrac{1}{\sqrt{2\pi}}(x-\theta)\exp\left(-\tfrac{1}{2}(x-\theta)^2\right)dx\\ &= -\tfrac{1}{\sqrt{2\pi}}\exp\left(-\tfrac{1}{2}\theta^2\right).\end{aligned}$$

Therefore

$$\sqrt{n}\left(\widehat{p}_1 - p_1(\theta), \overline{Y} - \theta\right)^T \xrightarrow[n\to\infty]{L} N\left(\begin{pmatrix} 0 \\ 0 \end{pmatrix}, \begin{pmatrix} p_1(\theta)\,p_2(\theta) & -\frac{\exp\left(-\theta^2/2\right)}{\sqrt{2\pi}} \\ -\frac{\exp\left(-\theta^2/2\right)}{\sqrt{2\pi}} & 1 \end{pmatrix}\right).$$

On the other hand $\sqrt{n}\left(\widehat{p}_1 - p_1(\theta), p_1\left(\overline{Y}\right) - p_1(\theta)\right)^T$ *can be written as*

$$\sqrt{n}\begin{pmatrix} 1 & 0 \\ 0 & -\frac{\exp\left(-\theta^2/2\right)}{\sqrt{2\pi}} \end{pmatrix}\begin{pmatrix} \widehat{p}_1 - p_1(\theta) \\ \overline{Y} - \theta \end{pmatrix} + o_P(1),$$

and denoting by

$$\boldsymbol{\Sigma} = \begin{pmatrix} p_1(\theta)\,p_2(\theta) & -\frac{\exp\left(-\theta^2/2\right)}{\sqrt{2\pi}} \\ -\frac{\exp\left(-\theta^2/2\right)}{\sqrt{2\pi}} & 1 \end{pmatrix} \text{ and } \boldsymbol{A} = \begin{pmatrix} 1 & 0 \\ 0 & -\frac{\exp\left(-\theta^2/2\right)}{\sqrt{2\pi}} \end{pmatrix},$$

we have

$$\boldsymbol{\Sigma}^* = \boldsymbol{A}\boldsymbol{\Sigma}\boldsymbol{A}^T = \begin{pmatrix} p_1(\theta)\,p_2(\theta) & \frac{\exp\left(-\theta^2\right)}{2\pi} \\ \frac{\exp\left(-\theta^2\right)}{2\pi} & \frac{\exp\left(-\theta^2\right)}{2\pi} \end{pmatrix}.$$

Then

$$\sqrt{n}\left(\widehat{p}_1 - p_1(\theta), p_1\left(\overline{Y}\right) - p_1(\theta)\right)^T \xrightarrow[n\to\infty]{L} N\left(\begin{pmatrix} 0 \\ 0 \end{pmatrix}, \boldsymbol{\Sigma}^*\right),$$

$$\sqrt{n}\left(\widehat{p}_1 - p_1\left(\overline{Y}\right)\right) \xrightarrow[n\to\infty]{L} N\left(0, p_1(\theta)\,p_2(\theta) - \frac{\exp\left(-\theta^2\right)}{2\pi}\right)$$

and

$$\frac{\sqrt{n}\left(\widehat{p}_1 - p_1\left(\overline{Y}\right)\right)}{\sqrt{p_1(\theta)\,p_2(\theta) - \frac{\exp\left(-\theta^2\right)}{2\pi}}} \xrightarrow[n\to\infty]{L} N(0,1).$$

Finally

$$n\left(\widehat{p}_1 - p_1\left(\overline{Y}\right)\right)^2 \xrightarrow[n\to\infty]{L} \left(p_1(\theta)\,p_2(\theta) - \frac{\exp\left(-\theta^2\right)}{2\pi}\right)\chi_1^2$$

and

$$\frac{n\left(\widehat{p}_1 - p_1\left(\overline{Y}\right)\right)^2}{p_1(\theta)\,p_2(\theta)} \xrightarrow[n\to\infty]{L} \lambda\chi_1^2$$

with $\lambda = 1 - \dfrac{\exp\left(-\theta^2\right)}{2\pi p_1(\theta)\,p_2(\theta)}.$

Let us prove that $\lambda \geq 0$. *As the matrix* $\mathbf{\Sigma}^*$ *is nonnegative definite, then*

$$\begin{aligned} 0 \leq \det\left(\mathbf{\Sigma}^*\right) &= \frac{\exp\left(-\theta^2\right)}{2\pi}\left(p_1\left(\theta\right)p_2\left(\theta\right) - \frac{\exp\left(-\theta^2\right)}{2\pi}\right) \\ &= \frac{p_1\left(\theta\right)p_2\left(\theta\right)}{2\pi}\exp\left(-\theta^2\right)\left(1 - \frac{\exp\left(-\theta^2\right)}{2\pi p_1\left(\theta\right)p_2\left(\theta\right)}\right). \end{aligned}$$

Therefore $0 \leq \lambda \leq 1$.

Now we present the theorem that states the asymptotic distribution of the ϕ-divergence test statistic $T_n^{\phi}\left(\widehat{\boldsymbol{\theta}}\right)$. The proof can be seen in Morales et *al*. (1995) for ϕ-divergence measures in general and in Chernoff and Lehman (1954) for the particular case $\phi_1\left(x\right) = \frac{1}{2}\left(x-1\right)^2$.

Theorem 6.4

Under the conditions given in Morales et al. (1995) the asymptotic distribution of the ϕ*-divergence test statistic* $T_n^{\phi}\left(\widehat{\boldsymbol{\theta}}\right)$, *where* $\widehat{\boldsymbol{\theta}}$ *is the maximum likelihood estimator based on the original data, verifies*

$$\frac{2n}{\phi''\left(1\right)}D_{\phi}(\widehat{\mathbf{p}},\mathbf{p}(\widehat{\boldsymbol{\theta}})) \underset{n\to\infty}{\overset{L}{\longrightarrow}} \chi^2_{M-M_0-1} + \sum_{j=1}^{M_0}\left(1-\lambda_j\right)Z_j^2,$$

where Z_j *are independent and normally distributed random variables with mean zero and unit variance, and the* λ_j, $0 \leq \lambda_j \leq 1$, *are the roots of the equation*

$$\det\left(\boldsymbol{I}_F\left(\boldsymbol{\theta}_0\right) - \lambda\mathcal{I}_{\mathcal{F}}\left(\boldsymbol{\theta}_0\right)\right) = 0,$$

where $\mathcal{I}_{\mathcal{F}}\left(\boldsymbol{\theta}_0\right)$ *and* $\boldsymbol{I}_F\left(\boldsymbol{\theta}_0\right)$ *are the Fisher information matrices from the original and discretized models, respectively.*

Remark 6.1

Let F_1 $\left(F_1 \equiv \chi^2_{M-M_0-1}\right)$ *be the asymptotic distribution function of the* ϕ*-divergence test statistic*

$$T_n^{\phi_1}\left(\widehat{\boldsymbol{\theta}}_{\phi_2}\right) = \frac{2n}{\phi_1''\left(1\right)}D_{\phi_1}\left(\widehat{\mathbf{p}},\mathbf{p}(\widehat{\boldsymbol{\theta}}_{\phi_2})\right)$$

and F_2 $\left(F_2 \equiv \chi^2_{M-M_0-1} + \sum\limits_{j=1}^{M_0}\left(1-\lambda_j\right)\chi^2_{1j}\right)$ *be the asymptotic distribution function of the* ϕ*-divergence test statistic*

$$T_n^{\phi}\left(\widehat{\boldsymbol{\theta}}\right) = \frac{2n}{\phi''\left(1\right)}D_{\phi}(\widehat{\mathbf{p}},\mathbf{p}(\widehat{\boldsymbol{\theta}})).$$

We have

$$F_2(x) \leq F_1(x) \qquad \forall x \geq 0.$$

This result indicates that the decision rule "reject H_0, with a significance level α, if $T_n^{\phi}\left(\widehat{\boldsymbol{\theta}}\right) > \chi^2_{M-M_0-1,\alpha}$", will lead to a probability of rejection greater than the desired level of significance when the hypothesis is true. In symbols

$$\begin{aligned}
\alpha^* &= \Pr_{H_0}\left(\frac{2n}{\phi_1''(1)} D_{\phi_1}(\widehat{\mathbf{p}}, \mathbf{p}(\widehat{\boldsymbol{\theta}})) > \chi^2_{M-M_0-1,\alpha}\right) \\
&= 1 - F_2\left(\chi^2_{M-M_0-1,\alpha}\right) \geq 1 - F_1\left(\chi^2_{M-M_0-1,\alpha}\right) \\
&= \Pr_{H_0}\left(\frac{2n}{\phi_1''(1)} D_{\phi_1}\left(\widehat{\mathbf{p}}, \mathbf{p}(\widehat{\boldsymbol{\theta}}_{\phi_2})\right) > \chi^2_{M-M_0-1,\alpha}\right) \\
&= \alpha.
\end{aligned}$$

Then when we consider the approximation $F_1 \simeq F_2$, the probability of rejecting the null hypothesis increases and then we are raising the probability of type I error. However, a numerical investigation of a few special cases indicates that, at least in the Poisson models (see Chernoff and Lehman (1954)) this excess of probability of type I error will be so small as not to be serious. The situation appears to be not quite so favorable in the normal case.

6.3.2. Goodness-of-fit with Quantile Characterization

In this section we consider the problem studied in Section 3.4.1, but here we assume that $F \in \{F_{\boldsymbol{\theta}}\}_{\boldsymbol{\theta}\in\Theta}$ and Θ is an open set in $\mathbb{R}^{M_0}$. We consider the values π_i defined in (3.17) and the vector

$$\boldsymbol{q}^0 = (\pi_j - \pi_{j-1} : 1 \leq j \leq M)^T.$$

We are going to test $H_0 : F = F_{\boldsymbol{\theta}}$ by testing

$$H_0 : \boldsymbol{q} = \mathbf{p}(\boldsymbol{Y}_n, \boldsymbol{\theta}), \tag{6.11}$$

where

$$\mathbf{p}(\boldsymbol{Y}_n, \boldsymbol{\theta}) = (p_1(\boldsymbol{Y}_n, \boldsymbol{\theta}), ..., p_M(\boldsymbol{Y}_n, \boldsymbol{\theta}))^T = \left(F_{\boldsymbol{\theta}}(Y_{n_j}) - F_{\boldsymbol{\theta}}(Y_{n_{j-1}}) : 1 \leq j \leq M\right)^T,$$

$n_0 = 0$, $n_M = +\infty$, $Y_{n_i} = Y_{(n_i)}$ ($n_i = [n\pi_i] + 1$) is the n_ith-order statistic, $Y_{n_0} = -\infty$ and $Y_{n_M} = +\infty$.

If we consider $G \notin \{F_{\boldsymbol{\theta}}\}_{\boldsymbol{\theta}\in\Theta}$ an alternative hypothesis to the null hypothesis given in (6.11) is

$$\boldsymbol{q}(\boldsymbol{Y}_n) = (q_1(\boldsymbol{Y}_n), ..., q_M(\boldsymbol{Y}_n))^T = \left(G(Y_{n_j}) - G(Y_{n_{j-1}}) : 1 \leq j \leq M\right)^T. \tag{6.12}$$

For testing (6.11) we can consider the ϕ-divergence test statistic defined by

$$T_n^{\phi_1}\left(\widehat{\boldsymbol{\theta}}_{\phi_2}\right) = \frac{2n}{\phi_1''(1)} D_{\phi_1}\left(\boldsymbol{p}(\boldsymbol{Y}_n, \widehat{\boldsymbol{\theta}}_{\phi_2}), \boldsymbol{q}^0\right), \tag{6.13}$$

where

$$\widehat{\boldsymbol{\theta}}_{\phi_2} = \arg \inf_{\theta \in \Theta \subset R^{M_0}} D_{\phi_2}\left(\boldsymbol{p}(\boldsymbol{Y}_n, \boldsymbol{\theta}), \boldsymbol{q}^0\right).$$

Under some regularity assumptions Menéndez et *al.* (1998a) established that $\widehat{\boldsymbol{\theta}}_{\phi_2} \underset{n\to\infty}{\overset{P}{\longrightarrow}} \boldsymbol{\theta}_0$, $\sqrt{n}(\widehat{\boldsymbol{\theta}}_{\phi_2} - \boldsymbol{\theta}_0) \underset{n\to\infty}{\overset{L}{\longrightarrow}} N(\boldsymbol{0}, \boldsymbol{I}_F(\boldsymbol{\theta}_0)^{-1})$ and $T_n^{\phi_1}\left(\widehat{\boldsymbol{\theta}}_{\phi_2}\right) \underset{n\to\infty}{\overset{L}{\longrightarrow}} \chi^2_{M-M_0-1}$. We can consider the decision rule: "Reject the null hypothesis given in (6.11), with significance level α, if

$$T_n^{\phi_1}(\widehat{\boldsymbol{\theta}}_{\phi_2}) > \chi^2_{M-M_0-1,\alpha}\text{"}. \tag{6.14}$$

Let $\boldsymbol{\theta}_a$ a point in the parameter space verifying, under the alternative hypothesis given in (6.12), $\widehat{\boldsymbol{\theta}}_{\phi_2} = \boldsymbol{\theta}_a + o_P(1)$. We denote

$$c_j^* = G^{-1}(\pi_j),\ j = 1, ..., M,$$

and

$$\boldsymbol{p}(\boldsymbol{c}^*, \boldsymbol{\theta}_a) = \left(F_{\boldsymbol{\theta}_a}\left(G^{-1}(\pi_j)\right) - F_{\boldsymbol{\theta}_a}\left(G^{-1}(\pi_{j-1})\right) : j = 1, ..., M\right).$$

Under the assumption

$$\sqrt{n}\left(\boldsymbol{p}(\boldsymbol{Y}_n, \widehat{\boldsymbol{\theta}}_{\phi_2}) - \boldsymbol{p}(\boldsymbol{c}^*, \boldsymbol{\theta}_a)\right) \underset{n\to\infty}{\overset{L}{\longrightarrow}} N(\boldsymbol{0}, \boldsymbol{\Sigma}),$$

for some matrix $\boldsymbol{\Sigma}$, and if $\boldsymbol{p}(\boldsymbol{c}^*, \boldsymbol{\theta}_a) \neq \boldsymbol{q}^0$, the power of the ϕ-divergence test statistic (6.13) for testing (6.11), under the alternative hypothesis given in (6.12), satisfies

$$\beta_{n,\phi_1}(\boldsymbol{q}(\boldsymbol{Y}_n)) = 1 - \Phi_n\left(\frac{\sqrt{n}}{\sigma_{\phi_1}(\boldsymbol{q}(\boldsymbol{Y}_n))}\left(\frac{\phi_1''(1)}{2n}\chi^2_{M-M_0-1,\alpha} - D_{\phi_1}\left(\boldsymbol{p}(\boldsymbol{c}^*, \boldsymbol{\theta}_a), \boldsymbol{q}^0\right)\right)\right),$$

where $\Phi_n(x)$ is a sequence of distribution functions tending uniformly to the standard normal distribution $\Phi(x)$ and

$$\sigma^2_{\phi_1}(\boldsymbol{q}(\boldsymbol{Y}_n)) = \sum_{j,k=1}^{M} \sigma_{jk} \phi_1'\left(\frac{p_j(\boldsymbol{c}^*, \boldsymbol{\theta}_a)}{q_j^0}\right) \phi_1'\left(\frac{p_k(\boldsymbol{c}^*, \boldsymbol{\theta}_a)}{q_k^0}\right) \frac{1}{q_j^0}\frac{1}{q_k^0},$$

for $\boldsymbol{\Sigma} = (\sigma_{ij})_{i,j=1,...,M}$.

Other interesting results in this direction can be seen in Menéndez et *al.* (2001a, b, c).

6.3.3. Estimation from an Independent Sample

Consider the problem in which we wish to obtain the limiting distribution of $T_n^{\phi_1}$ when $\boldsymbol{\theta}$ is estimated using information from a second multinomial sample that is independent of $\mathbf{Y}_n = (Y_1, \ldots, Y_n)$. Then in addition to $\mathbf{Y}_n = (Y_1, \ldots, Y_n)$ we consider the samples $\mathbf{Y}^*_{n^*} = (Y^*_1, \ldots, Y^*_{n^*}) = (Y_{n+1}, ..., Y_{n+n^*})$ and $\mathbf{Y}^{**}_{n^{**}} = (Y^{**}_1, \ldots, Y^{**}_{n^{**}}) = (Y_1, \ldots, Y_{n+n^*})$. Obviously, the samples $\mathbf{Y}_n$ and $\mathbf{Y}^*_{n^*}$ are independent and $\mathbf{Y}_n, \mathbf{Y}^*_{n^*}$ are subsamples of $\mathbf{Y}^{**}_{n^{**}}$. We assume that n^* depends on n in such a way that $\tau = \lim_{n\to\infty} \frac{n_*}{n}$ exists. Let $\widehat{\boldsymbol{\theta}}^*_{\phi_2}, \widehat{\boldsymbol{\theta}}^{**}_{\phi_2}$ be the minimum ϕ_2-divergence estimators based on the random samples $\mathbf{Y}^*_{n^*}$ and $\mathbf{Y}^{**}_{n^{**}}$, respectively and let $\boldsymbol{p}(\widehat{\boldsymbol{\theta}}^*_{\phi_2})$ and $\boldsymbol{p}(\widehat{\boldsymbol{\theta}}^{**}_{\phi_2})$ be the parametric distribution estimators. Then we have for every $\tau > 0$ that the asymptotic distributions of the ϕ-divergence test statistics

$$\frac{2n}{\phi_1''(1)} D_{\phi_1}\left(\widehat{\mathbf{p}}, \boldsymbol{p}(\widehat{\boldsymbol{\theta}}^*_{\phi_2})\right) \quad \text{and} \quad \frac{2n}{\phi_1''(1)} D_{\phi_1}\left(\widehat{\mathbf{p}}, \boldsymbol{p}(\widehat{\boldsymbol{\theta}}^{**}_{\phi_2})\right)$$

are

$$\chi^2_{M-M_0-1} + \frac{1+\tau}{\tau}\chi^2_{M_0} \quad \text{and} \quad \chi^2_{M-M_0-1} + \frac{\tau}{1+\tau}\chi^2_{M_0}$$

respectively, where $\chi^2_{M-M_0-1}$ and $\chi^2_{M_0}$ are independent chi-squared distributed random variables with $M - M_0 - 1$ and M_0 degrees of freedom, respectively. This result has been established in Morales et *al*. (1995).

This is an important generalization of the classical result with $\phi_2(x) = x\log x - x + 1$ and $\phi_1(x) = \frac{1}{2}(1-x)^2$, that is to say the classical chi-square test statistic. In that case the first result was given by Murthy and Gafarian (1970) and the second one Chase (1972).

Now we denote by $\hat{\boldsymbol{\theta}}^*$ and $\hat{\boldsymbol{\theta}}^{**}$ the MLE based on the original data $\mathbf{Y}^*_{n^*} = (Y^*_1, ..., Y^*_{n^*})$ and $\mathbf{Y}^{**}_{n^{**}} = (Y^{**}_1, ..., Y^{**}_{n^{**}})$ and by $\widehat{\boldsymbol{p}}$ the relative frequency vector based on the sample $\mathbf{Y}_n = (Y_1, \ldots, Y_n)$. Under some regularity conditions given in Morales et *al*. (1995) the ϕ-divergence test statistics

$$\frac{2n}{\phi_1''(1)} D_{\phi_1}\left(\widehat{\mathbf{p}}, \boldsymbol{p}(\widehat{\boldsymbol{\theta}}^*)\right) \quad \text{and} \quad \frac{2n}{\phi_1''(1)} D_{\phi_1}\left(\widehat{\mathbf{p}}, \boldsymbol{p}(\widehat{\boldsymbol{\theta}}^{**})\right)$$

are asymptotically distributed as

$$\chi^2_{M-M_0-1} + \sum_{j=1}^{M_0}(1+\lambda_j/\tau)Z_j^2 \text{ and } \chi^2_{M-M_0-1} + \sum_{j=1}^{M_0}(1-\lambda_j/(1+\tau))Z_j^2,$$

respectively, where Z_j, $j = 1, ..., M_0$, are independent standard normal variables and the λ_j, $0 \leq \lambda_j \leq 1$, are the roots of the equation

$$\det\left(\boldsymbol{I}_F(\boldsymbol{\theta}_0) - \lambda \mathcal{I}_{\mathcal{F}}(\boldsymbol{\theta}_0)\right) = 0,$$

where $\mathcal{I}_{\mathcal{F}}(\boldsymbol{\theta}_0)$ and $\boldsymbol{I}_F(\boldsymbol{\theta}_0)$ are the Fisher information matrices of the original and discretized models, respectively.

6.3.4. Goodness-of-fit with Dependent Observations

In this Section we extend some results considered in Chapter 3, related to stationary irreducible aperiodic Markov chains $\boldsymbol{Y} = \{Y_k,\ k \geq 0\}$ with state space $\{1, ..., M\}$, to the situation in which the stationary distribution depend on an unknown parameter $\boldsymbol{\theta} \in \Theta$ and Θ is an open subset of $\mathbb{R}^{M_0}$. We denote by $\boldsymbol{p}(\boldsymbol{\theta})$ the stationary distribution and by $\boldsymbol{P}_{\boldsymbol{\theta}}$ the set of all matrices $\boldsymbol{P}$ such that their stationary distribution $\boldsymbol{p}$ coincides with $\boldsymbol{p}(\boldsymbol{\theta})$ and each element of $\boldsymbol{P}_{\boldsymbol{\theta}}$ by $\boldsymbol{P}(\boldsymbol{\theta})$. A basic statistical problem is how to estimate in a consistent and asymptotically normal way the unknown true parameter $\boldsymbol{\theta}_0 \in \Theta$ by using a random sample of size n from $\boldsymbol{Y} = \{Y_k,\ k \geq 0\}$ about the states of the chain, i.e., how to find a measurable mapping $\widetilde{\boldsymbol{\theta}}$ taking on values in Θ such that

i) $\widetilde{\boldsymbol{\theta}} \underset{n\to\infty}{\overset{P}{\longrightarrow}} \boldsymbol{\theta}_0$

ii) $\sqrt{n}(\widetilde{\boldsymbol{\theta}} - \boldsymbol{\theta}_0) \underset{n\to\infty}{\overset{L}{\longrightarrow}} N(\boldsymbol{0}, \boldsymbol{V}_0)$

and how to evaluate the $M_0 \times M_0$ matrix $\boldsymbol{V}_0$. In this context the minimum ϕ-divergence estimator was obtained in Menéndez et *al*. (1999a) and is given by

$$\widehat{\boldsymbol{\theta}}_{\phi_2} = \arg \inf_{\theta \in \Theta \subset R^{M_0}} D_{\phi_2}(\widehat{\boldsymbol{p}}, \boldsymbol{p}(\boldsymbol{\theta})),$$

where $\widehat{\boldsymbol{p}}$ is the nonparametric estimator based on the sample of size n.

Under some regularity assumptions in the cited paper of Menéndez et *al*. the following statements were established:

a) $\widehat{\boldsymbol{\theta}}_{\phi_2} \underset{n\to\infty}{\overset{a.s.}{\longrightarrow}} \boldsymbol{\theta}_0$

b) $\widehat{\boldsymbol{\theta}}_{\phi_2} = \boldsymbol{\theta}_0 + \left(\boldsymbol{A}(\boldsymbol{\theta}_0)^T \boldsymbol{A}(\boldsymbol{\theta}_0)\right)^{-1} \boldsymbol{A}(\boldsymbol{\theta}_0)^T diag\left(\boldsymbol{p}(\boldsymbol{\theta}_0)^{-1/2}\right)(\widehat{\boldsymbol{p}} - \boldsymbol{p}(\boldsymbol{\theta}_0))$
$+o(\| \widehat{\boldsymbol{p}} - \boldsymbol{p}(\boldsymbol{\theta}_0) \|)$

c) $\sqrt{n}(\widehat{\boldsymbol{\theta}}_{\phi_2} - \boldsymbol{\theta}_0) \xrightarrow[n\to\infty]{L} N\left(\mathbf{0}, \boldsymbol{\Delta}_0^T \boldsymbol{B}_0 \boldsymbol{\Delta}_0\right),$

where

- $\boldsymbol{\Delta}_0 = \boldsymbol{A}(\boldsymbol{\theta}_0)\left(\boldsymbol{A}(\boldsymbol{\theta}_0)^T \boldsymbol{A}(\boldsymbol{\theta}_0)\right)^{-1}$
- $\boldsymbol{B}_0 = diag\left(\boldsymbol{p}(\boldsymbol{\theta}_0)^{-1/2}\right) \boldsymbol{S}_0 \, diag\left(\boldsymbol{p}(\boldsymbol{\theta}_0)^{-1/2}\right)$

and

- $\boldsymbol{S}_0 = diag(\boldsymbol{p}(\boldsymbol{\theta}_0))\, \boldsymbol{C}_0 + \boldsymbol{C}_0^T diag(\boldsymbol{p}(\boldsymbol{\theta}_0)) - diag(\boldsymbol{p}(\boldsymbol{\theta}_0)) - \boldsymbol{p}(\boldsymbol{\theta}_0)\,\boldsymbol{p}(\boldsymbol{\theta}_0)^T,$

where $\boldsymbol{C}_0 = \left(\boldsymbol{I}_{M\times M} - \boldsymbol{P}(\boldsymbol{\theta}_0) + \mathbf{1}\boldsymbol{p}(\boldsymbol{\theta}_0)^T\right)^{-1}$ and $\mathbf{1} = (1, ..., 1)^T$ is the column vector of M units.

After estimating the unknown parameter $\boldsymbol{\theta}_0$ we are interested in testing $H_0: \boldsymbol{p} = \boldsymbol{p}(\boldsymbol{\theta}_0)$. Regarding this problem of testing, in the cited paper of Menéndez et *al*. (1999a), it was established that

$$\frac{2n}{\phi_1''(1)} D_{\phi_1}\left(\widehat{\boldsymbol{p}}, \boldsymbol{p}(\widehat{\boldsymbol{\theta}}_{\phi_2})\right) \xrightarrow[n\to\infty]{L} \sum_{j=1}^{M} \rho_i Z_i^2,$$

where Z_i are independently normally distributed random variables with mean zero and unit variance and ρ_i are the eigenvalues of the matrix

$$\boldsymbol{L}_0 = diag\left(\boldsymbol{p}(\boldsymbol{\theta}_0)^{-1/2}\right)(\boldsymbol{I} - \boldsymbol{W}_+)\, \boldsymbol{S}_0\, (\boldsymbol{I} - \boldsymbol{W}_-)\, diag\left(\boldsymbol{p}(\boldsymbol{\theta}_0)^{-1/2}\right),$$

where

$$\boldsymbol{W}_+ = diag\left(\boldsymbol{p}(\boldsymbol{\theta}_0)^{1/2}\right) \boldsymbol{\Sigma}_0\, diag\left(\boldsymbol{p}(\boldsymbol{\theta}_0)^{-1/2}\right),$$

$$\boldsymbol{W}_- = diag\left(\boldsymbol{p}(\boldsymbol{\theta}_0)^{-1/2}\right) \boldsymbol{\Sigma}_0\, diag\left(\boldsymbol{p}(\boldsymbol{\theta}_0)^{1/2}\right),$$

and

$$\boldsymbol{\Sigma}_0 = \boldsymbol{A}(\boldsymbol{\theta}_0)^T \left(\boldsymbol{A}(\boldsymbol{\theta}_0)^T \boldsymbol{A}(\boldsymbol{\theta}_0)\right)^{-1} \boldsymbol{A}(\boldsymbol{\theta}_0).$$

Another interesting problem is to test the transition matrix of the chain, i.e., to test

$$H_0: \boldsymbol{P} = \boldsymbol{P}(\boldsymbol{\theta}_0) = (p_{ij}(\boldsymbol{\theta}_0))_{i,j=1,\dots,M},$$

for some unknown $\boldsymbol{\theta}_0 \in \Theta$. Menéndez et *al*. (1999b) considered the ϕ-divergence test statistic

$$T_n^{\phi_1}\left(\widehat{\boldsymbol{\theta}}_{\phi_2}\right) = \frac{2n}{\phi''(1)} \sum_{i=1}^{M} \frac{v_{i*}}{n} \sum_{j=1}^{M} p_{ij}(\widehat{\boldsymbol{\theta}}_{\phi_2})\phi_1\left(\frac{\widehat{p}(i,j)}{p_{ij}(\widehat{\boldsymbol{\theta}}_{\phi_2})}\right),$$

where $\widehat{p}(i,j)$ was defined in Section 3.4.2, and

$$\widehat{\boldsymbol{\theta}}_{\phi_2} = \arg \inf_{\boldsymbol{\theta} \in \Theta \subset \mathbb{R}^{M_0}} D_{\phi_2}(\widehat{\boldsymbol{P}}, \boldsymbol{P}(\boldsymbol{\theta})) = \arg \inf_{\boldsymbol{\theta} \in \Theta \subset \mathbb{R}^{M_0}} \sum_{i=1}^{M} \frac{v_{i*}}{n} \sum_{j=1}^{M} p_{ij}(\boldsymbol{\theta}) \phi_2 \left(\frac{\widehat{p}(i,j)}{p_{ij}(\boldsymbol{\theta})} \right).$$

Its asymptotic distribution is chi-square with $c - M - M_0$ degrees of freedom, where c is the number of elements of the set,

$$C = \{(i,j) : p_{ij}(\boldsymbol{\theta}_0) > 0\}.$$

The results presented in this section extend the ideas of Glesser and Moore (1983a, 1983b).

6.3.5. Goodness-of-fit with Constraints

With the notation introduced in Section 5.5, Pardo, J. A. et *al*. (2002), established that given ν ($\nu < M_0$) real valued functions $f_1(\boldsymbol{\theta}), ..., f_\nu(\boldsymbol{\theta})$ that constrain the parameter $\boldsymbol{\theta} \in \Theta \subset \mathbb{R}^{M_0}$, $f_m(\boldsymbol{\theta}) = 0$, $m = 1, ..., \nu$ and under the null hypothesis given in (6.2) it holds

$$\frac{2n}{\phi_1''(1)} D_{\phi_1}\left(\widehat{\mathbf{p}}, \mathbf{p}(\widehat{\boldsymbol{\theta}}_{\phi_2}^{(r)})\right) \underset{n\to\infty}{\overset{L}{\longrightarrow}} \chi^2_{M-M_0+\nu-1}.$$

The importance of this result will be seen in Chapter 8.

6.4. Exercises

1. Consider the following random sample of size $n = 20$; 1, 1, 1, 2, 2, 2, 2, 1, 1, 1, 1, 3, 3, 3, 4, 4, 4, 1, 1, 1. We wish to test if these observations fit the distribution given in Exercise 3 of Chapter 5, using $\phi_1(x) = \phi_2(x) = x \log x - x + 1$ and significance level $\alpha = 0.05$.

2. Consider the following random sample of size $n = 20$; -1, 1, 1,-1, 1, 1, 1, 1, 1, 0, 0, 1, 0, -1, -1, 0, 0, 1, 0, 1. We wish to test if these observations fit the distribution given in Exercise 9 of Chapter 5, using $\phi_1(x) = \frac{1}{2}(1-x)^2$, $\phi_2(x) = x \log x - x + 1$ and significance level $\alpha = 0.05$.

3. A genetic model indicates that the distributions of the population between men or women and colour-blind or normal have the probabilities

	Men	*Women*
Normal	$\frac{\theta}{2}$	$\theta\left(1-\frac{\theta}{2}\right)$
Colour-blind	$\frac{(1-\theta)}{2}$	$\frac{(1-\theta)^2}{2}$

   a) Find the minimum ϕ_2-divergence estimator with $\phi_2(x) = x \log x - x + 1$.

   b) Obtain its asymptotic distribution.

   c) We consider a random sample of size $n = 2000$ and we obtain the following results

	Men	*Women*
Normal	894	1015
Colour-blind	81	10

   We want to know if the previous random sample is from the genetic model considered using $\phi_2(x) = \frac{1}{2}(x-1)^2$, $\phi_1(x) = x \log x - x + 1$ and significance level $\alpha = 0.05$.

4. Solve Example 6.1 using Theorem 6.4.

5. Let $Y_1, ..., Y_n$ be a random sample from a exponential population of parameter θ. We consider the partition

$$E_1 = (0,1) \text{ and } E_2 = [1, \infty)$$

and define $p_i(\theta) = \Pr_\theta(E_i)\,, i = 1, 2$.

a) Find the asymptotic distribution of the test statistic

$$T_n^{\phi}\left(\widehat{\theta}\right) = \frac{2n}{\phi''(1)} D_{\phi}(\widehat{\mathbf{p}}, \mathbf{p}(\widehat{\theta})),$$

where $\widehat{\theta}$ is the maximum likelihood estimator based on the original data and $\phi(x) = \frac{1}{2}(x-1)^2$.

b) Prove that the eigenvalue λ associated with the asymptotic distribution of the test statistic $T_n^{\phi}\left(\widehat{\theta}\right)$ verifies $0 \leq \lambda \leq 1$.

6. Solve Exercise 5 using Theorem 6.4.

7. Let X be a random variable with probability mass function

$$\Pr_{\theta}(X=1) = 0.5 - 2\theta, \ \Pr_{\theta}(X=2) = 0.5 + \theta \text{ and } \Pr_{\theta}(X = x_3) = \theta,$$

with $0 < \theta < \frac{1}{4}$.

a) Find the minimum power divergence estimator of θ for $\lambda = -2$.

b) Find the asymptotic distribution of the estimator obtained in part *a)*.

c) In 8000 independent trials the events $\{i\}$, $i = 1, 2, 3$, have occurred 2014, 5012 and 974 times respectively. Given the significance level $\alpha = 0.05$, test the hypothesis that the data are from the population described by the random variable X, using the power-divergence test statistic for $\lambda = -2, -1, -1/2, 0, 2/3$ and 1.

8. Let X be a random variable with probability mass function

$$\begin{aligned}
\Pr(X=1) &= p_1(\theta_1,\theta_2) = \theta_1\theta_2 \\
\Pr(X=2) &= p_2(\theta_1,\theta_2) = \theta_1(1-\theta_2) \\
\Pr(X=3) &= p_3(\theta_1,\theta_2) = \theta_2(1-\theta_1) \\
\Pr(X=4) &= p_4(\theta_1,\theta_2) = (1-\theta_2)(1-\theta_1),
\end{aligned}$$

with $0 \leq \theta_1, \ \theta_2 \leq 1$.

a) Find Fisher information matrix as well as the asymptotic distribution of the minimum ϕ-divergence estimator.

b) Find the minimum ϕ-divergence estimator with $\phi(x) = x\log x - x + 1$ as well as its asymptotic distribution.

c) Find the expression of the test statistic $T_n^{\phi}\left(\widehat{\boldsymbol{\theta}}_{(0)}\right)$ for $\phi(x) = \frac{1}{2}(x-1)^2$ for testing $H_0 : \boldsymbol{p}(\boldsymbol{\theta})$, where

$$\begin{aligned}\boldsymbol{p}(\boldsymbol{\theta}) &= (p_{11}(\boldsymbol{\theta}), p_{12}(\boldsymbol{\theta}), p_{21}(\boldsymbol{\theta}), p_{22}(\boldsymbol{\theta}))^T \\ &= (\theta_1\theta_2, \theta_1(1-\theta_2), \theta_2(1-\theta_1), (1-\theta_2)(1-\theta_1))^T.\end{aligned}$$

9. Let X be a random variable with probability density function given by

$$f_\theta(x) = \begin{cases} \dfrac{1-\theta\cos x}{2\pi} & x \in [0, 2\pi) \\ 0 & \text{otherwise} \end{cases},$$

$\theta \in (-1, 1)$ and we consider the discretized model obtained on dividing the interval $[0, 2\pi)$ into M intervals of equal size.

a) Find the Fisher information matrix in the discretized model.

b) Find the minimum power-divergence estimator for $\lambda = -2$.

c) Find its asymptotic distribution.

10. Given the model

$$\boldsymbol{P}(\theta) = \begin{pmatrix} 1-\theta & \theta \\ 1 & 0 \end{pmatrix} \in \boldsymbol{P}_\theta,$$

and the stationary distribution given by

$$\boldsymbol{p}(\theta) = \left(\frac{1}{1+\theta}, \frac{\theta}{1+\theta}\right)^T, \quad \theta \in \Theta = (0, 1):$$

a) Find the minimum ϕ-divergence estimator as well as its asymptotic properties.

b) Find the expression of the minimum power-divergence estimator for $\lambda \neq -1$.

6.5. Answers to Exercises

1. In our case we have

$$\widehat{p}_1 = 0.5, \ \widehat{p}_2 = 0.2, \ \widehat{p}_3 = 0.15 \text{ and } \widehat{p}_4 = 0.15,$$

then, taking into account Exercise 3 in Chapter 5, we have

$$\widehat{\theta}_{\phi_2}(y_1,...,y_n) = \frac{(-1+2\widehat{p}_1-\widehat{p}_2-\widehat{p}_3)+\left((-1+2\widehat{p}_1-\widehat{p}_2-\widehat{p}_3)^2+8\widehat{p}_4\right)^{1/2}}{2} = 0.4,$$

and

$$\begin{aligned}
p_1(\widehat{\theta}_{\phi_2}) &= \frac{1}{4}(2+0.4) = 0.6 \\
p_2(\widehat{\theta}_{\phi_2}) &= \frac{1}{4}(1-0.4) = 0.15 \\
p_3(\widehat{\theta}_{\phi_2}) &= 0.15 \\
p_4(\widehat{\theta}_{\phi_2}) &= 0.1.
\end{aligned}$$

The expression of the test statistic is given by

$$\begin{aligned}
\frac{2n}{\phi_2''(1)} D_{\phi_2}\left(\widehat{\boldsymbol{p}}, \boldsymbol{p}(\widehat{\theta}_{\phi_2})\right) &= 2n\left(\sum_{i=1}^{4} \widehat{p}_i \log \frac{\widehat{p}_i}{p_i(\widehat{\theta}_{\phi_2})}\right) \\
&= 40\left(0.5\log\frac{0.5}{0.6} + 0.2\log\frac{0.2}{0.15} + 0.15\log\frac{0.15}{0.15}\right. \\
&+ \left. 0.15\log\frac{0.15}{0.1}\right) = 1.0878.
\end{aligned}$$

On the other hand $\chi^2_{4-1-1,\,0.05} = 5.991$ and we should not reject the null hypothesis.

2. In this case we have

$$\widehat{\boldsymbol{p}} = (\widehat{p}_1, \widehat{p}_2, \widehat{p}_3)^T = (4/20, 6/20, 10/20)^T$$

and

$$\widehat{\theta}_{\phi_2}(y_1,...,y_n) = \frac{2\widehat{p}_1 + \widehat{p}_3}{2} = 0.45.$$

Then

$$p_1(\widehat{\theta}_{\phi_2}) = 0.2025,\ p_2(\widehat{\theta}_{\phi_2}) = 0.3025,\ p_3(\widehat{\theta}_{\phi_2}) = 0.4950.$$

Therefore

$$\begin{aligned}
\frac{2n}{\phi_1''(1)} D_{\phi_1}\left(\widehat{\boldsymbol{p}}, \boldsymbol{p}(\widehat{\theta}_{\phi_2})\right) &= n\sum_{i=1}^{3} p_i(\widehat{\theta}_{\phi_2})\left(\frac{\widehat{p}_i}{p_i(\widehat{\theta}_{\phi_2})} - 1\right)^2 \\
&= n\sum_{i=1}^{3} \frac{(\widehat{p}_i - p_i(\widehat{\theta}_{\phi_2}))^2}{p_i(\widehat{\theta}_{\phi_2})} = 2.0406 \times 10^{-3}.
\end{aligned}$$

On the other hand $\chi^2_{3-1-1,\,0.05} = 3.841$ and we should not reject the null hypothesis.

3. *a)* It is necessary to minimize in θ the function

$$\begin{aligned} g(\theta) &= \widehat{p}_1 \log \frac{2\widehat{p}_1}{\theta} + \widehat{p}_2 \log \frac{2\widehat{p}_2}{1-\theta} + \widehat{p}_3 \log \frac{2\widehat{p}_3}{\theta(2-\theta)} + \widehat{p}_4 \log \frac{2\widehat{p}_4}{(1-\theta)^2} \\ &= -\widehat{p}_1 \log \theta - \widehat{p}_2 \log(1-\theta) - \widehat{p}_3 \log(2-\theta) - \widehat{p}_3 \log \theta \\ &- \widehat{p}_4 2 \log(1-\theta) + c. \end{aligned}$$

Differentiating with respect to θ and equating to zero we have

$$\theta^2 (-1 - \widehat{p}_4 - \widehat{p}_3) + \theta (4 - \widehat{p}_1 - 2\widehat{p}_2) - 2(\widehat{p}_1 + \widehat{p}_3) = 0,$$

then

$$\widehat{\theta}_{\phi_2} = \frac{-(4-\widehat{p}_1-2\widehat{p}_2)+\left((4-\widehat{p}_1-2\widehat{p}_2)^2+8(-1-\widehat{p}_4-\widehat{p}_3)(\widehat{p}_1+\widehat{p}_3)\right)^{1/2}}{2(-1-\widehat{p}_4-\widehat{p}_3)}.$$

b) We have

$$\frac{\partial p_1(\theta)}{\partial \theta} = \frac{1}{2}, \; \frac{\partial p_2(\theta)}{\partial \theta} = -\frac{1}{2}, \; \frac{\partial p_3(\theta)}{\partial \theta} = 1 - \theta \text{ and } \frac{\partial p_4(\theta)}{\partial \theta} = -(1-\theta),$$

then

$$\begin{aligned} \boldsymbol{A}(\theta) &= diag\left(\boldsymbol{p}(\theta)^{-1/2}\right) \boldsymbol{J}(\theta) \\ &= \left(\tfrac{\sqrt{2}}{2}\theta^{-1/2}, -\tfrac{\sqrt{2}}{2}(1-\theta)^{-1/2}, \theta^{-1/2}\left(\tfrac{2-\theta}{2}\right)^{-1/2}(1-\theta), -\sqrt{2}\right) \end{aligned}$$

and

$$\boldsymbol{A}(\theta)^T \boldsymbol{A}(\theta) = \frac{1}{2}\theta^{-1} + \frac{1}{2}(1-\theta)^{-1} + \theta^{-1}\left(\frac{2-\theta}{2}\right)^{-1}(1-\theta)^2 + 2.$$

Therefore,

$$\sqrt{n}(\widehat{\theta}_{\phi_2}(Y_1, ..., Y_n) - \theta_0) \underset{n\to\infty}{\overset{L}{\longrightarrow}} N\left(0, \left(\boldsymbol{A}(\theta_0)^T \boldsymbol{A}(\theta_0)\right)^{-1}\right).$$

c) In our case

$$\widehat{p}_1 = 0.447, \; \widehat{p}_2 = 0.0405, \; \widehat{p}_3 = 0.5075, \; \widehat{p}_4 = 0.005,$$

then

$$\widehat{\theta}_{\phi_2}(y_1, ..., y_n) = 0.9129.$$

We have

$$p_1(\widehat{\theta}_{\phi_2}) = 0.4564, \; p_2(\widehat{\theta}_{\phi_2}) = 0.0435, \; p_3(\widehat{\theta}_{\phi_2}) = 0.4962$$

and

$$p_4(\widehat{\theta}_{\phi_2}) = 0.0039,$$

and the value of the ϕ-divergence test statistic is

$$\begin{aligned}
\frac{2n}{\phi_1''(1)} D_{\phi_1}\left(\widehat{\boldsymbol{p}}, \boldsymbol{p}(\widehat{\theta}_{\phi_2})\right) &= n \sum_{i=1}^{4} p_i(\widehat{\theta}_{\phi_2}) \left(\frac{\widehat{p}_i}{p_i(\widehat{\theta}_{\phi_2})} - 1 \right)^2 \\
&= n \sum_{i=1}^{4} \frac{\left(\widehat{p}_i - p_i(\widehat{\theta}_{\phi_2})\right)^2}{p_i(\widehat{\theta}_{\phi_2})} \\
&= 1.9361.
\end{aligned}$$

But $\chi^2_{M-M_0-1,\alpha} = \chi^2_{2,0.05} = 5.991$ and we should not reject the null hypothesis.

4. The Fisher information associated with the original model is

$$\mathcal{I}_{\mathcal{F}}(\theta) = \frac{1}{\sigma^2} = 1$$

and the Fisher information associated with the discretized model is given by

$$\boldsymbol{I}_F(\theta) = \boldsymbol{A}(\theta)^T \boldsymbol{A}(\theta)$$

where

$$\begin{aligned}
\boldsymbol{A}(\theta) &= \begin{pmatrix} p_1(\theta)^{-1/2} & 0 \\ 0 & p_2(\theta)^{-1/2} \end{pmatrix} \left(-\frac{1}{\sqrt{2\pi}} e^{-\theta^2/2}, \frac{1}{\sqrt{2\pi}} e^{-\theta^2/2} \right)^T \\
&= \left(-\frac{p_1(\theta)^{-1/2}}{\sqrt{2\pi}} e^{-\theta^2/2}, \frac{p_2(\theta)^{-1/2}}{\sqrt{2\pi}} e^{-\theta^2/2} \right)^T,
\end{aligned}$$

then

$$\boldsymbol{I}_F(\theta) = \boldsymbol{A}(\theta)^T \boldsymbol{A}(\theta) = \frac{e^{-\theta^2}}{p_1(\theta)\, p_2(\theta)\, 2\pi}.$$

We have $M = 2$, $M_0 = 1$ and $\phi(x) = \frac{1}{2}(1-x)^2$, then

$$\frac{2n}{\phi''(1)} D_\phi(\widehat{\boldsymbol{p}}, \boldsymbol{p}(\widehat{\theta})) = \frac{n\left(\widehat{p}_1 - p_1(\overline{Y})\right)^2}{p_1(\theta)\, p_2(\theta)} \underset{n\to\infty}{\overset{L}{\longrightarrow}} (1-\lambda_1)\, \chi^2_1$$

being λ_1 the solution of the equation

$$\det\left(\boldsymbol{I}_F(\theta) - \lambda \mathcal{I}_{\mathcal{F}}(\theta)\right) = \det\left(\frac{e^{-\theta^2}}{p_1(\theta)\, p_2(\theta)\, 2\pi} - \lambda \right) = 0,$$

i.e.,

$$\lambda_1 = \frac{e^{-\theta^2}}{p_1(\theta)\, p_2(\theta)\, 2\pi}.$$

5. We have

$$\Pr_\theta(E_1) = p_1(\theta) = 1 - \exp(-1/\theta) \text{ and } \Pr_\theta(E_2) = p_2(\theta) = \exp(-1/\theta),$$

and the maximum likelihood estimator of θ, based on the original data, is

$$\widehat{\theta} = \overline{Y}.$$

The test statistic $T_n^\phi\left(\widehat{\theta}\right)$ has the expression

$$\begin{aligned} T_n^\phi\left(\widehat{\theta}\right) &= n\sum_{i=1}^{2} \frac{(\widehat{p}_i - p_i(\overline{Y}))^2}{p_i(\overline{Y})} = n\left(\frac{(\widehat{p}_1 - p_1(\overline{Y}))^2}{p_1(\overline{Y})} + \frac{(\widehat{p}_2 - p_2(\overline{Y}))^2}{p_2(\overline{Y})}\right) \\ &= n\left(\frac{(\widehat{p}_1 - p_1(\overline{Y}))^2}{p_1(\overline{Y})} + \frac{(\widehat{p}_1 - p_1(\overline{Y}))^2}{p_2(\overline{Y})}\right) = n\left(\frac{(\widehat{p}_1 - p_1(\overline{Y}))^2}{p_1(\overline{Y})\, p_2(\overline{Y})}\right), \end{aligned}$$

but

$$p_1(\overline{Y})\, p_2(\overline{Y}) \underset{n\to\infty}{\overset{P}{\longrightarrow}} p_1(\theta)\, p_2(\theta).$$

Therefore the asymptotic distribution of $T_n^\phi\left(\widehat{\theta}\right)$ coincides with the asymptotic distribution of the random variable

$$R_n^\phi\left(\widehat{\theta}\right) = n\frac{(\widehat{p}_1 - p_1(\overline{Y}))^2}{p_1(\theta)\, p_2(\theta)}.$$

We know that

$$p_1(\overline{Y}) = p_1(\theta) + \frac{\partial p_1(\theta)}{\partial \theta}(\overline{Y} - \theta) + o_P\left(n^{-\frac{1}{2}}\right),$$

but

$$\frac{\partial p_1(\theta)}{\partial \theta} = -\frac{1}{\theta^2}\exp(-1/\theta),$$

then we have

$$\sqrt{n}(p_1(\overline{Y}) - p_1(\theta)) = -\frac{\sqrt{n}}{\theta^2}\exp\left(-\frac{1}{\theta}\right)(\overline{Y} - \theta) + o_P(1).$$

The asymptotic distribution of the random variable

$$\sqrt{n}(\widehat{p}_1 - p_1(\theta), \overline{Y} - \theta)^T$$

is bivariate normal with mean vector $(0,0)^T$ and variance-covariance matrix given by

$$\begin{aligned} Var\left[\sqrt{n}(\widehat{p}_1 - p_1(\theta))\right] &= nVar\left[\widehat{p}_1\right] = n\frac{1}{n^2}Var\left[N_i\right] \\ &= n\frac{1}{n^2}np_1(\theta)\,p_2(\theta) = p_1(\theta)\,p_2(\theta) \end{aligned}$$

$$\begin{aligned} Var\left[\sqrt{n}\left(\overline{Y}-\theta\right)\right] &= nVar\left[\overline{Y}-\theta\right] = nVar\left[\overline{Y}\right] \\ &= n\frac{1}{n}Var\left[Y_i\right] = \theta^2. \end{aligned}$$

Denoting by $T = \frac{1}{n}\sum_{i=1}^{n}(I_{(0,1)}(Y_i) - p_1(\theta))$, we have

$$\begin{aligned} Cov\left[\sqrt{n}\left(\widehat{p}_1 - p_1(\theta), \overline{Y}-\theta\right)\right] &= nCov\left[T, \frac{1}{n}\sum_{i=1}^{n}(Y_i-\theta)\right] \\ &= Cov\left[I_{(0,1)}(Y_i) - p_1(\theta), Y_i - \theta\right] \\ &= E\left[I_{(0,1)}(Y_i)(Y_i-\theta)\right] \\ &= \int_0^1 (x-\theta)\tfrac{1}{\theta}\exp(-x/\theta)\,dx \\ &= -\exp(-1/\theta). \end{aligned}$$

Therefore

$$\sqrt{n}\left(\widehat{p}_1 - p_1(\theta), \overline{Y}-\theta\right)^T \xrightarrow[n\to\infty]{L} N\left(\begin{pmatrix}0\\0\end{pmatrix}, \begin{pmatrix} p_1(\theta)\,p_2(\theta) & -\exp(-\frac{1}{\theta}) \\ -\exp(-\frac{1}{\theta}) & \theta^2 \end{pmatrix}\right).$$

On the other hand

$$\sqrt{n}\left(\widehat{p}_1\text{-}p_1(\theta), p_1\left(\overline{Y}\right)\text{-}p_1(\theta)\right)^T = \sqrt{n}\begin{pmatrix} 1 & 0 \\ 0 & -\exp(-\frac{1}{\theta}) \end{pmatrix}\begin{pmatrix} \widehat{p}_1 - p_1(\theta) \\ \overline{Y}-\theta \end{pmatrix},$$

and denoting by

$$\boldsymbol{\Sigma} = \begin{pmatrix} p_1(\theta)\,p_2(\theta) & -\exp(-1/\theta) \\ -\exp(-1/\theta) & \theta^2 \end{pmatrix} \text{ and } \boldsymbol{A} = \begin{pmatrix} 1 & 0 \\ 0 & -\exp(-1/\theta) \end{pmatrix}$$

we have

$$\boldsymbol{\Sigma}^* = \boldsymbol{A}\boldsymbol{\Sigma}\boldsymbol{A}^T = \begin{pmatrix} p_1(\theta)\,p_2(\theta) & \frac{1}{\theta^2}\exp(-2/\theta) \\ \frac{1}{\theta^2}\exp(-2/\theta) & \frac{1}{\theta^2}\exp(-2/\theta) \end{pmatrix},$$

then

$$\sqrt{n}\left(\widehat{p}_1 - p_1(\theta), p_1\left(\overline{Y}\right) - p_1(\theta)\right)^T \xrightarrow[n\to\infty]{L} N\left(\begin{pmatrix} 0 \\ 0 \end{pmatrix}, \boldsymbol{\Sigma}^*\right),$$

and

$$\sqrt{n}\left(\widehat{p}_1 - p_1\left(\overline{Y}\right)\right) \xrightarrow[n\to\infty]{L} N\left(0, p_1(\theta)\, p_2(\theta) - \theta^{-2}\exp(-2/\theta)\right)$$

or

$$\frac{\sqrt{n}\left(\widehat{p}_1 - p_1\left(\overline{Y}\right)\right)}{\sqrt{p_1(\theta)\, p_2(\theta) - \theta^{-2}\exp(-2/\theta)}} \xrightarrow[n\to\infty]{L} N(0,1).$$

Finally

$$n\left(\widehat{p}_1 - p_1\left(\overline{Y}\right)\right)^2 \xrightarrow[n\to\infty]{L} \left(p_1(\theta)\, p_2(\theta) - \theta^{-2}\exp(-2/\theta)\right)\chi_1^2$$

and

$$\frac{n\left(\widehat{p}_1 - p_1\left(\overline{Y}\right)\right)^2}{p_1(\theta)\, p_2(\theta)} \xrightarrow[n\to\infty]{L} \lambda\chi_1^2$$

being $\lambda = 1 - \dfrac{\exp(-2/\theta)}{\theta^2 p_1(\theta)\, p_2(\theta)}$.

Now we are going to see that $\lambda \geq 0$. We know that $\boldsymbol{\Sigma}^*$ is nonnegative definite, then

$$\begin{aligned} 0 \leq \det(\boldsymbol{\Sigma}^*) &= \frac{\exp(-2/\theta)}{\theta^2}\left(p_1(\theta)\, p_2(\theta) - \frac{\exp(-2/\theta)}{\theta^2}\right) \\ &= \frac{p_1(\theta)\, p_2(\theta)}{\theta^2}\exp(-2/\theta)\left(1 - \frac{\exp(-2/\theta)}{\theta^2 p_1(\theta)\, p_2(\theta)}\right), \end{aligned}$$

and $0 \leq \lambda \leq 1$.

6. The Fisher information associated with the original model is given by

$$\mathcal{I}_{\mathcal{F}}(\theta) = \frac{1}{\theta^2},$$

and the Fisher information associated with the discretized model is

$$\boldsymbol{I}_F(\theta) = \boldsymbol{A}(\theta)^T \boldsymbol{A}(\theta)$$

being

$$\begin{aligned} \boldsymbol{A}(\theta) &= \begin{pmatrix} p_1(\theta)^{-1/2} & 0 \\ 0 & p_2(\theta)^{-1/2} \end{pmatrix}\left(-\tfrac{1}{\theta^2}e^{-1/\theta}, \tfrac{1}{\theta^2}e^{-1/\theta}\right)^T \\ &= \left(-\tfrac{p_1(\theta)^{-1/2}}{\theta^2}e^{-1/\theta}, \tfrac{p_2(\theta)^{-1/2}}{\theta^2}e^{-1/\theta}\right)^T, \end{aligned}$$

then

$$\boldsymbol{I}_F(\theta) = \boldsymbol{A}(\theta)^T \boldsymbol{A}(\theta) = \frac{1}{\theta^4} \frac{e^{-2/\theta}}{p_1(\theta)\, p_2(\theta)}.$$

We have $M = 2$, $M_0 = 1$ and $\phi(x) = \frac{1}{2}(1-x)^2$, then

$$\frac{n\left(\widehat{p}_1 - p_1\left(\overline{Y}\right)\right)^2}{p_1(\theta)\, p_2(\theta)} \xrightarrow[n\to\infty]{L} (1-\lambda_1)\chi_1^2$$

being λ_1 the root of the equation

$$\det\left(\boldsymbol{I}_F(\theta) - \lambda_1 \mathcal{I}_{\mathcal{F}}(\theta)\right) = 0.$$

It is easy to get

$$\lambda_1 = \frac{\exp(-2/\theta)}{\theta^2 p_1(\theta)\, p_2(\theta)}.$$

7. *a)* The power-divergence for $\lambda = -2$ has the expression

$$\begin{aligned} D_{\phi_{(-2)}}(\widehat{\mathbf{p}}, \mathbf{p}(\theta)) = & \frac{1}{-2(-2+1)}\left(\sum_{j=1}^{3} \frac{p_j(\theta)^2}{\widehat{p}_j} - 1\right) \\ = & \frac{1}{2}\left(\frac{(0.5-2\theta)^2}{\widehat{p}_1} + \frac{(0.5+\theta)^2}{\widehat{p}_2} + \frac{\theta^2}{\widehat{p}_3} - 1\right). \end{aligned}$$

Differentiating

$$g(\theta) \equiv D_{\phi_{(-2)}}(\widehat{\mathbf{p}}, \mathbf{p}(\theta)),$$

with respect to θ and equating to zero we get

$$\theta(8\widehat{p}_2\widehat{p}_3 + 2\widehat{p}_1\widehat{p}_3 + 2\widehat{p}_1\widehat{p}_2) - 2\widehat{p}_2\widehat{p}_3 + \widehat{p}_1\widehat{p}_3 = 0,$$

then

$$\widehat{\theta}_{\phi_{(-2)}}(y_1, ..., y_n) = \frac{2\widehat{p}_2\widehat{p}_3 - \widehat{p}_1\widehat{p}_3}{8\widehat{p}_2\widehat{p}_3 + 2\widehat{p}_1\widehat{p}_3 + 2\widehat{p}_1\widehat{p}_2}.$$

b) We have

$$\frac{\partial p_1(\theta)}{\partial\theta} = -2, \quad \frac{\partial p_2(\theta)}{\partial\theta} = 1 \quad \text{and} \quad \frac{\partial p_3(\theta)}{\partial\theta} = 1,$$

then

$$\begin{aligned} \boldsymbol{A}(\theta) &= diag\left(\mathbf{p}(\theta)^{-1/2}\right) \boldsymbol{J}(\theta) \\ &= (-2(0.5-2\theta)^{-1/2}, (0.5+\theta)^{-1/2}, \theta^{-1/2})^T, \end{aligned}$$

and

$$\boldsymbol{A}(\theta)^T \boldsymbol{A}(\theta) = \frac{(2\theta + 0.5^2)}{(0.5 - 2\theta)(0.5 + \theta)\theta}.$$

Therefore

$$\sqrt{n}\left(\widehat{\theta}_{\phi_{(-2)}}(Y_1, ..., Y_n) - \theta_0\right) \underset{n\to\infty}{\overset{L}{\longrightarrow}} N\left(0, \frac{(0.5 - 2\theta_0)(0.5 + \theta_0)\theta_0}{(2\theta_0 + 0.5^2)}\right).$$

c) We have

$$D_{\phi_{(\lambda)}}(\widehat{\boldsymbol{p}}, \boldsymbol{p}(\theta)) = \frac{1}{\lambda(\lambda+1)}\left(\sum_{j=1}^{3} \frac{p_j(\theta)^{\lambda+1}}{\widehat{p}_j^{\lambda}} - 1\right).$$

It is immediate to get

$$\widehat{\theta}_{\phi_{(-2)}} = \frac{2\widehat{p}_2\widehat{p}_3 - \widehat{p}_1\widehat{p}_3}{8\widehat{p}_2\widehat{p}_3 + 2\widehat{p}_1\widehat{p}_3 + 2\widehat{p}_1\widehat{p}_2} = 0.1057,$$

then

$$p_1(\widehat{\theta}_{\phi_{(-2)}}) = .2886,\ p_2(\widehat{\theta}_{\phi_{(-2)}}) = 0.6057 \text{ and } p_3(\widehat{\theta}_{\phi_{(-2)}}) = 0.1057,$$

and

λ	-2	-1	-0.5	0	2/3	1
$T_n^{\phi(\lambda)}\left(\widehat{\theta}_{\phi_{(-2)}}\right)$	62.853	63.583	64.017	64.498	65.213	65.603

On the other hand $\chi^2_{3-1-1,0.05} = 3.841$. Then we should reject the null hypothesis for all the values of λ considered.

8. *a)* The matrix $\boldsymbol{A}(\boldsymbol{\theta})$ is given by

$$\boldsymbol{A}(\boldsymbol{\theta}) = \begin{pmatrix} \sqrt{\frac{\theta_2}{\theta_1}} & \sqrt{\frac{\theta_1}{\theta_2}} \\ \sqrt{\frac{1-\theta_2}{\theta_1}} & -\sqrt{\frac{\theta_1}{1-\theta_2}} \\ -\sqrt{\frac{\theta_2}{1-\theta_1}} & \sqrt{\frac{1-\theta_1}{\theta_2}} \\ -\sqrt{\frac{1-\theta_2}{1-\theta_1}} & -\sqrt{\frac{1-\theta_1}{1-\theta_2}} \end{pmatrix};$$

then we have

$$\boldsymbol{I}_F(\boldsymbol{\theta}) = \boldsymbol{A}(\boldsymbol{\theta})^T \boldsymbol{A}(\boldsymbol{\theta}) = \begin{pmatrix} (\theta_1(1-\theta_1))^{-1} & 0 \\ 0 & (\theta_2(1-\theta_2))^{-1} \end{pmatrix}.$$

Therefore,

$$\sqrt{n}\left((\widehat{\theta}^1_{\phi_2},\widehat{\theta}^2_{\phi_2})^T-(\theta_{10},\theta_{20})^T\right)\underset{n\to\infty}{\overset{L}{\longrightarrow}}N\left(\mathbf{0},\boldsymbol{I}_F\left(\boldsymbol{\theta}_0\right)^{-1}\right).$$

b) It is necessary to minimize in θ_1 and θ_2 the function

$$\begin{aligned} g\left(\theta_1,\theta_2\right) &= \widehat{p}_1\log\frac{\widehat{p}_1}{\theta_1\theta_2}+\widehat{p}_2\log\frac{\widehat{p}_2}{\theta_1\left(1-\theta_2\right)}+\widehat{p}_3\log\frac{\widehat{p}_3}{\theta_2\left(1-\theta_1\right)} \\ &+ \widehat{p}_4\log\frac{\widehat{p}_4}{\left(1-\theta_1\right)\left(1-\theta_2\right)} \\ &= -\widehat{p}_1\log\theta_1\theta_2-\widehat{p}_2\log\theta_1\left(1-\theta_2\right)-\widehat{p}_3\log\theta_2\left(1-\theta_1\right) \\ &- \widehat{p}_4\log\left(1-\theta_1\right)\left(1-\theta_2\right)+c. \end{aligned}$$

Differentiating with respect to θ_1 and θ_2 and equating to zero we have

$$\widehat{\theta}_1=\widehat{p}_1+\widehat{p}_2 \text{ and } \widehat{\theta}_2=\widehat{p}_1+\widehat{p}_3.$$

The asymptotic distribution has been obtained, in general, for any ϕ in *a)*.

c) If we denote $\widehat{\boldsymbol{p}}=(\widehat{p}_1,\widehat{p}_2,\widehat{p}_3,\widehat{p}_4)^T$ by $\widehat{\boldsymbol{p}}=(\widehat{p}_{11},\widehat{p}_{12},\widehat{p}_{21},\widehat{p}_{22})^T$, we have

$$\begin{aligned} \widehat{\theta}_1=\widehat{p}_1+\widehat{p}_2=\widehat{p}_{11}+\widehat{p}_{12}=\widehat{p}_{1*}=\frac{n_{11}+n_{12}}{n}\equiv\frac{n_{1*}}{n} \\ \widehat{\theta}_2=\widehat{p}_1+\widehat{p}_3=\widehat{p}_{11}+\widehat{p}_{21}=\widehat{p}_{*1}=\frac{n_{11}+n_{21}}{n}\equiv\frac{n_{*1}}{n}. \end{aligned}$$

In a similar way

$$\begin{aligned} 1-\widehat{\theta}_1=1-\widehat{p}_{1*}=\widehat{p}_{2*}=\frac{n_{21}+n_{22}}{n}\equiv\frac{n_{2*}}{n} \\ 1-\widehat{\theta}_2=1-\widehat{p}_{*1}=\widehat{p}_{*2}=\frac{n_{12}+n_{22}}{n}\equiv\frac{n_{*2}}{n}. \end{aligned}$$

Then we get

$$T_n^{\phi}\left(\widehat{\theta}_{\phi(0)}\right)=n\sum_{i=1}^{2}\sum_{j=1}^{2}p_{ij}(\widehat{\theta}_{\phi(0)})\left(\frac{\widehat{p}_{ij}}{p_{ij}(\widehat{\theta}_{\phi(0)})}-1\right)^2=\sum_{i=1}^{2}\sum_{j=1}^{2}\frac{(nn_{ij}-n_{i*}n_{*j})^2}{n_{i*}n_{*j}}.$$

9. *a)* The partition of the interval $[0,2\pi)$ is given by $E_j=\left\{[j-1\frac{2\pi}{M},j\frac{2\pi}{M})\right\}$, $j=1,...,M$. It is clear that

$$p_j\left(\theta\right)=\Pr_{\theta}(E_j)=F_{\theta}\left(j\frac{2\pi}{M}\right)-F_{\theta}\left((j-1)\frac{2\pi}{M}\right)=\frac{1}{M}+\theta\ c_j,$$

where $c_j=\frac{1}{2\pi}(\text{sen}\,(j-1)\frac{2\pi}{M}-\text{sen}\,j\frac{2\pi}{M})$.

Then,

$$\boldsymbol{p}(\theta) = (\frac{1}{M} + \theta\, c_1, ..., \frac{1}{M} + \theta\, c_M)^T$$

and

$$\boldsymbol{A}(\theta) = \begin{pmatrix} c_1 \left(\frac{1}{M} + \theta\, c_1\right)^{-1/2} \\ \cdot \\ \cdot \\ \cdot \\ c_M \left(\frac{1}{M} + \theta\, c_M\right)^{-1/2} \end{pmatrix}.$$

Then,

$$\boldsymbol{A}(\theta)^T \boldsymbol{A}(\theta) = \sum_{j=1}^{M} \frac{c_j^2}{\left(\frac{1}{M} + \theta\, c_j\right)} = M \sum_{j=1}^{M} \frac{c_j^2}{1 + M\theta\, c_j}$$

and

$$\boldsymbol{I}_F(\theta) = M \sum_{j=1}^{M} \frac{c_j^2}{1 + M\theta\, c_j}.$$

b) The power-divergence for $\lambda = -2$ is

$$\begin{aligned} D_{\phi_{(-2)}}(\widehat{\boldsymbol{p}}, \boldsymbol{p}(\theta)) &= \frac{1}{-2(-2+1)} \left(\sum_{j=1}^{M} \frac{p_j(\theta)^2}{\widehat{p}_j} - 1 \right) \\ &= \frac{1}{2} \left(\sum_{j=1}^{M} \frac{1}{\widehat{p}_j} \left(\tfrac{1}{M} + \theta c_j\right)^2 - 1 \right). \end{aligned}$$

Differentiating and equating to zero we have

$$\widehat{\theta}_{\phi_{(-2)}} = -\left(\frac{1}{M} \sum_{j=1}^{M} \frac{c_j}{\widehat{p}_j} \right) \left(\sum_{j=1}^{M} \frac{c_j^2}{\widehat{p}_j} \right)^{-1}.$$

c) The asymptotic distribution is given by

$$\sqrt{n}(\widehat{\theta}_{\phi_{(-2)}}(Y_1, ..., Y_n) - \theta_0) \underset{n\to\infty}{\overset{L}{\longrightarrow}} N\left(0, \left(M \sum_{j=1}^{M} \frac{c_j^2}{1 + M\theta_0 c_j} \right)^{-1} \right).$$

10. *a)* First we are going to obtain the asymptotic properties of the minimum ϕ-divergence estimator. We have

$$\cdot\ \boldsymbol{A}(\theta_0) = \frac{1}{(1+\theta_0)^{-1/2}} \begin{pmatrix} 1 & 0 \\ 0 & \theta_0^{-1/2} \end{pmatrix} \begin{pmatrix} -\dfrac{1}{(1+\theta_0)^2} \\ \dfrac{1}{(1+\theta_0)^2} \end{pmatrix}$$

$$= \frac{1}{(1+\theta_0)^{3/2}} \begin{pmatrix} -1 \\ \theta_0^{-1/2} \end{pmatrix}$$

$$\cdot\ \boldsymbol{A}(\theta_0)^T \boldsymbol{A}(\theta_0) = (1+\theta_0)^{-2}\, \theta_0^{-1}$$

$$\cdot\ \boldsymbol{\Delta}_0^T = ((1+\theta_0)\,\theta_0)^{1/2}\,(-\theta_0^{1/2}, 1)$$

$$\cdot\ \boldsymbol{C}_0^{-1} = \frac{1}{(1+\theta_0)} \begin{pmatrix} \theta_0^2+\theta_0+1 & -\theta_0^2 \\ -\theta_0 & 2\theta_0+1 \end{pmatrix}.$$

Therefore,

$$\boldsymbol{S}_0 = \tfrac{(1-\theta_0)\theta_0}{(1+\theta_0)^3} \begin{pmatrix} 1 & -1 \\ -1 & 1 \end{pmatrix} \text{ and } \boldsymbol{B}_0 = \tfrac{(1-\theta_0)}{(1+\theta_0)^2} \begin{pmatrix} \theta_0 & -\theta_0^{1/2} \\ -\theta_0^{1/2} & 1 \end{pmatrix}.$$

Then,

$$\sqrt{n}\left(\widehat{\theta}_{\phi_2}(Y_1,...,Y_n) - \theta_0\right) \underset{n\to\infty}{\overset{L}{\longrightarrow}} N\left(0, \boldsymbol{\Delta}_0^T \boldsymbol{B}_0 \boldsymbol{\Delta}_0\right),$$

and

$$\boldsymbol{\Delta}_0^T \boldsymbol{B}_0 \boldsymbol{\Delta}_0 = \theta_0(1-\theta_0^2).$$

The minimum ϕ-divergence estimator is obtained minimizing, in $\theta \in (0,1)$, the function

$$g(\theta) = D_{\phi_2}(\widehat{\boldsymbol{p}}, \boldsymbol{p}(\theta)) = \frac{1}{1+\theta}\left(\phi_2\left((1+\theta)\,\widehat{p}_1\right) + \theta\phi_2\left(\theta^{-1}\widehat{p}_2\,(1+\theta)\right)\right).$$

b) If we consider the power-divergence measure we obtain

$$\begin{aligned} g(\theta) &= \tfrac{1}{\lambda(\lambda+1)}\left(\widehat{p}_1^{\lambda+1}(1+\theta)^\lambda + (1+\theta)^\lambda\,\theta^{-\lambda}\widehat{p}_2^{\lambda+1} - 1\right) \\ &= \tfrac{1}{\lambda(\lambda+1)}\left((1+\theta)^\lambda\left(\widehat{p}_1^{\lambda+1} + \theta^{-\lambda}\widehat{p}_2^{\lambda+1}\right) - 1\right). \end{aligned}$$

Then,

$$\begin{aligned} g'(\theta) &= \tfrac{1}{\lambda(\lambda+1)}\left((1+\theta)^{\lambda-1}\lambda\left(\widehat{p}_1^{\lambda+1} + \theta^{-\lambda}\widehat{p}_2^{\lambda+1}\right) - (1+\theta)^\lambda\,\lambda\tfrac{1}{\theta^{\lambda+1}}\widehat{p}_2^{\lambda+1}\right) \\ &= \tfrac{(1+\theta)^{\lambda-1}\lambda}{\lambda(\lambda+1)}\left(\widehat{p}_1^{\lambda+1} + \frac{\widehat{p}_2^{\lambda+1}}{\theta^\lambda} - \frac{\widehat{p}_2^{\lambda+1}}{\theta^{\lambda+1}} - \frac{\widehat{p}_2^{\lambda+1}}{\theta^\lambda}\right) \\ &= \tfrac{(1+\theta)^{\lambda-1}}{(\lambda+1)}\left(\widehat{p}_1^{\lambda+1} - \frac{\widehat{p}_2^{\lambda+1}}{\theta^\lambda}\right), \end{aligned}$$

and therefore $g'(\theta) = 0$ implies $\theta = \widehat{p}_2/\widehat{p}_1$. Then we can conclude that

$$\widehat{\theta}_{\phi_{(\lambda)}} = \widehat{p}_2/\widehat{p}_1 \quad \text{if} \quad 0 < \widehat{p}_2/\widehat{p}_1 < 1.$$

7

Testing Loglinear Models Using Phi-divergence Test Statistics

7.1. Introduction

One basic and straightforward method for analyzing categorical data is via crosstabulation. For example, a medical researcher may tabulate the frequency of different symptoms by patient's age and gender; an educational researcher may tabulate the number of high school drop-outs by age, gender, and ethnic background; an economist may tabulate the number of business failures by industry, region, and initial capitalization; a market researcher may tabulate consumer preferences by product, age, and gender, etc. In all of these cases, the most interesting results can be summarized in a multiway frequency table, that is, in a crosstabulation table with two or more factors. Loglinear models provide a more "sophisticated" way of looking at crosstabulation tables. Specifically, one can test the different factors that are used in the crosstabulation (e.g., gender, region, etc.) and their interactions for statistical significance. In this introductory section we present an intuitive approach to loglinear models and in the remaining sections a systematic study of them.

Example 7.1

The data in Table 7.1 for 205 married persons, reported initially by Galton, give the number of cases in which a tall, medium or short man was married to a tall, medium or short woman.

		Wife			
		Tall	Medium	Short	Totals
	Tall	18	28	14	60
Husband	Medium	20	51	28	99
	Short	12	25	9	46
	Totals	50	104	51	205

Table 7.1

Source: Christensen, R. (1997, p. 67).

Are the heights of husband and wife independent ?

The answer to the previous question could be given using the classical chi-square test statistic for independence. However, we are going to deal with the problem considering a loglinear model for the data assuming independence. Let X and Y denote two categorical response variables, X and Y having I and J levels, respectively. The responses (X, Y) of a subject randomly chosen from some population have a probability distribution. Let $p_{ij} = \Pr(X = i, Y = j)$, with $p_{ij} > 0$, $i = 1, ..., I$, $j = 1, ..., J$. We display this distribution in a rectangular table having I rows for the categories of X and J columns for the categories of Y. The corresponding matrix $I \times J$ is called a contingency table. Consider a random sample of size n on (X, Y) and we denote by n_{ij} the observed frequency in the (i, j)th-cell for $(i, j) \in I \times J$ with $n = \sum_{i=1}^{I} \sum_{j=1}^{J} n_{ij}$ and the totals for the ith-row and jth-column by $n_{i*} = \sum_{j=1}^{J} n_{ij}$ and $\sum_{i=1}^{I} n_{ij} = n_{*j}$, $i = 1, ..., I, j = 1, ..., J$, respectively.

In the following we assume that n_{ij} is the observed value corresponding to a random variable N_{ij}, $i = 1, ..., I, j = 1, ..., J$, in such a way that the random variable $(N_{11}, ..., N_{IJ})$ is multinomially distributed with parameters n and $(p_{11}, ..., p_{IJ})$. We denote $m_{ij} = E[N_{ij}] = np_{ij}$, $i = 1, ..., I$, $j = 1, ..., J$. Under the independence assumption we have

$$H_0 : p_{ij} = p_{i*}p_{*j}, \ i = 1, ..., I, j = 1, ..., J \Leftrightarrow m_{ij} = np_{i*}p_{*j}, \tag{7.1}$$

where $p_{i*} = \sum_{j=1}^{J} p_{ij}$ and $p_{*j} = \sum_{i=1}^{I} p_{ij}$. The hypothesis (7.1) can be written as

$$H_0 \colon \log m_{ij} = \log n + \log p_{i*} + \log p_{*j}, \tag{7.2}$$

or equivalently

$$\begin{aligned}\log m_{ij} =& \log n + \log p_{i*} + \log p_{*j} \\ =& \log p_{i*} - \left(\sum_{h=1}^{I} \log p_{h*}\right)/I + \log p_{*j} - \left(\sum_{h=1}^{J} \log p_{*h}\right)/J \\ +& \log n + \left(\sum_{h=1}^{I} \log p_{h*}\right)/I + \left(\sum_{h=1}^{J} \log p_{*h}\right)/J.\end{aligned}$$

If we denote

$$\begin{aligned}\theta_{1(i)} =& \log p_{i*} - \left(\sum_{h=1}^{I} \log p_{h*}\right)/I \\ \theta_{2(j)} =& \log p_{*j} - \left(\sum_{h=1}^{J} \log p_{*h}\right)/J \\ u =& \log n + \left(\sum_{h=1}^{I} \log p_{h*}\right)/I + \left(\sum_{h=1}^{J} \log p_{*h}\right)/J,\end{aligned}$$

we have $\log m_{ij}(\boldsymbol{\theta}) = u+\theta_{1(i)}+\theta_{2(j)}, i = 1, ..., I, j = 1, ..., J$, where the parameters $\{\theta_{1(i)}\}$ and $\{\theta_{2(j)}\}$ verify $\sum_{i=1}^{I} \theta_{1(i)} = \sum_{j=1}^{J} \theta_{2(j)} = 0$.

Then the hypothesis of independence given in a two-way contingency table can be specified by the model

$$\log m_{ij}(\boldsymbol{\theta}) = u + \theta_{1(i)} + \theta_{2(j)}, \ i = 1, ..., I, \ j = 1, ..., J,$$

where the parameters verify $\sum_{i=1}^{I} \theta_{1(i)} = \sum_{j=1}^{J} \theta_{2(j)} = 0$.

The number of parameters of the model, initially, is

$$\left.\begin{array}{ll} u & 1 \\ \theta_{1(i)} & I-1 \\ \theta_{2(j)} & J-1 \end{array}\right\} I+J-1,$$

but we are assuming that our data are from a multinomial population with $n = \sum_{i=1}^{I}\sum_{j=1}^{J} m_{ij}(\boldsymbol{\theta})$; therefore actually we have $I+J-2$ parameters because u is a function of $\theta_{1(i)}$ and $\theta_{2(j)}$.

The loglinear model of independence which justifies the data from Example 7.1 is

$$\log m_{ij}(\boldsymbol{\theta}) = u + \theta_{1(i)} + \theta_{2(j)}, \ i = 1, ..., 3, \ j = 1, ..., 3,$$

with $\sum_{i=1}^{3} \theta_{1(i)} = \sum_{j=1}^{3} \theta_{2(j)} = 0$.

Since it may be useful to express the loglinear model in a matrix notation, we obtain the corresponding matrix form for the model of Example 7.1,

$$\begin{aligned}
\log m_{11}(\boldsymbol{\theta}) &= u+\theta_{1(1)}+\theta_{2(1)} \\
\log m_{12}(\boldsymbol{\theta}) &= u+\theta_{1(1)}+\theta_{2(2)} \\
\log m_{13}(\boldsymbol{\theta}) &= u+\theta_{1(1)}+\theta_{2(3)} = u+\theta_{1(1)}-\theta_{2(1)}-\theta_{2(2)} \\
\log m_{21}(\boldsymbol{\theta}) &= u+\theta_{1(2)}+\theta_{2(1)} \\
\log m_{22}(\boldsymbol{\theta}) &= u+\theta_{1(2)}+\theta_{2(2)} \\
\log m_{23}(\boldsymbol{\theta}) &= u+\theta_{1(2)}+\theta_{2(3)} = u+\theta_{1(2)}-\theta_{2(1)}-\theta_{2(2)} \\
\log m_{31}(\boldsymbol{\theta}) &= u+\theta_{1(3)}+\theta_{2(1)} = u-\theta_{1(1)}-\theta_{1(2)}+\theta_{2(1)} \\
\log m_{32}(\boldsymbol{\theta}) &= u+\theta_{1(3)}+\theta_{2(2)} = u-\theta_{1(1)}-\theta_{1(2)}+\theta_{2(2)} \\
\log m_{33}(\boldsymbol{\theta}) &= u+\theta_{1(3)}+\theta_{2(3)} = u-\theta_{1(1)}-\theta_{1(2)}-\theta_{2(1)}-\theta_{2(2)}.
\end{aligned}$$

If we denote by

$$\boldsymbol{X} = \begin{pmatrix} 1 & 1 & 0 & 1 & 0 \\ 1 & 1 & 0 & 0 & 1 \\ 1 & 1 & 0 & -1 & -1 \\ 1 & 0 & 1 & 1 & 0 \\ 1 & 0 & 1 & 0 & 1 \\ 1 & 0 & 1 & -1 & -1 \\ 1 & -1 & -1 & 1 & 0 \\ 1 & -1 & -1 & 0 & 1 \\ 1 & -1 & -1 & -1 & -1 \end{pmatrix},$$

$\boldsymbol{m}(\boldsymbol{\theta}) = (m_{11}(\boldsymbol{\theta}), ..., m_{13}(\boldsymbol{\theta}), ..., m_{31}(\boldsymbol{\theta}), ..., m_{33}(\boldsymbol{\theta}))^T$ and

$$\boldsymbol{\theta} = \left(u, \theta_{1(1)}, \theta_{1(2)}, \theta_{2(1)}, \theta_{2(2)}\right)^T,$$

we have $\log \boldsymbol{m}(\boldsymbol{\theta}) = \boldsymbol{X}\boldsymbol{\theta}$.

In a two-way contingency table the most general loglinear model is

$$\log m_{ij}(\boldsymbol{\theta}) = u+\theta_{1(i)}+\theta_{2(j)}+\theta_{12(ij)} \tag{7.3}$$

where

$$\begin{aligned}
u &= \tfrac{1}{IJ}\textstyle\sum_{i=1}^{I}\sum_{j=1}^{J}\log m_{ij}(\boldsymbol{\theta}) & \theta_{1(i)} &= \tfrac{1}{J}\textstyle\sum_{j=1}^{J}\log m_{ij}(\boldsymbol{\theta})-u \\
\theta_{2(j)} &= \tfrac{1}{I}\textstyle\sum_{i=1}^{I}\log m_{ij}(\boldsymbol{\theta})-u & \theta_{12(ij)} &= \log m_{ij}(\boldsymbol{\theta})-u-\theta_{1(i)}-\theta_{2(j)}.
\end{aligned}$$

The term $\theta_{12(ij)}$ represents the interaction between the two random variables X and Y. It is easy to verify $\sum_{i=1}^{I}\theta_{12(ij)} = \sum_{j=1}^{J}\theta_{12(ij)} = 0$. The number of parameters in this model is $IJ-1$. A loglinear model, with multinomial sampling, in which the number of cells minus one is equal to the number of parameters is called saturated model. A saturated model is one which attempts to estimate parameters for all single-variable and all interaction effects. That is, saturated models include all possible terms, including all interaction effects. They provide an exact fit for the observed cell counts. Since observed and expected are the same, there are too many unknown parameters to compute goodness-of-fit test statistics.

If we assume $I = J = 3$, denoting

$$\boldsymbol{X} = \begin{pmatrix} 1 & 1 & 0 & 1 & 0 & 1 & 0 & 0 & 0 \\ 1 & 1 & 0 & 0 & 1 & 0 & 1 & 0 & 0 \\ 1 & 1 & 0 & -1 & -1 & -1 & -1 & 0 & 0 \\ 1 & 0 & 1 & 1 & 0 & 0 & 0 & 1 & 0 \\ 1 & 0 & 1 & 0 & 1 & 0 & 0 & 0 & 1 \\ 1 & 0 & 1 & 1 & -1 & 0 & 0 & -1 & -1 \\ 1 & -1 & -1 & 1 & 0 & -1 & 0 & -1 & 0 \\ 1 & -1 & -1 & 0 & 1 & 0 & -1 & 0 & -1 \\ 1 & -1 & -1 & -1 & -1 & 1 & 1 & 1 & 1 \end{pmatrix},$$

$$\boldsymbol{m}(\boldsymbol{\theta}) = \left(m_{11}(\boldsymbol{\theta}), ..., m_{13}(\boldsymbol{\theta}), ..., m_{31}(\boldsymbol{\theta}), ..., m_{33}(\boldsymbol{\theta})\right)^T$$

and

$$\boldsymbol{\theta} = \left(u, \theta_{1(1)}, \theta_{1(2)}, \theta_{2(1)}, \theta_{2(2)}, \theta_{12(11)}, \theta_{12(12)}, \theta_{12(21)}, \theta_{12(22)}\right)^T$$

we have that the matrix form of the saturated model is $\log \boldsymbol{m}(\boldsymbol{\theta}) = \boldsymbol{X\theta}$. Before introducing the loglinear models in a three-way contingency table we present an example.

Example 7.2

Worchester (1971) describes a case-control study of women diagnosed with thromboembolism for the purpose of studying the risks associated with smoking and oral contraceptive use. Their data are summarized in Table 7.2.

Contraceptive	Cases		Controls	
use?	Smoker	nonsmoker	Smoker	nonsmoker
Yes	14	12	2	8
No	7	25	22	84

Table 7.2

Source: Worchester, J. (1971).

Loglinear models are useful methods to describe the inter-relationships between these three factors.

Let X, Y and Z denote three categorical response variables: X having I levels, Y having J levels and Z having K levels. When subjects are classified based on the three variables, there are IJK possible combinations of classification. The responses (X, Y, Z) of a subject randomly chosen from some population have a probability distribution. Let $p_{ijk} = \Pr(X = i, Y = j, Z = k)$, with $p_{ijk} > 0$, $i = 1, ..., I, j = 1, ..., J$ and $k = 1, ..., K$, and let $\boldsymbol{p} = (p_{111}, ..., p_{IJK})^T$ be the joint distribution of X, Y and Z. This distribution is displayed in a three-way contingency table. Consider a random sample of size n on (X, Y, Z) and let n_{ijk} be the observed frequency in the (i, j, k)th-cell for $(i, j, k) \in I \times J \times K$ with $\sum_{i=1}^{I} \sum_{j=1}^{J} \sum_{k=1}^{K} n_{ijk} = n$. In the following we assume that n_{ijk} is the observed value corresponding to a random variable N_{ijk}, $i = 1, ..., I, j = 1, ..., J$, $k = 1, ..., K$, in such a way that the random variable $(N_{111}, ..., N_{IJK})$ is multinomially distributed with parameters n and $(p_{111}, ..., p_{IJK})$. We denote $m_{ijk} = E[N_{ijk}] = np_{ijk}$, $i = 1, ..., I$, $j = 1, ..., J$, $k = 1, ..., K$. We also denote

$$
\begin{aligned}
u &= \sum_{i=1}^{I} \sum_{j=1}^{J} \sum_{k=1}^{K} \log m_{ijk}/IJK \\
\theta_{1(i)} &= \sum_{j=1}^{J} \sum_{k=1}^{K} \log m_{ijk}/JK - u \\
\theta_{2(j)} &= \sum_{i=1}^{I} \sum_{k=1}^{K} \log m_{ijk}/IK - u \\
\theta_{3(k)} &= \sum_{i=1}^{I} \sum_{j=1}^{J} \log m_{ijk}/IJ - u \\
\theta_{12(ij)} &= \sum_{k=1}^{K} \log m_{ijk}/K - \sum_{j=1}^{J} \sum_{k=1}^{K} \log m_{ijk}/JK - \sum_{i=1}^{I} \sum_{k=1}^{K} \log m_{ijk}/IK + u \\
\theta_{13(ik)} &= \sum_{j=1}^{J} \log m_{ijk}/J - \sum_{j=1}^{J} \sum_{k=1}^{K} \log m_{ijk}/JK - \sum_{i=1}^{I} \sum_{j=1}^{J} \log m_{ijk}/IJ + u \\
\theta_{23(jk)} &= \sum_{i=1}^{I} \log m_{ijk}/I - \sum_{i=1}^{I} \sum_{k=1}^{K} \log m_{ijk}/IK - \sum_{i=1}^{I} \sum_{j=1}^{J} \log m_{ijk}/IJ + u \\
\theta_{123(ijk)} &= \log m_{ijk} - \sum_{k=1}^{K} \log m_{ijk}/K - \sum_{j=1}^{J} \log m_{ijk}/J - \sum_{i=1}^{I} \log m_{ijk}/I \\
&\quad + \sum_{j=1}^{J} \sum_{k=1}^{K} \log m_{ijk}/JK + \sum_{i=1}^{I} \sum_{k=1}^{K} \log m_{ijk}/IK + \sum_{i=1}^{I} \sum_{j=1}^{J} \log m_{ijk}/IJ + u.
\end{aligned}
$$

Then we have

$$\log m_{ijk}(\boldsymbol{\theta}) = u + \theta_{1(i)} + \theta_{2(j)} + \theta_{3(k)} + \theta_{12(ij)} + \theta_{13(ik)} + \theta_{23(jk)} + \theta_{123(ijk)}. \quad (7.4)$$

It is clear that the parameters verify the following constraints,

$$\sum_{i=1}^{I} \theta_{1(i)} = \sum_{j=1}^{J} \theta_{2(j)} = \sum_{k=1}^{K} \theta_{3(k)} = 0$$
$$\sum_{i=1}^{I} \theta_{12(ij)} = \sum_{j=1}^{J} \theta_{12(ij)} = \sum_{i=1}^{I} \theta_{13(ik)} = \sum_{k=1}^{K} \theta_{13(ik)} = \sum_{j=1}^{J} \theta_{23(jk)} = \sum_{k=1}^{K} \theta_{23(jk)} = 0$$
$$\sum_{i=1}^{I} \theta_{123(ijk)} = \sum_{j=1}^{J} \theta_{123(ijk)} = \sum_{k=1}^{K} \theta_{123(ijk)} = 0.$$

The number of parameters, initially, in this model is

$$\left.\begin{array}{ll} u & 1 \\ \theta_{1(i)} & I-1 \\ \theta_{2(j)} & J-1 \\ \theta_{3(k)} & K-1 \\ \theta_{12(ij)} & (I-1)(J-1) \\ \theta_{13(ik)} & (I-1)(K-1) \\ \theta_{23(jk)} & (J-1)(K-1) \\ \theta_{123(ijk)} & (I-1)(J-1)(K-1) \end{array}\right\} IJK,$$

but under the assumption of multinomial sampling we have $IJK-1$ parameters because $\sum_{i=1}^{I}\sum_{j=1}^{J}\sum_{k=1}^{K} m_{ijk}(\boldsymbol{\theta}) = n$. We have the saturated model, in a three-way contingency table, because the number of parameters coincide with the number of cells minus one.

A simpler model in a three-way contingency table is the model

$$\log m_{ijk}(\boldsymbol{\theta}) = u + \theta_{1(i)} + \theta_{2(j)} + \theta_{3(k)}. \quad (7.5)$$

This is the independence model in a three-way contingency table, i.e., $p_{ijk} = p_{i**}p_{*j*}p_{**k}$ with $p_{i**} = \sum_{j=1}^{J}\sum_{k=1}^{K} p_{ijk}$, $p_{*j*} = \sum_{i=1}^{I}\sum_{k=1}^{K} p_{ijk}$ and $p_{**k} = \sum_{i=1}^{I}\sum_{j=1}^{J} p_{ijk}$.

Between the models (7.4) and (7.5) there are different loglinear models that can be considered:

i) Variable X is jointly independent of Y and Z. In this case we have

$$\log m_{ijk}(\boldsymbol{\theta}) = u + \theta_{1(i)} + \theta_{2(j)} + \theta_{3(k)} + \theta_{23(jk)},$$

i.e., $p_{ijk} = p_{*jk}p_{i**}$. The number of parameters in this model is

$$(I-1)+(J-1)+(K-1)+(K-1)(J-1).$$

ii) The random variables X and Z are independent given the random variable Y. In this case we have

$$\log m_{ijk}(\boldsymbol{\theta}) = u + \theta_{1(i)} + \theta_{2(j)} + \theta_{3(k)} + \theta_{12(ij)} + \theta_{23(jk)},$$

i.e., $p_{ijk}/p_{*j*} = (p_{ij*}/p_{*j*})(p_{*jk}/p_{*j*})$. The number of parameters of this model is

$$(I-1)+(J-1)+(K-1)+(I-1)(J-1)+(J-1)(K-1).$$

iii) The random variables X, Y and Z are pairwise dependent but the three random variables X, Y and Z are jointly independent. In this case we have

$$\log m_{ijk}(\boldsymbol{\theta}) = u + \theta_{1(i)} + \theta_{2(j)} + \theta_{3(k)} + \theta_{12(ij)} + \theta_{13(ik)} + \theta_{23(jk)}.$$

This model does not admit a representation in terms of probabilities. The number of parameters is given by

$$(I-1)+(J-1)+(K-1)+(I-1)[(J-1)+(K-1)]+(J-1)(K-1).$$

Hierarchical loglinear models require that high order interactions are always accompanied by all of their lower order interactions. All hierarchical models for a three-way contingency table are the following:

H_1: $\log m_{ijk}(\boldsymbol{\theta}) = u + \theta_{1(i)} + \theta_{2(j)} + \theta_{3(k)} + \theta_{12(ij)} + \theta_{13(ik)} + \theta_{23(jk)}$

H_2 : $\log m_{ijk}(\boldsymbol{\theta}) = u + \theta_{1(i)} + \theta_{2(j)} + \theta_{3(k)} + \theta_{13(ik)} + \theta_{23(jk)}$

H_2^* : $\log m_{ijk}(\boldsymbol{\theta}) = u + \theta_{1(i)} + \theta_{2(j)} + \theta_{3(k)} + \theta_{12(ij)} + \theta_{13(ik)}$

H_2^{**} : $\log m_{ijk}(\boldsymbol{\theta}) = u + \theta_{1(i)} + \theta_{2(j)} + \theta_{3(k)} + \theta_{12(ij)} + \theta_{23(jk)}$

H_3 : $\log m_{ijk}(\boldsymbol{\theta}) = u + \theta_{1(i)} + \theta_{2(j)} + \theta_{3(k)} + \theta_{23(jk)}$

H_3^*: $\log m_{ijk}(\boldsymbol{\theta}) = u + \theta_{1(i)} + \theta_{2(j)} + \theta_{3(k)} + \theta_{12(ij)}$

H_3^{**}: $\log m_{ijk}(\boldsymbol{\theta}) = u + \theta_{1(i)} + \theta_{2(j)} + \theta_{3(k)} + \theta_{13(ik)}$

H_4 : $\log m_{ijk}(\boldsymbol{\theta}) = u + \theta_{1(i)} + \theta_{2(j)} + \theta_{3(k)}$

H_4^* : $\log m_{ijk}(\boldsymbol{\theta}) = u + \theta_{2(j)} + \theta_{3(k)} + \theta_{23(jk)}$

H_4^{**}: $\log m_{ijk}(\boldsymbol{\theta}) = u + \theta_{1(i)} + \theta_{2(j)} + \theta_{12(ij)}$

H_4^{***}: $\log m_{ijk}(\boldsymbol{\theta}) = u + \theta_{1(i)} + \theta_{3(k)} + \theta_{13(ik)}$

H_5 : $\log m_{ijk}(\boldsymbol{\theta}) = u + \theta_{2(j)} + \theta_{3(k)}$

H_5^*: $\log m_{ijk}(\boldsymbol{\theta}) = u + \theta_{1(i)} + \theta_{2(j)}$

H_5^{**}: $\log m_{ijk}(\boldsymbol{\theta}) = u + \theta_{1(i)} + \theta_{3(k)}$

H_6 : $\log m_{ijk}(\boldsymbol{\theta}) = u + \theta_{3(k)}$

H_6^*: $\log m_{ijk}(\boldsymbol{\theta}) = u + \theta_{1(i)}$

H_6^{**}: $\log m_{ijk}(\boldsymbol{\theta}) = u + \theta_{2(j)}$

H_7 : $\log m_{ijk}(\boldsymbol{\theta}) = u.$

It is clear that, from a practical point of view, the problem consists of obtaining the model that presents a better fit to our data. In relation with Example 7.2 the problem is to find the model, among the previous considered models, that is able to explain more clearly the given data. In the procedure that follows in the next Sections it will be necessary to give a method to estimate the parameters of the model and then to choose the best model. In order to do this second step we first choose a nested sequence of loglinear models (two loglinear models are said to be nested when one contains a subset of the parameters in the other) in the way which is explained in Section 7.4. and then we shall give some procedure to choose the best model among the models considered in a nested sequence.

7.2. Loglinear Models: Definition

Let $Y_1, Y_2, ..., Y_n$ be a sample of size $n \geq 1$, with independent realizations in the statistical space $\mathcal{X} = \{1, 2, ..., M\}$, which are identically distributed according to a probability distribution $\boldsymbol{p}(\boldsymbol{\theta}_0)$. For a two-way contingency table we have $M = IJ$ and for a three-way contingency table we have $M = IJK$. This distribution is assumed to be unknown, but belonging to a known family

$$T = \left\{ \boldsymbol{p}(\boldsymbol{\theta}) = (p_1(\boldsymbol{\theta}), ..., p_M(\boldsymbol{\theta}))^T : \boldsymbol{\theta} \in \Theta \right\},$$

of distributions on $\mathcal{X}$ with $\Theta \subseteq \mathbb{R}^{M_0}$ $(M_0 < M - 1)$. In other words, the true value $\boldsymbol{\theta}_0$ of the parameter $\boldsymbol{\theta} = (\theta_1, ..., \theta_{M_0})^T \in \Theta \subseteq \mathbb{R}^{M_0}$ is assumed to be fixed but unknown. We denote $\boldsymbol{p} = (p_1, ..., p_M)^T$ and $\widehat{\boldsymbol{p}} = (\widehat{p}_1, ..., \widehat{p}_M)^T$ with

$$\widehat{p}_j = \frac{N_j}{n} \quad \text{and} \quad N_j = \sum_{i=1}^{n} I_{\{j\}}(Y_i);\ j = 1, ..., M. \tag{7.6}$$

The statistic $(N_1, ..., N_M)$ is obviously sufficient for the statistical model under consideration and is multinomially distributed; that is,

$$\Pr(N_1 = n_1, ..., N_M = n_M) = \frac{n!}{n_1! ... n_M!} p_1(\boldsymbol{\theta})^{n_1} ... p_M(\boldsymbol{\theta})^{n_M}, \tag{7.7}$$

for integers $n_1, ..., n_M \geq 0$ such that $n_1 + ... + n_M = n$.

In what follows, we assume that $\boldsymbol{p}(\boldsymbol{\theta})$ belongs to the general class of loglinear models. That is, we assume:

$$p_u(\boldsymbol{\theta}) = \exp\left(\boldsymbol{w}_u^T \boldsymbol{\theta}\right) / \sum_{v=1}^{M} \exp\left(\boldsymbol{w}_v^T \boldsymbol{\theta}\right);\ u = 1, ..., M, \tag{7.8}$$

where $\boldsymbol{w}_u^T = (w_{u1}, ..., w_{uM_0})$. The $M \times M_0$ matrix $\boldsymbol{W} = (\boldsymbol{w}_1, ..., \boldsymbol{w}_M)^T$ is assumed to have full column rank M_0 $(M_0 < M - 1)$ and columns linearly independent of the $M \times 1$ column vector $(1, ..., 1)^T$. This is the model that we shall consider for the theoretical results in the next sections.

We restrict ourselves to multinomial random sampling but it is possible to give the results presented in the next sections under the assumptions of either Poisson, multinomial, or product-multinomial sampling jointly. For more details see Cressie and Pardo (2002b).

7.3. Asymptotic Results for Minimum Phi-divergence Estimators in Loglinear Models

In this Section we present some asymptotic results for the minimum ϕ-divergence estimator under the loglinear model (7.8). These results are obtained particularizing the results obtained in Chapter 5 for a general multinomial model. In the cited Chapter 5 we establish that the Fisher information matrix in a multinomial model has the expression

$$\boldsymbol{I}_F(\boldsymbol{\theta}) = \boldsymbol{A}(\boldsymbol{\theta})^T \boldsymbol{A}(\boldsymbol{\theta}),$$

where $\boldsymbol{A}(\boldsymbol{\theta})$ is the $M \times M_0$ matrix defined in (5.7).

For a loglinear model we have

$$\frac{\partial p_j(\boldsymbol{\theta})}{\partial \theta_r} = p_j(\boldsymbol{\theta}) w_{jr} - p_j(\boldsymbol{\theta}) \sum_{v=1}^{M} w_{vr} p_v(\boldsymbol{\theta}).$$

Then

$$\frac{\partial \boldsymbol{p}(\boldsymbol{\theta})}{\partial \boldsymbol{\theta}} = \left(diag(\boldsymbol{p}(\boldsymbol{\theta})) - \boldsymbol{p}(\boldsymbol{\theta}) \boldsymbol{p}(\boldsymbol{\theta})^T \right) \boldsymbol{W} = \boldsymbol{\Sigma}_{\boldsymbol{p}(\boldsymbol{\theta})} \boldsymbol{W}$$

and hence

$$\boldsymbol{A}(\boldsymbol{\theta}) = diag\left(\boldsymbol{p}(\boldsymbol{\theta})^{-1/2} \right) \boldsymbol{\Sigma}_{\boldsymbol{p}(\boldsymbol{\theta})} \boldsymbol{W}.$$

Then the Fisher information matrix for a loglinear model is given by

$$\boldsymbol{I}_F(\boldsymbol{\theta}) = \boldsymbol{W}^T \boldsymbol{\Sigma}_{\boldsymbol{p}(\boldsymbol{\theta})} \boldsymbol{W}.$$

By Theorem 5.2, if $\widehat{\boldsymbol{\theta}}_\phi$ is the minimum ϕ-divergence estimator for the loglinear model given in (7.8), then

$$\sqrt{n}(\widehat{\boldsymbol{\theta}}_\phi - \boldsymbol{\theta}_0) \underset{n\to\infty}{\overset{L}{\longrightarrow}} N\left(\boldsymbol{0}, \left(\boldsymbol{W}^T \boldsymbol{\Sigma}_{\boldsymbol{p}(\boldsymbol{\theta}_0)} \boldsymbol{W} \right)^{-1} \right),$$

where

$$\widehat{\boldsymbol{\theta}}_\phi = \arg \inf_{\boldsymbol{\theta} \in \Theta} D_\phi(\widehat{\boldsymbol{p}}, \boldsymbol{p}(\boldsymbol{\theta})), \tag{7.9}$$

and by Theorem 5.1 verifies

$$\widehat{\boldsymbol{\theta}}_\phi = \boldsymbol{\theta}_0 + \boldsymbol{I}_F(\boldsymbol{\theta}_0)^{-1} \boldsymbol{W}^T \boldsymbol{\Sigma}_{\boldsymbol{p}(\boldsymbol{\theta}_0)} diag\left(\boldsymbol{p}(\boldsymbol{\theta}_0)^{-1}\right) (\widehat{\boldsymbol{p}} - \boldsymbol{p}(\boldsymbol{\theta}_0)) + o(\| \widehat{\boldsymbol{p}} - \boldsymbol{p}(\boldsymbol{\theta}_0) \|). \tag{7.10}$$

Another interesting result, useful later, is the following

$$\sqrt{n}\left(\boldsymbol{p}(\widehat{\boldsymbol{\theta}}_{\phi})-\boldsymbol{p}(\boldsymbol{\theta}_0)\right) \underset{n\to\infty}{\overset{L}{\longrightarrow}} N\left(\mathbf{0},\, \boldsymbol{\Sigma}_{\boldsymbol{p}(\boldsymbol{\theta}_0)}\boldsymbol{W}\left(\boldsymbol{W}^T\boldsymbol{\Sigma}_{\boldsymbol{p}(\boldsymbol{\theta}_0)}\boldsymbol{W}\right)^{-1}\boldsymbol{W}^T\boldsymbol{\Sigma}_{\boldsymbol{p}(\boldsymbol{\theta}_0)}\right). \tag{7.11}$$

From a practical point of view in order to find the minimum ϕ-divergence estimator $\widehat{\boldsymbol{\theta}}_{\phi}$ we must solve the following system of equations

$$\begin{cases} \dfrac{\partial D_{\phi}(\widehat{\mathbf{p}},\boldsymbol{p}(\boldsymbol{\theta}))}{\partial\theta_i}=0 \\ i=1,...,M_0 \end{cases},$$

with the condition that $\boldsymbol{p}(\boldsymbol{\theta})$ verifies (7.8).

These equations are nonlinear functions of the minimum ϕ-divergence estimator, $\widehat{\boldsymbol{\theta}}_{\phi}$. In order to solve these equations numerically the Newton-Raphson method is used. We have,

$$\begin{aligned}\frac{\partial D_{\phi}(\widehat{\mathbf{p}},\boldsymbol{p}(\boldsymbol{\theta}))}{\partial\theta_j} &= \sum_{l=1}^{M}\left(\phi\left(\frac{\widehat{p}_l}{p_l(\boldsymbol{\theta})}\right)-\phi'\left(\frac{\widehat{p}_l}{p_l(\boldsymbol{\theta})}\right)\frac{\widehat{p}_l}{p_l(\boldsymbol{\theta})}\right)\\ &\times\left(p_l(\boldsymbol{\theta})w_{lj}-p_l(\boldsymbol{\theta})\sum_{u=1}^{M}w_{uj}p_u(\boldsymbol{\theta})\right),\end{aligned}$$

and

$$\begin{aligned}\frac{\partial}{\partial\theta_r}\left(\frac{\partial D_{\phi}(\widehat{\mathbf{p}},\boldsymbol{p}(\boldsymbol{\theta}))}{\partial\theta_j}\right) &= \sum_{l=1}^{M}\phi''\left(\frac{\widehat{p}_l}{p_l(\boldsymbol{\theta})}\right)\frac{\widehat{p}_l}{p_l(\boldsymbol{\theta})^2}\frac{\partial p_l(\boldsymbol{\theta})}{\partial\theta_r}\frac{\partial p_l(\boldsymbol{\theta})}{\partial\theta_j}\frac{\widehat{p}_l}{p_l(\boldsymbol{\theta}^{(t)})}\\ &+\sum_{l=1}^{k}\frac{\partial^2 p_l(\boldsymbol{\theta})}{\partial\theta_j\partial\theta_r}\left(\phi\left(\frac{\widehat{p}_l}{p_l(\boldsymbol{\theta})}\right)-\phi'\left(\frac{\widehat{p}_l}{p_l(\boldsymbol{\theta})}\right)\frac{\widehat{p}_l}{p_l(\boldsymbol{\theta})}\right).\end{aligned} \tag{7.12}$$

Therefore the $(t+1)$th-step estimate, $\widehat{\boldsymbol{\theta}}^{(t+1)}$, in a Newton-Raphson procedure is obtained from $\widehat{\boldsymbol{\theta}}^{(t)}$ as

$$\widehat{\boldsymbol{\theta}}^{(t+1)}=\widehat{\boldsymbol{\theta}}^{(t)}-\boldsymbol{G}(\widehat{\boldsymbol{\theta}}^{(t)})^{-1}\left(\frac{\partial D_{\phi}(\widehat{\mathbf{p}},\boldsymbol{p}(\widehat{\boldsymbol{\theta}}^{(t)}))}{\partial\theta_j}\right)^T_{j=1,...,M_0},$$

where $\boldsymbol{G}(\boldsymbol{\theta}^{(t)})$ is the matrix whose elements are given in (7.12).

An interesting simulation study to analyze the behavior of the minimum power-divergence estimator, defined by

$$\widehat{\boldsymbol{\theta}}_{(\lambda)}\equiv\arg\min_{\boldsymbol{\theta}\in\Theta}D_{\phi_{(\lambda)}}(\widehat{\mathbf{p}},\boldsymbol{p}(\boldsymbol{\theta})), \tag{7.13}$$

in a three-way contingency table, has been considered in Pardo, L. and Pardo, M. C. (2003). Notice that $\widehat{\boldsymbol{\theta}}_{(0)}$ is the maximum likelihood estimator.

7.4. Testing in Loglinear Models

We denote by $\boldsymbol{p}(\widehat{\boldsymbol{\theta}}_{\phi_2})$ the parametric estimator, based on the minimum $\widehat{\boldsymbol{\theta}}_{\phi_2}$-divergence estimator, of the loglinear model defined in (7.8). For testing if our data are from a loglinear model we can use the family of ϕ-divergence test statistics

$$T_n^{\phi_1}\left(\widehat{\boldsymbol{\theta}}_{\phi_2}\right) = \frac{2n}{\phi_1''(1)} D_{\phi_1}\left(\widehat{\boldsymbol{p}}, \boldsymbol{p}(\widehat{\boldsymbol{\theta}}_{\phi_2})\right) \tag{7.14}$$

which by Theorem 6.1 are asymptotically distributed chi-squared with $M-M_0-1$ degrees of freedom. In Example 7.1 we have $M = 9$ and $M_0 = 4$, therefore $M - M_0 - 1 = 4 = (I-1)(J-1)$. In the previous expression $\widehat{\boldsymbol{\theta}}_{\phi_2}$ is the minimum ϕ_2-divergence estimator for the parameters of the considered loglinear model. In the rest of the chapter we shall assume the conditions given for the function ϕ_2 in Theorem 5.1 as well as that $\phi_i(x)$ is twice continuously differentiable in a neighborhood of 1 with the second derivative $\phi_i''(1) \neq 0$, $i = 1, 2$.

Based on this result it is possible to select a nested sequence of loglinear models. In order to fix ideas we consider all the possible loglinear models associated with a $I \times J \times K$ contingency table where the first element of the sequence is given by

$$H_1 : \log m_{ijk}(\boldsymbol{\theta}) = u + \theta_{1(i)} + \theta_{2(j)} + \theta_{3(k)} + \theta_{12(ij)} + \theta_{13(ik)} + \theta_{23(jk)}.$$

The second element should be chosen between the models

$$H_2 : \log m_{ijk}(\boldsymbol{\theta}) = u + \theta_{1(i)} + \theta_{2(j)} + \theta_{3(k)} + \theta_{13(ik)} + \theta_{23(jk)}$$

$$H_2^* : \log m_{ijk}(\boldsymbol{\theta}) = u + \theta_{1(i)} + \theta_{2(j)} + \theta_{3(k)} + \theta_{12(ij)} + \theta_{13(ik)}$$

$$H_2^{**} : \log m_{ijk}(\boldsymbol{\theta}) = u + \theta_{1(i)} + \theta_{2(j)} + \theta_{3(k)} + \theta_{12(ij)} + \theta_{23(jk)}.$$

We consider the three tests of hypotheses

$$H_{Null} : H_2 \text{ versus } H_{Alt} : \text{The saturated model}$$
$$H_{Null} : H_2^* \text{ versus } H_{Alt} : \text{The saturated model}$$
$$H_{Null} : H_2^{**} \text{ versus } H_{Alt} : \text{The saturated model}$$

and we shall choose the model that better support the data. To do the previous tests we consider the ϕ-divergence test statistic $T_n^{\phi_1}\left(\widehat{\boldsymbol{\theta}}_{\phi_2}^{(l)}\right)$, given in (7.14), where by $\widehat{\boldsymbol{\theta}}_{\phi_2}^{(l)}$ we indicate the minimum ϕ_2-divergence estimator for the parameters of the loglinear model l, being $l = H_2$, H_2^* or H_2^{**}.

If we assume that we have chosen the model H_2^*, the third element in the sequence of nested sequence of loglinear models should be chosen between the models

H_3^*: $\log m_{ijk}(\boldsymbol{\theta}) = u + \theta_{1(i)} + \theta_{2(j)} + \theta_{3(k)} + \theta_{12(ij)}$

H_3^{**}: $\log m_{ijk}(\boldsymbol{\theta}) = u + \theta_{1(i)} + \theta_{2(j)} + \theta_{3(k)} + \theta_{13(ik)}$.

We consider the two tests of hypotheses

$$\begin{array}{l} H_{Null} : H_3^* \text{ versus } H_{Alt} : \text{The saturated model} \\ H_{Null} : H_3^{**} \text{ versus } H_{Alt} : \text{The saturated model} \end{array}$$

and we choose the model that better support the data. To do the previous tests we consider the ϕ-divergence test statistic $T_n^{\phi_1}\left(\widehat{\boldsymbol{\theta}}_{\phi_2}^{(l)}\right)$, given in (7.14), where by $\widehat{\boldsymbol{\theta}}_{\phi_2}^{(l)}$ we indicate the minimum ϕ_2-divergence estimator for the parameters of the loglinear model l, being $l = H_3^*$ or H_3^{**}. We can continue in the same way until getting a convenient nested sequence of loglinear models.

One of the main problems in loglinear models is to test a nested sequence of hypotheses,

$$H_l : \mathbf{p} = \mathbf{p}(\boldsymbol{\theta})\,;\ \boldsymbol{\theta} \in \Theta_l;\ l = 1, ..., m,\ m \le M_0 < M - 1, \tag{7.15}$$

where $\Theta_m \subset \Theta_{m-1} \subset ... \subset \Theta_1 \subset \mathbb{R}^{M_0}$; $M_0 < M-1$ and $\dim(\Theta_l) = d_l$; $l = 1, ..., m$, with

$$d_m < d_{m-1} < ... < d_1 \le M_0. \tag{7.16}$$

Our strategy will be to test successively the hypotheses

$$H_{l+1} \text{ against } H_l; \quad l = 1, ..., m-1, \tag{7.17}$$

as null and alternative hypotheses respectively. We continue to test as long as the null hypothesis is accepted and choose the loglinear model H_l according to the first l for which H_{l+1} is rejected (as a null hypothesis) in favor of H_l (as an alternative hypothesis). This strategy is quite standard for nested models (Read and Cressie, 1988, p. 42). The nesting occurs naturally because of the hierarchical principle, which says that interactions should not be fitted unless the corresponding main effects are present (e.g., Collett, 1994, p. 78).

The results presented in this Section were obtained in Cressie, N. and Pardo, L. (2000) and Cressie, N. et *al.* (2003).

Theorem 7.1

Suppose that data $(N_1, ..., N_M)$ are multinomially distributed according to (7.7) and (7.8). Consider the nested sequence of hypotheses given by (7.15) and (7.16). Choose two functions $\phi_1, \phi_2 \in \Phi^*$. Then, for testing hypotheses,

$$H_{Null} : H_{l+1} \text{ against } H_{Alt} : H_l,$$

the asymptotic null distribution of the ϕ-divergence test statistic,

$$\widetilde{T}^{(l)}_{\phi_1,\phi_2} \equiv \frac{2n}{\phi_1''(1)} D_{\phi_1}(\boldsymbol{p}(\widehat{\boldsymbol{\theta}}^{(l+1)}_{\phi_2}), \boldsymbol{p}(\widehat{\boldsymbol{\theta}}^{(l)}_{\phi_2})), \tag{7.18}$$

is chi-square with $d_l - d_{l+1}$ degrees of freedom; $l = 1, ..., m-1$. In (7.18), $\widehat{\theta}^{(l)}_{\phi_2}$ and $\widehat{\theta}^{(l+1)}_{\phi_2}$ are the minimum ϕ_2-divergence estimators under the models H_l and H_{l+1}, respectively, where the minimum ϕ-divergence estimators are defined by (7.9).

Proof. The second-order expansion of $D_{\phi_1,\phi_2} \equiv D_{\phi_1}\left(\boldsymbol{p}(\widehat{\boldsymbol{\theta}}^{(l+1)}_{\phi_2}), \boldsymbol{p}(\widehat{\boldsymbol{\theta}}^{(l)}_{\phi_2})\right)$ about $(\boldsymbol{p}(\boldsymbol{\theta}_0), \boldsymbol{p}(\boldsymbol{\theta}_0))$ gives

$$\begin{aligned}
D_{\phi_1,\phi_2} = {} & \frac{1}{2}\sum_{j=1}^{M}\left(\frac{\partial^2 D_{\phi_1}(\boldsymbol{p},\boldsymbol{q})}{\partial p_j^2}\right)_{(\boldsymbol{p}(\boldsymbol{\theta}_0),\boldsymbol{p}(\boldsymbol{\theta}_0))}\left(p_j(\widehat{\boldsymbol{\theta}}^{(l+1)}_{\phi_2}) - p_j(\boldsymbol{\theta}_0)\right)^2 \\
& + \frac{1}{2}\sum_{j=1}^{M}\left(\frac{\partial^2 D_{\phi_1}(\boldsymbol{p},\boldsymbol{q})}{\partial q_j^2}\right)_{(\boldsymbol{p}(\boldsymbol{\theta}_0),\boldsymbol{p}(\boldsymbol{\theta}_0))}\left(p_j(\widehat{\boldsymbol{\theta}}^{(l)}_{\phi_2}) - p_j(\boldsymbol{\theta}_0)\right)^2 \\
& + \sum_{i=1}^{M}\sum_{j=1}^{M}\left(\frac{\partial^2 D_{\phi_1}(\boldsymbol{p},\boldsymbol{q})}{\partial p_i \partial q_j}\right)_{(\boldsymbol{p}(\boldsymbol{\theta}_0),\boldsymbol{p}(\boldsymbol{\theta}_0))}\left(p_i(\widehat{\boldsymbol{\theta}}^{(l+1)}_{\phi_2}) - p_i(\boldsymbol{\theta}_0)\right) \\
& \times \left(p_j(\widehat{\boldsymbol{\theta}}^{(l)}_{\phi_2}) - p_j(\boldsymbol{\theta}_0)\right) \\
& + o\left(\left\|\boldsymbol{p}(\widehat{\boldsymbol{\theta}}^{(l+1)}_{\phi_2}) - \boldsymbol{p}(\boldsymbol{\theta}_0)\right\|^2 + \left\|\boldsymbol{p}(\widehat{\boldsymbol{\theta}}^{(l)}_{\phi_2}) - \boldsymbol{p}(\boldsymbol{\theta}_0)\right\|^2\right).
\end{aligned}$$

We have used in the previous Taylor's expansion that $D_{\phi_1}(\boldsymbol{p}(\boldsymbol{\theta}_0), \boldsymbol{p}(\boldsymbol{\theta}_0)) = 0$ and the first order term in Taylor expansion is also zero. Thus, $\widetilde{T}^{(l)}_{\phi_1,\phi_2}$ can be written as

$$
\begin{aligned}
\widetilde{T}^{(l)}_{\phi_1,\phi_2} &= n\sum_{j=1}^{M}\frac{1}{p_j(\boldsymbol{\theta}_0)}\left(p_j(\widehat{\boldsymbol{\theta}}^{(l+1)}_{\phi_2}) - p_j(\boldsymbol{\theta}_0)\right)^2 \\
&+ n\sum_{j=1}^{M}\frac{1}{p_j(\boldsymbol{\theta}_0)}\left(p_j(\widehat{\boldsymbol{\theta}}^{(l)}_{\phi_2}) - p_j(\boldsymbol{\theta}_0)\right)^2 \\
&- 2n\sum_{j=1}^{M}\frac{1}{p_j(\boldsymbol{\theta}_0)}\left(p_j(\widehat{\boldsymbol{\theta}}^{(l+1)}_{\phi_2}) - p_j(\boldsymbol{\theta}_0)\right)\left(p_j(\widehat{\boldsymbol{\theta}}^{(l)}_{\phi_2}) - p_j(\boldsymbol{\theta}_0)\right) \\
&+ n\, o\left(\left\|\boldsymbol{p}(\widehat{\boldsymbol{\theta}}^{(l+1)}_{\phi_2}) - \boldsymbol{p}(\boldsymbol{\theta}_0)\right\|^2 + \left\|\boldsymbol{p}(\widehat{\boldsymbol{\theta}}^{(l)}_{\phi_2}) - \boldsymbol{p}(\boldsymbol{\theta}_0)\right\|^2\right) \\
&= \left(n^{1/2} diag\left(\boldsymbol{p}(\boldsymbol{\theta}_0)^{-1/2}\right)\left(\boldsymbol{p}(\widehat{\boldsymbol{\theta}}^{(l+1)}_{\phi_2}) - \boldsymbol{p}(\widehat{\boldsymbol{\theta}}^{(l)}_{\phi_2})\right)\right)^T \\
&\times \left(n^{1/2} diag\left(\boldsymbol{p}(\boldsymbol{\theta}_0)^{-1/2}\right)\left(\boldsymbol{p}(\widehat{\boldsymbol{\theta}}^{(l+1)}_{\phi_2}) - \boldsymbol{p}(\widehat{\boldsymbol{\theta}}^{(l)}_{\phi_2})\right)\right) \\
&+ n\, o\left(\left\|\boldsymbol{p}(\widehat{\boldsymbol{\theta}}^{(l+1)}_{\phi_2}) - \boldsymbol{p}(\boldsymbol{\theta}_0)\right\|^2 + \left\|\boldsymbol{p}(\widehat{\boldsymbol{\theta}}^{(l)}_{\phi_2}) - \boldsymbol{p}(\boldsymbol{\theta}_0)\right\|^2\right).
\end{aligned}
$$

We know from (7.11) that, under the loglinear model (7.8) and the null hypothesis H_{l+1},

$$\sqrt{n}\left(\boldsymbol{p}(\widehat{\boldsymbol{\theta}}^{(l+1)}_{\phi_2}) - \boldsymbol{p}(\boldsymbol{\theta}_0)\right) \underset{n\to\infty}{\overset{L}{\longrightarrow}} N\left(\boldsymbol{0}, \boldsymbol{\Sigma}^*_{(l+1)}\right),$$

with $\boldsymbol{\Sigma}^*_{(l+1)} = \boldsymbol{\Sigma}_{\boldsymbol{p}(\boldsymbol{\theta}_0)}\boldsymbol{W}_{(l+1)}\left(\boldsymbol{W}^T_{(l+1)}\boldsymbol{\Sigma}_{\boldsymbol{p}(\boldsymbol{\theta}_0)}\boldsymbol{W}_{(l+1)}\right)^{-1}\boldsymbol{W}^T_{(l+1)}\boldsymbol{\Sigma}_{\boldsymbol{p}(\boldsymbol{\theta}_0)}$, and $\boldsymbol{W}_{(l+1)}$ is the loglinear model matrix of explanatory variables under the null hypothesis H_{l+1}.

Then $\left\|\boldsymbol{p}(\widehat{\boldsymbol{\theta}}^{(l+1)}_{\phi_2}) - \boldsymbol{p}(\boldsymbol{\theta}_0)\right\|^2 = O_P\left(n^{-1}\right)$, and because it is assumed that $\boldsymbol{\theta}_0 \in \Theta_{l+1} \subset \Theta_l$, we also have that $\left\|\boldsymbol{p}(\widehat{\boldsymbol{\theta}}^{(l)}_{\phi_2}) - \boldsymbol{p}(\boldsymbol{\theta}_0)\right\|^2 = O_P\left(n^{-1}\right)$. Consequently,

$$n\, o\left(\left\|\boldsymbol{p}(\widehat{\boldsymbol{\theta}}^{(l+1)}_{\phi_2}) - \boldsymbol{p}(\boldsymbol{\theta}_0)\right\|^2 + \left\|\boldsymbol{p}(\widehat{\boldsymbol{\theta}}^{(l)}_{\phi_2}) - \boldsymbol{p}(\boldsymbol{\theta}_0)\right\|^2\right) = o_P(1),$$

and hence the asymptotic distribution of the test statistic $\widetilde{T}^{(l)}_{\phi_1,\phi_2}$ is the same as the asymptotic distribution of the random variable $\boldsymbol{Z}^T\boldsymbol{Z}$, where

$$\boldsymbol{Z} \equiv \sqrt{n} diag\left(\boldsymbol{p}(\boldsymbol{\theta}_0)^{-1/2}\right)\left(\boldsymbol{p}(\widehat{\boldsymbol{\theta}}^{(l+1)}_{\phi_2}) - \boldsymbol{p}(\widehat{\boldsymbol{\theta}}^{(l)}_{\phi_2})\right).$$

Then, using (7.10), we obtain

$$\begin{aligned}
\boldsymbol{p}(\widehat{\boldsymbol{\theta}}_{\phi_2}^{(l+1)}) - \boldsymbol{p}(\widehat{\boldsymbol{\theta}}_{\phi_2}^{(l)}) = & \left(\boldsymbol{\Sigma}_{\boldsymbol{p}(\boldsymbol{\theta}_0)} \boldsymbol{W}_{(l+1)} \left(\boldsymbol{W}_{(l+1)}^T \boldsymbol{\Sigma}_{\boldsymbol{p}(\boldsymbol{\theta}_0)} \boldsymbol{W}_{(l+1)} \right)^{-1} \boldsymbol{W}_{(l+1)}^T \boldsymbol{\Sigma}_{\boldsymbol{p}(\boldsymbol{\theta}_0)} \right. \\
& - \left. \boldsymbol{\Sigma}_{\boldsymbol{p}(\boldsymbol{\theta}_0)} \boldsymbol{W}_{(l)} \left(\boldsymbol{W}_{(l)}^T \boldsymbol{\Sigma}_{\boldsymbol{p}(\boldsymbol{\theta}_0)} \boldsymbol{W}_{(l)} \right)^{-1} \boldsymbol{W}_{(l)}^T \boldsymbol{\Sigma}_{\boldsymbol{p}(\boldsymbol{\theta}_0)} \right) \\
& \times \ diag \left(\boldsymbol{p}(\boldsymbol{\theta}_0)^{-1} \right) (\widehat{\boldsymbol{p}} - \boldsymbol{p}(\boldsymbol{\theta}_0)) \\
& + \ o\left(\left\| \widehat{\boldsymbol{\theta}}_{\phi_2}^{(l+1)} - \boldsymbol{\theta}_0 \right\| \right) - o\left(\left\| \widehat{\boldsymbol{\theta}}_{\phi_2}^{(l)} - \boldsymbol{\theta}_0 \right\| \right).
\end{aligned}$$

If we denote

$$\begin{aligned}
\boldsymbol{A}_{(i)} \equiv & \ diag \left(\boldsymbol{p}(\boldsymbol{\theta}_0)^{-1/2} \right) \boldsymbol{\Sigma}_{\boldsymbol{p}(\boldsymbol{\theta}_0)} \boldsymbol{W}_{(i)} \left(\boldsymbol{W}_{(i)}^T \boldsymbol{\Sigma}_{\boldsymbol{p}(\boldsymbol{\theta}_0)} \boldsymbol{W}_{(i)} \right)^{-1} \\
\times & \ \boldsymbol{W}_{(i)}^T \boldsymbol{\Sigma}_{\boldsymbol{p}(\boldsymbol{\theta}_0)} diag \left(\boldsymbol{p}(\boldsymbol{\theta}_0)^{-1/2} \right); \ i = l, \ l+1,
\end{aligned} \tag{7.19}$$

which is a symmetric matrix, the random vector $\boldsymbol{Z}$ can be written as

$$\begin{aligned}
\boldsymbol{Z} = & \ \left(\boldsymbol{A}_{(l+1)} - \boldsymbol{A}_{(l)} \right) diag \left(\boldsymbol{p}(\boldsymbol{\theta}_0)^{-1/2} \right) \sqrt{n} \left(\widehat{\boldsymbol{p}} - \boldsymbol{p}(\boldsymbol{\theta}_0) \right) \\
& + \ o\left(\left\| \widehat{\boldsymbol{\theta}}_{\phi_2}^{(l+1)} - \boldsymbol{\theta}_0 \right\| \right) - o\left(\left\| \widehat{\boldsymbol{\theta}}_{\phi_2}^{(l)} - \boldsymbol{\theta}_0 \right\| \right).
\end{aligned}$$

Thus,

$$\sqrt{n} diag \left(\boldsymbol{p}(\boldsymbol{\theta}_0)^{-1/2} \right) \left(\boldsymbol{p}(\widehat{\boldsymbol{\theta}}_{\phi_2}^{(l+1)}) - \boldsymbol{p}(\widehat{\boldsymbol{\theta}}_{\phi_2}^{(l)}) \right) \underset{n \to \infty}{\overset{L}{\longrightarrow}} N\left(\boldsymbol{0}, \boldsymbol{\Sigma}^* \right),$$

where

$$\begin{aligned}
\boldsymbol{\Sigma}^* = & \ \left(\boldsymbol{A}_{(l+1)} - \boldsymbol{A}_{(l)} \right) diag \left(\boldsymbol{p}(\boldsymbol{\theta}_0)^{-1/2} \right) \boldsymbol{\Sigma}_{\boldsymbol{p}(\boldsymbol{\theta}_0)} diag \left(\boldsymbol{p}(\boldsymbol{\theta}_0)^{-1/2} \right) \\
\times & \ \left(\boldsymbol{A}_{(l+1)} - \boldsymbol{A}_{(l)} \right) \\
= & \ \left(\boldsymbol{A}_{(l+1)} - \boldsymbol{A}_{(l)} \right) \left(\boldsymbol{I} - \boldsymbol{p}(\boldsymbol{\theta}_0)^{1/2} \left(\boldsymbol{p}(\boldsymbol{\theta}_0)^{1/2} \right)^T \right) \left(\boldsymbol{A}_{(l+1)} - \boldsymbol{A}_{(l)} \right),
\end{aligned}$$

and $\boldsymbol{p}(\boldsymbol{\theta}_0)^{1/2} = \left(p_1(\boldsymbol{\theta}_0)^{1/2}, ..., p_M(\boldsymbol{\theta}_0)^{1/2} \right)^T$. Then, because

$$\left(\boldsymbol{p}(\boldsymbol{\theta}_0)^{1/2} \right)^T diag \left(\boldsymbol{p}(\boldsymbol{\theta}_0)^{-1/2} \right) \boldsymbol{\Sigma}_{\boldsymbol{p}(\boldsymbol{\theta}_0)}$$

can be written as

$$\boldsymbol{1}^T diag \left(\boldsymbol{p}(\boldsymbol{\theta}_0) \right) - \boldsymbol{1}^T \boldsymbol{p}(\boldsymbol{\theta}_0) \boldsymbol{p}(\boldsymbol{\theta}_0)^T = \boldsymbol{0},$$

where $\boldsymbol{1} = (1, ..., 1)^T$, we have, using Exercise 10, that

$$\boldsymbol{\Sigma}^* = \left(\boldsymbol{A}_{(l+1)} - \boldsymbol{A}_{(l)} \right) \left(\boldsymbol{A}_{(l+1)} - \boldsymbol{A}_{(l)} \right) = \boldsymbol{A}_{(l)} - \boldsymbol{A}_{(l+1)}.$$

Therefore the matrix $\left(\boldsymbol{A}_{(l)} - \boldsymbol{A}_{(l+1)}\right)$ is symmetric and idempotent with

$$trace\left(\boldsymbol{A}_{(l)} - \boldsymbol{A}_{(l+1)}\right) = d_l - d_{l+1}$$

and $\boldsymbol{Z}$ is asymptotically normal with mean vector $\mathbf{0}$ and variance-covariance matrix $\left(\boldsymbol{A}_{(l)} - \boldsymbol{A}_{(l+1)}\right)$. Applying Lemma 3, p. 57, in Ferguson (1996) (see Remark 2.6), we have that $\boldsymbol{Z}^T\boldsymbol{Z}$ is asymptotically chi-squared distributed with $d_l - d_{l+1}$ degrees of freedom.

Finally,

$$\widetilde{T}^{(l)}_{\phi_1,\phi_2} \equiv \frac{2n}{\phi_1''(1)} D_{\phi_1}\left(\boldsymbol{p}(\widehat{\boldsymbol{\theta}}^{(l+1)}_{\phi_2}), \boldsymbol{p}(\widehat{\boldsymbol{\theta}}^{(l)}_{\phi_2})\right) \xrightarrow[n\to\infty]{L} \chi^2_{d_l - d_{l+1}}. \quad \blacksquare$$

For the (h,ϕ)-divergence measures we have the following result:

Theorem 7.2

Under the assumptions given in Theorem 7.1, the asymptotic null distribution of the (h,ϕ)-divergence test statistic,

$$\widetilde{T}^{(l)}_{\phi_1,\phi_2,h} \equiv \frac{2n}{\phi_1''(1)\, h'(0)} h\left(D_{\phi_1}\left(\boldsymbol{p}(\widehat{\boldsymbol{\theta}}^{(l+1)}_{\phi_2}), \boldsymbol{p}(\widehat{\boldsymbol{\theta}}^{(l)}_{\phi_2})\right)\right),$$

is chi-square with $d_l - d_{l+1}$ degrees of freedom; $l = 1, ..., m-1$, where h is a differentiable function mapping from $[0,\infty)$ onto $[0,\infty)$, with $h(0) = 0$ and $h'(0) > 0$.

Proof. We know

$$D_{\phi_1}\left(\boldsymbol{p}(\widehat{\boldsymbol{\theta}}^{(l+1)}_{\phi_2}), \boldsymbol{p}(\widehat{\boldsymbol{\theta}}^{(l)}_{\phi_2})\right) = \frac{\phi_1''(1)}{2n} \boldsymbol{Z}^T\boldsymbol{Z} + o_P\left(n^{-1}\right),$$

where $\boldsymbol{Z}^T\boldsymbol{Z}$ is asymptotically chi-squared distributed with $d_l - d_{l+1}$ degrees of freedom. Further, because $h(x) = h(0) + h'(0)\,x + o(x)$, we have

$$\widetilde{T}^{(l)}_{\phi_1,\phi_2,h} = \boldsymbol{Z}^T\boldsymbol{Z} + o_P(1) \xrightarrow[n\to\infty]{L} \chi^2_{d_l - d_{l+1}}. \quad \blacksquare$$

We can observe that in the two previous theorems the estimated model in the null hypothesis appears in the left argument of the ϕ-divergence. It is usual to consider the null hypothesis in the right argument of the ϕ-divergence; for this reason we present the following theorem.

Theorem 7.3

Under the assumptions of Theorems 7.1 and 7.2, the asymptotic null distribution of the ϕ-divergence test statistics,

$$T_{\phi_1,\phi_2}^{(l)} = \frac{2n}{\phi_1''(1)} D_{\phi_1}\left(\boldsymbol{p}(\widehat{\boldsymbol{\theta}}_{\phi_2}^{(l)}), \boldsymbol{p}(\widehat{\boldsymbol{\theta}}_{\phi_2}^{(l+1)})\right) \tag{7.20}$$

and

$$T_{\phi_1,\phi_2,h}^{(l)} = \frac{2n}{\phi_1''(1)\, h'(0)} h\left(D_{\phi_1}\left(\boldsymbol{p}(\widehat{\boldsymbol{\theta}}_{\phi_2}^{(l)}), \boldsymbol{p}(\widehat{\boldsymbol{\theta}}_{\phi_2}^{(l+1)})\right)\right), \tag{7.21}$$

is chi-square with $d_l - d_{l+1}$ degrees of freedom; $l = 1, ..., m-1$, where h is a differentiable function mapping from $[0, \infty)$ onto $[0, \infty)$, with $h(0) = 0$ and $h'(0) > 0$.

Proof. We consider the function $\varphi(x) = x\phi_1(x^{-1})$. It is clear that $\varphi(x) \in \Phi^*$, $\widetilde{T}_{\varphi,\phi_2}^{(l)} = T_{\phi_1,\phi_2}^{(l)}$ and $\widetilde{T}_{\varphi,\phi_2,h}^{(l)} = T_{\phi_1,\phi_2,h}^{(l)}$. Then the result follows directly from Theorems 7.1 and 7.2. ■

Remark 7.1

A well known test statistic appears when we choose

$$\phi_1(x) = \phi_2(x) = x\log x - x + 1$$

in the test statistic $T_{\phi_1,\phi_2}^{(l)}$ given in (7.20). It is the classical likelihood ratio test statistic (Agresti, 1996, p. 197; Christensen, 1997, p. 322) and it holds the well known result,

$$G^2 \equiv 2n \sum_{j=1}^{k} p_j(\widehat{\boldsymbol{\theta}}^{(l)}) \log \frac{p_j(\widehat{\boldsymbol{\theta}}^{(l)})}{p_j(\widehat{\boldsymbol{\theta}}^{(l+1)})} \xrightarrow[n\to\infty]{L} \chi^2_{d_l - d_{l+1}},$$

where $\widehat{\boldsymbol{\theta}}^{(i)}$ is the maximum likelihood estimator of $\boldsymbol{\theta}$ under the model H_i $(\boldsymbol{\theta} \in \Theta_i)$; $i = l, l+1$.

Another important test statistic appears when we put $\phi_1(x) = \frac{1}{2}(1-x)^2$, $\phi_2(x) = x\log x - x + 1$ in the test statistic $T_{\phi_1,\phi_2}^{(l)}$ given in (7.20). Then we obtain the chi-square test statistic given in Agresti (1996, p. 197), as well as the result,

$$X^2 \equiv n \sum_{j=1}^{k} \frac{\left(p_j(\widehat{\boldsymbol{\theta}}^{(l)}) - p_j(\widehat{\boldsymbol{\theta}}^{(l+1)})\right)^2}{p_j(\widehat{\boldsymbol{\theta}}^{(l+1)})} \xrightarrow[n\to\infty]{L} \chi^2_{d_l - d_{l+1}}.$$

But this is not the only family of test statistics for testing nested sequences of loglinear models given in (7.17) based on ϕ-divergence measures. In the following theorem another family of test statistics based on this measure is proposed.

Theorem 7.4

Suppose that data $(N_1, ..., N_k)$ are multinomially distributed according to (7.7) and (7.8). Consider the nested sequence of hypotheses given by (7.15) and (7.16). Choose the two functions $\phi_1, \phi_2 \in \Phi^*$. Then, for testing hypotheses,

$$H_{Null} : H_{l+1} \text{ against } H_{Alt} : H_l; \quad l = 1, ..., m-1,$$

the ϕ-divergence test statistics,

$$S_\phi^{(l)} = \frac{2n}{\phi''(1)} \left(D_\phi \left(\widehat{\mathbf{p}}, \mathbf{p}(\widehat{\boldsymbol{\theta}}_\phi^{(l+1)}) \right) - D_\phi \left(\widehat{\mathbf{p}}, \mathbf{p}(\widehat{\boldsymbol{\theta}}_\phi^{(l)}) \right) \right), \tag{7.22}$$

and

$$S_{\phi,h}^{(l)} = \frac{2n}{\phi''(1) h'(0)} \left(h \left(D_\phi \left(\widehat{\mathbf{p}}, \mathbf{p}(\widehat{\boldsymbol{\theta}}_\phi^{(l+1)}) \right) \right) - h \left(D_\phi \left(\widehat{\mathbf{p}}, \mathbf{p}(\widehat{\boldsymbol{\theta}}_\phi^{(l)}) \right) \right) \right) \tag{7.23}$$

are nonnegative and their asymptotic null distribution is chi-square with $d_l - d_{l+1}$ degrees of freedom; $l = 1, ..., m-1$, where h is a differentiable increasing function mapping from $[0, \infty)$ onto $[0, \infty)$, with $h(0) = 0$ and $h'(0) > 0$.

Proof. It is clear that $S_\phi^{(l)} \geq 0$; $l = 1, ..., m-1$, because

$$D_\phi \left(\widehat{\mathbf{p}}, \mathbf{p}(\widehat{\boldsymbol{\theta}}_\phi^{(l+1)}) \right) = \inf_{\boldsymbol{\theta} \in \Theta_{l+1}} D_\phi(\widehat{\mathbf{p}}, \mathbf{p}(\boldsymbol{\theta})) \geq \inf_{\boldsymbol{\theta} \in \Theta_l} D_\phi(\widehat{\mathbf{p}}, \mathbf{p}(\boldsymbol{\theta})) = D_\phi \left(\widehat{\mathbf{p}}, \mathbf{p}(\widehat{\boldsymbol{\theta}}_\phi^{(l)}) \right).$$

The proof of the asymptotic distribution of the test statistics $S_\phi^{(l)}$ and $S_{\phi,h}^{(l)}$ follows the same steps as the proof given in Theorems 7.1 and 7.2. ∎

Remark 7.2

The asymptotic result of Theorem 7.4 can be generalized further to include a ϕ_1 for divergence D_{ϕ_1}, and a ϕ_2 for estimation $\widehat{\boldsymbol{\theta}}_{\phi_2}^{(i)}$. That is, the statistic

$$S_{\phi_1,\phi_2}^{(l)} \equiv \frac{2n}{\phi_1''(1)} \left(D_{\phi_1} \left(\widehat{\mathbf{p}}, \mathbf{p}(\widehat{\boldsymbol{\theta}}_{\phi_2}^{(l+1)}) \right) - D_{\phi_1} \left(\widehat{\mathbf{p}}, \mathbf{p}(\widehat{\boldsymbol{\theta}}_{\phi_2}^{(l)}) \right) \right) \xrightarrow[n\to\infty]{L} \chi^2_{d_l - d_{l+1}}$$

under H_{l+1}.

The special case of $\phi_1(x) = \frac{1}{2}(1-x)^2$, $\phi_2(x) = x \log x - x + 1$ yields a statistic based on the difference of chi-square test statistic with maximum likelihood *estimation used to obtain the expected frequencies (e.g., Agresti, 1996, p. 197), namely*

$$n\left(D_{\frac{1}{2}(1-x)^2}\left(\widehat{\boldsymbol{p}},\boldsymbol{p}(\widehat{\boldsymbol{\theta}}^{(l+1)})\right)-D_{\frac{1}{2}(1-x)^2}\left(\widehat{\boldsymbol{p}},\boldsymbol{p}(\widehat{\boldsymbol{\theta}}^{(l)})\right)\right).$$

However, the nonnegativity of $S^{(l)}_{\phi_1,\phi_2}$ does not hold when $\phi_1 \neq \phi_2$. Thus, for the case above, considered by Agresti, the difference of the chi-square test statistics is not necessarily nonnegative. Since it is common to use maximum likelihood estimation (that is, $\phi_2(x) = x\log x - x + 1$), the test statistic $S^{(l)}_{\phi_1,\phi_2}$; $\phi_1 \neq \phi_2$ is not all that interesting to us. In the following we shall concentrate on the statistics $T^{(l)}_{\phi_1,\phi_2}$.

For testing the nested hypotheses $\{H_l : l = 1, ..., m\}$ given by (7.15), we test $H_{Null} : H_{l+1}$ against $H_{Alt} : H_l$, using the ϕ-divergence test statistic $T^{(l)}_{\phi_1,\phi_2}$ given by (7.20); if it is too large, H_{Null} is rejected. When $T^{(l)}_{\phi_1,\phi_2} > c$, we reject H_{Null} in (7.17), where c is specified so that the significance level of the test is α :

$$\Pr\left(T^{(l)}_{\phi_1,\phi_2} > c \mid H_{l+1}\right) = \alpha;\ \alpha \in (0,1). \tag{7.24}$$

Theorem 7.1 was shown that under (7.7), (7.8) and (7.15), and $H_{Null} : H_{l+1}$, the test statistic $T^{(l)}_{\phi_1,\phi_2}$ converges in distribution to a chi-square distribution with $d_l - d_{l+1}$ degrees of freedom; $l = 1, ..., m-1$. Therefore, c could be chosen as the $100(1-\alpha)$ percentile of a chi-square distribution with $d_l - d_{l+1}$ degrees of freedom,

$$c = \chi^2_{d_l - d_{l+1},\alpha}. \tag{7.25}$$

The choice of (7.25) in (7.24) only guarantees an asymptotic size-α test. Here we use (7.25) but ask, in the finite-sample simulations given in Section 5, for what choices of λ in $T^{(l)}_{\phi_1(\lambda),\phi_2(\lambda)}$ is the relation (7.24) most accurately attained?

The asymptotic chi-square approximation, $c = \chi^2_{d_l - d_{l+1},\alpha}$, is checked for a sequence of loglinear models in the simulation study given in Section 7.5. We give a small illustration of those results now. Figures 7.1 and 7.2 show *departures* of the exact simulated size from the nominal size of $\alpha = 0.05$ for one particular choice (specified in Section 7.5) of H_{l+1} and H_l, for various choices of λ in $\phi_1 = \phi_{(\lambda)}$, and for small to large sample sizes (n = 15, 20, 25, 35, 50, 100, 200). Figure 7.1 represents nonpositive choices of λ, and Figure 7.2 represents nonnegative choices of λ. The positive values of λ perform the best.

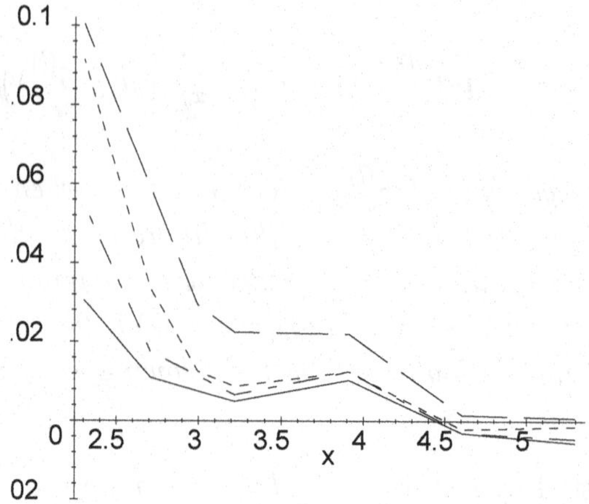

Figure 7.1. (Exact size – Nominal size of 0.05) as a function of $x = \log n$. Shown are $\lambda = -2$ (dashed line), $\lambda = -1$ (dotted line), $\lambda = -1/2$ (dash-dotted line) and $\lambda = 0$ (solid line).

Source: Cressie, N., Pardo, L. and Pardo, M.C. (2003).

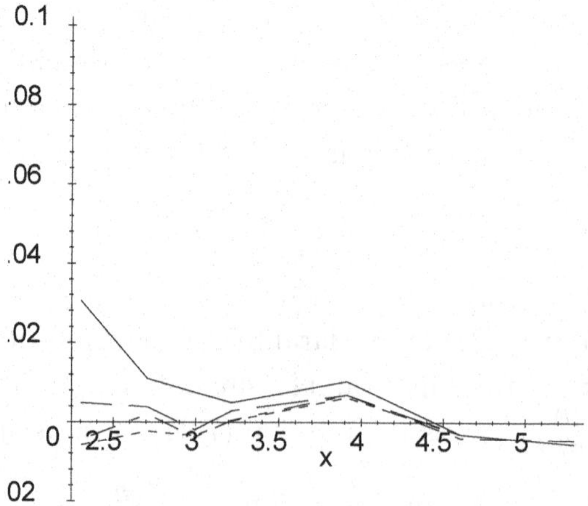

Figure 7.2. (Exact size – Nominal size of 0.05) as a function of $x = \log n$. Shown are $\lambda = 0$ (solid line), $\lambda = 2/3$ (dashed line), $\lambda = 1$ (dotted line) and $\lambda = 2$ (dash-dotted line).

Source: Cressie, N., Pardo, L. and Pardo, M.C. (2003).

To test the nested sequence of hypotheses $\{H_l : l = 1, ..., m\}$ effectively, we need an asymptotic independence result for the sequence of test statistics $T^{(1)}_{\phi_1,\phi_2}$, $T^{(2)}_{\phi_1,\phi_2}, ..., T^{(m^*)}_{\phi_1,\phi_2}$, where m^* is the integer $1 \leq m^* \leq m$ for which H_{m^*} is true but H_{m^*+1} is not true. This result is given in the theorem below. Notice that our hypothesis-testing strategy is an attempt to find this value of m^*; we infer it to be l^*, the first in the sequence of hypothesis tests for which H_{l^*+1} is rejected as a null hypothesis.

Theorem 7.5

Suppose that data $(N_1, ..., N_M)$ *are multinomially distributed according to (7.7) and (7.8). Suppose we wish to test first*

$$H_{Null} : H_l \text{ against } H_{Alt} : H_{l-1},$$

followed by

$$H_{Null} : H_{l+1} \text{ against } H_{Alt} : H_l.$$

Then, under the null hypothesis H_l, *the* ϕ*-divergence test statistics*

$$T^{(l-1)}_{\phi_1,\phi_2} \quad \text{and} \quad T^{(l)}_{\phi_1,\phi_2}$$

are asymptotically independent and chi-squared distributed with $d_{l-1} - d_l$ *and* $d_l - d_{l+1}$ *degrees of freedom, respectively.*

Proof. A similar development to the one presented in Theorem 7.1 gives

$$T^{(l)}_{\phi_1,\phi_2} = \sqrt{n}\left(\widehat{\mathbf{p}} - \mathbf{p}\left(\boldsymbol{\theta}_0\right)\right)^T \boldsymbol{M}_l^T \boldsymbol{M}_l \sqrt{n}\left(\widehat{\mathbf{p}} - \mathbf{p}\left(\boldsymbol{\theta}_0\right)\right) + o_P\left(1\right)$$

and

$$T^{(l-1)}_{\phi_1,\phi_2} = \sqrt{n}\left(\widehat{\mathbf{p}} - \mathbf{p}\left(\boldsymbol{\theta}_0\right)\right)^T \boldsymbol{M}_{l-1}^T \boldsymbol{M}_{l-1} \sqrt{n}\left(\widehat{\mathbf{p}} - \mathbf{p}\left(\boldsymbol{\theta}_0\right)\right) + o_P\left(1\right),$$

where

$$\boldsymbol{M}_i = \left(\boldsymbol{A}_{(i+1)} - \boldsymbol{A}_{(i)}\right) diag\left(\mathbf{p}\left(\boldsymbol{\theta}_0\right)^{-1/2}\right); \; i = l-1, l,$$

$\boldsymbol{A}_{(i)}$ is given in (7.19) and $\boldsymbol{W}_{(i)}$ is the loglinear model matrix of explanatory variables under the ith-loglinear model; $i = l-1,\, l,\, l+1$.

Now because

$$\sqrt{n}\left(\widehat{\mathbf{p}} - \mathbf{p}\left(\boldsymbol{\theta}_0\right)\right) \xrightarrow[n\to\infty]{L} N\left(\mathbf{0}, \boldsymbol{\Sigma}_{\mathbf{p}(\boldsymbol{\theta}_0)}\right),$$

from Theorem 4 in Searle (1971, p. 59), $T^{(l)}_{\phi_1,\phi_2}$ and $T^{(l-1)}_{\phi_1,\phi_2}$ are asymptotically independent if

$$\boldsymbol{P}_{(l)} \equiv \boldsymbol{M}_{l-1}^T \boldsymbol{M}_{l-1} \boldsymbol{\Sigma}_{\mathbf{p}(\boldsymbol{\theta}_0)} \boldsymbol{M}_l^T \boldsymbol{M}_l = \mathbf{0}.$$

We have

$$\begin{aligned}\boldsymbol{P}_{(l)} &= \boldsymbol{M}_{l-1}^T\left(\boldsymbol{A}_{(l)}-\boldsymbol{A}_{(l-1)}\right)\left(\boldsymbol{I}-\sqrt{\boldsymbol{p}(\boldsymbol{\theta}_0)}\sqrt{\boldsymbol{p}(\boldsymbol{\theta}_0)^T}\right)\left(\boldsymbol{A}_{(l+1)}-\boldsymbol{A}_{(l)}\right)\boldsymbol{M}_l\\ &= \boldsymbol{M}_{l-1}^T\left(\boldsymbol{A}_{(l)}-\boldsymbol{A}_{(l-1)}\right)\left(\boldsymbol{A}_{(l+1)}-\boldsymbol{A}_{(l)}\right)\boldsymbol{M}_l,\end{aligned}$$

since $\boldsymbol{A}_{(i)}\sqrt{\boldsymbol{p}(\boldsymbol{\theta}_0)}=\boldsymbol{0}$; $i=l-1,l,l+1$.

Applying Exercise 10, we have

$$\boldsymbol{A}_{(i)}\boldsymbol{A}_{(i+1)}=\boldsymbol{A}_{(i+1)}\boldsymbol{A}_{(i)}=\boldsymbol{A}_{(i+1)};\ \boldsymbol{A}_{(i)}\boldsymbol{A}_{(i)}=\boldsymbol{A}_{(i)}; i=l-1,l,l+1,$$

therefore $\boldsymbol{M}_{l-1}^T\boldsymbol{M}_{l-1}\boldsymbol{\Sigma}_{\boldsymbol{p}(\boldsymbol{\theta}_0)}\boldsymbol{M}_l^T\boldsymbol{M}_l=\boldsymbol{0}$. ■

Similar results to one obtained in this theorem can be obtained for the ϕ-divergence test statistics $\widetilde{T}^{(l)}_{\phi_1,\phi_2}$, $T^{(l)}_{\phi_1,\phi_2,h}$, $\widetilde{T}^{(l)}_{\phi_1,\phi_2,h}$, $S^{(l)}_{\phi}$, $S^{(l)}_{\phi,h}$ and $S^{(l)}_{\phi_1,\phi_2}$.

In general, theoretical results for the test statistic $T^{(l)}_{\phi_1,\phi_2}$ under alternative hypotheses are not easy to obtain. An exception to this is when there is a contiguous sequence of alternatives that approaches the null hypothesis H_{l+1} at the rate of $O\left(n^{-1/2}\right)$. Regarding the alternative, Haberman (1974) was the first to study the asymptotic distribution of the chi-square test statistic and likelihood ratio test statistic under contiguous alternative hypotheses, establishing that the asymptotic distribution is noncentral chi-square with d_l-d_{l+1} degrees of freedom. Oler (1985) presented a systematic study of the contiguous alternative hypotheses in multinomial populations, obtaining as a particular case the asymptotic distribution for the loglinear models. Through simulations, she also studied how closely the noncentral chi-square distributions agree with the exact sampling distributions. Fenech and Westfall (1988) presented an interesting analytic study of the noncentrality parameter in the case of loglinear models. Now we generalize their results to tests based on the ϕ-divergence test statistic $T^{(l)}_{\phi_1,\phi_2}$ given by (7.20).

Consider the multinomial probability vector

$$\boldsymbol{p}_n\equiv\boldsymbol{p}(\boldsymbol{\theta}_0)+\boldsymbol{d}/\sqrt{n};\ \boldsymbol{\theta}_0\in\Theta_{l+1}, \tag{7.26}$$

where $\boldsymbol{d}\equiv(d_1,...,d_M)^T$ is a fixed $M\times 1$ vector such that $\sum_{j=1}^M d_j=0$, and recall that n is the total-count parameter of the multinomial distribution. As $n\to\infty$, the sequence of multinomial probabilities $\{\boldsymbol{p}_n\}_{n\in N}$ converges to a multinomial probability in H_{l+1} at the rate of $O\left(n^{-1/2}\right)$. We call

$$H_{l+1,n}:\boldsymbol{p}_n=\boldsymbol{p}(\boldsymbol{\theta}_0)+\boldsymbol{d}/\sqrt{n};\ \boldsymbol{\theta}_0\in\Theta_{l+1} \tag{7.27}$$

a sequence of contiguous alternative hypotheses, here contiguous to the null hypothesis H_{l+1}.

Now consider the problem of testing

$$H_{Null} : H_{l+1} \text{ against } H_{Alt} : H_{l+1,n},$$

using the ϕ-divergence test statistics $T^{(l)}_{\phi_1,\phi_2}$ given by (7.20). The power of this test is

$$\pi^{(l)}_n \equiv \Pr\left(T^{(l)}_{\phi_1,\phi_2} > c\,|H_{l+1,n}\right). \tag{7.28}$$

In what follows, we show that under the alternative $H_{l+1,n}$, and as $n \to \infty$, $T^{(l)}_{\phi_1,\phi_2}$ converges in distribution to a noncentral chi-square random variable with noncentrality parameter δ, where δ is given in Theorem 7.6, and $d_l - d_{l+1}$ degrees of freedom $(\chi^2_{d_l-d_{l+1}}(\delta))$. Consequently, as $n \to \infty$,

$$\pi^{(l)}_n \to \Pr\left(\chi^2_{d_l-d_{l+1}}(\delta) > c\right). \tag{7.29}$$

In (7.10), it was established that the asymptotic expansion of the minimum ϕ-divergence estimator about $\boldsymbol{\theta}_0 \in \Theta_{l+1}$ is given by

$$\begin{aligned}\widehat{\boldsymbol{\theta}}^{(l+1)}_{\phi} &= \boldsymbol{\theta}_0 + \left(\boldsymbol{W}^T_{(l+1)}\boldsymbol{\Sigma}_{\boldsymbol{p}(\boldsymbol{\theta}_0)}\boldsymbol{W}_{(l+1)}\right)^{-1}\boldsymbol{W}^T_{(l+1)}\boldsymbol{\Sigma}_{\boldsymbol{p}(\boldsymbol{\theta}_0)} diag\left(\boldsymbol{p}(\boldsymbol{\theta}_0)^{-1}\right)\\ &\times \left(\widehat{\boldsymbol{p}} - \boldsymbol{p}(\boldsymbol{\theta}_0)\right) + o\left(\|\widehat{\boldsymbol{p}} - \boldsymbol{p}(\boldsymbol{\theta}_0)\|\right)\end{aligned} \tag{7.30}$$

where $\boldsymbol{W}_{l+1}$ is the loglinear-model matrix of explanatory variables under the null hypothesis H_{l+1}.

Under the hypothesis given in (7.27), we have

$$\sqrt{n}\left(\widehat{\boldsymbol{p}} - \boldsymbol{p}(\boldsymbol{\theta}_0)\right) = \sqrt{n}\left(\widehat{\boldsymbol{p}} - \boldsymbol{p}_n\right) + \boldsymbol{d},$$

and hence

$$\sqrt{n}\left(\widehat{\boldsymbol{p}} - \boldsymbol{p}(\boldsymbol{\theta}_0)\right) \xrightarrow[n\to\infty]{L} N\left(\boldsymbol{d}, \boldsymbol{\Sigma}_{\boldsymbol{p}(\boldsymbol{\theta}_0)}\right), \tag{7.31}$$

so

$$o\left(\|\widehat{\boldsymbol{p}} - \boldsymbol{p}(\boldsymbol{\theta}_0)\|\right) = o\left(O_P\left(n^{-1/2}\right)\right) = o_P\left(n^{-1/2}\right).$$

Therefore, we have established that under the contiguous alternative hypotheses given in (7.27), and for $\boldsymbol{\theta}_0 \in \Theta_{l+1}$,

$$\begin{aligned}\widehat{\boldsymbol{\theta}}^{(l+1)}_{\phi} &= \boldsymbol{\theta}_0 + \left(\boldsymbol{W}^T_{(l+1)}\boldsymbol{\Sigma}_{\boldsymbol{p}(\boldsymbol{\theta}_0)}\boldsymbol{W}_{(l+1)}\right)^{-1}\boldsymbol{W}^T_{(l+1)}\boldsymbol{\Sigma}_{\boldsymbol{p}(\boldsymbol{\theta}_0)} diag\left(\boldsymbol{p}(\boldsymbol{\theta}_0)^{-1}\right)\\ &\times \left(\widehat{\boldsymbol{p}} - \boldsymbol{p}(\boldsymbol{\theta}_0)\right) + o_P\left(n^{-1/2}\right).\end{aligned} \tag{7.32}$$

This result will be important in the following theorem.

Theorem 7.6

Suppose that data $(N_1, ..., N_M)$ *are multinomially distributed according to (7.7) and (7.8). The asymptotic distribution of the* ϕ*-divergence test statistic* $T^{(l)}_{\phi_1,\phi_2}$*, under the contiguous alternative hypotheses (7.27), is chi-square with* $d_l - d_{l+1}$ *degrees of freedom and noncentrality parameter* δ *given by*

$$\delta = \boldsymbol{d}^T diag\left(\boldsymbol{p}\left(\boldsymbol{\theta}_0\right)^{-1/2}\right)\left(\boldsymbol{A}_{(l)} - \boldsymbol{A}_{(l+1)}\right) diag\left(\boldsymbol{p}\left(\boldsymbol{\theta}_0\right)^{-1/2}\right)\boldsymbol{d},$$

where $\boldsymbol{d} = (d_1, ..., d_M)^T$ *is defined in (7.26) and satisfies* $\sum_{i=1}^M d_i = 0$*, and* $\boldsymbol{A}_{(i)}$*,* $i = l, l+1$*, is given in (7.19).*

Proof. By Theorem 7.1, we know that

$$T^{(l)}_{\phi_1,\phi_2} = \boldsymbol{Z}^T\boldsymbol{Z} + n\ o\left(\left\|\boldsymbol{p}(\widehat{\boldsymbol{\theta}}^{(l+1)}_{\phi_2}) - \boldsymbol{p}\left(\boldsymbol{\theta}_0\right)\right\|^2 + \left\|\boldsymbol{p}(\widehat{\boldsymbol{\theta}}^{(l)}_{\phi_2}) - \boldsymbol{p}\left(\boldsymbol{\theta}_0\right)\right\|^2\right),$$

where

$$\boldsymbol{Z} = \sqrt{n}\ diag\left(\boldsymbol{p}\left(\boldsymbol{\theta}_0\right)^{-1/2}\right)\left(\boldsymbol{p}(\widehat{\boldsymbol{\theta}}^{(l+1)}_{\phi_2}) - \boldsymbol{p}(\widehat{\boldsymbol{\theta}}^{(l)}_{\phi_2})\right).$$

But

$$\boldsymbol{p}(\widehat{\boldsymbol{\theta}}^{(l+1)}_{\phi_2}) - \boldsymbol{p}\left(\boldsymbol{\theta}_0\right) = \frac{\partial \boldsymbol{p}\left(\boldsymbol{\theta}_0\right)}{\partial \boldsymbol{\theta}}(\widehat{\boldsymbol{\theta}}^{(l+1)}_{\phi_2} - \boldsymbol{\theta}_0) + o\left(\left\|\boldsymbol{p}(\widehat{\boldsymbol{\theta}}^{(l+1)}_{\phi_2}) - \boldsymbol{p}\left(\boldsymbol{\theta}_0\right)\right\|\right),$$

and (7.32) we have

$$\boldsymbol{p}(\widehat{\boldsymbol{\theta}}^{(l+1)}_{\phi_2}) - \boldsymbol{p}\left(\boldsymbol{\theta}_0\right) = O_P\left(n^{-1/2}\right)$$

and

$$\left\|\boldsymbol{p}(\widehat{\boldsymbol{\theta}}^{(l+1)}_{\phi_2}) - \boldsymbol{p}\left(\boldsymbol{\theta}_0\right)\right\|^2 = O_P\left(n^{-1}\right).$$

In a similar way and taking into account that $\boldsymbol{\theta}_0 \in \Theta_{l+1} \subset \Theta_l$, it can be obtained that

$$\left\|\boldsymbol{p}(\widehat{\boldsymbol{\theta}}^{(l)}_{\phi_2}) - \boldsymbol{p}\left(\boldsymbol{\theta}_0\right)\right\|^2 = O_P\left(n^{-1}\right).$$

Then

$$T^{(l)}_{\phi_1,\phi_2} = \boldsymbol{Z}^T\boldsymbol{Z} + o_P\left(1\right).$$

From (7.32), we have, under the contiguous alternative hypotheses, that

$$\boldsymbol{Z} = \sqrt{n}\left(\boldsymbol{A}_{(l+1)} - \boldsymbol{A}_{(l)}\right) diag\left(\boldsymbol{p}\left(\boldsymbol{\theta}_0\right)^{-1/2}\right)\left(\widehat{\boldsymbol{p}} - \boldsymbol{p}\left(\boldsymbol{\theta}_0\right)\right) + o_P\left(1\right).$$

By (7.31)

$$\sqrt{n}\left(\widehat{\boldsymbol{p}}-\boldsymbol{p}\left(\boldsymbol{\theta}_0\right)\right) \xrightarrow[n\to\infty]{L} N\left(\boldsymbol{d}, \boldsymbol{\Sigma}_{\boldsymbol{p}(\boldsymbol{\theta}_0)}\right),$$

and hence

$$\boldsymbol{Z} \xrightarrow[n\to\infty]{L} N\left(\boldsymbol{\delta}, \boldsymbol{\Sigma}^*\right),$$

where $\boldsymbol{\delta} = \left(\boldsymbol{A}_{(l+1)} - \boldsymbol{A}_{(l)}\right) diag\left(\boldsymbol{p}\left(\boldsymbol{\theta}_0\right)^{-1/2}\right) \boldsymbol{d}$ and

$$\begin{aligned}
\boldsymbol{\Sigma}^* = & \left(\boldsymbol{A}_{(l+1)} - \boldsymbol{A}_{(l)}\right) diag\left(\boldsymbol{p}\left(\boldsymbol{\theta}_0\right)^{-1/2}\right) \boldsymbol{\Sigma}_{\boldsymbol{p}(\boldsymbol{\theta}_0)} diag\left(\boldsymbol{p}\left(\boldsymbol{\theta}_0\right)^{-1/2}\right) \\
\times & \left(\boldsymbol{A}_{(l+1)} - \boldsymbol{A}_{(l)}\right) \\
= & \left(\boldsymbol{A}_{(l+1)} - \boldsymbol{A}_{(l)}\right) \left(\boldsymbol{I} - \sqrt{\boldsymbol{p}\left(\boldsymbol{\theta}_0\right)}\sqrt{\boldsymbol{p}\left(\boldsymbol{\theta}_0\right)^T}\right) \left(\boldsymbol{A}_{(l+1)} - \boldsymbol{A}_{(l)}\right).
\end{aligned}$$

Using the results in the proof of Theorem 7.1, it can be shown that

$$\boldsymbol{\Sigma}^* = \left(\boldsymbol{A}_{(l)} - \boldsymbol{A}_{(l+1)}\right),$$

and it is an idempotent matrix of rank $(d_l - d_{l+1})$.

If we establish that $\boldsymbol{\Sigma}^*\boldsymbol{\mu} = \boldsymbol{\mu}$, the theorem follows from the Lemma on page 63 in Ferguson (1996), because in this case the noncentrality parameter is given by $\delta = \boldsymbol{\mu}^T\boldsymbol{\mu}$.

Applying the results obtained in Exercise 10, we have

$$\begin{aligned}
\boldsymbol{\Sigma}^*\boldsymbol{\mu} = & \left(\boldsymbol{A}_{(l)} - \boldsymbol{A}_{(l+1)}\right)\boldsymbol{\mu} = \boldsymbol{A}_{(l)}\boldsymbol{\mu} - \boldsymbol{A}_{(l+1)}\boldsymbol{\mu} \\
= & \boldsymbol{A}_{(l)}\left(\boldsymbol{A}_{(l+1)} - \boldsymbol{A}_{(l)}\right) diag\left(\boldsymbol{p}\left(\boldsymbol{\theta}_0\right)^{-1/2}\right)\boldsymbol{d} - \boldsymbol{A}_{(l+1)}\left(\boldsymbol{A}_{(l+1)} - \boldsymbol{A}_{(l)}\right) \\
\times & diag\left(\boldsymbol{p}\left(\boldsymbol{\theta}_0\right)^{-1/2}\right)\boldsymbol{d} \\
= & \boldsymbol{\mu}.
\end{aligned}$$

Then the noncentrality parameter δ is given by

$$\delta = \boldsymbol{\mu}^T\boldsymbol{\mu} = \boldsymbol{d}^T diag\left(\boldsymbol{p}\left(\boldsymbol{\theta}_0\right)^{-1/2}\right)\left(\boldsymbol{A}_{(l)} - \boldsymbol{A}_{(l+1)}\right) diag\left(\boldsymbol{p}\left(\boldsymbol{\theta}_0\right)^{-1/2}\right)\boldsymbol{d}. \qquad \blacksquare$$

Remark 7.3

Theorem 7.6 can be used to obtain an approximation to the power function of the test (7.17), as follows. Write

$$\boldsymbol{p}(\boldsymbol{\theta}^{(l)}) = \boldsymbol{p}(\boldsymbol{\theta}_0^{(l+1)}) + \frac{1}{\sqrt{n}}\left(\sqrt{n}\left(\boldsymbol{p}(\boldsymbol{\theta}^{(l)}) - \boldsymbol{p}(\boldsymbol{\theta}_0^{(l+1)})\right)\right),$$

and define

$$\boldsymbol{p}_n(\boldsymbol{\theta}^{(l)}) \equiv \boldsymbol{p}(\boldsymbol{\theta}_0^{(l+1)}) + \frac{1}{\sqrt{n}}\mathbf{d},$$

where $\mathbf{d} = \sqrt{n}\left(\boldsymbol{p}(\boldsymbol{\theta}^{(l)}) - \boldsymbol{p}(\boldsymbol{\theta}_0^{(l+1)})\right)$. *Then substitute* $\mathbf{d}$ *into the definition of and finally* δ *into the right hand side of (7.29).*

The asymptotic noncentral chi-square approximation for power is checked f finite samples in the simulation given in Section 7.5. Figures 7.3 and 7.4 shc departures of the exact power from the asymptotic power for one particular choi (specified in Section 7.5) of H_{l+1} and H_l, for various choices of λ in $\phi_1 = \phi_{(\lambda}$ and for small to large sample sizes (n = 15, 20, 25, 35, 50, 100, 200). Figu 7.3 represents nonpositive choices of λ and Figure 7.4. represents nonnegati choices of λ. These figures need to be interpreted in light of associated exact size see Section 7.5. However, it is immediately apparent that from an asymptoti approximation point of view, $\lambda = 2/3$ seems to perform the best, particularly fo small and moderate sample sizes.

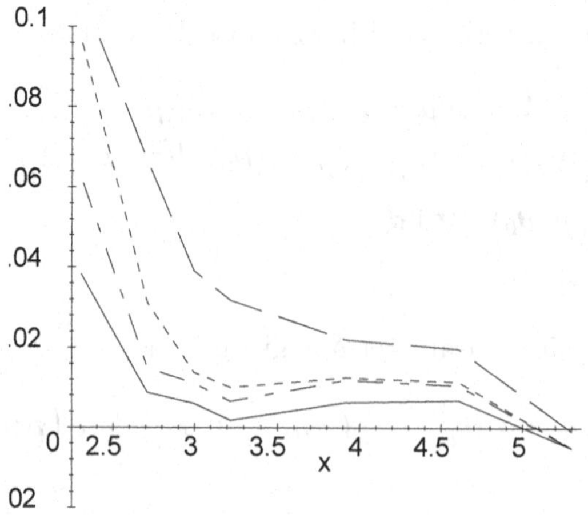

Figure 7.3. (Exact Power – Asymptotic power) as a function of $x = \log n$. Shown are $\lambda = -2$ (dashed line), $\lambda = -1$ (dotted line), $\lambda = -1/2$ (dash-dotted line) and $\lambda = 0$ (solid line).

Source: Cressie, N., Pardo, L. and Pardo, M.C. (2003).

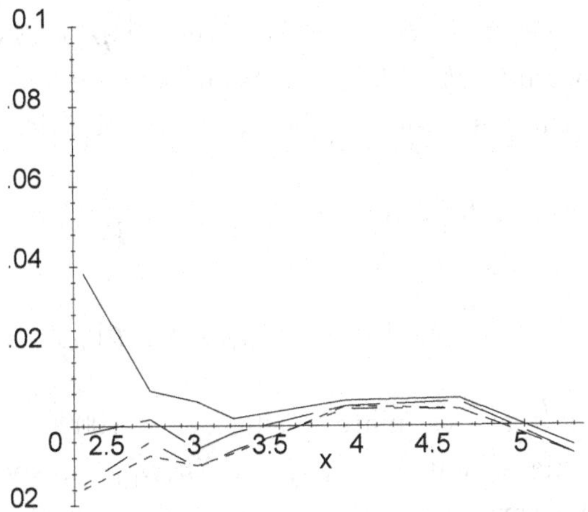

Figure 7.4. (Exact Power – Asymptotic power) as a function of $x = \log n$. Shown are $\lambda = 0$ (solid line), $\lambda = 2/3$ (dashed line), $\lambda = 1$ (dotted line) and $\lambda = 2$ (dash-dotted line).

Source: Cressie, N., Pardo, L. and Pardo, M.C. (2003).

7.5. Simulation Study

Section 7.4 presents theoretical results for testing hypotheses in loglinear models. The results give asymptotic distribution theory for the ϕ-divergence test statistic $T^{(l)}_{\phi_1,\phi_2}$ under the null hypothesis and a sequence of contiguous alternative hypotheses. The appropriateness of these results in finite samples is demonstrated in Figures 7.1, 7.2, 7.3 and 7.4. We now describe the finite-sample simulation study from which these figures were obtained, and we give new results that compare the powers of tests based on $T^{(l)}_{\phi_{(\lambda)},\phi_{(0)}}$ for $\lambda = -2, -1, -1/2, 0, 2/3, 1, 2$.

We consider for the estimation problem the maximum likelihood estimator and for the testing problem the power-divergence test statistics with the values of λ given previously.

Consider a $2 \times 2 \times 2$ contingency table, so $M = 8$. We simulate data $N_1, ..., N_M$ from a multinomial distribution with sample size n and probability vector $\boldsymbol{p} = (p_1, ..., p_M)^T$, where n and $\boldsymbol{p}$ are specified. The motivation for our simulation study comes from a similar one carried out by Oler (1985). The simulation study, presented in this Section, was carried out in Cressie et *al.* (2003). For

the moment, fix l and consider the power-divergence test statistics $T^{(l)}_{\phi_{(\lambda)},\phi_{(0)}}$ f
testing $H_{Null} : H_{l+1}$ against $H_{Alt} : H_{l+1,n}$, and let $\boldsymbol{p}_0 \in H_{Null}$ and $\boldsymbol{p}_{1,n} \in H_{A}$
where $\boldsymbol{p}_{1,n}$ is a subscript with n because its entries may depend on n. The essen
of our simulation study is to obtain the exact probabilities,

$$\begin{aligned} \alpha_n^{(l)} &\equiv \Pr\left(T^{(l)}_{\phi_{(\lambda)},\phi_{(0)}} > c \mid \boldsymbol{p}_0\right) \\ \pi_n^{(l)} &\equiv \Pr\left(T^{(l)}_{\phi_{(\lambda)},\phi_{(0)}} > c \mid \boldsymbol{p}_{1,n}\right). \end{aligned} \qquad (7.3$$

In fact, $\alpha_n^{(l)}$ and $\pi_n^{(l)}$ are estimated using $N = 100000$ simulations from t
multinomial sampling schemes $(n, \boldsymbol{p}_0)$ and $(n, \boldsymbol{p}_{1,n})$, respectively. For a giv
$\boldsymbol{p}_0$ (see below), the various choices of n and $\boldsymbol{p}_{1,n}$ represent the design of o
simulation study. We choose

$$n = 15, 20, 25, 35, 50, 100, 200$$

to represent small, moderate and large sample sizes.

We simulate multinomial random vectors $(N_1, ..., N_M)$ and compute prob
bilities $\alpha_n^{(l)}$ for $(n, \boldsymbol{p}_0)$ and $\pi_n^{(l)}$ for $(n, \boldsymbol{p}_{1,n})$. To see what happens for contiguo
alternatives, we fix $\boldsymbol{p}_1 \in H_l$ (see below) and define

$$\boldsymbol{p}^*_{1,n} \equiv \boldsymbol{p}_0 + (25/n)^{1/2}(\boldsymbol{p}_1 - \boldsymbol{p}_0). \qquad (7.3$$

Notice that $\boldsymbol{p}^*_{1,25} = \boldsymbol{p}_1$ and, as n increases, $\boldsymbol{p}^*_{1,n}$ converges to $\boldsymbol{p}_0$ at the ra
$n^{-1/2}$; that is, $\left\{\boldsymbol{p}^*_{1,n}\right\}$ is a sequence of contiguous alternatives. Our design f
the simulation study is to choose $(n, \boldsymbol{p}_{1,n})$ as fixed and contiguous alternative
which we now give.

Contiguous alternatives:

$$\left\{(n, \boldsymbol{p}^*_{1,n}) : n = 15, 20, 25, 35, 50, 100, 200\right\},$$

where $\boldsymbol{p}^*_{1,n}$ is given by (7.34) and $\boldsymbol{p}_1$ is specified below.

Fixed alternatives:

$$\left\{(n, \boldsymbol{p}_1) : n = 15, 20, 25, 35, 50, 100, 200\right\},$$

where $\boldsymbol{p}_1$ is specified below.

Notice that for $n < 25$, the contiguous alternatives are further from H_{Null} than are the fixed alternatives and that the two sequences share the alternative $(25, \boldsymbol{p}_1)$. These choices allow reasonable coverage of the space of alternatives.

In the simulation study, we consider the following nested sequence of loglinear models for the $2 \times 2 \times 2$ table:

$$
\begin{aligned}
H_1 &: p_{ijk}(\boldsymbol{\theta}) = \exp\left\{u + \theta_{1(i)} + \theta_{2(j)} + \theta_{3(k)} + \theta_{12(ij)} + \theta_{13(ik)} + \theta_{23(jk)}\right\} \\
H_2 &: p_{ijk}(\boldsymbol{\theta}) = \exp\left\{u + \theta_{1(i)} + \theta_{2(j)} + \theta_{3(k)} + \theta_{12(ij)} + \theta_{13(ik)}\right\} \\
H_3 &: p_{ijk}(\boldsymbol{\theta}) = \exp\left\{u + \theta_{1(i)} + \theta_{2(j)} + \theta_{3(k)} + \theta_{12(ij)}\right\} \\
H_4 &: p_{ijk}(\boldsymbol{\theta}) = \exp\left\{u + \theta_{1(i)} + \theta_{2(j)} + \theta_{3(k)}\right\}.
\end{aligned}
$$

Here, $\exp(-u)$ is the normalizing constant and the subscript θ-terms add to zero over each of their indices. Based on Oler's study, we used a moderate value for each main effect and a small value for the interactions. That is, we used

$$\exp\left(\theta_{1(1)}\right) = \exp\left(\theta_{2(1)}\right) = \exp\left(\theta_{3(1)}\right) = 5/6$$

$$\exp\left(\theta_{12(11)}\right) = \exp\left(\theta_{13(11)}\right) = \exp\left(\theta_{23(11)}\right) = 9/10.$$

Then the simulation experiment is designed so that $\boldsymbol{p}_1 \in H_l$; $l = 1, 2, 3$.

In Section 7.4, we showed Figures 7.1 and 7.2 (Exact size-Nominal size) for $H_{Null} : H_4$, using the test statistic $T^{(3)}_{\phi(\lambda),\phi(0)}$ and $c = \chi^2_{1,0.05}$ as well as Figure 7.3 and 7.4 (Exact power − Asymptotic power) for $H_{Null} : \boldsymbol{p}_0 \in H_4$ and $H_{Alt} : \boldsymbol{p}^*_{1,n}$; with $\boldsymbol{p}_1 \in H_3$, using the test statistic $T^{(3)}_{\phi(\lambda),\phi(0)}$ and $c = \chi^2_{1,0.05}$.

In the simulation study, we shall compare members of the power-divergence family of test statistics; our criteria for a good performance are:

i) good exact power and size for small to moderate sample sizes. For this, we consider the following three hypothesis tests with fixed alternatives:

$H_{Null} : \boldsymbol{p}_0 \in H_{l+1}$ versus $H_{Alt} : (n, \boldsymbol{p}_1)$, where $\boldsymbol{p}_1 \in H_l$ and $n = 15, 20, 25, 35$; $l = 1, 2, 3$.

ii) good agreement of exact and asymptotic probabilities for reasonably sm and moderate sample sizes. For this we consider the following three h pothesis tests with contiguous alternatives:

$H_{Null}: \boldsymbol{p}_0 \in H_{l+1}$ versus $H_{Alt}: \left(n, \boldsymbol{p}_{1,n}^*\right)$, where $\boldsymbol{p}_{1,n}^*$ is given by (7.34), $\boldsymbol{p}_1 \in H_l$ and $n = 15, 20, 25, 35;\ l = 1, 2, 3$.

First of all, we study the closeness of the exact size for $H_{Null}: H_2$, $H_{Null}:$ and $H_{Null}: H_4$ to the nominal size $\alpha = 0.05$. Following Dale (1986), we consid the inequality,

$$\left|\text{logit}(1-\alpha_n^{(l)}) - \text{logit}(1-\alpha)\right| \leq e,$$ (7.3

where $\text{logit}(p) \equiv \ln(p/(1-p))$. The two probabilities are considered to be "clos if they satisfy (7.35) with $e = 0.35$ and "fairly close" if they satisfy (7.35) wi $e = 0.7$. Note that for $\alpha = 0.05$, $e = 0.35$ corresponds to $\alpha_n^{(l)} \in [0.0357, 0.0695$ and $e = 0.7$ corresponds to $\alpha_n^{(l)} \in [0.0254, 0.0959]$. From the calculations th yield Figures 7.5-7.7, the test statistics that satisfy (7.35) for $e = 0.35$ are tho corresponding to $\lambda = 2/3, 1, 2$. For $e = 0.7$, only one extra test statistic, th corresponding to $\lambda = 0$, is added.

Figures 7.8-7.10 give a similar comparison of exact to asymptotic, this tin for power under a contiguous alternative. In Figures 7.8 and 7.9, it is clear th the test statistic corresponding to $\lambda = 2/3$ has the best behavior. In Figure 7.1 $\lambda = 1, 2$ are the best but $\lambda = 2/3$ is still competitive.

Figures 7.11-7.13 show (Exact power-Exact size) for the three hypothesis test H_4 versus H_3, H_3 versus H_2, and H_2 versus H_1. This is a measure of how quick the power curve increases from its probability of type I error. We see from th figures that the increase in power is a little more for tests based on negative than for positive λ. This should be tempered with the fact that for negative the exact size is considered not even "fairly close". This trade-off between si behavior and power behavior is a classical problem in hypothesis testing.

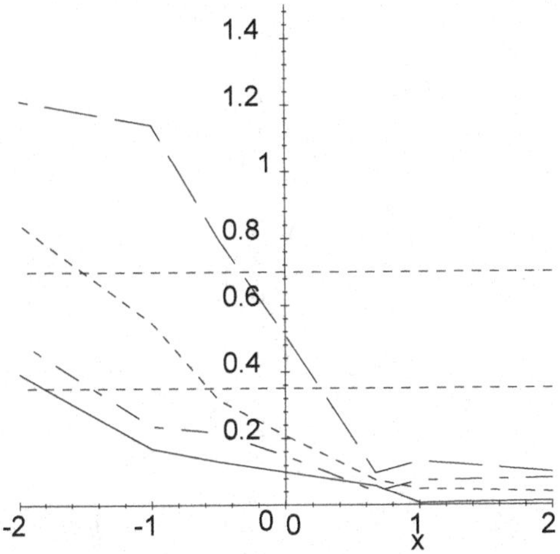

Figure 7.5. $|\text{logit}(1 - \text{Exact size}) - \text{logit}(1 - \text{Nominal size of } .05)|$ as a function of λ for model H_4. Shown are $n = 15$ (dashed line), $n = 20$ (dotted line), $n = 25$ (dash-dotted line) and $n = 35$ (solid line). The two horizontal lines correspond to Dale's bounds of $e = .35$ and $e = .7$ in (7.35).

Source: Cressie, N., Pardo, L. and Pardo, M.C. (2001).

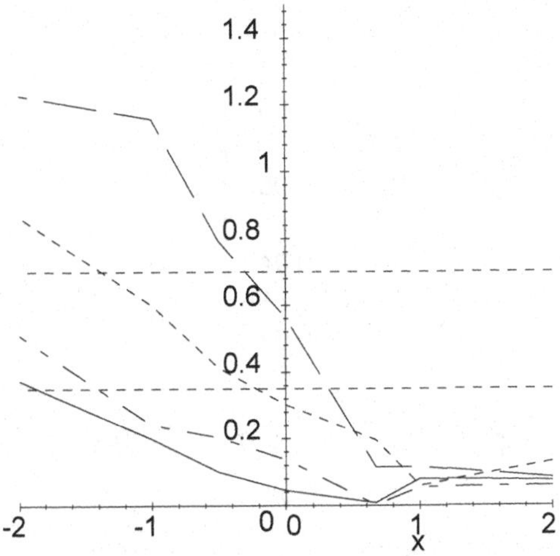

Figure 7.6. $|\text{logit}(1 - \text{Exact size}) - \text{logit}(1 - \text{Nominal size of } .0.5)|$ as a function of λ for model H_3. Shown are $n = 15$ (dashed line), $n = 20$ (dotted line), $n = 25$ (dash-dotted line) and $n = 35$ (solid line). The two horizontal lines correspond to Dale's bounds of $e = .35$ and $e = .7$ in (7.35).

Source: Cressie, N., Pardo, L. and Pardo, M.C. (2001).

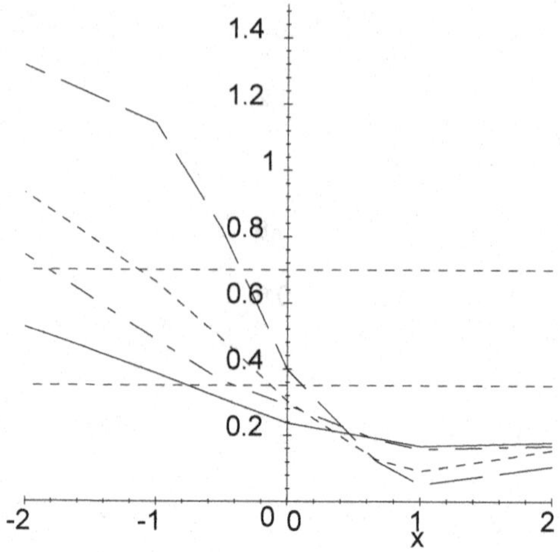

Figure 7.7. |logit(1 – Exact size) – logit(1 – Nominal size of .05)| as a function of λ for model H_2. Shown are $n = 15$ (dashed line), $n = 20$ (dotted line), $n = 25$ (dash-dotted line) and $n = 35$ (solid line). The two horizontal lines correspond to Dale's bounds $e = .35$ and $e = .7$ in (7.35).

Source: Cressie, N., Pardo, L. and Pardo, M.C. (2001).

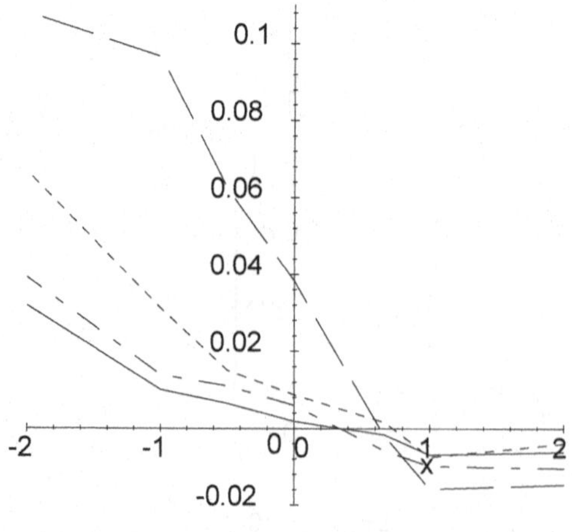

Figure 7.8. (Exact Power – Asymptotic power) as a function of λ for testing $\boldsymbol{p}_0 \in H_4$ versus $\boldsymbol{p}^*_{1,n}$ with $\boldsymbol{p}_1 \in H_3$. Shown are $n = 15$ (dashed line), $n = 20$ (dotted line), $n = 25$ (dash-dotted line) and $n = 35$ (solid line).

Source: Cressie, N., Pardo, L. and Pardo, M.C. (2001).

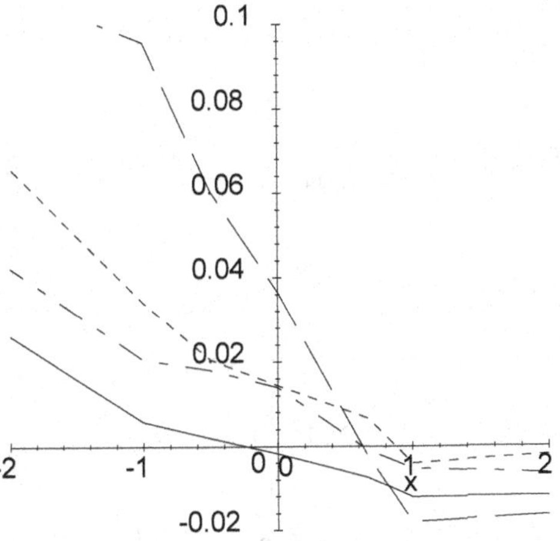

Figure 7.9. (Exact Power – Asymptotic power) as a function of λ for testing $\boldsymbol{p}_0 \in H_3$ versus $\boldsymbol{p}_{1,n}^*$ with $\boldsymbol{p}_1 \in H_2$. Shown are $n = 15$ (dashed line), $n = 20$ (dotted line), $n = 25$ (dash-dotted line) and $n = 35$ (solid line).

Source: Cressie, N., Pardo, L. and Pardo, M.C. (2001).

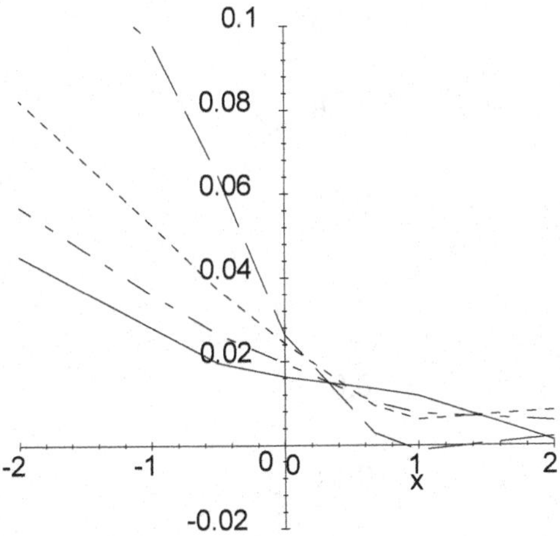

Figure 7.10. (Exact Power – Asymptotic power) as a function of λ for testing $\boldsymbol{p}_0 \in H_2$ versus $\boldsymbol{p}_{1,n}^*$ with $\boldsymbol{p}_1 \in H_1$. Shown are $n = 15$ (dashed line), $n = 20$ (dotted line), $n = 25$ (dash-dotted) and $n = 35$ (solid line).

Source: Cressie, N., Pardo, L. and Pardo, M.C. (2001).

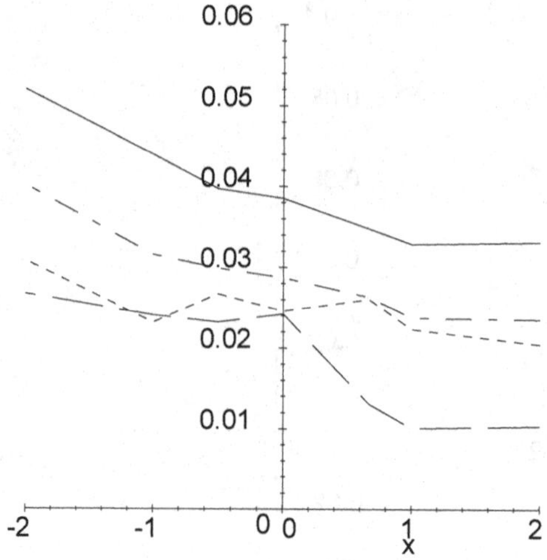

Figure 7.11. (Exact Power – Exact size) as a function of λ for testing H_4 versus H_3. Shown are $n = 15$ (dashed line), $n = 20$ (dotted line), $n = 25$ (dash-dotted line) and $n = 35$ (solid line).

Source: Cressie, N., Pardo, L. and Pardo, M.C. (2001).

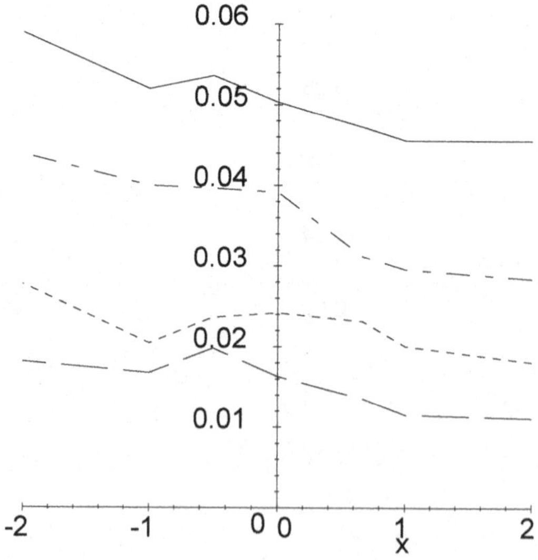

Figure 7.12. (Exact Power – Exact size) as a function of λ for testing H_3 versus H_2. Shown are $n = 15$ (dashed line), $n = 20$ (dotted line), $n = 25$ (dash-dotted line) and $n = 35$ (solid line).

Source: Cressie, N., Pardo, L. and Pardo, M.C. (2001).

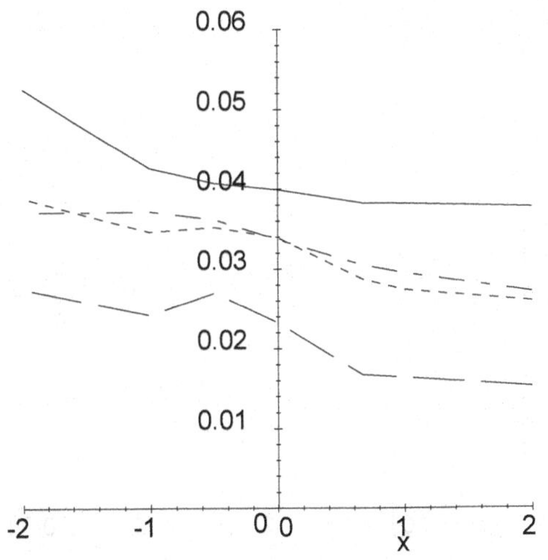

Figure 7.13. (Exact Power – Exact size) as a function of λ for testing H_2 versus H_1. Shown are $n = 15$ (dashed line), $n = 20$ (dotted line), $n = 25$ (dash-dotted line) and $n = 35$ (solid line).

Source: Cressie, N., Pardo, L. and Pardo, M.C. (2001).

In what follows, we consider only the test power-divergence test statistic that satisfy (7.34) with $e = 0.7$, and to discriminate between them we calculate:

$$g_1(\lambda) \equiv \left| AP_{i,n}^{(l)}(\lambda) - SEP_{i,n}^{(l)}(\lambda) \right|$$

and

$$g_2(\lambda) \equiv \left(SEP_{i,n}^{(l)}(\lambda) - STS_{i,n}^{(l)}(\lambda) \right)^{-1},$$

where $AP_{i,n}^{(l)}(\lambda)$ is the asymptotic power, $SEP_{i,n}^{(l)}(\lambda)$ is the simulated exact power, and $STS_{i,n}^{(l)}(\lambda)$ is the simulated test size of the test statistic $T_{\phi(\lambda),\phi(0)}^{(l)}$; $l = 1, 2, 3$, under the alternative $i = F$ (fixed), C (contiguous) and $n = 15, 20, 25, 35$. Then, for a given l, we consider a test statistic $T_{\phi(\lambda_1),\phi(0)}^{(l)}$ to be better than a test statistic $T_{\phi(\lambda_2),\phi(0)}^{(l)}$ iff

$$g_1(\lambda_1) < g_1(\lambda_2) \text{ and } g_2(\lambda_1) < g_2(\lambda_2). \tag{7.36}$$

In Figures 7.14-7.16, we plot $y = g_2(\lambda)$ versus $x = g_1(\lambda)$; from (7.36), we look for values of λ that are as close to $(0, 0)$ as possible in the (x, y) plane.

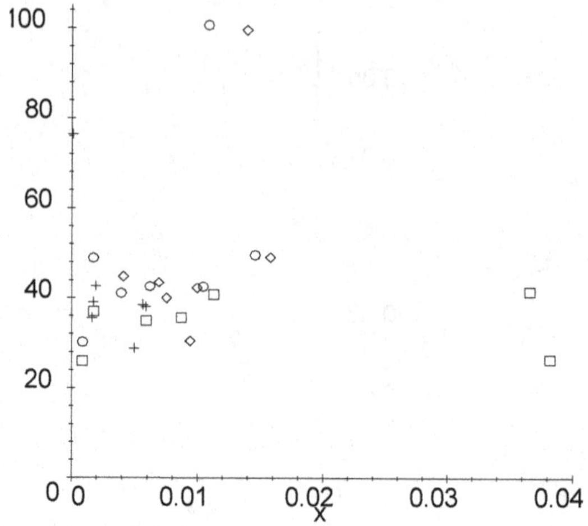

Figure 7.14. $y = g_2(\lambda)$ versus $x = g_1(\lambda)$ for $T^{(3)}_{\phi(\lambda),\phi(0)}$. Shown are $\lambda = 0$ (Square), $\lambda = 2/3$ (Cross), $\lambda = 1$ (Diamond) and $\lambda = 2$ (Circle).

Source: Cressie, N., Pardo, L. and Pardo, M.C. (2001).

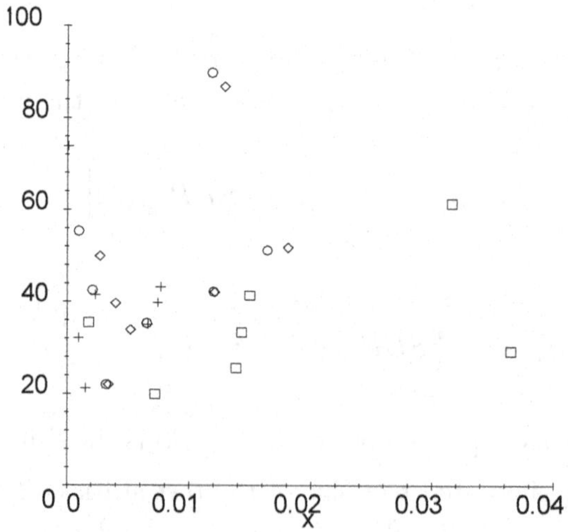

Figure 7.15. $y = g_2(\lambda)$ versus $x = g_1(\lambda)$ for $T^{(2)}_{\phi(\lambda),\phi(0)}$. Shown are $\lambda = 0$ (Square), $\lambda = 2/3$ (Cross), $\lambda = 1$ (Diamond) and $\lambda = 2$ (Circle).

Source: Cressie, N., Pardo, L. and Pardo, M.C. (2001).

The points $(g_1(\lambda), g_2(\lambda))$ far away from $(0,0)$ are those corresponding to smallest sample size $n = 15$, as expected. For this sample size, Figures 7.14-7.16 show that the exact size of the tests based on $T^{(l)}_{\phi(0),\phi(0)}$ is too large in relation to that of $T^{(l)}_{\phi(\lambda),\phi(0)}$,

$\lambda = 2/3, 1, 2$. For $n = 15$, the points $(g_1(2/3), g_2(2/3))$ are closer to $(0,0)$ than the points $(g_1(1), g_2(1))$ and $(g_1(2), g_2(2))$. Thus, according to the criterion (7.36), the test based on $T^{(l)}_{\phi(2/3),\phi(0)}$ is the best for $n = 15$.

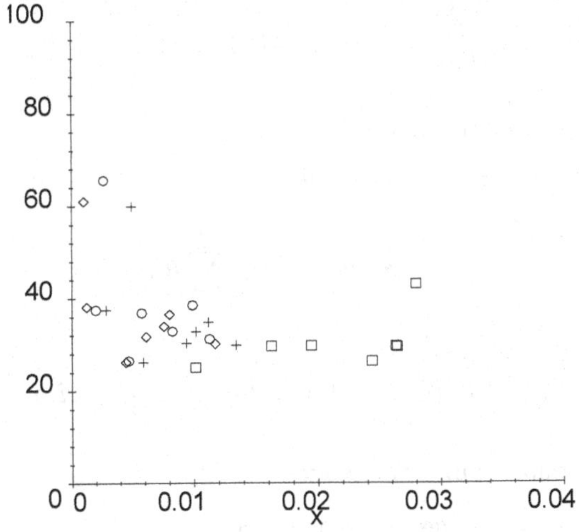

Figure 7.16. $y = g_2(\lambda)$ versus $x = g_1(\lambda)$ for $T^{(1)}_{\phi(\lambda),\phi(0)}$. Shown are $\lambda = 0$ (Square), $\lambda = 2/3$ (Cross), $\lambda = 1$ (Diamond) and $\lambda = 2$ (Circle).
Source: Cressie, N., Pardo, L. and Pardo, M.C. (2001).

For $n = 20, 25, 35$, it can be seen that $T^{(l)}_{\phi(2/3),\phi(0)}$ is better than $T^{(l)}_{\phi(\lambda),\phi(0)}$, for $\lambda = 1, 2$, according to (7.36). However, $T^{(l)}_{\phi(2/3),\phi(0)}$ is not obviously better than $T^{(l)}_{\phi(0),\phi(0)}$, since $g_1(2/3) < g_1(0)$ but $g_2(0) < g_2(2/3)$. The choice of $T^{(l)}_{\phi(2/3),\phi(0)}$ or $T^{(l)}_{\phi(0),\phi(0)}$ for $n = 20, 25, 35$, is going to depend on whether we need to make a very good approximation of the exact power (g_1 small) or whether we want to use a test statistic with as much exact power as possible (g_2 small). In the first instance, we should choose $T^{(l)}_{\phi(2/3),\phi(0)}$ and in the second instance, $T^{(l)}_{\phi(0),\phi(0)}$.

From all the simulation studies we have carried out, our conclusion is that the test based on $T^{(l)}_{\phi(2/3),\phi(0)}$ is a very good and often better alternative to the tests based on the classical statistics $T^{(l)}_{\phi(\lambda),\phi(0)}$ with $\lambda = 0, 1$.

7.6. Exercises

1. Find the matrix form of the loglinear models,

   a) $\log m_{ij}(\boldsymbol{\theta}) = u + \theta_i + \theta_j + \theta_{ij}$, $i, j = 1, 2, 3$, with

   $$\sum_{i=1}^{3} \theta_i = 0, \ \sum_{i=1}^{3} \theta_{ij} = 0, \ j = 1, 2, 3, \text{ and } \theta_{ij} = \theta_{ji}, \ i, j = 1, 2, 3.$$

   This is the symmetry model.

   b) $\log m_{ij}(\boldsymbol{\theta}) = u + \theta_{1(i)} + \theta_{2(j)} + \theta_{12(ij)}$, $i, j = 1, 2, 3$, with

   $$\sum_{i=1}^{3} \theta_{1(i)} = 0, \ \sum_{j=1}^{3} \theta_{2(j)} = 0$$

   $$\sum_{i=1}^{3} \theta_{12(ij)} = 0, \ j = 1, 2, 3 \text{ and } \theta_{12(ij)} = \theta_{12(ji)}, i, j = 1, 2, 3, \ i \neq j.$$

   This is the quasi-symmetry model.

2. Find the matrix form for the loglinear models

   a) $\log m_{ij}(\boldsymbol{\theta}) = u + \theta_{1(i)} + \theta_{2(j)} + \delta_i c_{ij}$, $i, j = 1, 2$, with

   $$\sum_{i=1}^{2} \theta_{1(i)} = \sum_{j=1}^{2} \theta_{2(j)} = \sum_{i=1}^{2} \delta_i = 0$$

   and c_{ij} known constants.

   b) $\log m_{ij}(\boldsymbol{\theta}) = u + \theta_{1(i)} + \theta_{2(j)} + \rho_j d_{ij}$, $i, j = 1, 2$, with

   $$\sum_{i=1}^{2} \theta_{1(i)} = \sum_{j=1}^{2} \theta_{2(j)} = \sum_{j=1}^{2} \rho_j = 0$$

   and d_{ij} known constants.

   c) $\log m_{ij}(\boldsymbol{\theta}) = u + \theta_{1(i)} + \theta_{2(j)} + \lambda (l_{ij})^2$, $i, j = 1, 2$, with

   $$\sum_{i=1}^{2} \theta_{1(i)} = \sum_{j=1}^{2} \theta_{2(j)} = 0$$

   and l_{ij} known constants.

3. We consider the loglinear model of quasi-symmetry given in Exercise 1. Find the maximum likelihood estimator of the expected values, $\boldsymbol{m}(\boldsymbol{\theta})$.

4. Find the expression of the minimum power-divergence estimator for the expected values, $\boldsymbol{m}(\boldsymbol{\theta})$, in the symmetry model.

5. We consider the loglinear model,

$$\log m_{ij}(\boldsymbol{\theta}) = u + \theta_{1(i)} + \theta_{2(j)} + \theta_{12(ij)}$$

with

$$\begin{aligned}
u &= \tfrac{1}{IJ} \sum_{i=1}^{I} \sum_{j=1}^{J} \log m_{ij} \\
\theta_{1(i)} &= \tfrac{1}{J} \sum_{j=1}^{J} \log m_{ij} - u \\
\theta_{2(j)} &= \tfrac{1}{I} \sum_{i=1}^{I} \log m_{ij} - u \\
\theta_{12(ij)} &= \log m_{ij} - u - \theta_{1(i)} - \theta_{2(j)}.
\end{aligned}$$

Show that

$$\theta_{12(ij)} = \frac{1}{IJ} \sum_{s=1}^{I} \sum_{t=1}^{J} \log \frac{m_{ij}(\boldsymbol{\theta})\, m_{st}(\boldsymbol{\theta})}{m_{it}(\boldsymbol{\theta})\, m_{sj}(\boldsymbol{\theta})}.$$

6. The maximum likelihood estimators for $m_{ijk}(\boldsymbol{\theta})$, $i, j, k = 1, 2$, in the model

$$\log m_{ijk}(\boldsymbol{\theta}) = u + \theta_{1(i)} + \theta_{2(j)} + \theta_{3(k)} + \theta_{12(ij)} + \theta_{13(ik)} + \theta_{23(jk)},$$

based on a random sample of size 820, are given in the following table

		Variable (C)			
		C_1		C_2	
	Variable (B)	B_1	B_2	B_1	B_2
	A_1	350.5	149.5	59.51	112.5
Variable (A)	A_2	25.51	23.49	19.49	79.51

Find the maximum likelihood estimators for the parameters $\theta_{1(i)}$ and $\theta_{12(ij)}$.

7. In 1968, 715 blue collar workers, selected from Danish Industry, were asked a number of questions concerning their job satisfaction. Some of these questions were summarized in a measure of job satisfaction. Based on similar questions the job satisfaction of the supervisors was measured. Also included in the investigation there was an external evaluation of the quality of management for each factory. The following table shows 715 workers distributed on the three variables: A: Own job satisfaction, B: Supervisors job satisfaction and C: Quality of management.

			A	
		B	Low	High
C	Bad	Low	103	87
		High	32	42
	Good	Low	59	109
		High	78	205

Source: Andersen, E. B. (1990, p. 156).

We consider the nested sequence of loglinear models H_1, H_2^{**}, H_3, H_4, H_5 and H_6 given in Section 7.1.

a) Find the maximum likelihood estimators of $\boldsymbol{m}(\boldsymbol{\theta})$ for all the models.

b) Using as significance level $\alpha = 0.05$, test the null hypothesis that data are from model H_4 based on the maximum likelihood estimator as well as on the power-divergence test statistic with $\lambda = -2, -1, -1/2, 0, 2/3$ and 1.

c) Find the best loglinear model among the models: H_1, H_2^{**}, H_3, H_4, H_5 and H_6 using the maximum likelihood estimator as well as the power-divergence test statistics $T^{(l)}_{\phi(\lambda),\phi(0)}$ with $\lambda = 0, 1, -1$ and taking as significance level $\alpha = 0.01$.

d) The same as in the previous part but with the power-divergence test statistic $S^{(l)}_{\phi(\lambda),\phi(0)}$.

8. The Swedish traffic authorities investigated in 1961 and 1962 on a trial basis the possible effects of speed limitations. In certain weeks a speed limit of 90 km/hour was enforced, while in other weeks no limits were enforced. The following table shows for two periods of the same length, one in 1961 and one in 1962, the observed number of killed persons in traffic accidents on main roads and on secondary roads.

		Type of Roads (C)			
		Main		Secondaries	
	Speed limit (B)	90Km (Hour)	Free	90Km (Hour)	Free
Year (A)	1961	8	57	42	106
	1962	11	45	37	69

Source: Andersen, E. B. (1990, p. 158).

We consider the nested sequence of loglinear-models H_1, H_2, H_3, H_4, H_5 and H_6 given in Section 7.1.

a) Find the maximum likelihood estimators of $\boldsymbol{m}(\boldsymbol{\theta})$ for all the models.

b) We consider the model H_3. Find the maximum likelihood estimators for the parameters $\theta_{1(i)}, \theta_{2(j)}, \theta_{3(k)}, \theta_{23(jk)}$.

c) Using as significance level $\alpha = 0.05$, test the null hypothesis that data are from model H_3 based on the maximum likelihood estimator as well as on the power-divergence test statistic with $\lambda = -2, -1, -1/2, 0, 2/3, 1$ and 2.

d) Find the best loglinear model among the models: H_1, H_2, H_3, H_4, H_5 and H_6 using the maximum likelihood estimator as well as the power-divergence test statistics $T^{(l)}_{\phi(\lambda),\phi(0)}$ with $\lambda = -2, -1/2, 0, 2/3, 1$ and 2 assuming that for testing

$$H_{Null} : H_3 \text{ versus } H_{Alt} : H_2$$

the null hypothesis was accepted and taking as significance level $\alpha = 0.05$.

e) The same as in the previous part but with the power-divergence test statistic $S^{(l)}_{\phi(\lambda),\phi(0)}$ with $\lambda = 1$.

9. In 1973 a retrospective study of cancer of the ovary was carried out. Information was obtained from 299 women, who were operated on for cancer of the ovary 10 years before. The following four dichotomous variables were observed:

A: Whether $X-$ray treatment was received or not.

B: Whether the woman had survived the operation by 10 years or not.

C: Whether the operation was radical or limited.

D: Whether the cancer at the time of operation was in an early or an advanced stage.

The observed number of women are shown in the following table:

			A : X-ray	
D (*Stage*)	*C* (*Operation*)	*B* (*Survival*)	NO	YES
Early	Radical	No	10	17
		YES	41	64
	Limited	No	1	3
		YES	13	9
Advanced	Radical	NO	38	64
		YES	6	11
	Limited	NO	3	13
		YES	1	5

Source: Andersen, E. B. (1998, p. 121).

We consider the following nested sequence of loglinear models:

$$\begin{aligned} H_1 : \log m_{ijkl}(\boldsymbol{\theta}) &= u + \theta_{1(i)} + \theta_{2(j)} + \theta_{3(k)} + \theta_{4(l)} \\ &+ \theta_{12(ij)} + \theta_{13(ik)} + \theta_{14(il)} + \theta_{23(jk)} + \theta_{24(jl)} + \theta_{34(kl)} \\ &+ \theta_{123(ijk)} + \theta_{124(ijl)} + \theta_{134(ikl)} + \theta_{234(jkl)}. \end{aligned}$$

$$\begin{aligned} H_2 : \log m_{ijkl}(\boldsymbol{\theta}) &= u + \theta_{1(i)} + \theta_{2(j)} + \theta_{3(k)} + \theta_{4(l)} \\ &+ \theta_{12(ij)} + \theta_{13(ik)} + \theta_{14(il)} + \theta_{23(jk)} + \theta_{24(jl)} + \theta_{34(kl)} \\ &+ \theta_{124(ijl)} + \theta_{134(ikl)} + \theta_{234(jkl)}. \end{aligned}$$

$$\begin{aligned} H_3 : \log m_{ijkl}(\boldsymbol{\theta}) &= u + \theta_{1(i)} + \theta_{2(j)} + \theta_{3(k)} + \theta_{4(l)} \\ &+ \theta_{12(ij)} + \theta_{13(ik)} + \theta_{14(il)} + \theta_{23(jk)} + \theta_{24(jl)} + \theta_{34(kl)} \\ &+ \theta_{134(ikl)} + \theta_{234(jkl)}. \end{aligned}$$

$$\begin{aligned} H_4 : \log m_{ijkl}(\boldsymbol{\theta}) &= u + \theta_{1(i)} + \theta_{2(j)} + \theta_{3(k)} + \theta_{4(l)} \\ &+ \theta_{12(ij)} + \theta_{13(ik)} + \theta_{14(il)} + \theta_{23(jk)} + \theta_{24(jl)} + \theta_{34(kl)} \\ &+ \theta_{134(ikl)}. \end{aligned}$$

$$\begin{aligned} H_5: \log m_{ijkl}(\boldsymbol{\theta}) = & \; u + \theta_{1(i)} + \theta_{2(j)} + \theta_{3(k)} + \theta_{4(l)} \\ + & \; \theta_{13(ik)} + \theta_{14(il)} + \theta_{23(jk)} + \theta_{24(jl)} + \theta_{34(kl)} + \theta_{134(ikl)}. \end{aligned}$$

$$\begin{aligned} H_6: \log m_{ijkl}(\boldsymbol{\theta}) = & \; u + \theta_{1(i)} + \theta_{2(j)} + \theta_{3(k)} \\ + & \; \theta_{4(l)} + \theta_{13(ik)} + \theta_{14(il)} + \theta_{24(jl)} + \theta_{34(kl)} + \theta_{134(ikl)}. \end{aligned}$$

$$\begin{aligned} H_7: \log m_{ijkl}(\boldsymbol{\theta}) = & \; u + \theta_{1(i)} + \theta_{2(j)} + \theta_{3(k)} \\ + & \; \theta_{4(l)} + \theta_{13(ik)} + \theta_{14(il)} + \theta_{34(kl)} + \theta_{134(ikl)}. \end{aligned}$$

a) Using as significance level $\alpha = 0.05$, test the null hypothesis that data are from model H_5 based on the maximum likelihood estimator as well as on the power-divergence test statistic with $\lambda = -2, -1, -1/2, 0, 2/3, 1$ and 2.

b) Find the best loglinear models among the models $H_1, H_2, H_3, H_4, H_5, H_6$ and H_7 considered above based on $S^{(l)}_{\phi(0),\phi(0)}$, using as significance level $\alpha = 0.05$

10. Show that $\boldsymbol{A}_{(l)}\boldsymbol{A}_{(l+1)} = \boldsymbol{A}_{(l+1)}\boldsymbol{A}_{(l)} = \boldsymbol{A}_{(l+1)}$ and $\boldsymbol{A}_{(l)}\boldsymbol{A}_{(l)} = \boldsymbol{A}_{(l)}$, where $\boldsymbol{A}_{(i)}$ is defined in (7.19) for $i = l, l+1$.

7.7. Answers to Exercises

1. Under the given restrictions it is easy to verify that the matrix form is given by

$$\begin{pmatrix} \log m_{11}(\boldsymbol{\theta}) \\ \log m_{12}(\boldsymbol{\theta}) \\ \log m_{13}(\boldsymbol{\theta}) \\ \log m_{21}(\boldsymbol{\theta}) \\ \log m_{22}(\boldsymbol{\theta}) \\ \log m_{23}(\boldsymbol{\theta}) \\ \log m_{31}(\boldsymbol{\theta}) \\ \log m_{32}(\boldsymbol{\theta}) \\ \log m_{33}(\boldsymbol{\theta}) \end{pmatrix} = \begin{pmatrix} 1 & 2 & 0 & 1 & 0 & 0 \\ 1 & 1 & 1 & 0 & 1 & 0 \\ 1 & 0 & -1 & -1 & -1 & 0 \\ 1 & 1 & 1 & 0 & 1 & 0 \\ 1 & 0 & 2 & 0 & 0 & 1 \\ 1 & -1 & 0 & 0 & -1 & -1 \\ 1 & 0 & -1 & -1 & -1 & 0 \\ 1 & -1 & 0 & 0 & -1 & -1 \\ 1 & -2 & -2 & 1 & 2 & 1 \end{pmatrix} \begin{pmatrix} u \\ \theta_1 \\ \theta_2 \\ \theta_{11} \\ \theta_{12} \\ \theta_{22} \end{pmatrix}$$

in the case *a)* and by

$$\begin{pmatrix} \log m_{11}(\boldsymbol{\theta}) \\ \log m_{12}(\boldsymbol{\theta}) \\ \log m_{13}(\boldsymbol{\theta}) \\ \log m_{21}(\boldsymbol{\theta}) \\ \log m_{22}(\boldsymbol{\theta}) \\ \log m_{23}(\boldsymbol{\theta}) \\ \log m_{31}(\boldsymbol{\theta}) \\ \log m_{32}(\boldsymbol{\theta}) \\ \log m_{33}(\boldsymbol{\theta}) \end{pmatrix} = \begin{pmatrix} 1 & 1 & 0 & 1 & 0 & 1 & 0 & 0 \\ 1 & 1 & 0 & 0 & 1 & 0 & 1 & 0 \\ 1 & 1 & 0 & -1 & -1 & -1 & -1 & 0 \\ 1 & 0 & 1 & 1 & 0 & 0 & 1 & 0 \\ 1 & 0 & 1 & 0 & 1 & 0 & 0 & 1 \\ 1 & 0 & 1 & -1 & -1 & 0 & -1 & -1 \\ 1 & -1 & -1 & 1 & 0 & -1 & -1 & 0 \\ 1 & -1 & -1 & 0 & 1 & 0 & -1 & -1 \\ 1 & -1 & -1 & -1 & -1 & 1 & 2 & 1 \end{pmatrix} \begin{pmatrix} u \\ \theta_{1(1)} \\ \theta_{1(2)} \\ \theta_{2(1)} \\ \theta_{2(2)} \\ \theta_{12(11)} \\ \theta_{12(12)} \\ \theta_{12(22)} \end{pmatrix}$$

in the second case.

2. *a)* The matrix $\boldsymbol{X}$ is given by

$$\boldsymbol{X} = \begin{pmatrix} 1 & 1 & 1 & c_{11} \\ 1 & 1 & -1 & c_{12} \\ 1 & -1 & 1 & -c_{21} \\ 1 & -1 & -1 & -c_{22} \end{pmatrix}.$$

b) The matrix $\boldsymbol{X}$ is given by

$$\boldsymbol{X} = \begin{pmatrix} 1 & 1 & 1 & d_{11} \\ 1 & 1 & -1 & -d_{12} \\ 1 & -1 & 1 & d_{21} \\ 1 & -1 & -1 & -d_{22} \end{pmatrix}.$$

c) The matrix $\boldsymbol{X}$ is given by

$$\boldsymbol{X} = \begin{pmatrix} 1 & 1 & 1 & l_{11}^2 \\ 1 & 1 & -1 & l_{12}^2 \\ 1 & -1 & 1 & l_{21}^2 \\ 1 & -1 & -1 & l_{22}^2 \end{pmatrix}.$$

3. To find the maximum likelihood estimator of $\boldsymbol{m}(\boldsymbol{\theta})$ we must maximize the expression, $\sum_{i=1}^{I}\sum_{j=1}^{I} n_{ij} \log m_{ij}(\boldsymbol{\theta})$. We have

$$\begin{aligned}\sum_{i=1}^{I}\sum_{j=1}^{I} n_{ij} \log m_{ij}(\boldsymbol{\theta}) = \; & nu + \sum_{i=1}^{I}\sum_{j=1}^{I} n_{ij}\theta_{1(i)} + \sum_{i=1}^{I}\sum_{j=1}^{I} n_{ij}\theta_{2(j)} \\ + \; & \sum_{i=1}^{I}\sum_{j=1}^{I} \left(\frac{n_{ij}+n_{ji}}{2}\right)\theta_{12(ij)},\end{aligned}$$

and

$$\sum_{i=1}^{I}\sum_{j=1}^{I} n_{ij} \log m_{ij}(\boldsymbol{\theta}) = \sum_{i=1}^{I}\sum_{j=1}^{I} n_{ij} \log np_{ij}(\boldsymbol{\theta}) = n\log n + \sum_{i=1}^{I}\sum_{j=1}^{I} n_{ij} \log p_{ij}(\boldsymbol{\theta}).$$

We denote

$$L(\boldsymbol{\theta}) \equiv \log \Pr{}_{\boldsymbol{\theta}}\left(N_{11} = n_{11}, ..., N_{II} = n_{II}\right);$$

then,

$$\begin{aligned}
L(\boldsymbol{\theta}) = & \log \tfrac{n!}{n_{11}!...n_{II}!} + \sum_{i=1}^{I}\sum_{j=1}^{I} n_{ij} \log p_{ij}(\boldsymbol{\theta}) \\
= & \log \tfrac{n!}{n_{11}!...n_{II}!} - n\log n + \sum_{i=1}^{I}\sum_{j=1}^{I} n_{ij} \log m_{ij}(\boldsymbol{\theta}) \\
= & \log \tfrac{n!}{n_{11}!...n_{II}!} - n\log n + nu + \sum_{i=1}^{I} n_{i*}\theta_{1(i)} + \sum_{j=1}^{I} n_{*j}\theta_{2(j)} \\
+ & \sum_{i=1}^{I}\sum_{j=1}^{I} \left(\tfrac{n_{ij}+n_{ji}}{2}\right)\theta_{12(ij)}
\end{aligned}$$

and

$$\begin{aligned}
\Pr_{\boldsymbol{\theta}}(N_{11} = n_{11}, ..., N_{II} = n_{II}) = & \exp(-n\log n)\tfrac{n!}{n_{11}!...n_{II}!} \exp\left\{ nu + \sum_{i=1}^{I} n_{i*}\theta_{1(i)} \right. \\
+ & \left. \sum_{j=1}^{I} n_{*j}\theta_{2(j)} + \sum_{i=1}^{I}\sum_{j=1}^{J} \left(\tfrac{n_{ij}+n_{ji}}{2}\right)\theta_{12(ij)} \right\}.
\end{aligned}$$

Therefore,

$$\begin{aligned}
\exp(-nu) = & \sum_{n_{11},...,n_{II}} \exp(-n\log n)\tfrac{n!}{n_{11}!...n_{II}!} \exp\left\{ \sum_{i=1}^{I} n_{i*}\theta_{1(i)} + \right. \\
+ & \left. \sum_{j=1}^{I} n_{*j}\theta_{2(j)} + \sum_{i=1}^{I}\sum_{j=1}^{I} \left(\tfrac{n_{ij}+n_{ji}}{2}\right)\theta_{12(ij)} \right\}.
\end{aligned}$$

Differentiating in the two members of the previous equality with respect to $\theta_{12(ij)}$, we have

$$\begin{aligned}
-n\exp(-nu)\frac{\partial u}{\partial \theta_{12(ij)}} = & \sum_{n_{11},...,n_{II}} \exp(-n\log n)\tfrac{n!}{n_{11}!...n_{II}!} \exp\left\{ \sum_{i=1}^{I} n_{i*}\theta_{1(i)} + \right. \\
+ & \left. \sum_{j=1}^{I} n_{*j}\theta_{2(j)} + \sum_{i=1}^{I}\sum_{j=1}^{I} \left(\tfrac{n_{ij}+n_{ji}}{2}\right)\theta_{12(ij)} \right\} \left(\tfrac{n_{ij}+n_{ji}}{2}\right).
\end{aligned}$$

Then

$$\begin{aligned}
-n\frac{\partial u}{\partial \theta_{12(ij)}} = & \sum_{n_{11},...,n_{II}} \left(\tfrac{n_{ij}+n_{ji}}{2}\right) \Pr_{\boldsymbol{\theta}}(N_{11} = n_{11}, ..., N_{II} = n_{II}) \\
= & E\left[\tfrac{N_{ij}+N_{ji}}{2}\right] \\
= & \tfrac{m_{ij}(\boldsymbol{\theta})+m_{ji}(\boldsymbol{\theta})}{2}.
\end{aligned}$$

On the other hand

$$\begin{aligned}\frac{\partial L(\boldsymbol{\theta})}{\partial \theta_{12(ij)}} &= n\frac{\partial u}{\partial \theta_{12(ij)}} + \frac{n_{ij}+n_{ji}}{2} \\ &= -\frac{m_{ij}(\boldsymbol{\theta}) + m_{ji}(\boldsymbol{\theta})}{2} + \frac{n_{ij}+n_{ji}}{2}.\end{aligned}$$

Therefore

$$\frac{\partial L(\boldsymbol{\theta})}{\partial \theta_{12(ij)}} = 0,$$

or equivalently

$$0 = -\frac{m_{ij}(\boldsymbol{\theta}) + m_{ji}(\boldsymbol{\theta})}{2} + \frac{n_{ij}+n_{ji}}{2}.$$

Then

$$m_{ij}(\widehat{\boldsymbol{\theta}}) + m_{ji}(\widehat{\boldsymbol{\theta}}) = n_{ij} + n_{ji}, \; i \neq j.$$

In the same way differentiating with respect to $\theta_{1(i)}$ and $\theta_{2(j)}$ we have

$$m_{i*}(\widehat{\boldsymbol{\theta}}) = n_{i*} \text{ and } m_{*j}(\widehat{\boldsymbol{\theta}}) = n_{*j}.$$

Finally, the maximum likelihood estimators, $m_{ij}(\widehat{\boldsymbol{\theta}})$, are obtained as a solution of the system of equations,

$$\begin{cases} m_{ij}(\widehat{\boldsymbol{\theta}}) + m_{ji}(\widehat{\boldsymbol{\theta}}) = n_{ij} + n_{ji}, \; i \neq j \\ m_{i*}(\widehat{\boldsymbol{\theta}}) = n_{i*} \\ m_{*j}(\widehat{\boldsymbol{\theta}}) = n_{*j} \end{cases}.$$

The first set of equations jointly with one of the other two is enough to find the maximum likelihood estimator of $m_{ij}(\widehat{\boldsymbol{\theta}})$.

4. To find the minimum power-divergence estimator we must minimize in $\boldsymbol{\theta}$ the expression

$$D_{\phi_{(\lambda)}}(\widehat{\boldsymbol{p}}, \boldsymbol{p}(\boldsymbol{\theta})) + \mu\left(\sum_{i=1}^{I}\sum_{j=1}^{I} p_{ij}(\boldsymbol{\theta}) - 1\right) = 0, \tag{7.37}$$

with

$$\begin{aligned} D_{\phi_{(\lambda)}}(\widehat{\boldsymbol{p}}, \boldsymbol{p}(\boldsymbol{\theta})) &= \frac{1}{\lambda(\lambda+1)}\left(\sum_{i=1}^{I}\sum_{j=1}^{I}\frac{\widehat{p}_{ij}^{\lambda+1}}{p_{ij}(\boldsymbol{\theta})^{\lambda}} - 1\right) \\ &= \frac{1}{\lambda(\lambda+1)}\sum_{i=1}^{I}\sum_{j=1}^{I}\left(\frac{\widehat{p}_{ij}^{\lambda+1}+\widehat{p}_{ji}^{\lambda+1}}{2}\exp\left(-\lambda\log p_{ij}(\boldsymbol{\theta})\right) - 1\right) \end{aligned} \tag{7.38}$$

and

$$\log p_{ij}(\boldsymbol{\theta}) = \log n + \log m_{ij}(\boldsymbol{\theta}) = \log n + u + \theta_i + \theta_j + \theta_{ij}.$$

It is clear that

$$\frac{\partial \log p_{ij}(\boldsymbol{\theta})}{\partial \theta_{ij}} = \frac{\partial u}{\partial \theta_{ij}} + 1,$$

and in a similar way to Exercise 3 we can find that

$$n\frac{\partial u}{\partial \theta_{ij}} = -m_{ij}(\boldsymbol{\theta}),$$

then

$$\frac{\partial \log p_{ij}(\boldsymbol{\theta})}{\partial \theta_{ij}} = -p_{ij}(\boldsymbol{\theta}) + 1.$$

Also,

$$\frac{\partial \sum_{i=1}^{I}\sum_{j=1}^{I} p_{ij}(\boldsymbol{\theta})}{\partial \theta_{ij}} = p_{ij}(\boldsymbol{\theta})\left(-p_{ij}(\boldsymbol{\theta}) + 1\right).$$

Differentiating in (7.37) with respect to θ_{ij}, taking into account the expression (7.38),

$$\frac{1}{\lambda(\lambda+1)}\left(\frac{\widehat{p}_{ij}^{\lambda+1}+\widehat{p}_{ji}^{\lambda+1}}{2} p_{ij}(\boldsymbol{\theta})^{-\lambda}(-\lambda)\left(-p_{ij}(\boldsymbol{\theta})+1\right)\right) + \mu p_{ij}(\boldsymbol{\theta})\left(-p_{ij}(\boldsymbol{\theta})+1\right) = 0,$$

i.e.,

$$\frac{1}{\lambda+1}\frac{\widehat{p}_{ij}^{\lambda+1}+\widehat{p}_{ji}^{\lambda+1}}{2} = \mu p_{ij}(\boldsymbol{\theta})^{\lambda+1},$$

then

$$\left(\frac{1}{\lambda+1}\right)^{\frac{1}{\lambda+1}}\left(\frac{\widehat{p}_{ij}^{\lambda+1}+\widehat{p}_{ji}^{\lambda+1}}{2}\right)^{\frac{1}{\lambda+1}} = \mu^{\frac{1}{\lambda+1}} p_{ij}(\boldsymbol{\theta}),$$

and

$$\mu^{\frac{1}{\lambda+1}} = \left(\frac{1}{\lambda+1}\right)^{\frac{1}{\lambda+1}}\sum_{i=1}^{I}\sum_{j=1}^{I}\left(\frac{\widehat{p}_{ij}^{\lambda+1}+\widehat{p}_{ji}^{\lambda+1}}{2}\right)^{\frac{1}{\lambda+1}}.$$

Therefore

$$p_{ij}(\widehat{\boldsymbol{\theta}}^{(\lambda)}) = \frac{\left(\frac{\widehat{p}_{ij}^{\lambda+1}+\widehat{p}_{ji}^{\lambda+1}}{2}\right)^{\frac{1}{\lambda+1}}}{\sum_{i=1}^{I}\sum_{j=1}^{I}\left(\frac{\widehat{p}_{ij}^{\lambda+1}+\widehat{p}_{ji}^{\lambda+1}}{2}\right)^{\frac{1}{\lambda+1}}} = \frac{\left(\widehat{p}_{ij}^{\lambda+1}+\widehat{p}_{ji}^{\lambda+1}\right)^{\frac{1}{\lambda+1}}}{\sum_{i=1}^{I}\sum_{j=1}^{I}\left(\widehat{p}_{ij}^{\lambda+1}+\widehat{p}_{ji}^{\lambda+1}\right)^{\frac{1}{\lambda+1}}}.$$

5. We have

$$\theta_{12(ij)} = \log m_{ij}(\boldsymbol{\theta}) - u - \theta_{1(i)} - \theta_{2(j)},$$

then

$$\begin{aligned}
\theta_{12(ij)} &= \log m_{ij}(\boldsymbol{\theta}) - u - \tfrac{1}{J}\sum_{j=1}^{J}\log m_{ij}(\boldsymbol{\theta}) + u - \tfrac{1}{I}\sum_{i=1}^{I}\log m_{ij}(\boldsymbol{\theta}) + u \\
&= \log m_{ij}(\boldsymbol{\theta}) - \tfrac{1}{J}\sum_{j=1}^{J}\log m_{ij}(\boldsymbol{\theta}) - \tfrac{1}{I}\sum_{i=1}^{I}\log m_{ij}(\boldsymbol{\theta}) + \tfrac{1}{IJ}\sum_{i=1}^{I}\sum_{j=1}^{J}\log m_{ij}(\boldsymbol{\theta}) \\
&= \tfrac{1}{IJ}\sum_{s=1}^{I}\sum_{t=1}^{J}\log\frac{m_{ij}(\boldsymbol{\theta})\, m_{st}(\boldsymbol{\theta})}{m_{it}(\boldsymbol{\theta})\, m_{sj}(\boldsymbol{\theta})}.
\end{aligned}$$

6. For a three-way contingency table we know

$$\begin{aligned}
\widehat{u} &= \sum_{i=1}^{I}\sum_{j=1}^{J}\sum_{k=1}^{K}\log m_{ijk}(\widehat{\boldsymbol{\theta}})/IJK \\
\widehat{\theta}_{1(i)} &= \sum_{j=1}^{J}\sum_{k=1}^{K}\log m_{ijk}(\widehat{\boldsymbol{\theta}})/JK - \widehat{u}
\end{aligned}$$

and

$$\begin{aligned}
\widehat{\theta}_{12(ij)} &= \sum_{k=1}^{K}\log m_{ijk}(\widehat{\boldsymbol{\theta}})/K - \sum_{j=1}^{J}\sum_{k=1}^{K}\log m_{ijk}(\widehat{\boldsymbol{\theta}})/JK \\
&- \sum_{i=1}^{I}\sum_{k=1}^{K}\log m_{ijk}(\widehat{\boldsymbol{\theta}})/IK + \widehat{u}.
\end{aligned}$$

Substituting we obtain

$$\begin{aligned}
\widehat{u} &= \tfrac{1}{8}\sum_{i=1}^{I}\sum_{j=1}^{J}\sum_{k=1}^{K}\log m_{ijk}(\widehat{\boldsymbol{\theta}}) && = 4.1771 \\
\widehat{\theta}_{1(1)} &= \sum_{j=1}^{J}\sum_{k=1}^{K}\log m_{1jk}(\widehat{\boldsymbol{\theta}})/JK - \widehat{u} && = 0.74818 \\
\widehat{\theta}_{1(2)} &= \sum_{j=1}^{J}\sum_{k=1}^{K}\log m_{2jk}(\widehat{\boldsymbol{\theta}})/JK - \widehat{u} && = -0.7418.
\end{aligned}$$

Only we have to calculate $\widehat{\theta}_{12(11)}$ because

$$\begin{aligned}
\widehat{\theta}_{12(11)} + \widehat{\theta}_{12(12)} &= 0 \\
\widehat{\theta}_{12(11)} + \widehat{\theta}_{12(21)} &= 0.
\end{aligned}$$

Then we get

$$\widehat{\theta}_{12(11)} = 4.9727 - 4.9189 - 4.0386 + 4.1771 = 0.1923.$$

7. *a)* For the models H_2^{**}, H_3, H_4, H_5 and H_6, the maximum likelihood estimators are

H_2^{**}	H_3	H_4	H_5	H_6
$\frac{n_{ij*}n_{*jk}}{nn_{*j*}}$	$\frac{n_{*jk}n_{i**}}{n}$	$\frac{n_{i**}n_{*j*}n_{**k}}{n^2}$	$\frac{n_{*j*}n_{**k}}{nI}$	$\frac{n_{**k}}{IJ}$.

This result follows because, for instance, the model H_2^{**} can be characterized by

$$p_{ijk} = \frac{p_{ij*}\, p_{*jk}}{p_{*j*}}.$$

For the model H_1 we do not have an explicit expression and we must use a statistical package (Statgraphics, SAS, SPSS, etc.). In the following table we present the maximum likelihood estimators for $m_{ijk}(\boldsymbol{\theta})$ for the different models.

(i,j,k)	n_{ijk}	H_1	H_2^{**}	H_3	H_4	H_5	H_6
$(1,1,1)$	103	102.3	86	72.3	50.3	66.1	66
$(1,1,2)$	59	59.7	76	63.9	85.9	112.9	112.8
$(1,2,1)$	32	32.7	22.8	28.2	50.1	65.9	66
$(1,2,2)$	78	77.3	87.2	107.7	85.7	112.6	112.8
$(2,1,1)$	87	87.7	104	117.7	81.9	66.1	66
$(2,1,2)$	109	108.3	92	104.1	139.9	112.9	112.8
$(2,2,1)$	42	41.3	51.2	45.8	81.7	65.9	66
$(2,2,2)$	205	205.7	195.8	175.3	139.5	112.6	112.8

b) We are going to establish the goodness-of-fit to the data of the model H_4. We consider the power-divergence test statistics

$$T^{(4)}_{\phi_{(\lambda)}\phi_{(0)}} = \frac{2}{\lambda(\lambda+1)} \sum_{i=1}^{2}\sum_{j=1}^{2}\sum_{k=1}^{2} \frac{(n_{ijk})^{\lambda+1}}{\left(m_{ijk}(\widehat{\boldsymbol{\theta}}^{(4)})\right)^{\lambda}} - \frac{2n}{\lambda(\lambda+1)},$$

whose asymptotic distribution is chi-square with 4 degrees of freedom. In the following table the values of the test statistics are presented for some choices of λ.

λ	-2	-1	-1/2	0	2/3	1
$T^{(4)}_{\phi_{(\lambda)},\phi_{(0)}}$	117.738	114.746	115.57	117.98	123.85	128.057

On other hand we have $\chi^2_{4;\,0.05} = 9.49$. Then we should reject the null hypothesis.

c) For $\lambda = 1$, we have

	$d_l - d_{l+1}$	$T^{(l)}_{\phi_{(1)},\phi_{(0)}}$	$\chi^2_{d_l-d_{l+1},0.01}$
H_2^{**} versus H_1	1	19.9	6.635

For $\lambda = 0$, we have

	$d_l - d_{l+1}$	$T^{(l)}_{\phi_{(0)},\phi_{(0)}}$	$\chi^2_{d_l-d_{l+1},0.01}$
H_2^{**} versus H_1	1	19.647	6.635

Finally, for $\lambda = -1$,

	$d_l - d_{l+1}$	$T^{(l)}_{\phi_{(-1)},\phi_{(0)}}$	$\chi^2_{d_l-d_{l+1},0.01}$
H_2^{**} versus H_1	1	19.419	6.635

Independently of the value of λ we should choose the model H_1.

d) If we consider the test statistic

$$S^{(l)}_{\phi(\lambda),\phi(0)} = 2n\left(D_{\phi(\lambda)}\left(\widehat{\mathbf{p}},\mathbf{p}(\widehat{\boldsymbol{\theta}}^{(l+1)})\right) - D_{\phi(\lambda)}\left(\widehat{\mathbf{p}},\mathbf{p}(\widehat{\boldsymbol{\theta}}^{(l)})\right)\right).$$

For $l = 1$, we have

$$\begin{aligned} S^{(l)}_{\phi(1),\phi(0)} &= 19.884 - 0.0648 = 19.819 \\ S^{(l)}_{\phi(0),\phi(0)} &= 19.712 - 0.0646 = 19.647 \\ S^{(l)}_{\phi(-1),\phi(0)} &= 19.568 - 0.0851 = 19.483, \end{aligned}$$

and the conclusion is the same as in *c*).

8. *a)* In a similar way to the previous exercise we get

(i,j,k)	n_{ijk}	H_1	H_2	H_3	H_4	H_5	H_6
$(1,1,1)$	8	8.8	10.21	10.8	18	15.8	36.3
$(1,1,2)$	42	41.2	46	44.9	37.7	33.2	63.5
$(1,2,1)$	57	56.2	54.80	57.9	50.8	44.7	30.3
$(1,2,2)$	106	106.8	102	99.4	106.6	93.8	63.5
$(2,1,1)$	11	10.2	8.8	8.2	13.7	15.8	30.3
$(2,1,2)$	37	37.8	33	34.1	28.7	33.2	63.5
$(2,2,1)$	45	45.8	47	44.1	38.6	44.7	30.3
$(2,2,2)$	69	68.2	73	75.6	81.1	93.8	63.5

b) The maximum likelihood estimators for the parameters of the model H_3 are given by

$\widehat{\theta}_{1(1)}$	$\widehat{\theta}_{2(1)}$	$\widehat{\theta}_{3(1)}$	$\widehat{\theta}_{23(12)}$
0.1368	-0.619	-0.4912	-0.2213

c) We have to check the goodness-of-fit of the data to the model H_3,

$$T^{(3)}_{\phi(\lambda)\phi(0)} = \frac{2}{\lambda(\lambda+1)}\sum_{i=1}^{2}\sum_{j=1}^{2}\sum_{k=1}^{2}\frac{(n_{ijk})^{\lambda+1}}{\left(m_{ijk}(\widehat{\boldsymbol{\theta}}^{(3)})\right)^{\lambda}} - \frac{2n}{\lambda(\lambda+1)}.$$

In the following table we present the values of the power-divergence test statistic for the different values of λ,

λ	-2	$-1/2$	-1	0	$2/3$	1	2
$T^{(3)}_{\phi(\lambda)\phi(0)}$	3.1795	3.1342	3.1431	3.132	3.1391	3.147	3.1882

On the other hand $\chi^2_{3;\ 0.05} = 7.81$. Then we should not reject the null hypothesis.

d) First we present the test

$$H_{Null} : H_4 \text{ versus } H_{Alt} : H_3,$$

using the power-divergence test statistic

$$T^{(3)}_{\phi_{(\lambda)}\phi_{(0)}} = \frac{2}{\lambda(\lambda+1)} \sum_{i=1}^{2}\sum_{j=1}^{2}\sum_{k=1}^{2} m_{ijk}(\widehat{\boldsymbol{\theta}}^{(4)}) \left(\left(\frac{m_{ijk}(\widehat{\boldsymbol{\theta}}^{(4)})}{m_{ijk}(\widehat{\boldsymbol{\theta}}^{(3)})} \right)^{\lambda} - 1 \right).$$

We have

λ	-2	-0.5	-1	0	2/3	1	2
$T^{(3)}_{\phi_{(\lambda)}\phi_{(0)}}$	10.069	11.136	10.714	11.626	12.413	12.873	14.579

and $\chi^2_{1;\,0.05} = 3.84$. Then we should reject the null hypothesis and to choose the model H_3, i.e.,

$$H_3 : \log p_{ijk}(\boldsymbol{\theta}) = u + \theta_{1(i)} + \theta_{2(j)} + \theta_{3(k)} + \theta_{23(ij)}.$$

e) The results using the test statistics $S^{(l)}_{\phi_{(\lambda)}\phi_{(0)}}$ are given in the following table

Model	$2nD_{\phi_{(0)}}(\widehat{\boldsymbol{p}}, \boldsymbol{p}(\widehat{\boldsymbol{\theta}}^{(l)}))$	$2nD_{\phi_{(1)}}(\widehat{\boldsymbol{p}}, \boldsymbol{p}(\widehat{\boldsymbol{\theta}}^{(l)}))$	$S^{(l)}_{\phi_{(0)}\phi_{(0)}}$	$S^{(l)}_{\phi_{(1)}\phi_{(0)}}$
H$_1$	0.1935	0.1928		
H$_2$	2.4427	2.4508	2.2492	2.258
H$_3$	3.142	3.147	0.6993	0.6962
H$_4$	13.851	12.566	10.709	9.419
H$_5$	20.809	19.638	6.958	7.072
H$_6$	109.83	106.72	89.021	87.082

On the other hand $\chi^2_{d_l - d_{l+1};\alpha} = \chi^2_{1;\,0.05} = 3.84$, and we should choose the model H_3.

9. *a)* The predicted values for the frequencies associated with the model H_5 are:

(10.69911, 37.45849, 1.781327, 3.034881, 40.30089, 6.541514, 12.21867, 0.0965124, 16.99271, 63.84969, 1.526851, 13.65695, 64.00729, 11.15031, 10.47315, 4.343055).

Based on this vector we have

λ	-2	-1	-0.5	0	2/3	1	2
$T^{(5)}_{\phi_{(\lambda)}\phi_{(0)}}$	1.8629	1.8986	1.9479	2.0242	2.171	2.2667	2.6674

and $\chi^2_{5;\,0.05} = 11.1$. Therefore we should not reject the null hypothesis.

b) The results obtained with $S^{(l)}_{\phi(0),\phi(0)}$ are given in the following table,

Model	$2nD_{\phi_{(0)}}(\widehat{\boldsymbol{p}},\boldsymbol{p}(\widehat{\boldsymbol{\theta}}^{(l)}))$	$S^{(l)}_{\phi(0)\phi(0)}$	$d_l - d_{l+1}$
H_1	0.60		
H_2	1.23	0.63	1
H_3	1.55	0.32	1
H_4	1.93	0.38	1
H_5	2.02	0.09	1
H_6	4.12	2.10	1
H_7	136.73	132.61	1

On the basis of this table we should choose the model H_6 because $\chi^2_{1;\ 0.05} = 3.84$.

10. We have

$$\begin{aligned}
\boldsymbol{A}_{(l)}\boldsymbol{A}_{(l)} = & \; diag\left(\boldsymbol{p}\left(\boldsymbol{\theta}_0\right)^{-1/2}\right)\boldsymbol{\Sigma}_{\boldsymbol{p}(\boldsymbol{\theta}_0)}\boldsymbol{W}_{(l)}\left(\boldsymbol{W}^T_{(l)}\boldsymbol{\Sigma}_{\boldsymbol{p}(\boldsymbol{\theta}_0)}\boldsymbol{W}_{(l)}\right)^{-1}\boldsymbol{W}^T_{(l)} \\
\times & \; \boldsymbol{\Sigma}_{\boldsymbol{p}(\boldsymbol{\theta}_0)}diag\left(\boldsymbol{p}\left(\boldsymbol{\theta}_0\right)^{-1/2}\right)diag\left(\boldsymbol{p}\left(\boldsymbol{\theta}_0\right)^{-1/2}\right)\boldsymbol{\Sigma}_{\boldsymbol{p}(\boldsymbol{\theta}_0)}\boldsymbol{W}_{(l)} \\
\times & \; \left(\boldsymbol{W}^T_{(l)}\boldsymbol{\Sigma}_{\boldsymbol{p}(\boldsymbol{\theta}_0)}\boldsymbol{W}_{(l)}\right)^{-1}\boldsymbol{W}^T_{(l)}\boldsymbol{\Sigma}_{\boldsymbol{p}(\boldsymbol{\theta}_0)}diag\left(\boldsymbol{p}\left(\boldsymbol{\theta}_0\right)^{-1/2}\right) \\
= & \; diag\left(\boldsymbol{p}\left(\boldsymbol{\theta}_0\right)^{-1/2}\right)\boldsymbol{\Sigma}_{\boldsymbol{p}(\boldsymbol{\theta}_0)}\boldsymbol{W}_{(l)}\left(\boldsymbol{W}^T_{(l)}\boldsymbol{\Sigma}_{\boldsymbol{p}(\boldsymbol{\theta}_0)}\boldsymbol{W}_{(l)}\right)^{-1} \\
\times & \; \boldsymbol{W}^T_{(l)}\boldsymbol{\Sigma}_{\boldsymbol{p}(\boldsymbol{\theta}_0)}diag\left(\boldsymbol{p}\left(\boldsymbol{\theta}_0\right)^{-1/2}\right).
\end{aligned}$$

The last equality follows because

$$\boldsymbol{\Sigma}_{\boldsymbol{p}(\boldsymbol{\theta}_0)}diag\left(\boldsymbol{p}\left(\boldsymbol{\theta}_0\right)^{-1/2}\right)diag\left(\boldsymbol{p}\left(\boldsymbol{\theta}_0\right)^{-1/2}\right)\boldsymbol{\Sigma}_{\boldsymbol{p}(\boldsymbol{\theta}_0)} = \boldsymbol{\Sigma}_{\boldsymbol{p}(\boldsymbol{\theta}_0)}.$$

We know that $\boldsymbol{W}_{(l+1)}$ is a submatrix of $\boldsymbol{W}_{(l)}$; therefore there exists a matrix $\boldsymbol{B}$ verifying $\boldsymbol{W}_{(l+1)} = \boldsymbol{W}_{(l)}\boldsymbol{B}$. Therefore,

$$\begin{aligned}
\boldsymbol{A}_{(l)}\boldsymbol{A}_{(l+1)} = & \; diag\left(\boldsymbol{p}\left(\boldsymbol{\theta}_0\right)^{-1/2}\right)\boldsymbol{\Sigma}_{\boldsymbol{p}(\boldsymbol{\theta}_0)}\boldsymbol{W}_{(l)}\left(\boldsymbol{W}^T_{(l)}\boldsymbol{\Sigma}_{\boldsymbol{p}(\boldsymbol{\theta}_0)}\boldsymbol{W}_{(l)}\right)^{-1}\boldsymbol{W}^T_{(l)}\boldsymbol{\Sigma}_{\boldsymbol{p}(\boldsymbol{\theta}_0)} \\
\times & \; diag\left(\boldsymbol{p}\left(\boldsymbol{\theta}_0\right)^{-1/2}\right)diag\left(\boldsymbol{p}\left(\boldsymbol{\theta}_0\right)^{-1/2}\right)\boldsymbol{\Sigma}_{\boldsymbol{p}(\boldsymbol{\theta}_0)}\boldsymbol{W}_{(l+1)} \\
\times & \; \left(\boldsymbol{W}^T_{(l+1)}\boldsymbol{\Sigma}_{\boldsymbol{p}(\boldsymbol{\theta}_0)}\boldsymbol{W}_{(l+1)}\right)^{-1}\boldsymbol{W}^T_{(l+1)}\boldsymbol{\Sigma}_{\boldsymbol{p}(\boldsymbol{\theta}_0)}diag\left(\boldsymbol{p}\left(\boldsymbol{\theta}_0\right)^{-1/2}\right) \\
= & \; diag\left(\boldsymbol{p}\left(\boldsymbol{\theta}_0\right)^{-1/2}\right)\boldsymbol{\Sigma}_{\boldsymbol{p}(\boldsymbol{\theta}_0)}\boldsymbol{W}_{(l)}\boldsymbol{B}\left(\boldsymbol{W}^T_{(l+1)}\boldsymbol{\Sigma}_{\boldsymbol{p}(\boldsymbol{\theta}_0)}\boldsymbol{W}_{(l+1)}\right)^{-1} \\
\times & \; \boldsymbol{W}^T_{(l+1)}\boldsymbol{\Sigma}_{\boldsymbol{p}(\boldsymbol{\theta}_0)}diag\left(\boldsymbol{p}\left(\boldsymbol{\theta}_0\right)^{-1/2}\right) \\
= & \; diag\left(\boldsymbol{p}\left(\boldsymbol{\theta}_0\right)^{-1/2}\right)\boldsymbol{\Sigma}_{\boldsymbol{p}(\boldsymbol{\theta}_0)}\boldsymbol{W}_{(l+1)}\left(\boldsymbol{W}^T_{(l+1)}\boldsymbol{\Sigma}_{\boldsymbol{p}(\boldsymbol{\theta}_0)}\boldsymbol{W}_{(l+1)}\right)^{-1} \\
\times & \; \boldsymbol{W}^T_{(l+1)}\boldsymbol{\Sigma}_{\boldsymbol{p}(\boldsymbol{\theta}_0)}diag\left(\boldsymbol{p}\left(\boldsymbol{\theta}_0\right)^{-1/2}\right)\boldsymbol{W}^T_{(l+1)}\boldsymbol{\Sigma}_{\boldsymbol{p}(\boldsymbol{\theta}_0)}diag\left(\boldsymbol{p}\left(\boldsymbol{\theta}_0\right)^{-1/2}\right) \\
= & \; \boldsymbol{A}_{(l+1)}.
\end{aligned}$$

We know $\left(\boldsymbol{A}_{(l)}\boldsymbol{A}_{(l+1)}\right)^T = \boldsymbol{A}^T_{(l+1)}$ and $\boldsymbol{A}^T_{(l+1)}\boldsymbol{A}^T_{(l)} = \boldsymbol{A}^T_{(l+1)}$. But $\boldsymbol{A}_{(i)}$, $i = l, l+1$, is a symmetric matrix, therefore $\boldsymbol{A}_{(l+1)}\boldsymbol{A}_{(l)} = \boldsymbol{A}_{(l+1)}$.

8

Phi-divergence Measures in Contingency Tables

8.1. Introduction

In this chapter we study the importance of the estimators based on ϕ-divergences as well as the test statistics based on ϕ-divergences in some classical problems of contingency tables. First, we study the problems of independence, symmetry, marginal homogeneity and quasi-symmetry in a two-way contingency table and then the classical problem of homogeneity. There are different approaches to these problems but we study them using the result given in Section 5.5 in Chapter 5 regarding the minimum ϕ-divergence estimator with constraints as well as the result given in Section 6.3.5 in Chapter 6 regarding the testing problem when the null hypothesis can be written in terms of some constraints on the parameter space.

Throughout the chapter, the cited results of Chapters 5 and 6 will be very useful. Due to this we reproduce them now.

We suppose that we have ν $(\nu < M_0)$ real-valued functions $f_1(\boldsymbol{\theta}), ..., f_\nu(\boldsymbol{\theta})$ that constrain the parameter $\boldsymbol{\theta} \in \Theta \subset \mathbb{R}^{M_0}$, $f_m(\boldsymbol{\theta}) = 0$, $m = 1, ..., \nu$, such that they verify the conditions *i)* and *ii)* given in Section 5.5, and we denote by $\boldsymbol{p}(\boldsymbol{\theta}_0) = (p_1(\boldsymbol{\theta}_0), ..., p_M(\boldsymbol{\theta}_0))^T$, where $\boldsymbol{\theta}_0$ is unknown, the probability vector associated with the multinomial model under consideration. If we denote by

$$\Theta_0 = \{\boldsymbol{\theta} \in \Theta : f_m(\boldsymbol{\theta}) = 0, m = 1, ..., \nu\},$$

the restricted minimum ϕ-divergence estimator of $\boldsymbol{\theta}_0$ or the minimum ϕ-divergence estimator of $\boldsymbol{\theta}_0$ in Θ_0, $\widehat{\boldsymbol{\theta}}_{\phi}^{(r)}$ verifies

$$\widehat{\boldsymbol{\theta}}_{\phi}^{(r)} = \boldsymbol{\theta}_0 + \boldsymbol{H}(\boldsymbol{\theta}_0)\boldsymbol{I}_F(\boldsymbol{\theta}_0)^{-1}\boldsymbol{A}(\boldsymbol{\theta}_0)^T diag\left(\mathbf{p}(\boldsymbol{\theta}_0)^{-1/2}\right)(\widehat{\mathbf{p}}-\mathbf{p}(\boldsymbol{\theta}_0)) + o(\|\widehat{\mathbf{p}}-\mathbf{p}(\boldsymbol{\theta}_0)\|) \tag{8.1}$$

where the $M_0 \times M_0$ matrix $\boldsymbol{H}(\boldsymbol{\theta}_0)$ is defined by

$$\boldsymbol{H}(\boldsymbol{\theta}_0) = \boldsymbol{I} - \boldsymbol{I}_F(\boldsymbol{\theta}_0)^{-1}\boldsymbol{B}(\boldsymbol{\theta}_0)^T\left(\boldsymbol{B}(\boldsymbol{\theta}_0)\boldsymbol{I}_F(\boldsymbol{\theta}_0)^{-1}\boldsymbol{B}(\boldsymbol{\theta}_0)^T\right)^{-1}\boldsymbol{B}(\boldsymbol{\theta}_0); \tag{8.2}$$

$\boldsymbol{I}$ denotes the $M_0 \times M_0$ identity matrix and also

$$\sqrt{n}\left(\widehat{\boldsymbol{\theta}}_{\phi}^{(r)} - \boldsymbol{\theta}_0\right) \underset{n\to\infty}{\overset{L}{\longrightarrow}} N(\mathbf{0}, \boldsymbol{W}(\boldsymbol{\theta}_0)), \tag{8.3}$$

where the $M_0 \times M_0$ matrix $\boldsymbol{W}(\boldsymbol{\theta}_0)$ is given by

$$\boldsymbol{I}_F(\boldsymbol{\theta}_0)^{-1}\left(\boldsymbol{I} - \boldsymbol{B}(\boldsymbol{\theta}_0)^T\left(\boldsymbol{B}(\boldsymbol{\theta}_0)\boldsymbol{I}_F(\boldsymbol{\theta}_0)^{-1}\boldsymbol{B}(\boldsymbol{\theta}_0)^T\right)^{-1}\boldsymbol{B}(\boldsymbol{\theta}_0)\boldsymbol{I}_F(\boldsymbol{\theta}_0)^{-1}\right).$$

$\boldsymbol{I}_F(\boldsymbol{\theta}_0)$ is the Fisher information matrix associated with the multinomial model and $\boldsymbol{B}(\boldsymbol{\theta}_0)$ is the $\nu \times M_0$ matrix of partial derivatives

$$\boldsymbol{B}(\boldsymbol{\theta}_0) = \left(\frac{\partial f_m(\boldsymbol{\theta}_0)}{\partial \theta_k}\right)_{\substack{m=1,\ldots,\nu \\ k=1,\ldots,M_0}}.$$

For testing,

$$H_0 : \mathbf{p}{=}\mathbf{p}(\boldsymbol{\theta}_0), \text{ for some unknown } \boldsymbol{\theta}_0 \in \Theta_0 \subset \Theta \subset \mathbb{R}^{M_0}, \tag{8.4}$$

we consider the ϕ-divergence test statistic

$$\frac{2n}{\phi_1''(1)}D_{\phi_1}\left(\widehat{\mathbf{p}}, \mathbf{p}(\widehat{\boldsymbol{\theta}}_{\phi_2}^{(r)})\right) \underset{n\to\infty}{\overset{L}{\longrightarrow}} \chi^2_{M-M_0+\nu-1}, \tag{8.5}$$

where $\widehat{\boldsymbol{\theta}}_{\phi_2}^{(r)}$ is the minimum ϕ_2-divergence estimator of $\boldsymbol{\theta}_0$ in Θ_0.

We shall assume that $\phi_1, \phi_2 \in \Phi^*$ are twice continuously differentiable in $x > 0$ with $\phi_1''(1) > 0$ and $\phi_2''(1) > 0$.

8.2. Independence

One of the most interesting models in a two-way contingency table consists of testing the independence between two random variables X and Y.

We consider a two-way contingency table and let $\widehat{\mathbf{p}} = (\widehat{p}_{11}, ..., \widehat{p}_{IJ})^T$ be the nonparametric estimator of the unknown probability vector $\mathbf{p} = (p_{11}, ..., p_{IJ})^T$, where $p_{ij} =$

Pr $(X = i, Y = j)$, with $p_{ij} > 0$, $i = 1, ..., I$, $j = 1, ..., J$. The hypothesis of independence is given by

$$H_0 : p_{ij} = p_{i*}p_{*j}, \; i = 1, ..., I, j = 1, ..., J, \tag{8.6}$$

where $p_{*j} = \sum_{i=1}^{I} p_{ij}$ and $p_{i*} = \sum_{j=1}^{J} p_{ij}$.

We consider the parameter space

$$\Theta = \left\{ \boldsymbol{\theta} : \boldsymbol{\theta} = (p_{ij}; \; i = 1, ..., I, \; j = 1, ..., J, (i,j) \neq (I,J))^T \right\} \tag{8.7}$$

and we denote by

$$\boldsymbol{p}(\boldsymbol{\theta}) = (p_{11}, ..., p_{IJ})^T = \boldsymbol{p}, \tag{8.8}$$

the probability vector characterizing our model with $p_{IJ} = 1 - \sum\limits_{i=1}^{I} \sum\limits_{\substack{j=1 \\ (i,j)\neq(I,J)}}^{J} p_{ij}$. The hypothesis of independence given in (8.6) can be formulated again using the $(I-1)(J-1)$ constraints

$$h_{ij}(\boldsymbol{\theta}) = p_{ij} - p_{i*}p_{*j} = 0, \; i = 1, ..., I-1, \; j = 1, ..., J-1. \tag{8.9}$$

Also considering the parameter $\boldsymbol{\beta} = (p_{1*}, ..., p_{I-1*}, p_{*1}, ..., p_{*J-1})^T$ and the set

$$B = \{(a_1, ..., a_{I-1}, b_1, ..., b_{J-1})^T \in \mathbb{R}^{I+J-2} / \textstyle\sum_{i=1}^{I-1} a_i < 1, \sum_{j=1}^{J-1} b_j < 1,$$
$$a_i > 0, \; b_j > 0, \; i = 1, ..., I-1, \; j = 1, ..., J-1\},$$

the hypothesis (8.6) can be expressed for some unknown $\boldsymbol{\theta}_0 \in \Theta$ with $\boldsymbol{p}(\boldsymbol{\theta}_0) = \boldsymbol{p}_0$ as

$$H_0 : \boldsymbol{p}_0 = \boldsymbol{p}(\boldsymbol{g}(\boldsymbol{\beta}_0)), \quad \boldsymbol{\beta}_0 \in B \text{ and } \boldsymbol{g}(\boldsymbol{\beta}_0) = \boldsymbol{\theta}_0, \tag{8.10}$$

where the function $\boldsymbol{g}$ is defined by $\boldsymbol{g} = (g_{ij}; \; i = 1, ..., I, \; j = 1, ..., J, (i,j) \neq (I,J))$ with

$$g_{ij}(\boldsymbol{\beta}) = p_{i*}p_{*j}, \quad i = 1, ..., I-1, \; j = 1, ..., J-1$$

and

$$g_{Ij}(\boldsymbol{\beta}) = \left(1 - \sum_{i=1}^{I-1} p_{i*}\right) p_{*j}, \; j = 1, ..., J-1$$
$$g_{iJ}(\boldsymbol{\beta}) = \left(1 - \sum_{j=1}^{J-1} p_{*j}\right) p_{i*}, \; i = 1, ..., I-1.$$

It is important to remark by (8.8) that $\boldsymbol{p}(\boldsymbol{g}(\boldsymbol{\beta})) = (g_{ij}(\boldsymbol{\beta}), i = 1, ..., I, \; j = 1, ..., J)^T$, where

$$g_{IJ}(\boldsymbol{\beta}) = 1 - \sum_{i=1}^{I} \sum_{\substack{j=1 \\ (i,j)\neq(I,J)}}^{J} g_{ij}(\boldsymbol{\beta}).$$

With this approach the hypothesis of independence can be formulated in terms of the results presented in Chapter 6 (see Exercise 8).

In this Chapter we consider the approach given in (8.9). We can observe that with this approach the hypothesis of independence can be written as

$$H_0 : \mathbf{p}{=}\mathbf{p}\left(\boldsymbol{\theta}_0\right), \text{ for some unknown } \boldsymbol{\theta}_0{\in}\Theta_0,$$

where

$$\Theta_0 = \left\{\boldsymbol{\theta} \in \Theta : h_{ij}\left(\boldsymbol{\theta}\right) = 0,\ i = 1, ..., I-1,\ j = 1, ..., J-1\right\}, \tag{8.11}$$

with h_{ij} defined in (8.9) and $\mathbf{p}\left(\boldsymbol{\theta}\right)$ defined in (8.8).

For $I = J = 2$, we only have a constraint which is given by

$$h_{11}\left(\boldsymbol{\theta}\right) = p_{11} - p_{1*}p_{*1} = 0,$$

and for $I = J = 3$, we have four constraints given by

$$\begin{aligned} h_{11}\left(\boldsymbol{\theta}\right) &= \ p_{11} - p_{1*}p_{*1} = 0 \\ h_{12}\left(\boldsymbol{\theta}\right) &= \ p_{12} - p_{1*}p_{*2} = 0 \\ h_{21}\left(\boldsymbol{\theta}\right) &= \ p_{21} - p_{2*}p_{*1} = 0 \\ h_{22}\left(\boldsymbol{\theta}\right) &= \ p_{22} - p_{2*}p_{*2} = 0. \end{aligned}$$

8.2.1. Restricted Minimum Phi-divergence Estimator

In the following theorem we present the expression of the restricted minimum ϕ–divergence estimator for the problem of independence as well as its asymptotic properties. We can observe that this estimator is the minimum ϕ-divergence estimator under the independence hypothesis.

Theorem 8.1

The minimum ϕ-divergence estimator,

$$\widehat{\boldsymbol{\theta}}^{I,\phi} = \left(p_{i,j}^{I,\phi};\ i = 1, ..., I,\ j = 1, ..., J \text{ and } (i,j) \neq (I,J)\right)^T,$$

of $\boldsymbol{\theta}_0{\in}\Theta_0$ (i.e., under the hypothesis of independence) is obtained as a solution of the system of equations

$$\left\{\begin{array}{ll} \sum\limits_{j=1}^{J}\left(p_{*j}\phi\left(\dfrac{n_{ij}}{np_{i*}p_{*j}}\right) - \dfrac{n_{ij}}{np_{i*}}\phi'\left(\dfrac{n_{ij}}{np_{i*}p_{*j}}\right)\right) - \mu & = 0,\quad i = 1, ..., I \\ \sum\limits_{i=1}^{I}\left(p_{i*}\phi\left(\dfrac{n_{ij}}{np_{i*}p_{*j}}\right) - \dfrac{n_{ij}}{np_{*j}}\phi'\left(\dfrac{n_{ij}}{np_{i*}p_{*j}}\right)\right) - \mu & = 0,\quad j = 1, ..., J, \end{array}\right. \tag{8.12}$$

where μ is given by

$$\mu = \sum_{i=1}^{I}\sum_{j=1}^{J}\left(p_{i*}p_{*j}\phi\left(\frac{n_{ij}}{np_{i*}p_{*j}}\right) - \frac{n_{ij}}{n}\phi'\left(\frac{n_{ij}}{np_{i*}p_{*j}}\right)\right).$$

Its asymptotic distribution is

$$\sqrt{n}\left(\widehat{\boldsymbol{\theta}}^{I,\phi}-\boldsymbol{\theta}_0\right)\underset{n\to\infty}{\overset{L}{\longrightarrow}}N(\mathbf{0},\boldsymbol{W}_I(\boldsymbol{\theta}_0)) \tag{8.13}$$

where the $(IJ-1)\times(IJ-1)$ *matrix* $\boldsymbol{W}_I(\boldsymbol{\theta}_0)$ *is given by*

$$\boldsymbol{\Sigma}_{\boldsymbol{\theta}_0}\left(\boldsymbol{I}_{(IJ-1)\times(IJ-1)}-\boldsymbol{B}_I(\boldsymbol{\theta}_0)^T\left(\boldsymbol{B}_I(\boldsymbol{\theta}_0)\boldsymbol{\Sigma}_{\boldsymbol{\theta}_0}\boldsymbol{B}_I(\boldsymbol{\theta}_0)^T\right)^{-1}\boldsymbol{B}_I(\boldsymbol{\theta}_0)\boldsymbol{\Sigma}_{\boldsymbol{\theta}_0}\right), \tag{8.14}$$

the $(I-1)(J-1)\times(IJ-1)$ *matrix* $\boldsymbol{B}_I(\boldsymbol{\theta}_0)$ *is defined by*

$$\boldsymbol{B}_I(\boldsymbol{\theta}_0)=\left(\frac{\partial h_{ij}(\boldsymbol{\theta}_0)}{\partial\boldsymbol{\theta}}\right)_{i=1,\ldots,I-1,j=1,\ldots,J-1},$$

and $\boldsymbol{\Sigma}_{\boldsymbol{\theta}_0}=\boldsymbol{I}_F(\boldsymbol{\theta}_0)^{-1}$ *is the* $(IJ-1)\times(IJ-1)$ *matrix verifying* $\boldsymbol{\Sigma}_{\boldsymbol{\theta}_0}=diag(\boldsymbol{\theta}_0)-\boldsymbol{\theta}_0\boldsymbol{\theta}_0^T$.

Proof. Instead of getting

$$\widehat{\boldsymbol{\theta}}^{I,\phi}=\left(p_{ij}^{I,\phi};\ i=1,\ldots,I,\ j=1,\ldots,J\text{ and }(i,j)\neq(I,J)\right)^T$$

we shall obtain

$$\boldsymbol{p}(\widehat{\boldsymbol{\theta}}^{I,\phi})=\left(p_{ij}^{I,\phi};\ i=1,\ldots,I,\ j=1,\ldots,J\right)^T.$$

The p_{ij}'s minimizing the ϕ-divergence

$$D_\phi(\widehat{\mathbf{p}},\boldsymbol{p}(\boldsymbol{\theta}))=\sum_{i=1}^{I}\sum_{j=1}^{J}p_{ij}\phi\left(\frac{n_{ij}}{np_{ij}}\right)$$

subject to the hypothesis of independence (or the constraints about $\boldsymbol{\theta}$ given in (8.9)) may be obtained minimizing the Lagrangian function

$$\sum_{i=1}^{I}\sum_{j=1}^{J}p_{ij}\phi\left(\frac{n_{ij}}{np_{ij}}\right)+\sum_{i=1}^{I-1}\sum_{j=1}^{J-1}\lambda_{ij}h_{ij}(\boldsymbol{\theta})+\mu\left(1-\sum_{i=1}^{I}\sum_{j=1}^{J}p_{ij}\right) \tag{8.15}$$

where λ_{ij} and μ are undetermined Lagrangian multipliers. Taking into account the constraints given in (8.9), minimizing the expression (8.15) is equivalent to minimizing the expression

$$\sum_{i=1}^{I}\sum_{j=1}^{J}p_{i*}p_{*j}\phi\left(\frac{n_{ij}}{np_{i*}p_{*j}}\right)+\mu_1\left(1-\sum_{j=1}^{J}p_{*j}\right)+\mu_2\left(1-\sum_{i=1}^{I}p_{i*}\right).$$

Finally, differentiating with respect to p_{i*} and p_{*j} we get the system of equations given in (8.12), whose solutions provide the minimum ϕ-divergence estimator, $\widehat{\boldsymbol{\theta}}^{I,\phi}$.

The asymptotic distribution follows from (8.3) because $\boldsymbol{I}_F(\boldsymbol{\theta}_0)^{-1}=\boldsymbol{\Sigma}_{\boldsymbol{\theta}_0}$, under the considered parametric multinomial model. ∎

Corollary 8.1
The minimum power-divergence estimator

$$\boldsymbol{p}(\widehat{\boldsymbol{\theta}}^{I,\phi(\lambda)}) = \left(p_{ij}^{I,\phi(\lambda)};\ i=1,...,I,\ j=1,...,J\right)^T$$

of $\boldsymbol{p}(\boldsymbol{\theta}_0)$ *(i.e., under the hypothesis of independence (8.9)) is given by*

$$p_{ij}^{I,\phi(\lambda)} = p_{i*}^{I,\phi(\lambda)} \times p_{*j}^{I,\phi(\lambda)},\ i=1,...,I,\ j=1,...,J$$

where $p_{i*}^{I,\phi(\lambda)}$ *and* $p_{*j}^{I,\phi(\lambda)}$ *are the solutions of the system of equations*

$$\left\{\begin{array}{l} p_{i*}^{I,\phi(\lambda)} = \dfrac{\left(\displaystyle\sum_{j=1}^{J} \frac{n_{ij}^{\lambda+1}}{\left(\widehat{p}_{*j}^{I,\phi(\lambda)}\right)^{\lambda}}\right)^{\frac{1}{\lambda+1}}}{\displaystyle\sum_{i=1}^{I}\left(\sum_{j=1}^{J} \frac{n_{ij}^{\lambda+1}}{\left(\widehat{p}_{*j}^{I,\phi(\lambda)}\right)^{\lambda}}\right)^{\frac{1}{\lambda+1}}} \quad i=1,...,I \\ \\ p_{*j}^{I,\phi(\lambda)} = \dfrac{\left(\displaystyle\sum_{i=1}^{I} \frac{n_{ij}^{\lambda+1}}{\left(\widehat{p}_{i*}^{I,\phi(\lambda)}\right)^{\lambda}}\right)^{\frac{1}{\lambda+1}}}{\displaystyle\sum_{j=1}^{J}\left(\sum_{i=1}^{I} \frac{n_{ij}^{\lambda+1}}{\left(\widehat{p}_{i*}^{I,\phi(\lambda)}\right)^{\lambda}}\right)^{\frac{1}{\lambda+1}}} \quad j=1,...,J \end{array}\right. . \tag{8.16}$$

Proof. In this case the system of the equations (8.12) can be written as

$$\left\{\begin{array}{l} \displaystyle\sum_{j=1}^{J} \frac{n_{ij}^{\lambda+1}}{n^{\lambda+1}\left(p_{i*}^{I,\phi(\lambda)}\right)^{\lambda+1}\left(p_{*j}^{I,\phi(\lambda)}\right)^{\lambda}} = \left(\sum_{i=1}^{I}\left(\sum_{j=1}^{J} \frac{n_{ij}^{\lambda+1}}{n^{\lambda+1}\left(p_{*j}^{I,\phi(\lambda)}\right)^{\lambda}}\right)^{\frac{1}{\lambda+1}}\right)^{\lambda+1} \\ \\ \displaystyle\sum_{i=1}^{I} \frac{n_{ij}^{\lambda+1}}{n^{\lambda+1}\left(p_{*j}^{I,\phi(\lambda)}\right)^{\lambda+1}\left(p_{i*}^{I,\phi(\lambda)}\right)^{\lambda}} = \left(\sum_{i=1}^{I}\left(\sum_{j=1}^{J} \frac{n_{ij}^{\lambda+1}}{n^{\lambda+1}\left(p_{i*}^{I,\phi(\lambda)}\right)^{\lambda}}\right)^{\frac{1}{\lambda+1}}\right)^{\lambda+1} \end{array}\right. .$$

It is clear that the solution of this system is given by (8.16).

For $\lambda = 0$,

$$p_{i*}^{I,\phi(0)} = \frac{n_{i*}}{n} \quad \text{and} \quad p_{*j}^{I,\phi(0)} = \frac{n_{*j}}{n}, i=1,...,I,\ j=1,...,J,$$

hence the maximum likelihood estimator, under the hypothesis of independence, is obtained.

If $\lambda \to -1$ the minimum modified likelihood estimator under independence is obtained as the solution of the system of equations

$$\begin{cases} \widehat{p}_{i*}^{I,\phi(-1)} = \dfrac{\prod_{j=1}^{J}\left(\dfrac{n_{ij}}{\widehat{p}_{*j}^{I,\phi(-1)}}\right)^{\widehat{p}_{*j}^{I,\phi(-1)}}}{\sum_{i=1}^{I}\prod_{j=1}^{J}\left(\dfrac{n_{ij}}{\widehat{p}_{*j}^{I,\phi(-1)}}\right)^{\widehat{p}_{*j}^{I,\phi(-1)}}} & i=1,...,I \\ \widehat{p}_{*j}^{I,\phi(-1)} = \dfrac{\prod_{i=1}^{I}\left(\dfrac{n_{ij}}{\widehat{p}_{i*}^{I,\phi(-1)}}\right)^{\widehat{p}_{i*}^{I,\phi(-1)}}}{\sum_{j=1}^{J}\prod_{i=1}^{I}\left(\dfrac{n_{ij}}{\widehat{p}_{i*}^{I,\phi(-1)}}\right)^{\widehat{p}_{i*}^{I,\phi(-1)}}} & j=1,...,J \end{cases}.$$

We call the resulting estimators for $\lambda = -2$, $-1/2$, $2/3$ and 1 the minimum modified chi-square estimator, Freeman-Tukey estimator, Cressie-Read estimator and the minimum chi-square estimator, respectively.

Remark 8.1

For $I = J = 2$, given $\boldsymbol{\theta}_0 = (p_{*1}p_{1*},\ (1-p_{*1})\,p_{1*},\ p_{*1}\,(1-p_{1*}))^T \in \Theta_0$, it is easy to check that

- $\boldsymbol{B}_I(\boldsymbol{\theta}_0) = (1 - p_{1*} - p_{*1}, -p_{*1}, -p_{1*})$,
- $\boldsymbol{\Sigma}_{\boldsymbol{\theta}_0} = diag(\boldsymbol{\theta}_0) - \boldsymbol{\theta}_0\boldsymbol{\theta}_0^T$,
- $\left(\boldsymbol{B}_I(\boldsymbol{\theta}_0)\,\boldsymbol{\Sigma}_{\boldsymbol{\theta}_0}\boldsymbol{B}_I(\boldsymbol{\theta}_0)^T\right)^{-1} = (p_{1*}p_{*1}(1 - p_{1*} - p_{*1} + p_{1*}p_{*1}))^{-1}$,

and it is not difficult to establish that the matrix $\boldsymbol{W}_I(\boldsymbol{\theta}_0)$ is given by

$$\begin{pmatrix} p_{*1} & p_{1*} \\ 1-p_{*1} & -p_{1*} \\ -p_{*1} & 1-p_{1*} \end{pmatrix}\begin{pmatrix} p_{1*}(1-p_{1*}) & 0 \\ 0 & p_{*1}(1-p_{*1}) \end{pmatrix}\begin{pmatrix} p_{*1} & 1-p_{*1} & -p_{*1} \\ p_{1*} & -p_{1*} & 1-p_{1*} \end{pmatrix},$$

where:

- *The matrix*

$$\begin{pmatrix} p_{1*}(1-p_{1*}) & 0 \\ 0 & p_{*1}(1-p_{*1}) \end{pmatrix}$$

is the inverse of the Fisher information matrix corresponding to the paramete[r] $\boldsymbol{\beta}_0 = (p_{1*},\ p_{*1})^T$ *, i.e.,* $\boldsymbol{I}_F(\boldsymbol{\beta}_0)^{-1}$

· *The matrix*

$$\begin{pmatrix} p_{*1} & p_{1*} \\ 1 - p_{*1} & -p_{1*} \\ -p_{*1} & 1 - p_{1*} \end{pmatrix}$$

is the matrix of partial derivatives $\left(\frac{\partial \boldsymbol{g}(\boldsymbol{\beta}_0)}{\partial \boldsymbol{\beta}} \right)^T$.

Then we have

$$\boldsymbol{W}_I(\boldsymbol{\theta}_0) = \left(\frac{\partial \boldsymbol{g}(\boldsymbol{\beta}_0)}{\partial \boldsymbol{\beta}} \right)^T \boldsymbol{I}_F(\boldsymbol{\beta}_0)^{-1} \left(\frac{\partial \boldsymbol{g}(\boldsymbol{\beta}_0)}{\partial \boldsymbol{\beta}} \right).$$

This relation is true in general.

Theorem 8.2

Under the independence model, we have

$$\boldsymbol{W}_I(\boldsymbol{\theta}_0) = \boldsymbol{M}_{\boldsymbol{\beta}_0}^T \boldsymbol{I}_F(\boldsymbol{\beta}_0)^{-1} \boldsymbol{M}_{\boldsymbol{\beta}_0}, \tag{8.17}$$

where

$$\boldsymbol{M}_{\boldsymbol{\beta}_0} = \left(\frac{\partial \boldsymbol{g}(\boldsymbol{\beta}_0)}{\partial \boldsymbol{\beta}} \right),$$

and $\boldsymbol{g}$ *was defined in (8.10).*

Proof.

It is not difficult to establish that

$$\boldsymbol{I}_F(\boldsymbol{\beta}_0) = \boldsymbol{M}_{\boldsymbol{\beta}_0} \boldsymbol{\Sigma}_{\boldsymbol{\theta}_0}^{-1} \boldsymbol{M}_{\boldsymbol{\beta}_0}^T.$$

Then we have

$$\boldsymbol{M}_{\boldsymbol{\beta}_0}^T \boldsymbol{I}_F(\boldsymbol{\beta}_0)^{-1} \boldsymbol{M}_{\boldsymbol{\beta}_0} = \boldsymbol{M}_{\boldsymbol{\beta}_0}^T \left(\boldsymbol{M}_{\boldsymbol{\beta}_0} \boldsymbol{\Sigma}_{\boldsymbol{\theta}_0}^{-1} \boldsymbol{M}_{\boldsymbol{\beta}_0}^T \right)^{-1} \boldsymbol{M}_{\boldsymbol{\beta}_0}. \tag{8.18}$$

Multiplying (8.17) by $\boldsymbol{\Sigma}_{\boldsymbol{\theta}_0}^{-1} \boldsymbol{M}_{\boldsymbol{\beta}_0}^T$ on the right and by $\boldsymbol{M}_{\boldsymbol{\beta}_0} \boldsymbol{\Sigma}_{\boldsymbol{\theta}_0}^{-1}$ on the left and taking into account (8.18) and the expression of $\boldsymbol{W}_I(\boldsymbol{\theta}_0)$ given in (8.14), we have

$$\begin{aligned} \boldsymbol{M}_{\boldsymbol{\beta}_0} \boldsymbol{\Sigma}_{\boldsymbol{\theta}_0}^{-1} \boldsymbol{M}_{\boldsymbol{\beta}_0}^T = & \ \boldsymbol{M}_{\boldsymbol{\beta}_0} \boldsymbol{\Sigma}_{\boldsymbol{\theta}_0}^{-1} \boldsymbol{M}_{\boldsymbol{\beta}_0}^T \\ & - \boldsymbol{M}_{\boldsymbol{\beta}_0} \boldsymbol{B}_I(\boldsymbol{\theta}_0)^T \left(\boldsymbol{B}_I(\boldsymbol{\theta}_0) \boldsymbol{\Sigma}_{\boldsymbol{\theta}} \boldsymbol{B}_I(\boldsymbol{\theta}_0)^T \right)^{-1} \boldsymbol{B}_I(\boldsymbol{\theta}_0) \boldsymbol{M}_{\boldsymbol{\beta}_0}^T. \end{aligned}$$

But

$$\boldsymbol{M}_{\boldsymbol{\beta}_0}\boldsymbol{B}_I(\boldsymbol{\theta}_0)^T = \boldsymbol{0} \text{ and } \boldsymbol{B}_I(\boldsymbol{\theta}_0)\boldsymbol{M}_{\boldsymbol{\beta}_0}^T = \boldsymbol{0}.$$

Then we have established that

$$\boldsymbol{A}^T\boldsymbol{X}\boldsymbol{A} = \boldsymbol{A}^T\boldsymbol{Y}\boldsymbol{A},$$

where $\boldsymbol{X} = \boldsymbol{M}_{\boldsymbol{\beta}_0}^T\boldsymbol{I}_F(\boldsymbol{\beta}_0)^{-1}\boldsymbol{M}_{\boldsymbol{\beta}_0}$, $\boldsymbol{Y} = \boldsymbol{W}_I(\boldsymbol{\theta}_0)$ and $\boldsymbol{A} = \boldsymbol{\Sigma}_{\boldsymbol{\theta}_0}^{-1}\boldsymbol{M}_{\boldsymbol{\beta}_0}^T$. But the matrix $\boldsymbol{A}$ has full rank, then $\boldsymbol{X} = \boldsymbol{Y}$. ■

Remark 8.2

The Fisher information matrix associated with the independence model is given by

$$\left(\boldsymbol{M}_{\boldsymbol{\beta}_0}^T\boldsymbol{I}_F(\boldsymbol{\beta}_0)^{-1}\boldsymbol{M}_{\boldsymbol{\beta}_0}\right)^{-1},$$

since if we formulate the problem of independence based on (8.10) we have

$$\sqrt{n}\left(\widehat{\boldsymbol{\beta}} - \boldsymbol{\beta}_0\right) \xrightarrow[n\to\infty]{L} N\left(\boldsymbol{0}, \boldsymbol{I}_F(\boldsymbol{\beta}_0)^{-1}\right),$$

where $\widehat{\boldsymbol{\beta}}$ is the maximum likelihood estimator of $\boldsymbol{\beta}_0$, and

$$\sqrt{n}\left(\boldsymbol{g}(\widehat{\boldsymbol{\beta}}) - \boldsymbol{g}(\boldsymbol{\beta}_0)\right) \xrightarrow[n\to\infty]{L} N\left(\boldsymbol{0},\ \boldsymbol{M}_{\boldsymbol{\beta}_0}^T\boldsymbol{I}_F(\boldsymbol{\beta}_0)^{-1}\boldsymbol{M}_{\boldsymbol{\beta}_0}\right).$$

In Theorem 8.1 we have established that

$$\sqrt{n}\left(\widehat{\boldsymbol{\theta}}^{I,\phi} - \boldsymbol{\theta}_0\right) \xrightarrow[n\to\infty]{L} N(\boldsymbol{0}, \boldsymbol{W}_I(\boldsymbol{\theta}_0)).$$

Then the restricted minimum ϕ-divergence estimator will be BAN if and only if

$$\boldsymbol{W}_I(\boldsymbol{\theta}_0) = \boldsymbol{M}_{\boldsymbol{\beta}_0}^T\boldsymbol{I}_F(\boldsymbol{\beta}_0)^{-1}\boldsymbol{M}_{\boldsymbol{\beta}_0}.$$

Therefore, $\widehat{\boldsymbol{\theta}}^{I,\phi}$ is BAN by Theorem 8.2.

Now we are going to present a result that establishes sufficient conditions for uniqueness of the minimum ϕ-divergence estimator in the independence problem.

Theorem 8.3

The minimum ϕ-divergence estimator under the hypothesis of independence, $\widehat{\boldsymbol{\theta}}^{I,\phi} = \left(p_{ij}^{I,\phi};\ i = 1, ..., I,\ j = 1, ..., J \text{ and } (i,j) \neq (I,J)\right)^T$, is unique if

$$k_2(i,j) > -k_1(i,j) > 0,\ i = 1, ..., I;\ j = 1, ..., J$$

where

$$k_1(i,j) = \phi\left(\frac{n_{ij}}{np_{i*}p_{*j}}\right) - \frac{n_{ij}}{np_{i*}p_{*j}}\phi'\left(\frac{n_{ij}}{np_{i*}p_{*j}}\right),\ i = 1, ..., I;\ j = 1, ..., J$$

$$k_2(i,j) = 2\frac{n_{ij}^2}{n^2p_{i*}^2p_{*j}^2}\phi''\left(\frac{n_{ij}}{np_{i*}p_{*j}}\right),\ i = 1, ..., I;\ j = 1, ..., J.$$

Proof. It is necessary to establish that the function

$$f(p_{1*},...,p_{I*},p_{*1},...,p_{*J}) = \sum_{i=1}^{I}\sum_{j=1}^{J} p_{i*}p_{*j}\phi\left(\frac{n_{ij}}{np_{i*}p_{*j}}\right)$$

is strictly convex. Consider the functions

$$f_{ij}(p_{1*},...,p_{I*},p_{*1},...,p_{*J}) = p_{i*}p_{*j}\phi\left(\frac{n_{ij}}{np_{i*}p_{*j}}\right),\ i=1,...,I,\ j=1,...,J.$$

We will use the fact that if f_{ij} is strictly convex for all i and j, then $\sum_{i=1}^{I}\sum_{j=1}^{J} f_{ij}$ is strictly convex.

For the sake of brevity, we forget indexes and write

$$f(x,y) = xy\phi(\frac{k}{xy}),$$

where k is a positive constant. The determinant of the Hessian matrix of the function f is

$$\frac{k^4}{x^4y^4}\phi''\left(\frac{k}{xy}\right)^2 - \left(\phi\left(\frac{k}{xy}\right) - \frac{k}{xy}\phi'\left(\frac{k}{xy}\right) + \frac{k^2}{x^2y^2}\phi''\left(\frac{k}{xy}\right)\right)^2$$

which can be written as follows

$$-\left(\phi\left(\frac{k}{xy}\right) - \frac{k}{xy}\phi'\left(\frac{k}{xy}\right)\right)\left(\phi\left(\frac{k}{xy}\right) - \frac{k}{xy}\phi'\left(\frac{k}{xy}\right) + 2\frac{k^2}{x^2y^2}\phi''\left(\frac{k}{xy}\right)\right).$$

Hence the result holds. ■

It is not difficult to establish that in the case of the power-divergence family the result is verified for $\lambda > -1/2$.

8.2.2. Test of Independence

Based on the ϕ-divergence test statistic given in (8.5) we should reject the hypothesis of independence if

$$I_n^{\phi_1}\left(\widehat{\boldsymbol{\theta}}^{I,\phi_2}\right) \equiv \frac{2n}{\phi_1''(1)} D_{\phi_1}\left(\widehat{\mathbf{p}}, \mathbf{p}(\widehat{\boldsymbol{\theta}}^{I,\phi_2})\right) > c,$$

where c is a positive constant. In the following theorem we establish the asymptotic distribution of the test statistic $I_n^{\phi_1}\left(\widehat{\boldsymbol{\theta}}^{I,\phi_2}\right)$.

Theorem 8.4

The asymptotic distribution of the ϕ-divergence test statistics

$$I_n^{\phi_1}\left(\widehat{\boldsymbol{\theta}}^{I,\phi_2}\right) \equiv \frac{2n}{\phi_1''(1)} D_{\phi_1}\left(\widehat{\mathbf{p}}, \mathbf{p}(\widehat{\boldsymbol{\theta}}^{I,\phi_2})\right) \tag{8.19}$$

and

$$I_n^{\phi_1,h}\left(\widehat{\boldsymbol{\theta}}^{I,\phi_2}\right) \equiv \frac{2n}{h'(0)\,\phi_1''(1)} h\left(D_{\phi_1}\left(\widehat{\mathbf{p}},\mathbf{p}(\widehat{\boldsymbol{\theta}}^{I,\phi_2})\right)\right), \tag{8.20}$$

for testing the hypothesis of independence, is chi-square with $(I-1)(J-1)$ degrees of freedom.

Proof. In our case we have IJ cells and by (8.7) $IJ-1$ parameters that are necessary to estimate. Using relation (8.9) the number of constraints is $(I-1)\times(J-1)$. Then by (8.5) the asymptotic distribution of the family of test statistics (8.19) is chi-square with

$$\underbrace{IJ}_{\text{Cells}} - \underbrace{IJ-1}_{\substack{\text{Estimated}\\ \text{Parameters}}} + \underbrace{(I-1)(J-1)}_{\text{Constraints}} -1 = \underbrace{(I-1)(J-1)}_{\substack{\text{Degrees of}\\ \text{freedom}}}.$$

In relation with the family of (h,ϕ)-divergence test statistics given in (8.20), we have

$$h(x) = h(0) + h'(0)\,x + o(x),$$

then

$$h\left(D_{\phi_1}\left(\widehat{\mathbf{p}},\mathbf{p}(\widehat{\boldsymbol{\theta}}^{I,\phi_2})\right)\right) = h'(0)\,D_{\phi_1}\left(\widehat{\mathbf{p}},\mathbf{p}(\widehat{\boldsymbol{\theta}}^{I,\phi_2})\right) + o_P(1)$$

and we get that the asymptotic distribution of the family of test statistics given in (8.20) is also chi-square with $(I-1)(J-1)$ degrees of freedom.

Remark 8.3

For $\phi_1(x) = \phi_2(x) = x\log x - x + 1 = \phi_{(0)}(x)$ we get that $I_n^{\phi_1}\left(\widehat{\boldsymbol{\theta}}^{I,\phi_2}\right)$ coincides with the classical likelihood ratio test statistic for testing independence given by

$$G^2 \equiv I_n^{\phi_{(0)}}\left(\widehat{\boldsymbol{\theta}}^{I,\phi_{(0)}}\right) = 2n\sum_{i=1}^{I}\sum_{j=1}^{J}\widehat{p}_{ij}\log\frac{\widehat{p}_{ij}}{\widehat{p}_{i*}\widehat{p}_{*j}},$$

and for $\phi_2(x) = x\log x - x + 1 = \phi_{(0)}(x)$ and $\phi_1(x) = \frac{1}{2}(x-1)^2 = \phi_{(1)}(x)$ the classical chi-square test statistic given by

$$X^2 \equiv I_n^{\phi_{(1)}}\left(\widehat{\boldsymbol{\theta}}^{I,\phi_{(0)}}\right) = n\sum_{i=1}^{I}\sum_{j=1}^{J}\frac{\left(\widehat{p}_{ij} - \widehat{p}_{i*}\widehat{p}_{*j}\right)^2}{\widehat{p}_{i*}\widehat{p}_{*j}}.$$

It is important to note that for $\phi_2(x) = x\log x - x + 1$, we get the family of test statistics studied by Zografos (1993) and Pardo, L. et al. (1993a), which is

$$I_n^{\phi_1}\left(\widehat{\boldsymbol{\theta}}\right) \equiv I_n^{\phi_1}\left(\widehat{\boldsymbol{\theta}}^{I,\phi_{(0)}}\right) = \frac{2n}{\phi_1''(1)}D_{\phi_1}(\widehat{\mathbf{p}},\mathbf{p}(\widehat{\boldsymbol{\theta}})), \tag{8.21}$$

where $\widehat{\boldsymbol{\theta}} \equiv \widehat{\boldsymbol{\theta}}^{I,\phi_{(0)}}$ is the maximum likelihood estimator,

$$\widehat{\boldsymbol{\theta}} = \left(\frac{n_{i*}}{n}\times\frac{n_{*j}}{n},\ i=1,...,I,\ j=1,..,J,(i,j)\neq(I,J)\right)^T.$$

Based on the previous test we should reject the hypothesis of independence with significance level α if

$$I_n^{\phi_1}\left(\widehat{\boldsymbol{\theta}}^{I,\phi_2}\right) \text{ or } \left(I_n^{\phi_1,h}\left(\widehat{\boldsymbol{\theta}}^{I,\phi_2}\right)\right) \geq \chi^2_{(I-1)(J-1),\alpha}. \tag{8.22}$$

Power of the Test

Let $\boldsymbol{q} = (q_{11}, ..., q_{IJ})^T$ be a point at the alternative hypothesis, i.e., there exist at least two indexes i and j for which $q_{ij} \neq q_{i*} \times q_{*j}$. We denote by $\boldsymbol{\theta}_a^{\phi_2}$ the value of $\boldsymbol{\theta}$ verifying

$$\boldsymbol{\theta}_a^{\phi_2} = \arg\min_{\boldsymbol{\theta}\in\Theta_0} D_{\phi_2}(\boldsymbol{q}, \boldsymbol{p}(\boldsymbol{\theta})),$$

where Θ_0 was defined in (8.11).

It is clear that

$$\boldsymbol{\theta}_a^{\phi_2} = (f_{ij}(\boldsymbol{q});\ i = 1, ..., I,\ j = 1, ..., J, (i,j) \neq (I,J)\ )^T$$

and

$$\boldsymbol{p}(\boldsymbol{\theta}_a^{\phi_2}) = (f_{ij}(\boldsymbol{q});\ i = 1, ..., I,\ j = 1, ..., J\ )^T \equiv \boldsymbol{f}(\boldsymbol{q}),$$

with

$$f_{IJ}(\boldsymbol{q}) = 1 - \sum_{i=1}^{I} \sum_{\substack{j=1 \\ (i,j)\neq(I,J)}}^{J} f_{ij}(\boldsymbol{q}).$$

The notation $f_{ij}(\boldsymbol{q})$ indicates that the elements of the vector $\boldsymbol{\theta}_a^{\phi_2}$ depend on $\boldsymbol{q}$. For instance, for $\phi_2(x) = x\log x - x + 1$,

$$\boldsymbol{p}(\boldsymbol{\theta}_a^{\phi_2}) = (q_{i*} \times q_{*j};\ i = 1, ..., I,\ j = 1, ..., J\ )^T,$$

and $f_{ij}(\boldsymbol{q}) = q_{i*} \times q_{*j}$.

We also denote

$$\widehat{\boldsymbol{\theta}}^{I,\phi_2} = \left(p_{ij}^{I,\phi_2};\ i = 1, ..., I,\ j = 1, ..., J, (i,j) \neq (I,J)\ \right)^T$$

and then

$$\boldsymbol{p}(\widehat{\boldsymbol{\theta}}^{I,\phi_2}) = \left(p_{ij}^{I,\phi_2};\ i = 1, ..., I,\ j = 1, ..., J\ \right)^T \equiv \boldsymbol{f}(\widehat{\boldsymbol{p}}),$$

where $\boldsymbol{f} = (f_{ij};\ i = 1, ..., I,\ j = 1, ..., J)^T$. If the alternative $\boldsymbol{q}$ is true we have that $\widehat{\boldsymbol{p}}$ tends to $\boldsymbol{q}$ and $\boldsymbol{p}(\widehat{\boldsymbol{\theta}}^{I,\phi_2})$ to $\boldsymbol{p}(\boldsymbol{\theta}_a^{\phi_2})$ in probability.

If we define the function

$$\Psi_{\phi_1}(\boldsymbol{q}) = D_{\phi_1}(\boldsymbol{q}, \boldsymbol{f}(\boldsymbol{q})),$$

we have

$$\Psi_{\phi_1}(\widehat{\boldsymbol{p}}) = \Psi_{\phi_1}(\boldsymbol{q}) + \sum_{i=1}^{I}\sum_{j=1}^{J} \frac{\partial D_{\phi_1}(\boldsymbol{q}, \boldsymbol{f}(\boldsymbol{q}))}{\partial q_{ij}} (\widehat{p}_{ij} - q_{ij}) + o(\|\widehat{\boldsymbol{p}} - \boldsymbol{q}\|).$$

Hence the random variables

$$\sqrt{n}\,(D_{\phi_1}(\widehat{\boldsymbol{p}}, \boldsymbol{f}(\widehat{\boldsymbol{p}})) - D_{\phi_1}(\boldsymbol{q}, \boldsymbol{f}(\boldsymbol{q})))$$

and

$$\sqrt{n}\sum_{i=1}^{I}\sum_{j=1}^{J} \frac{\partial D_{\phi_1}(\boldsymbol{q}, \boldsymbol{f}(\boldsymbol{q}))}{\partial q_{ij}} (\widehat{p}_{ij} - q_{ij})$$

have the same asymptotic distribution. If we denote by

$$l_{ij} = \frac{\partial D_{\phi_1}(\boldsymbol{q}, \boldsymbol{f}(\boldsymbol{q}))}{\partial q_{ij}} \tag{8.23}$$

and by $\boldsymbol{l} = (l_{ij};\ i = 1, ..., I,\ j = 1, ..., J\ )^T$ we have

$$\sqrt{n}\,(D_{\phi_1}(\widehat{\boldsymbol{p}}, \boldsymbol{f}(\widehat{\boldsymbol{p}})) - D_{\phi_1}(\boldsymbol{q}, \boldsymbol{f}(\boldsymbol{q}))) \xrightarrow[n\to\infty]{L} N\left(\boldsymbol{0}, \boldsymbol{l}^T\boldsymbol{\Sigma}_{\boldsymbol{q}}\boldsymbol{l}\right), \tag{8.24}$$

where

$$\boldsymbol{\Sigma}_{\boldsymbol{q}} = \left(q_{i_1j_1}\left(\delta_{(i_1j_1)(i_2j_2)} - q_{i_2j_2}\right)\right)_{\substack{i_1,i_2=1,...,I\\ j_1,j_2=1,...,J}}.$$

In the following theorem we present the asymptotic distribution using the maximum likelihood estimator. In this case we have:

Theorem 8.5

Let $\boldsymbol{p}(\widehat{\boldsymbol{\theta}}) = (\widehat{p}_{i*} \times \widehat{p}_{*j},\ i = 1, ..., I,\ j = 1, ..., J)^T$ *the maximum likelihood estimator of* $\boldsymbol{p}(\boldsymbol{\theta}) = (p_{11}, ..., p_{IJ})^T$, *under the independence hypothesis, and* $\boldsymbol{q}$ *a point at the alternative hypothesis. Then,*

$$\sqrt{n}\left(D_{\phi_1}(\widehat{\boldsymbol{p}}, \boldsymbol{p}(\widehat{\boldsymbol{\theta}})) - D_{\phi_1}(\boldsymbol{q}, \boldsymbol{q}_{I\times J})\right) \xrightarrow[n\to\infty]{L} N\left(0, \sigma^2_{\phi_1}(\boldsymbol{q})\right)$$

where

$$\sigma^2_{\phi_1}(\boldsymbol{q}) = \sum_{i=1}^{I}\sum_{j=1}^{J} q_{ij}l_{ij}^2 - \left(\sum_{i=1}^{I}\sum_{j=1}^{J} q_{ij}l_{ij}\right)^2,$$

$$\begin{aligned} l_{ij} &= \sum_{r=1}^{I}\left(q_{r*}\phi_1\left(\frac{q_{rj}}{q_{r*}q_{*j}}\right) - \frac{q_{rj}}{q_{*j}}\phi_1'\left(\frac{q_{rj}}{q_{r*}q_{*j}}\right)\right) \\ &+ \sum_{s=1}^{J}\left(q_{*s}\phi_1\left(\frac{q_{is}}{q_{i*}q_{*s}}\right) - \frac{q_{is}}{q_{i*}}\phi_1'\left(\frac{q_{is}}{q_{i*}q_{*s}}\right)\right) + \phi_1'\left(\frac{q_{ij}}{q_{i*}q_{*j}}\right)\end{aligned}$$

and $\boldsymbol{q}_{I\times J} = (q_{i*} \times q_{*j},\ i = 1, ..., I,\ j = 1, ..., J)^T$ *whenever* $\sigma^2_{\phi_1}(\boldsymbol{q}) > 0$.

Proof.

We have to calculate the elements l_{ij}, given in (8.23) taking into account that in this case $f_{ij}(\boldsymbol{q}) = q_{i*} \times q_{*j}$. We can write the function $\Psi_{\phi_1}(\boldsymbol{q})$ as

$$\begin{aligned}
\Psi_{\phi_1}(\boldsymbol{q}) = & \sum_{i=1}^{I}\sum_{j=1}^{J} q_{i*}q_{*j}\phi_1\left(\frac{q_{ij}}{q_{i*}q_{*j}}\right) = q_{i*}q_{*j}\phi_1\left(\frac{q_{ij}}{q_{i*}q_{*j}}\right) \\
+ & \sum_{\substack{s=1\\ s\neq j}}^{J} q_{i*}q_{*s}\phi_1\left(\frac{q_{is}}{q_{i*}q_{*s}}\right) + \sum_{\substack{r=1\\ r\neq i}}^{I} q_{r*}q_{*j}\phi_1\left(\frac{q_{rj}}{q_{r*}q_{*j}}\right) \\
+ & \sum_{\substack{r=1\\ r\neq i}}^{I}\sum_{\substack{s=1\\ s\neq j}}^{J} q_{r*}q_{*s}\phi_1\left(\frac{q_{rs}}{q_{r*}q_{*s}}\right) = G_1 + G_2 + G_3 + G_4.
\end{aligned}$$

Then we have

$$\begin{aligned}
\frac{\partial G_1}{\partial q_{ij}} &= \phi_1\left(\frac{q_{ij}}{q_{i*}q_{*j}}\right)(q_{*j} + q_{i*}) + \phi_1'\left(\frac{q_{ij}}{q_{i*}q_{*j}}\right)\frac{q_{i*}q_{*j} - q_{ij}\left(q_{*j} + q_{i*}\right)}{q_{i*}q_{*j}} \\
&= \phi_1\left(\frac{q_{ij}}{q_{i*}q_{*j}}\right)(q_{*j} + q_{i*}) + \phi_1'\left(\frac{q_{ij}}{q_{i*}q_{*j}}\right)\left(1 - \frac{q_{ij}}{q_{i*}} - \frac{q_{ij}}{q_{*j}}\right)
\end{aligned}$$

$$\begin{aligned}
\frac{\partial G_2}{\partial q_{ij}} &= \sum_{\substack{s=1\\ s\neq j}}^{J}\left\{q_{*s}\phi_1\left(\frac{q_{is}}{q_{i*}q_{*s}}\right) + q_{i*}q_{*s}\phi_1'\left(\frac{q_{is}}{q_{i*}q_{*s}}\right)\frac{q_{is}}{q_{*s}}(-1)\frac{1}{q_{i*}^2}\right\} \\
&= \sum_{\substack{s=1\\ s\neq j}}^{J} q_{*s}\phi_1\left(\frac{q_{is}}{q_{i*}q_{*s}}\right) - \sum_{\substack{s=1\\ s\neq j}}^{J}\phi_1'\left(\frac{q_{is}}{q_{i*}q_{*s}}\right)\frac{q_{is}}{q_{i*}}
\end{aligned}$$

$$\frac{\partial G_3}{\partial q_{ij}} = \sum_{\substack{r=1\\ r\neq i}}^{I} q_{r*}\phi_1\left(\frac{q_{rj}}{q_{r*}q_{*j}}\right) - \sum_{\substack{r=1\\ r\neq i}}^{I}\frac{q_{rj}}{q_{*j}}\phi_1'\left(\frac{q_{rj}}{q_{r*}q_{*j}}\right)$$

$$\frac{\partial G_4}{\partial q_{ij}} = 0.$$

Therefore,

$$\begin{aligned}
l_{ij} = \frac{\partial \Psi_{\phi_1}(\boldsymbol{q})}{\partial q_{ij}} &= \sum_{r=1}^{I}\left(q_{r*}\phi_1\left(\frac{q_{rj}}{q_{r*}q_{*j}}\right) - \frac{q_{rj}}{q_{*j}}\phi_1'\left(\frac{q_{rj}}{q_{r*}q_{*j}}\right)\right) \\
&+ \sum_{s=1}^{J}\left(q_{*s}\phi_1\left(\frac{q_{is}}{q_{i*}q_{*s}}\right) - \frac{q_{is}}{q_{i*}}\phi_1'\left(\frac{q_{is}}{q_{i*}q_{*s}}\right)\right) \\
&+ \phi_1'\left(\frac{q_{ij}}{q_{i*}q_{*j}}\right)
\end{aligned}$$

and based on (8.24) we have

$$\sigma_{\phi_1}^2(\boldsymbol{q}) = \boldsymbol{l}^T \boldsymbol{\Sigma}_{\boldsymbol{q}} \boldsymbol{l} = \sum_{i=1}^{I}\sum_{j=1}^{J} q_{ij} l_{ij}^2 - \left(\sum_{i=1}^{I}\sum_{j=1}^{J} q_{ij} l_{ij}\right)^2 .$$

■

Using this result it is possible to get the power of the test of independence. This is given by

$$\beta_{n,\phi_1}(\boldsymbol{q}) = 1 - \Phi_n\left(\frac{\sqrt{n}}{\sigma_{\phi_1}(\boldsymbol{q})}\left(\frac{\phi_1''(1)}{2n}\chi^2_{(I-1)(J-1),\alpha} - D_{\phi_1}(\boldsymbol{q}, \boldsymbol{q}_{I\times J})\right)\right),$$

where $\Phi_n(x)$ is a sequence of distribution functions leading uniformly to the standard normal distribution $\Phi(x)$.

Remark 8.4

For the Kullback-Leibler divergence we have

$$\sigma^2(\boldsymbol{q}) = \sum_{i=1}^{I}\sum_{j=1}^{J} q_{ij} \log^2 \frac{q_{ij}}{q_{i*}q_{*j}} - \left(\sum_{i=1}^{I}\sum_{j=1}^{J} q_{ij} \log \frac{q_{ij}}{q_{i*}q_{*j}}\right)^2 .$$

In general, theoretical results for the test statistics $I_n^{\phi_1}\left(\widehat{\boldsymbol{\theta}}^{I,\phi_2}\right)$ or $I_n^{\phi_1,h}\left(\widehat{\boldsymbol{\theta}}^{I,\phi_2}\right)$, under alternative hypotheses, with $\phi_2(x) \neq x\log x - x + 1$, are not easy to obtain. An exception to this fact is when there is a contiguous sequence of alternatives that approaches the null hypothesis $H_0 : \boldsymbol{p} = \boldsymbol{p}(\boldsymbol{\theta}_0)$, for some unknown $\boldsymbol{\theta}_0 \in \Theta_0$, at the rate $O\left(n^{-1/2}\right)$. Consider the multinomial probability vector

$$p_{n,ij} = \boldsymbol{p}_{ij}(\boldsymbol{\theta}_0) + \frac{d_{ij}}{\sqrt{n}}, \; i = 1, ..., I, \; j = 1, ..., J,$$

where $\boldsymbol{d} = (d_{11}, ..., d_{IJ})^T$ is a fixed $IJ \times 1$ vector such that $\sum_{i=1}^{I}\sum_{j=1}^{J} d_{ij} = 0$; recall that n is the total count parameter of the multinomial distribution and $\boldsymbol{\theta}_0$ is unknown but belonging to Θ_0. As $n \to \infty$, the sequence of multinomial probabilities $\{\boldsymbol{p}_n\}_{n\in N}$ with

$$\boldsymbol{p}_n = (p_{n,ij}, \; i = 1, ..., I, \; j = 1, ..., J)^T$$

converges to a multinomial probability in H_0 at the rate of $O\left(n^{-1/2}\right)$. Let

$$H_{1,n} : \boldsymbol{p}_n = \boldsymbol{p}(\boldsymbol{\theta}_0) + \frac{\boldsymbol{d}}{\sqrt{n}}, \qquad \boldsymbol{\theta}_0 \in \Theta_0 \tag{8.25}$$

be a sequence of contiguous alternative hypotheses, here contiguous to the null hypothesis $H_0 : \boldsymbol{p}=\boldsymbol{p}(\boldsymbol{\theta}_0)$, for some unknown $\boldsymbol{\theta}_0 \in \Theta_0$. We are interested in studying the asymptotic behavior of the power of $I_n^{\phi_1}\left(\widehat{\boldsymbol{\theta}}^{I,\phi_2}\right)$ $\left(\text{or } I_n^{\phi_1,h}\left(\widehat{\boldsymbol{\theta}}^{I,\phi_2}\right)\right)$ under contiguous alternative hypotheses given in (8.25).

The power of this test is

$$\pi_n \equiv \Pr\left(I_n^{\phi_1}\left(\widehat{\boldsymbol{\theta}}^{I,\phi_2}\right)\left(\text{or } I_n^{\phi_1,h}\left(\widehat{\boldsymbol{\theta}}^{I,\phi_2}\right)\right) > \chi^2_{(I-1)(J-1),\alpha}/H_{1,n}\right).$$

In what follows, we prove that under the alternative $H_{1,n}$, and as $n \to \infty$,

$$I_n^{\phi_1}\left(\widehat{\boldsymbol{\theta}}^{I,\phi_2}\right)\left(\text{or } I_n^{\phi_1,h}\left(\widehat{\boldsymbol{\theta}}^{I,\phi_2}\right)\right)$$

converge in distribution to a noncentral chi-square random variable with noncentrality parameter δ given in Theorem 8.6, and $(I-1)(J-1)$ degrees of freedom $\left(\chi^2_{(I-1)(J-1)}(\delta)\right)$. Consequently, as $n \to \infty$

$$\pi_n \to \Pr\left(\chi^2_{(I-1)(J-1)}(\delta) > \chi^2_{(I-1)(J-1),\alpha}\right).$$

Theorem 8.6

Under $H_{1,n} : \boldsymbol{p}_n = \boldsymbol{p}(\boldsymbol{\theta}_0) + \frac{\boldsymbol{d}}{\sqrt{n}}$, $\boldsymbol{\theta}_0$ some unknown value in Θ_0, the ϕ-divergence test statistics

$$I_n^{\phi_1}\left(\widehat{\boldsymbol{\theta}}^{I,\phi_2}\right)\left(\text{or } I_n^{\phi_1,h}\left(\widehat{\boldsymbol{\theta}}^{I,\phi_2}\right)\right) \tag{8.26}$$

are asymptotically noncentrally chi-squared distributed with $(I-1)(J-1)$ degrees of freedom and noncentrality parameter

$$\delta = \sum_{i=1}^{I}\sum_{j=1}^{J}\frac{d_{ij}^2}{p_{i*}p_{*j}} - \sum_{i=1}^{I}\frac{d_{i*}^2}{p_{i*}} - \sum_{j=1}^{J}\frac{d_{*j}^2}{p_{*j}}.$$

Proof. Pardo, J. A. et *al.* (2002) established that for testing $H_0 : \boldsymbol{p} = \boldsymbol{p}(\boldsymbol{\theta}_0)$, for some unknown $\boldsymbol{\theta}_0 \in \Theta_0 \subset \Theta \subset \mathbb{R}^{M_0}$ with

$$\Theta_0 = \{\boldsymbol{\theta} \in \Theta : f_m(\boldsymbol{\theta}) = 0,\ m = 1, ..., \nu\},$$

versus $H_{1,n} : \boldsymbol{p} = \boldsymbol{p}(\boldsymbol{\theta}_0) + \boldsymbol{d}/\sqrt{n}$, the asymptotic distribution of the test statistic given in (8.26), under $H_{1,n}$, is noncentral chi-square with $M - M_0 + \nu - 1$ degrees of freedom and noncentrality parameter given by $\delta = \boldsymbol{\mu}^T\boldsymbol{\mu}$, where

$$\boldsymbol{\mu} = diag\left(\boldsymbol{p}(\boldsymbol{\theta}_0)^{-1/2}\right)(\boldsymbol{I} - \boldsymbol{L}(\boldsymbol{\theta}_0))\,\boldsymbol{d},$$

$$\boldsymbol{L}(\boldsymbol{\theta}_0) = diag\left(\boldsymbol{p}(\boldsymbol{\theta}_0)^{-1/2}\right)\boldsymbol{A}(\boldsymbol{\theta}_0)\,\boldsymbol{H}(\boldsymbol{\theta}_0)\,\boldsymbol{\Sigma}_{\boldsymbol{\theta}_0}\boldsymbol{A}(\boldsymbol{\theta}_0)^T diag\left(\boldsymbol{p}(\boldsymbol{\theta}_0)^{-1/2}\right),$$

and $\boldsymbol{H}(\boldsymbol{\theta}_0)$ was given in (8.2).

In our case it is not difficult to establish that

$$\begin{aligned}\boldsymbol{\mu}^T\boldsymbol{\mu} = & \ \boldsymbol{d}^T diag\left(\boldsymbol{p}(\boldsymbol{\theta}_0)^{-1/2}\right)\left\{\left(\boldsymbol{I}-\boldsymbol{A}(\boldsymbol{\theta}_0)\boldsymbol{\Sigma}_{\boldsymbol{\theta}_0}\boldsymbol{A}(\boldsymbol{\theta}_0)^T\right.\right.\\ + & \left.\left.\boldsymbol{A}(\boldsymbol{\theta}_0)\boldsymbol{\Sigma}_{\boldsymbol{\theta}_0}\boldsymbol{B}_I(\boldsymbol{\theta}_0)^T\left(\boldsymbol{B}_I(\boldsymbol{\theta}_0)\boldsymbol{\Sigma}_{\boldsymbol{\theta}_0}\boldsymbol{B}_I(\boldsymbol{\theta}_0)^T\right)^{-1}\boldsymbol{B}_I(\boldsymbol{\theta}_0)\boldsymbol{\Sigma}_{\boldsymbol{\theta}_0}\boldsymbol{A}(\boldsymbol{\theta}_0)^T\right)\right\}\\ \times & \ diag\left(\boldsymbol{p}(\boldsymbol{\theta}_0)^{-1/2}\right)\boldsymbol{d}.\end{aligned}$$

Applying (8.17) we have

$$\boldsymbol{\Sigma}_{\boldsymbol{\theta}_0}-\boldsymbol{\Sigma}_{\boldsymbol{\theta}_0}\boldsymbol{B}_I(\boldsymbol{\theta}_0)^T\left(\boldsymbol{B}_I(\boldsymbol{\theta}_0)\boldsymbol{\Sigma}_{\boldsymbol{\theta}_0}\boldsymbol{B}_I(\boldsymbol{\theta}_0)^T\right)^{-1}\boldsymbol{B}_I(\boldsymbol{\theta}_0)\boldsymbol{\Sigma}_{\boldsymbol{\theta}_0}=\boldsymbol{M}_{\boldsymbol{\beta}_0}^T\boldsymbol{I}_F(\boldsymbol{\beta}_0)^{-1}\boldsymbol{M}_{\boldsymbol{\beta}_0}\ .$$

Multiplying, in the previous expression, on the right by $\boldsymbol{A}(\boldsymbol{\theta}_0)$ and on the left by $\boldsymbol{A}(\boldsymbol{\theta}_0)^T$ we have

$$\begin{aligned}\boldsymbol{\mu}^T\boldsymbol{\mu} = & \ \boldsymbol{d}^T diag\left(\boldsymbol{p}(\boldsymbol{\theta}_0)^{-1/2}\right)\left(\boldsymbol{I}-\boldsymbol{A}(\boldsymbol{\theta}_0)\boldsymbol{M}_{\boldsymbol{\beta}_0}^T\boldsymbol{I}_F(\boldsymbol{\beta}_0)^{-1}\boldsymbol{M}_{\boldsymbol{\beta}_0}\boldsymbol{A}(\boldsymbol{\theta}_0)^T\right)\\ \times & \ diag\left(\boldsymbol{p}(\boldsymbol{\theta}_0)^{-1/2}\right)\boldsymbol{d}\end{aligned}$$

and

$$\boldsymbol{p}(\boldsymbol{\theta}_0)=\boldsymbol{p}(\boldsymbol{g}(\boldsymbol{\beta}_0)),\ \boldsymbol{A}(\boldsymbol{\theta}_0)\boldsymbol{M}_{\boldsymbol{\beta}_0}=\boldsymbol{A}(\boldsymbol{g}(\boldsymbol{\beta}_0)),$$

then

$$\begin{aligned}\boldsymbol{\mu}^T\boldsymbol{\mu} & = \boldsymbol{d}^T diag\left(\boldsymbol{p}(\boldsymbol{g}(\boldsymbol{\beta}_0)^{-1})\right)\left(\boldsymbol{I}-\boldsymbol{A}(\boldsymbol{g}(\boldsymbol{\beta}_0))\boldsymbol{I}_F(\boldsymbol{g}(\boldsymbol{\beta}_0))^{-1}\boldsymbol{A}(\boldsymbol{g}(\boldsymbol{\beta}_0))^T\right)\\ & \times diag\left(\boldsymbol{p}(\boldsymbol{g}(\boldsymbol{\beta}_0))^{-1/2}\right)\boldsymbol{d}.\end{aligned}$$

It is not difficult to establish that the matrix $\boldsymbol{d}^T diag\left(\boldsymbol{p}(\boldsymbol{g}(\boldsymbol{\beta}_0))^{-1/2}\right)$ is given by

$$\left(\frac{d_{11}}{\sqrt{p_{1*}p_{*1}}},\frac{d_{12}}{\sqrt{p_{1*}p_{*2}}},...,\frac{d_{1J}}{\sqrt{p_{1*}p_{*J}}},......,\frac{d_{I1}}{\sqrt{p_{I*}p_{*1}}},\frac{d_{I2}}{\sqrt{p_{I*}p_{*2}}},...,\frac{d_{IJ}}{\sqrt{p_{I*}p_{*J}}}\right)$$

and the matrix $\boldsymbol{I}_F(\boldsymbol{\beta}_0)^{-1}$ by

$$\begin{pmatrix}\Sigma_{\boldsymbol{p}_I} & 0\\ 0 & \Sigma_{\boldsymbol{p}_J}\end{pmatrix},$$

where $\Sigma_{\boldsymbol{p}_I}=diag(\boldsymbol{p}_I)-\boldsymbol{p}_I\boldsymbol{p}_I^T$ and $\Sigma_{\boldsymbol{p}_J}=diag(\boldsymbol{p}_J)-\boldsymbol{p}_J\boldsymbol{p}_J^T$, being

$$\boldsymbol{p}_I=(p_{1*},...,p_{I-1*})^T \text{ and } \boldsymbol{p}_J=(p_{*1},...,p_{*J-1})^T.$$

With these expressions one gets

$$\delta=\boldsymbol{\mu}^T\boldsymbol{\mu}=\sum_{i=1}^{I}\sum_{j=1}^{J}\frac{d_{ij}^2}{p_{i*}p_{*j}}-\sum_{i=1}^{I}\frac{d_{i*}^2}{p_{i*}}-\sum_{j=1}^{J}\frac{d_{*j}^2}{p_{*j}}.$$

■

For more details about the problem of independence in a two-way contingency table see Menéndez et *al.* (2005b).

8.2.3. Multiway Contingency Tables

The parameter space associated with a three-way contingency table is

$$\Theta=\left\{\boldsymbol{\theta}:\boldsymbol{\theta}=(p_{ijk};\ i=1,...,I,\ j=1,...,J,\ k=1,...,K,\ (i,j,k)\neq(I,J,K))^T\right\},$$

where $p_{ijk}=\Pr\left(X=i,Y=j,Z=k\right)$. The hypothesis of independence,

$$H_0: p_{ijk}=p_{i**}p_{*j*}p_{**k},\ i=1,...,I,\ j=1,...,J,\ k=1,...,K,$$

can be formulated by using the $IJK-I-J-K+2$ constraints,

$$h_{ijk}\left(\boldsymbol{\theta}\right)=p_{ijk}-p_{i**}p_{*j*}p_{**k}=0,$$

where $i,j,k\in D$, being D the set,

$$\{(i,j,k)/i=1,...,I,j=1,...,J,k=1,...,K,(i,j,k)\neq(i_1,J,K),i_1=1,...,I,$$
$$(i,j,k)\neq(I,j_1,K),\ j_1=1,...,J-1,\ (i,j,k)\neq(I,J,k_1),\ k_1=1,...,K-1\}$$

In this situation similar results to those of the previous subsections can be obtained (see Pardo, J. A. (2004)).

8.3. Symmetry

An interesting problem in a two-way contingency table is to investigate whether there are symmetric patterns in the data: Cell probabilities on one side of the main diagonal are a mirror image of those on the other side. This problem was first discussed by Bowker (1948) who gave the maximum likelihood estimator as well as a large sample chi-square type test for the null hypothesis of symmetry. In Ireland et *al*. (1969) it was proposed the minimum discrimination information estimator and in Quade and Salama (1975) the minimum chi-squared estimator. Based on the maximum likelihood estimator and on the family of ϕ-divergence measures, in Menéndez et *al*. (2001e) a new family of test statistics was introduced. This family contains as a particular case the test statistic given by Bowker (1948) as well as the likelihood ratio test. The state-of-the art in relation with the symmetry problem can be seen in Bishop et *al*. (1975), Agresti (2002), Andersen (1998) and references therein.

We consider a two-way contingency table with $I=J$ and let $\widehat{\boldsymbol{p}}=(\widehat{p}_{11},...,\widehat{p}_{II})^T$ be the nonparametric estimator of the unknown probability vector $\boldsymbol{p}$ with components $p_{ij}=\Pr(X=i,Y=j)$, with $p_{ij}>0$ and $i,j=1,...,I$.

The hypothesis of symmetry is

$$H_0: p_{ij}=p_{ji},\ ,\ i,j=1,...,I. \tag{8.27}$$

We consider the parameter space

$$\Theta = \left\{ \boldsymbol{\theta} : \boldsymbol{\theta} = (p_{ij};\ i,j = 1,...,I,\ (i,j) \neq (I,I))^T \right\}, \tag{8.28}$$

and we denote by

$$\boldsymbol{p}(\boldsymbol{\theta}) = (p_{11},...,p_{II})^T = \boldsymbol{p}, \tag{8.29}$$

the probability vector characterizing our model, with $p_{II} = 1 - \sum_{i=1}^{I}\sum_{j=1}^{I} p_{ij}$ with $(i,j) \neq (I,I)$. The hypothesis of symmetry introduced in (8.27) can be formulated using the following $I(I-1)/2$ constraints

$$h_{ij}(\boldsymbol{\theta}) = p_{ij} - p_{ji} = 0,\ i < j,\ i = 1,...,I-1, j = 1,...,I. \tag{8.30}$$

Also considering the parameter $\boldsymbol{\beta} = (p_{11},...,p_{1I}, p_{22},...,p_{2I},...,p_{I-1I-1}, p_{I-1I})^T$ and the set

$$\begin{aligned} B = \ & \{(a_{11},...,a_{1I},\ a_{22},...,a_{2I},...,a_{I-1I-1}, a_{I-1I})^T \in \mathbb{R}^{\frac{I(I+1)}{2}-1} : \\ & \textstyle\sum_{i\leq j} a_{ij} < 1,\ 0 < a_{ij},\ i,j = 1,..,I\}, \end{aligned}$$

the hypothesis (8.27) can be expressed for some unknown $\boldsymbol{\theta}_0 \in \Theta$ with $\boldsymbol{p}(\boldsymbol{\theta}_0) = \boldsymbol{p}_0$ as

$$H_0 : \boldsymbol{p}_0 = \boldsymbol{p}(\boldsymbol{g}(\boldsymbol{\beta}_0)), \quad \boldsymbol{\beta}_0 \in B \text{ and } \boldsymbol{g}(\boldsymbol{\beta}_0) = \boldsymbol{\theta}_0, \tag{8.31}$$

where the function $\boldsymbol{g}$ is defined by $\boldsymbol{g} = (g_{ij};\ i,j = 1,...,I,\ (i,j) \neq (I,I))$ with

$$g_{ij}(\boldsymbol{\beta}) = \begin{cases} p_{ij} & i \leq j \\ p_{ji} & i > j \end{cases}, \quad i,j = 1,...,I-1,$$

and

$$\begin{aligned} g_{Ij}(\boldsymbol{\beta}) = &\ \ p_{jI},\ j = 1,...,I-1 \\ g_{iI}(\boldsymbol{\beta}) = &\ \ p_{iI},\ i = 1,...,I-1. \end{aligned}$$

Note that $\boldsymbol{p}(\boldsymbol{g}(\boldsymbol{\beta})) = (g_{ij}(\boldsymbol{\beta});\ i,j = 1,...,I)^T$, where

$$g_{II}(\boldsymbol{\beta}) = 1 - \sum_{\substack{i,j=1 \\ (i,j)\neq(I,I)}}^{I} g_{ij}(\boldsymbol{\beta}).$$

This approach can be formulated in terms of the results presented in Chapter 6.

In this Chapter we consider the approach given in (8.30). We can observe that with this approach the hypothesis of symmetry can be written as

$$H_0 : \boldsymbol{p} = \boldsymbol{p}(\boldsymbol{\theta}_0),\ \text{for some unknown } \boldsymbol{\theta}_0 \in \Theta_0,$$

with

$$\Theta_0 = \{\boldsymbol{\theta} \in \Theta : h_{ij}(\boldsymbol{\theta}) = 0,\ i < j,\ i = 1,...,I-1, j = 1,...,I\}, \tag{8.32}$$

$\boldsymbol{p}(\boldsymbol{\theta})$ defined in (8.29) and h_{ij} in (8.30).

It is clear that the $\frac{I(I-1)}{2} \times (I^2-1)$ matrix $\boldsymbol{B}_S(\boldsymbol{\theta}_0)$ given by

$$\boldsymbol{B}_S(\boldsymbol{\theta}_0) = \left(\frac{\partial h_{ij}(\boldsymbol{\theta}_0)}{\partial \theta_{ij}}\right)_{\frac{I(I-1)}{2}\times(I^2-1)} \tag{8.33}$$

has rank $I(I-1)/2$, because the matrix of partial derivatives $\boldsymbol{B}_S(\boldsymbol{\theta}_0)$ has the same rank as the matrix

$$\left(\boldsymbol{I}_{\frac{I(I-1)}{2}\times\frac{I(I-1)}{2}}, -\boldsymbol{I}_{\frac{I(I-1)}{2}\times\frac{I(I-1)}{2}}, \boldsymbol{O}_{\frac{I(I-1)}{2}\times I}\right)$$

where $\boldsymbol{I}_{\frac{I(I-1)}{2}\times\frac{I(I-1)}{2}}$ is the identity matrix of order $I(I-1)/2$ and $\boldsymbol{O}_{\frac{I(I-1)}{2}\times I}$ is the matrix with $I(I-1)/2$ rows and I columns whose elements are all zero.

For $I = 2$, we only have a constraint which is given by

$$h_{12}(\boldsymbol{\theta}) = p_{12} - p_{21} = 0,$$

and for $I = 3$, we have three constraints given by

$$\begin{aligned} h_{12}(\boldsymbol{\theta}) &= \; p_{12} - p_{21} = 0 \\ h_{13}(\boldsymbol{\theta}) &= \; p_{13} - p_{31} = 0 \\ h_{23}(\boldsymbol{\theta}) &= \; p_{23} - p_{32} = 0. \end{aligned}$$

In the following we obtain the expression of the minimum ϕ-divergence estimator of $\boldsymbol{\theta}_0 \in \Theta_0$, i.e., the minimum ϕ-divergence estimator under the hypothesis of symmetry (8.30).

Theorem 8.7

The minimum ϕ-divergence estimator,

$$\widehat{\boldsymbol{\theta}}^{S,\phi} = \left(p_{ij}^{S,\phi};\ i,j = 1,...,I \text{ and } (i,j) \neq (I,I)\right)^T,$$

of $\boldsymbol{\theta}_0$ in Θ_0 (i.e., under the hypothesis of symmetry) is obtained as a solution of the system of equations

$$\tfrac{1}{2}\left(\phi\left(\frac{\widehat{p}_{ij}}{p_{ij}}\right) + \phi\left(\frac{\widehat{p}_{ji}}{p_{ij}}\right) - \frac{\widehat{p}_{ij}}{p_{ij}}\phi'\left(\frac{\widehat{p}_{ij}}{p_{ij}}\right) - \frac{\widehat{p}_{ji}}{p_{ij}}\phi'\left(\frac{\widehat{p}_{ji}}{p_{ij}}\right)\right) - \mu \;= 0, \quad i,j = 1,...,I, \tag{8.34}$$

where μ has the expression

$$\mu = \frac{1}{2}\sum_{i=1}^{I}\sum_{j=1}^{J}\left\{p_{ij}\left[\phi\left(\frac{\widehat{p}_{ij}}{p_{ij}}\right) + \phi\left(\frac{\widehat{p}_{ji}}{p_{ij}}\right)\right] - \widehat{p}_{ij}\phi'\left(\frac{\widehat{p}_{ij}}{p_{ij}}\right) - \widehat{p}_{ji}\phi'\left(\frac{\widehat{p}_{ji}}{p_{ij}}\right)\right\}.$$

Its asymptotic distribution is given by

$$\sqrt{n}\left(\widehat{\boldsymbol{\theta}}^{S,\phi} - \boldsymbol{\theta}_0\right) \underset{n\to\infty}{\overset{L}{\to}} N(\boldsymbol{0}, \boldsymbol{W}_S(\boldsymbol{\theta}_0)), \tag{8.35}$$

where the $(I^2-1)\times(I^2-1)$ *matrix* $\boldsymbol{W}_S(\boldsymbol{\theta}_0)$ *is*

$$\boldsymbol{W}_S(\boldsymbol{\theta}_0)=\boldsymbol{\Sigma}_{\boldsymbol{\theta}_0}-\boldsymbol{\Sigma}_{\boldsymbol{\theta}_0}\boldsymbol{B}_S(\boldsymbol{\theta}_0)^T\left(\boldsymbol{B}_S(\boldsymbol{\theta}_0)\boldsymbol{\Sigma}_{\boldsymbol{\theta}_0}\boldsymbol{B}_S(\boldsymbol{\theta}_0)^T\right)^{-1}\boldsymbol{B}_S(\boldsymbol{\theta}_0)\boldsymbol{\Sigma}_{\boldsymbol{\theta}_0},$$

and $\boldsymbol{B}_S(\boldsymbol{\theta}_0)$ *is the matrix defined in (8.33).*

Proof. Instead of getting $\widehat{\boldsymbol{\theta}}^{S,\phi}=\left(p_{ij}^{S,\phi};\ i,j=1,...,I \text{ and } (i,j)\neq(I,I)\right)^T$ we shall obtain $\boldsymbol{p}(\widehat{\boldsymbol{\theta}}^{S,\phi})=\left(p_{ij}^{S,\phi};\ i,j=1,...,I\right)^T$. The p_{ij}'s, $i,j=1,...,I$, which minimize the ϕ-divergence

$$D_\phi(\widehat{\boldsymbol{p}},\boldsymbol{p}(\boldsymbol{\theta}))$$

subject to the null hypothesis of symmetry (or the constraints about $\boldsymbol{\theta}$ given in (8.30)), may be obtained by minimizing

$$\sum_{i=1}^{I}\sum_{j=1}^{I}p_{ij}\phi\left(\frac{\widehat{p}_{ij}}{p_{ij}}\right)+\sum_{\substack{i=1\\i<j}}^{I}\sum_{j=1}^{I}\lambda_{ij}\left(p_{ij}-p_{ji}\right)+\mu\left(1-\sum_{i=1}^{I}\sum_{j=1}^{I}p_{ij}\right) \tag{8.36}$$

with respect to the p_{ij}, where μ and λ_{ij} are undetermined Lagrangian multipliers. Minimizing the expression (8.36) is equivalent to minimizing the expression

$$\frac{1}{2}\sum_{i=1}^{I}\sum_{j=1}^{I}p_{ij}\left(\phi\left(\frac{\widehat{p}_{ij}}{p_{ij}}\right)+\phi\left(\frac{\widehat{p}_{ji}}{p_{ij}}\right)\right)+\mu\left(1-\sum_{i=1}^{I}\sum_{j=1}^{I}p_{ij}\right).$$

Differentiating with respect to p_{ij}, $i,j=1,...,I$ we get the system of equations given in (8.34), whose solutions provide the minimum ϕ-divergence estimator $\widehat{\boldsymbol{\theta}}^{S,\phi}$.

The asymptotic distribution follows from (8.3) because in our model $\boldsymbol{I}_F(\boldsymbol{\theta}_0)^{-1}=\boldsymbol{\Sigma}_{\boldsymbol{\theta}_0}$. ∎

Corollary 8.2

The minimum power-divergence estimator, $\boldsymbol{p}(\widehat{\boldsymbol{\theta}}^{S,\phi_{(\lambda)}})=\left(p_{ij}^{S,\phi_{(\lambda)}};\ i,j=1,...,I\right)^T$ *of* $\boldsymbol{p}(\boldsymbol{\theta}_0)$, *under the hypothesis of symmetry (8.27), is given by*

$$p_{ij}^{S,\phi_{(\lambda)}}=\frac{\left(\widehat{p}_{ij}^{\lambda+1}+\widehat{p}_{ji}^{\lambda+1}\right)^{\frac{1}{\lambda+1}}}{\sum_{i=1}^{I}\sum_{j=1}^{I}\left(\widehat{p}_{ij}^{\lambda+1}+\widehat{p}_{ji}^{\lambda+1}\right)^{\frac{1}{\lambda+1}}},\qquad i,j=1,...,I. \tag{8.37}$$

Proof. In this case the system of equations (8.34) can be written as

$$\frac{1}{p_{ij}^{\lambda+1}}\left(\frac{\widehat{p}_{ij}^{\lambda+1}+\widehat{p}_{ji}^{\lambda+1}}{2}\right)=\left(\sum_{i=1}^{I}\sum_{j=1}^{I}\left(\frac{\widehat{p}_{ij}^{\lambda+1}+\widehat{p}_{ji}^{\lambda+1}}{2}\right)^{\frac{1}{\lambda+1}}\right)^{\lambda+1},\quad i,j=1,...,I$$

and it is clear that the solution of this system of equations is given by (8.37).

For $\lambda = 0$,

$$p_{ij}^{S,\phi_{(0)}} = \frac{\widehat{p}_{ij} + \widehat{p}_{ji}}{2}, \ i,j = 1,...,I,$$

so we obtain the maximum likelihood estimator for symmetry introduced by Bowker (1948).

For $\lambda \to -1$, we obtain as a limit case,

$$p_{ij}^{S,\phi_{(-1)}} = \frac{\left(\widehat{p}_{ij}\widehat{p}_{ji}\right)^{1/2}}{\sum_{i=1}^{I}\sum_{j=1}^{I}\left(\widehat{p}_{ij}\widehat{p}_{ji}\right)^{1/2}}, \ i,j = 1,...,I$$

i.e., the minimum modified likelihood estimator for symmetry introduced and studied in Ireland et *al.* (1969).

For $\lambda = 1$,

$$p_{ij}^{S,\phi_{(1)}} = \frac{\left(\widehat{p}_{ij}^{2} + \widehat{p}_{ji}^{2}\right)^{1/2}}{\sum_{i=1}^{I}\sum_{j=1}^{I}\left(\widehat{p}_{ij}^{2} + \widehat{p}_{ji}^{2}\right)^{1/2}}$$

we get the minimum chi-square estimator for symmetry introduced in Quade and Salama (1975).

Other interesting estimators for symmetry may be: For $\lambda = -2$ the minimum modified chi-square estimator, for $\lambda = -1/2$ the Freeman-Tukey estimator and finally for $\lambda = 2/3$ the Cressie-Read estimator.

Remark 8.5

For $I = 2$, given $\boldsymbol{\theta}_0 = (p_{11},\ p_{12},\ p_{21})^T \in \Theta_0$, it is easy to check that

- $\boldsymbol{B}_S(\boldsymbol{\theta}_0) = (0, 1, -1)$
- $\boldsymbol{\Sigma}_{\boldsymbol{\theta}_0} = diag(\boldsymbol{\theta}_0) - \boldsymbol{\theta}_0\boldsymbol{\theta}_0^T,$
- $\left(\boldsymbol{B}_S(\boldsymbol{\theta}_0)\boldsymbol{\Sigma}_{\boldsymbol{\theta}_0}\boldsymbol{B}_S(\boldsymbol{\theta}_0)^T\right)^{-1} = \left(-(p_{12} - p_{21})^2 + p_{21} + p_{12}\right)^{-1},$

and it is not difficult to establish that the matrix $\boldsymbol{W}_S(\boldsymbol{\theta}_0)$ has the expression

$$\begin{pmatrix} p_{11}(1-p_{11}) & -p_{11}p_{12} & -p_{11}p_{12} \\ -p_{11}p_{12} & -\frac{1}{2}p_{12}(2p_{12}-1) & -\frac{1}{2}p_{12}(2p_{12}-1) \\ -p_{11}p_{12} & -\frac{1}{2}p_{12}(2p_{12}-1) & -\frac{1}{2}p_{12}(2p_{12}-1) \end{pmatrix}.$$

This matrix can be written as

$$\begin{pmatrix} 1 & 0 \\ 0 & 1 \\ 0 & 1 \end{pmatrix} \begin{pmatrix} p_{11}(1-p_{11}) & -p_{11}p_{12} \\ -p_{11}p_{12} & \frac{1}{2}p_{12}(1-2p_{12}) \end{pmatrix} \begin{pmatrix} 1 & 0 & 0 \\ 0 & 1 & 1 \end{pmatrix},$$

where:

- *The matrix*

$$\begin{pmatrix} p_{11}(1-p_{11}) & -p_{11}p_{12} \\ -p_{11}p_{12} & \frac{1}{2}p_{12}(1-2p_{12}) \end{pmatrix}$$

is the inverse of the Fisher information matrix corresponding to the parameter $\boldsymbol{\beta}_0 = (p_{11}, p_{12})^T$, *i.e.,* $\boldsymbol{I}_F(\boldsymbol{\beta}_0)^{-1}$.

- *The matrix*

$$\begin{pmatrix} 1 & 0 \\ 0 & 1 \\ 0 & 1 \end{pmatrix}$$

is the matrix of partial derivatives $\left(\frac{\partial \mathbf{g}(\boldsymbol{\beta}_0)}{\partial \boldsymbol{\beta}}\right)^T$.

Then we have

$$\boldsymbol{W}_S(\boldsymbol{\theta}_0) = \left(\frac{\partial \mathbf{g}(\boldsymbol{\beta}_0)}{\partial \boldsymbol{\beta}}\right)^T \boldsymbol{I}_F(\boldsymbol{\beta}_0)^{-1} \left(\frac{\partial \mathbf{g}(\boldsymbol{\beta}_0)}{\partial \boldsymbol{\beta}}\right).$$

This result can be proved in general in the same way as the one corresponding to the problem of independence in Section 8.2 .

Theorem 8.8
Under the hypothesis of symmetry, we have

$$\boldsymbol{W}_S(\boldsymbol{\theta}_0) = \boldsymbol{M}_{\boldsymbol{\beta}_0}^T \boldsymbol{I}_F(\boldsymbol{\beta}_0) \boldsymbol{M}_{\boldsymbol{\beta}_0}.$$

Proof. The proof follows the same steps as the proof of Theorem 8.2.

From Theorem 8.8, it can be seen that $\widehat{\boldsymbol{\theta}}^{S,\phi_2}$ is BAN.

8.3.1. Test of Symmetry

Based on the test statistic given in (8.5) we should reject the hypothesis of symmetry if

$$S_n^{\phi_1}\left(\widehat{\boldsymbol{\theta}}^{S,\phi_2}\right) \equiv \frac{2n}{\phi_1''(1)} D_{\phi_1}\left(\widehat{\mathbf{p}}, \mathbf{p}(\widehat{\boldsymbol{\theta}}^{S,\phi_2})\right) > c,$$

where c is a positive constant. In the following theorem we establish the asymptotic distribution of the test statistic $S_n^{\phi_1}\left(\widehat{\boldsymbol{\theta}}^{S,\phi_2}\right)$.

Theorem 8.9

The asymptotic distribution of the ϕ-divergence test statistics

$$S_n^{\phi_1}\left(\widehat{\boldsymbol{\theta}}^{S,\phi_2}\right)=\frac{2n}{\phi_1''(1)}D_{\phi_1}\left(\widehat{\mathbf{p}},\mathbf{p}(\widehat{\boldsymbol{\theta}}^{S,\phi_2})\right) \tag{8.38}$$

and

$$S_n^{\phi_1,h}\left(\widehat{\boldsymbol{\theta}}^{S,\phi_2}\right)\equiv\frac{2n}{h'(0)\,\phi_1''(1)}h\left(D_{\phi_1}\left(\widehat{\mathbf{p}},\mathbf{p}(\widehat{\boldsymbol{\theta}}^{S,\phi_2})\right)\right), \tag{8.39}$$

for testing the hypothesis of symmetry, given in (8.27), is chi-square with $\frac{I(I-1)}{2}$ degrees of freedom.

Proof. In our case we have I^2 cells and by (8.28) I^2-1 parameters that are necessary to estimate. Using the relation (8.30) the number of constraints is $I(I-1)/2$. Then by (8.5) the asymptotic distribution of the family of ϕ-divergence test statistics (8.38) is chi-square with

$$\underbrace{I^2}_{\text{Cells}} \quad - \quad \underbrace{I^2-1}_{\substack{\text{Estimated}\\ \text{Parameters}}} \quad + \quad \underbrace{I(I-1)/2}_{\text{Constraints}} \quad -1 \quad = \quad \underbrace{I(I-1)/2}_{\substack{\text{Degrees of}\\ \text{freedom}}}.$$

In relation with the family of test statistics given in (8.39), we have

$$h(x)=h(0)+h'(0)\,x+o(x),$$

then

$$h\left(D_{\phi_1}\left(\widehat{\mathbf{p}},\mathbf{p}(\widehat{\boldsymbol{\theta}}^{S,\phi_2})\right)\right)=h'(0)\,D_{\phi_1}\left(\widehat{\mathbf{p}},\mathbf{p}(\widehat{\boldsymbol{\theta}}^{S,\phi_2})\right)+o_P(1)$$

and we get that the asymptotic distribution of the family of test statistics given in (8.39) is also chi-square with $I(I-1)/2$ degrees of freedom. ■

Based on this result we should reject the hypothesis of symmetry if

$$S_n^{\phi_1}\left(\widehat{\boldsymbol{\theta}}^{S,\phi_2}\right)>\chi^2_{I(I-1)/2,\alpha}\quad\left(\text{or } S_n^{\phi_1,h}\left(\widehat{\boldsymbol{\theta}}^{S,\phi_2}\right)\right)>\chi^2_{I(I-1)/2,\alpha}. \tag{8.40}$$

Remark 8.6

For $\phi_1(x)=\phi_2(x)=\phi_{(0)}(x)=x\log x-x+1$ we get that $S_n^{\phi_1}\left(\widehat{\boldsymbol{\theta}}^{S,\phi_2}\right)$ coincides with the classical likelihood ratio test statistic for symmetry

$$G^2\equiv S_n^{\phi_{(0)}}\left(\widehat{\boldsymbol{\theta}}^{S,\phi_{(0)}}\right)=2\sum_{\substack{i,j=1\\ i\neq j}}^{I}n_{ij}\log\frac{2n_{ij}}{n_{ij}+n_{ji}}.$$

For $\phi_2(x) = \phi_{(0)}(x) = x\log x - x + 1$ *and* $\phi_1(x) = \phi_{(1)}(x) = \frac{1}{2}(x-1)^2$, $S_n^{\phi_1}\left(\widehat{\boldsymbol{\theta}}^{S,\phi_2}\right)$ *coincides with the classical chi-square test statistic of Bowker (1948) and is given by*

$$X^2 \equiv S_n^{\phi_{(1)}}\left(\widehat{\boldsymbol{\theta}}^{S,\phi_{(0)}}\right) = \sum_{\substack{i,j=1\\ i<j}}^{I} \frac{(n_{ij}-n_{ji})^2}{n_{ij}+n_{ji}}.$$

It is also important to note that for $\phi_1(x) = \phi_2(x) = \phi_{(1)}(x) = \frac{1}{2}(x-1)^2$ *we get the test statistic given by Quade and Salama (1975) and for* $\phi_2(x) = x\log x - x + 1$ *and whatever* $\phi_1(x)$ *we get the family of test statistics introduced and studied in Menéndez et al. (2001c).*

Power of the Test

Let $\boldsymbol{q} = (q_{11}, ..., q_{II})^T$ be a point at the alternative hypothesis, i.e., there exist at least two indexes i and j for which $q_{ij} \neq q_{ji}$. We denote by $\boldsymbol{\theta}_a^{\phi_2}$ the point on Θ verifying

$$\boldsymbol{\theta}_a^{\phi_2} = \arg\min_{\boldsymbol{\theta}\in\Theta_0} D_{\phi_2}(\boldsymbol{q}, \boldsymbol{p}(\boldsymbol{\theta})),$$

where Θ_0 was defined in (8.32).

It is clear that

$$\boldsymbol{\theta}_a^{\phi_2} = (f_{ij}(\boldsymbol{q});\ i,j = 1,...,I,\ \ (i,j) \neq (I,I))^T$$

and

$$\boldsymbol{p}(\boldsymbol{\theta}_a^{\phi_2}) = (f_{ij}(\boldsymbol{q});\ i,j = 1,...,I,\ )^T \equiv \boldsymbol{f}(\boldsymbol{q}),$$

with

$$f_{II}(\boldsymbol{q}) = 1 - \sum_{\substack{i,j=1\\ (i,j)\neq(I,I)}}^{I} f_{ij}(\boldsymbol{q}).$$

The notation $f_{ij}(\boldsymbol{q})$ indicates that the elements of the vector $\boldsymbol{\theta}_a^{\phi_2}$ depend on $\boldsymbol{q}$. For instance, for the power-divergence family $\phi_{(\lambda)}(x)$ we have

$$f_{ij}(\boldsymbol{q}) = \frac{\left(q_{ij}^{\lambda+1} + q_{ji}^{\lambda+1}\right)^{\frac{1}{\lambda+1}}}{\sum_{i=1}^{I}\sum_{j=1}^{I}\left(q_{ij}^{\lambda+1} + q_{ji}^{\lambda+1}\right)^{\frac{1}{\lambda+1}}}, \qquad i,j = 1,...,I.$$

We also denote

$$\widehat{\boldsymbol{\theta}}^{S,\phi_2} = \left(p_{ij}^{S,\phi_2};\ i,j = 1,...,I,\ \ (i,j) \neq (I,I)\ \right)^T$$

and then

$$\boldsymbol{p}(\widehat{\boldsymbol{\theta}}^{S,\phi_2}) = \left(p_{ij}^{S,\phi_2};\ i,j = 1,...,I\right)^T \equiv \boldsymbol{f}(\widehat{\mathbf{p}}),$$

where $\boldsymbol{f} = (f_{ij};\ i,j = 1,...,I)^T$. If the alternative $\boldsymbol{q}$ is true we have that $\widehat{\mathbf{p}}$ tends to $\boldsymbol{q}$ and $\boldsymbol{p}(\widehat{\boldsymbol{\theta}}^{S,\phi_2})$ to $\boldsymbol{p}\left(\boldsymbol{\theta}_a^{\phi_2}\right)$ in probability.

If we define the function

$$\Psi_{\phi_1}(\boldsymbol{q}) = D_{\phi_1}(\boldsymbol{q}, \boldsymbol{f}(\boldsymbol{q})),$$

we have

$$\Psi_{\phi_1}(\widehat{\mathbf{p}}) = \Psi_{\phi_1}(\boldsymbol{q}) + \sum_{i=1}^{I}\sum_{j=1}^{J} \frac{\partial D_{\phi_1}(\boldsymbol{q}, \boldsymbol{f}(\boldsymbol{q}))}{\partial q_{ij}} (\widehat{p}_{ij} - q_{ij}) + o\left(\|\widehat{\mathbf{p}} - \boldsymbol{q}\|\right).$$

Then the random variables

$$\sqrt{n}\left(D_{\phi_1}(\widehat{\mathbf{p}}, \boldsymbol{f}(\widehat{\mathbf{p}})) - D_{\phi_1}(\boldsymbol{q}, \boldsymbol{f}(\boldsymbol{q}))\right)$$

and

$$\sqrt{n} \sum_{i=1}^{I}\sum_{j=1}^{J} \frac{\partial D_{\phi_1}(\boldsymbol{q}, \boldsymbol{f}(\boldsymbol{q}))}{\partial q_{ij}} (\widehat{p}_{ij} - q_{ij})$$

have the same asymptotic distribution. If we define

$$l_{ij} = \frac{\partial D_{\phi_1}(\boldsymbol{q}, \boldsymbol{f}(\boldsymbol{q}))}{\partial q_{ij}} \tag{8.41}$$

and $\boldsymbol{l} = (l_{ij};\ i,j = 1,...,I)^T$, we have

$$\sqrt{n}\left(D_{\phi_1}(\widehat{\mathbf{p}}, \boldsymbol{f}(\widehat{\mathbf{p}})) - D_{\phi_1}(\boldsymbol{q}, \boldsymbol{f}(\boldsymbol{q}))\right) \underset{n\to\infty}{\overset{L}{\longrightarrow}} N\left(\mathbf{0}, \boldsymbol{l}^T \boldsymbol{\Sigma}_{\boldsymbol{q}} \boldsymbol{l}\right), \tag{8.42}$$

where $\boldsymbol{\Sigma}_{\boldsymbol{q}} = diag(\boldsymbol{q}) - \boldsymbol{q}\boldsymbol{q}^T$.

In the following theorem we present the asymptotic distribution using the maximum likelihood estimator. In this case we have:

Theorem 8.10

Let $\mathbf{p}(\widehat{\boldsymbol{\theta}}) = \left(\frac{\widehat{p}_{ij}+\widehat{p}_{ji}}{2},\ i,j = 1,...,I\right)^T$ *the maximum likelihood estimator, under the symmetry hypothesis, of* $\mathbf{p}(\boldsymbol{\theta}_0) = (p_{11},...,p_{II})^T$ *and let* $\boldsymbol{q}$ *be a point at the alternative hypothesis* $(\boldsymbol{q} \neq \mathbf{p}(\boldsymbol{\theta}_0))$. *Then,*

$$\sqrt{n}\left(D_{\phi_1}(\widehat{\mathbf{p}}, \mathbf{p}(\widehat{\boldsymbol{\theta}})) - D_{\phi_1}(\boldsymbol{q}, \boldsymbol{f}(\boldsymbol{q}))\right) \underset{n\to\infty}{\overset{L}{\longrightarrow}} N\left(0, \sigma_{\phi_1}^2(\boldsymbol{q})\right),$$

where

$$\sigma_{\phi_1}^2(\boldsymbol{q}) = \sum_{i=1}^{I}\sum_{\substack{j=1\\ i\neq j}}^{I} q_{ij} m_{ij}^2 - \left(\sum_{i=1}^{I}\sum_{\substack{j=1\\ i\neq j}}^{I} q_{ij} m_{ij}\right)^2,$$

and

$$m_{ij} = \frac{1}{2}\phi_1\left(\frac{2q_{ij}}{q_{ij}+q_{ji}}\right) + \frac{1}{2}\phi_1\left(\frac{2q_{ji}}{q_{ij}+q_{ji}}\right) + \frac{q_{ij}}{q_{ij}+q_{ji}}\left(\phi_1'\left(\frac{2q_{ij}}{q_{ij}+q_{ji}}\right) - \phi_1'\left(\frac{2q_{ji}}{q_{ij}+q_{ji}}\right)\right).$$

Proof. The result follows from (8.41) and (8.42). ∎

In the same way as in the problem of independence it is not easy to get the expression of the power function of the test statistics given in (8.40) when $\phi_2(x) \neq x\log x - x + 1$; however we can consider a contiguous sequence of alternative hypotheses that approach the null hypothesis $H_0 : p_{ij} = p_{ji}$ at the rate $O\left(n^{-1/2}\right)$. Consider the multinomial probability vector

$$p_{n,ij} = p_{ij}(\boldsymbol{\theta}_0) + \frac{d_{ij}}{\sqrt{n}},\ i,j = 1,...,I,$$

where $\boldsymbol{d} = (d_{11},...,d_{II})^T$ is a fixed vector such that $\sum_{i,j=1}^{I} d_{ij} = 0$ and $\boldsymbol{\theta}_0$ is unknown but belonging to Θ_0. As $n \to \infty$, the sequence of multinomial probabilities $\{\boldsymbol{p}_n\}_{n\in N}$ with

$$\boldsymbol{p}_n = \left(p_{n,ij} + \frac{d_{ij}}{\sqrt{n}}, i,j = 1,...,I\right)^T$$

converges to a multinomial probability in H_0 at the rate of $O\left(n^{-1/2}\right)$. We name

$$H_{1,n} : \boldsymbol{p}_n = \boldsymbol{p}(\boldsymbol{\theta}) + \frac{\boldsymbol{d}}{\sqrt{n}},\ \boldsymbol{\theta} \in \Theta_0.$$

Then, we have the following result:

Theorem 8.11
Under $H_{1,n} : \boldsymbol{p}_n = \boldsymbol{p}(\boldsymbol{\theta}_0) + \frac{\boldsymbol{d}}{\sqrt{n}}$, $\boldsymbol{\theta}_0$ *some unknown value in* Θ_0, *the test statistics*

$$S_n^{\phi_1}\left(\widehat{\boldsymbol{\theta}}^{S,\phi_2}\right) \quad \text{and} \quad S_n^{\phi_1,h}\left(\widehat{\boldsymbol{\theta}}^{S,\phi_2}\right)$$

are asymptotically noncentrally chi-squared distributed with $I(I-1)/2$ *degrees of freedom and noncentrality parameter*

$$\delta = \frac{1}{2}\sum_{\substack{i,j=1\\ i\neq j}}^{I} \frac{d_{ij}^2}{p_{ij}} - \sum_{\substack{i,j=1\\ i<j}}^{I} \frac{d_{ij}d_{ji}}{p_{ij}}.$$

∎

An interesting analysis of the problem of symmetry as well as a simulation study can be seen in Menéndez et *al.* (2005a).

In some real problems (i.e., medicine, psychology, sociology, etc.) the categorica response variables (X, Y) represent the measure after or before a treatment. In such situations our interest is to determine the treatment effect, i.e., if $X \geq Y$ (we assume that X represents the measure after the treatment and Y before the treatment). In the following we understand that X is preferred or indifferent to Y, according to joint likelihood ratio ordering, if and only if $p_{ij} \geq p_{ji}$ $\forall i \geq j$. In this situation the alternative hypothesis is

$$H_1 : p_{ij} \geq p_{ji}, \text{ for all } i \geq j.$$

This problem was first considered by Barmi & Kochar (1995) who presented the likelihood ratio test statistic for the problem of testing

$$H_0 : p_{ij} = p_{ji} \text{ against } H_1 : p_{ij} \geq p_{ji}, \ \forall i \geq j, \tag{8.43}$$

and considered the application of it to a real life problem: He tested if the vision of both the eyes, for 7477 women, is the same against the alternative that the right eye has better vision than the left eye. Menéndez et *al*. (2003c) considered the three hypotheses, H_0 and H_1 given in (8.43) and H_2 no restriction over the p_{ij}'s and studied some ϕ-divergence test statistics for testing H_0 against H_1 and H_1 against H_2.

8.3.2. Symmetry in a Three-way Contingence Table

If we consider a three-way contingency table, the parameter space, given in (8.7), is

$$\Theta = \left\{ \boldsymbol{\theta} : \boldsymbol{\theta} = (p_{ijk};\ i, j, k = 1, ..., I,\ (i, j, k) \neq (I, I, I))^T \right\},$$

where $p_{ijk} = \Pr(X = i, Y = j, Z = k)$, $i, j, k = 1, ..., I$, $(i, j, k) \neq (I, I, I)$ and

$$p_{III} = 1 - \sum_{\substack{i,j,k=1 \\ (i,j,k)\neq(I,I,I)}}^{I} p_{ijk}.$$

The hypothesis of symmetry is given by

$$H_0 : p_{ijk} = p_{i'j'k'} \quad \forall (i, j, k) \quad \text{and} \quad \forall \text{ permutations } (i', j', k') \text{ of } (i, j, k). \tag{8.44}$$

We denote by $\boldsymbol{p}(\boldsymbol{\theta}) = (p_{111}, ..., p_{III})^T$ the probability vector characterizing our model. The problem of symmetry formulated in (8.44) can be formulated using the following $I(I-1)(5I+2)/6$ constraints about $\boldsymbol{\theta}$

$$h_{ijk}(\boldsymbol{\theta}) = p_{ijk} - p_{i'j'k'} = 0, \forall \text{ permutations } (i', j', k') \text{ of } (i, j, k). \tag{8.45}$$

We denote

$$\Theta_0 = \{\theta \in \Theta / h_{ijk}(\boldsymbol{\theta}) = 0,\ \forall \text{ permutations } (i', j', k') \text{ of } (i, j, k)\}.$$

The matrix

$$\boldsymbol{B}_S^*(\boldsymbol{\theta}_0) = \left(\frac{\partial h_{ijk}(\boldsymbol{\theta}_0)}{\partial \theta_{ijk}}\right)_{\frac{I(I-1)(5I+2)}{6} \times I^3} \tag{8.46}$$

has rank $I(I-1)(5I+2)/6$, because the matrix $\boldsymbol{B}_S^*(\boldsymbol{\theta}_0)$ has the same rank as the matrix

$$\left(-\boldsymbol{I}_{\frac{I(I-1)(5I+2)}{6} \times \frac{I(I-1)(5I+2)}{6}}, \boldsymbol{C}_{\frac{I(I-1)(5I+2)}{6} \times \frac{I(I-1)(I+4)}{6}}, \boldsymbol{0}_{\frac{I(I-1)(5I+2)}{6} \times I}\right),$$

where $\boldsymbol{I}_{\frac{I(I-1)(5I+2)}{6}}$ is the identity matrix of order $I(I-1)(5I+2)/6$, $\boldsymbol{0}_{\frac{I(I-1)(5I+2)}{6} \times I}$ is the matrix with $I(I-1)(5I+2)/6$ rows and I columns whose elements are all zero and $\boldsymbol{C}_{\frac{I(I-1)(5I+2)}{6} \times \frac{I(I-1)(I+4)}{6}}$ is the matrix with $\frac{I(I-1)(5I+2)}{6}$ rows and $\frac{I(I-1)(I+4)}{6}$ columns whose rank is the same as the rank of the matrix $\left(\boldsymbol{I}_{\frac{I(I-1)(I+4)}{6}}, \boldsymbol{I}_{\frac{I(I-1)(I+4)}{6}}\right)^T$.

Now we present in the following theorem the expression of the minimum ϕ-divergence estimator of $\boldsymbol{\theta}_0$ in Θ_0.

Theorem 8.12

The minimum ϕ-divergence estimator,

$$\widehat{\boldsymbol{\theta}}^{S,\phi} = \left(p_{ijk}^{S,\phi};\ i,j,k = 1,...,I,\ (i,j,k) \neq (I,I,I)\right)^T,$$

of $\boldsymbol{\theta}_0$ in Θ_0 is obtained as a solution of the system of equations

$$\left\{\begin{array}{r} 6\left(\phi\left(\frac{\widehat{p}_{ijk}}{p_{ijk}}\right)+\phi\left(\frac{\widehat{p}_{ikj}}{p_{ijk}}\right)+\phi\left(\frac{\widehat{p}_{jik}}{p_{ijk}}\right)+\phi\left(\frac{\widehat{p}_{jki}}{p_{ijk}}\right)+\phi\left(\frac{\widehat{p}_{kij}}{p_{ijk}}\right)+\phi\left(\frac{\widehat{p}_{kji}}{p_{ijk}}\right)\right)- \\ -\left(\frac{\widehat{p}_{ijk}}{p_{ijk}}\phi'\left(\frac{\widehat{p}_{ijk}}{p_{ijk}}\right)+\frac{\widehat{p}_{ikj}}{p_{ijk}}\phi'\left(\frac{\widehat{p}_{ikj}}{p_{ijk}}\right)+\frac{\widehat{p}_{jik}}{p_{ijk}}\phi'\left(\frac{\widehat{p}_{jik}}{p_{ijk}}\right)\right) \\ -\left(\frac{\widehat{p}_{jki}}{p_{ijk}}\phi'\left(\frac{\widehat{p}_{jki}}{p_{ijk}}\right)+\frac{\widehat{p}_{kij}}{p_{ijk}}\phi'\left(\frac{\widehat{p}_{kij}}{p_{ijk}}\right)+\frac{\widehat{p}_{kji}}{p_{ijk}}\phi'\left(\frac{\widehat{p}_{kji}}{p_{ijk}}\right)\right)-6\mu \quad =0 \\ i,j,k=1,...,I \end{array}\right. \tag{8.47}$$

where μ has the expression

$$\sum_{i,j,k=1}^{I}\left(p_{ijk}\left(\phi\left(\frac{\widehat{p}_{ijk}}{p_{ijk}}\right)+\phi\left(\frac{\widehat{p}_{ikj}}{p_{ijk}}\right)+\phi\left(\frac{\widehat{p}_{jik}}{p_{ijk}}\right)+\phi\left(\frac{\widehat{p}_{jki}}{p_{ijk}}\right)+\phi\left(\frac{\widehat{p}_{kij}}{p_{ijk}}\right)+\phi\left(\frac{\widehat{p}_{kji}}{p_{ijk}}\right)\right)\right.$$
$$-\frac{1}{6}\left(\widehat{p}_{ijk}\phi'\left(\frac{\widehat{p}_{ijk}}{p_{ijk}}\right)+\widehat{p}_{ikj}\phi'\left(\frac{\widehat{p}_{ikj}}{p_{ijk}}\right)+\widehat{p}_{jik}\phi'\left(\frac{\widehat{p}_{jik}}{p_{ijk}}\right)\right)$$
$$\left.-\frac{1}{6}\left(\widehat{p}_{jki}\phi'\left(\frac{\widehat{p}_{jki}}{p_{ijk}}\right)+\widehat{p}_{kij}\phi'\left(\frac{\widehat{p}_{kij}}{p_{ijk}}\right)+\widehat{p}_{kji}\phi'\left(\frac{\widehat{p}_{kji}}{p_{ijk}}\right)\right)\right)$$

and its asymptotic distribution is given by

$$\sqrt{n}(\widehat{\boldsymbol{\theta}}^{S,\phi} - \boldsymbol{\theta}_0) \underset{n\to\infty}{\overset{L}{\longrightarrow}} N(\boldsymbol{0}, \boldsymbol{W}_S(\boldsymbol{\theta}_0)) \tag{8.48}$$

where the matrix $\boldsymbol{W}_S^*(\boldsymbol{\theta}_0)$ is

$$\boldsymbol{\Sigma}_{\boldsymbol{\theta}_0}\left(\boldsymbol{I}_{(I^3-1)\times(I^3-1)} - \boldsymbol{B}_S^*(\boldsymbol{\theta}_0)^T\left(\boldsymbol{B}_S^*(\boldsymbol{\theta}_0)\boldsymbol{\Sigma}_{\boldsymbol{\theta}_0}\boldsymbol{B}_S^*(\boldsymbol{\theta}_0)^T\right)^{-1}\boldsymbol{B}_S(\boldsymbol{\theta}_0)\right)\boldsymbol{\Sigma}_{\boldsymbol{\theta}_0},$$

and $\boldsymbol{B}_S^*(\boldsymbol{\theta}_0)$ *is the matrix given in (8.46).*

Proof. Instead of getting

$$\widehat{\boldsymbol{\theta}}^{S,\phi} = \left(p_{ijk}^{S,\phi};\ i,j,k=1,...,I,\ (i,j,k)\neq(I,I,I)\right)^T$$

we shall obtain $\boldsymbol{p}(\widehat{\boldsymbol{\theta}}^{S,\phi}) = \left(p_{ijk}^{S,\phi};\ i,j,k=1,...,I\right)^T$. The p_{ijk}, $i,j,k=1,...,I$, which minimize the ϕ-divergence $D_\phi(\widehat{\boldsymbol{p}},\boldsymbol{p}(\boldsymbol{\theta}))$ subject to the null hypothesis of symmetry may be obtained minimizing

$$\sum_{i,j,k=1}^{I} p_{ijk}\phi\left(\frac{\widehat{p}_{ijk}}{p_{ijk}}\right) + \sum_{i,j,k=1}^{I} \lambda_{ijk}\left(p_{ijk}-p_{i'j'k'}\right) + \mu\left(1-\sum_{i,j,k=1}^{I} p_{ijk}\right) \tag{8.49}$$

with respect to p_{ijk}, where μ and λ_{ijk} are undetermined Lagrangian multipliers. Minimizing the expression (8.49) is equivalent to minimizing the expression

$$\sum_{i,j,k=1}^{I} p_{ijk}\left(\phi\left(\frac{\widehat{p}_{ijk}}{p_{ijk}}\right)+\phi\left(\frac{\widehat{p}_{ikj}}{p_{ijk}}\right)+\phi\left(\frac{\widehat{p}_{jik}}{p_{ijk}}\right)+\phi\left(\frac{\widehat{p}_{jki}}{p_{ijk}}\right)+\phi\left(\frac{\widehat{p}_{kij}}{p_{ijk}}\right)\right.$$

$$\left.+\phi\left(\frac{\widehat{p}_{kji}}{p_{ijk}}\right)+\phi\left(\frac{\widehat{p}_{kji}}{p_{ijk}}\right)\right)+\mu\left(1-\sum_{i,j,k=1}^{I} p_{ijk}\right).$$

Differentiating with respect to p_{ijk}, $i,j,k=1,...,I$, we get the system of equations given in (8.47) whose solutions provide the minimum ϕ-divergence estimator $\widehat{\boldsymbol{\theta}}^{S,\phi}$.

The asymptotic distribution follows from (8.3) because in our model $\boldsymbol{I}_F(\boldsymbol{\theta}_0)^{-1} = \boldsymbol{\Sigma}_{\boldsymbol{\theta}_0}$. ■

Corollary 8.3

The minimum power-divergence estimator

$$\boldsymbol{p}(\widehat{\boldsymbol{\theta}}^{S,\phi_{(\lambda)}}) = \left(p_{ijk}^{S,\phi};\ i,j,k=1,...,I\right),$$

under the hypothesis of symmetry, is given by

$$p_{ijk}^{S,\phi_{(\lambda)}} = \frac{\left(\widehat{p}_{ijk}^{\lambda+1}+\widehat{p}_{ikj}^{\lambda+1}+\widehat{p}_{jik}^{\lambda+1}+\widehat{p}_{jki}^{\lambda+1}+\widehat{p}_{kij}^{\lambda+1}+\widehat{p}_{kji}^{\lambda+1}\right)^{\frac{1}{\lambda+1}}}{\sum_{i,j,k=1}^{I}\left(\widehat{p}_{ijk}^{\lambda+1}+\widehat{p}_{ikj}^{\lambda+1}+\widehat{p}_{jik}^{\lambda+1}+\widehat{p}_{jki}^{\lambda+1}+\widehat{p}_{kij}^{\lambda+1}+\widehat{p}_{kji}^{\lambda+1}\right)^{\frac{1}{\lambda+1}}},\quad i,j,k=1,...,I. \tag{8.50}$$

For $\lambda=0$,

$$p_{ijk}^{S,\phi_{(0)}} = \frac{\widehat{p}_{ijk}+\widehat{p}_{ikj}+\widehat{p}_{jik}+\widehat{p}_{jki}+\widehat{p}_{kij}+\widehat{p}_{kji}}{6},\quad i,j,k=1,...,I,$$

ence, we obtain the maximum likelihood estimator. This estimator was introduced and studied by Bowker (1948).

For $\lambda \to -1$, we have

$$p_{ijk}^{S,\phi_{(-1)}} = \frac{\left(\widehat{p}_{ijk} \times \widehat{p}_{ikj} \times \widehat{p}_{jik} \times \widehat{p}_{jki} \times \widehat{p}_{kij} \times \widehat{p}_{kji}\right)^{1/6}}{\sum\limits_{i,j,k=1}^{I} \left(\widehat{p}_{ijk} \times \widehat{p}_{ikj} \times \widehat{p}_{jik} \times \widehat{p}_{jki} \times \widehat{p}_{kij} \times \widehat{p}_{kji}\right)^{1/6}}, \quad i,j,k = 1,...,I,$$

he minimum modified likelihood estimator and finally for $\lambda = 1$,

$$p_{ijk}^{S,\phi_{(1)}} = \frac{\left(\widehat{p}^2_{ijk} + \widehat{p}^2_{ikj} + \widehat{p}^2_{jik} + \widehat{p}^2_{jki} + \widehat{p}^2_{kij} + \widehat{p}^2_{kji}\right)^{1/2}}{\sum\limits_{i,j,k=1}^{I} \left(\widehat{p}^2_{ijk} + \widehat{p}^2_{ikj} + \widehat{p}^2_{jik} + \widehat{p}^2_{jki} + \widehat{p}^2_{kij} + \widehat{p}^2_{kji}\right)^{1/2}},$$

ve get the minimum chi-square estimator for symmetry.

Based on the test statistic given in (8.5) we should reject the hypothesis of symmetry f

$$S_n^{\phi_1}\left(\widehat{\boldsymbol{\theta}}^{S,\phi_2}\right) \equiv \frac{2n}{\phi_1''(1)} D_{\phi_1}\left(\widehat{\mathbf{p}}, \mathbf{p}(\widehat{\boldsymbol{\theta}}^{S,\phi_2})\right) > c,$$

vhere c is a positive constant.

In the following theorem we establish the asymptotic distribution of the ϕ-divergence est statistic $S_n^{\phi_1}\left(\widehat{\boldsymbol{\theta}}^{S,\phi_2}\right)$.

Theorem 8.13

The asymptotic distribution of the ϕ-divergence test statistics

$$S_n^{\phi_1}\left(\widehat{\boldsymbol{\theta}}^{S,\phi_2}\right) = \frac{2n}{\phi_1''(1)} D_{\phi_1}\left(\widehat{\mathbf{p}}, \mathbf{p}(\widehat{\boldsymbol{\theta}}^{S,\phi_2})\right) \quad (8.51)$$

ınd

$$S_n^{\phi_1,h}\left(\widehat{\boldsymbol{\theta}}^{S,\phi_2}\right) \equiv \frac{2n}{h'(0)\,\phi_1''(1)} h\left(D_{\phi_1}\left(\widehat{\mathbf{p}}, \mathbf{p}(\widehat{\boldsymbol{\theta}}^{S,\phi_2})\right)\right), \quad (8.52)$$

for testing the hypothesis of symmetry, given in (8.44), is chi-square with $I(I-1)(5I+2)/6$ degrees of freedom.

Proof. In our case we have I^3 cells and by (8.28) I^3-1 parameters that are necessary to estimate. Using the relation (8.45) the number of constraints is $I(I-1)(5I+2)/6$. Then by (8.5) the asymptotic distribution of the family of test statistics (8.51) is a chi-square distribution with

$$\underbrace{I^3}_{\text{Cells}} - \underbrace{I^3-1}_{\substack{\text{Estimated}\\ \text{Parameters}}} + \underbrace{I(I-1)(5I+2)/6}_{\text{Constraints}} - 1 = \underbrace{I(I-1)(5I+2)/6}_{\substack{\text{Degrees of}\\ \text{freedom}}}.$$

In relation with the family of test statistics given in (8.52), we have

$$h(x) = h(0) + h'(0)x + o(x),$$

then

$$h\left(D_{\phi_1}\left(\widehat{\boldsymbol{p}}, \boldsymbol{p}(\widehat{\boldsymbol{\theta}}^{S,\phi_2})\right)\right) = h'(0)\, D_{\phi_1}\left(\widehat{\boldsymbol{p}}, \boldsymbol{p}(\widehat{\boldsymbol{\theta}}^{S,\phi_2})\right) + o_P(1)$$

and we get that the asymptotic distribution of the family of test statistics given in (8.52) is also chi-square with $I(I-1)(5I+2)/6$ degrees of freedom.

A study of this problem with maximum likelihood estimator and ϕ-divergence test statistics can be seen in Menéndez et *al.* (2004b).

8.4. Marginal Homogeneity

In case the pattern is not completely symmetric, one likes to check whether, at least, the two sets of marginal totals have the same distribution: marginal symmetry (or marginal homogeneity) or the local odd ratios are symmetric: quasi-symmetry. The problem of marginal homogeneity was first discussed by Stuart (1955), who defined a test statistic which is a quadratic form in the differences of the corresponding marginal values, whose matrix is the inverse of a consistent estimate of the covariance matrix of the differences under the null hypothesis and its asymptotic distribution is chi-square with $I-1$ degrees of freedom under the null hypothesis of marginal homogeneity. This hypothesis has been discussed by several authors (e.g., Bhapkar 1966, 1979, Ireland et *al.* 1969, Bishop et *al.* 1975, Agresti 1983, Bhapkar and Darroch 1990, Kullback 1971).

We consider a two-way contingency table with $I = J$ and let $\widehat{\boldsymbol{p}} = (\widehat{p}_{11}, ..., \widehat{p}_{II})^T$ be the nonparametric estimator of the unknown probability vector $\boldsymbol{p} = (p_{11}, ..., p_{II})^T$, where $p_{ij} = \Pr(X = i, Y = j)$, with $p_{ij} > 0$, and $i, j = 1, ..., I$.

The hypothesis of marginal homogeneity is given by

$$H_0 : \sum_{i=1}^{I} p_{ji} = \sum_{i=1}^{I} p_{ij}, \quad j = 1, ..., I-1, \tag{8.53}$$

and the parameter space is

$$\Theta = \left\{\boldsymbol{\theta} : \boldsymbol{\theta} = (p_{ij};\ i, j = 1, ..., I,\ (i,j) \neq (I,I))^T\right\}. \tag{8.54}$$

We denote by

$$\boldsymbol{p}(\boldsymbol{\theta}) = (p_{11}, ..., p_{II})^T = \boldsymbol{p} \tag{8.55}$$

the probability vector characterizing the marginal homogeneity model with $p_{II} = 1 - \sum_{i=1}^{I}\sum_{j=1}^{I} p_{ij}$ with $(i,j) \neq (I,I)$.

The hypothesis of marginal homogeneity formulated in (8.53) can be formulated using he following $I-1$ constraints about the parameter $\boldsymbol{\theta}$,

$$h_j(\boldsymbol{\theta}) = \sum_{i=1}^{I} p_{ji} - \sum_{i=1}^{I} p_{ij} = 0, \quad j = 1, ..., I-1, \tag{8.56}$$

or considering the parameter

$$\boldsymbol{\beta} = (p_{11}, p_{21}, .., p_{2I}, ..., p_{I1}, .., p_{II-1})^T$$

and the set

$$B = \left\{ (a_{11}, a_{21}, .., a_{2I}, ..., a_{I1}, .., a_{II-1}) : \ a_{ij} > 0, \ \sum_{(i,j)\in L} a_{ij} < 1 \right\}$$

where

$$L = \{(i,j) : i,j = 1, ..., I, (i,j) \neq (I,I), (i,j) \neq (1,j), \ j = 2, ..., I\}.$$

The hypothesis (8.53) can be expressed for some unknown $\boldsymbol{\theta}_0 \in \Theta$ with $\boldsymbol{p}(\boldsymbol{\theta}_0) = \boldsymbol{p}_0$, by

$$H_0 : \boldsymbol{p}_0 = \boldsymbol{p}(\boldsymbol{g}(\boldsymbol{\beta}_0)), \ \boldsymbol{\beta}_0 \in B \text{ and } \boldsymbol{g}(\boldsymbol{\beta}_0) = \boldsymbol{\theta}_0, \tag{8.57}$$

where the function $\boldsymbol{g}$ is defined by

$$\begin{aligned} g_{ij}(\boldsymbol{\beta}) &= -\sum_{i\neq 1,j}^{I} p_{ij} + \sum_{l\neq j}^{I} p_{jl} && i = 1, j = 2, ..., I \\ g_{ij}(\boldsymbol{\beta}) &= p_{ij} && i \neq 1, j = 1, ..., I, (i,j) \neq (I,I) \\ g_{11}(\boldsymbol{\beta}) &= p_{11}. \end{aligned}$$

We recall $\boldsymbol{p}(\boldsymbol{g}(\boldsymbol{\beta})) = (g_{ij}(\boldsymbol{\beta}), \ i,j = 1, ..., I)^T$, where

$$g_{II}(\boldsymbol{\beta}) = 1 - \sum_{\substack{i,j=1 \\ (i,j)\neq(I,I)}}^{I} g_{ij}(\boldsymbol{\beta}).$$

This approach can be formulated and solved in terms of the results presented in Chapter 5.

In this Chapter we consider the approach given in (8.56). We can observe that with this approach the problem of marginal homogeneity can be written as

$$H_0 : \boldsymbol{p} = \boldsymbol{p}(\boldsymbol{\theta}_0), \text{ for some unknown } \boldsymbol{\theta}_0 \in \Theta_0,$$

with

$$\Theta_0 = \{\boldsymbol{\theta} \in \Theta : h_j(\boldsymbol{\theta}) = 0, \ j = 1, ..., I-1\},$$

$\boldsymbol{p}(\boldsymbol{\theta})$ defined in (8.55) and h_j, $j = 1, ..., I-1$, in (8.56).

It is clear that the matrix

$$\boldsymbol{B}_{MH}(\boldsymbol{\theta}_0) = \left(\frac{\partial h_j(\boldsymbol{\theta}_0)}{\partial \theta_{ij}} \right)_{(I-1)\times(I^2-1)} \tag{8.58}$$

has rank $I-1$ because the matrix $\boldsymbol{I}_{(I-1)\times(I-1)}$ is a submatrix of it.

For $I = 2$, we only have a constraint and it is given by

$$h_1(\boldsymbol{\theta}) = p_{12} - p_{21} = 0,$$

so that the problem of marginal homogeneity coincides with the problem of symmetry.

For $I = 3$, we have the two following constraints,

$$\begin{aligned} h_1(\boldsymbol{\theta}) = &\ p_{12} + p_{13} - p_{21} - p_{31} = 0 \\ h_2(\boldsymbol{\theta}) = &\ p_{21} + p_{23} - p_{12} - p_{32} = 0. \end{aligned}$$

In the following we present the expression of the minimum ϕ-divergence estimator of $\boldsymbol{\theta}_0$ under the constraints given in (8.56).

Theorem 8.14

The minimum ϕ-divergence estimator,

$$\widehat{\boldsymbol{\theta}}^{MH,\phi} = \left(p_{ij}^{MH,\phi};\ i,j = 1, ..., I,\ \text{and}\ (i,j) \neq (I,I) \right)^T,$$

of $\boldsymbol{\theta}_0$ in Θ_0 (i.e., under the hypothesis of marginal homogeneity) is obtained as a solution of the system of equations

$$\left\{ \begin{array}{rl} \phi\left(\frac{\widehat{p}_{ij}}{p_{ij}}\right) - \frac{\widehat{p}_{ij}}{p_{ij}}\phi'\left(\frac{\widehat{p}_{ij}}{p_{ij}}\right) - \mu + \lambda_i - \lambda_j & = 0, \quad i,j = 1, ..., I,\ i \neq j \\ \sum_{i=1}^{I} p_{ji} - \sum_{i=1}^{I} p_{ij} & = 0, \quad j = 1, ..., I \end{array} \right. \tag{8.59}$$

where μ has the expression

$$\mu = \sum_{i=1}^{I}\sum_{j=1}^{J} \left(p_{ij} \left(\phi\left(\frac{\widehat{p}_{ij}}{p_{ij}}\right) + \lambda_i - \lambda_j \right) - \widehat{p}_{ij}\phi'\left(\frac{\widehat{p}_{ij}}{p_{ij}}\right) \right).$$

Its asymptotic distribution is

$$\sqrt{n}(\widehat{\boldsymbol{\theta}}^{MH,\phi} - \boldsymbol{\theta}_0) \underset{n\to\infty}{\overset{L}{\to}} N(\boldsymbol{0}, \boldsymbol{W}_{MH}(\boldsymbol{\theta}_0)), \tag{8.60}$$

where the $(I^2-1)\times(I^2-1)$ matrix $\boldsymbol{W}_{MH}(\boldsymbol{\theta}_0)$ is given by

$$\boldsymbol{\Sigma}_{\boldsymbol{\theta}_0} - \boldsymbol{\Sigma}_{\boldsymbol{\theta}_0}\boldsymbol{B}_{MH}(\boldsymbol{\theta}_0)^T \left(\boldsymbol{B}_{MH}(\boldsymbol{\theta}_0)\boldsymbol{\Sigma}_{\boldsymbol{\theta}_0}\boldsymbol{B}_{MH}(\boldsymbol{\theta}_0)^T \right)^{-1} \boldsymbol{B}_{MH}(\boldsymbol{\theta}_0)\boldsymbol{\Sigma}_{\boldsymbol{\theta}_0}.$$

Proof. Instead of getting $\widehat{\boldsymbol{\theta}}^{MH,\phi} = \left(p_{ij}^{MH,\phi};\ i,j=1,...,I, \text{ and } (i,j) \neq (I,I)\right)^T$ we shall obtain $\boldsymbol{p}\left(\widehat{\boldsymbol{\theta}}^{MH,\phi}\right) = \left(p_{ij}^{MH,\phi};\ i,j=1,...,I\right)^T$. The p'_{ij}s, $i,j=1,...,I$, which minimize the ϕ-divergence

$$D_\phi(\widehat{\boldsymbol{p}}, \boldsymbol{p}(\boldsymbol{\theta}))$$

subject to the null hypothesis of marginal homogeneity (or the constraints about $\boldsymbol{\theta}$ given in (8.56)), may be obtained minimizing

$$\sum_{i=1}^{I}\sum_{j=1}^{I} p_{ij}\phi\left(\frac{\widehat{p}_{ij}}{p_{ij}}\right) + \sum_{i=1}^{I}\lambda_i\left(\sum_{j=1}^{I}p_{ij} - \sum_{j=1}^{I}p_{ji}\right) + \mu\left(1 - \sum_{i=1}^{I}\sum_{j=1}^{I}p_{ij}\right) \tag{8.61}$$

with respect to the p_{ij}, where μ and λ_i are undetermined Lagrangian multipliers. Minimizing the expression (8.61), is equivalent to solving the system of equations given in (8.59). The asymptotic distribution is obtained from (8.3) because in our case $\boldsymbol{I}_F(\boldsymbol{\theta}_0)^{-1} = \boldsymbol{\Sigma}_{\boldsymbol{\theta}_0}$. ■

Corollary 8.4

The minimum power-divergence estimator, $\mathbf{p}(\widehat{\boldsymbol{\theta}}^{MH,\phi}) = \left(p_{ij}^{MH,\phi};\ i,j=1,...,I\right)^T$ *of* $\boldsymbol{\theta}_0$ *under the hypothesis of marginal homogeneity (8.56), is given by*

$$p_{ij}^{MH,\phi_{(\lambda)}} = \frac{\widehat{p}_{ij}}{\left((\lambda+1)\left(-\mu+\lambda_i-\lambda_j\right)\right)^{\frac{1}{\lambda+1}}} \qquad i,j=1,...,I$$

where μ *and* λ_i, $i=1,...,I$, *must satisfy the system of equations*

$$\begin{cases} \displaystyle\sum_{i=1}^{I}\sum_{j=1}^{I}\frac{\widehat{p}_{ij}}{((\lambda+1)(-\mu+\lambda_i-\lambda_j))^{\frac{1}{\lambda+1}}} = 1 \\ \displaystyle\sum_{j=1}^{I}\frac{\widehat{p}_{ij}}{((\lambda+1)(-\mu+\lambda_i-\lambda_j))^{\frac{1}{\lambda+1}}} = \sum_{l=1}^{I}\frac{\widehat{p}_{li}}{((\lambda+1)(-\mu+\lambda_l-\lambda_i))^{\frac{1}{\lambda+1}}}, \quad i=1,...,I \end{cases}.$$

The maximum likelihood estimator, obtained for $\lambda = 0$, *is*

$$p_{ij}^{MH,\phi_{(0)}} = \frac{\widehat{p}_{ij}}{1+\lambda_i-\lambda_j}, \qquad i,j=1,...,I$$

where the λ_i *holds*

$$\sum_{j=1}^{I}\frac{\widehat{p}_{ij}}{1+\lambda_i-\lambda_j} = \sum_{j=1}^{I}\frac{\widehat{p}_{ji}}{1+\lambda_j-\lambda_i}, \quad i=1,...,I.$$

The minimum modified likelihood estimator, obtained for $\lambda \to -1$, *is*

$$p_{ij}^{MH,\phi_{(-1)}} = \frac{\widehat{p}_{ij}\frac{a_j}{a_i}}{\sum_{i=1}^{I}\sum_{j=1}^{I}\widehat{p}_{ij}\frac{a_j}{a_i}}, \qquad i,j=1,...,I,$$

where the a_i holds

$$a_i \sum_{j=1}^{I} \frac{\widehat{p}_{ji}}{a_j} = \frac{1}{a_i} \sum_{l=1}^{I} \widehat{p}_{il} a_l, \; i = 1, ..., I.$$

Remark 8.7

For $I = 3$, given $\boldsymbol{\theta}_0 = (p_{11}, p_{21} + p_{23} - p_{32}, p_{31} + p_{32} - p_{23}, p_{21}, p_{22}, p_{23}, p_{31}, p_{32})^T \in \Theta_0$, it is easy to check that

- $\boldsymbol{B}_{MH}(\boldsymbol{\theta}_0) = \begin{pmatrix} 0 & 1 & 1 & -1 & 0 & 0 & -1 & 0 \\ 0 & -1 & 0 & 1 & 0 & 1 & 0 & -1 \end{pmatrix},$
- $\boldsymbol{\Sigma}_{\boldsymbol{\theta}_0} = diag(\boldsymbol{\theta}_0) - \boldsymbol{\theta}_0 \boldsymbol{\theta}_0^T,$
- *The matrix* $\left(\boldsymbol{B}_{MH}(\boldsymbol{\theta}_0) \boldsymbol{\Sigma}_{\boldsymbol{\theta}_0} \boldsymbol{B}_{MH}^T(\boldsymbol{\theta}_0)\right)^{-1}$ *is*

$$\frac{1}{4p_{21}p_{31} + 4p_{23}p_{31} + 4p_{21}p_{32} - (p_{23} - p_{32})^2} \begin{pmatrix} 2(p_{23} + p_{21}) & 2p_{21} + p_{23} - p_{32} \\ 2p_{21} + p_{23} - p_{32} & 2(p_{31} + p_{21}) \end{pmatrix}$$

and it is not difficult to establish that the matrix $\boldsymbol{W}_{MH}(\boldsymbol{\theta}_0)$ is

$$\boldsymbol{W}_{MH}(\boldsymbol{\theta}_0) = \left(\frac{\partial \boldsymbol{g}(\boldsymbol{\beta}_0)}{\partial \boldsymbol{\beta}}\right)^T \boldsymbol{I}_F(\boldsymbol{\beta}_0)^{-1} \left(\frac{\partial \boldsymbol{g}(\boldsymbol{\beta}_0)}{\partial \boldsymbol{\beta}}\right)$$

where

$$\left(\frac{\partial \boldsymbol{g}(\boldsymbol{\beta}_0)}{\partial \boldsymbol{\beta}}\right)^T = \begin{pmatrix} 1 & 0 & 0 & 0 & 0 & 0 \\ 0 & 1 & 0 & 1 & 0 & -1 \\ 0 & 0 & 0 & -1 & 1 & 1 \\ 0 & 1 & 0 & 0 & 0 & 0 \\ 0 & 0 & 1 & 0 & 0 & 0 \\ 0 & 0 & 0 & 1 & 0 & 0 \\ 0 & 0 & 0 & 0 & 1 & 0 \\ 0 & 0 & 0 & 0 & 0 & 1 \end{pmatrix}$$

and $\boldsymbol{I}_F(\boldsymbol{\beta}_0)^{-1}$ is the inverse of the Fisher information matrix for the parameter $\boldsymbol{\beta}_0 = (p_{11}, p_{21}, p_{22}, p_{23}, p_{31}, p_{32})^T$.

This result can be proved in general in the same way as the results in Sections 2 and 3 corresponding to hypotheses of independence and symmetry, respectively. From this result, $\widehat{\boldsymbol{\theta}}^{MH,\phi_2}$ is also BAN.

Theorem 8.15

Under the marginal homogeneity model, we have

$$\boldsymbol{W}_{MH}(\boldsymbol{\theta}_0) = \boldsymbol{M}_{\boldsymbol{\beta}_0}^T \boldsymbol{I}_F(\boldsymbol{\beta}_0)^{-1} \boldsymbol{M}_{\boldsymbol{\beta}_0}.$$

Proof. The proof follows the same steps as the proof of Theorem 8.2 ■

In the following theorem we present two families of test statistics for testing marginal homogeneity based on the minimum ϕ-divergence estimator.

Theorem 8.16
The asymptotic distribution of the ϕ-divergence test statistics

$$MH_n^{\phi_1}\left(\widehat{\boldsymbol{\theta}}^{MH,\phi_2}\right) \equiv \frac{2n}{\phi_1''(1)} D_{\phi_1}\left(\widehat{\mathbf{p}}, \mathbf{p}(\widehat{\boldsymbol{\theta}}^{MH,\phi_2})\right) \tag{8.62}$$

and

$$MH_n^{\phi_1,h}\left(\widehat{\boldsymbol{\theta}}^{MH,\phi_2}\right) \equiv \frac{2n}{h'(0)\,\phi_1''(1)} h\left(D_{\phi_1}\left(\widehat{\mathbf{p}}, \mathbf{p}(\widehat{\boldsymbol{\theta}}^{MH,\phi_2})\right)\right) \tag{8.63}$$

for testing the hypothesis of marginal homogeneity given in (8.53) is chi-square with $I-1$ degrees of freedom.

Proof. In our case we have I^2 cells and by (8.54) I^2-1 parameters that are necessary to estimate. Using the relation (8.56) the number of constraints is $I-1$. Then by (8.5), the asymptotic distribution of the family of ϕ-divergence test statistics (8.62) is chi-square with

$$\underbrace{I^2}_{\text{Cells}} - \underbrace{I^2-1}_{\text{Estimated Parameters}} + \underbrace{(I-1)}_{\text{Constraints}} - 1 = \underbrace{(I-1)}_{\text{Degrees of freedom}}.$$

In relation with the family of test statistics given in (8.63), we have

$$h(x) = h(0) + h'(0)\,x + o(x),$$

therefore

$$h\left(D_{\phi_1}\left(\widehat{\mathbf{p}}, \mathbf{p}(\widehat{\boldsymbol{\theta}}^{MH,\phi})\right)\right) = h'(0)\, D_{\phi_1}\left(\widehat{\mathbf{p}}, \mathbf{p}(\widehat{\boldsymbol{\theta}}^{MH,\phi})\right) + o_P(1)$$

and we get that the asymptotic distribution of the family of test statistics given in (8.63) is also chi-square with $I-1$ degrees of freedom. ■

If we use the test statistics $MH_n^{\phi_1}\left(\widehat{\boldsymbol{\theta}}^{MH,\phi_2}\right)$ $\left(MH_n^{\phi_1,h}\left(\widehat{\boldsymbol{\theta}}^{MH,\phi_2}\right)\right)$ for testing the marginal homogeneity we should reject the null hypothesis, i.e., the hypothesis of marginal homogeneity if $MH_n^{\phi_1}\left(\widehat{\boldsymbol{\theta}}^{MH,\phi_2}\right)$ $\left(MH_n^{\phi_1,h}\left(\widehat{\boldsymbol{\theta}}^{MH,\phi_2}\right)\right)$ is too large. Based on the previous theorem we should reject the hypothesis of marginal homogeneity, with significance level α, if

$$MH_n^{\phi_1}\left(\widehat{\boldsymbol{\theta}}^{MH,\phi_2}\right) > \chi^2_{I-1,\alpha} \text{ or } MH_n^{\phi_1,h}\left(\widehat{\boldsymbol{\theta}}^{MH,\phi_2}\right) > \chi^2_{I-1,\alpha}.$$

8.5. Quasi-symmetry

The symmetry concept is stronger than that of marginal symmetry in the sense that the latter is implied by the former. The hypothesis of quasi-symmetry was introduced by Caussinus (1965) who introduced the maximum likelihood estimator for quasi-symmetry as well as a chi-square type statistic for testing this hypothesis. For additional discussion of quasi-symmetry, see Darroch (1981, 1986), McCullagh (1982), Darroch and McCullagh (1986) and Agresti (2002). Recently, Matthews and Growther (1997) have studied quasi-symmetry and independence for cross-classified data in a two-way contingency table. These models are expressed in terms of the cross-product ratios and a maximum likelihood estimation procedure is proposed for estimating the expected frequencies subject to the constraints imposed on the frequencies through the cross-product ratios.

We consider a two-way contingency table with $I = J$ and let $\widehat{\boldsymbol{p}} = (\widehat{p}_{11}, ..., \widehat{p}_{II})^T$ be the nonparametric estimator of the unknown probability vector $\boldsymbol{p} = (p_{11}, ..., p_{II})^T$, where $p_{ij} = \Pr(X = i, Y = j)$, with $p_{ij} > 0$ and $i, j = 1, ..., I$.

The hypothesis of quasi-symmetry is given by

$$H_0 : p_{ij}p_{jk}p_{ki} = p_{ik}p_{kj}p_{ji}, \tag{8.64}$$

for all $i, j, k = 1, ..., I$.

We consider the parameter space

$$\Theta = \left\{ \boldsymbol{\theta} : \boldsymbol{\theta} = (p_{ij};\ i, j = 1, ..., I,\ (i, j) \neq (I, I))^T \right\}, \tag{8.65}$$

and we denote by

$$\boldsymbol{p}(\boldsymbol{\theta}) = (p_{11}, ..., p_{II})^T = \boldsymbol{p} \tag{8.66}$$

the probability vector characterizing our model with $p_{II} = 1 - \sum_{i=1}^{I} \sum_{j=1}^{I} p_{ij}$ and $(i, j) \neq (I, I)$.

There are other two equivalent ways to characterize the quasi-symmetry showed by Caussinus (1965). The hypothesis (8.64) is equivalent to

$$H_0 : s_{ij} = s_{ji},\ i \neq j$$

with

$$s_{ij} = \frac{p_{ij}p_{II}}{p_{iI}p_{Ij}}, i, j = 1, ..., I - 1, \tag{8.67}$$

as well as to

$$H_0 : c_{ij} = c_{ji},\ i \neq j,\ c_{ij} > 0 \tag{8.68}$$

with

$$p_{ij} = a_i c_{ij},\ a_i > 0.$$

The hypothesis of quasi-symmetry in the way given by (8.67) could be formulated using the following $(I-1)(I-2)/2$ constraints about $\boldsymbol{\theta}$,

$$h_{ij}(\boldsymbol{\theta}) = p_{ij}p_{jI}p_{Ii} - p_{iI}p_{Ij}p_{ji} = 0, \tag{8.69}$$

for all $i,j = 1,...,I-1$, $i<j$. Also considering the parameter

$$\boldsymbol{\beta} = (p_{11}, p_{1I}, p_{21}, p_{22}, p_{2I}, ..., p_{I1}, p_{I2}, ..., p_{II-1})^T$$

and the set

$$\begin{aligned} B = \ & \{(a_{11}, a_{1I}, a_{21}, a_{22}, a_{2I}, ..., a_{I1}, a_{I2}, ..., a_{II-1})^T \in \mathbb{R}^{\frac{(I-1)(I+4)}{2}} : \\ & \sum_{i,j} a_{ij} < 1,\ a_{ij} > 0,\ i,j = 1,...,I\}, \end{aligned}$$

the hypothesis (8.69) can be expressed as

$$H_0 : \boldsymbol{p}{=}\boldsymbol{p}\left(\boldsymbol{g}(\boldsymbol{\beta}_0)\right),\ \boldsymbol{\beta}_0{\in}B \text{ and } \boldsymbol{g}(\boldsymbol{\beta}_0) = \boldsymbol{\theta}_0,$$

where the function $\boldsymbol{g}$ is defined by

$$\begin{array}{ll} g_{ij}(\boldsymbol{\beta}) = \dfrac{p_{iI}p_{Ij}p_{ji}}{p_{jI}p_{I1}} & i,j = 1,...,I-1,\ i<j \\ g_{ij}(\boldsymbol{\beta}) = p_{ij} & i = 1,...,I, j = 1,...,I-1,\ i \geq j \\ g_{iI}(\boldsymbol{\beta}) = p_{iI} & i = 1,...,I-1. \end{array}$$

We observe that $\boldsymbol{p}\left(\boldsymbol{g}(\boldsymbol{\beta})\right) = \left(g_{ij}(\boldsymbol{\beta}),\ i,j = 1,...,I\right)^T$, where

$$g_{II}(\boldsymbol{\beta}) = 1 - \sum_{\substack{i,j=1 \\ (i,j)\neq(I,I)}}^{I} g_{ij}(\boldsymbol{\beta}).$$

This approach can be formulated and solved in terms of the results presented in Chapter 6.

In this Section we consider the approach given in (8.67). We can observe that with this approach the problem of quasi-symmetry can be written as

$$H_0 : \boldsymbol{p} = \boldsymbol{p}\left(\boldsymbol{\theta}_0\right), \qquad \text{for some unknown } \boldsymbol{\theta}_0 \in \Theta_0,$$

with

$$\Theta_0 = \left\{\theta \in \Theta :\ h_{ij}(\boldsymbol{\theta}) = 0,\ i,j = 1,...,I-1,\ i<j\right\},$$

$\boldsymbol{p}(\boldsymbol{\theta})$ defined in (8.66) and $h_{ij}(\boldsymbol{\theta})$ in (8.69).

In this case the matrix of partial derivatives

$$\boldsymbol{B}_{QS}(\boldsymbol{\theta}_0) = \left(\frac{\partial h_{ij}(\boldsymbol{\theta}_0)}{\partial \theta_{ij}}\right)_{(I-1)(I-2)/2\times(I^2-1)} \tag{8.70}$$

has rank $(I-1)(I-2)/2$. For more details about quasi-symmetry hypothesis see Bhapka (1979).

For $I=3$, we have a constraint given by

$$h_{12}(\boldsymbol{\theta}) = p_{12}p_{23}p_{31} - p_{13}p_{32}p_{21} = 0.$$

Next theorem presents the expression of the minimum ϕ-divergence estimator of $\boldsymbol{\theta}_0$ under the constraints given in (8.69). We shall denote $p_{i*} = \sum_{j=1}^{I} p_{ij}$.

Theorem 8.17

The minimum ϕ-divergence estimator,

$$\widehat{\boldsymbol{\theta}}^{QS,\phi} = \left(p_{ij}^{QS}, i,j=1,...,I,\ (i,j) \neq (I,I)\right)^T,$$

of $\boldsymbol{\theta}_0 \in \Theta_0$ (i.e., under the hypothesis of quasi-symmetry) is obtained as a solution of the system of equations

$$\begin{cases} p_{ij}\phi\left(\frac{\widehat{p}_{ij}}{p_{ij}}\right) + p_{ji}\phi\left(\frac{\widehat{p}_{ji}}{p_{ji}}\right) - \widehat{p}_{ij}\phi'\left(\frac{\widehat{p}_{ij}}{p_{ij}}\right) - \widehat{p}_{ji}\phi'\left(\frac{\widehat{p}_{ji}}{p_{ji}}\right) & = \mu\left(p_{ij}+p_{ji}\right),\ i,j=1,...,I \\ \sum_{j=1}^{I} p_{ij}\phi\left(\frac{\widehat{p}_{ij}}{p_{ij}}\right) - \sum_{j=1}^{I} \widehat{p}_{ij}\phi'\left(\frac{\widehat{p}_{ij}}{p_{ij}}\right) & = \mu p_{i*},\ i=1,...,I \end{cases} \tag{8.71}$$

with

$$\mu = \sum_{i=1}^{I}\sum_{j=1}^{I}\left(p_{ij}\phi\left(\frac{\widehat{p}_{ij}}{p_{ij}}\right) - \widehat{p}_{ij}\phi'\left(\frac{\widehat{p}_{ij}}{p_{ij}}\right)\right).$$

Its asymptotic distribution is

$$\sqrt{n}(\widehat{\boldsymbol{\theta}}^{QS,\phi} - \boldsymbol{\theta}_0) \underset{n\to\infty}{\overset{L}{\to}} N(\mathbf{0}, \boldsymbol{W}_{QS}(\boldsymbol{\theta}_0)), \tag{8.72}$$

where the $(I^2-1)\times(I^2-1)$ matrix $\boldsymbol{W}_{QS}(\boldsymbol{\theta}_0)$ has the expression

$$\boldsymbol{\Sigma}_{\boldsymbol{\theta}_0} - \boldsymbol{\Sigma}_{\boldsymbol{\theta}_0}\boldsymbol{B}_{QS}(\boldsymbol{\theta}_0)^T\left(\boldsymbol{B}_{QS}(\boldsymbol{\theta}_0)\boldsymbol{\Sigma}_{\boldsymbol{\theta}_0}\boldsymbol{B}_{QS}(\boldsymbol{\theta}_0)^T\right)^{-1}\boldsymbol{B}_{QS}(\boldsymbol{\theta}_0)\boldsymbol{\Sigma}_{\boldsymbol{\theta}_0},$$

and the matrix $\boldsymbol{B}_{QS}(\boldsymbol{\theta}_0)$ is defined in (8.70).

Proof. Instead of getting $\widehat{\boldsymbol{\theta}}^{QS,\phi} = \left(p_{ij}^{QS,\phi};\ i,j=1,...,I, \text{ and } (i,j)\neq(I,I)\right)^T$ we shall obtain

$$\boldsymbol{p}(\widehat{\boldsymbol{\theta}}^{QS,\phi}) = \left(p_{ij}^{QS,\phi};\ i,j=1,...,I\right)^T.$$

The p_{ij}'s which minimize the ϕ-divergence

$$D_\phi(\widehat{\boldsymbol{p}}, \boldsymbol{p}(\boldsymbol{\theta}))$$

subject to the null hypothesis of quasi-symmetry may be obtained minimizing

$$\sum_{i=1}^{I}\sum_{j=1}^{I} p_{ij}\phi\left(\frac{\widehat{p}_{ij}}{p_{ij}}\right) + \sum_{(i,j)\in C} \lambda_{ij}\left(p_{ij}p_{jI}p_{Ii} - p_{iI}p_{Ij}p_{ji}\right) + \mu\left(1 - \sum_{i=1}^{I}\sum_{j=1}^{I} p_{ij}\right)$$

where $C = \{(i,j) : i,j = 1,...,I-1, i<j\}$. By using the characterization given in (8.68), we must obtain the minimum of the function

$$\sum_{i=1}^{I}\sum_{j=1}^{I} a_i c_{ij}\phi\left(\frac{\widehat{p}_{ij}}{a_i c_{ij}}\right) + \sum_{(i,j)\in C} \lambda_{ij}\left(c_{ij} - c_{ji}\right) + \mu\left(1 - \sum_{i=1}^{I}\sum_{j=1}^{I} a_i c_{ij}\right) \qquad (8.73)$$

with respect to a_i and c_{ij}, where μ and λ_{ij} are undetermined Lagrangian multipliers. Minimizing the expression (8.73) is equivalent to minimizing the expression

$$\frac{1}{2}\sum_{i=1}^{I}\sum_{j=1}^{I}\left(a_i c_{ij}\phi\left(\frac{\widehat{p}_{ij}}{a_i c_{ij}}\right) + a_j c_{ij}\phi\left(\frac{\widehat{p}_{ji}}{a_j c_{ij}}\right)\right) + \mu\left(1 - \frac{1}{2}\sum_{i=1}^{I}\sum_{j=1}^{I}\left(a_i + a_j\right)c_{ij}\right)$$

and the values of a_i and c_{ij} are obtained as a solution of the system of equations given in (8.71). The asymptotic distribution is obtained from (8.3) because in this case $\boldsymbol{I}_F(\boldsymbol{\theta}_0)^{-1} = \boldsymbol{\Sigma}_{\boldsymbol{\theta}_0}$. ■

Corollary 8.5

The minimum power-divergence estimator, $\mathbf{p}\left(\widehat{\boldsymbol{\theta}}^{QS,\phi_{(\lambda)}}\right) = \left(p_{ij}^{QS,\phi_{(\lambda)}};\ i,j = 1,...,I\right)^T$ *of* $\mathbf{p}(\boldsymbol{\theta}_0)$ *under the hypothesis of quasi-symmetry (8.64) is given by the solution of the system of equations*

$$\begin{cases} \frac{1}{\lambda+1}\left(p_{ij} - \frac{\widehat{p}_{ij}^{\lambda+1}}{p_{ij}^{\lambda}} + p_{ji} - \frac{\widehat{p}_{ji}^{\lambda+1}}{p_{ji}^{\lambda}}\right) & = \mu\left(p_{ij} + p_{ji}\right), i,j = 1,...,I, i \neq j \\ \frac{1}{\lambda+1}\left(p_{i*} - \sum_{j=1}^{I} \frac{\widehat{p}_{ij}^{\lambda+1}}{p_{ij}^{\lambda}}\right) & = \mu p_{i*}, i = 1,...,I, \end{cases}$$

where μ *is*

$$\mu = \frac{1}{\lambda+1}\left(1 - \sum_{i=1}^{I}\sum_{j=1}^{I} \frac{\widehat{p}_{ij}^{\lambda+1}}{p_{ij}^{\lambda}}\right).$$

It is clear that for $\lambda \to 0$, *we obtain the maximum likelihood estimator introduced by Caussinus (1965), i.e.,*

$$\begin{cases} p_{ij} + p_{ji} & = \widehat{p}_{ij} + \widehat{p}_{ji} \\ p_{i*} & = \widehat{p}_{i*} \end{cases}.$$

Remark 8.8

For $I = 3$, given

$$\boldsymbol{\theta}_0 = \left(p_{11}, \frac{p_{13}p_{32}p_{31}}{p_{21}p_{31}}, p_{13}, p_{21}, p_{22}, p_{23}, p_{31}, p_{32}\right)^T \in \Theta_0$$

it is easy to check that

- $\boldsymbol{B}_{QS}(\boldsymbol{\theta}_0) = \left(\, 0, \; p_{23}p_{31}, \; -p_{32}p_{21}, \; -p_{13}p_{32}, \; 0, \; p_{12}p_{31}, \; p_{12}p_{23}, \; -p_{13}p_{21}\right.$
- $\boldsymbol{\Sigma}_{\boldsymbol{\theta}_0} = diag(\boldsymbol{\theta}_0) - \boldsymbol{\theta}_0\boldsymbol{\theta}_0^T,$
- *The matrix* $\left(\boldsymbol{B}_{QS}(\boldsymbol{\theta}_0)\boldsymbol{\Sigma}_{\boldsymbol{\theta}_0}\boldsymbol{B}_{QS}(\boldsymbol{\theta}_0)^T\right)^{-1}$ *is given by*

$$\frac{p_{23}p_{31}}{p_{13}p_{32}p_{21}\left(p_{23}p_{31}\left(p_{23}p_{31} + p_{21}p_{32} + p_{13}p_{32} + p_{13}p_{21}\right) + p_{13}p_{32}p_{21}\left(p_{23} + p_{31}\right)\right)}$$

and it is not difficult to establish that the matrix $\boldsymbol{W}_{QS}(\boldsymbol{\theta}_0)$ *is*

$$\boldsymbol{W}_{QS}(\boldsymbol{\theta}_0) = \left(\frac{\partial \boldsymbol{g}(\boldsymbol{\beta}_0)}{\partial \boldsymbol{\beta}}\right)^T \boldsymbol{I}_F(\boldsymbol{\beta}_0)^{-1}\left(\frac{\partial \boldsymbol{g}(\boldsymbol{\beta}_0)}{\partial \boldsymbol{\beta}}\right),$$

where

$$\left(\frac{\partial \boldsymbol{g}(\boldsymbol{\beta}_0)}{\partial \boldsymbol{\beta}}\right)^T = \begin{pmatrix} 1 & 0 & 0 & 0 & 0 & 0 & 0 \\ 0 & \frac{p_{21}p_{32}}{p_{23}p_{31}} & \frac{p_{13}p_{32}}{p_{23}p_{31}} & 0 & -\frac{p_{13}p_{32}p_{21}}{p_{23}^2p_{31}} & -\frac{p_{13}p_{32}p_{21}}{p_{23}p_{31}^2} & \frac{p_{13}p_{21}}{p_{23}p_{31}} \\ 0 & 1 & 0 & 0 & 0 & 0 & 0 \\ 0 & 0 & 1 & 0 & 0 & 0 & 0 \\ 0 & 0 & 0 & 1 & 0 & 0 & 0 \\ 0 & 0 & 0 & 0 & 1 & 0 & 0 \\ 0 & 0 & 0 & 0 & 0 & 1 & 0 \\ 0 & 0 & 0 & 0 & 0 & 0 & 1 \end{pmatrix}$$

and $\boldsymbol{I}_F(\boldsymbol{\beta}_0)^{-1}$ *is the inverse of the Fisher information matrix for the parameter* $\boldsymbol{\beta}_0 = (p_{11}, p_{13}, p_{21}, p_{22}, p_{23}, p_{31}, p_{32})^T$. *This result can be obtained in general in the same way as in the previous problems. Therefore,* $\widehat{\boldsymbol{\theta}}^{QS,\phi_2}$ *is BAN.*

Now we shall establish a theorem to get the asymptotic distribution of two new families of test statistics introduced for testing the hypothesis of quasi-symmetry given in (8.64).

Theorem 8.18

The asymptotic distribution of the ϕ-divergence test statistics

$$QS_n^{\phi_1}\left(\widehat{\boldsymbol{\theta}}^{QS,\phi_2}\right) \equiv \frac{2n}{\phi_1''(1)} D_{\phi_1}\left(\widehat{\boldsymbol{p}}, \boldsymbol{p}(\widehat{\boldsymbol{\theta}}^{QS,\phi_2})\right) \qquad (8.74)$$

and

$$QS_n^{\phi_1}\left(\widehat{\boldsymbol{\theta}}^{QS,\phi_2,h}\right) \equiv \frac{2n}{h'(0)\,\phi_1''(1)} h\left(D_{\phi_1}\left(\widehat{\mathbf{p}}, \mathbf{p}(\widehat{\boldsymbol{\theta}}^{QS,\phi})\right)\right) \tag{8.75}$$

for testing the hypothesis of quasi-symmetry is chi-square with $(I-1)(I-2)/2$ degrees of freedom.

Proof. In our case we have I^2 cells and I^2-1 parameters that are necessary to estimate. Using the relation (8.69) the number of constraints is $(I-1)(I-2)/2$. Then the asymptotic distribution of the family of statistics (8.74) is chi-square with

$$\underbrace{I^2}_{\text{Cells}} - \underbrace{I^2-1}_{\text{Estimated Parameters}} + \underbrace{(I-1)(I-2)/2}_{\text{Constraints}} - 1 = \underbrace{(I-1)(I-2)/2}_{\text{Degrees of freedom}}.$$

In relation with the family of tests statistics given in (8.75), we have

$$h(x) = h(0) + h'(0)\,x + o(x),$$

then

$$h\left(D_{\phi_1}\left(\widehat{\mathbf{p}}, \mathbf{p}(\widehat{\boldsymbol{\theta}}^{QS,\phi_2})\right)\right) = h'(0)\,D_{\phi_1}\left(\widehat{\mathbf{p}}, \mathbf{p}(\widehat{\boldsymbol{\theta}}^{QS,\phi_2})\right) + o_P(1)$$

and we get that the asymptotic distribution of the family of tests statistics given in (8.75) is also chi-square with $(I-1)(I-2)/2$ degrees of freedom. ■

Caussinus (1965) showed that symmetry is equivalent to quasi-symmetry and marginal homogeneity simultaneously holding; then we have

$$\text{Quasi-Symmetry} + \text{Marginal homogeneity} = \text{Symmetry}. \tag{8.76}$$

From this idea we consider the two following families of ϕ-divergences test statistics

$$W_{\phi_1,\phi_2}^{MH} = \frac{2n}{\phi_1''(1)}\left(D_{\phi_1}\left(\widehat{\mathbf{p}}, \mathbf{p}(\widehat{\boldsymbol{\theta}}^{S,\phi_2})\right) - D_{\phi_1}\left(\widehat{\mathbf{p}}, \mathbf{p}(\widehat{\boldsymbol{\theta}}^{QS,\phi_2})\right)\right) \tag{8.77}$$

and

$$S_{\phi_1,\phi_2}^{MH} = \frac{2n}{\phi_2''(1)} D_{\phi_1}\left(\mathbf{p}(\widehat{\boldsymbol{\theta}}^{QS,\phi_2}), \mathbf{p}(\widehat{\boldsymbol{\theta}}^{S,\phi_2})\right). \tag{8.78}$$

In the following theorem, its proof can be seen in Menéndez et *al.* (2005c), we present their asymptotic distribution.

Theorem 8.19

For testing hypotheses,

$$H_0: \textit{Symmetry versus } H_1: \textit{Quasi-Symmetry},$$

the asymptotic null distribution of the test statistics W_{ϕ_1,ϕ_2}^{MH} and S_{ϕ_1,ϕ_2}^{MH} given in (8.77) and (8.78) respectively is chi-square with $I-1$ degrees of freedom.

8.6. Homogeneity

Suppose we have ν independent random samples and we are interested in testing the null hypothesis that the samples are homogeneous, i.e., are from the same unknown population. We denote the samples by $\boldsymbol{X}^{(1)} = \left(X_1^{(1)}, ..., X_{n_1}^{(1)}\right)^T, ..., \boldsymbol{X}^{(\nu)} = \left(X_1^{(\nu)}, ..., X_{n_\nu}^{(\nu)}\right)^T$, of sizes $n_1, ..., n_\nu$ respectively, and we are interested in deciding if the samples $\boldsymbol{X}^{(1)}, ..., \boldsymbol{X}^{(\nu)}$ are all derived from the same distribution function $F(x) = Q(X \leq x)$, $x \in \mathbb{R}$, where Q is a probability measure on the real line. In this direction let $\mathcal{P} = \{E_i\}_{i=1,...,M}$ be a partition of the real line into M mutually exclusive and exhaustive intervals, where $\Pr\left(X_k^{(i)} \in E_j\right) = p_{ij}$ for $i = 1, ..., \nu$, $j = 1, ..., M$ and $k = 1, ..., n_i$. We denote $\boldsymbol{\theta}_i = (p_{i1}, ..., p_{iM-1})^T$, where $0 < p_{ij} < 1$, $\sum_{j=1}^{M} p_{ij} = 1$, $\forall i = 1, ..., \nu$, and $\boldsymbol{p}(\boldsymbol{\theta}_i) = (p_{i1}, ..., p_{iM})^T$. If $\boldsymbol{X}^{(1)}, ..., \boldsymbol{X}^{(\nu)}$ are all drawn, at random, from the same distribution function F, then it is expected that $p_{ij} = Q(E_j)$, for every $i = 1, ..., \nu$ and $j = 1, ..., M$ and therefore the problem is now reduced to a problem of testing homogeneity in multinomial populations, i.e., the null hypothesis,

$$H_0 : p_{1j} = ... = p_{\nu j} = Q(E_j) = p_j^* \; , j = 1, ..., M, \tag{8.79}$$

where $0 < p_j^* < 1$ and $\sum_{j=1}^{M} p_j^* = 1$.

We can observe that our parameter space is given by

$$\Theta = \left\{\boldsymbol{\theta} : \boldsymbol{\theta} = (p_{11}, ..., p_{1M-1}, ..., p_{\nu 1}, ..., p_{\nu M-1})^T\right\}, \tag{8.80}$$

that is, $\boldsymbol{\theta} = \left(\boldsymbol{\theta}_1^T, ..., \boldsymbol{\theta}_\nu^T\right)^T$ and its dimension is $(M-1)\,\nu$.

The usual test statistics for testing (8.79), if $\boldsymbol{p}^* = (p_1^*, ..., p_M^*)^T$ is completely unknown, are the chi-square test statistic and the likelihood ratio test given, respectively by

$$X^2 = \sum_{i=1}^{\nu}\sum_{j=1}^{M} \frac{\left(n_{ij} - \dfrac{n_{i*}n_{*j}}{n}\right)^2}{\dfrac{n_{i*}n_{*j}}{n}} \tag{8.81}$$

and

$$G^2 = 2\sum_{i=1}^{\nu}\sum_{j=1}^{M} n_{ij}\left(\log\frac{n_{ij}}{n_{i*}} - \log\frac{n_{*j}}{n}\right). \tag{8.82}$$

Their asymptotic distribution under the null hypothesis H_0 given in (8.79) is chi-square with $(M-1)(\nu-1)$ degrees of freedom. In the expressions (8.81) and (8.82), n_{ij} is the observed number of the components of $\boldsymbol{X}^{(i)}$, $i = 1, .., \nu$, belonging to the interval E_j, $j = 1, ..., M$, $n_{*j} = \sum_{i=1}^{\nu} n_{ij}$, $j = 1, ..., M$, $n_{i*} = \sum_{j=1}^{M} n_{ij}$, $i = 1, ..., \nu$ and

$n = \sum_{i=1}^{\nu}\sum_{j=1}^{M} n_{ij} = \sum_{j=1}^{M} n_{*j} = \sum_{i=1}^{\nu} n_{i*}$. In the two previous test statistics the unknown parameters have been estimated using the maximum likelihood estimator in all the parameter space (n_{ij}/n_{i*}) and the maximum likelihood estimator under the null hypothesis (n_{*j}/n). The test statistics given in (8.81) and (8.82) were extended by Pardo, L. et *al.* (1999) by considering test statistics based on ϕ-divergence measures and using the maximum likelihood estimator. An extension of that result when the probability distribution $\boldsymbol{p}^* = (p_1^*, ..., p_M^*)^T$ depends on some unknown parameters can be seen in Pardo L. et *al.* (2001).

In the following we shall denote by

$$\boldsymbol{p}(\widehat{\boldsymbol{\theta}}_i) = \left(\frac{n_{i1}}{n_{i*}}, ..., \frac{n_{iM}}{n_{i*}}\right)^T,$$

the maximum likelihood estimator of $\boldsymbol{p}(\boldsymbol{\theta}_i)$, i.e., the probabilities corresponding to the ith-population. We also consider the two following probability vectors

$$\widehat{\boldsymbol{p}} = \left(\frac{n_{1*}}{n}\boldsymbol{p}(\widehat{\boldsymbol{\theta}}_1)^T, ..., \frac{n_{\nu*}}{n}\boldsymbol{p}(\widehat{\boldsymbol{\theta}}_\nu)^T\right)^T = \left(\frac{n_{11}}{n}, ..., \frac{n_{1M}}{n}, ..., \frac{n_{\nu 1}}{n}, ..., \frac{n_{\nu M}}{n}\right)^T$$

and

$$\begin{aligned}\boldsymbol{p}^*(\boldsymbol{\theta}) &= \left(\frac{n_{1*}}{n}p_{11}, ..., \frac{n_{1*}}{n}p_{1M}, ..., \frac{n_{\nu*}}{n}p_{\nu 1}, ..., \frac{n_{\nu*}}{n}p_{\nu M}\right)^T \\ &= \left(\frac{n_{1*}}{n}\boldsymbol{p}(\boldsymbol{\theta}_1)^T, ..., \frac{n_{\nu*}}{n}\boldsymbol{p}(\boldsymbol{\theta}_\nu)^T\right)^T.\end{aligned}$$

The ϕ-divergence between the vectors $\widehat{\boldsymbol{p}}$ and $\boldsymbol{p}^*(\boldsymbol{\theta})$ is given by

$$D_\phi(\widehat{\boldsymbol{p}}, \boldsymbol{p}^*(\boldsymbol{\theta})) = \sum_{i=1}^{\nu}\sum_{j=1}^{M}\frac{n_{i*}}{n}p_{ij}\phi\left(\frac{n_{ij}}{n_{i*}p_{ij}}\right).$$

It is clear that

$$H_n^\phi\left(\widehat{\boldsymbol{\theta}}\right) \equiv \frac{2n}{\phi''(1)}D_\phi(\widehat{\boldsymbol{p}}, \boldsymbol{p}^*(\widehat{\boldsymbol{\theta}})) = \frac{2n}{\phi''(1)}\sum_{i=1}^{\nu}\sum_{j=1}^{M}\frac{n_{i*}}{n}\frac{n_{*j}}{n}\phi\left(\frac{n_{ij}n}{n_{i*}n_{*j}}\right)$$

coincides with the test statistic X^2, given in (8.81), for $\phi(x) = \frac{1}{2}(x-1)^2$, where $\widehat{\boldsymbol{\theta}}$ is the maximum likelihood estimator of $\boldsymbol{\theta} \in \Theta$, under the null hypothesis, given by

$$\widehat{\boldsymbol{\theta}} = \left(\frac{n_{*1}}{n}, ..., \frac{n_{*M-1}}{n}, \overset{(\nu}{...}, \frac{n_{*1}}{n}, ..., \frac{n_{*M-1}}{n}\right)^T, \tag{8.83}$$

and $\boldsymbol{p}^*(\widehat{\boldsymbol{\theta}})$ is obtained from $\boldsymbol{p}^*(\boldsymbol{\theta})$ replacing $\boldsymbol{\theta}$ by $\widehat{\boldsymbol{\theta}}$. The likelihood ratio test statistic G^2, given in (8.82), is obtained for $\phi(x) = x\log x - x + 1$.

Pardo, L. et *al.* (1999) established that the asymptotic distribution of the test statistic $H_n^\phi\left(\widehat{\boldsymbol{\theta}}\right)$ is chi-square with $(M-1)(\nu-1)$ degrees of freedom. In this section instead of considering the maximum likelihood estimator we consider the minimum ϕ-divergence estimator.

The hypothesis of homogeneity given in (8.79) can be formulated using the $(\nu-1)(M-1)$ constraints,

$$h_{ij}(\boldsymbol{\theta}) = p_{ij} - p_{\nu j} = 0, \ i = 1, ..., \nu-1, \ j = 1, ..., M-1. \tag{8.84}$$

In this case

$$\Theta_0 = \{\boldsymbol{\theta} \in \Theta : h_{ij}(\boldsymbol{\theta}) = 0\},$$

therefore if $\boldsymbol{\theta}_0 \in \Theta_0$, $\boldsymbol{\theta}_0$ is given by $\boldsymbol{\theta}_0 = \left(p_1, ..., p_{M-1}, \overset{(\nu}{...}, p_1, ..., p_{M-1}\right)^T$. We denote $\boldsymbol{\theta}_1^0 = (p_1, ..., p_{M-1})^T$ and

$$\boldsymbol{I}_\lambda(\boldsymbol{\theta}_0) = diag(\lambda) \otimes \mathcal{I}_\mathcal{F}(\boldsymbol{\theta}_1^0),$$

where $\lambda_i = \lim_{n\to\infty} n_{i*}/n$, $i = 1, ..., \nu$, $\lambda = (\lambda_1, ..., \lambda_\nu)^T$ and

$$\mathcal{I}_\mathcal{F}(\boldsymbol{\theta}_1^0) = \begin{cases} \frac{1}{p_i} + \frac{1}{p_M} & \text{if} \quad i = j \\ \frac{1}{p_M} & \text{if} \quad i \neq j \end{cases} \quad i, j = 1, ..., M-1.$$

By $\otimes$ we are denoting the Kronecker product of the matrices $diag(\lambda)$ and $\mathcal{I}_\mathcal{F}(\boldsymbol{\theta}_1^0)$.

In the following theorem, its proof can be seen in Menéndez et *al*. (2003b), we present the expression of the minimum ϕ–divergence estimator of $\boldsymbol{\theta}$ under the constraints given in (8.84).

Theorem 8.20

The minimum ϕ-divergence estimator, $\widehat{\boldsymbol{\theta}}^{H,\phi} = \left(p_j^{H,\phi}, j = 1, ..., M-1\right)^T$, *of* $\boldsymbol{\theta}_0$ *under the hypothesis of homogeneity, is obtained as a solution of the system of the equations*

$$\sum_{i=1}^{\nu} \frac{n_{i*}}{n} \left(\phi\left(\frac{n_{ij}}{n_{i*}p_j^*}\right) - \frac{n_{ij}}{n_{i*}p_j^*}\phi'\left(\frac{n_{ij}}{n_{i*}p_j^*}\right)\right) - \mu = 0, \ j = 1, ..., M, \tag{8.85}$$

where

$$\mu = \sum_{j=1}^{M}\sum_{i=1}^{\nu} p_j^* \frac{n_{i*}}{n}\left(\phi\left(\frac{n_{ij}}{n_{i*}p_j^*}\right) - \frac{n_{ij}}{n_{i*}p_j^*}\phi'\left(\frac{n_{ij}}{n_{i*}p_j^*}\right)\right).$$

Its asymptotic distribution is

$$\sqrt{n}\left(\widehat{\boldsymbol{\theta}}^{H,\phi} - \boldsymbol{\theta}_0\right) \underset{n\to\infty}{\overset{L}{\to}} N(\boldsymbol{0}, \boldsymbol{W}_H(\boldsymbol{\theta}_0)), \tag{8.86}$$

where the matrix $\boldsymbol{W}_H(\boldsymbol{\theta}_0)$ *has the expression*

$$\boldsymbol{I}_\lambda^{-1}(\boldsymbol{\theta}_0)\left(\boldsymbol{I} - \boldsymbol{B}_H(\boldsymbol{\theta}_0)^T\left(\boldsymbol{B}_H(\boldsymbol{\theta}_0)\boldsymbol{I}_\lambda^{-1}(\boldsymbol{\theta}_0)\boldsymbol{B}_H(\boldsymbol{\theta}_0)^T\right)^{-1}\boldsymbol{B}_H(\boldsymbol{\theta}_0)\boldsymbol{I}_\lambda^{-1}(\boldsymbol{\theta}_0)\right),$$

$$\boldsymbol{B}_H(\boldsymbol{\theta}_0)_{(\nu-1)(M-1)\times\nu(M-1)} = \left(\left(\frac{\partial h_{ij}(\boldsymbol{\theta})}{\partial\theta_{st}}\right)_{\boldsymbol{\theta}=\boldsymbol{\theta}_0}\right)_{\substack{i=1,...,\nu-1;j=1,...,M-1\\ s=1,...,\nu;\ t=1,...,M-1}},$$
$$= (\boldsymbol{I}, \boldsymbol{M}_H)$$

I denotes the $(M-1)(\nu-1)\times(M-1)(\nu-1)$ identity matrix, $\boldsymbol{M}_H$ is the matrix given by

$$\boldsymbol{M}_H = \left(-\boldsymbol{I}_{(M-1)\times(M-1)}, \overset{(\nu-1}{\ldots}, -\boldsymbol{I}_{(M-1)\times(M-1)}\right)^T$$

and the rank of $\boldsymbol{B}_H(\boldsymbol{\theta}_0)$ is $(M-1)\times(\nu-1)$.

Corollary 8.6

The minimum power-divergence estimator, $\mathbf{p}(\widehat{\boldsymbol{\theta}}^{H,\phi_{(\lambda)}}) = \left(p_j^{H,\phi}, j=1,...,M\right)^T$, of $\mathbf{p}(\boldsymbol{\theta}_0)$ under the hypothesis of homogeneity (8.84) is given by

$$p_j^{H,\phi_{(\lambda)}} = \frac{\left(\sum\limits_{i=1}^{\nu}\frac{n_{ij}^{\lambda+1}}{n_{i*}^{\lambda}}\right)^{1/(\lambda+1)}}{\sum\limits_{j=1}^{M}\left(\sum\limits_{i=1}^{\nu}\frac{n_{ij}^{\lambda+1}}{n_{i*}^{\lambda}}\right)^{1/(\lambda+1)}},\ j=1,...,M. \tag{8.87}$$

The proof follows from (8.85) and taking into account the expression of $\phi_{(\lambda)}$. It is interesting to observe that for $\lambda=0$,

$$p_j^{H,\phi_{(0)}} = \frac{n_{*j}}{n},\ j=1,...,M,$$

i.e., we obtain the classical maximum likelihood estimator under homogeneity and for $\lambda=1$,

$$p_j^{H,\phi_{(1)}} = \frac{\left(\sum\limits_{i=1}^{\nu}\frac{n_{ij}^{2}}{n_{i*}}\right)^{1/2}}{\sum\limits_{j=1}^{M}\left(\sum\limits_{i=1}^{\nu}\frac{n_{ij}^{2}}{n_{i*}}\right)^{1/2}},\ j=1,...,M$$

the minimum chi-square estimator under homogeneity. This estimator was obtained, for the first time, in a different way from the one presented here, by Quade and Salama (1975).

Other interesting estimators for homogeneity are: For $\lambda=-2$ the minimum modified chi-square estimator; for $\lambda\to-1$, the minimum modified likelihood estimator; for $\lambda=-1/2$ Freeman-Tukey estimator and finally for $\lambda=2/3$ Cressie-Read estimator.

The asymptotic distribution of the ϕ-divergence test statistic for testing homogeneity, based on the restricted minimum ϕ-divergence estimator, is given in the following theorem. Its proof can be seen in Menéndez et *al.* (2003b).

Theorem 8.21

The asymptotic distribution of the ϕ-divergence test statistic

$$H_n^{\phi_1}\left(\widehat{\boldsymbol{\theta}}^{H,\phi_2}\right) \equiv \frac{2n}{\phi_1''(1)} D_{\phi_1}\left(\widehat{\mathbf{p}}, \mathbf{p}^*(\widehat{\boldsymbol{\theta}}^{H,\phi_2})\right)$$

for testing the hypothesis of homogeneity is chi-square with $(\nu-1)(M-1)$ degrees o freedom.

Based on this theorem we should reject the null hypothesis of homogeneity given in (8.79), with significance level α, iff

$$H_n^{\phi_1}\left(\widehat{\boldsymbol{\theta}}^{H,\phi_2}\right) > \chi^2_{(\nu-1)(M-1),\alpha}. \tag{8.88}$$

Remark 8.9

For $\phi_1(x) = \phi_2(x) = x\log x - x + 1$ we get that $H_n^{\phi_1}\left(\widehat{\boldsymbol{\theta}}^{H,\phi_2}\right)$ coincides with the classical likelihood ratio test statistic for homogeneity and for $\phi_2(x) = x\log x - x + 1$ and $\phi_1(x) = \frac{1}{2}(x-1)^2$ coincides with the chi-square test statistic.

8.7. Exercises

1. Find the power function of the test statistic given in (8.88) using the maximum likelihood estimator and the alternative hypothesis

$$\boldsymbol{p}^* = \left(\frac{n_{1*}}{n}\left(\boldsymbol{p}_1^*\right)^T, ..., \frac{n_{\nu*}}{n}\left(\boldsymbol{p}_\nu^*\right)^T\right)^T,$$

where

$$\boldsymbol{p}_i^* = \left(p_{i1}^*, ..., p_{iM}^*\right)^T, \ 0 < p_{ij}^* < 1,$$

for $i = 1, ..., \nu$ and $j = 1, ..., M$ and there exists an index j with $1 \leq j \leq M$ and two indexes i and k, $1 \leq i, k \leq \nu$ and $i \neq k$ such that $p_{ij}^* \neq p_{kj}^*$.

2. The following data represent the blood types and the predisposition to suffer from diabetes ($A \equiv Low$, $B \equiv Average$ and $C \equiv High$)

	Low	*Average*	*High*	Total
O	137	86	35	258
A	42	23	11	76
B	19	17	7	43
AB	14	7	2	23
	212	133	55	400

Is there evidence to conclude that blood type is independent of predisposition to suffer from diabetes? Use the power-divergence family with $\lambda = -1 - 1/2, 0, 2/3$ and 1 for testing and $\lambda = 0$ for estimation, taking as significance level $\alpha = 0.05$.

3. Fifteen 3-year-old boys and fifteen 4-year-old girls were observed during 30 minutes play sessions, and each child's play during these two periods was scored as follows for incidence and degree of aggression:

Boys	96, 65, 74, 82, 121, 68, 79, 111, 48, 53, 92, 81, 31, 48
Girls	12, 47, 32, 59, 83, 14, 32, 15, 17, 82, 21, 34, 9, 15, 50

Test the hypothesis that there were not sex differences in the amount of aggression shown, using the power-divergence test statistic with $\lambda = -1, -1/2, 0, 2/3$ and 1, taking as significance level $\alpha = 0.05$.

4. Let $S_n = n^{1/2} \sum_{i=1}^{I} \sum_{j=1}^{J} l_{ij}\left(\widehat{p}_{ij} - q_{ij}\right)$ be the first term in the Taylor expansion of $D_{\phi_1}(\widehat{\boldsymbol{p}}, \boldsymbol{p}(\widehat{\boldsymbol{\theta}}))$ around $\left(\boldsymbol{q}, \boldsymbol{q}_{I\times J}\right)$ in Theorem 8.5. Prove that $S_n = 0 \ \forall n$ if and only if $\sigma_{\phi_1}^2(\boldsymbol{q}) = 0$.

5. If $q_{ij} = q_{i*} \times q_{*j}$ then $\sigma_{\phi_1}^2(\boldsymbol{q}) = 0$ in Theorem 8.5 because $l_{ij} = 0 \ \forall i = 1, ..., I, j = 1, ..., J$. Find a counter example, using the Kullback-Leibler divergence, in which $\sigma^2(\boldsymbol{q}) = 0$ but $q_{ij} \neq q_{i*} \times q_{*j}$.

6. Find the expression of the test statistics: Likelihood ratio test, chi-square test statistic, modified likelihood ratio test statistic, Freeman-Tukey and modified chi-square test statistic for the problem of symmetry.

7. Find the expression of the power for Pearson test statistic as well as for the likelihood ratio test in the problem of symmetry using the maximum likelihood estimator.

8. The data in the following table report the relative heights of 205 married couples

		Women		
		Tall	Medium	Short
	Tall	18	28	14
Men	Medium	20	51	28
	Short	12	25	9

Source: Christensen, R. (1997, p. 67).

Find the minimum power-divergence estimators, under the hypothesis of symmetry, for $\lambda = -2, -1, -0.5, 0$, and 1.

9. Find the expression of $\sigma_{\phi}^2(\boldsymbol{q})$ in Theorem 8.5 for the ϕ-divergences defined by $\phi_1(x) = .\frac{1}{2}(x-1)^2$; $\phi_2(x) = (1-\sqrt{x})^2$; $\phi_3(x) = (s-1)^{-1}(x^s - x)$; $\phi_4(x) = (\lambda(\lambda+1))^{-1}(x^{\lambda+1} - x)$ and $\phi_5(x) = \frac{(1-a)(1-x)}{a+(1-a)x}$.

8.8. Answers to Exercises

1. We denote

$$q_s^* = \sum_{l=1}^{\nu} \frac{n_{l*}}{n} p_{ls}^*,\ s = 1, ..., M,\ \boldsymbol{q} = (q_1^*, ..., q_M^*)^T$$

and

$$\boldsymbol{q}^* = \left(\frac{n_{1*}}{n}\boldsymbol{q}^T, ..., \frac{n_{\nu*}}{n}\boldsymbol{q}^T\right)^T.$$

Under the alternative hypothesis

$$\widehat{\boldsymbol{p}} \xrightarrow[n\to\infty]{P} \boldsymbol{p}^*$$

and

$$\boldsymbol{p}^*(\widehat{\boldsymbol{\theta}}) \xrightarrow[n\to\infty]{P} \boldsymbol{q}^*.$$

It is clear that in this case we have:

$$2nD_{\phi_1}(\widehat{\boldsymbol{p}}, \boldsymbol{p}^*(\widehat{\boldsymbol{\theta}})) = 2n \sum_{l=1}^{\nu} \frac{n_{i*}}{n} \sum_{j=1}^{M} \frac{n_{*j}}{n} \phi\left(\frac{\frac{n_{ij}}{n}}{\frac{n_{i*}}{n}\frac{n_{*j}}{n}}\right)$$

$$\cdot\ 2nD_{\phi_1}(\boldsymbol{p}^*,\boldsymbol{q}^*) = 2n\sum_{l=1}^{\nu}\sum_{j=1}^{M}\frac{n_{i*}}{n}q_j^*\phi\left(\frac{p_{ij}^*}{q_j^*}\right)$$

and

$$\sqrt{n}\left(2nD_{\phi_1}(\widehat{\boldsymbol{p}},\boldsymbol{p}^*(\widehat{\boldsymbol{\theta}})) - 2nD_{\phi_1}(\boldsymbol{p}^*,\boldsymbol{q}^*)\right)$$

takes the expression

$$\sqrt{n}\left(\sum_{i=1}^{\nu}\sum_{j=1}^{M}\left(\frac{\partial L}{\partial p_{ij}^*}\right)(\widehat{p}_{ij}-p_{ij}^*)\right) + o\left(\|\widehat{\boldsymbol{p}}-\boldsymbol{p}^*\|\right),$$

where

$$L = 2\sum_{l=1}^{\nu}\sum_{j=1}^{M}n_{i*}q_j^*\phi\left(\frac{p_{ij}^*}{q_j^*}\right).$$

It is not difficult to establish that

$$\begin{aligned} L = \; & 2\sum_{\substack{h=1\\h\neq i}}^{\upsilon} n_{h*}\left(q_1^*\phi\left(\frac{p_{h1}^*}{q_1^*}\right) + \ldots\ldots + q_s^*\phi\left(\frac{p_{hs}^*}{q_s^*}\right) + \ldots + q_M^*\phi\left(\frac{p_{hM}^*}{q_M^*}\right)\right) \\ & + 2n_{i*}q_j^*\phi\left(\frac{p_{ij}^*}{q_j^*}\right) + k, \end{aligned}$$

where k is independent of p_{ij}^*. Taking into account that

$$\frac{\partial q_s^*}{\partial p_{ij}^*} = \begin{cases} \dfrac{n_{i*}}{n} & s = j \\ 0 & s \neq j \end{cases},$$

we have

$$\frac{\partial L}{\partial p_{ij}^*} = 2\frac{n_{i*}}{n}\sum_{h=1}^{\upsilon}n_{h*}\left(\phi\left(\frac{p_{hj}^*}{q_j^*}\right) - \frac{p_{hj}^*}{q_j^*}\phi'\left(\frac{p_{hj}^*}{q_j^*}\right)\right) + 2n_{i*}\phi'\left(\frac{p_{ij}^*}{q_j^*}\right).$$

Then, if we denote

$$A_{ij}^{\phi} = \frac{\partial L}{\partial p_{ij}^*},$$

we have that the random variables

$$\sqrt{n}\left(\sum_{i=1}^{\upsilon}\sum_{j=1}^{M}\left(\frac{\partial L}{\partial p_{ij}^*}\right)(\widehat{p}_{ij}-p_{ij}^*)\right) = \left(A^{\phi}\right)^T\sqrt{n}\left(\widehat{\boldsymbol{p}}-\boldsymbol{p}^*\right),$$

being $A^{\phi} = \left(A_{11}^{\phi},\ldots,A_{1M}^{\phi},\ldots,A_{\nu 1}^{\phi},\ldots,A_{\upsilon M}^{\phi}\right)^T$, and

$$\sqrt{n}\left(2nD_{\phi_1}(\widehat{\boldsymbol{p}},\boldsymbol{p}^*(\widehat{\boldsymbol{\theta}})) - 2nD_{\phi_1}(\boldsymbol{p}^*,\boldsymbol{q}^*)\right)$$

have the same asymptotic distribution. But

$$\sqrt{n}\left(\widehat{\boldsymbol{p}}-\boldsymbol{p}^*\right) \xrightarrow[n\to\infty]{L} N\left(\mathbf{0},\boldsymbol{\Sigma}^*\right),$$

where $\boldsymbol{\Sigma}^* = diag\left(\lambda_1^{-1}\boldsymbol{\Sigma}_{\boldsymbol{p}_1^*},...,\lambda_\upsilon^{-1}\boldsymbol{\Sigma}_{\boldsymbol{p}_\upsilon^*}\right)$, $\boldsymbol{\Sigma}_{\boldsymbol{p}_i^*} = diag\left(\boldsymbol{p}_i^*\right) - \boldsymbol{p}_i^*\left(\boldsymbol{p}_i^*\right)^T$, $i = 1,...,$ and $\lambda_i = \lim_{n\to\infty}\dfrac{n_{i*}}{n}$.

Therefore,

$$\sqrt{n}\left(2nD_{\phi_1}(\widehat{\boldsymbol{p}},\boldsymbol{p}^*(\widehat{\boldsymbol{\theta}})) - 2nD_{\phi_1}(\boldsymbol{p}^*,\boldsymbol{q}^*)\right) \xrightarrow[n\to\infty]{L} N\left(0,\sigma_{\phi_1}^2(\boldsymbol{p}^*)\right),$$

where

$$\sigma_{\phi_1}^2(\boldsymbol{p}^*) = \sum_{i=1}^{\upsilon}\lambda_i^{-1}\left(\sum_{j=1}^{M}\left(A_{ij}^{\phi}\right)^2 p_{ij}^* - \left(\sum_{j=1}^{M}A_{ij}^{\phi}p_{ij}^*\right)^2\right).$$

Based on this result, we have that the power of the test statistic is given by

$$\beta_{n,\phi_1}(\boldsymbol{p}^*) = 1-\Phi_n\left(\frac{\sqrt{n}}{\sigma_{\phi_1}(\boldsymbol{p}^*)}\left(\frac{\chi^2_{(\nu-1)(M-1),\alpha}}{2n} - D_{\phi_1}(\boldsymbol{p}^*,\boldsymbol{q}^*)\right)\right),$$

where $\Phi_n(x)$ is a sequence of distribution functions leading uniformly to the standard normal distribution $\Phi(x)$.

2. If we consider the maximum likelihood estimator, the power-divergence test statistic, $I_n^{\phi(\lambda)}(\widehat{\boldsymbol{\theta}}^{I,\phi(0)})$ has the expression

$$I_n^{\phi(\lambda)}(\widehat{\boldsymbol{\theta}}^{I,\phi(0)}) = \frac{2n}{\lambda(\lambda+1)}\left(n^{\lambda-1}\sum_{i=1}^{I}\sum_{j=1}^{J}\frac{n_{ij}^{\lambda+1}}{n_{i*}^{\lambda}n_{*j}^{\lambda}} - 1\right).$$

In our case,

$$n_{1*} = 258,\ n_{2*} = 76,\ n_{3*} = 43 \text{ and } n_{4*} = 23$$
$$n_{*1} = 212,\ n_{*2} = 133 \text{ and } n_{*3} = 55.$$

Then we have

λ	-1	-1/2	0	2/3	1
$I_n^{\phi(\lambda)}(\widehat{\boldsymbol{\theta}}^{I,\phi(0)})$	2.543	2.499	2.462	2.422	2.405

On the other hand $\chi^2_{(I-1)(J-1),\alpha} = \chi^2_{6,\ 0.05} = 12.59$, then we should not reject the hypothesis that the two variables are independent.

3. If we consider the maximum likelihood estimator, the power-divergence test statistic $H_n^{\phi(\lambda)}(\widehat{\boldsymbol{\theta}}^{H,\phi(0)})$ has the expression

$$H_n^{\phi(\lambda)}(\widehat{\boldsymbol{\theta}}^{H,\phi(0)}) = \frac{2}{\lambda(\lambda+1)}\left(\sum_{i=1}^{m}\sum_{j=1}^{k}n_{i*}\frac{\left(\frac{n_{ij}}{n}\right)^{\lambda+1}}{\left(\frac{n_{*j}}{n}\right)^{\lambda}} - n\right).$$

If we consider the following partition of the interval $[8, 122]$,

$$E_1 = [8, 43), \ E_2 = [43, 78) \text{ and } E_3 = [78, 122],$$

we have

	E_1	E_2	E_3	n_{i*}
Boys	1	6	8	15
Girls	10	3	2	15
n_{*j}	11	9	10	

Then,

$$H_n^{\phi(\lambda)}(\widehat{\boldsymbol{\theta}}^{H,\phi(0)}) = \frac{2}{\lambda(\lambda+1)} \left\{ n_{1*} \left(\left(\frac{\left(\frac{1}{15}\right)^{\lambda+1}}{\left(\frac{11}{30}\right)^{\lambda}} \right) + \left(\frac{\left(\frac{6}{15}\right)^{\lambda+1}}{\left(\frac{9}{30}\right)^{\lambda}} \right) + \left(\frac{\left(\frac{8}{15}\right)^{\lambda+1}}{\left(\frac{10}{30}\right)^{\lambda}} \right) \right) \right.$$
$$\left. + \ n_{2*} \left(\left(\frac{\left(\frac{10}{15}\right)^{\lambda+1}}{\left(\frac{11}{30}\right)^{\lambda}} \right) + \left(\frac{\left(\frac{3}{15}\right)^{\lambda+1}}{\left(\frac{9}{30}\right)^{\lambda}} \right) + \left(\frac{\left(\frac{2}{15}\right)^{\lambda+1}}{\left(\frac{10}{30}\right)^{\lambda}} \right) \right) - n \right\}$$

and

λ	-1	-1/2	0	2/3	1
$H_n^{\phi(\lambda)}(\widehat{\boldsymbol{\theta}}^{H,\phi(0)})$	17.69	15.051	13.42	12.24	11.96

On the other hand $\chi^2_{2,0.05} = 5.991$, then we should reject the null hypothesis.

4. If $S_n = 0$ a.s. $\forall n$, we have $Var(S_n) = 0 \ \forall n$ and then

$$\lim_{n\to\infty} Var(S_n) = \sigma^2_{\phi_1}(\boldsymbol{q}) = 0.$$

Suppose that $\sigma^2_{\phi_1}(\boldsymbol{q}) = 0$. We have

$$S_n = \sqrt{n} \sum_{i=1}^{I} \sum_{j=1}^{J} l_{ij} (\widehat{p}_{ij} - q_{ij}),$$

then

$$E[S_n] = \sqrt{n} \sum_{i=1}^{I} \sum_{j=1}^{J} l_{ij} E[\widehat{p}_{ij} - q_{ij}] = 0$$

$$\begin{aligned} Var[S_n] &= E\left[n \sum_{i_1,i_2=1}^{I} \sum_{j_1,j_2=1}^{J} l_{i_1j_1} l_{i_2j_2} (\widehat{p}_{i_1j_1} - q_{i_1j_1})(\widehat{p}_{i_2j_2} - q_{i_2j_2}) \right] \\ &= n \sum_{i_1=1}^{I} \sum_{j_1=1}^{J} l^2_{i_1j_1} E\left[(\widehat{p}_{i_1j_1} - q_{i_1j_1})^2 \right] \\ &+ n \sum_{\substack{i_1,i_2=1 \\ i_1\neq i_2}}^{I} \sum_{\substack{j_1,j_2=1 \\ j_1\neq j_2}}^{J} l_{i_1j_1} l_{i_2j_2} E\left[(\widehat{p}_{i_1j_1} - q_{i_1j_1})(\widehat{p}_{i_2j_2} - q_{i_2j_2}) \right]. \end{aligned}$$

But,

$$\begin{aligned} E\left[(\widehat{p}_{i_1j_1} - q_{i_1j_1})^2 \right] &= V[\widehat{p}_{i_1j_1}] = \frac{q_{i_1j_1}(1 - p_{i_1j_1})}{n} \\ E\left[(\widehat{q}_{i_1j_1} - q_{i_1j_1})(\widehat{p}_{i_2j_2} - q_{i_2j_2}) \right] &= Cov[\widehat{p}_{i_1j_1}, \widehat{p}_{i_2j_2}] = -\frac{q_{i_1j_1} q_{i_2j_2}}{n}, \end{aligned}$$

therefore

$$\begin{aligned} Var\left[S_n\right] &= \sum_{i_1=1}^{I}\sum_{j_1=1}^{J} l_{i_1j_1}^2 q_{i_1j_1}\left(1-q_{i_1j_1}\right) - \sum_{\substack{i_1,i_2=1\\ i_1\neq i_2}}^{I}\sum_{\substack{j_1,j_2=1\\ j_1\neq j_2}}^{J} l_{i_1j_1}\, l_{\,i_2j_2} q_{i_1j_1} q_{i_2j_2} \\ &= \sum_{i_1=1}^{I}\sum_{j_1=1}^{J} l_{i_1j_1}^2 q_{i_1j_1} - \left(\sum_{i_1=1}^{I}\sum_{j_1=1}^{J} l_{i_1j_1} q_{i_1j_1}\right)^2 = \sigma_{\phi_1}^2\left(\boldsymbol{q}\right) = 0. \end{aligned}$$

Then $S_n = 0$.

5. We consider the bivariate random variable (X, Y) with probability distribution

$$\begin{aligned} q_{11} &= \Pr\left(X = x_1, Y = y_1\right) = 1/2 \qquad & q_{12} &= \Pr\left(X = x_1, Y = y_2\right) = 0 \\ q_{21} &= \Pr\left(X = x_2, Y = y_1\right) = 0 & q_{22} &= \Pr\left(X = x_2, Y = y_2\right) = 1/2. \end{aligned}$$

We have

$$l_{11} = l_{22} = -\log 2,\ l_{12} = l_{21} = 0$$

with

$$\sigma^2\left(\boldsymbol{q}\right) = \sum_{i=1}^{2}\sum_{j=1}^{2} q_{ij} l_{ij}^2 - \left(\sum_{i=1}^{2}\sum_{j=1}^{2} q_{ij} l_{ij}\right)^2 = 0.$$

On the other hand

$$q_{1*} = 1/2,\ q_{2*} = 1/2,\ q_{*1} = 1/2 \text{ and } q_{*2} = 1/2.$$

Then $q_{ij} \neq q_{i*} \times q_{*j}$ and $\sigma^2\left(\boldsymbol{q}\right) = 0$.

6. It is immediate to get for the power-divergence family and for the maximum likelihood estimator that

$$S_n^{\phi_{(\lambda)}}(\widehat{\boldsymbol{\theta}}^{S,\phi_{(0)}}) = \frac{2}{\lambda(\lambda+1)} \sum_{\substack{i,j\\ i\neq j}} n_{ij} \left(\left(\frac{2n_{ij}}{n_{ij}+n_{ji}}\right)^{\lambda} - 1\right).$$

Then for $\lambda \to 0$ (likelihood ratio test statistic), we have

$$S_n^{\phi_{(0)}}(\widehat{\boldsymbol{\theta}}^{S,\phi_{(0)}}) \equiv G^2 = 2\sum_{\substack{i,j\\ i\neq j}} n_{ij} \log \frac{2n_{ij}}{n_{ij}+n_{ji}},$$

for $\lambda \to -1$ (modified likelihood ratio test statistic),

$$S_n^{\phi_{(-1)}}(\widehat{\boldsymbol{\theta}}^{S,\phi_{(0)}}) = \sum_{\substack{i,j\\ i\neq j}} (n_{ij}+n_{ji}) \log \frac{n_{ij}+n_{ji}}{2n_{ij}},$$

for $\lambda = 1$ (chi-square test statistic),

$$S_n^{\phi_{(1)}}(\widehat{\boldsymbol{\theta}}^{S,\phi_{(0)}}) \equiv X^2 = \sum_{\substack{i,j\\ i<j}} \frac{\left(n_{ij}-n_{ji}\right)^2}{n_{ij}+n_{ji}},$$

for $\lambda = -1/2$ (Freeman-Tukey test statistic),

$$S_n^{\phi(-1/2)}(\widehat{\boldsymbol{\theta}}^{S,\phi(0)}) = 8\sum_{\substack{i,j\\ i\neq j}} n_{ij}\left(1 - \left(\frac{n_{ij}+n_{ji}}{2n_{ij}}\right)^{1/2}\right)$$

and finally for $\lambda = -2$ (modified chi-square test statistic)

$$S_n^{\phi(-2)}(\widehat{\boldsymbol{\theta}}^{S,\phi(0)}) = \sum_{\substack{i,j\\ i\neq j}} n_{ij}\left(\left(\frac{n_{ij}+n_{ji}}{2n_{ij}}\right)^{2} - 1\right).$$

7. The expression of the m'_{ij}s given in Theorem 8.10 for the power-divergence family is as follows

$$\begin{aligned} m_{ij}^{(\lambda)} &= \frac{1}{2\lambda(\lambda+1)}\left(\left(\frac{2q_{ij}}{q_{ij}+q_{ji}}\right)^{\lambda+1} + \left(\frac{2q_{ji}}{q_{ij}+q_{ji}}\right)^{\lambda+1} - 2\right) \\ &+ \frac{1}{\lambda}\frac{q_{ji}}{q_{ij}+q_{ji}}\left(\left(\frac{2q_{ij}}{q_{ij}+q_{ji}}\right)^{\lambda} - \left(\frac{2q_{ji}}{q_{ij}+q_{ji}}\right)^{\lambda}\right). \end{aligned}$$

Then for $\lambda \to 0$ and $\lambda = 1$ we get

$$m_{ij}^{(0)} = \log\frac{2q_{ji}}{q_{ij}+q_{ji}},\quad m_{ij}^{(1)} = \frac{q_{ij}^2 - 3q_{ji}^2 + 2q_{ij}q_{ji}}{2(q_{ij}+q_{ji})^2}.$$

Then we have

$$\sigma^2_{\phi_{(0)}}(\boldsymbol{q}) = \sum_{\substack{i,j\\ i\neq j}} q_{ij}\left(\log\frac{2q_{ij}}{q_{ij}+q_{ji}}\right)^2 - \left(\sum_{\substack{i,j\\ i\neq j}} q_{ij}\log\frac{2q_{ij}}{q_{ij}+q_{ji}}\right)^2$$

and

$$\sigma^2_{\phi_{(1)}}(\boldsymbol{q}) = \sum_{\substack{i,j\\ i\neq j}} q_{ij}\left(\frac{q_{ij}^2 - 3q_{ji}^2 + 2q_{ij}q_{ji}}{2(q_{ij}+q_{ji})^2}\right)^2 - \left(\sum_{\substack{i,j\\ i\neq j}} q_{ij}\frac{q_{ij}^2 - 3q_{ji}^2 + 2q_{ij}q_{ji}}{2(q_{ij}+q_{ji})^2}\right)^2.$$

According to Theorem 8.10

$$\beta_{n,\phi_{(\lambda)}}(\boldsymbol{q}) = 1 - \Phi_n\left(\frac{\sqrt{n}}{\sigma_{\phi_{(\lambda)}}(\boldsymbol{q})}\left(\frac{\chi^2_{I(I-1)/2,\alpha}}{2n} - D_{\phi_{(\lambda)}}(\boldsymbol{q},\boldsymbol{f}(\boldsymbol{q}))\right)\right),$$

then it is easy to obtain $\beta_{n,\phi_{(0)}}(\boldsymbol{q})$ and $\beta_{n,\phi_{(1)}}(\boldsymbol{q})$.

8. In (8.37) it was established that

$$p_{ij}^{S,\phi_{(\lambda)}} = \frac{\left(\widehat{p}_{ij}^{\lambda+1} + \widehat{p}_{ji}^{\lambda+1}\right)^{\frac{1}{\lambda+1}}}{\sum_{i=1}^{I}\sum_{j=1}^{I}\left(\widehat{p}_{ij}^{\lambda+1} + \widehat{p}_{ji}^{\lambda+1}\right)^{\frac{1}{\lambda+1}}}, \qquad i,j = 1,...,I.$$

In the following table we present the minimum power-divergence estimators

	$p_{11}^{S,\phi_{(\lambda)}}$	$p_{12}^{S,\phi_{(\lambda)}}$	$p_{13}^{S,\phi_{(\lambda)}}$	$p_{22}^{S,\phi_{(\lambda)}}$	$p_{23}^{S,\phi_{(\lambda)}}$	$p_{33}^{S,\phi_{(\lambda)}}$
$\lambda = -2$	0.0885	0.1147	0.0635	0.2508	0.1299	0.0442
$\lambda = -1$	0.0881	0.1159	0.0634	0.2498	0.1295	0.0440
$\lambda = -1/2$	0.0879	0.1165	0.0634	0.2493	0.1294	0.0439
$\lambda = 0$	0.0878	0.1171	0.0634	0.2487	0.1293	0.0439
$\lambda = 1$	0.0874	0.1182	0.0633	0.2477	0.1290	0.0437

9. Taking into account the expression of $\sigma^2(\boldsymbol{q})$ in Theorem 8.5, we get the following table

Divergence	l_{ij}
$\phi_1(x)$	$\frac{q_{ij}}{q_{i*}q_{*j}} - \frac{1}{2}\sum_{r=1}^{I}\frac{q_{rj}^2}{q_{r*}q_{*j}^2} - \frac{1}{2}\sum_{s=1}^{J}\frac{q_{is}^2}{q_{i*}q_{*s}^2}$
$\phi_2(x)$	$3 - \sum_{r=1}^{I}\frac{q_{rj}^{1/2}q_{r*}^{1/2}}{q_{*j}^{1/2}} - \sum_{s=1}^{J}\frac{q_{*s}^{1/2}q_{is}^{1/2}}{q_{i*}^{1/2}} - \frac{q_{i*}^{1/2}q_{*j}^{1/2}}{q_{ij}^{1/2}}$
$\phi_3(x)$	$-\sum_{r=1}^{I}\frac{q_{rj}^s}{q_{r*}^{s-1}q_{*j}^s} - \sum_{s=1}^{J}\frac{q_{is}^s}{q_{*s}^{s-1}q_{i*}^s} + \frac{1}{1-s}\left(1 - s\frac{q_{ij}^{s-1}}{q_{i*}^{s-1}q_{*j}^{s-1}}\right)$
$\phi_4(x)$	$\frac{1}{\lambda+1}\left(-\sum_{r=1}^{I}\frac{q_{rj}^{\lambda+1}}{q_{r*}^{\lambda}q_{*j}^{\lambda+1}} - \sum_{s=1}^{J}\frac{q_{is}^{\lambda+1}}{q_{*s}^{\lambda}q_{i*}^{\lambda+1}} + 2\right) + \frac{1}{\lambda}\left(\frac{q_{ij}^{\lambda}}{q_{i*}^{\lambda}q_{*j}^{\lambda}} - 1\right)$
$\phi_5(x)$	$(1-a)\left\{\sum_{r=1}^{I}\left(\frac{q_{rj}}{q_{*j}\left(a+(1-a)\frac{q_{rj}}{q_{r*}q_{*j}}\right)^2} + \frac{q_{r*}}{(1-a)\left(a+(1-a)\frac{q_{rj}}{q_{r*}q_{*j}}\right)}\right) + \sum_{s=1}^{J}\left(\frac{q_{is}}{q_{*j}\left(a+(1-a)\frac{q_{is}}{q_{i*}q_{*s}}\right)^2} + \frac{q_{*s}}{(1-a)\left(a+(1-a)\frac{q_{rj}}{q_{i*}q_{*s}}\right)}\right) - \frac{1}{\left(a+(1-a)\frac{q_{ij}}{q_{*}q_{*s}}\right)^2}\right\}.$

The previous ϕ-divergences correspond to Pearson, Matusita (a=1/2), Rathie-Kanappan, Cressie-Read and Rukhin. The expressions of Rathie-Kanappan and Rukhin ϕ-divergences presented in Chapter 1 are obtained from here with

$$\phi_3(x) - \phi_3'(1)(x-1) \text{ and } \phi_5(x) - \phi_5'(1)(x-1),$$

respectively.

9

Testing in General Populations

9.1. Introduction

The domains of application of ϕ-divergence test statistics go far beyond that of multinomial testing presented in previous chapters. Thus in this chapter the ϕ-divergence test statistics are introduced and studied in general populations.

Let $(\mathcal{X}, \beta_{\mathcal{X}}, P_{\boldsymbol{\theta}})_{\boldsymbol{\theta}\in\Theta}$ be the statistical space associated with the random variable X, where $\beta_{\mathcal{X}}$ is the σ-field of Borel subsets $A \subset \mathcal{X}$ and $\{P_{\boldsymbol{\theta}}\}_{\boldsymbol{\theta}\in\Theta}$ is a family of probability distributions defined on the measurable space $(\mathcal{X}, \beta_{\mathcal{X}})$ where Θ is an open subset of $\mathbb{R}^{M_0}$, with $M_0 \geq 1$. Probability measures $P_{\boldsymbol{\theta}}$ are assumed to be described by densities $f_{\boldsymbol{\theta}}(x) = dP_{\boldsymbol{\theta}}/d\mu(x)$, where μ is a σ-finite measure on $(\mathcal{X}, \beta_{\mathcal{X}})$. Let $Y_1, \ldots, Y_n$ be a random sample from a population described by the random variable X. For testing $H_0 : \boldsymbol{\theta} = \boldsymbol{\theta}_0$ against $H_1 : \boldsymbol{\theta} = \boldsymbol{\theta}_1$ (simple null hypothesis against simple alternative), the uniformly most powerful test is given by the Neyman-Pearson criterion. If μ is the Lebesgue measure, i.e., X is a continuous random variable, the criterion establishes: reject H_0 if

$$h_n = h_n(Y_1, \ldots, Y_n) = L(\boldsymbol{\theta}_1; Y_1, \ldots, Y_n)/L(\boldsymbol{\theta}_0; Y_1, \ldots, Y_n) \geq k_\alpha,$$

with

$$L(\boldsymbol{\theta}; Y_1, \ldots, Y_n) = \prod_{i=1}^{n} f_{\boldsymbol{\theta}}(Y_i),$$

and accept H_0 otherwise, where $k_\alpha = k_\alpha(n, \alpha, \boldsymbol{\theta}_0, \boldsymbol{\theta}_1)$ is determined by

$$\Pr_{\boldsymbol{\theta}_0}(h_n \geq k_\alpha) = \alpha,$$

where α $(0 < \alpha < 1)$ is the significance level.

More generally, for testing $H_0 : \boldsymbol{\theta} = \boldsymbol{\theta}_0$ against $H_1 : \boldsymbol{\theta} \neq \boldsymbol{\theta}_0$, the most powerful or uniformly most powerful test does not exist and we have to rely on other criteria for the choice of an appropriate test statistic. In such situations the classical solutions are Wald test statistic, Rao test statistic, likelihood ratio test statistic and more recently the test statistics based on ϕ-divergence measures: ϕ-divergence test statistics. The same problem appears with composite null hypotheses of the type $H_0 : \boldsymbol{\theta} \in \Theta_0 \subset \Theta$ and again the previous test statistics provide good solutions. In this chapter we study the properties of ϕ-divergence test statistics for testing simple and composite null hypotheses.

We assume that the statistical model $(\mathcal{X}, \beta_{\mathcal{X}}, P_{\boldsymbol{\theta}})_{\boldsymbol{\theta}\in\Theta}$ satisfies the standard regularity assumptions considered in parametric asymptotic statistics, i.e., conditions *i)-v)* in Section 2 of Chapter 2, as well as the following assumptions in relation to the function ϕ involved in the definition of general ϕ-divergence test statistics:

(Φ1) The function $\phi \in \Phi^*$ is twice continuously differentiable, with $\phi''(1) > 0$;

(Φ2) For each $\boldsymbol{\theta}_0 \in \Theta$ there exists an open neighborhood $N(\boldsymbol{\theta}_0)$ such that for all $\boldsymbol{\theta} \in N(\boldsymbol{\theta}_0)$ and $1 \leq i, j \leq M_0$ it holds:

$$\frac{\partial}{\partial \theta_i} \int_{\mathcal{X}} f_{\boldsymbol{\theta}_0}(x)\phi\left(\frac{f_{\boldsymbol{\theta}}(x)}{f_{\boldsymbol{\theta}_0}(x)}\right) d\mu(x) = \int_{\mathcal{X}} \frac{\partial}{\partial \theta_i}\left(f_{\boldsymbol{\theta}_0}(x)\phi\left(\frac{f_{\boldsymbol{\theta}}(x)}{f_{\boldsymbol{\theta}_0}(x)}\right)\right) d\mu(x),$$

$$\frac{\partial^2}{\partial \theta_i \partial \theta_j} \int_{\mathcal{X}} f_{\boldsymbol{\theta}_0}(x)\phi\left(\frac{f_{\boldsymbol{\theta}}(x)}{f_{\boldsymbol{\theta}_0}(x)}\right) d\mu(x) = \int_{\mathcal{X}} \frac{\partial^2}{\partial \theta_i \partial \theta_j}\left(f_{\boldsymbol{\theta}_0}(x)\phi\left(\frac{f_{\boldsymbol{\theta}}(x)}{f_{\boldsymbol{\theta}_0}(x)}\right)\right) d\mu(x),$$

and these expressions are continuous on $N(\boldsymbol{\theta}_0)$.

9.2. Simple Null Hypotheses: Wald, Rao, Wilks and Phi-divergence Test Statistics

The classical test statistics for testing $H_0 : \boldsymbol{\theta} = \boldsymbol{\theta}_0$ against $H_1 : \boldsymbol{\theta} \neq \boldsymbol{\theta}_0$ (simple null hypothesis against composite alternative hypothesis) are the following:

- *Wald test statistic*

$$W_n^0 = n(\widehat{\boldsymbol{\theta}} - \boldsymbol{\theta}_0)^T \mathcal{I}_{\mathcal{F}}(\widehat{\boldsymbol{\theta}})(\widehat{\boldsymbol{\theta}} - \boldsymbol{\theta}_0), \qquad (9.1)$$

where $\widehat{\boldsymbol{\theta}}$ is the maximum likelihood estimator of $\boldsymbol{\theta}$ in Θ obtained from the sample $Y_1, ..., Y_n$ and $\mathcal{I}_{\mathcal{F}}(\boldsymbol{\theta})$ is the Fisher information matrix for the original model.

- *Likelihood ratio test statistic*

$$L_n^0 = 2n\left(\lambda_n(\widehat{\boldsymbol{\theta}}) - \lambda_n(\boldsymbol{\theta}_0)\right), \qquad (9.2)$$

where

$$\lambda_n(\boldsymbol{\theta}) = \frac{1}{n}\sum_{i=1}^{n} \log f_{\boldsymbol{\theta}}(Y_i).$$

· *Rao test statistic*

$$R_n^0 = \frac{1}{n}\boldsymbol{U}_n(\boldsymbol{\theta}_0)^T \mathcal{I}_{\mathcal{F}}(\boldsymbol{\theta}_0)^{-1}\boldsymbol{U}_n(\boldsymbol{\theta}_0) \tag{9.3}$$

being

$$\boldsymbol{U}_n(\boldsymbol{\theta}_0) = \left(\sum_{i=1}^{n}\frac{\partial \log f_{\boldsymbol{\theta}}(Y_i)}{\partial\theta_1}, ..., \sum_{i=1}^{n}\frac{\partial \log f_{\boldsymbol{\theta}}(Y_i)}{\partial\theta_{M_0}}\right)^T_{\boldsymbol{\theta}=\boldsymbol{\theta}_0}.$$

Kupperman (1957, 1958) established that the test statistic based on the Kullback-Leibler divergence measure

$$T_n^{Kull}(\widehat{\boldsymbol{\theta}}, \boldsymbol{\theta}_0) \equiv 2nD_{Kull}(\widehat{\boldsymbol{\theta}}, \boldsymbol{\theta}_0) = 2n\int_{\mathcal{X}} f_{\widehat{\boldsymbol{\theta}}}(x)\log\frac{f_{\widehat{\boldsymbol{\theta}}}(x)}{f_{\boldsymbol{\theta}_0}(x)}d\mu(x)$$

is asymptotically chi-squared distributed with M_0 degrees of freedom. This result allows to test the null hypothesis $H_0 : \boldsymbol{\theta} = \boldsymbol{\theta}_0$ by means of the Kullback-Leibler divergence measure.

More recently, in the line of Kupperman, Salicrú et *al.* (1994) introduced the ϕ-divergence test statistic and studied its properties including its asymptotic behavior.

· *ϕ-divergence test statistic*

$$T_n^{\phi}(\widehat{\boldsymbol{\theta}}, \boldsymbol{\theta}_0) = \frac{2n}{\phi''(1)}D_{\phi}(\widehat{\boldsymbol{\theta}}, \boldsymbol{\theta}_0). \tag{9.4}$$

In the following theorem we present the asymptotic distribution of the test statistics given in (9.1), (9.2), (9.3) and (9.4).

Theorem 9.1

Let the model $(\mathcal{X}, \beta_{\mathcal{X}}, P_{\boldsymbol{\theta}})_{\boldsymbol{\theta}\in\Theta}$. Suppose ϕ satisfy the assumptions i)-v) considered in Section 2 of Chapter 2 and (Φ1)-(Φ2) respectively. Under the null hypothesis

$$H_0 : \boldsymbol{\theta} = \boldsymbol{\theta}_0 \tag{9.5}$$

the asymptotic distribution of the test statistics given in (9.1), (9.2), (9.3) and (9.4) is chi-square with M_0 degrees of freedom.

Proof. We start with Wald test statistic. Since under $H_0 : \boldsymbol{\theta} = \boldsymbol{\theta}_0$,

$$\widehat{\boldsymbol{\theta}} \xrightarrow[n\to\infty]{P} \boldsymbol{\theta}_0,$$

and by assumption v), the elements of $\mathcal{I}_{\mathcal{F}}(\boldsymbol{\theta})$ are continuous in $\boldsymbol{\theta}$, it holds

$$\mathcal{I}_{\mathcal{F}}(\widehat{\boldsymbol{\theta}}) \xrightarrow[n\to\infty]{P} \mathcal{I}_{\mathcal{F}}(\boldsymbol{\theta}_0).$$

Then, the asymptotic distribution of the quadratic form

$$n(\widehat{\boldsymbol{\theta}}-\boldsymbol{\theta}_0)^T\mathcal{I}_{\mathcal{F}}(\widehat{\boldsymbol{\theta}})(\widehat{\boldsymbol{\theta}}-\boldsymbol{\theta}_0)$$

coincides with that of the quadratic form

$$\boldsymbol{X}^T\boldsymbol{X} \equiv n(\widehat{\boldsymbol{\theta}}-\boldsymbol{\theta}_0)^T\mathcal{I}_{\mathcal{F}}(\boldsymbol{\theta}_0)(\widehat{\boldsymbol{\theta}}-\boldsymbol{\theta}_0),$$

where $\boldsymbol{X} = \sqrt{n}\mathcal{I}_{\mathcal{F}}(\boldsymbol{\theta}_0)^{1/2}(\widehat{\boldsymbol{\theta}}-\boldsymbol{\theta}_0)$.

We also know that under H_0,

$$\sqrt{n}(\widehat{\boldsymbol{\theta}}-\boldsymbol{\theta}_0) \xrightarrow[n\to\infty]{L} N\left(\mathbf{0}, \mathcal{I}_{\mathcal{F}}(\boldsymbol{\theta}_0)^{-1}\right)$$

which implies

$$\boldsymbol{X} \xrightarrow[n\to\infty]{L} N(\mathbf{0}, \boldsymbol{I}_{M_0\times M_0}).$$

Therefore,

$$n(\widehat{\boldsymbol{\theta}}-\boldsymbol{\theta}_0)^T\mathcal{I}_{\mathcal{F}}(\widehat{\boldsymbol{\theta}})(\widehat{\boldsymbol{\theta}}-\boldsymbol{\theta}_0) \xrightarrow[n\to\infty]{L} \chi^2_{M_0},$$

and the Wald test statistic should reject the null hypothesis whenever

$$n(\widehat{\boldsymbol{\theta}}-\boldsymbol{\theta}_0)^T\mathcal{I}_{\mathcal{F}}(\widehat{\boldsymbol{\theta}})(\widehat{\boldsymbol{\theta}}-\boldsymbol{\theta}_0) > \chi^2_{M_0,\alpha}.$$

As to Rao efficient score test statistic by assumption iv), we know that

$$E_{\boldsymbol{\theta}}\left[\frac{\partial \log f_{\boldsymbol{\theta}}(X)}{\partial \theta_i}\right] = 0, \quad i = 1, ..., M_0.$$

Further, by the Central Limit Theorem, under H_0 the random vector

$$\frac{1}{n}\boldsymbol{U}_n(\boldsymbol{\theta}_0) = \left(\frac{1}{n}\sum_{i=1}^n \frac{\partial \log f_{\boldsymbol{\theta}_0}(Y_i)}{\partial \theta_1}, ..., \frac{1}{n}\sum_{i=1}^n \frac{\partial \log f_{\boldsymbol{\theta}_0}(Y_i)}{\partial \theta_{M_0}}\right)^T$$

satisfies

$$\sqrt{n}\frac{1}{n}\boldsymbol{U}_n(\boldsymbol{\theta}_0) \xrightarrow[n\to\infty]{L} N(\mathbf{0}, \boldsymbol{\Sigma}),$$

where $\boldsymbol{\Sigma}$ is the variance-covariance matrix of the random vector

$$\left(\frac{\partial \log f_{\boldsymbol{\theta}_0}(X)}{\partial \theta_1}, ..., \frac{\partial \log f_{\boldsymbol{\theta}_0}(X)}{\partial \theta_{M_0}}\right)^T,$$

which is obviously $\mathcal{I}_{\mathcal{F}}(\boldsymbol{\theta}_0)$. Therefore

$$\frac{1}{\sqrt{n}}\boldsymbol{U}_n(\boldsymbol{\theta}_0) \xrightarrow[n\to\infty]{L} N(\mathbf{0}, \mathcal{I}_{\mathcal{F}}(\boldsymbol{\theta}_0))$$

and consequently

$$R_n = \frac{1}{n}\boldsymbol{U}_n(\boldsymbol{\theta}_0)^T \mathcal{I}_{\mathcal{F}}(\boldsymbol{\theta}_0)^{-1}\boldsymbol{U}_n(\boldsymbol{\theta}_0) \underset{n\to\infty}{\overset{L}{\longrightarrow}} \chi^2_{M_0}.$$

Now let us establish the asymptotic distribution of the ϕ-divergence test statistic,

$$T_n^{\phi}(\widehat{\boldsymbol{\theta}}, \boldsymbol{\theta}_0) = \frac{2n}{\phi''(1)} D_{\phi}(\widehat{\boldsymbol{\theta}}, \boldsymbol{\theta}_0).$$

A second order Taylor expansion of $D_\phi(\boldsymbol{\theta}, \boldsymbol{\theta}_0)$ around $\boldsymbol{\theta} = \boldsymbol{\theta}_0$ at $\boldsymbol{\theta} = \widehat{\boldsymbol{\theta}}$ gives

$$\begin{aligned} D_\phi(\widehat{\boldsymbol{\theta}}, \boldsymbol{\theta}_0) &= D_\phi(\boldsymbol{\theta}_0, \boldsymbol{\theta}_0) + \sum_{i=1}^{M_0} \left(\frac{\partial D_\phi(\boldsymbol{\theta}, \boldsymbol{\theta}_0)}{\partial \theta_i}\right)_{\boldsymbol{\theta}=\boldsymbol{\theta}_0} (\widehat{\theta}_i - \theta_{0i}) \\ &+ \frac{1}{2}\sum_{i=1}^{M_0}\sum_{j=1}^{M_0} \left(\frac{\partial D^2_\phi(\boldsymbol{\theta}, \boldsymbol{\theta}_0)}{\partial \theta_i \partial \theta_j}\right)_{\boldsymbol{\theta}=\boldsymbol{\theta}_0} (\widehat{\theta}_i - \theta_{0i})(\widehat{\theta}_j - \theta_{0j}) + o\left(\left\|\widehat{\boldsymbol{\theta}} - \boldsymbol{\theta}_0\right\|^2\right), \end{aligned}$$

where by (1.6), it is clear that

$$D_\phi(\boldsymbol{\theta}_0, \boldsymbol{\theta}_0) = \phi(1) = 0.$$

Now we prove that the second term in the previous expansion is zero.

Indeed,

$$\frac{\partial D_\phi(\boldsymbol{\theta}, \boldsymbol{\theta}_0)}{\partial \theta_i} = \int_{\mathcal{X}} \phi'\left(\frac{f_{\boldsymbol{\theta}}(x)}{f_{\boldsymbol{\theta}_0}(x)}\right) \frac{\partial f_{\boldsymbol{\theta}}(x)}{\partial \theta_i} d\mu(x),$$

and for $\boldsymbol{\theta} = \boldsymbol{\theta}_0$

$$\left(\frac{\partial D_\phi(\boldsymbol{\theta}, \boldsymbol{\theta}_0)}{\partial \theta_i}\right)_{\boldsymbol{\theta}=\boldsymbol{\theta}_0} = \phi'(1) \int_{\mathcal{X}} \frac{\partial f_{\boldsymbol{\theta}}(x)}{\partial \theta_i} d\mu(x) = \phi'(1) \frac{\partial}{\partial \theta_i} \int_{\mathcal{X}} f_{\boldsymbol{\theta}}(x) d\mu(x) = 0.$$

This means that the random variables

$$2nD_\phi(\widehat{\boldsymbol{\theta}}, \boldsymbol{\theta}_0)$$

and

$$n\sum_{i=1}^{M_0}\sum_{j=1}^{M_0} \left(\frac{\partial^2 D_\phi(\boldsymbol{\theta}, \boldsymbol{\theta}_0)}{\partial \theta_i \partial \theta_j}\right)_{\boldsymbol{\theta}=\boldsymbol{\theta}_0} (\widehat{\theta}_i - \theta_{0i})(\widehat{\theta}_j - \theta_{0j})$$

have the same asymptotic distribution, since under H_0 $o\left(\left\|\widehat{\boldsymbol{\theta}} - \boldsymbol{\theta}_0\right\|^2\right) = o_P(n^{-1})$.

The second order derivatives are

$$\begin{aligned} \frac{\partial^2 D_\phi(\boldsymbol{\theta}, \boldsymbol{\theta}_0)}{\partial \theta_i \partial \theta_j} &= \int_{\mathcal{X}} \phi''\left(\frac{f_{\boldsymbol{\theta}}(x)}{f_{\boldsymbol{\theta}_0}(x)}\right) \frac{1}{f_{\boldsymbol{\theta}_0}(x)} \frac{\partial f_{\boldsymbol{\theta}}(x)}{\partial \theta_i} \frac{\partial f_{\boldsymbol{\theta}}(x)}{\partial \theta_j} d\mu(x) \\ &+ \int_{\mathcal{X}} \phi'\left(\frac{f_{\boldsymbol{\theta}}(x)}{f_{\boldsymbol{\theta}_0}(x)}\right) \frac{\partial^2 f_{\boldsymbol{\theta}}(x)}{\partial \theta_i \partial \theta_j} d\mu(x), \end{aligned}$$

and then

$$\begin{aligned}\left(\frac{\partial^2 D_\phi(\boldsymbol{\theta},\boldsymbol{\theta}_0)}{\partial\theta_i\partial\theta_j}\right)_{\boldsymbol{\theta}=\boldsymbol{\theta}_0} &= \phi''(1)\int_{\mathcal{X}}\frac{1}{f_{\boldsymbol{\theta}_0}(x)}\frac{\partial f_{\boldsymbol{\theta}_0}(x)}{\partial\theta_i}\frac{\partial f_{\boldsymbol{\theta}_0}(x)}{\partial\theta_j}d\mu(x)\\ &= \phi''(1)\mathcal{I}_{\mathcal{F}}^{ij}(\boldsymbol{\theta}_0),\end{aligned}$$

where $\mathcal{I}_{\mathcal{F}}^{ij}(\boldsymbol{\theta}_0)$ is the (i,j)th-element of the Fisher information matrix. Therefore,

$$T_n^\phi(\widehat{\boldsymbol{\theta}},\boldsymbol{\theta}_0)=\frac{2n}{\phi''(1)}D_\phi(\widehat{\boldsymbol{\theta}},\boldsymbol{\theta}_0) = n(\widehat{\boldsymbol{\theta}}-\boldsymbol{\theta}_0)^T\mathcal{I}_{\mathcal{F}}(\boldsymbol{\theta}_0)(\widehat{\boldsymbol{\theta}}-\boldsymbol{\theta}_0)+o_P(1),$$

and finally, this implies

$$T_n^\phi(\widehat{\boldsymbol{\theta}},\boldsymbol{\theta}_0)=\frac{2n}{\phi''(1)}D_\phi(\widehat{\boldsymbol{\theta}},\boldsymbol{\theta}_0)\underset{n\to\infty}{\overset{L}{\longrightarrow}}\chi^2_{M_0}.$$

In order to obtain the asymptotic distribution of L_n^0 under H_0, we consider the following Taylor expansion

$$\begin{aligned}\lambda_n(\widehat{\boldsymbol{\theta}}) =\ & \lambda_n(\boldsymbol{\theta}_0)+\sum_{i=1}^{M_0}\left(\frac{\partial\lambda_n(\boldsymbol{\theta})}{\partial\theta_i}\right)_{\boldsymbol{\theta}=\boldsymbol{\theta}_0}(\widehat{\theta}_i-\theta_{0i})\\ +\ & \tfrac{1}{2}\sum_{i=1}^{M_0}\sum_{j=1}^{M_0}\left(\frac{\partial^2\lambda_n(\boldsymbol{\theta})}{\partial\theta_i\partial\theta_j}\right)_{\boldsymbol{\theta}=\boldsymbol{\theta}_0}(\widehat{\theta}_i-\theta_{0i})(\widehat{\theta}_j-\theta_{0j})+o\left(\left\|\widehat{\boldsymbol{\theta}}-\boldsymbol{\theta}_0\right\|^2\right).\end{aligned}$$

But

$$\left(\frac{\partial\lambda_n(\boldsymbol{\theta})}{\partial\theta_i}\right)_{\boldsymbol{\theta}=\boldsymbol{\theta}_0}=\frac{1}{n}\sum_{l=1}^{n}\frac{\partial\log f_{\boldsymbol{\theta}_0}(Y_l)}{\partial\theta_i}.$$

Then, applying Khintchine Law of Large Numbers we get

$$\left(\frac{\partial\lambda_n(\boldsymbol{\theta})}{\partial\theta_i}\right)_{\boldsymbol{\theta}=\boldsymbol{\theta}_0}\underset{n\to\infty}{\overset{a.s.}{\longrightarrow}}\int_{\mathcal{X}}\frac{\partial\log f_{\boldsymbol{\theta}_0}(x)}{\partial\theta_i}f_{\boldsymbol{\theta}_0}(x)\,d\mu(x)=0.$$

On the other hand, the second derivatives are

$$\left(\frac{\partial^2\lambda_n(\boldsymbol{\theta})}{\partial\theta_i\partial\theta_j}\right)_{\boldsymbol{\theta}=\boldsymbol{\theta}_0}=-\frac{1}{n}\sum_{l=1}^{n}\frac{1}{f_{\boldsymbol{\theta}_0}(Y_l)^2}\frac{\partial f_{\boldsymbol{\theta}_0}(Y_l)}{\partial\theta_i}\frac{\partial f_{\boldsymbol{\theta}_0}(Y_l)}{\partial\theta_j}+\frac{1}{n}\sum_{l=1}^{n}\frac{\partial^2 f_{\boldsymbol{\theta}_0}(Y_l)}{\partial\theta_i\partial\theta_j}\frac{1}{f_{\boldsymbol{\theta}_0}(Y_l)}.$$

But by Khintchine Law of Large Numbers,

$$-\frac{1}{n}\sum_{l=1}^{n}\frac{1}{f_{\boldsymbol{\theta}_0}(Y_l)^2}\frac{\partial f_{\boldsymbol{\theta}_0}(Y_l)}{\partial\theta_i}\frac{\partial f_{\boldsymbol{\theta}_0}(Y_l)}{\partial\theta_j}\underset{n\to\infty}{\overset{a.s.}{\longrightarrow}}\mathcal{I}_{\mathcal{F}}^{ij}(\boldsymbol{\theta}_0)$$

and

$$\frac{1}{n}\sum_{l=1}^{n}\frac{\partial^2 f_{\boldsymbol{\theta}_0}(Y_l)}{\partial\theta_i\partial\theta_j}\frac{1}{f_{\boldsymbol{\theta}_0}(Y_l)}\underset{n\to\infty}{\overset{a.s.}{\longrightarrow}}\int_{\mathcal{X}}\frac{\partial^2 f_{\boldsymbol{\theta}_0}(x)}{\partial\theta_i\partial\theta_j}d\mu(x)=0.$$

Therefore,

$$2n\left(\lambda_n(\widehat{\boldsymbol{\theta}}) - \lambda_n(\boldsymbol{\theta}_0)\right) = \sqrt{n}(\widehat{\boldsymbol{\theta}} - \boldsymbol{\theta}_0)^T \mathcal{I}_{\mathcal{F}}(\boldsymbol{\theta}_0)\sqrt{n}(\widehat{\boldsymbol{\theta}} - \boldsymbol{\theta}_0) + o_P(1),$$

and this implies

$$L_n^0 \xrightarrow[n\to\infty]{L} \chi^2_{M_0}.$$

Remark 9.1

The (h,ϕ)-divergence test statistics associated with the (h,ϕ)-divergence measures satisfy

$$T_n^{\phi,h}(\widehat{\boldsymbol{\theta}},\boldsymbol{\theta}_0) \equiv \frac{2n}{h'(0)\,\phi''(1)} D_\phi^h(\widehat{\boldsymbol{\theta}},\boldsymbol{\theta}_0) \xrightarrow[n\to\infty]{L} \chi^2_{M_0}.$$

This result is immediate from the relation

$$D_\phi^h(\widehat{\boldsymbol{\theta}},\boldsymbol{\theta}_0) = h'(0)\,D_\phi(\widehat{\boldsymbol{\theta}},\boldsymbol{\theta}_0) + o\left(D_\phi(\widehat{\boldsymbol{\theta}},\boldsymbol{\theta}_0)\right).$$

Based on Theorem 9.1 and Remark 9.1, we should reject the null hypothesis given in (9.5), with significance level α, if $T_n^\phi(\widehat{\boldsymbol{\theta}},\boldsymbol{\theta}_0) > \chi^2_{M_0,\alpha}$ (or if $T_n^{\phi,h}(\widehat{\boldsymbol{\theta}},\boldsymbol{\theta}_0) > \chi^2_{M_0,\alpha}$). The rejection rule is analogous for W_n^0, L_n^0 and R_n^0.

In most cases, the power function of this testing procedure can not be calculated explicitly. In the following theorem we present a useful asymptotic result for approximating the power function.

Theorem 9.2

Let the model and $\phi \in \Phi^*$ satisfy the assumptions i)-v) considered in Section 2 of Chapter 2 and (Φ1)-(Φ2) respectively. Let $\boldsymbol{\theta}^*$ be the true parameter, with $\boldsymbol{\theta}^* \neq \boldsymbol{\theta}_0$. Then it holds

$$\sqrt{n}\left(D_\phi(\widehat{\boldsymbol{\theta}},\boldsymbol{\theta}_0) - D_\phi(\boldsymbol{\theta}^*,\boldsymbol{\theta}_0)\right) \xrightarrow[n\to\infty]{L} N\left(0,\sigma_\phi^2(\boldsymbol{\theta}^*)\right),$$

where $\sigma_\phi^2(\boldsymbol{\theta}^*) = \boldsymbol{T}^T\mathcal{I}_{\mathcal{F}}(\boldsymbol{\theta}^*)^{-1}\boldsymbol{T}$, $\boldsymbol{T} = (t_1,...,t_M)^T$ and $t_j = \left(\frac{\partial D_\phi(\boldsymbol{\theta},\boldsymbol{\theta}_0)}{\partial\theta_j}\right)_{\boldsymbol{\theta}=\boldsymbol{\theta}^*}$, $j = 1,...,M$.

Proof. A first order Taylor expansion gives

$$D_\phi(\widehat{\boldsymbol{\theta}},\boldsymbol{\theta}_0) = D_\phi(\boldsymbol{\theta}^*,\boldsymbol{\theta}_0) + \boldsymbol{T}^T(\widehat{\boldsymbol{\theta}} - \boldsymbol{\theta}^*) + o\left(\left\|\widehat{\boldsymbol{\theta}} - \boldsymbol{\theta}^*\right\|\right).$$

By the Central Limit Theorem,

$$\sqrt{n}(\widehat{\boldsymbol{\theta}} - \boldsymbol{\theta}^*) \xrightarrow[n\to\infty]{L} N\left(\mathbf{0},\mathcal{I}_{\mathcal{F}}(\boldsymbol{\theta}^*)^{-1}\right)$$

and

$$\sqrt{n}\,o\left(\left\|\widehat{\boldsymbol{\theta}} - \boldsymbol{\theta}^*\right\|\right) = o_P(1).$$

Then, it is clear that the random variables,

$$\sqrt{n}\left(D_\phi(\widehat{\boldsymbol{\theta}}, \boldsymbol{\theta}_0) - D_\phi\left(\boldsymbol{\theta}^*, \boldsymbol{\theta}_0\right)\right) \quad \text{and} \quad \boldsymbol{T}^T\sqrt{n}(\widehat{\boldsymbol{\theta}} - \boldsymbol{\theta}^*),$$

have the same asymptotic distribution.

Remark 9.2

Through Theorem 9.2, a first approximation to the power function, at $\boldsymbol{\theta}^* \neq \boldsymbol{\theta}_0$, i given by

$$\beta^1_{n,\phi}(\boldsymbol{\theta}^*) = 1 - \Phi\left(\frac{\sqrt{n}}{\sigma_\phi\left(\boldsymbol{\theta}^*\right)}\left(\frac{\phi''\left(1\right)\ \chi^2_{M_0,\alpha}}{2n} - D_\phi\left(\boldsymbol{\theta}^*, \boldsymbol{\theta}_0\right)\right)\right), \qquad (9.6$$

where $\Phi\left(x\right)$ is the standard normal distribution function. If some alternative $\boldsymbol{\theta}^* \neq \boldsymbol{\theta}_0$ is the true parameter, then the probability of rejecting $\boldsymbol{\theta}_0$ with the rejection rul $T^\phi_n\left(\widehat{\boldsymbol{\theta}}, \boldsymbol{\theta}_0\right) > \chi^2_{M_0,\alpha}$, for fixed significance level α, tends to one as $n \to \infty$. The tes is consistent in Fraser's sense.

In order to produce some less trivial asymptotic powers that are not all equal to 1 we can use a Pitman-type local analysis, as developed by LeCam (1960), confining the attention to $n^{-1/2}$-neighborhoods of the true parameter values. More specifically, we consider the power at contiguous alternative hypotheses described by

$$H_{1,n} : \boldsymbol{\theta}_n = \boldsymbol{\theta}_0 + n^{-1/2}\boldsymbol{d},$$

where $\boldsymbol{d}$ is a fixed vector in $\mathbb{R}^{M_0}$ such that $\boldsymbol{\theta}_n \in \Theta \subset \mathbb{R}^{M_0}$.

A fundamental tool to get the asymptotic distribution of the ϕ-divergence test statistic $T^\phi_n(\widehat{\boldsymbol{\theta}}, \boldsymbol{\theta}_0)$ under the contiguous alternative hypotheses is LeCam's third lemma, as presented in Hájek and Sidàk (1967). Instead, in the following theorem we present a simpler proof.

Theorem 9.3

Let the model and $\phi \in \Phi^*$ satisfy the assumptions i)–v) considered in Section 2 of Chapter 2 and (Φ1)–(Φ2) respectively. Under the contiguous alternative hypotheses

$$H_{1,n} : \boldsymbol{\theta}_n = \boldsymbol{\theta}_0 + n^{-1/2}\boldsymbol{d},$$

where $\boldsymbol{d}$ is a fixed vector in $\mathbb{R}^{M_0}$ such that $\boldsymbol{\theta}_n \in \Theta \subset \mathbb{R}^{M_0}$, the asymptotic distribution of the ϕ-divergence test statistic $T^\phi_n(\widehat{\boldsymbol{\theta}}, \boldsymbol{\theta}_0)$ is noncentral chi-square, with M_0 degrees of freedom and noncentrality parameter $\delta = \boldsymbol{d}^T\mathcal{I}_\mathcal{F}\left(\boldsymbol{\theta}_0\right)\boldsymbol{d}$.

Proof. We can write

$$\sqrt{n}(\widehat{\boldsymbol{\theta}} - \boldsymbol{\theta}_0) = \sqrt{n}(\widehat{\boldsymbol{\theta}} - \boldsymbol{\theta}_n) + \sqrt{n}\left(\boldsymbol{\theta}_n - \boldsymbol{\theta}_0\right) = \sqrt{n}(\widehat{\boldsymbol{\theta}} - \boldsymbol{\theta}_n) + \boldsymbol{d}.$$

Under $H_{1,n}$, it holds

$$\sqrt{n}(\widehat{\boldsymbol{\theta}} - \boldsymbol{\theta}_n) \underset{n\to\infty}{\overset{L}{\longrightarrow}} N\left(\boldsymbol{0}, \mathcal{I}_\mathcal{F}\left(\boldsymbol{\theta}_0\right)^{-1}\right) \quad \text{and} \quad \sqrt{n}(\widehat{\boldsymbol{\theta}} - \boldsymbol{\theta}_0) \underset{n\to\infty}{\overset{L}{\longrightarrow}} N\left(\boldsymbol{d}, \mathcal{I}_\mathcal{F}\left(\boldsymbol{\theta}_0\right)^{-1}\right).$$

By applying the delta method, it is not difficult to establish that

$$T_n^{\phi}(\widehat{\boldsymbol{\theta}},\boldsymbol{\theta}_0) = n(\widehat{\boldsymbol{\theta}}-\boldsymbol{\theta}_0)\mathcal{I}_{\mathcal{F}}(\boldsymbol{\theta}_0)(\widehat{\boldsymbol{\theta}}-\boldsymbol{\theta}_0)^T + n\, o\left(\left\|\widehat{\boldsymbol{\theta}}-\boldsymbol{\theta}_0\right\|^2\right).$$

Then, $T_n^{\phi}(\widehat{\boldsymbol{\theta}},\boldsymbol{\theta}_0)$ has the same asymptotic distribution as the quadratic form

$$\boldsymbol{X}^T\boldsymbol{X} \equiv \left(\mathcal{I}_{\mathcal{F}}(\boldsymbol{\theta})^{1/2}\sqrt{n}(\widehat{\boldsymbol{\theta}}-\boldsymbol{\theta}_0)\right)^T\left(\mathcal{I}_{\mathcal{F}}(\boldsymbol{\theta})^{1/2}\sqrt{n}(\widehat{\boldsymbol{\theta}}-\boldsymbol{\theta}_0)\right).$$

On the other hand

$$\boldsymbol{X} \equiv \mathcal{I}_{\mathcal{F}}(\boldsymbol{\theta}_0)^{1/2}\sqrt{n}(\widehat{\boldsymbol{\theta}}-\boldsymbol{\theta}_0) \xrightarrow[n\to\infty]{L} N\left(\mathcal{I}_{\mathcal{F}}(\boldsymbol{\theta}_0)^{1/2}\mathbf{d}, \boldsymbol{I}_{M_0\times M_0}\right).$$

Therefore

$$\left(\mathcal{I}_{\mathcal{F}}(\boldsymbol{\theta}_0)^{1/2}\sqrt{n}(\widehat{\boldsymbol{\theta}}-\boldsymbol{\theta}_0)\right)^T\left(\mathcal{I}_{\mathcal{F}}(\boldsymbol{\theta}_0)^{1/2}\sqrt{n}(\widehat{\boldsymbol{\theta}}-\boldsymbol{\theta}_0)\right) \xrightarrow[n\to\infty]{L} \chi^2_{M_0}(\delta),$$

being $\delta = \mathbf{d}^T\,\mathcal{I}_{\mathcal{F}}(\boldsymbol{\theta}_0)\,\mathbf{d}$.

Remark 9.3

Using Theorem 9.3, we get a second approximation to the power function, at $\boldsymbol{\theta}_n = \boldsymbol{\theta}_0 + n^{-1/2}\mathbf{d}$, *by means of*

$$\beta_n^2(\boldsymbol{\theta}_n) = 1 - G_{\chi^2_{M_0}(\delta)}\left(\chi^2_{M_0,\alpha}\right),$$

where $G_{\chi^2_{M_0}(\delta)}$ *is the distribution function of a noncentral chi-square random variable with* M_0 *degrees of freedom and noncentrality parameter* $\delta = \mathbf{d}^T\mathcal{I}_{\mathcal{F}}(\boldsymbol{\theta}_0)\,\mathbf{d}$. *If we want to approximate the power at the alternative* $\boldsymbol{\theta} \neq \boldsymbol{\theta}_0$, *then we can take* $\mathbf{d} = \mathbf{d}(n,\boldsymbol{\theta},\boldsymbol{\theta}_0) = \sqrt{n}(\boldsymbol{\theta}-\boldsymbol{\theta}_0)$. *We can observe that this approximation is independent of* ϕ.

The same result can be obtained for the test statistics W_n^0, L_n^0 and R_n^0 (see, e.g., Sen and Singer (1993) or Serfling (1980)).

Example 9.1

Let $Y_1, ..., Y_n$ *be a random sample from a normal population with mean* μ *and variance* σ^2. *Consider the parameter* $\boldsymbol{\theta} = (\mu,\sigma)$. *Find the expression of the test statistics given in (9.1), (9.2) and (9.3), as well as the expression of Rényi test statistic, for testing*

$$H_0 : \boldsymbol{\theta} = \boldsymbol{\theta}_0 = (\mu_0,\sigma_0) \qquad \textit{versus} \qquad H_1 : \boldsymbol{\theta} \neq \boldsymbol{\theta}_0.$$

It is well known that the maximum likelihood estimators of μ *and* σ *are given by*

$$\widehat{\mu} = \overline{Y} \quad \textit{and} \quad \widehat{\sigma} = \sqrt{\frac{1}{n}\sum_{i=1}^{n}(Y_i-\overline{Y})^2}.$$

The likelihood ratio test statistic becomes

$$L_n^0 = 2\left(\log\left(\frac{1}{\widehat{\sigma}^n\left(\sqrt{2\pi}\right)^n}\exp(-\frac{n}{2})\right) - \log\left(\frac{1}{\sigma_0^n\left(\sqrt{2\pi}\right)^n}\exp\left(-\frac{n}{2}\frac{\widetilde{\sigma}^2}{\sigma_0^2}\right)\right)\right),$$

where

$$\widetilde{\sigma} = \sqrt{\frac{1}{n}\sum_{i=1}^{n}\left(Y_i - \mu_0\right)^2},$$

i.e.,

$$L_n^0 = 2n\log\frac{\sigma_0}{\widehat{\sigma}} + n\left(\frac{\widetilde{\sigma}^2}{\sigma_0^2} - 1\right).$$

The Fisher information matrix is

$$\mathcal{I}_{\mathcal{F}}\left(\mu,\sigma\right) = \begin{pmatrix} \frac{1}{\sigma^2} & 0 \\ 0 & \frac{2}{\sigma^2} \end{pmatrix}.$$

Therefore, the Wald test statistic has the expression

$$\begin{aligned} W_n^0 &= n\left(\overline{Y} - \mu_0, \widehat{\sigma} - \sigma_0\right)\begin{pmatrix} \frac{1}{\widehat{\sigma}^2} & 0 \\ 0 & \frac{2}{\widehat{\sigma}^2} \end{pmatrix}\begin{pmatrix} \overline{Y} - \mu_0 \\ \widehat{\sigma} - \sigma_0 \end{pmatrix} \\ &= n\left(\overline{Y} - \mu_0\right)^2/\widehat{\sigma}^2 + 2n\left(\widehat{\sigma} - \sigma_0\right)^2/\widehat{\sigma}^2. \end{aligned}$$

Concerning the Rao test statistic, the score vector is

$$\begin{aligned} \boldsymbol{U}_n\left(\mu_0,\sigma_0\right) &= \left(\sum_{i=1}^{n}\frac{\partial\log f_{\mu,\sigma}\left(Y_i\right)}{\partial\mu}, \sum_{i=1}^{n}\frac{\partial\log f_{\mu,\sigma}\left(Y_i\right)}{\partial\sigma}\right)^T_{(\mu,\sigma)=(\mu_0,\sigma_0)} \\ &= \left(n\left(\frac{\overline{Y} - \mu_0}{\sigma_0^2}\right), n\left(\frac{\widetilde{\sigma}^2 - \sigma_0^2}{\sigma_0^3}\right)\right)^T \end{aligned}$$

and then

$$R_n^0 = n\frac{\left(\overline{Y} - \mu_0\right)^2}{\sigma_0^2} + \frac{n}{2}\left(\frac{\widetilde{\sigma}^2 - \sigma_0^2}{\sigma_0^2}\right)^2.$$

Finally, we obtain the expression of Rényi test statistic.

The Rényi divergence may be regarded as a (h,ϕ)-divergence; that is,

$$D_r^1(\widehat{\boldsymbol{\theta}},\boldsymbol{\theta}_0) = h\left(D_\phi(\widehat{\boldsymbol{\theta}},\boldsymbol{\theta}_0)\right),$$

with

$$h\left(x\right) = \frac{1}{r\left(r-1\right)}\log\left(r\left(r-1\right)x + 1\right) \tag{9.7}$$

and

$$\phi\left(x\right) = \frac{1}{r\left(r-1\right)}\left(x^r - r\left(x-1\right) - 1\right). \tag{9.8}$$

Rényi test statistic has the expression

$$T_n^r(\widehat{\boldsymbol{\theta}},\boldsymbol{\theta}_0)=\frac{2n}{h'(0)\,\phi''(1)}D_r^1(\widehat{\boldsymbol{\theta}},\boldsymbol{\theta}_0).$$

It is easy to see that $h'(0)=1$ *and* $\phi''(1)=1$. *Therefore,*

$$T_n^r(\widehat{\boldsymbol{\theta}},\boldsymbol{\theta}_0)=\frac{2n}{r(r-1)}\log\int_{\mathcal{X}} f_{\widehat{\boldsymbol{\theta}}}(x)^r\, f_{\boldsymbol{\theta}_0}(x)^{1-r}\,d\mu(x)\,,\;\; r\neq 1,0.$$

In Chapter 1 it was established that Rényi's divergence measure between two normal populations is

$$D_r^1\left((\mu,\sigma),(\mu_0,\sigma_0)\right)=\frac{1}{2}\frac{\left(\mu-\mu_0\right)^2}{r\sigma_0^2+(1-r)\,\sigma^2}-\frac{1}{2r\,(r-1)}\log\frac{r\sigma_0^2+(1-r)\,\sigma^2}{\left(\sigma^2\right)^{1-r}\left(\sigma_0^2\right)^r}. \tag{9.9}$$

Using this formula, we get

$$T_n^r(\widehat{\boldsymbol{\theta}},\boldsymbol{\theta}_0)=n\left(\frac{\left(\overline{Y}-\mu_0\right)^2}{r\sigma_0^2+(1-r)\,\widehat{\sigma}^2}-\frac{1}{r\,(r-1)}\log\frac{r\sigma_0^2+(1-r)\,\widehat{\sigma}^2}{\left(\widehat{\sigma}^2\right)^{1-r}\left(\sigma_0^2\right)^r}\right),$$

where $\widehat{\boldsymbol{\theta}}=(\overline{Y},\widehat{\sigma}^2)$ *and* $\boldsymbol{\theta}_0=(\mu_0,\sigma_0)$.

When $r\to 1$, *we get the test statistic based on Kullback-Leibler divergence,*

$$\lim_{r\to 1}T_n^r(\widehat{\boldsymbol{\theta}},\boldsymbol{\theta}_0)=2nD_{Kull}(\widehat{\boldsymbol{\theta}},\boldsymbol{\theta}_0)=2n\log\frac{\sigma_0}{\widehat{\sigma}}+n\left(\frac{\widehat{\sigma}^2}{\sigma_0^2}-1\right)\equiv T_n^{Kull}(\widehat{\boldsymbol{\theta}},\boldsymbol{\theta}_0).$$

Observe that in this case $T_n^1(\widehat{\boldsymbol{\theta}},\boldsymbol{\theta}_0)=T_n^{Kull}(\widehat{\boldsymbol{\theta}},\boldsymbol{\theta}_0)$ *coincides with* L_n^0. *The following Remark provides a sufficient condition for the equality of these test statistics.*

Remark 9.4

When the density $f_{\boldsymbol{\theta}}(x)$ *belongs to the exponential family, that is,*

$$f_{\boldsymbol{\theta}}(x)=q(\boldsymbol{\theta})\,t(x)\exp\left(\sum_{i=1}^{M_0}S_i(\boldsymbol{\theta})\,t_i(x)\right)\quad x\in\mathcal{X},$$

likelihood ratio test statistic coincides with the test statistic based on Kullback-Leibler divergence measure.

For the exponential family, likelihood ratio test statistic is given by

$$
\begin{aligned}
L_n^0 &= 2n\left(\lambda_n(\widehat{\boldsymbol{\theta}}) - \lambda_n(\boldsymbol{\theta}_0)\right) = 2\log\frac{f_{\widehat{\boldsymbol{\theta}}}(Y_1,...,Y_n)}{f_{\boldsymbol{\theta}_0}(Y_1,...,Y_n)} \\
&= 2\log\frac{q(\widehat{\boldsymbol{\theta}})^n \prod\limits_{i=1}^{n} t(Y_i)\exp\left(\sum\limits_{j=1}^{n}\sum\limits_{i=1}^{M_0} S_i(\widehat{\boldsymbol{\theta}}) t_i(Y_j)\right)}{q(\boldsymbol{\theta}_0)^n \prod\limits_{i=1}^{n} t(Y_i)\exp\left(\sum\limits_{j=1}^{n}\sum\limits_{i=1}^{M_0} S_i(\boldsymbol{\theta}_0) t_i(Y_j)\right)} \\
&= 2\left(\log\frac{q(\widehat{\boldsymbol{\theta}})^n}{q(\boldsymbol{\theta}_0)^n} + \sum_{j=1}^{n}\sum_{i=1}^{M_0}\left(S_i(\widehat{\boldsymbol{\theta}}) - S_i(\boldsymbol{\theta}_0)\right) t_i(Y_j)\right) \\
&= 2n\left(\log\frac{q(\widehat{\boldsymbol{\theta}})}{q(\boldsymbol{\theta}_0)} + \sum_{i=1}^{M_0}\left(S_i(\widehat{\boldsymbol{\theta}}) - S_i(\boldsymbol{\theta}_0)\right)\left(\frac{1}{n}\sum_{j=1}^{n} t_i(Y_j)\right)\right),
\end{aligned}
$$

while the test statistic based on Kullback-Leibler divergence takes the form

$$
T_n^{Kull}(\widehat{\boldsymbol{\theta}},\boldsymbol{\theta}_0) = 2n\log\frac{q(\widehat{\boldsymbol{\theta}})}{q(\boldsymbol{\theta}_0)} + 2n\sum_{i=1}^{M_0}\left(S_i(\widehat{\boldsymbol{\theta}}) - S_i(\boldsymbol{\theta}_0)\right) E_{\widehat{\boldsymbol{\theta}}}\left[t_i(X)\right].
$$

It is clear that the two statistics coincide asymptotically since

$$
\frac{1}{n}\sum_{j=1}^{n} t_i(Y_j) \underset{n\to\infty}{\overset{a.s.}{\longrightarrow}} E_{\boldsymbol{\theta}_0}\left[t_i(X)\right]
$$

and

$$
E_{\widehat{\boldsymbol{\theta}}}\left[t_i(X)\right] \underset{n\to\infty}{\overset{a.s.}{\longrightarrow}} E_{\boldsymbol{\theta}_0}\left[t_i(X)\right].
$$

The following question then arises: Is it possible to find assumptions under which the two exact tests coincide? Thus, we need to find the conditions under which

$$
E_{\widehat{\boldsymbol{\theta}}}\left[t_i(X)\right] = \frac{1}{n}\sum_{j=1}^{n} t_i(Y_j).
$$

The likelihood function is

$$
L(\boldsymbol{\theta},Y_1,...,Y_n) = q(\boldsymbol{\theta})^n\prod_{i=1}^{n} t(Y_i)\exp\left(\sum_{i=1}^{M_0} S_i(\boldsymbol{\theta})\sum_{j=1}^{n} t_i(Y_j)\right),
$$

and the loglikelihood function is

$$
\log L(\boldsymbol{\theta},Y_1,...,Y_n) = n\log q(\boldsymbol{\theta}) + \log\prod_{i=1}^{n} t(Y_i) + \sum_{i=1}^{M_0} S_i(\boldsymbol{\theta})\sum_{j=1}^{n} t_i(Y_j).
$$

Therefore, the maximum likelihood estimator, $\widehat{\boldsymbol{\theta}}$, is a solution of the system of equation

$$
\begin{cases}
\dfrac{\partial\log L(\widehat{\boldsymbol{\theta}},Y_1,...,Y_n)}{\partial\theta_j} = \dfrac{n}{q(\boldsymbol{\theta})}\dfrac{\partial q(\widehat{\boldsymbol{\theta}})}{\partial\theta_j} + \sum\limits_{i=1}^{M_0}\dfrac{\partial S_i(\widehat{\boldsymbol{\theta}})}{\partial\theta_j}\sum\limits_{k=1}^{n} t_i(Y_k) = 0, \\
j = 1,...,M_0.
\end{cases}
$$

On the other hand, we know that

$$\int_{\mathcal{X}} q(\boldsymbol{\theta})\, t(x) \exp\left(\sum_{i=1}^{M_0} S_i(\boldsymbol{\theta})\, t_i(x)\right) d\mu(x) = 1.$$

Taking derivatives with respect to θ_j in both sides we get

$$\begin{aligned}&\int_{\mathcal{X}} \frac{\partial q(\boldsymbol{\theta})}{\partial\theta_j} t(x) \exp\left(\sum_{i=1}^{M_0} S_i(\boldsymbol{\theta})\, t_i(x)\right) d\mu(x)\\ &+\int_{\mathcal{X}} q(\boldsymbol{\theta})\, t(x) \exp\left(\sum_{i=1}^{M_0} S_i(\boldsymbol{\theta})\, t_i(x)\right) \sum_{i=1}^{M_0} \frac{\partial S_i(\boldsymbol{\theta})}{\partial\theta_j} t_i(x)\, d\mu(x) = 0,\end{aligned}$$

which is equal to

$$\begin{cases} \dfrac{n}{q(\boldsymbol{\theta})}\dfrac{\partial q(\boldsymbol{\theta})}{\partial\theta_j} + n\sum\limits_{i=1}^{M_0} \dfrac{\partial S_i(\boldsymbol{\theta})}{\partial\theta_j} E_{\boldsymbol{\theta}}\left[t_i(X)\right] = 0, \\ j = 1, ..., M_0. \end{cases}$$

Thus, for the exponential family it holds

$$\begin{cases} \sum\limits_{i=1}^{M_0} \dfrac{\partial S_i(\widehat{\boldsymbol{\theta}})}{\partial\theta_j} \dfrac{1}{n}\sum\limits_{k=1}^{n} t_i(Y_k) = \sum\limits_{i=1}^{M_0} \dfrac{\partial S_i(\widehat{\boldsymbol{\theta}})}{\partial\theta_j} E_{\widehat{\boldsymbol{\theta}}}\left[t_i(X)\right], \\ j = 1, ..., M_0, \end{cases}$$

and this can be written as a homogeneous system

$$\begin{cases} \sum\limits_{i=1}^{M_0} a_{ij} z_i = 0 \\ j = 1, ..., M_0 \end{cases}$$

with

$$a_{ij} = \frac{\partial S_i(\widehat{\boldsymbol{\theta}})}{\partial\theta_j} \text{ and } z_i = \frac{1}{n}\sum_{k=1}^{n} t_i(Y_k) - E_{\widehat{\boldsymbol{\theta}}}\left[t_i(X)\right].$$

This system has the unique solution $z_i = 0$, $i = 1, ..., M_0$, if and only if the determinant of the matrix

$$\boldsymbol{A} = (a_{ij})_{i,j=1,...,M_0}$$

is different from zero.

Therefore, the exact likelihood ratio test statistic coincides with the exact test based on the Kullback-Leibler divergence measure if and only if the determinant of the matrix

$$\left(\frac{\partial S_i(\widehat{\boldsymbol{\theta}})}{\partial\theta_j}\right)_{i,j=1,...,M_0}$$

is different from zero. This condition means that the equality is obtained when the expo nential family is not overparametrized.

Buse (1982), by means of simple diagrams, gave an intuitive meaning of the likelihoo ratio test statistic, Rao test statistic and Wald test statistic. Interesting survey article about these test procedures were given by Breusch and Pagan (1980) and Engle (1981)

9.2.1. Confidence Regions

We can easily construct families of confidence regions with a prescribed asymptoti confidence coefficient $1-\alpha$ for $\boldsymbol{\theta} \in \Theta \subset \mathbb{R}^{M_0}$. We consider the problem of testing H_0 $\boldsymbol{\theta} = \boldsymbol{\theta}_0$ based on the ϕ-divergence test statistic $T_n^{\phi}(\widehat{\boldsymbol{\theta}}, \boldsymbol{\theta}_0)$. Let $A_\phi(\boldsymbol{\theta}_0)$, $\boldsymbol{\theta}_0 \in \Theta \subset \mathbb{R}^{M_0}$ the region of acceptance associated with the test statistic "reject the null hypothesis H_0 with significance level α, if $T_n^{\phi}(\widehat{\boldsymbol{\theta}}, \boldsymbol{\theta}_0) > \chi^2_{M_0,\alpha}$", i.e.,

$$A_\phi(\boldsymbol{\theta}_0) = \left\{ (y_1, ..., y_n) : T_n^{\phi}(\widehat{\boldsymbol{\theta}}, \boldsymbol{\theta}_0) < \chi^2_{M_0,\alpha} \right\}.$$

For each observation $(y_1, ..., y_n)$ let $S_\phi(y_1, ..., y_n)$ denote the set

$$S_\phi(y_1, ..., y_n) = \{ \boldsymbol{\theta} : (y_1, ..., y_n) \in A_\phi(\boldsymbol{\theta}), \ \boldsymbol{\theta} \in \Theta \}.$$

Then $S_\phi(y_1, ..., y_n)$ is a family of confidence regions for $\boldsymbol{\theta} \in \Theta \subset \mathbb{R}^{M_0}$ with asymptoti confidence coefficient $1-\alpha$.

9.3. Composite Null Hypothesis

We assume that the statistical model $(\mathcal{X}, \beta_{\mathcal{X}}, P_{\boldsymbol{\theta}})_{\boldsymbol{\theta} \in \Theta}$, satisfies the standard regu larity assumptions *i)-v)* given in Section 2 of Chapter 2 (see also Serfling (1980)). As t the composite null hypothesis

$$H_0 : \boldsymbol{\theta} \in \Theta_0 \subset \Theta, \tag{9.10}$$

we assume the following:

(H1) Θ_0 is a subset of $\mathbb{R}^{M_0}$, and there exist $1 \leq d_0 \leq M_0$, an open subset $B \subset \mathbb{R}^{M_0-d}$ and mappings

$$\boldsymbol{g} : \Theta \longrightarrow \mathbb{R}^{d_0} \qquad \text{and} \qquad \boldsymbol{h} : B \longrightarrow \Theta$$

such that $\Theta_0 = \{ \boldsymbol{h}(\boldsymbol{\beta}) : \boldsymbol{\beta} \in B \}$ and $\boldsymbol{g}(\boldsymbol{\theta}) = \mathbf{0}$ on Θ_0.

(H2) The $M_0 \times d_0$ matrix

$$\boldsymbol{B}(\boldsymbol{\theta}) = \left(\frac{\partial g_j(\boldsymbol{\theta})}{\partial \theta_s} \right)_{\substack{j=1,...,d_0 \\ s=1,...,M_0}}$$

exists and is of rank d_0 for all $\boldsymbol{\theta} \in \Theta_0$, with all elements continuous on Θ_0.

H3) The $M_0 \times (M_0 - d_0)$ matrix

$$\boldsymbol{M}_{\boldsymbol{\beta}} = \left(\frac{\partial h_j(\boldsymbol{\theta})}{\partial \beta_s} \right)_{\substack{j=1,\ldots,M_0 \\ s=1,\ldots,M_0-d_0}}$$

exists and is of rank $M_0 - d_0$ for all $\boldsymbol{\beta} \in B$, with all elements continuous on B.

H4) The statistical submodel

$$\left((\mathcal{X},\ \beta_{\mathcal{X}}) , \left\{ p_{\boldsymbol{\beta}} = f_{\boldsymbol{h}(\boldsymbol{\beta})} : \boldsymbol{\beta} \in B \right\}, \mu \right)$$

satisfies regularity conditions *i)-v)*.

In order to solve the testing problem

$$H_0 : \boldsymbol{\theta} \in \Theta_0 \subset \Theta \text{ versus } H_1 : \boldsymbol{\theta} \in \Theta - \Theta_0$$

ve shall consider the test statistics given in (9.1), (9.2), (9.3) and (9.4), but adapted to urrent context:

- *Wald test statistic*

$$W_n = n\boldsymbol{g}(\widehat{\boldsymbol{\theta}}) \left(\boldsymbol{B}(\widehat{\boldsymbol{\theta}}) \mathcal{I}_{\mathcal{F}}(\widehat{\boldsymbol{\theta}})^{-1} \boldsymbol{B}^T(\widehat{\boldsymbol{\theta}}) \right)^{-1} \boldsymbol{g}(\widehat{\boldsymbol{\theta}})^T, \tag{9.11}$$

where $\widehat{\boldsymbol{\theta}}$ is the maximum likelihood estimator of $\boldsymbol{\theta}$ in Θ, $\mathcal{I}_{\mathcal{F}}(\widehat{\boldsymbol{\theta}})^{-1}$ denotes the inverse of the Fisher information matrix, $\boldsymbol{g}(\boldsymbol{\theta})$, the mapping defined in (H1) and $\boldsymbol{B}(\boldsymbol{\theta})$ the matrix in (H2).

- *Likelihood ratio test statistic*

$$L_n = 2n \left(\lambda_n(\widehat{\boldsymbol{\theta}}) - \lambda_n(\widetilde{\boldsymbol{\theta}}) \right) \tag{9.12}$$

where $\widetilde{\boldsymbol{\theta}} = \boldsymbol{h}(\widehat{\boldsymbol{\beta}})$ is the maximum likelihood estimator restricted to the null hypothesis Θ_0, and

$$\lambda_n(\boldsymbol{\theta}) = \frac{1}{n} \sum_{i=1}^{n} \log f_{\boldsymbol{\theta}}(Y_i).$$

- *Rao test statistic*

$$R_n = \frac{1}{n} \boldsymbol{U}_n(\widetilde{\boldsymbol{\theta}}) \mathcal{I}_{\mathcal{F}}(\widetilde{\boldsymbol{\theta}})^{-1} \boldsymbol{U}_n(\widetilde{\boldsymbol{\theta}})^T, \tag{9.13}$$

where

$$\boldsymbol{U}_n(\widetilde{\boldsymbol{\theta}}) = \left(\sum_{i=1}^{n} \frac{\partial \log f_{\boldsymbol{\theta}}(Y_i)}{\partial \theta_1}, \ldots, \sum_{i=1}^{n} \frac{\partial \log f_{\boldsymbol{\theta}}(Y_i)}{\partial \theta_M} \right)_{\boldsymbol{\theta} = \widetilde{\boldsymbol{\theta}}}.$$

For testing the special composite null hypotheses $H \equiv \Theta_0 = \Theta_1 \times \{\boldsymbol{\theta}_{20}\}$ and $H^* \equiv \Theta_0^* = \{\boldsymbol{\theta}_{10}\} \times \Theta_2$ in models with $\Theta = \Theta_1 \times \Theta_2$, Salicrú et *al.* (1994) proposed the ϕ-divergence test statistics

$$T_n^{\phi}\left((\widehat{\boldsymbol{\theta}}_1, \widehat{\boldsymbol{\theta}}_2), (\widehat{\boldsymbol{\theta}}_1, \boldsymbol{\theta}_{20})\right) = \frac{2n}{\phi''(1)} D_{\phi}\left((\widehat{\boldsymbol{\theta}}_1, \widehat{\boldsymbol{\theta}}_2), (\widehat{\boldsymbol{\theta}}_1, \boldsymbol{\theta}_{20})\right),$$

$$T_n^{\phi}\left((\widehat{\boldsymbol{\theta}}_1, \widehat{\boldsymbol{\theta}}_2), (\boldsymbol{\theta}_{10}, \widehat{\boldsymbol{\theta}}_2)\right) = \frac{2n}{\phi''(1)} D_{\phi}\left((\widehat{\boldsymbol{\theta}}_1, \widehat{\boldsymbol{\theta}}_2), (\boldsymbol{\theta}_{10}, \widehat{\boldsymbol{\theta}}_2)\right)$$

using the maximum likelihood estimator $\widehat{\boldsymbol{\theta}} = (\widehat{\boldsymbol{\theta}}_1, \widehat{\boldsymbol{\theta}}_2)$.

Later Morales et *al.* (1997), using the ϕ-divergence test statistics, studied the problem under any hypothesis H with the properties (H1) – (H4). They dealt with the following family of test statistics:

· *ϕ-divergence test statistic*

$$T_n^{\phi}(\widehat{\boldsymbol{\theta}}, \widetilde{\boldsymbol{\theta}}) \equiv \frac{2n}{\phi''(1)} D_{\phi}(\widehat{\boldsymbol{\theta}}, \widetilde{\boldsymbol{\theta}}), \tag{9.14}$$

where $\widehat{\boldsymbol{\theta}}$ is the maximum likelihood estimator of $\boldsymbol{\theta}$ in Θ and $\widetilde{\boldsymbol{\theta}}$ is the maximum likelihood estimator in the null hypothesis Θ_0.

Theorem 9.4

Let the model and ϕ satisfy the assumptions i) – v) considered in Section 2 of Chapter 2 and $(\Phi 1), (\Phi 2)$. Then, under any hypothesis H with properties (H1) – (H4), the asymptotic distribution of the test statistics given in (9.11), (9.12), (9.13) and (9.14) is chi-square with d_0 degrees of freedom.

The proof of this theorem for the ϕ-divergence test statistic given in (9.14) was established by Morales et *al.* (1997). For the rest of test statistics, see Serfling (1980) or Sen and Singer (1993). For the test statistic based on the (h, ϕ)-divergence measures $T_n^{\phi,h}(\widehat{\boldsymbol{\theta}}, \widetilde{\boldsymbol{\theta}})$, we have a similar result.

For composite null hypothesis a result analogous to Remark 9.4 can be established.

Consider the null hypothesis $H_0 : \boldsymbol{\theta} \in \Theta_0 \subset \Theta$ and assume that conditions (H1) – (H4) hold. By Theorem 9.4, the null hypothesis should be rejected if $T_n^{\phi}(\widehat{\boldsymbol{\theta}}, \widetilde{\boldsymbol{\theta}}) \geq \chi^2_{d_0,\alpha}$. The following theorem can be used to approximate the power function. Assume that $\boldsymbol{\theta}^* \notin \Theta_0$ is the true value of the parameter so that $\widehat{\boldsymbol{\theta}} \xrightarrow[n\to\infty]{a.s.} \boldsymbol{\theta}^*$ and that there exists $\boldsymbol{\theta}_0 \in \Theta_0$ such that the restricted maximum likelihood estimator satisfies $\widetilde{\boldsymbol{\theta}} \xrightarrow[n\to\infty]{a.s.} \boldsymbol{\theta}_0$. Then, it holds

$$\sqrt{n}\left((\widehat{\boldsymbol{\theta}}, \widetilde{\boldsymbol{\theta}}) - (\boldsymbol{\theta}^*, \boldsymbol{\theta}_0)\right) \xrightarrow[n\to\infty]{L} N\left(\begin{pmatrix} \mathbf{0} \\ \mathbf{0} \end{pmatrix}, \begin{pmatrix} \mathcal{I}_{\mathcal{F}}(\boldsymbol{\theta}^*)^{-1} & \mathbf{A}_{12} \\ \mathbf{A}_{21} & \mathbf{A}_{22} \end{pmatrix}\right),$$

where $\boldsymbol{A}_{12} = \boldsymbol{A}_{12}(\boldsymbol{\theta}^*, \boldsymbol{\theta}_0)$, $\boldsymbol{A}_{21} = \boldsymbol{A}_{12}^T$ and $\boldsymbol{A}_{22} = \boldsymbol{A}_{22}(\boldsymbol{\theta}^*, \boldsymbol{\theta}_0)$ are $d_0 \times d_0$ matrices. We have the following result.

Theorem 9.5

Let the model $\phi \in \Phi^*$ and Θ_0 satisfy the assumptions i)–v), (ϕ1)–(ϕ2) and (H1)–(H4) respectively. Then

$$\sqrt{n}\left(D_\phi(\widehat{\boldsymbol{\theta}}, \widetilde{\boldsymbol{\theta}}) - D_\phi(\boldsymbol{\theta}^*, \boldsymbol{\theta}_0)\right) \underset{n\to\infty}{\overset{L}{\longrightarrow}} N\left(0, \sigma_\phi^2(\boldsymbol{\theta}^*)\right),$$

where

$$\sigma_\phi^2(\boldsymbol{\theta}^*) = \boldsymbol{T}^T \mathcal{I}_{\mathcal{F}}(\boldsymbol{\theta}^*)^{-1} \boldsymbol{T} + \boldsymbol{T}^T \boldsymbol{A}_{12} \boldsymbol{S} + \boldsymbol{S}^T \boldsymbol{A}_{21} \boldsymbol{T} + \boldsymbol{S}^T \boldsymbol{A}_{22} \boldsymbol{S},$$

$$\boldsymbol{T} = \left(\frac{\partial D_\phi(\boldsymbol{\theta}_1, \boldsymbol{\theta}_0)}{\partial \boldsymbol{\theta}_1}\right)^T_{\boldsymbol{\theta}_1 = \boldsymbol{\theta}^*} \quad \textit{and } \boldsymbol{S} = \left(\frac{\partial D_\phi(\boldsymbol{\theta}^*, \boldsymbol{\theta}_2)}{\partial \boldsymbol{\theta}_2}\right)^T_{\boldsymbol{\theta}_2 = \boldsymbol{\theta}_0}.$$

Proof. The result follows straightforward by making a first order Taylor expansion of $D_\phi(\widehat{\boldsymbol{\theta}}, \widetilde{\boldsymbol{\theta}})$

$$D_\phi(\widehat{\boldsymbol{\theta}}, \widetilde{\boldsymbol{\theta}}) = D_\phi(\boldsymbol{\theta}^*, \boldsymbol{\theta}_0) + \boldsymbol{T}^T(\widehat{\boldsymbol{\theta}} - \boldsymbol{\theta}^*) + \boldsymbol{S}^T(\widetilde{\boldsymbol{\theta}} - \boldsymbol{\theta}_0) + o\left(\left\|\widehat{\boldsymbol{\theta}} - \boldsymbol{\theta}^*\right\| + \left\|\widetilde{\boldsymbol{\theta}} - \boldsymbol{\theta}_0\right\|\right).$$

Remark 9.5

On the basis of this theorem we can get an approximation of the power function at $\boldsymbol{\theta}^*$, $\beta_{n,\phi}(\boldsymbol{\theta}^*) = \Pr_{\boldsymbol{\theta}^*}\left(T_n^\phi(\widehat{\boldsymbol{\theta}}, \widetilde{\boldsymbol{\theta}}) > \chi^2_{d_0,\alpha}\right)$, in the following way:

$$\beta_{n,\phi}(\boldsymbol{\theta}^*) = 1 - \Phi\left(\frac{1}{\sigma_\phi(\boldsymbol{\theta}^*)}\left(\frac{\phi''(1)}{2\sqrt{n}}\chi^2_{d_0,\alpha} - \sqrt{n} D_\phi(\boldsymbol{\theta}^*, \boldsymbol{\theta}_0)\right)\right),$$

where $\Phi(x)$ is the standard normal distribution function.

We may also find an approximation of the power of $T_n^\phi(\widehat{\boldsymbol{\theta}}, \widetilde{\boldsymbol{\theta}})$ at an alternative close to the null hypothesis. Let $\boldsymbol{\theta}_n \in \Theta - \Theta_0$ be a given alternative and let $\boldsymbol{\theta}_0$ be the element in Θ_0 closest to $\boldsymbol{\theta}_n$ in the Euclidean distance sense. A first possibility to introduce contiguous alternative hypotheses is to consider a fixed $\boldsymbol{d} \in \mathbb{R}^{M_0}$ and to permit $\boldsymbol{\theta}_n$ moving towards $\boldsymbol{\theta}_0$ as n increases in the following way

$$H_{1,n} : \boldsymbol{\theta}_n = \boldsymbol{\theta}_0 + n^{-1/2}\boldsymbol{d}.$$

A second approach is to relax the condition $\boldsymbol{g}(\boldsymbol{\theta}) = \boldsymbol{0}$ defining Θ_0. Let $\delta \in \mathbb{R}^{M_0}$ and consider the following sequence, $\boldsymbol{\theta}_n$, of parameters moving towards $\boldsymbol{\theta}_0$ according to

$$H_{1,n}^* : \boldsymbol{g}(\boldsymbol{\theta}_n) = n^{-1/2}\delta.$$

Note that a Taylor series expansion of $\boldsymbol{g}(\boldsymbol{\theta}_n)$ around $\boldsymbol{\theta}_0$ yields

$$\boldsymbol{g}(\boldsymbol{\theta}_n) = \boldsymbol{g}(\boldsymbol{\theta}_0) + \boldsymbol{B}(\boldsymbol{\theta}_0)^T(\boldsymbol{\theta}_n - \boldsymbol{\theta}_0) + o(\|\boldsymbol{\theta}_n - \boldsymbol{\theta}_0\|). \tag{9.15}$$

By substituting $\boldsymbol{\theta}_n = \boldsymbol{\theta}_0 + n^{-1/2}\boldsymbol{d}$ in (9.15) and taking into account that $\boldsymbol{g}(\boldsymbol{\theta}_0) = \boldsymbol{0}$, we get

$$\boldsymbol{g}(\boldsymbol{\theta}_n) = n^{-1/2}\boldsymbol{B}(\boldsymbol{\theta}_0)^T \boldsymbol{d} + o(\|\boldsymbol{\theta}_n - \boldsymbol{\theta}_0\|),$$

so that the equivalence in the limit is obtained for $\delta = \boldsymbol{B}(\boldsymbol{\theta}_0)^T \boldsymbol{d}$. In Morales and Pardo (2001), the following result was established:

Theorem 9.6

Under some regularity conditions, the asymptotic distribution of $T_n^{\phi}(\widehat{\boldsymbol{\theta}}, \widetilde{\boldsymbol{\theta}})$ under $H_{1,n}^*$ is noncentral chi-square with d_0 degrees of freedom and noncentrality parameter $\delta = \boldsymbol{d}^T \mathcal{I}_{\mathcal{F}}(\boldsymbol{\theta}_0)^{-1} \boldsymbol{d}$ and noncentral chi-square with d_0 degrees of freedom and noncentrality parameter $\delta^* = \boldsymbol{d}^T \left(\boldsymbol{B}(\boldsymbol{\theta}_0) \mathcal{I}_{\mathcal{F}}(\boldsymbol{\theta}_0)^{-1} \boldsymbol{B}^T(\boldsymbol{\theta}_0)\right)^{-1} \boldsymbol{d}$ under $H_{1,n}$.

Davidson and Lever (1970) obtained the same result for the test statistic L_n.

Example 9.2 *(Morales et al. 1997)*

Let $Y_1, ..., Y_n$ be a random sample from a normal population with mean μ and variance σ^2, $\boldsymbol{\theta} = (\mu, \sigma) \in \Theta = (-\infty, \infty) \times (0, \infty)$. We shall test composite hypotheses

$$H_0 : \sigma = \mu/3 \qquad \text{versus} \qquad H_1 : \sigma \neq \mu/3.$$

The maximum likelihood estimators of μ and σ are $\widehat{\mu} = \overline{Y} = \frac{1}{n}\sum_{i=1}^n Y_i$ and $\widehat{\sigma} = (\frac{1}{n}\sum_{i=1}^n (Y_i - \overline{Y})^2)^{1/2}$, respectively. In this case,

$$\Theta_0 = \{(\mu, \sigma) \in \Theta : \sigma = \mu/3\}.$$

Now we are going to calculate the maximum likelihood estimator, $\widetilde{\boldsymbol{\theta}}$, of $\boldsymbol{\theta}$ in Θ_0. The density function is

$$f_{N(\mu,\sigma=\mu/3)}(x) = \frac{3}{\mu(2\pi)^{1/2}} \exp\left\{-\frac{1}{2}\left(\frac{x-\mu}{\mu/3}\right)^2\right\} = \frac{1}{\mu}\frac{3}{(2\pi)^{1/2}} \exp\left\{-\frac{9}{2}\left(\frac{x}{\mu} - 1\right)^2\right\}.$$

Then, the likelihood function is

$$L(\mu; Y_1, ..., Y_n) = \left(\frac{3}{(2\pi)^{1/2}}\right)^n \frac{1}{\mu^n} \exp\left\{-\frac{9}{2}\left(\sum_{i=1}^n \frac{Y_i^2}{\mu^2} + n - \frac{2}{\mu}\sum_{i=1}^n Y_i\right)\right\}$$

and the loglikelihood function

$$\log L(\mu; Y_1, ..., Y_n) = n\left(-\log\mu + \log\frac{3}{\sqrt{2\pi}}\right) - \frac{9}{2}\left(\sum_{i=1}^n \frac{Y_i^2}{\mu^2} + n - \frac{2}{\mu}\sum_{i=1}^n Y_i\right).$$

Therefore

$$\frac{\partial \log L(\mu; Y_1, ..., Y_n)}{\partial\mu} = \frac{1}{\mu}\left(-n + \frac{9}{\mu^2}\sum_{i=1}^n Y_i^2 - \frac{9}{\mu}\sum_{i=1}^n Y_i\right) = 0$$

and $\widetilde{\mu}$ is a solution of the equation

$$-n\mu^2+9\sum_{i=1}^{n}Y_i^2-9\mu\sum_{i=1}^{n}Y_i=0\Leftrightarrow \mu^2+9\overline{Y}\mu-9\frac{1}{n}\sum_{i=1}^{n}Y_i^2=0.$$

This solution is

$$\mu=\frac{1}{2}\left(-9\overline{Y}\pm\sqrt{(9\overline{Y})^2+4\times 9\frac{1}{n}\sum_{i=1}^{n}Y_i^2}\right)$$

and the maximum likelihood estimator of $\boldsymbol{\theta}$ in Θ_0 is $\widetilde{\boldsymbol{\theta}}=(\widetilde{\mu},\widetilde{\mu}/3)$, where

$$\widetilde{\mu}=\frac{3}{2}\left(-3\overline{Y}+\sqrt{13\overline{Y}^2+4\widehat{\sigma}^2}\right).$$

Now, taking into account the expression of Rényi divergence for two normal populations given in (9.9), we get the expression of the Rényi test statistic

$$\begin{aligned} T_n^r(\widehat{\boldsymbol{\theta}},\widetilde{\boldsymbol{\theta}}) &= \frac{2n}{\phi''(1)\,h'(0)}D_r^1((\widehat{\mu},\widehat{\sigma}),(\widetilde{\mu},\tfrac{\widetilde{\mu}}{3}))\\ &= n\left(\frac{1}{r(r-1)}\log\frac{(\widehat{\sigma}^2)^{1-r}(\widetilde{\mu}/3)^{2r}}{r(\widetilde{\mu}/3)^2+(1-r)\widehat{\sigma}^2}+\frac{(\widehat{\mu}-\widetilde{\mu})^2}{(r(\widetilde{\mu}/3)^2+(1-r)\widehat{\sigma}^2)}\right). \end{aligned}$$

The likelihood ratio test statistic is given by

$$L_n=T_n^{Kull}(\widehat{\boldsymbol{\theta}},\widetilde{\boldsymbol{\theta}})=\lim_{r\to 1}D_r^1(\widehat{\boldsymbol{\theta}},\widetilde{\boldsymbol{\theta}})=n\left(\frac{(\widehat{\mu}-\widetilde{\mu})^2}{\widetilde{\mu}^2}9+9\frac{\widehat{\sigma}^2}{\widetilde{\mu}^2}-1+\log\frac{\widetilde{\mu}^2}{9\widehat{\sigma}^2}\right)$$

and for $r\to 0$, we have

$$T_n^0(\widehat{\boldsymbol{\theta}},\widetilde{\boldsymbol{\theta}})=\lim_{r\to 0}T_n^r(\widehat{\boldsymbol{\theta}},\widetilde{\boldsymbol{\theta}})=n\left(\frac{(\widehat{\mu}-\widetilde{\mu})^2}{\widehat{\sigma}^2}+\frac{\widetilde{\mu}^2}{9}\frac{1}{\widehat{\sigma}^2}-1+\log\frac{9\widehat{\sigma}^2}{\widetilde{\mu}^2}\right).$$

Now we consider the test statistics W_n and R_n. For the Wald test statistic W_n, we need the Fisher information matrix, the function $\boldsymbol{g}$ and the matrix $\boldsymbol{B}(\boldsymbol{\theta})$. The estimated Fisher information matrix is

$$\mathcal{I}_{\mathcal{F}}(\widehat{\mu},\widehat{\sigma})=\begin{pmatrix}\widehat{\sigma}^{-2} & 0\\ 0 & 2\widehat{\sigma}^{-2}\end{pmatrix},$$

the function $\boldsymbol{g}$ and the matrix $\boldsymbol{B}(\boldsymbol{\theta})$ are $\boldsymbol{g}(\widehat{\boldsymbol{\theta}})=\boldsymbol{g}(\widehat{\mu},\widehat{\sigma})=\widehat{\mu}-3\widehat{\sigma}$ and $\boldsymbol{B}(\widehat{\boldsymbol{\theta}})=(1,-3)$, respectively.

Then, the Wald test statistic is

$$W_n= n\boldsymbol{g}(\widehat{\boldsymbol{\theta}})\left(\boldsymbol{B}(\widehat{\boldsymbol{\theta}})\mathcal{I}_{\mathcal{F}}(\widehat{\boldsymbol{\theta}})^{-1}\boldsymbol{B}^T(\widehat{\boldsymbol{\theta}})\right)^{-1}\boldsymbol{g}(\widehat{\boldsymbol{\theta}})^T=n\frac{2(\widehat{\mu}-3\widehat{\sigma})^2}{11\widehat{\sigma}^2}.$$

Regarding Rao test statistic, it is necessary to calculate

$$\boldsymbol{U}_n(\widetilde{\boldsymbol{\theta}}) = \left(\sum_{i=1}^{n} \frac{\partial \log f_{\mu,\sigma}(Y_i)}{\partial \mu}, \sum_{i=1}^{n} \frac{\partial \log f_{\mu,\sigma}(Y_i)}{\partial \sigma} \right)_{(\mu,\sigma)=(\widetilde{\mu},\widetilde{\mu}/3)}.$$

It holds

$$\left(\sum_{i=1}^{n} \frac{\partial \log f_{\mu,\sigma}(Y_i)}{\partial \mu} \right)_{(\widetilde{\mu},\widetilde{\mu}/3)} = n\frac{9}{\widetilde{\mu}^2}(\overline{Y} - \widetilde{\mu}),$$

and

$$\sum_{i=1}^{n} \left(\frac{\partial \log f_{\mu,\sigma}(Y_i)}{\partial \sigma} \right)_{(\widetilde{\mu},\widetilde{\mu}/3)} = n\left(-\frac{3}{\widetilde{\mu}} + \frac{27}{\widetilde{\mu}^3}\frac{1}{n}\sum_{i=1}^{n}(Y_i - \widetilde{\mu})^2 \right).$$

Therefore

$$\begin{aligned} R_n &= \tfrac{1}{n}\boldsymbol{U}_n(\widetilde{\boldsymbol{\theta}})^T \mathcal{I}_{\mathcal{F}}(\widetilde{\boldsymbol{\theta}})^{-1}\boldsymbol{U}_n(\widetilde{\boldsymbol{\theta}}) \\ &= n\left(\frac{9}{\widetilde{\mu}^2}(\overline{Y} - \widetilde{\mu})^2 + \frac{\widetilde{\mu}^2}{18}\left(-\frac{3}{\widetilde{\mu}} + \frac{27}{\widetilde{\mu}^3}\frac{1}{n}\sum_{i=1}^{n}(Y_i - \widetilde{\mu})^2 \right)^2 \right). \end{aligned}$$

In the following table we summarize the final expressions of the different test statistic presented here

Statistic	Expression
W_n	$n\dfrac{2(\widehat{\mu} - 3\widehat{\sigma})^2}{11\widehat{\sigma}^2}$
R_n	$n\left(\dfrac{9}{\widetilde{\mu}^2}(\overline{Y} - \widetilde{\mu})^2 + \dfrac{\widetilde{\mu}^2}{18}\left(-\dfrac{3}{\widetilde{\mu}} + \dfrac{27}{\widetilde{\mu}^3}\dfrac{1}{n}\sum_{i=1}^{n}(Y_i - \widetilde{\mu})^2 \right)^2 \right)$
$L_n \equiv T_n^{Kull}$	$n\left(\dfrac{(\widehat{\mu} - \widetilde{\mu})^2}{\widetilde{\mu}^2}9 + 9\dfrac{\widehat{\sigma}^2}{\widetilde{\mu}^2} - 1 + \log\dfrac{\widetilde{\mu}^2}{9\widehat{\sigma}^2} \right)$
T_n^0	$n\left(\dfrac{(\widehat{\mu} - \widetilde{\mu})^2}{\widehat{\sigma}^2} + \dfrac{\widetilde{\mu}^2}{9}\dfrac{1}{\widehat{\sigma}^2} - 1 + \log\dfrac{9\widehat{\sigma}^2}{\widetilde{\mu}^2} \right)$
T_n^r	$n\left(\dfrac{1}{r(r-1)}\log\dfrac{(\widehat{\sigma}^2)^{1-r}(\widetilde{\mu}/3)^{2r}}{r(\widetilde{\mu}/3)^2 + (1-r)\widehat{\sigma}^2} + \dfrac{(\widehat{\mu} - \widetilde{\mu})^2}{r(\widetilde{\mu}/3)^2 + (1-r)\widehat{\sigma}^2} \right).$

The formula for T_n^r *holds for* $r \neq 0$, $r \neq 1$ *when* $r(\widetilde{\mu}/3)^2 + (1-r)\widehat{\sigma}^2 > 0$. *When* $r(\widetilde{\mu}/3)^2 + (1-r)\widehat{\sigma}^2 < 0$, *then* $T_n^r = \infty$.

The null hypothesis, Θ_0, *is given by the line* $\sigma = \mu/3$. *Now we consider the perpendicular line to the null hypothesis across the point* $(0, c)$. *This line is given by* $\sigma = -3\mu + c$

The intersection of both lines is a point at the null hypothesis $P_5 = (3c/10, c/10)$. *We consider 10 points on the line* $\sigma = -3\mu + c$, *four of them under* P_5 *and six over* P_5, *namely*

$$P_j = \frac{c}{16}\left(\frac{101}{20}, \frac{17}{20}\right) + j\frac{c}{320}(-1,3),\ j = 1, 2, ..., 11.$$

For $c = 1/2$ *we have the following* 11 *points:*

$$\begin{aligned}
&P_1 = (.1562, .0312)\,, && P_2 = (.1547, .0359)\,, && P_3 = (.1531, .0400)\,,\\
&P_4 = (.1516, .0453)\,, && P_5 = (.1500, .0500)\,, && P_6 = (.1484, .0547)\,,\\
&P_7 = (.1469, .0594)\,, && P_8 = (.1453, .0641)\,, && P_9 = (.1437, .0687),\\
&P_{10} = (.1422, .0734), && P_{11} = (.1406, .0781).
\end{aligned}$$

Given, for instance, $P_2 = (.1547, .0359)$, *we calculate the power of the test statistics simulating random normal variables with mean 0.1547 and standard deviation* 0.0359.

For any statistic

$$T \in S = \{W_n, R_n, T_n^r,\ r = -1, -.6, -.3, 0, .3, .5, .7, 1, 1.3, 1.6, 2\}$$

and any point P_j, *the power is*

$$\beta_T(P_j) = \Pr(T > \chi^2_{1,0.05}/P_j) = \Pr(T > 3.84/P_j).$$

This power will be obtained as follows:

- *For each* P_j, $j = 1, ..., 11$, *repeat* $N = 1000$ *times:*
  - *Generate* $n = 50$ *normal random variables with parameter* P_j *and obtain the maximum likelihood estimators.*
  - *Evaluate* $T_{j,i}$ *(value of* T *for the random sample* i $(i = 1, ..., 1000)$ *of parameter* P_j $(j = 1, ..., 11)$ *with* $c = 1/2$*).*
- *Estimate*

$$\widehat{\beta}_T(j) = \frac{Number\ of\ T_{j,i} > 3.84\ (i = 1, ..., 1000)}{1000}.$$

In Table 9.1 we present the powers corresponding to the test statistics W_n, R_n *and* T_n^r, $r = -1, -.6, -.3, 0, .3, .5, .7, 1, 1.3, 1.6, 2$. *For* $r = 1$ *we have the likelihood ratio test statistic. Let us denote*

$$S = \{W_n, R_n, T_n^r,\ r = -1, -.6, -.3, 0, .3, .5, .7, 1, 1.3, 1.6, 2\}$$

the set of test statistics considered here. We define $\beta_{MAX}(P_j) = \sup_{T\in S}\beta_T(P_j)$, *for* $j = 1, ..., 11$, *the maximum power achieved in the class* S *for each alternative hypothesis* P_j. *Then*

$$i_T(P_j) = \beta_{MAX}(P_j) - \beta_T(P_j)$$

is the inefficiency on the set of alternative hypotheses for each $T \in S$. Thus, the quantit

$$\eta(T) = \max_{j \neq 5} \{ i_T (P_j) \}$$

can be regarded as a measure of inefficiency of the test statistic T. The minimax criterio chooses the test statistic $T \in S$ minimizing $\eta(.)$.

	Powers								
			T_n^r						
P_j	W_n	R_n	$r=-1$	$r=-.6$	$r=-.3$	$r=0$	$r=.3$	$r=.5$	$r=.7$
1	1.00	.997	1.00	1.00	1.00	1.00	1.00	1.00	1.00
2	.965	.853	.979	.975	.970	.960	.951	.946	.936
3	.675	.431	.742	.710	.689	.664	.640	.615	.602
4	.252	.099	.312	.280	.260	.238	.215	.205	.191
5	.068	.044	.095	.078	.074	.067	.059	.056	.054
6	.083	.194	.067	.069	.082	.095	.109	.119	.133
7	.301	.477	.222	.264	.294	.326	.349	.363	.377
8	.544	.715	.459	.504	.536	.563	.596	.620	.633
9	.811	.907	.754	.786	.806	.825	.841	.850	.861
10	.910	.962	.872	.895	.907	.915	.931	.933	.943
11	.972	.989	.964	.967	.969	.974	.977	.979	.980

Table 9.1

	Powers				
	T_n^r				
P_j	$r=1$	$r=1.3$	$r=1.6$	$r=2$	$\beta_{MAX}(j)$
1	1.00	.999	.997	.997	1.00
2	.927	.913	.894	.880	.979
3	.573	.540	.516	.477	.742
4	.176	.159	.147	.126	.312
5	.055	.052	.049	.050	.050
6	.147	.162	.181	.214	.214
7	.404	.437	.463	.496	.496
8	.659	.677	.701	.730	.730
9	.874	.888	.897	.915	.915
10	.950	.958	.960	.966	.966
11	.981	.986	.989	.992	.992

Table 9.1 (Continuation)

In Table 9.2 we present the inefficiencies $\eta(T)$ of the test statistics T_n^r.

			Inefficiencies							
			T_n^r							
P_j	W_n	R_n	$r=-1$	$r=-.6$	$r=-.3$	$r=0$	$r=.3$	$r=.5$	$r=.7$
1	.000	.003	.000	.000	.000	.000	.000	.000	.000
2	.014	.126	.000	.004	.009	.019	.028	.033	.043
3	.067	.311	.000	.032	.053	.078	.102	.127	.140
4	.060	.213	.000	.032	.052	.074	.097	.107	.121
5	-.018	.006	-.045	-.028	.024	-.017	-.009	-.006	-.004
6	.131	.020	.147	.145	.132	.119	.105	.095	.081
7	.195	.019	.274	.232	.202	.170	.147	.133	.119
8	.186	.015	.271	.226	.194	.167	.134	.110	.097
9	.104	.008	.161	.129	.109	.090	.074	.065	.054
10	.056	.004	.094	.071	.059	.051	.035	.033	.023
11	.020	.003	.028	.025	.023	.018	.015	.013	.012
	.195	.311	.274	.232	.202	.170	.147	.133	.140

Table 9.2

	Inefficiencies			
	T_n^r			
j	$r=1$	$r=1.3$	$r=1.6$	$r=2$
1	.003	.001	.003	.003
2	.052	.066	.085	.099
3	.169	.202	.226	.265
4	.136	.153	.165	.186
5	-.005	-.002	.001	.000
6	.067	.052	.033	.000
7	.092	.059	.033	.000
8	.071	.053	.029	.000
9	.041	.027	.018	.000
10	.016	.008	.006	.000
11	.011	.006	.003	.000
	.169	.202	.226	.265

Table 9.2 (Continuation)

We can see from the last line in Table 9.2 that the test statistic $T_n^{0.5}$ presents the minimum relative inefficiency. The relative inefficiency of the test statistic using, e.g., the Rao test statistic, is more than 100% higher.

Observe that $T_n^{-0.5} = -8n \log(1 - T_n^{Fre}/8n)$, *where* T_n^{Fre} *is the Freeman-Tukey tes statistic given by*

$$T_n^{Fre} = 4n \int_{\mathbb{R}} \left(\sqrt{f_{\widehat{\boldsymbol{\theta}}}(x)} - \sqrt{f_{\widetilde{\boldsymbol{\theta}}}(x)} \right)^2 dx.$$

Furthermore, $T_n^2 = n \log \left(1 + \frac{X_n^2}{n}\right)$ *where*

$$X_n^2 = n \int_{\mathbb{R}} \frac{\left(f_{\widehat{\boldsymbol{\theta}}}(x) - f_{\widetilde{\boldsymbol{\theta}}}(x)\right)^2}{f_{\widetilde{\boldsymbol{\theta}}}(x)} dx$$

is the chi-square test statistic.

9.4. Multi-sample Problem

This section deals with testing a composite null hypothesis H_0 about parameter from s populations whose distributional structure differs just in the value of a parame ter. From each population i, a sample of size n_i is drawn at random, $i = 1, ..., s$. Le $\widehat{\boldsymbol{\theta}}_1, \ldots, \widehat{\boldsymbol{\theta}}_s$ denote the maximum likelihood estimators and $\widetilde{\boldsymbol{\theta}}_1, \ldots, \widetilde{\boldsymbol{\theta}}_s$ the maximum like lihood estimators under H_0. When $n_1 = \ldots = n_s$, Morales et *al.* (1997) developed a testing procedure based on the ϕ-divergence test statistic

$$T_n^{\phi} \left((\widehat{\boldsymbol{\theta}}_1, \ldots, \widehat{\boldsymbol{\theta}}_s), (\widetilde{\boldsymbol{\theta}}_1, \ldots, \widetilde{\boldsymbol{\theta}}_s) \right), \tag{9.16}$$

which is obtained by calculating a ϕ–divergence between the joint densities

$$\prod_{i=1}^{s} f_{\widehat{\boldsymbol{\theta}}_i}(x_i) \quad \text{and} \quad \prod_{i=1}^{s} f_{\widetilde{\boldsymbol{\theta}}_i}(x_i).$$

When the sample sizes are different, the ϕ-divergence test statistics given in (9.16 cannot be used unless they were generalized in some sense. In the literature of Statistica Information Theory, problems related to s samples have been treated by using familie of divergences between s populations (see Menéndez et *al.* (1997e), Zografos (1998a) Morales et *al.* (1998)). This is a nice possibility, but it is not the natural extension o the likelihood ratio test statistic, and in some situations the asymptotic distribution o these test statistics are based on a linear combination of chi-square distributions instea of on a chi-square distribution.

The likelihood ratio test statistic uses the ratio between

$$\prod_{i=1}^{s} \prod_{j=1}^{n_i} f_{\widehat{\boldsymbol{\theta}}_i}(Y_{ij}) \quad \text{and} \quad \prod_{i=1}^{s} \prod_{j=1}^{n_i} f_{\widetilde{\boldsymbol{\theta}}_i}(Y_{ij}) \tag{9.17}$$

or introducing the decision rule. As a parallel approach, in this section we consider divergence between the two estimated likelihood functions appearing in (9.17) and provide a decision rule on the basis of the resulting statistic.

Let $(\mathcal{X}_1, \beta_{\mathcal{X}_1}, P_{1,\boldsymbol{\theta}_1})_{\boldsymbol{\theta}_1 \in \Theta_1}, \dots, (\mathcal{X}_s, \beta_{\mathcal{X}_s}, P_{s,\boldsymbol{\theta}_s})_{\boldsymbol{\theta}_s \in \Theta_s}$ be statistical spaces associated with independent populations. For $i = 1, \cdots, s$, $\mathcal{X}_i \subset \mathbb{R}^{n_i}$ is the sample space, $\beta_{\mathcal{X}_i}$ is the Borel σ-field of subsets of $\mathcal{X}_i$, $\Theta_i \subset \mathbb{R}^{k_i}$ is an open set and $f_{i,\boldsymbol{\theta}_i}$ is the probability density function of $P_{i,\boldsymbol{\theta}_i}$ with respect to a σ-finite measure μ_i, $i = 1, ..., s$. In this section, we deal with the product statistical space

$$(\mathcal{X}_1 \times \cdots \times \mathcal{X}_s, \beta_{\mathcal{X}_1} \times \cdots \times \beta_{\mathcal{X}_s}, P_{1,\boldsymbol{\theta}_1} \times \cdots \times P_{s,\boldsymbol{\theta}_s}),$$

with $(\boldsymbol{\theta}_1, \dots, \boldsymbol{\theta}_s) \in \Theta_1 \times \cdots \times \Theta_s$. Let $\mu \triangleq \mu_1^{n_1} \times \cdots \times \mu_s^{n_s}$ be the product measure and $\mathcal{Y} \triangleq \mathcal{X}_1^{n_1} \times \cdots \times \mathcal{X}_s^{n_s}$ be the product sample space. Consider s independent random samples, $Y_{i1}, Y_{i2}, \dots, Y_{in_i}$, $i = 1, ..., s$, from independent and identically distributed random variables with common probability density function $f_{i,\boldsymbol{\theta}_i}$, $\boldsymbol{\theta}_i \in \Theta_i$, $i = 1, \dots, s$. Assume that the sample sizes n_i tend to infinity with the same rate, that is, if $n = \sum_{i=1}^{s} n_i$, then

$$\frac{n_i}{n} \underset{n_i \to \infty}{\longrightarrow} \lambda_i \in (0,1), \; i = 1, ..., s \tag{9.18}$$

where $\sum_{i=1}^{s} \lambda_i = 1$. Further, assume that the parameters $\boldsymbol{\theta}_i = (\theta_{i,1}, \dots, \theta_{i,k_i})$, $i = 1, \dots, s$, have the same k first components, that is,

$$\begin{array}{ccccccc} \theta_{1,1} & = & \theta_{2,1} & = & \dots & = & \theta_{s,1} \\ \theta_{1,2} & = & \theta_{2,2} & = & \dots & = & \theta_{s,2} \\ \vdots & & \vdots & & & & \vdots \\ \theta_{1,k} & = & \theta_{2,k} & = & \dots & = & \theta_{s,k}, \end{array}$$

where $k \le \min\{k_1, \cdots, k_s\}$. Let us consider the joint sample

$$Y = (Y_{11}, \dots, Y_{1n_1}; Y_{21}, \dots, Y_{2n_2}; \dots; Y_{s1}, \dots, Y_{sn_s}),$$

and the joint parameter

$$\gamma = (\theta_{1,1}, \dots, \theta_{1,k_1}; \theta_{2,k+1}, \dots, \theta_{2,k_2}; \dots; \theta_{s,k+1}, \dots, \theta_{s,k_s}),$$

with $\gamma \in \Gamma$, where Γ is an open subset of $\mathbb{R}^M$ and $M = \sum_{i=1}^{s} k_i - (s-1)k$. Let

$$f_{i,\boldsymbol{\theta}_i}(\boldsymbol{Y}_i) = \prod_{j=1}^{n_i} f_{i,\boldsymbol{\theta}_i}(Y_{ij}) \quad \text{and} \quad L(\boldsymbol{\theta}_i; Y_{i1}, ..., Y_{in_i}) = \sum_{j=1}^{n_i} \log f_{i,\boldsymbol{\theta}_i}(Y_{ij})$$

be the likelihood and the log-likelihood function of $\boldsymbol{\theta}_i$ based on the i-th sample. The likelihood and the log-likelihood function of $\boldsymbol{\gamma} = (\gamma_1, \dots, \gamma_M) \in \Gamma$ based on the joint sample are

$$f_{\gamma}(\boldsymbol{Y}) = \prod_{i=1}^{s} f_{i,\boldsymbol{\theta}_i}(Y_i) \quad \text{and} \quad l(\gamma) = \sum_{i=1}^{s} L(\boldsymbol{\theta}_i; Y_{i1}, ..., Y_{in_i})$$

respectively. The likelihood equations are

$$\frac{\partial l(\gamma)}{\partial \gamma_p} = \sum_{i=1}^{s} \frac{\partial L(\boldsymbol{\theta}_i; Y_{i1}, ..., Y_{in_i})}{\partial \gamma_p} = 0, \quad p = 1, \ldots, M,$$

where, for each i,

$$\frac{\partial L(\boldsymbol{\theta}_i; Y_{i1}, ..., Y_{in_i})}{\partial \gamma_p} = \sum_{j=1}^{n_i} \frac{\partial}{\partial \gamma_p} \log f_{i,\boldsymbol{\theta}_i}(Y_{ij}).$$

Let $\mathcal{I}^i(\boldsymbol{\theta}_i)$ be the Fisher information matrix of the parameter $\boldsymbol{\theta}_i$ associated with the ith-population with density function $f_{i,\boldsymbol{\theta}_i}$ and let $I^i_{p,q}(\boldsymbol{\theta}_i)$ the (p,q)th-element of $\mathcal{I}^i(\boldsymbol{\theta}_i)$. We split $\mathcal{I}^i(\boldsymbol{\theta}_i)$ into blocks as follows

$$\mathcal{I}^i(\boldsymbol{\theta}_i) = \left(\begin{array}{c|c} \mathcal{I}^i_{k,k}(\boldsymbol{\theta}_i) & \mathcal{I}^i_{k,k_i}(\boldsymbol{\theta}_i) \\ \hline \mathcal{I}^i_{k_i,k}(\boldsymbol{\theta}_i) & \mathcal{I}^i_{k_i,k_i}(\boldsymbol{\theta}_i) \end{array} \right)_{k_i \times k_i},$$

where $\mathcal{I}^i_{k,k}(\boldsymbol{\theta}_i)$, $\mathcal{I}^i_{k,k_i}(\boldsymbol{\theta}_i)$, $\mathcal{I}^i_{k_i,k}(\boldsymbol{\theta}_i)$ and $\mathcal{I}^i_{k_i,k_i}(\boldsymbol{\theta}_i)$ are the submatrices whose lower-right corner elements are respectively $I^i_{k,k}(\boldsymbol{\theta}_i)$, $I^i_{k,k_i}(\boldsymbol{\theta}_i)$, $I^i_{k_i,k}(\boldsymbol{\theta}_i)$ and $I^i_{k_i,k_i}(\boldsymbol{\theta}_i)$, and whose sizes are $k \times k$, $k \times (k_i - k)$, $(k_i - k) \times k$ and $(k_i - k) \times (k_i - k)$, respectively.

The following $M \times M$ matrix, denoted by $V(\gamma)$, plays the fundamental role of Fisher information matrix,

$$\boldsymbol{V}(\gamma) = \left(\begin{array}{c|c|c|c} \sum_{i=1}^{s} \lambda_i \mathcal{I}^i_{k,k}(\boldsymbol{\theta}_i) & \lambda_1 \mathcal{I}^1_{k,k_1}(\boldsymbol{\theta}_1) & \cdots & \lambda_s \mathcal{I}^s_{k,k_s}(\boldsymbol{\theta}_s) \\ \hline \lambda_1 \mathcal{I}^1_{k_1,k}(\boldsymbol{\theta}_1) & \lambda_1 \mathcal{I}^1_{k_1,k_1}(\boldsymbol{\theta}_1) & \cdots & 0 \\ \hline \vdots & \vdots & \ddots & \vdots \\ \hline \lambda_s \mathcal{I}^s_{k_s,k}(\boldsymbol{\theta}_s) & 0 & \cdots & \lambda_s \mathcal{I}^s_{k_s,k_s}(\boldsymbol{\theta}_s) \end{array} \right).$$

The regularity assumption *(vi)* is needed to derive the asymptotic normality of the maximum likelihood estimators.

(vi) The matrix $\boldsymbol{V}(\gamma)$ is positive definite.

A composite null hypothesis H_0 can be usually described by a subset Γ_0 of the parametric space and, consequently, the alternative hypothesis is associated with $\Gamma_1 = \Gamma - \Gamma_0$. Suppose that Γ_0 can be expressed as

$$\Gamma_0 = \{\gamma \in \Gamma : \gamma_i = h_i(\boldsymbol{\beta}),\ i = 1, \ldots, M\} = \{\gamma \in \Gamma : \boldsymbol{h}(\boldsymbol{\beta}) = \gamma\},$$

where $\boldsymbol{\beta} = (\beta_1, \ldots, \beta_{M_0})^T \in B$, $\boldsymbol{h} = (h_1, ..., h_M)$ and $B \subset \mathbb{R}^{M_0}$ is an open subset. Suppose that Γ_0 can be also described by the $M - M_0$ restrictions

$$g_i(\gamma) = 0, \quad i = 1, \ldots, M - M_0,$$

where the functions h_i and g_j have continuous first order partial derivatives and the ranks of the matrices

$$\boldsymbol{T}_\gamma = \left(\frac{\partial g_i(\boldsymbol{\gamma})}{\partial \gamma_j}\right)_{\substack{i=1,\ldots,M-M_0\\ j=1,\ldots,M}}, \quad \boldsymbol{M}_\beta = \left(\frac{\partial h_i(\boldsymbol{\beta})}{\partial \beta_j}\right)_{\substack{i=1,\ldots,M\\ j=1,\ldots,M_0}}$$

are $M - M_0$ and M_0 respectively.

Further, suppose that the s submodels

$$(\mathcal{X}_1, \beta_{\mathcal{X}_1}, P_{1,\boldsymbol{\theta}_1})_{\boldsymbol{\theta}_1 \in \Theta_1}, \ldots, (\mathcal{X}_s, \beta_{\mathcal{X}_s}, P_{s,\boldsymbol{\theta}_s})_{\boldsymbol{\theta}_s \in \Theta_s}$$

restricted to null hypothesis (i.e., with $\boldsymbol{\gamma} \in \Gamma_0$) satisfy *i)–v)*, with derivatives taken with respect to the new parameter $\boldsymbol{\beta}$.

For testing the null hypothesis $H_0 : \boldsymbol{\gamma} = \boldsymbol{h}(\boldsymbol{\beta})$, we shall use the family of ϕ-divergence test statistics

$$T_n^\phi(\widehat{\boldsymbol{\gamma}}, \boldsymbol{h}(\widehat{\boldsymbol{\beta}})) \equiv \frac{2}{\phi''(1)} D_\phi(\widehat{\boldsymbol{\gamma}}, \boldsymbol{h}(\widehat{\boldsymbol{\beta}})),$$

where $D_\phi(\widehat{\boldsymbol{\gamma}}, \boldsymbol{h}(\widehat{\boldsymbol{\beta}}))$ is the ϕ-divergence between $f_{\widehat{\boldsymbol{\gamma}}}(x)$ and $f_{\boldsymbol{h}(\widehat{\boldsymbol{\beta}})}(x)$.

The following theorem was proved by Morales et *al.* (2001) for $k = 0$ and by Hobza et *al.* (2001) for $k > 1$.

Theorem 9.7

Let $H_0 : \boldsymbol{\gamma} = \boldsymbol{h}(\boldsymbol{\beta})$ be true. For each $i = 1, \ldots, s$, let $Y_{i1}, \ldots, Y_{in_i}$ be independent and identically distributed random variables with common probability density function $f_{i,\boldsymbol{\theta}_i}(x)$ satisfying the regularity assumptions (9.18) and i)-v). Suppose also that the necessary conditions for differentiating inside the integrals hold. Then

$$T_n^\phi(\widehat{\boldsymbol{\gamma}}, \boldsymbol{h}(\widehat{\boldsymbol{\beta}})) = \frac{2}{\phi''(1)} D_\phi(\widehat{\boldsymbol{\gamma}}, \boldsymbol{h}(\widehat{\boldsymbol{\beta}})) \underset{n\to\infty}{\overset{L}{\longrightarrow}} \chi^2_{M-M_0}.$$

In the multi-sample case with exponential models, the Kullback-Leibler and the likelihood ratio test statistics coincide for $k = 0$; however this result does not hold for $k > 1$.

In the case of the (h, ϕ)-divergence measure the analogous result holds, i.e.,

$$T_n^{\phi,h}(\widehat{\boldsymbol{\gamma}}, \boldsymbol{h}(\widehat{\boldsymbol{\beta}})) = \frac{2}{\phi''(1)\, h'(0)} D_\phi^h(\widehat{\boldsymbol{\gamma}}, \boldsymbol{h}(\widehat{\boldsymbol{\beta}})) \underset{n\to\infty}{\overset{L}{\longrightarrow}} \chi^2_{M-M_0}.$$

For instance, in the special case of Rényi's divergence, the functions h and ϕ were given in (9.7) and (9.8) and Rényi test statistic is

$$T_n^r(\widehat{\boldsymbol{\gamma}}, \boldsymbol{h}(\widehat{\boldsymbol{\beta}})) \equiv 2D_1^r(\widehat{\boldsymbol{\gamma}}, \boldsymbol{h}(\widehat{\boldsymbol{\beta}})) = \frac{1}{r(r-1)} \log \int_{\mathcal{X}} f_{\widehat{\boldsymbol{\gamma}}}^r(x) f_{\boldsymbol{h}(\widehat{\boldsymbol{\beta}})}^{1-r}(x) d\mu(x) \underset{n\to\infty}{\overset{L}{\longrightarrow}} \chi^2_{M-M_0}.$$

The example below, due to Morales et *al.* (2001), shows the behaviour of Rényi test statistic with normal populations.

Example 9.3

Let $(Y_{11},...,Y_{1n_1}),...,(Y_{s1},...,Y_{sn_s})$ *be s independent random samples from normal populations with unknown parameters* $(\mu_1,\sigma_1),...,(\mu_s,\sigma_s)$ *respectively. We are interested in testing*

$$H_0:\sigma_1^2=...=\sigma_s^2 \quad versus \quad H_1:\exists i\neq j \text{ with } \sigma_i^2\neq\sigma_j^2.$$

In this case, the joint parameter space is

$$\Gamma=\left\{(\mu_1,...,\mu_s,\sigma_1^2,...,\sigma_s^2)\,/\,\mu_i\in\mathbb{R},\ \sigma_i^2>0, i=1,...,s\right\},$$

and its restriction to H_0 *is*

$$\Gamma_0=\left\{(\mu_1,...,\mu_s,\sigma_1^2,...,\sigma_s^2)\in\Gamma\,/\,\sigma_1^2=...=\sigma_s^2>0\right\}.$$

Using the functions h_i *and* g_i *given previously, the null hypothesis* Γ_0 *can be written in the following alternative forms:*

i) Consider the set B defined by

$$B=\left\{\boldsymbol{\beta}=(\mu_1,...,\mu_s,\sigma,...\sigma)\in\mathbb{R}^{2s}:\mu_i\in\mathbb{R} \text{ and } \sigma\in\mathbb{R}^+\right\}$$

and the functions,

$$h_i(\boldsymbol{\beta})=\begin{cases}\mu_i & i=1,...,s\\ \sigma^2 & i=s+1,...,2s\end{cases}.$$

Then

$$\Gamma_0=\{\gamma\in\Gamma:\gamma_i=h_i(\boldsymbol{\beta}),\ i=1,...,2s\}.$$

ii) Consider the function $g_j:\Theta\to\mathbb{R}^{s-1}$, $j=1,...,s-1$, *defined by*

$$g_j(\mu_1,...,\mu_s,\sigma_1,...,\sigma_s)=\sigma_1-\sigma_j,\ j=1,...,s-1.$$

Obviously $g_j(\mu_1,...,\mu_s,\sigma_1,...,\sigma_s)=0$ *for* $(\mu_1,...,\mu_s,\sigma_1,...,\sigma_s)\in\Gamma_0$. *Therefore,*

$$\Gamma_0=\{\gamma\in\Gamma:g_j(\gamma)=0,\ j=1,...,s-1\}.$$

The maximum likelihood estimator is $\widehat{\gamma}=\left(\overline{Y}_{1*},\ldots,\overline{Y}_{s*},\widehat{\sigma}_1^2,\ldots,\widehat{\sigma}_s^2\right)$ *and the maximum likelihood estimator under the null hypothesis is* $\boldsymbol{h}(\widehat{\boldsymbol{\beta}})=\left(\overline{Y}_{1*},\ldots,\overline{Y}_{s*},\widehat{\sigma}^2,\ldots,\widehat{\sigma}^2\right)$ *where* $\widehat{\sigma}^2=\frac{1}{n}\sum_{i=1}^s\widehat{\sigma}_i^2$. *The Rényi test statistic for testing* $H_0:\sigma_1^2=\ldots=\sigma_s^2$ *is given by*

$$T_n^r(\widehat{\gamma},\boldsymbol{h}(\widehat{\boldsymbol{\beta}}))=\frac{2}{\phi''(1)h'(0)}D_r^1(\widehat{\gamma},\boldsymbol{h}(\widehat{\boldsymbol{\beta}}))=2D_r^1((\widehat{\boldsymbol{\mu}},\widehat{\boldsymbol{\Sigma}}_1),(\widetilde{\boldsymbol{\mu}},\widetilde{\boldsymbol{\Sigma}}_2)),$$

with

$$\widehat{\boldsymbol{\mu}} = \left(\widehat{\boldsymbol{\mu}}_1^T, ..., \widehat{\boldsymbol{\mu}}_s^T\right)^T, \qquad \widehat{\boldsymbol{\mu}}_i^T = \left(\overline{Y}_{i*}, \ldots, \overline{Y}_{i*}\right)^T, \ i = 1, \ldots, s,$$
$$\widehat{\boldsymbol{\Sigma}}_1 = diag\left(\widehat{\boldsymbol{\Sigma}}_1^*, \ldots, \widehat{\boldsymbol{\Sigma}}_s^*\right), \quad \widehat{\boldsymbol{\Sigma}}_i^* = diag\left(\widehat{\sigma}_i^2, \ldots, \widehat{\sigma}_i^2\right), \ i = 1, \ldots, s,$$
$$\widetilde{\boldsymbol{\mu}} = \left(\widetilde{\boldsymbol{\mu}}_1^T, ..., \widetilde{\boldsymbol{\mu}}_s^T\right)^T, \qquad \widetilde{\boldsymbol{\mu}}_i^T = \left(\overline{Y}_{i*}, \ldots, \overline{Y}_{i*}\right)^T, \ i = 1, \ldots, s,$$
$$\widetilde{\boldsymbol{\Sigma}}_2 = diag\left(\widetilde{\boldsymbol{\Sigma}}_1^*, \ldots, \widetilde{\boldsymbol{\Sigma}}_s^*\right), \quad \widetilde{\boldsymbol{\Sigma}}_i^* = diag\left(\widehat{\sigma}^2, \ldots, \widehat{\sigma}^2\right), \ i = 1, \ldots, s.$$

After straightforward algebra, we get

$$T_n^r(\widehat{\gamma}, \boldsymbol{h}(\widehat{\boldsymbol{\beta}})) = \begin{cases} \frac{1}{r(1-r)} \sum_{i=1}^{s} n_i \log \frac{(r\widehat{\sigma}^2 + (1-r)\widehat{\sigma}_i^2)}{\widehat{\sigma}^{2r}\widehat{\sigma}_i^{2(1-r)}} & \text{if} \quad r \neq 0, r \neq 1 \\ \sum_{i=1}^{s} n_i \log \frac{\widehat{\sigma}^2}{\widehat{\sigma}_i^2} & \text{if} \quad r = 1 \\ \sum_{i=1}^{s} n_i \left(\frac{\widehat{\sigma}^2}{\widehat{\sigma}_i^2} - 1 + \log \frac{\widehat{\sigma}_i^2}{\widehat{\sigma}^2}\right) & \text{if} \quad r = 0. \end{cases}$$

By Theorem 9.7, the asymptotic distribution of $T_n^r(\widehat{\gamma}, \boldsymbol{h}(\widehat{\boldsymbol{\beta}}))$ *is chi-square with* $s-1$ *degrees of freedom. Therefore, an asymptotically test, with significance level* α*, for the problem of testing the equality of variances should reject* H_0 *when* $T_n^r(\widehat{\gamma}, \boldsymbol{h}(\widehat{\boldsymbol{\beta}})) > \chi^2_{s-1,\alpha}$.

In Morales et al. (2001), a power simulation study was carried out for comparing several members of the family of Rényi test statistics.

As a result of this Monte Carlo simulation experiment, they recommended any Rényi divergence test statistic with $r \in [5/4, 3/2]$*. They also emphasize that* $r = 5/4$ *emerges as a good alternative since it presents the best numerical results.*

Under the assumption $\sigma_1^2 = ... = \sigma_s^2$, it is not difficult to establish that Rényi test statistic for testing

$$H_0 : \mu_1 = ... = \mu_s \text{ versus } H_1 : \exists\, i \neq j \text{ with } \mu_i \neq \mu_j$$

rejects the null hypothesis if

$$\frac{\frac{1}{s}\sum_{i=1}^{s} n_i \left(\overline{Y}_{i*} - \overline{Y}\right)^2}{\frac{1}{n-s}\sum_{i=1}^{s}\sum_{j=1}^{n_i} \left(Y_{ij} - \overline{Y}_{i*}\right)^2} > F_{s-1,n-s,\alpha},$$

where $F_{s-1,n-s,\alpha}$ is the $100\,(1-\alpha)$ percentile of the F distribution with $s-1$ and $n-s$ degrees of freedom.

9.5. Some Topics in Multivariate Analysis

A considerable part of this book is devoted to the use of the ϕ-divergence concept developing methods of estimation and testing for the analysis of categorical data in severa contexts like analysis of cross-classified data, log-linear models, etc. From this point c view, the ϕ−divergence is exploited here for the analysis of discrete multivariate data.

The use of divergences in order to meet and study problems of Multivariate Analysi is not new. It starts in the early 1950, when Kullback (1959), in his pioneer book dedicated five chapters to the use of the minimum information discrimination principle fo the study of several problems in multivariate analysis. The bridge which links Statistica Information Theory and Multivariate Analysis is founded on the fact that multivariat analysis methods are mainly created on the notion of the distance between observation or distance among their respective distributions, while on the other hand, Statistica Information Theory is mainly concerned with the definition of statistical distances o divergences between distributions and on the development of metric geometries base mainly on the Fisher information matrix.

The operational link between these two statistical areas has received great atten tion over the last four decades. Since Kullback's pioneer work, there is a vast amoun of contributions based on the information theoretic formulation of multivariate statis tical topics, like distribution theory, statistical inference using multivariate continuous categorical or mixed data, concepts of multivariate dependence, discrimination and clas sification etc. An indicatory, nonexhaustive literature of the subject is the following: i Construction of Multivariate Distributions (Kapur (1989), Cuadras (1992a), Cuadras e *al.* (1997a), Zografos (1999)), ii) Statistical Inference (Krzanowski (1983), Bar-Hen an Daudin (1995), Morales et *al.* (1997, 1998), Zografos (1998a), Garren (2000)), iii) Mea sures of Multivariate Association (Kent (1983), Inaba and Shirahata (1986), Joe (1989) Zografos (1998b, 2000)), iv) Discriminant Analysis (Matusita (1966, 1973), Krzanowsk (1982), Cacoullos and Koutras (1985), Cuadras (1992b), Koutras (1992), Bar-Hen, A (1996), Cuadras et *al.* (1997b), Menéndez et *al.* (2004, 2005d)). In Sy and Gupta (2004 it can be seen that Information Theory is a useful tool for data mining.

).6. Exercises

1. Find the asymptotic distribution of the statistic $D_{\phi}^{h}(\widehat{\boldsymbol{\theta}},\boldsymbol{\theta}_0)$, where $\widehat{\boldsymbol{\theta}}$ is the maximum likelihood estimator obtained from a population with parameter $\boldsymbol{\theta} \neq \boldsymbol{\theta}_0$.

2. Find the asymptotic distribution of the (h,ϕ)-divergence test statistic $T_n^{\phi,h}(\widehat{\boldsymbol{\theta}},\boldsymbol{\theta}_0)$ under the contiguous alternative hypotheses

$$H_{1,n} : \boldsymbol{\theta}_n = \boldsymbol{\theta}_0 + n^{-1/2}\boldsymbol{d},$$

where $\boldsymbol{d}$ is a fixed vector in $\mathbb{R}^{M_0}$ such that $\boldsymbol{\theta}_n \in \Theta \subset \mathbb{R}^{M_0}$.

3. We consider the divergence measure of order r and degree s.

   a) Find the asymptotic distribution of the statistic $D_r^s(\widehat{\boldsymbol{\theta}},\boldsymbol{\theta}_0)$ under the hypothesis that the observations are from a population with $\boldsymbol{\theta} \neq \boldsymbol{\theta}_0$.

   b) We consider the random sample

0.0789	0.1887	0.0828	0.0086	0.0572	0.0041	0.3551	0.0783
0.0732	0.1839	0.1439	0.1681	0.0115	0.1155	0.0566	

   from a exponential population of unknown parameter θ. Using the (h,ϕ)-divergence test statistic based on the divergence measure of order r and degree s ($s = 2$ and $r = 0.5$) test if $\theta = 10$ versus $\theta \neq 10$ using as significance level $\alpha = 0.05$.

4. We consider the ϕ-divergence test statistic $T_n^{\phi}(\widehat{\boldsymbol{\theta}},\boldsymbol{\theta}_0)$ for testing $H_0 : \boldsymbol{\theta} = \boldsymbol{\theta}_0$. Obtain the approximate size n, guaranteeing a power β at a given alternative $\boldsymbol{\theta} \neq \boldsymbol{\theta}_0$.

5. We consider two random samples from two populations with parameters $\boldsymbol{\theta}_1$ and $\boldsymbol{\theta}_2$ of sizes n and m respectively, and the corresponding maximum likelihood estimators, $\widehat{\boldsymbol{\theta}}_1 = (\widehat{\theta}_{11},...,\widehat{\theta}_{1M_0})^T$ and $\widehat{\boldsymbol{\theta}}_2 = (\widehat{\theta}_{21},...,\widehat{\theta}_{2M_0})^T$, associated with them. Find the asymptotic distribution of the test statistic $D_{\phi}(\widehat{\boldsymbol{\theta}}_1,\widehat{\boldsymbol{\theta}}_2)$ under the two following assumptions: *a)* $\boldsymbol{\theta}_1 \neq \boldsymbol{\theta}_2$ and *b)* $\boldsymbol{\theta}_1 = \boldsymbol{\theta}_2$.

6. Use the result obtained in Exercise 5 for testing $H_0 : \boldsymbol{\theta}_1 = \boldsymbol{\theta}_2$ versus $H_1 : \boldsymbol{\theta}_1 \neq \boldsymbol{\theta}_2$.

7. Let $Y_1,...,Y_n$ be a random sample from a exponential distribution of parameter θ.

   a) Find the expression of Cressie-Read test statistic, $D_{\phi_{(\lambda)}}(\widehat{\theta},\theta_0)$, for $\lambda = 2$ where $\widehat{\theta}$ is the maximum likelihood estimator of θ.

   b) Test $H_0 : \theta = 3/2$ versus $H_1 : \theta \neq 3/2$, on the basis of the exact test based on $D_{\phi_{(\lambda)}}(\widehat{\theta},\theta_0)$ and using as significance level α.

8. Let $\boldsymbol{X}$ be a M-variate normal population with unknown mean vector $\boldsymbol{\mu}$ and known variance-covariance matrix $\boldsymbol{\Sigma}_0$.

   a) Find a test statistic for testing $\boldsymbol{\mu} = \boldsymbol{\mu}_0$ versus $\boldsymbol{\mu} \neq \boldsymbol{\mu}_0$ using Kullback-Leibler divergence measure as well as the $(1-\alpha)$ confidence region in $\mathbb{R}^2$ for $\boldsymbol{\mu}$.

b) The measurements on the first and second adult sons in a sample of 25 families are the following:

First son (X_1)	191	195	181	183	176	163	195	186	181
	208	189	197	188	192	175	192	174	176
	179	183	174	190	188	197	190		
Second son (X_2)	179	201	185	188	177	161	183	173	182
	192	190	189	197	187	165	185	178	176
	186	174	185	195	187	200	187		

Source: Mardia, K. V., Kent, J. T. and Bibby, J. M. (1979, p. 121).

We assume that X_1 and X_2 are independent and that each one is normally distributed with variance 100, i.e.,

$$\boldsymbol{\Sigma} = \boldsymbol{\Sigma}_0 = \begin{pmatrix} 100 & 0 \\ 0 & 100 \end{pmatrix}.$$

Test using as significance level α,

$$H_0 : \boldsymbol{\mu} = \boldsymbol{\mu}_0 = (\mu_{01}, \mu_{02}) = (182, 182)^T \text{ against } H_1 : \boldsymbol{\mu} \neq \boldsymbol{\mu}_0$$

and get a 95% confidence region for the mean of X_1 and X_2.

9. Let $\boldsymbol{X}$ be a M-variate normal population with known vector mean $\boldsymbol{\mu}_0$ and unknown variance-covariance matrix $\boldsymbol{\Sigma}$.

   a) Find a test statistic for testing $\boldsymbol{\Sigma} = \boldsymbol{\Sigma}_0$ versus $\boldsymbol{\Sigma} \neq \boldsymbol{\Sigma}_0$ using the Kullback-Leibler divergence measure as well as the $100\,(1-\alpha)$ confidence region for $\boldsymbol{\Sigma}$ and $M = 2$.

   b) Test using as significance level $\alpha = 0.05$ for the data given in Exercise 8

$$H_0 : \boldsymbol{\Sigma} = \boldsymbol{\Sigma}_0 = \begin{pmatrix} 100 & 0 \\ 0 & 100 \end{pmatrix} \text{ assuming that } \boldsymbol{\mu} = (182, 182)^T.$$

10. Let $\boldsymbol{X}$ be a M-variate normal population with unknown mean vector $\boldsymbol{\mu}$ and unknown variance-covariance matrix $\boldsymbol{\Sigma}$. Find a test statistic for testing

$$H_0 : \boldsymbol{\mu} = \boldsymbol{\mu}_0 \qquad \text{versus } H_1 : \boldsymbol{\mu} \neq \boldsymbol{\mu}_0,$$

based on the Kullback-Leibler divergence measure.

11. Let $\boldsymbol{X}$ be a M-variate normal population with unknown mean vector $\boldsymbol{\mu}$ and unknown variance-covariance matrix $\boldsymbol{\Sigma}$. Find a test statistic for testing

$$H_0 : \boldsymbol{\Sigma} = \boldsymbol{\Sigma}_0 \qquad \text{versus } H_1 : \boldsymbol{\Sigma} \neq \boldsymbol{\Sigma}_0,$$

based on the Kullback-Leibler divergence measure.

12. Given the independent random samples:

0.4068 1.7698 1.7830 1.0186 1.5880 1.9616 1.1334 0.1288 0.8306

and

1.3863 2.5470 0.9480 0.0420 0.1449 0.7971 1.2858 2.9358 1.4829
0.7971 1.2858 2.5335 1.4829

from uniform distributions in the intervals $(0, \theta_1)$ and $(0, \theta_2)$ respectively, using as significance level $\alpha = 0.05$, find the exact test based on Kullback-Leibler divergence measure for testing $\theta_1 = \theta_2$ versus $\theta_1 \neq \theta_2$ $(\theta_1 \leq \theta_2)$.

13. Let $X_1, ..., X_n$ and $Y_1, ..., Y_m$ be two independent random samples from the distributions

$$f_{\theta_i}(x) = \exp(-(x - \theta_i)) \qquad \theta_i < x < \infty, \quad i = 1, 2.$$

Find the test statistic based on Kullback-Leibler divergence measure for testing

$$H_0 : \theta_1 = \theta_2 \qquad \text{versus} \qquad H_1 : \theta_1 > \theta_2.$$

14. We consider a population X with probability density function

$$f_\theta(x) = \frac{\theta 2^\theta}{x^{\theta+1}} \qquad x \geq 2, \ \theta > 0.$$

Find the expression of the test statistics of Wald, Rao, Likelihood ratio and Rényi for testing

$$H_0 : \theta = \theta_0 \qquad \text{versus} \qquad H_1 : \theta \neq \theta_0,$$

based on a sample of size n.

15. We consider a population with exponential distribution with parameter θ and we wish to test $H_0 : \theta = 1$ versus $H_1 : \theta \neq 1$ using as significance level $\alpha = 0.05$.

   a) Find Rényi test statistic for $r = 1/4, 3/4, 1, 5/4, 7/4$ and 2.

   b) Study the accuracy of powers approximations given for $n = 20, 40, 80$ and 200, $r = 1/4, 3/4, 1, 5/4, 7/4$, 2 and $\theta = 0.5$, 0.6, 0.65, 0.70, 0.75, 0.80, 0.85, 0.90, 0.95, 1.10, 1.20, 1.30, 1.40, 1.50, 1.60, 1.70, 1.80, 1.90, 2, 2.10, 2.20, 2.30, 2.40 and 2.50.

   c) Obtain the sample size to get a power of $\beta = 0.8$ for $\theta = 0.5$, 0.6, 0.65, 0.70, 0.75, 0.80, 0.85, 0.90, 0.95, 1.10, 1.20, 1.30, 1.40, 1.50, 1.60, 1.70, 1.80, 1.90, 2, 2.10, 2.20, 2.30, 2.40 and 2.50.

9.7. Answers to Exercises

1. We have, $D_\phi^h(\boldsymbol{\theta}_1,\boldsymbol{\theta}_2) = h(D_\phi(\boldsymbol{\theta}_1,\boldsymbol{\theta}_2))$, where h is a differentiable increasing function mapping from $[0,\infty)$ onto $[0,\infty)$, with $h(0)=0$ and $h'(0)>0$.
We know that $h(x) = h'(x_0)(x-x_0) + o(x-x_0)$, then
$$\begin{aligned}\sqrt{n}\left(D_\phi^h(\widehat{\boldsymbol{\theta}},\boldsymbol{\theta}_0) - D_\phi^h(\boldsymbol{\theta},\boldsymbol{\theta}_0)\right) =& \; h'(D_\phi(\boldsymbol{\theta},\boldsymbol{\theta}_0))\left[\sqrt{n}\left(D_\phi(\widehat{\boldsymbol{\theta}},\boldsymbol{\theta}_0) - D_\phi(\boldsymbol{\theta},\boldsymbol{\theta}_0)\right)\right]\\ &+ \; o_P(1),\end{aligned}$$
and
$$\sqrt{n}\left(D_\phi^h(\widehat{\boldsymbol{\theta}},\boldsymbol{\theta}_0) - D_\phi^h(\boldsymbol{\theta},\boldsymbol{\theta}_0)\right) \underset{n\to\infty}{\overset{L}{\longrightarrow}} N\left(0,\sigma_{h,\phi}^2(\boldsymbol{\theta})\right),$$
where $\sigma_{h,\phi}^2(\boldsymbol{\theta}) = \left(h'(D_\phi(\boldsymbol{\theta},\boldsymbol{\theta}_0))\right)^2\sigma_\phi^2(\boldsymbol{\theta})$ and $\sigma_\phi^2(\boldsymbol{\theta}) = \boldsymbol{T}^T\mathcal{I}_{\mathcal{F}}(\boldsymbol{\theta})^{-1}\boldsymbol{T}$ is given in Theorem 9.2.

2. In a similar way to the previous Exercise we have, under $H_{1,n}$, that the statistics
$$D_\phi^h(\widehat{\boldsymbol{\theta}},\boldsymbol{\theta}_0) \text{ and } h'(0)\,D_\phi(\widehat{\boldsymbol{\theta}},\boldsymbol{\theta}_0)$$
have the same asymptotic distribution. Therefore,
$$T_n^{\phi,h}(\widehat{\boldsymbol{\theta}},\boldsymbol{\theta}_0) = \frac{2n}{\phi''(1)\,h'(0)}D_\phi^h(\widehat{\boldsymbol{\theta}},\boldsymbol{\theta}_0) \underset{n\to\infty}{\overset{L}{\longrightarrow}} \chi_{M_0}^2(\delta),$$
being $\delta = \boldsymbol{d}^T\,\mathcal{I}_{\mathcal{F}}(\boldsymbol{\theta}_0)\,\boldsymbol{d}$.
If we want to approximate the power at some alternative $\boldsymbol{\theta}$, then $\boldsymbol{d} = \boldsymbol{d}(n,\boldsymbol{\theta},\boldsymbol{\theta}_0) = \sqrt{n}(\boldsymbol{\theta}-\boldsymbol{\theta}_0)$ should be used in the formula of the noncentrality parameter.

3. The divergence of order r and degree s is given by
$$D_r^s(\boldsymbol{\theta},\boldsymbol{\theta}_0) = \frac{1}{s-1}\left(\left(\int_{\mathcal{X}} f_{\boldsymbol{\theta}}(x)^r\, f_{\boldsymbol{\theta}_0}(x)^{1-r}\,d\mu(x)\right)^{\frac{s-1}{r-1}} - 1\right),$$
and it can be considered as a (h,ϕ)-divergence with
$$h(x) = \frac{1}{s-1}\left((1+r(r-1)x)^{\frac{s-1}{r-1}} - 1\right);\; s,r\neq 1$$
and
$$\phi(x) = \frac{x^r - r(x-1) - 1}{r(r-1)};\; r\neq 0,1.$$

   a) From Exercise 1 it is only necessary to get the elements of the vector $\boldsymbol{T}$. These are given by
$$\begin{aligned} t_i &= \frac{r}{r-1}\left(\int_{\mathcal{X}} f_{\boldsymbol{\theta}}(x)^r\, f_{\boldsymbol{\theta}_0}(x)^{1-r}\,d\mu(x)\right)^{\frac{s-r}{r-1}}\\ &\times \int_{\mathcal{X}} f_{\boldsymbol{\theta}_0}(x)^{1-r}\, f_{\boldsymbol{\theta}}(x)^{r-1}\,\frac{\partial f_{\boldsymbol{\theta}}(x)}{\partial\theta_i}\,d\mu(x),\; i=1,...,M.\end{aligned}$$

b) The divergence measure of order r and degree s between two exponential distributions with parameters λ and μ is given for $\lambda r + \mu(1-r) > 0$ by

$$D_r^s(\lambda,\mu) = (s-1)^{-1}\left(\left(\frac{\lambda^r \mu^{1-r}}{\lambda r + \mu(1-r)}\right)^{\frac{s-1}{r-1}} - 1\right),$$

and for $\lambda r + \mu(1-r) \leq 0$, $D_r^s(\lambda,\mu) = +\infty$. The maximum likelihood estimator in the exponential model is $\widehat{\lambda} = \overline{Y}^{-1}$ ($\overline{Y}$ is the sample mean) and for the given random sample takes on the value $\widehat{\lambda} = 9.3379$.
Now it is necessary to evaluate the test statistic

$$T_n^{r,s}(\widehat{\lambda},10) \equiv \frac{2n}{h'(0)\,\phi''(1)} D_r^s(\widehat{\lambda},10).$$

In our case $h'(0) = r$ and $\phi''(1) = 1$, then we have

$$T_n^{r,s}(\widehat{\lambda},10) = \frac{2n}{0.5} D_{0.5}^2(\widehat{\lambda},10) = 0.0704.$$

On the other hand $\chi^2_{1,\,0.05} = 3.841$, and we should not reject the null hypothesis.

4. If we consider the expression of the power given in (9.6) the problem will be solved if we consider the sample size n^* obtained as a solution of the equation

$$\beta^1_{n,\phi}(\boldsymbol{\theta}) = 1 - \Phi\left(\frac{\sqrt{n}}{\sigma_\phi(\boldsymbol{\theta})}\left(\frac{\phi''(1)\,\chi^2_{M_0,\alpha}}{2n} - D_\phi(\boldsymbol{\theta},\boldsymbol{\theta}_0)\right)\right),$$

i.e.,

$$n^* = \frac{A + B + \sqrt{A(A+2B)}}{2D_\phi(\boldsymbol{\theta},\boldsymbol{\theta}_0)^2}, \tag{9.19}$$

where $A = \sigma^2_\phi(\boldsymbol{\theta})\left(\Phi^{-1}(1-\beta^1_{n,\phi}(\boldsymbol{\theta}))\right)^2$ and $B = \phi''(1)\,\chi^2_{M_0,\alpha} D_\phi(\boldsymbol{\theta},\boldsymbol{\theta}_0)$. The required sample size is $n = [n^*] + 1$, where $[\cdot]$ is used to denote "integer part of".

5. First we consider part *a)*.

a) We have

$$\begin{aligned} D_\phi(\widehat{\boldsymbol{\theta}}_1,\widehat{\boldsymbol{\theta}}_2) &= D_\phi(\boldsymbol{\theta}_1,\boldsymbol{\theta}_2) + \sum_{i=1}^{M_0} \frac{\partial D_\phi(\boldsymbol{\theta}_1,\boldsymbol{\theta}_2)}{\partial \theta_{1i}}(\widehat{\theta}_{1i} - \theta_{1i}) \\ &+ \sum_{i=1}^{M_0} \frac{\partial D_\phi(\boldsymbol{\theta}_1,\boldsymbol{\theta}_2)}{\partial \theta_{2i}}(\widehat{\theta}_{2i} - \theta_{2i}) + o\left(\left\|\widehat{\boldsymbol{\theta}}_1 - \boldsymbol{\theta}_1\right\|\right) \\ &+ o\left(\left\|\widehat{\boldsymbol{\theta}}_2 - \boldsymbol{\theta}_2\right\|\right), \end{aligned}$$

then

$$\begin{aligned} \sqrt{\tfrac{nm}{n+m}}\left(D_\phi(\widehat{\boldsymbol{\theta}}_1,\widehat{\boldsymbol{\theta}}_2) - D_\phi(\boldsymbol{\theta}_1,\boldsymbol{\theta}_2)\right) &= \boldsymbol{T}^T(\widehat{\boldsymbol{\theta}}_1 - \boldsymbol{\theta}_1) + \boldsymbol{S}^T(\widehat{\boldsymbol{\theta}}_2 - \boldsymbol{\theta}_2) \\ &+ o\left(\left\|\widehat{\boldsymbol{\theta}}_1 - \boldsymbol{\theta}_1\right\|\right) + o\left(\left\|\widehat{\boldsymbol{\theta}}_2 - \boldsymbol{\theta}_2\right\|\right), \end{aligned}$$

where

$$\boldsymbol{T} = (t_1, ..., t_{M_0})^T \text{ and } \boldsymbol{S} = (s_1, ..., s_{M_0})^T$$

with

$$t_i = \int_{\mathcal{X}} \phi' \left(\frac{f_{\boldsymbol{\theta}_1}(x)}{f_{\boldsymbol{\theta}_2}(x)} \right) \frac{\partial f_{\boldsymbol{\theta}_1}(x)}{\partial \theta_{1i}} d\mu(x)$$

and

$$s_i = \int_{\mathcal{X}} \left(\frac{\partial f_{\boldsymbol{\theta}_2}(x)}{\partial \theta_{2i}} \phi \left(\frac{f_{\boldsymbol{\theta}_1}(x)}{f_{\boldsymbol{\theta}_2}(x)} \right) - \phi' \left(\frac{f_{\boldsymbol{\theta}_1}(x)}{f_{\boldsymbol{\theta}_2}(x)} \right) \frac{f_{\boldsymbol{\theta}_1}(x)}{f_{\boldsymbol{\theta}_2}(x)} \right) d\mu(x).$$

On the other hand

$$\begin{array}{lcl} \sqrt{n} T^T (\widehat{\boldsymbol{\theta}}_1 - \boldsymbol{\theta}_1) & \underset{n\to\infty}{\overset{L}{\longrightarrow}} & N\left(\mathbf{0}, \boldsymbol{T}^T \mathcal{I}_{\mathcal{F}}(\boldsymbol{\theta}_1)^{-1} \boldsymbol{T}\right) \\ \sqrt{m} S^T (\widehat{\boldsymbol{\theta}}_2 - \boldsymbol{\theta}_2) & \underset{n\to\infty}{\overset{L}{\longrightarrow}} & N\left(\mathbf{0}, \boldsymbol{S}^T \mathcal{I}_{\mathcal{F}}(\boldsymbol{\theta}_2)^{-1} \boldsymbol{S}\right). \end{array}$$

Therefore the test statistic

$$\sqrt{\frac{nm}{n+m}} \left(D_\phi(\widehat{\boldsymbol{\theta}}_1, \widehat{\boldsymbol{\theta}}_2) - D_\phi(\boldsymbol{\theta}_1, \boldsymbol{\theta}_2) \right)$$

is asymptotically distributed as a normal distribution with mean zero and variance

$$\lambda \boldsymbol{T}^T \mathcal{I}_{\mathcal{F}}(\boldsymbol{\theta}_1)^{-1} \boldsymbol{T} + (1-\lambda) \boldsymbol{S}^T \mathcal{I}_{\mathcal{F}}(\boldsymbol{\theta}_2)^{-1} \boldsymbol{S},$$

where

$$\lambda = \lim_{n,m\to\infty} \frac{m}{m+n},$$

because $o\left(\left\|\widehat{\boldsymbol{\theta}}_1 - \boldsymbol{\theta}_1\right\|\right) = o_P\left(n^{-1/2}\right)$ and $o\left(\left\|\widehat{\boldsymbol{\theta}}_2 - \boldsymbol{\theta}_2\right\|\right) = o_P\left(m^{-1/2}\right)$.

b) In this case we have

$$\begin{array}{rcl} D_\phi(\widehat{\boldsymbol{\theta}}_1, \widehat{\boldsymbol{\theta}}_2) & = & \frac{1}{2} \sum_{i,j=1}^{M_0} \left(\frac{\partial^2 D_\phi(\boldsymbol{\theta}_1, \boldsymbol{\theta}_2)}{\partial \theta_{1i} \partial \theta_{1j}} \right)_{\boldsymbol{\theta}_1 = \boldsymbol{\theta}_2} (\widehat{\theta}_{1i} - \theta_{1i})(\widehat{\theta}_{1j} - \theta_{1j}) \\ & + & \sum_{i,j=1}^{M_0} \left(\frac{\partial^2 D_\phi(\boldsymbol{\theta}_1, \boldsymbol{\theta}_2)}{\partial \theta_{1i} \partial \theta_{2j}} \right)_{\boldsymbol{\theta}_1 = \boldsymbol{\theta}_2} (\widehat{\theta}_{1i} - \theta_{1i})(\widehat{\theta}_{2j} - \theta_{2j}) \\ & + & \frac{1}{2} \sum_{i,j=1}^{M_0} \left(\frac{\partial^2 D_\phi(\boldsymbol{\theta}_1, \boldsymbol{\theta}_2)}{\partial \theta_{2i} \partial \theta_{2j}} \right)_{\boldsymbol{\theta}_1 = \boldsymbol{\theta}_2} (\widehat{\theta}_{2i} - \theta_{2i})(\widehat{\theta}_{2j} - \theta_{2j}) \\ & + & o\left(\left\| \widehat{\boldsymbol{\theta}}_1 - \boldsymbol{\theta}_1 \right\|^2 \right) + o\left(\left\| \widehat{\boldsymbol{\theta}}_2 - \boldsymbol{\theta}_2 \right\|^2 \right). \end{array}$$

We have

$$\frac{\partial D_\phi(\boldsymbol{\theta}_1, \boldsymbol{\theta}_2)}{\partial \theta_{1i}} = \int_{\mathcal{X}} \phi' \left(\frac{f_{\boldsymbol{\theta}_1}(x)}{f_{\boldsymbol{\theta}_2}(x)} \right) \frac{\partial f_{\boldsymbol{\theta}_1}(x)}{\partial \theta_{1i}} d\mu(x) \qquad i = 1, ..., M_0$$

then

$$\left(\frac{\partial^2 D_\phi(\boldsymbol{\theta}_1,\boldsymbol{\theta}_2)}{\partial\theta_{1i}\partial\theta_{1j}}\right)_{\boldsymbol{\theta}_1=\boldsymbol{\theta}_2} = \phi''(1)\int_{\mathcal{X}}\frac{1}{f_{\boldsymbol{\theta}_1}(x)}\frac{\partial f_{\boldsymbol{\theta}_1}(x)}{\partial\theta_{1i}}\frac{\partial f_{\boldsymbol{\theta}_1}(x)}{\partial\theta_{1j}}d\mu(x)$$

$$\left(\frac{\partial^2 D_\phi(\boldsymbol{\theta}_1,\boldsymbol{\theta}_2)}{\partial\theta_{1i}\partial\theta_{2j}}\right)_{\boldsymbol{\theta}_1=\boldsymbol{\theta}_2} = -\left(\frac{\partial^2 D_\phi(\boldsymbol{\theta}_1,\boldsymbol{\theta}_2)}{\partial\theta_{1i}\partial\theta_{1j}}\right)_{\boldsymbol{\theta}_1=\boldsymbol{\theta}_2}$$

$$\left(\frac{\partial^2 D_\phi(\boldsymbol{\theta}_1,\boldsymbol{\theta}_2)}{\partial\theta_{2i}\partial\theta_{2j}}\right)_{\boldsymbol{\theta}_1=\boldsymbol{\theta}_2} = \left(\frac{\partial^2 D_\phi(\boldsymbol{\theta}_1,\boldsymbol{\theta}_2)}{\partial\theta_{1i}\partial\theta_{1j}}\right)_{\boldsymbol{\theta}_1=\boldsymbol{\theta}_2}.$$

Therefore,

$$\begin{aligned}\frac{2}{\phi''(1)}D_\phi(\widehat{\boldsymbol{\theta}}_1,\widehat{\boldsymbol{\theta}}_2) &= (\widehat{\boldsymbol{\theta}}_1-\boldsymbol{\theta}_1)^T\mathcal{I}_{\mathcal{F}}(\boldsymbol{\theta}_1)(\widehat{\boldsymbol{\theta}}_1-\boldsymbol{\theta}_1)\\ &- 2(\widehat{\boldsymbol{\theta}}_1-\boldsymbol{\theta}_1)^T\mathcal{I}_{\mathcal{F}}(\boldsymbol{\theta}_1)(\widehat{\boldsymbol{\theta}}_2-\boldsymbol{\theta}_2)\\ &+ (\widehat{\boldsymbol{\theta}}_2-\boldsymbol{\theta}_1)^T\mathcal{I}_{\mathcal{F}}(\boldsymbol{\theta}_1)(\widehat{\boldsymbol{\theta}}_2-\boldsymbol{\theta}_1)+o\left(\left\|\widehat{\boldsymbol{\theta}}_1-\boldsymbol{\theta}_1\right\|^2\right)\\ &+ o\left(\left\|\widehat{\boldsymbol{\theta}}_2-\boldsymbol{\theta}_2\right\|^2\right)\\ &= (\widehat{\boldsymbol{\theta}}_1-\widehat{\boldsymbol{\theta}}_2)^T\mathcal{I}_{\mathcal{F}}(\boldsymbol{\theta}_1)(\widehat{\boldsymbol{\theta}}_1-\widehat{\boldsymbol{\theta}}_2)+o\left(\left\|\widehat{\boldsymbol{\theta}}_1-\boldsymbol{\theta}_1\right\|^2\right)\\ &+ o\left(\left\|\widehat{\boldsymbol{\theta}}_2-\boldsymbol{\theta}_2\right\|^2\right).\end{aligned}$$

On the other hand

$$\begin{aligned}\sqrt{n}(\widehat{\boldsymbol{\theta}}_1-\boldsymbol{\theta}_1) &\underset{n\to\infty}{\overset{L}{\longrightarrow}} N\left(\mathbf{0},\mathcal{I}_{\mathcal{F}}(\boldsymbol{\theta}_1)^{-1}\right)\\ \sqrt{m}(\widehat{\boldsymbol{\theta}}_2-\boldsymbol{\theta}_2) &\underset{n\to\infty}{\overset{L}{\longrightarrow}} N\left(\mathbf{0},\mathcal{I}_{\mathcal{F}}(\boldsymbol{\theta}_2)^{-1}\right)\end{aligned}$$

then

$$\begin{aligned}\sqrt{\tfrac{mn}{m+n}}(\widehat{\boldsymbol{\theta}}_1-\boldsymbol{\theta}_1) &\underset{n,m\to\infty}{\overset{L}{\longrightarrow}} N\left(\mathbf{0},\lambda\mathcal{I}_{\mathcal{F}}(\boldsymbol{\theta}_1)^{-1}\right)\\ \sqrt{\tfrac{mn}{m+n}}(\widehat{\boldsymbol{\theta}}_2-\boldsymbol{\theta}_2) &\underset{n,m\to\infty}{\overset{L}{\longrightarrow}} N\left(\mathbf{0},(1-\lambda)\mathcal{I}_{\mathcal{F}}(\boldsymbol{\theta}_2)^{-1}\right).\end{aligned}$$

Under the hypothesis $\boldsymbol{\theta}_1=\boldsymbol{\theta}_2$, we have

$$\sqrt{\tfrac{mn}{m+n}}(\widehat{\boldsymbol{\theta}}_1-\widehat{\boldsymbol{\theta}}_2) \underset{n,m\to\infty}{\overset{L}{\longrightarrow}} N\left(\mathbf{0},\mathcal{I}_{\mathcal{F}}(\boldsymbol{\theta}_1)^{-1}\right)$$

therefore

$$\frac{2}{\phi''(1)}\frac{mn}{m+n}D_\phi(\widehat{\boldsymbol{\theta}}_1,\widehat{\boldsymbol{\theta}}_2) \underset{n,m\to\infty}{\overset{L}{\longrightarrow}} \chi^2_{M_0},$$

because $o\left(\left\|\widehat{\boldsymbol{\theta}}_1-\boldsymbol{\theta}_1\right\|^2\right)=o_P\left(n^{-1}\right)$ and $o\left(\left\|\widehat{\boldsymbol{\theta}}_2-\boldsymbol{\theta}_2\right\|^2\right)=o_P\left(m^{-1}\right)$.

6. We should reject the null hypothesis if

$$D_\phi(\widehat{\boldsymbol{\theta}}_1, \widehat{\boldsymbol{\theta}}_2) > c,$$

where c is a positive constant. Now it is possible to find two situations:

i) The distribution of the test statistic $D_\phi(\widehat{\boldsymbol{\theta}}_1, \widehat{\boldsymbol{\theta}}_2)$ is known; then independentl of the regularity conditions, we choose c in such a way that the significanc level of the test is α,

$$\Pr\left(D_\phi(\widehat{\boldsymbol{\theta}}_1, \widehat{\boldsymbol{\theta}}_2) > c\right) = \alpha.$$

ii) The exact distribution of the test statistic $D_\phi(\widehat{\boldsymbol{\theta}}_1, \widehat{\boldsymbol{\theta}}_2)$ is unknown. In thi case we have to use the asymptotic distribution given in Exercise 5 and w should reject the null hypothesis if

$$\frac{2}{\phi''(1)} \frac{mn}{m+n} D_\phi(\widehat{\boldsymbol{\theta}}_1, \widehat{\boldsymbol{\theta}}_2) > \chi^2_{M_0,\alpha}.$$

7. *a)* The divergence of Cressie and Read for $\lambda = 2$ between two exponential distrib utions is given by

$$\begin{aligned} D_{\phi_{(2)}}(\theta, \theta_0) &= \tfrac{1}{6}\left(\int_0^\infty \frac{f_\theta(x)^3}{f_{\theta_0}(x)^2}dx - 1\right) \\ &= \tfrac{1}{6}\left(\frac{\theta^3}{\theta_0^2}\int_0^\infty \exp\left(-\left(3\theta - 2\theta_0\right)x\right)dx - 1\right) \\ &= \tfrac{1}{6}\left(\frac{\theta^3}{\theta_0^2}\frac{1}{3\theta - 2\theta_0} - 1\right) \end{aligned}$$

for $\theta > \frac{2}{3}\theta_0$. If $\theta \leq \frac{2}{3}\theta_0$ we have $D_{\phi_{(2)}}(\theta, \theta_0) = +\infty$.

b) We should reject the null hypothesis, $H_0 : \theta = \theta_0$, if

$$T_n^2(\widehat{\theta}, \theta_0) \equiv 2nD_{\phi_{(2)}}(\widehat{\theta}, \theta_0) > c.$$

The maximum likelihood estimator for θ is $\widehat{\theta} = \overline{Y}^{-1}$. Therefore, we can write

$$T_n^2(\widehat{\theta}, \theta_0) = \begin{cases} \frac{n}{3}\left(\dfrac{4}{27\overline{Y}^2\left(1 - \overline{Y}\right)} - 1\right) & \text{if } \ \overline{Y} < 1 \\ +\infty & \text{if } \ \overline{Y} \geq 1 \end{cases}.$$

We denote

$$g(y) = \begin{cases} \frac{n}{3}\left(\dfrac{4}{27y^2(1-y)} - 1\right) & \text{if } \ y < 1 \\ +\infty & \text{if } \ y \geq 1 \end{cases}$$

and it is clear that if $y \to 0$ or 1 then g tends to $+\infty$ and also the minimum i obtained for $y = 2/3$. Therefore,

$$g(y) > k$$

if and only if $y < k_1$ or $y > k_2$. The values k_1 and k_2 are obtained as a solution of the equation system

$$\begin{cases} g(k_1) = g(k_2) \\ \Pr_{\theta=3/2}\left(\overline{Y} \in (k_1, k_2)\right) = \alpha. \end{cases}$$

It is well known that if Y has an exponential distribution with parameter θ_0, then $\sum_{i=1}^{n} Y_i$ has a gamma distribution with parameters $a = \theta_0$ and $p = n$. Also $Z = 2n\theta_0\overline{Y}$ is chi-square with n degrees of freedom. One procedure to calculate the values k_1 and k_2 determining the critical region in an easier way is the following: Reject $H_0 : \theta = 3/2$ when

$$\overline{Y} < k_1 \text{ or } \overline{Y} > k_2$$

is equivalent to reject H_0 when $3n\overline{Y} < c_1$ or $3n\overline{Y} > c_2$, where c_1 and c_2 are obtained in such a way that

$$\begin{cases} 0.95 = F_{\chi_n^2}(c_2) - F_{\chi_n^2}(c_1) \\ g\left(\frac{c_1}{3n}\right) = g\left(\frac{c_2}{3n}\right) \end{cases}.$$

Then c_1 and c_2 are the solutions of the equation system

$$\begin{cases} 0.95 = \displaystyle\int_{c_1}^{c_2} \frac{1}{2^{n/2}\Gamma\left(\frac{n}{2}\right)} e^{-\frac{1}{2}x} x^{\frac{n}{2}-1} dx \\ c_2^{\frac{n}{2}}(3n - c_2) = c_1^{\frac{n}{2}}(3n - c_1) \end{cases}.$$

8. *a)* Given two M-variate normal distributions $N(\boldsymbol{\mu}_1, \boldsymbol{\Sigma}_1)$ and $N(\boldsymbol{\mu}_2, \boldsymbol{\Sigma}_2)$, Kullback-Leibler divergence (see Chapter 1) between them is

$$\begin{aligned} D_{Kull}(\boldsymbol{\theta}_1, \boldsymbol{\theta}_2) &= \tfrac{1}{2}\left((\boldsymbol{\mu}_1 - \boldsymbol{\mu}_2)^T \boldsymbol{\Sigma}_2^{-1} (\boldsymbol{\mu}_1 - \boldsymbol{\mu}_2) + trace\left(\boldsymbol{\Sigma}_2^{-1}\boldsymbol{\Sigma}_1 - I\right)\right) \\ &+ \tfrac{1}{2}\log\frac{|\boldsymbol{\Sigma}_2|}{|\boldsymbol{\Sigma}_1|}, \end{aligned}$$

with $\boldsymbol{\theta}_1 = (\boldsymbol{\mu}_1, \boldsymbol{\Sigma}_1)$ and $\boldsymbol{\theta}_2 = (\boldsymbol{\mu}_2, \boldsymbol{\Sigma}_2)$.

In our case $\boldsymbol{\theta}_1 = (\boldsymbol{\mu}, \boldsymbol{\Sigma}_0)$ and $\boldsymbol{\theta}_2 = (\boldsymbol{\mu}_0, \boldsymbol{\Sigma}_0)$ with $\boldsymbol{\Sigma}_0$ known. Given the random sample

$$\boldsymbol{Y}_1 = (Y_{11}, ..., Y_{1M})^T, \boldsymbol{Y}_2 = (Y_{21}, ..., Y_{2M})^T, ..., \boldsymbol{Y}_n = (Y_{n1}, ..., Y_{nM})^T$$

from the population distributed $N(\boldsymbol{\mu}, \boldsymbol{\Sigma}_0)$, the maximum likelihood estimator of $\boldsymbol{\mu}$ is $\overline{\mathbf{Y}} = \left(\overline{Y}_1, \overline{Y}_2, ..., \overline{Y}_M\right)^T$, where

$$\overline{Y}_i = \frac{1}{n}\sum_{l=1}^{n} Y_{li}, \; i = 1, ..., M.$$

Then the test statistic for testing $\boldsymbol{\mu} = \boldsymbol{\mu}_0$

$$T_n^{Kull}\left(\overline{\mathbf{Y}}, \boldsymbol{\mu}_0\right) \equiv 2nD_{Kull}\left(\overline{\mathbf{Y}}, \boldsymbol{\mu}_0\right) = n\left(\overline{\mathbf{Y}} - \boldsymbol{\mu}_0\right)^T \boldsymbol{\Sigma}_0^{-1}\left(\overline{\mathbf{Y}} - \boldsymbol{\mu}_0\right)$$

is asymptotically chi-squared distributed with M degrees of freedom. Then w should reject the null hypothesis if

$$T_n^{Kull}\left(\overline{\boldsymbol{Y}},\boldsymbol{\mu}_0\right)=n\left(\overline{\boldsymbol{Y}}-\boldsymbol{\mu}_0\right)^T\boldsymbol{\Sigma}_0^{-1}\left(\overline{\boldsymbol{Y}}-\boldsymbol{\mu}_0\right)>\chi^2_{M,\alpha}.$$

In this case the exact distribution of the random variable

$$n\left(\overline{\boldsymbol{Y}}-\boldsymbol{\mu}_0\right)^T\boldsymbol{\Sigma}_0^{-1}\left(\overline{\boldsymbol{Y}}-\boldsymbol{\mu}_0\right)$$

is also chi-square with M degrees of freedom. Then the exact and asymptotic test coincide.

A $(1-\alpha)\,100\%$ confidence region in $\mathbb{R}^2$ for the mean of X_1 and X_2 is given by th values $\boldsymbol{\mu}=(\mu_1,\mu_2)\in\mathbb{R}^2$ verifying

$$n\left(\overline{\boldsymbol{Y}}-\boldsymbol{\mu}\right)^T\boldsymbol{\Sigma}_0^{-1}\left(\overline{\boldsymbol{Y}}-\boldsymbol{\mu}\right)<\chi^2_{2,\alpha}.$$

b) Applying *a)* it is necessary to evaluate

$$A=n\left(\overline{\boldsymbol{Y}}-\boldsymbol{\mu}_0\right)^T\boldsymbol{\Sigma}_0^{-1}\left(\overline{\boldsymbol{Y}}-\boldsymbol{\mu}_0\right).$$

Hence

$$A=25\,(3.74,1.84)\begin{pmatrix}100&0\\0&100\end{pmatrix}^{-1}\begin{pmatrix}3.72\\1.84\end{pmatrix}=4.343,$$

and $\chi^2_{2,0.05}=5.99$. Then the null hypothesis should not be rejected.

The 95% confidence region for (μ_1,μ_2) is given by

$$\left\{(\mu_1,\mu_2)\in\mathbb{R}^2:(185.72-\mu_1)^2+(183.84-\mu_2)^2<23.96\right\}.$$

9. *a)* In this case $\theta_1=(\boldsymbol{\mu}_0,\boldsymbol{\Sigma})$ and $\theta_2=(\boldsymbol{\mu}_0,\boldsymbol{\Sigma}_0)$. Given the random sample

$$\boldsymbol{Y}_1=(Y_{11},...,Y_{1M})^T,\boldsymbol{Y}_2=(Y_{21},...,Y_{2M})^T,...,\boldsymbol{Y}_n=(Y_{n1},...,Y_{nM})^T$$

the maximum likelihood estimator of $\boldsymbol{\Sigma}$, with $\boldsymbol{\mu}=\boldsymbol{\mu}_0$ known (see Mardia et al 1979, p. 104), is given by

$$\widehat{\boldsymbol{\Sigma}}=\boldsymbol{S}+\boldsymbol{d}\boldsymbol{d}^T$$

where $\boldsymbol{d}=\overline{\boldsymbol{Y}}-\boldsymbol{\mu}_0$ and $\boldsymbol{S}$ is the sample variance covariance matrix given by

$$\boldsymbol{S}=(s_{ij})_{i,j=1,...,M}=\left(\frac{1}{n}\sum_{l=1}^{n}\left(Y_{il}-\overline{Y}_i\right)\left(Y_{jl}-\overline{Y}_j\right)\right)_{i,j=1,...,M}.$$

The expression of the test statistic based on Kullback-Leibler divergence is

$$T_n^{Kull}\left(\widehat{\boldsymbol{\Sigma}},\boldsymbol{\Sigma}_0\right)=2nD_{Kull}\left(\widehat{\boldsymbol{\Sigma}},\boldsymbol{\Sigma}_0\right)=n\left(trace\left(\boldsymbol{\Sigma}_0^{-1}\widehat{\boldsymbol{\Sigma}}-I\right)+\log\frac{|\boldsymbol{\Sigma}_0|}{\left|\widehat{\boldsymbol{\Sigma}}\right|}\right).$$

Therefore we should reject the null hypothesis if $2nD_{Kull}(\widehat{\Sigma}, \Sigma_0) > \chi^2_{\frac{M^2+M}{2},\alpha}$.

The confidence region for $\sigma = (\sigma_{11}, \sigma_{22}, \sigma_{12})$ is given by

$$\left\{ (\sigma_{11}, \sigma_{22}, \sigma_{12}) \ : n \left(trace \left(\Sigma^{-1}\widehat{\Sigma} - I \right) + \log \frac{|\Sigma|}{\left|\widehat{\Sigma}\right|} \right) < 7.82 \right\}.$$

b) With the data given in Exercise 8, we have

$$\widehat{\Sigma} = \begin{pmatrix} 93.48 & 66.87 \\ 66.87 & 96.77 \end{pmatrix} + \begin{pmatrix} 13.83 & 6.84 \\ 6.84 & 3.38 \end{pmatrix} = \begin{pmatrix} 107.31 & 73.71 \\ 73.71 & 100.11 \end{pmatrix},$$

i.e.,

$$\left|\widehat{\Sigma}\right| = 5308.96, \ |\Sigma_0| = 10000 \text{ and } trace \left(\Sigma_0^{-1}\widehat{\Sigma} - I \right) = 0.074.$$

Then

$$T_n^{Kull} \left(\widehat{\Sigma}, \Sigma_0 \right) = 17.679.$$

On the other hand $\chi^2_{3,0.05} = 7.82$ and we should reject the null hypothesis.

10. In this case the unknown parameter is $\theta = (\mu, \Sigma)$. Given the random sample

$$Y_1 = (Y_{11}, ..., Y_{1M})^T, Y_2 = (Y_{21}, ..., Y_{2M})^T, ..., Y_n = (Y_{n1}, ..., Y_{nM})^T,$$

from the normal population $N(\mu, \Sigma)$, the maximum likelihood estimator of θ, under the null hypothesis, is given by

$$\widetilde{\theta} = \left(\mu_0, S + dd^T \right),$$

where d and S were given in Exercise 9. The maximum likelihood estimator of θ in all the parameter space is given by

$$\widehat{\theta} = \left(\overline{Y}, S \right).$$

Therefore

$$\begin{aligned} D_{Kull}(\widehat{\theta}, \widetilde{\theta}) &= \tfrac{1}{2} \left(\overline{Y} - \mu_0 \right)^T \left(S + dd^T \right)^{-1} \left(\overline{Y} - \mu_0 \right) \\ &+ \tfrac{1}{2} \left(trace \left(\left(S + dd^T \right)^{-1} S - I_{M \times M} \right) + \log \frac{\left| S + dd^T \right|}{|S|} \right), \end{aligned}$$

and we must reject the null hypothesis if

$$T_n^{Kull}(\widehat{\theta}, \widetilde{\theta}) \equiv 2nD_{Kull}(\widehat{\theta}, \widetilde{\theta}) > c.$$

We are going to find the exact test statistic. We have

$$\begin{aligned} M = & \ trace(\boldsymbol{I}_{M\times M}) \\ = & \ trace\left(\left(\boldsymbol{S}+\boldsymbol{d}\boldsymbol{d}^T\right)^{-1}\left(\boldsymbol{S}+\boldsymbol{d}\boldsymbol{d}^T\right)\right) \\ = & \ trace\left(\left(\boldsymbol{S}+\boldsymbol{d}\boldsymbol{d}^T\right)^{-1}\boldsymbol{S}\right)+trace\left(\left(\boldsymbol{S}+\boldsymbol{d}\boldsymbol{d}^T\right)^{-1}\boldsymbol{d}\boldsymbol{d}^T\right) \\ = & \ trace\left(\left(\boldsymbol{S}+\boldsymbol{d}\boldsymbol{d}^T\right)^{-1}\boldsymbol{S}\right)+trace\left(\left(\overline{\mathbf{Y}}-\boldsymbol{\mu}_0\right)^T\left(\boldsymbol{S}+\boldsymbol{d}\boldsymbol{d}^T\right)^{-1}\left(\overline{\mathbf{Y}}-\boldsymbol{\mu}_0\right)\right) \\ = & \ trace\left(\left(\boldsymbol{S}+\boldsymbol{d}\boldsymbol{d}^T\right)^{-1}\boldsymbol{S}\right)+\left(\overline{\mathbf{Y}}-\boldsymbol{\mu}_0\right)^T\left(\boldsymbol{S}+\boldsymbol{d}\boldsymbol{d}^T\right)^{-1}\left(\overline{\mathbf{Y}}-\boldsymbol{\mu}_0\right), \end{aligned}$$

because $\left(\overline{\boldsymbol{Y}}-\boldsymbol{\mu}_0\right)^T\left(\boldsymbol{S}+\boldsymbol{d}\boldsymbol{d}^T\right)^{-1}\left(\overline{\mathbf{Y}}-\boldsymbol{\mu}_0\right)$ is a scalar.

Then we have

$$T_n^{Kull}(\widehat{\boldsymbol{\theta}},\widetilde{\boldsymbol{\theta}}) = n\log\frac{\left|\boldsymbol{S}+\boldsymbol{d}\boldsymbol{d}^T\right|}{|\boldsymbol{S}|}.$$

Taking into account that

$$\left|\boldsymbol{A}_{p\times p}+\boldsymbol{B}_{p\times n}\boldsymbol{C}_{n\times p}\right| = \left|\boldsymbol{A}_{p\times p}\right|\left|\boldsymbol{I}_{n\times n}+\boldsymbol{C}_{n\times p}\boldsymbol{A}_{p\times p}^{-1}\boldsymbol{B}_{p\times n}\right|,$$

we have for $\boldsymbol{A}_{p\times p}=\boldsymbol{S}$, $\boldsymbol{B}_{p\times n}=\boldsymbol{d}$ and $\boldsymbol{C}_{n\times p}=\boldsymbol{d}^T$ that

$$T_n^{Kull}(\widehat{\boldsymbol{\theta}},\widetilde{\boldsymbol{\theta}}) = n\log\frac{|\boldsymbol{S}|\left(1+\boldsymbol{d}^T\boldsymbol{S}^{-1}\boldsymbol{d}\right)}{|\boldsymbol{S}|} = n\log\left(1+\boldsymbol{d}^T\boldsymbol{S}^{-1}\boldsymbol{d}\right).$$

Hence $T_n^{Kull}(\widehat{\boldsymbol{\theta}},\widetilde{\boldsymbol{\theta}})>c$ is equivalent to

$$(n-1)\,\boldsymbol{d}^T\boldsymbol{S}^{-1}\boldsymbol{d} > T^2_{M,\,n-1,\,\alpha}, \qquad (9.20)$$

where $T^2_{M,\,n-1}$ is a Hotelling T^2 distribution with parameters M and $n-1$. For more details about this distribution see Mardia et *al*. (1979). Finally, the inequality given in (9.20) is equivalent to

$$\frac{n-M}{n}\boldsymbol{d}^T\boldsymbol{S}^{-1}\boldsymbol{d} > F_{M,\,n-M,\,\alpha}.$$

When we use the asymptotic distribution we have

$$T_n^{Kull}(\widehat{\boldsymbol{\theta}},\widetilde{\boldsymbol{\theta}}) \xrightarrow[n\to\infty]{L} \chi^2_{2M+\frac{M(M-1)}{2}-\left(M+\frac{M(M-1)}{2}\right)=M}.$$

11. In this case, $\boldsymbol{\theta}=(\boldsymbol{\mu},\boldsymbol{\Sigma})$, then

$$\widetilde{\boldsymbol{\theta}} = \left(\overline{\boldsymbol{Y}},\boldsymbol{\Sigma}_0,\right) \quad \text{and} \quad \widehat{\boldsymbol{\theta}} = \left(\overline{\boldsymbol{Y}},\boldsymbol{S}\right).$$

Therefore,

$$2nD_{Kull}(\widehat{\boldsymbol{\theta}},\widetilde{\boldsymbol{\theta}}) = n\left(trace\left(\boldsymbol{\Sigma}_0^{-1}\boldsymbol{S}-\boldsymbol{I}\right)+\log\frac{|\boldsymbol{\Sigma}_0|}{|\boldsymbol{S}|}\right)$$

whose asymptotic distribution is chi-square with degrees of freedom given by

$$\dim(\Theta)-\dim(\Theta_0) = M+M+\frac{1}{2}\left(M\left(M-1\right)\right)-M = \frac{M^2+M}{2}.$$

12. In this case we are interested in a exact test because the regularity assumptions are not verified and then it is not possible to use the asymptotic results. It is necessary to get the exact distribution of the test statistic

$$D_{Kull}(\widehat{\theta}_1,\widehat{\theta}_2).$$

For $\theta_1 > \theta_2$ we have

$$D_{Kull}\left(\theta_1,\theta_2\right) = \int_0^{\theta_1}\frac{1}{\theta_1}\log\frac{1/\theta_1}{1/\theta_2}dx + \int_{\theta_1}^{\theta_2} 0\log\frac{0}{1/\theta_2}dx = \log\frac{\theta_2}{\theta_1},$$

and

$$D_{Kull}(\widehat{\theta}_1,\widehat{\theta}_2) = \log\frac{\widehat{\theta}_2}{\widehat{\theta}_1},$$

with $\widehat{\theta}_1 = \max\left(X_1,...,X_n\right)$, $\widehat{\theta}_2 = \max\left(Y_1,...,Y_m\right)$, where $X_1,...,X_n$ is a random sample from the population $U\left(0,\theta_1\right)$ and $Y_1,...,Y_m$ is a random sample from the population $U\left(0,\theta_2\right)$. We should reject the null hypothesis if $\log(\widehat{\theta}_2/\widehat{\theta}_1) > c$, where c is obtained under the assumption that the test has a nominal size α. This is equivalent to reject H_0 if and only if $\widehat{\theta}_2/\widehat{\theta}_1 > k$.

The probability density functions of $\widehat{\theta}_1$ and $\widehat{\theta}_2$ are given by

$$\begin{aligned} f_{\widehat{\theta}_1}\left(t_1\right) &= nt_1^{n-1}/\theta_1^n \qquad t_1\in\left(0,\theta_1\right)\\ f_{\widehat{\theta}_2}\left(t_2\right) &= mt_2^{m-1}/\theta_2^m \qquad t_2\in\left(0,\theta_2\right), \end{aligned}$$

and the joint probability density function of the random variable $(\widehat{\theta}_1,\widehat{\theta}_2)$ by

$$f_{\widehat{\theta}_1\widehat{\theta}_2}\left(t_1,t_2\right) = \frac{mn}{\theta_1^n\theta_2^m}t_1^{n-1}t_2^{m-1} \qquad t_1\in\left(0,\theta_1\right),\ t_2\in\left(0,\theta_2\right).$$

Now we are going to get the distribution of the random variable $\widehat{\theta}_2/\widehat{\theta}_1$ under the null hypothesis $\theta_1 = \theta_2$. It is immediate to establish that the distribution of the random variable (W,T) where

$$W = \frac{\widehat{\theta}_2}{\widehat{\theta}_1} \quad \text{and} \quad T = \widehat{\theta}_1$$

is given by

$$g_{(W,T)}(w,t) = \frac{nm}{\theta_1^{n+m}} t^{n+m} w^{m-1} \qquad (w,t) \in A$$

where

$$A = \left\{(w,t) \in \mathbb{R}^2 : wt < \theta_2,\ 0 < t < \theta_1 \text{ and } w > 0\right\}.$$

Therefore the probability density function of the random variable W is

$$g_W(w) = \begin{cases} \frac{nm}{n+m} w^{m-1} & w \in (0,1] \\ \frac{nm}{n+m} w^{-n-1} & w \in (1,\infty) \end{cases}.$$

Finally, if we assume that $\alpha < m/(m+n)$, from the equation

$$\alpha = \Pr(W > k/\ H_0) = \int_k^\infty \frac{nm}{n+m} w^{-n-1} dw$$

we have

$$k = \left(\frac{\alpha}{m}(n+m)\right)^{-1/n}.$$

Then, the null hypothesis should be rejected if

$$\frac{\widehat{\theta}_2}{\widehat{\theta}_1} > \left(\frac{\alpha}{m}(n+m)\right)^{-1/n}.$$

In our case we have for $\alpha = 0.05$,

$$\widehat{\theta}_1 = 1.9616,\ \widehat{\theta}_2 = 2.9358,\ \widehat{\theta}_2/\widehat{\theta}_1 = 1.49 \text{ and } \left(\frac{\alpha}{m}(n+m)\right)^{-1/n} = 1.31;$$

hence we should reject the null hypothesis.

13. It is immediate to establish that

$$D_{Kull}(\theta_1,\theta_2) = \int_{\theta_1}^\infty \exp\left(-(x-\theta_1)\right)(\theta_1-\theta_2)\,dx = \theta_1 - \theta_2.$$

Therefore

$$D_{Kull}(\widehat{\theta}_1,\widehat{\theta}_2) = \widehat{\theta}_1 - \widehat{\theta}_2$$

with $\widehat{\theta}_1 = \min(X_1,...,X_n)$ and $\widehat{\theta}_2 = \min(Y_1,...,Y_m)$ and the null hypothesis should be rejected if

$$\widehat{\theta}_1 - \widehat{\theta}_2 > c$$

where c is obtained in such a way the test has size α. We know that

$$\begin{aligned} f_{\widehat{\theta}_1}(t_1) &= n\exp\left(-n(t_1-\theta_1)\right) \quad \theta_1 \le t_1 < \infty \\ f_{\widehat{\theta}_2}(t_2) &= m\exp\left(-m(t_2-\theta_2)\right) \quad \theta_2 \le t_2 < \infty \end{aligned}.$$

Then the probability density function of the random variable $(\widehat{\theta}_1, \widehat{\theta}_2)$ is given by

$$f_{\widehat{\theta}_1,\widehat{\theta}_2}(t_1, t_2) = nm \exp(-n(t_1 - \theta_1)) \exp(-m(t_2 - \theta_2))$$

where

$$\theta_1 \le t_1 < \infty, \ \theta_2 \le t_2 < \infty.$$

Under $\theta_1 = \theta_2$, the distribution of the bivariate random variable (W, T), with

$$W = \widehat{\theta}_1 - \widehat{\theta}_2 \quad \text{and} \quad T = \widehat{\theta}_2,$$

is given by

$$f_{W,T}(w, t) = mn \exp(-n(w + t - \theta_1) \exp(-m(t - \theta_1))) \qquad (w, t) \in A$$

where the domain A is

$$A = \left\{(w, t) \in \mathbb{R}^2 / t > \theta_2, \ w > \theta_1 - t\right\}.$$

The distribution of $W = \widehat{\theta}_1 - \widehat{\theta}_2$ under $\theta_1 = \theta_2$ is

$$f_{\widehat{\theta}_1 - \widehat{\theta}_2}(w) = \begin{cases} \dfrac{nm}{n+m} \exp(mw) & \text{if} \quad w \in (-\infty, 0) \\ \dfrac{nm}{n+m} \exp(-nw) & \text{if} \quad w \in (0, \infty) \end{cases}.$$

Therefore c is obtained, assuming that $\alpha \le m/(m+n)$, by solving the equation

$$\alpha = \int_c^\infty \frac{nm}{n+m} \exp(-nw)\, dw.$$

It is immediate to get

$$c = \frac{1}{n} \log\left(\frac{m}{(m+n)\alpha}\right)$$

and we should reject the null hypothesis if

$$\widehat{\theta}_1 - \widehat{\theta}_2 > \frac{1}{n} \log\left(\frac{m}{(m+n)\alpha}\right).$$

In our case

$$\frac{1}{n} \log\left(\frac{m}{(m+n)\alpha}\right) = 0.2349$$

and $\widehat{\theta}_1 - \widehat{\theta}_2 = 0.002$; then we should not reject the null hypothesis.

14. We have

$$L(\theta; Y_1, ..., Y_n) = \frac{\theta^n 2^{n\theta}}{\prod_{i=1}^n Y_i^{\theta+1}}$$

then

$$\frac{\partial \log L(\theta; Y_1, ..., Y_n)}{\partial \theta} = \frac{n}{\theta} + n \log 2 - \sum_{i=1}^{n} \log Y_i$$

and the maximum likelihood estimator is given by

$$\widehat{\theta} = \frac{n}{\sum\limits_{i=1}^{n} \log \frac{Y_i}{2}}.$$

We have

$$\begin{aligned} L_n^0 &= 2 \left(\log \left(\frac{(\widehat{\theta})^n 2^{n\widehat{\theta}}}{\prod\limits_{i=1}^{n} Y_i^{\widehat{\theta}+1}} \right) - \log \left(\frac{\theta_0^n 2^{n\theta_0}}{\prod\limits_{i=1}^{n} Y_i^{\theta_0+1}} \right) \right) \\ &= 2n \left(\frac{\theta_0 - \widehat{\theta}}{\widehat{\theta}} \right) + 2n \log \left(\frac{\widehat{\theta}}{\theta_0} \right). \end{aligned}$$

In order to obtain the expression of the test statistics W_n and R_n it is necessar to obtain the Fisher information. We have

$$\frac{\partial \log f_\theta(x)}{\partial \theta} = \frac{1}{\theta} + \log 2 - \log x$$

then

$$\mathcal{I}_{\mathcal{F}}(\theta) = E\left[-\left(\frac{\partial^2 \log f_\theta(x)}{\partial \theta^2} \right) \right] = \frac{1}{\theta^2},$$

and

$$W_n = n \left(\frac{\theta_0 - \widehat{\theta}}{\widehat{\theta}} \right)^2 \xrightarrow[n \to \infty]{L} \chi_1^2.$$

On the other hand,

$$U_n(\theta_0) = \left(\frac{\partial \log L(\theta; Y_1, ..., Y_n)}{\partial \theta} \right)_{\theta=\theta_0} = n \left(\frac{1}{\theta_0} + \log 2 - \frac{1}{n} \sum_{i=1}^{n} \log Y_i \right)$$

and then

$$R_n = n \left(-1 - \frac{\theta_0}{n} \sum_{i=1}^{n} \log \frac{Y_i}{2} \right)^2.$$

Now Rényi's divergence is given by

$$\begin{aligned} D_r^1(\theta_1, \theta_2) &= \frac{1}{r(r-1)} \log \int_2^\infty \left(\frac{2^{\theta_1} \theta_1}{x^{\theta_1+1}} \right)^r \left(\frac{2^{\theta_2} \theta_2}{x^{\theta_2+1}} \right)^{1-r} dx \\ &= \frac{1}{r(r-1)} \log \frac{\theta_1^r \theta_2^{1-r}}{r\theta_1 + (1-r)\theta_2} \end{aligned}$$

if $r\theta_1 + (1-r)\,\theta_2 > 0$. Therefore Rényi test statistic

$$T_n^r(\widehat{\theta}, \theta_0) = \frac{2n}{r\,(r-1)} \log \frac{\widehat{\theta}^r \theta_0^{1-r}}{r\widehat{\theta} + (1-r)\,\theta_0},$$

and for $r \to 1$ we have

$$\lim_{r\to 1} T_n^r(\widehat{\theta}, \theta_0) = T_n^{Kull}(\widehat{\theta}, \theta_0) = 2n\left(\frac{\theta_0 - \widehat{\theta}}{\widehat{\theta}}\right) + 2n \log\left(\frac{\widehat{\theta}}{\theta_0}\right) = L_n^0.$$

15. For the exponential model with

$$f_\theta(x) = \theta \exp(-\theta x)\, I_{(0,\infty)}(x), \quad \theta > 0,$$

we consider the problem of testing

$$H_0 : \theta = 1 \quad \text{versus} \quad H_1 : \theta \neq 1$$

based on Rényi test statistic $T_n^r(\widehat{\theta}, \theta_0)$. First, we study the accuracy of power approximations of Remarks 9.1–9.3. Rényi divergence between two exponential distributions is

$$D_r^1(\theta_1, \theta_2) = \begin{cases} \dfrac{1}{r\,(r-1)} \ln \dfrac{\theta_1^r\, \theta_2^{1-r}}{\theta_1 r + \theta_2\,(1-r)} & \text{if } \theta_1 r + \theta_2\,(1-r) > 0 \\ \infty & \text{if } \theta_1 r + \theta_2\,(1-r) \leq 0, \end{cases}$$

when $r \neq 0, 1$. Limiting cases are obtained for $r = 1$,

$$D_{Kull}(\theta_1, \theta_2) = \lim_{r\to 1} D_r^1(\theta_1, \theta_2) = \frac{\theta_2}{\theta_1} - 1 + \ln \frac{\theta_1}{\theta_2} \quad \text{(Kullback-Leibler)}$$

and for $r = 0$,

$$D_0^1(\theta_1, \theta_2) = D_{Kull}(\theta_2, \theta_1) = \frac{\theta_1}{\theta_2} - 1 + \ln \frac{\theta_2}{\theta_1}$$

the minimum discrimination information.

a) H_0 should be rejected if $T_n^r(\widehat{\theta}, 1) > c$, where $\widehat{\theta} = n / \sum_{i=1}^n Y_i = \overline{Y}^{-1}$ is the maximum likelihood estimator of θ. In this case $T_n^r(\widehat{\theta}, 1)$ is given by

$$\begin{cases} \frac{1}{r(r-1)} 2n \ln\left(\widehat{\theta}^r (\widehat{\theta} r + 1 - r)^{-1}\right) & \text{if } r \neq 0,1 \text{ and } \widehat{\theta} r + 1 - r > 0 \\ \infty & \text{if } r \neq 0,1 \text{ and } \widehat{\theta} r + 1 - r \leq 0 \\ 2n(\widehat{\theta}^{-1} - 1 + \ln \widehat{\theta}) & \text{if } r = 1 \\ 2n(\widehat{\theta} - 1 - \ln \widehat{\theta}) & \text{if } r = 0, \end{cases}$$

but $T_n^r(\widehat{\theta}, 1) > c$ is equivalent to

$$\begin{cases} (1-r)\overline{Y}^r + r\overline{Y}^{r-1} > c_1 & \text{if } \widehat{\theta} r + 1 - r > 0 \\ \overline{Y} - \ln \overline{Y} > c_2 & \text{if } r = 1 \\ \frac{1}{\overline{Y}} + \ln \overline{Y} > c_3 & \text{if } r = 0. \end{cases}$$

If we define the function

$$g_r(y) = \begin{cases} (1-r)y^r + ry^{r-1} & \text{if } r \neq 0,1 \\ y - \ln y & \text{if } r = 1 \\ \frac{1}{y} + \ln y & \text{if } r = 0, \end{cases}$$

we have

$$g_r'(y) = \begin{cases} > 0 & \Leftrightarrow \; y < 1 \\ < 0 & \Leftrightarrow \; y > 1 \end{cases}$$

and the rejection rule is $T_n^r(\widehat{\theta}, 1) > c \iff Z_n < c_{r,1}$ or $Z_n > c_{r,2}$ where $Z_n = 2n\overline{Y} \sim \chi^2_{2n}$ under H_0. Constants $c_{r,1}$ and $c_{r,2}$ are obtained by solving the equations

$$\begin{cases} 1-\alpha = & \Pr(c_{r,1} < Z_n < c_{r,2}) = F_{\chi^2_{2n}}(c_{r,2}) - F_{\chi^2_{2n}}(c_{r,1}), \\ g_r\left(\frac{c_{r,1}}{2n}\right) = & g_r\left(\frac{c_{r,2}}{2n}\right) \end{cases}$$

where

$$g_r(c) = \begin{cases} (1-r)c^r + rc^{r-1} & \text{if } r \neq 0,1 \\ c - \ln c & \text{if } r = 1 \\ \frac{1}{c} + \ln c & \text{if } r = 0. \end{cases}$$

The values of $c_{r,1}$ and $c_{r,2}$, for $n = 20, 40, 80, 200$ and $r = 1/4$, $3/4$, 1, $5/4$, $7/4$, 2, are presented in Table 9.3.

r	n	$n = 20$	$n = 40$	$n = 80$	$n = 200$
$1/4$	$c_{1/4,1}$	25.855	58.861	128.804	348.637
	$c_{1/4,2}$	64.033	110.541	200.373	460.411
$3/4$	$c_{3/4,1}$	25.253	58.100	127.914	347.618
	$c_{3/4,2}$	61.323	108.388	198.531	458.802
1	$c_{1,1}$	24.879	57.659	127.421	347.074
	$c_{1,2}$	60.275	107.479	197.709	458.051
$5/4$	$c_{5/4,1}$	24.449	57.173	126.892	346.505
	$c_{5/4,2}$	59.371	106.658	196.944	457.334
$7/4$	$c_{7/4,1}$	23.379	56.039	125.714	345.290
	$c_{7/4,2}$	57.898	105.239	195.565	455.990
2	$c_{2,1}$	22.700	55.374	125.057	344.639
	$c_{2,2}$	57.300	104.626	194.943	455.361

Table 9.3. Constants $c_{r,1}$ and $c_{r,2}$.

b) Exact powers of tests:

We know that under an alternative hypothesis $\theta \neq 1$, the random variable $2\theta\overline{Y}$ is chi-squared distributed with $2n$ degrees of freedom

Therefore the power function of Rényi test statistic, $\beta_{n,r}(\theta)$, is given by

$$\begin{aligned}\beta_{n,r}(\theta) &= 1 - \Pr{}_\theta\left(c_{r,1} < Z_n < c_{r,2}\right)\\ &= 1 - \Pr{}_\theta\left(\theta c_{r,1} < 2\theta\overline{Y} < \theta c_{r,2}\right)\\ &= 1 - \Pr{}_\theta\left(\theta c_{r,1} < \chi^2_{2n} < \theta c_{r,2}\right).\end{aligned}$$

For $n = 20$, $r = 0.25, 0.75, 1, 1.25, 1.75, 2$ and different values of θ in the interval $[0.5, 2.5]$, these exact powers are presented in Table 9.4.

θ	$\beta_{20,1/4}(\theta)$	$\beta_{20,3/4}(\theta)$	$\beta_{20,1}(\theta)$	$\beta_{20,5/4}(\theta)$	$\beta_{20,7/4}(\theta)$	$\beta_{20,2}(\theta)$
0.50	0.8117	0.8560	0.8714	0.8838	0.9025	0.9094
0.60	0.5417	0.6155	0.6437	0.6676	0.7057	0.7207
0.65	0.4005	0.4768	0.5074	0.5339	0.5774	0.5950
0.70	0.2778	0.3478	0.3771	0.4032	0.4473	0.4656
0.75	0.1821	0.2399	0.2652	0.2883	0.3284	0.3455
0.80	0.1149	0.1581	0.1779	0.1964	0.2295	0.2440
0.85	0.0729	0.1021	0.1160	0.1294	0.1539	0.1649
0.90	0.0511	0.0683	0.0768	0.0852	0.1010	0.1083
0.95	0.0446	0.0522	0.0561	0.0600	0.0676	0.0712
1.10	0.0879	0.0763	0.0699	0.0631	0.0490	0.0418
1.20	0.1557	0.1344	0.1221	0.1089	0.0803	0.0652
1.30	0.2481	0.2180	0.2002	0.1806	0.1366	0.1123
1.40	0.3578	0.3208	0.2983	0.2731	0.2143	0.1803
1.50	0.4750	0.4339	0.4084	0.3792	0.3085	0.2657
1.60	0.5892	0.5476	0.5211	0.4902	0.4124	0.3633
1.70	0.6922	0.6532	0.6278	0.5975	0.5184	0.4662
1.80	0.7788	0.7447	0.7219	0.6943	0.6192	0.5675
1.90	0.8473	0.8192	0.8000	0.7763	0.7092	0.6611
2.00	0.8984	0.8766	0.8613	0.8420	0.7854	0.7429
2.10	0.9348	0.9187	0.9071	0.8921	0.8466	0.8110
2.20	0.9595	0.9481	0.9397	0.9287	0.8937	0.8652
2.30	0.9756	0.9679	0.9621	0.9542	0.9285	0.9065
2.40	0.9857	0.9807	0.9768	0.9714	0.9532	0.9369
2.50	0.9919	0.9887	0.9862	0.9827	0.9701	0.9585

Table 9.4. Powers for $n = 20$

c) We know that Rényi divergence is a (h, ϕ)-divergence with h and ϕ given in (9.7) and (9.8). Therefore using Exercise 1 we have

$$\sqrt{n}\left(D_r^1(\widehat{\theta}, 1) - D_r^1(\theta, 1)\right) \underset{n\to\infty}{\overset{L}{\longrightarrow}} N(0, \sigma^2_{h,\phi}(\theta)),$$

where $\sigma^2_{h,\phi}(\theta) = \left(h'\left(D_\phi\left(\theta, 1\right)\right)\right)^2 \sigma^2_\phi\left(\theta\right)$ and $\sigma^2_\phi\left(\theta\right) = T^T \mathcal{I}_{\mathcal{F}}\left(\theta\right)^{-1} T$. In our case, $\mathcal{I}_{\mathcal{F}}(\theta) = \theta^{-2}$,

$$T = \frac{d}{d\theta} D_r^1(\theta, 1) = \begin{cases} \frac{1}{r-1}\left(\frac{1}{\theta} - \frac{1}{\theta r+1-r}\right) & \text{if } \ r \neq 0,1 \\ \frac{\theta-1}{\theta^2} & \text{if } \ r = 1 \\ \frac{\theta-1}{\theta} & \text{if } \ r = 0 \end{cases}$$

and

$$D_r^1(\theta, 1) = \begin{cases} \frac{1}{r(r-1)}\left(\frac{\theta^r}{\theta r+(1-r)} - 1\right) & \text{if } \ r \neq 0,1 \\ \theta - 1 - \ln\theta & \text{if } \ r = 0 \\ \ln\theta - 1 + \theta^{-1} & \text{if } \ r = 1 \end{cases} .$$

	$\beta_{20}^2(\theta)$	$\beta_{20,r}^1(\theta)$					
θ		r=1/4	r=3/4	r=1	r=5/4	r=7/4	r=2
.50	0.6088	0.8083	0.8165	0.8271	0.8436	0.8804	1.0000
.60	0.4320	0.5900	0.6342	0.6559	0.6787	0.7310	0.7588
.65	0.3463	0.4543	0.5116	0.5385	0.5651	0.6205	0.6505
.70	0.2677	0.3105	0.3714	0.4006	0.4291	0.4860	0.5155
.75	0.1989	0.1728	0.2236	0.2495	0.2756	0.3282	0.3552
.80	0.1412	0.0637	0.0920	0.1079	0.1249	0.1616	0.1812
.85	0.0944	0.0086	0.0147	0.0188	0.0235	0.0351	0.0421
.90	0.0571	0.0000	0.0001	0.0001	0.0002	0.0004	0.0006
.95	0.0268	0.0000	0.0000	0.0000	0.0000	0.0000	0.0000
1.1	0.0571	0.0000	0.0000	0.0000	0.0000	0.0000	0.0000
1.2	0.1412	0.0324	0.0202	0.0155	0.0116	0.0060	0.0042
1.3	0.2677	0.1692	0.1259	0.1059	0.0872	0.0551	0.0420
1.4	0.4320	0.3293	0.2697	0.2392	0.2087	0.1492	0.1215
1.5	0.6088	0.4666	0.4067	0.3742	0.3399	0.2667	0.2288
1.6	0.7653	0.5754	0.5238	0.4951	0.4637	0.3917	0.3510
1.7	0.8791	0.6600	0.6199	0.5979	0.5734	0.5136	0.4772
1.8	0.9471	0.7255	0.6973	0.6830	0.6669	0.6254	0.5980
1.9	0.9805	0.7766	0.7592	0.7521	0.7443	0.7222	0.7058
2.0	0.9940	0.8165	0.8083	0.8075	0.8069	0.8019	0.7956
2.1	0.9985	0.8480	0.8473	0.8514	0.8563	0.8642	0.8655
2.2	0.9997	0.8731	0.8781	0.8859	0.8947	0.9106	0.9163
2.3	0.9999	0.8931	0.9025	0.9128	0.9239	0.9435	0.9509
2.4	1.0000	0.9092	0.9218	0.9336	0.9457	0.9657	0.9728
2.5	1.0000	0.9223	0.9371	0.9497	0.9618	0.9800	0.9858

Table 9.5. Powers for $n = 20$.

θ	$r=1/4$	$r=3/4$	$r=1$	$r=5/4$	$r=7/4$	$r=2$
0.50	20	19	18	17	13	(1)
0.60	36	35	34	32	27	25
0.65	50	48	47	45	41	38
0.70	72	70	69	67	62	59
0.75	109	107	106	104	99	95
0.80	180	178	176	174	168	165
0.85	338	334	332	329	323	319
0.90	801	795	792	788	780	775
0.95	3370	3358	3352	3345	3331	3323
1.10	972	979	981	984	989	991
1.20	266	269	270	272	273	274
1.30	129	131	132	132	133	134
1.40	78	80	81	81	82	82
1.50	54	56	56	56	57	57
1.60	41	42	42	42	43	43
1.70	32	33	33	34	34	34
1.80	26	27	27	28	28	28
1.90	22	23	23	23	24	24
2.00	19	20	20	20	20	21
2.10	17	18	18	18	18	18
2.20	15	16	16	16	16	16
2.30	14	14	14	14	15	15
2.40	13	13	13	13	13	14
2.50	12	12	12	12	12	12

Table 9.6. Sample sizes $n=[n^*]+1$, where n^* is the root of the equation $0.8=\beta^1_{n^*,r}(\theta)$. (1)No value of n is obtained because approximate power is 1.

Therefore $\sigma^2_{h,\phi}(\theta)=\frac{|\theta-1|}{\theta^r}$.

First power approximation, $\beta^1_{n,r}(\theta)$, is given by

- $1-\Phi\left(\frac{\sqrt{n}\theta^r}{|\theta-1|}\left(\frac{3.84145}{2n}-\frac{1}{r(r-1)}\ln\frac{\theta^r}{\theta r+1-r}\right)\right)$ if $r\neq 0,1$
- $1-\Phi\left(\frac{\sqrt{n}}{|\theta-1|}\theta\left(\frac{3.84145}{2n}-\frac{1}{\theta}+1-\ln\theta\right)\right)$ if $r=1$

and

- $1-\Phi\left(\frac{\sqrt{n}}{|\theta-1|}\left(\frac{3.84145}{2n}-\theta+1+\ln\theta\right)\right)$ if $r=0$,

where $\Phi(\cdot)$ is the c.d.f. of the standard normal random variable.

Second power approximation (see Exercise 2) is

$$\beta^2_n(\theta)=1-G_{\chi^2_1(\delta)}\left(\chi^2_{1,\alpha}\right),$$

where $G_{\chi_1^2(\delta)}$ is the distribution function of a noncentral chi-squar random variable with 1 degree of freedom and noncentrality paramete $\delta = n(\theta - 1)^2 \mathcal{I}_{\mathcal{F}}(1) = n(\theta - 1)^2$. It is interesting to note that thi approximation does not depend on the functions h and ϕ considere in the test statistic T_n^r.

We are interested in the approximation of $\beta_{n,r}(\theta)$ by $\beta_{n,r}^1(\theta)$ an $\beta_n^2(\theta)$. In Table 9.5 we present their values for $n = 20$, $r = 0.25$, 0.75 1, 1.25, 1.75, 2 and several values of $\theta \neq 1$. We conclude that $\beta_{20,r}^1(\theta$ and $\beta_{20}^2(\theta)$ are good approximations for $\beta_{20,r}(\theta)$ in the present mode Approximations improve as n increases. For the sake of brevity, w do not present tables for $n > 20$. It is also interesting to note that fo values of θ near to 1, and more concretely if $\theta \in (0.8, 1.3)$, the approx imation $\beta_{20}^2(\theta)$ is better than the approximation $\beta_{20,r}^1(\theta)$. Otherwise the approximation $\beta_{20,r}^1(\theta)$ is better than $\beta_{20}^2(\theta)$. We conclude tha both approximations can be recommended for practical applications For a value of the power function equal to 0.8, $r = 0.25$, 0.75, 1, 1.25 1.75, 2 and several values of $\theta \neq 1$, in Table 9.6 we present the approx imate sample size $n = [n^*] + 1$, where n^* is the positive root of th equation $\beta = \beta_{n^*,r}^1(\theta)$. As expected, larger sample sizes are obtaine in the neighborhood of $\theta = 1$ (e.g., $\theta \in (0.8, 1.3)$). Observe that n^* ca be obtained from $\beta = \beta_{n^*,r}^1(\theta)$ in explicit form (cf. (9.19)); howeve this is not the case for n^{**} such that $\beta = \beta_{n^{**},r}(\theta)$. In the presen numerical example $\beta_{n,r}^1(\theta)$ and n^* are good and easy computable ap proximations for $\beta_{n,r}(\theta)$ and n^{**} respectively.

References

Adhikari, B. P. and Joshi, D. D. (1956). Distance, discrimination et résumé exhaustif. *Publications de l'Institut de Statistique de l'Université de Paris,* **5**, 57-74.

Agresti, A. (1983). Testing marginal homogeneity for ordinal categorical variables. *Biometrics,* **39**, 505-511.

Agresti, A. (1996). *An Introduction to Categorical Data Analysis.* John Wiley & Sons, New York.

Agresti, A. (2002). *Categorical Data Analysis* (Second Edition). John Wiley & Sons, New York.

Agresti, A. and Agresti, B. F. (1978). Statistical analysis of qualitative variation. *Sociological Methodology,* **9**, 204-237.

Aitchison, J. and Silvey, S. D. (1958). Maximum likelihood estimation of parameters subject to restraints. *Annals of Mathematical Statistics,* **29**, 813-828.

Ali, S. M. and Silvey, S. D. (1966). A general class of coefficient of divergence of one distribution from another. *Journal of the Royal Statistical Society,* **28**, 1, 131-142.

Andersen, E.B. (1990). *The Statistical Analysis of Categorical data.* Springer-Verlag, Berlin.

Andersen, E.B. (1998). *Introduction to the Statistical Analysis of Categorical Data.* Springer-Verlag, Berlin.

Arimoto, S. (1971). Information-theoretical considerations on estimation problems. *Information and Control,* **19**, 181-194.

Arndt, C. (2001). *Information Measures.* Springer-Verlag, Berlin.

Azalarov, T. A. and Narkhuzhaev, A. A. (1987). Asymptotic analysis of some chi-square type tests for Markov chains. *Doklady Akademii Nauk*, SSSR, **7**, 3-5.

Azalarov, T. A. and Narkhuzhaev, A. A. (1992). On asymptotic behavior of distributions of some statistics for Markov chains. *Theory of Probability and its Applications,* **37**, 117-119.

Bahadur, R. R. (1971). *Some Limit Theorems in Statistics.* Society for Industrial anc Applied Mathematics, Philadelphia.

Balakrishnan, V. and Sanghvi, L. D. (1968). Distance between populations on the basi of attribute. *Biometrics,* **24,** 859-865.

Bar-Hen, A. (1996). A preliminary test in discriminant analysis. *Journal of Multivariat Analysis,* **57**, 266-276.

Bar-Hen, A. and Daudin, J. J. (1995). Generalization of the Mahalanobis distance i the mixed case. *Journal of Multivariate Analysis,* **53**, 332-342.

Barmi, H. E. and Kochar, S. C. (1995). Likelihood ratio tests for bivariate symme try against ordered alternatives in a squared contingency table. *Statistics an Probability Letters,* **22,** 167-173.

Bartlett, M. S. (1951). The frequency goodness of fit test for probability chains. *Math ematical Proceedings of the Cambridge Philosophical Society,* **47,** 86-95.

Basawa, I. V. and Prakasa Rao, B. L. S. (1980). *Statistical Inference for Stochasti Processes.* Academic Press, London.

Basharin, G. P. (1959). On a statistical estimate for the entropy of a sequence c independent random variables. *Theory of Probability and its Applications,* **4,** 333 336.

Bednarski, T. and Ledwina, T. (1978). A note on a biasedness of test of fit. *Mathema tische Operationsforschung und Statistik, Series Statistics,* **9,** 191-193.

Behboodian, J. (1970). On a mixture of normal distributions. *Biometrika,* **57**, 215-217

Bhapkar, V. P. (1966). A note on the equivalence of two criteria for hypotheses i categorical data. *Journal of the American Statistical Association,* **61**, 228-235.

Bhapkar, V. P. (1979). On tests of symmetry when higher order interactions are absent *Journal Indian Statistical Association,* **17**, 17-26.

Bhapkar, V. P. and Darroch, J. N. (1990). Marginal Symmetry and Quasi Symmetr of General Order. *Journal of Multivariate Analysis,* **34,** 173-184.

Bhargava, T. N. and Doyle, P. H. (1974). A geometry study of diversity. *Journal c Theoretical Biology,* **43,** 241-251.

Bhargava, T. N. and Uppuluri, V. R. R. (1975). On an axiomatic derivation of Gin diversity with applications. *Metron,* **30,** 1-13.

Bhattacharyya, A. (1943). On a measure of divergence between two statistical popu lations defined by their probability distributions. *Bulletin Calcutta Mathematica Society,* **35**, 99-109.

Billingsley, P. (1961a). Statistical methods in Markov chains. *Annals of Mathematical Statistics,* **32,** 12-40.

Billingsley, P. (1961b). *Statistical Inference for Markov Processes.* Chicago Univ., Chicago Press.

Birch, M. W. (1964). A new proof of the Pearson-Fisher theorem. *Annals of Mathematical Statistics,* **35,** 817-824.

Bishop, M. M., Fienberg, S. E. and Holland, P. W. (1975). *Discrete Multivariate Analysis: Theory and Practice.* MIT Press, Cambridge, Mass.

Bofinger, E. (1973). Goodness of fit using sample quantiles. *Journal of the Royal Statistical Society,* Series B, **35**, 277-284.

Bonett, D. G. (1989). Pearson chi-square estimator and test for log-linear models with expected frequencies subject to linear constraints. *Statistics and Probability Letters,* **8,** 175-177.

Bowker, A. (1948). A test for symmetry in contingency tables. *Journal of the American Statistical Association,* **43**, 572-574.

Bregman, L. M. (1967). The relaxation method of finding the common point of convex sets and its application to the solution of problems in convex programming. *U.S.S.R. Computational Mathematics and Mathematical Physics,* **7,** 200-217.

Breusch, T. S. and Pagan, A. R. (1980). The Lagrange multiplier test and its applications to model specification in econometrics. *Review of Economic Studies,* **47**, 239-254.

Broffit, J. D. and Randles, R. H. (1977). A power approximation for the chi-square goodness-of-fit test: Simple hypothesis case. *Journal of the American Statistical Association,* **72,** 604-607.

Bross, I. (1954). Misclassification in 2x2 tables. *Biometrics,* **10**, 478-486.

Bryant, J. L. and Paulson, A. S. (1983). Estimation of mixing proportions via distance between characteristic functions. *Communications in Statistics,* **12**, 1009-1029.

Burbea, J. (1983). J-divergence and related concepts. *Encyclopedia of Statistical Sciences,* **4,** 290-296. John Wiley & Sons, New York.

Burbea, J. (1984). The Bose-Einstein entropy of degree α and its Jensen difference. *Utilitas Mathematica,* **25,** 225-240.

Burbea, J. and Rao, C. R. (1982a). On the convexity of some divergence measures based on entropy functions. *IEEE Transactions on Information Theory,* **28**, 489-495.

Burbea, J. and Rao, C. R. (1982b). On the convexity of higher order Jensen differences based on entropy functions. *IEEE Transactions on Information Theory,* **28**, 961-963.

Burbea, J. and Rao, C. R. (1982c). Entropy differential metric, distance and diver gence measures in probability spaces: A unified approach. *Journal of Multivariat Analysis,* **12,** 575-596.

Buse, A. (1982). The likelihood Ratio, Wald, and Lagrange multiplier tests: An expos itory note. *The American Statistician,* **36**, 153-157.

Cacoullos, T. and Koutras, M. (1985). Minimum distance discrimination for spherica distributions, in *Statistical Theory and Data Analysis.* (K. Matusita, Ed.), 91-101 Elsevier Science Publishers B. V. (North-Holland).

Caussinus, H. (1965). Contribution à l'analyse statistique des tableaux de correlation *Annales de la Faculté des Sciences de Toulouse,* **29,** 77-182.

Chacko, V. J. (1966). Modified chi-square test for ordered alternatives. *Sankhya,* Serie B, **28,** 185-190.

Chapman, J. W. (1976). A comparison of the χ^2, $-2\log R$, and the multinomial proba bility criteria for significance testing when expected frequencies are small. *Journa of the American Statistical Association,* **71,** 854-863.

Chase, G. R. (1972). On the chi-squared test when the parameters are estimate independently of the sample. *Journal of the American Statistical Association,* **67** 609-611.

Cheng, K. F., Hsueh, H. M. and Chien, T. H. (1998). Goodness of fit tests wit misclassified data. *Communications in Statistics (Theory and Methods)*, **27**, 1379 1393.

Chernoff, H. (1952). A measure of asymptotic efficiency for tests of a hypothesis base on the sum of obervations. *Annals of Mathematical Statistics,* **23,** 493-507.

Chernoff, H. and Lehmann, E. L. (1954). The use of maximum likelihood estimates i χ^2 tests for goodness of fit. *Annals of Mathematical Statistics,* **25**, 579-586.

Choi, K. and Bulgren, W. G. (1968). An estimation procedure for mixtures of distrib utions. *Journal of the Royal Statistical Society,* Series B, **30**, 444-460.

Christensen, R. (1997). *Log-Linear Model and Logistic Regression.* Springer-Verlag New York.

Cochran, W. G. (1952). The χ^2 test of goodness of fit. *Annals of Mathematical Statis tics,* **23**, 315-345.

Cohen, A. C. (1967). Estimation in mixtures of two normal distributions. *Technomet rics,* **9**, 15-28.

Cohen, A. and Sackrowitz, H. B. (1975). Unbiasedness of the chi-square, likelihoo ratio, and other goodness of fit tests for the equal cell case. *Annals of Statistic* **3**, 959-964.

Collett, D. (1994). *Modelling Survival Data in Medical Research.* Chapman and Hall, London.

Conover, W. J., Johnson, M. E. and Johnson M. M. (1981). A comparative study of tests for homogeneity of variances, with applications to the outer continental shelf bidding data. *Technometrics,* **23,** 351-361.

Cox, D. R. (2002). Karl Pearson and the chi-squared test, in *Goodness-of-Fit Tests and Model Validity.* (C. Huber-Carol, N. Balakrishnan, M. S. Nikulin and M. Mesbah, Eds.), Birkhäuser, Boston, 3-8.

Cressie, N. (1976). On the logarithms of high-order spacings. *Biometrika,* **63**, 343-355.

Cressie, N. (1979). An optimal statistic based on higher order gaps. *Biometrika,* **66**, 619-627.

Cressie, N. and Pardo, L. (2000). Minimum ϕ-divergence estimator and hierarchical testing in loglinear models. *Statistica Sinica,* **10,** 867-884.

Cressie, N. and Pardo, L. (2002a). Phi-divergence statistics, in *Encyclopedia of Environmetrics.* (A. H. ElShaarawi and W. W. Piegorich, Eds.). Volume 3, 1551-1555, John Wiley & Sons, New York.

Cressie, N. and Pardo, L. (2002b). Model checking in loglinear models using ϕ-divergences and MLEs. *Journal of Statistical Planning and Inference,* **103,** 437-453.

Cressie, N., Pardo, L. and Pardo, M. C. (2001). Size and power considerations for testing loglinear models using ϕ-divergence test statistics. Technical Report No. 680. Department of Statistics, The Ohio State University, Columbus, OH.

Cressie, N., Pardo, L. and Pardo, M. C. (2003). Size and power considerations for testing loglinear models using ϕ-divergence test statistics. *Statistica Sinica,* **13,** 550-570.

Cressie, N. and Read, T. R. C. (1984). Multinomial goodness-of-fit tests. *Journal of the Royal Statistical Society,* Series B, **46,** 440-464.

Csiszár, I. (1963). Eine Informationstheoretische Ungleichung und ihre Anwendung auf den Bewis der Ergodizität on Markhoffschen Ketten. *Publications of the Mathematical Institute of the Hungarian Academy of Sciences,* **8,** 84-108.

Csiszár, I. (1967). Information-type measures of difference of probability distributions and indirect observations. *Studia Scientiarum Mathematicarum Hungarica,* **2**, 299-318.

Cuadras, C. M. (1992a). Probability distributions with given multivariate marginals and given dependence structure. *Journal of Multivariate Analysis,* **42**, 51–66.

Cuadras, C. M. (1992b). Some examples of distance based discrimination. *Biometrical Letters,* **29**, 3-20.

Cuadras, C. M., Atkinson, R. A. and Fortiana, J. (1997a). Probability densities fron distances and discrimination. *Statistics and Probability Letters,* **33**, 405–411.

Cuadras, C. M., Fortiana, J. and Oliva, F. (1997b). The proximity of an individual t a population with applications in discriminant analysis. *Journal of Classification* **14,** 117–136.

D'Agostino, R. B. and Stephens, M. A. (1986). *Goodness-of-fit Techniques.* Marce Dekker, New York.

Dale, J. R. (1986). Asymptotic normality of goodness-of-fit statistics for sparse produc multinomials. *Journal of the Royal Statistical Society*, Series B, **41,** 48-59.

Dannenbring, D. G. (1997). Procedures for estimating optimal solution values for larg combinatorial problems. *Management Science,* **23**, 1273-1283.

Darling, D. A. (1953). On the class of problems related to the random division of a interval. *Annals of Mathematical Statistics*, **24**, 239-253.

Darroch, J. N. (1981). The Mantel-Haenszel test and tests of marginal symmetry fixed effects and mixed models for a categorical response. *International Statistica Review,* **49,** 285-307.

Darroch, J. N. (1986). Quasi-symmetry, in *Encyclopedia of Statistical Sciences.* (S Kotz and N. L. Johnson, Eds.), Volume 7, 469-473, John Wiley & Sons, New York.

Darroch, J. N. and McCullagh, P. (1986). Category distinguishability and observe agreement. *Australian Journal of Statistics,* **28,** 371-388.

Davidson, R. and Lever, W. (1970). The limiting distribution of the likelihood rati statistic under a class of local alternatives. *Sankhya,* Series A, **22**, 209-224.

Day, N. E. (1969). Estimating the components of a mixture of normal distributions *Biometrika,* **56**, 3, 463-474.

Del Pino, G. E. (1979). On the asymptotic distribution of k-spacings with application to goodness-of-fit tests. *The Annals of Statistics*, **7**, 1058-1065.

Diamond, E. L. and Lilienfeld, A. M. (1962). Effects of errors in classification an diagnosis in various types of epidemiological studies. *American Journal of Publi Health*, **10,** 2106-2110.

Diamond, E. L., Mitra, S. K. and Roy, S. N. (1960). Asymptotic power and asymp totic independence in the statistical analysis of categorical data. *Bulletin of th International Statistical Institute,* **37**, 3-23.

Dik, J. J. and de Gunst, M. C. M. (1985). The distribution of general quadratic form in normal variables. *Statistica Neerlandica,* **39,** 14-26.

Drost, F. C., Kallenberg, W. C. M., Moore, D. S. and Oosterhoff, J. (1989). Power approximations to multinomial tests of fit. *Journal of the American Statistical Association,* **84**, 130-141.

Dudewicz, F. J. and Van der Meulen, E. C. (1981). Entropy-based tests of uniformity. *Journal of the American Statistical Association*, **76**, 967-974.

Durbin, J. (1978). Goodness-of-tests based on order statistics, in *Transactions of the 7th Prague Conference on Information Theory,* Volume B, 109-127, Czech Academy of Sciences, Prague.

Ebraimi, N., Pflughoeft, K. and Soofi, E. S. (1994). Two measures of sample entropy. *Statistics and Probability Letters*, **20**, 225-234.

Engle, R. F. (1981). Wald, likelihood ratio, and Lagrange multiplier tests in econometrics, in *Handbook of Econometrics,* (Z. Griliches and M. Intriligator, Eds.), North Holland, Amsterdam.

Fenech, A. P. and Westfall, P. H. (1988). The power function of conditional log-linear model tests. *Journal of the American Statistical Association,* **83,** 198-203.

Ferguson, T. S. (1996). *A Course in Large Sample Theory.* Texts in Statistical Science. Chapman & Hall, New York.

Ferreri, C. (1980). Hypoentropy and related heterogeneity, divergence and information measures. *Statistica,* **2,** 155-167.

Fisher, R. A. (1924). The conditions under which χ^2measures the discrepancy between observation and hypothesis. *Journal of the Royal Statistical Society*, **87,** 442-450.

Fisher, R. A. (1925). *Statistical Methods for Research Workers.* Hafner Press, New York.

Fleming, W. (1977). *Functions of Several Variables* (Second Edition). Springer-Verlag, New York.

Fraser, D. A. S. (1957). *Nonparametric Methods in Statistics.* John Wiley & Sons, New York.

Fryer, J. G. and Roberson, C. A. (1972). A comparison of some methods for estimating mixed normal distributions. *Biometrika,* **59**, 3, 639-648.

Garren, Steven T. (2000). Asymptotic distribution of estimated affinity between multiparameter exponential families. *Annals of the Institute of Statistical Mathematica,* **52**, 426–437.

Gebert, J. R. and Kale, B. K. (1969). Goodness of fit tests based on discriminatory information. *Statistiche Hefte*, **10**, 192-200.

Gibbs, A. L. and Su, F. E. (2002). On choosing and bounding probability metrics. *International Statistical Review,* **70,** 419-435.

Gini, C. (1912). Variabilitá e Mutabilitá, Studi Economicoaguaridici della Facotta d Ginrisprudenza dell, Universite di Cagliari III, Parte II.

Gleser, L. J. and Moore, D. S. (1983a). The effect of dependence on chi-squared an empiric distribution tests of fit. *Annals of Statistics,* **11,** 1100-1108.

Gleser, L. J. and Moore, D. S. (1983b). The effect of positive dependence on chi-square tests for categorical data. *Journal of the Royal Statistical Society*, Series B, **47** 459-465.

Gokhale, D. V. (1973). Iterative maximum likelihood estimation for discrete distributions. *Sankhya,* **35,** 293-298.

Gokhale, D. V. and Kullback, S. (1978). *The Information in Contingency Tables* Marcel Dekker, New York.

Good, I. J., Gover, T. N. and Mitchell, G. J. (1970). Exact distributions for X^2 and fo the likelihood-ratio statistic for the equiprobable multinomial distribution. *Journa of the American Statistical Association,* **65**, 267-283.

Greenwood, M. (1946). The statistical study of infectious diseases. *Journal of the Roya Statistical Society*, Series A, **109**, 85-110.

Greenwood, P. E. and Nikulin, M. S. (1996). *A Guide to Chi-squared Testing.* Wile Series in Probability and Statistics, New York.

Gyorfi, L. and Nemetz, K. (1978). f-dissimilarity: A generalization of the affinity o several distributions. *Annals of the Institute of Statistical Mathematics,* **30**, 105 113.

Haber, M. (1980). A comparative simulation study of the small sample powers o several goodness-of-fit tests. *Journal of Statistical Computation and Simulation* **11,** 241-251.

Haber, M. (1984). A comparison of tests for the hypothesis of no three-factor interactio $2 \times 2 \times 2$ contingency tables. *Journal of Statistical Computation and Simulation* **20**, 205-215.

Haber, M. (1985). Maximum likelihood methods for linear and log-linear models i categorical data. *Computational Statistics & Data Analysis,* **3,** 1-10.

Haber, M. and Brown, M. B. (1986). Maximum likelihood estimation for log-linea models when expected frequencies are subject to linear constraints. *Journal of th American Statistical Association,* **81**, 477-482.

Haberman, S. J. (1974). *The Analysis of Frequency Data.* University of Chicago Press Chicago.

Haberman, S. J. (1978). *Analysis of Qualitative Data,* Volume 1. Academic Press, Ne York.

Hájek, J. and Sidák, Z. (1967). *Theory of Rank Tests.* Academic Press, New York.

Hall, P. (1986). On powerful distribution tests based on sample spacings. *Journal of Multivariate Analysis*, **19**, 201-224.

Hassenblad, V. (1966). Estimation of parameters for a mixture of normal distributions. *Technometrics,* **8**, 431-434.

Havrda, J. and Charvat, F. (1967). Concept of structural α-entropy. *Kybernetika,* **3**, 30-35.

Hobza, T., Molina, I. and Morales, D. (2001). Rényi statistics for testing hypotheses with s samples, with an application to familial data. *Trabajos I+D, I-2001-31*, Operations Research Center, Miguel Hernández University of Elche.

Hochberg, Y. (1977). On the use of double sampling schemes in analyzing categorical data with misclassification errors. *Journal of the American Statistical Association,* **72**, 914-921.

Hoeffding, W. (1965). Asymptotically optimal tests for multinomial distributions. *Annals of Mathematical Statistics,* **36**, 369-408.

Holst, L. (1972). Asymptotic normality and efficiency for certain goodness-of-fit tests. *Biometrika,* **59**, 137-145.

Hosmane, B. (1986). Improved likelihood ratio tests and Pearson chi-square tests for independence in two dimensional contingency tables. *Communications in Statistics (Theory and Methods),* **15**, 1875-1888.

Inaba, T. and Shirahata, S. (1986). Measures of dependence in normal models and exponential models by information gain. *Biometrika,* **73**, 345-352.

Ireland, C. T., Ku, H. H. and Kullback, S. (1969). Symmetry and marginal homogeneity of an $r \times r$ contingency table. *Journal of the American Statistical Association,* **64**, 1323-1341.

Ivchenko, G. and Medvedev, Y. (1990). *Mathematical Statistics.* Mir, Moscow.

Jammalamadaka, S. R., Zhou, X. and Tiwari, R. C. (1989): Asymptotic efficiencies of spacings tests for goodness of fit. *Metrika*, **36**, 355-377.

Jaynes, E. T. (1957). Information theory and statistical mechanics. *Physical Reviews,* **106**, 620-630.

Jaynes, E. T. (1957). Information theory and statistical mechanics. *Physical Reviews,* **108**, 171-197.

Jeffreys, H. (1946). An invariant form for the prior probability in estimation problems. *Proceedings of the Royal Society,* Series A, **186**, 453-561.

Jiménez, R. and Palacios, J. L. (1993). Shannon's measure of ordered samples: Som asymptotic results. Documento de trabajo. Universidad de Simón Bolívar, Caracas.

Joe, H., (1989). Relative entropy measures of multivariate dependence. *Journal of th American Statistical Association*, **84**, 157-164.

Kagan, A. M. (1963). On the theory of Fisher's amount of information. *Doklad Akademii Nauk SSSR,* **151,** 277-278.

Kailath, T. (1967). The Divergence and Bhatacharyya distance measures in signa selection. *IEEE Translation on Communication Technology,* **15,** 1, 52-60.

Kale, B. K. (1969). Unified derivation of test of goodness-of-fit based on spacings *Sankhya*, Series A, **31**, 43-48.

Kale, B. K. and Godambe, V. P. (1967). A test of goodness of fit. *Statistiche Hefte*, 8 165-172.

Kallenberg, W. C. M., Oosterhoff, J. and Schriever, B. F. (1985). The number of classe in chi-squared goodness-of-fit tests. *Journal of the American Statistical Association,* **80**, 959-968.

Kapur, J. N. (1972). Measures of uncertainty, mathematical programming and physics *Journal of the Indian Society of Agriculture and Statistics,* **24,** 47-66.

Kapur, J. N. (1982). Maximum-entropy probability distribution for a continuous random variable over a finite interval. *Journal of Mathematical Physics Sciences,* **16** 97-103.

Kapur, J. N. (1988). On measures of divergence based on Jensen difference. *Nation Academy Science Letters,* **11**, 1, 23-27.

Kapur, J. N. (1989). *Maximum-entropy Models in Science and Engineering.* John Wile & Sons, New York.

Kapur, J. N. and Kesaven, H. K. (1992). *Entropy Optimization Principles with Applications.* Academic Press, New York.

Kaufman, H. and Mathai, A. M. (1973). An axiomatic foundation for a multivariate measure of affinity among a number of distributions. *Journal of Multivariat Analysis,* **3,** 236-242.

Kent, J. T. (1983). Information gain and a general measure of correlation. *Biometrik* **70**, 163-173.

Kimball, B. F. (1947). Some basic theorems for developping tests of fit for the case o nonparametric probability distributions functions. *Annals of Mathematical Statistics*, **18**, 540-548.

Kirmani, S. N. U. A. and Alam, S. N. (1974). On goodness of fit tests based on spacings. *Sankhya*, Series A, **36**, 197-203.

Koehler, K. J. (1986). Goodness-of-fit statistics for log-linear models in sparse contingency tables. *Journal of the American Statistical Association,* **81**, 483-493.

Koehler, K. J. and Larntz, K. (1980). An empirical investigation of goodness-of-fit statistics for sparse multinomials. *Journal of the American Statistical Association,* **75**, 336-344.

Kolmogorov, A. N. (1933). Sulla determinazione empirica di una legge di distribuzione. *Giornale dell'Istituto Italiano degli Attuari,* **4**, 83-91.

Kolmogorov, A. N. (1963). On the approximations of distributions of sums of independent summands by infinitely divisible distributions. *Sankhya,* **25**, 159-174.

Kotze, T. J. V. W. and Gokhale, D. V. (1980). A comparison of the Pearson-X^2 and the log-likelihhod-ratio statistics for small samples by means of probability ordering. *Journal of Statistical Computation and Simulation,* **12**, 1-13.

Koutras, M. (1992). Minimum distance discrimination rules and success rates for elliptical normal mixtures. *Statistics and Probability Letters,* **13**, 259-268.

Krzanowski, W. J. (1982). Mixtures of continuous and categorical variables in discriminant analysis: A hypothesis-testing approach. *Biometrics,* **3**, 991-1002.

Kullback, S. (1959). *Information Theory and Statistics.* John Wiley & Sons, New York.

Kullback, S. (1971). Marginal homogeneity of multidimensional contingency tables. *The Annals of Mathematical Statistics,* **42**, 594-606.

Kullback, S. (1985). Kullback information, in *Encyclopedia of Statistical Sciences,* Volume 4 (S. Kotz and N. L. Johnson, Eds.), 421-425. John Wiley & Sons, New York.

Kullback, S. and Leibler, R. (1951). On information and sufficiency. *Annals of Mathematical Statistics*, **22**, 79-86.

Kumar, K. D., Nicklin, E. H. and Paulson, A. S. (1979). Comment on "An estimation procedure for mixtures of distributions" by Quandt and Ramsey, *Journal of the American Statistical Association*, **74**, 52-55.

Kuo, M. and Rao, J. S. (1984). Asymptotic results on the Greenwood statistic and some generalizations. *Journal of the Royal Statistical Association*, Series B, **46**, 228-237.

Kupperman, M. (1957). Further application to information theory to multivariate analysis and statistical inference. *Ph. D. Dissertation,* George Washington University.

Kupperman, M. (1958). Probability hypothesis and information statistics in samplin exponential class populations. *Annals of Mathematical Statistics,* **29,** 571-574.

Larntz, K. (1978). Small sample comparisons of exact levels for Chi-squared goodnes of fit statistics. *Journal of the American Statistical Association,* **73,** 253-263.

Lawal, H. B. (1984). Comparisons of X^2, Y^2, Freman-Tukey and Williams's improve G^2 test statistics in small samples of one-way multinomials. *Biometrika,* **71,** 415 458.

Lazo, A. C. G. V. and Rathie, P. N. (1978). On the entropy of continuous distribution *IEEE Transactions on Information Theory,* **IT-24,** 120-121.

Le Cam, L. (1960). *Locally Asymptotic Normal Families of Distribution.* Univ. Cali fornia Publications in Statistics. Univ. California Press, Berkeley.

Le Cam, L. (1990). Maximum likelihood: An introduction. *International Statistic Review,* **58,** 2, 153-171.

Lee, C. C. (1987). Chi-Squared tests for and against an order restriction on multinomia parameters. *Journal of the American Statistical Association,* **82,** 611-618.

Lehmann, E. L. (1959). *Testing Statistical Hypotheses.* John Wiley & Sons, New York

Lévy, P. (1925). *Calcul des Probabilités.* Gauthiers-Villars, Paris.

Lewontin, R. C. (1972). The apportionment of human diversity, *Evolutionary Biolog* **6,** 381-398.

Lieberson, S. (1969). Measuring population diversity. *American Sociology Review,* **34** 850-862.

Liese, F. and Vajda, I. (1987). *Convex Statistical Distances.* Teubner, Leipzig.

Light, R. J. and Margolin, B. H. (1971). An analysis of variance for categorical data *Journal of the American Statistical Association,* **66,** 534-544.

Lin, J. (1991). Divergence measures based on the Shannon entropy. *IEEE Transaction on Information Theory,* **37,** 145-151.

Lindley, D. V. (1956). On a measure of the information provided by an experiment *Annals of Mathematical Statistics,* **27,** 986-1005.

Lindsay, B. G. (1994). Efficiency versus robutness: The case for minimum Hellinge distance and other methods. *Annals of Statistics,* **22,** 1081-1114.

Lukacs, E. (1975). *Stochastic Convergence* (Second Edition). Academic Press, New York.

Lyons, N. I. and Hutcheson, K. (1979). Distributional properties of Simpson's index o diversity. *Communications in Statistics (Theory and Methods),* **6,** 569-574.

MacDonald, P. D. M. (1971). Comment on "Estimating mixtures of Normal distributions and switching distributions" by Choi and Bulgren, *Journal of the Royal Statistical Society*, Series B, **33**, 326-329.

Mahalanobis, P. C . (1936). On the generalized distance in statistics. *Proceedings of the National Academy of Sciences (India),* **2**, 49-55.

Mann, H. B. and Wald, A. (1942). On the choice of the number of class intervals in the application of the chi-square test. *Annals of Mathematical Statistics,* **13**, 306-317.

Mardia, K. V., Kent J. T. and Bibby, J. M. (1979). *Multivariate Analysis.* Academic Press, New York.

Margolin, B. H. and Light, R. L. (1974). An analysis of variance for categorical data, II: Small sample comparisons with chi square and other competitors. *Journal of the American Statistical Association,* **69**, 755-764.

Marhuenda, Y. (2003). *Contrastes de Uniformidad.* Ph. D. Dissertation. Universidad Miguel Hernández. Elche, Spain.

Marshall, A. W. and Olkin, I. (1979). *Inequalities: Theory of Majorization and Its Applications.* Academic Press, New York.

Mathai, A. and Rathie, P. N. (1975). *Basic Concepts in Information Theory.* John Wiley & Sons, New York.

Matthews, G. B. and Growther, N. A. S. (1997). A maximum likelihood estimation procedure when modelling categorical data in terms of cross-product ratios. *South African Statistics Journal,* **31**, 161-184.

Matusita, K. (1955). Decision rules, based on the distance for problems of fit, two samples, and estimation. *Annals of Mathematical Statistics,* **26**, 631-640.

Matusita, K. (1964). Distance and decision rules. *Annals of the Institute of Statistical Mathematics,* **16**, 305-320.

Matusita, K. (1966). A distance and related statistics in multivariate analysis. In *Multivariate Analysis.* (P. R. Krishnaiah, Ed.), 187-200, Academic Press, New York.

Matusita, K. (1967). On the notion of affinity of several distributions and some of its applications. *Annals of the Institute of Statistical Mathematics,* **19**, 181-192.

Matusita, K. (1973). Discrimination and the affinity of distributions, in *Discriminant Analysis and Applications.* (T. Cacoullus, Ed.), 213-223. Academic Press, New York.

McCullagh, P. (1982). Some applications of quasisymmetry. *Biometrika,* **69**, 303-308.

Menéndez, M. L., Pardo, L. and Taneja, I. J. (1992). On M-dimensional unified (r, s) Jensen difference divergence measures and their applications. *Kybernetika,* **28**, 309-324.

Menéndez, M. L., Morales, D., Pardo, L. and Salicrú, M. (1995). Asymptotic behaviour and statistical applications of divergence measures in multinomial populations: A unified study. *Statistical Papers,* **36**, 1-29.

Menéndez, M. L., Morales, D. and Pardo, L. (1997a). Maximum entropy principle and statistical inference on condensed ordered data. *Statistics and Probability Letters,* **34**, 85-93.

Menéndez, M. L., Morales, D., Pardo, L. and Vajda, I. (1997b). Testing in stationary models based on ϕ-divergences of observed and theoretical frequencies. *Kybernetika,* **33**, 465-475.

Menéndez, M. L., Pardo, J. A., Pardo, L. and Pardo, M. C. (1997c). Asymptotic approximations for the distributions of the (h, ϕ)-divergence goodness-of-fit statistics: Applications to Renyi's statistic. *Kybernetes,* **26**, 442-452.

Menéndez, M. L., Pardo, J. A., Pardo, L. and Pardo, M. C. (1997d). The Jensen-Shannon divergence. *Journal of the Franklin Institute,* **334**, 307-318

Menéndez, M. L., Morales, D., Pardo, L. and Salicrú, M. (1997e). Divergence measures between s populations: Statistical applications in the exponential family. *Communications in Statistics (Theory and Methods),* **26**, 1099-1117.

Menéndez, M. L., Morales, D., Pardo, L. and Vajda, I. (1998a). Two approaches to grouping of data and related disparity statistics. *Communications in Statistics (Theory and Methods),* **27**, 603-633.

Menéndez, M. L., Morales, D., Pardo, L. and Vajda, I. (1998b). Asymptotic distributions of ϕ-divergences of hypotetical and observed frequencies on refined partitions. *Statistica Neerlandica,* **52**, 1, 71-89.

Menéndez, M. L., Morales, D., Pardo, L. and Vajda, I. (1999a). Inference about stationary distributions of Markov chains based on divergences with observed frequencies. *Kybernetika,* **35**, 265-280

Menéndez, M. L., Morales, D., Pardo, L. and Zografos, K. (1999b). Statistical Inference for Markov chains based on divergences. *Statistics and Probability Letters,* **41**, 9-1

Menéndez, M. L., Morales, D., Pardo, L. and Vajda, I. (2001a). Approximations to powers of ϕ-disparity goodness-of-fit. *Communications in Statistics (Theory and Methods),* **30**, 105-134.

Menéndez, M. L., Morales, D., Pardo, L. and Vajda, I. (2001b). Minimum divergence estimators based on grouped data. *Annals of the Institute of Statistical Mathematics,* **53,2**, 277-288

Menéndez, M. L., Morales, D., Pardo, L. and Vajda, I. (2001c). Minimum disparity estimators for discrete and continuous models. *Applications on Mathematics,* **6,** 439-466

Menéndez, M. L., Pardo, J. A. and Pardo, L. (2001d). Csiszar's φ-divergence for testing the order in a Markov chain. *Statistical Papers,* **42,** 313-328.

Menéndez, M. L., Pardo, J. A. and Pardo, L. (2001e). Tests based on ϕ-divergences for bivariate symmetry. *Metrika,* **53**, 1, 13-30.

Menéndez, M. L., Pardo, L. and Zografos, K. (2002). Tests of hypotheses for and against order restrictions on multinomial parameters based on ϕ-divergences. *Utilitas Mathematica,* **61,** 209-223.

Menéndez, M. L., Morales, D. and Pardo, L. (2003a). Tests based on divergences for and against ordered alternatives in cubic contingency tables. *Applied Mathematics and Computation,* **134**, 207-216

Menéndez, M. L., Pardo, J.A., Pardo, L. and Zografos, K. (2003b). On tests of homogeneity based on minimum ϕ-divergence estimator with constraints. *Computational Statistics and Data Analysis,* **43,** 215-234.

Menéndez, M. L., Pardo, J. A. and Pardo, L. (2003c). Tests for bivariate symmetry against ordered alternatives in square contingency tables. *Australian and New Zealand Journal of Statistics,* **45,** 1, 115-124.

Menéndez, M. L., Pardo, L. and Zografos, K. (2004a). Minimum φ-dissimilarity approach in classification. Preprint of the Department of Mathematics, Probability, Statistics and Operational Research Section. University of Ioannina. Greece.

Menéndez, M. L., Pardo, J.A. and Pardo, L. (2004b). Tests of symmetry in three-dimensional contingency tables based on phi-divergence statistics. *Journal of Applied Statistics,* **31,** 9, 1095-1114.

Menéndez, M. L., Pardo, J.A., Pardo, L. and Zografos, K. (2005a). On tests of symmetry, marginal homogeneity and quasi-symmetry in two contingency tables based on minimum ϕ-divergence estimator with constraints. *Journal of Statistical Computation and Simulation,* **75,** 7, 555-580.

Menéndez, M. L., Pardo, J.A., Pardo, L. and Zografos, K. (2005b). On tests of independence based on minimum ϕ-divergence estimator with constraints: An application to modeling DNA. *Submitted.*

Menéndez, M. L., Pardo, J.A., Pardo, L. and Zografos, K. (2005c). A test for marginal homogeneity based on a ϕ-divergence statistic. *Submitted.*

Menéndez, M. L., Pardo, J.A., Pardo, L. and Zografos, K. (2005d). A prelimiary test in classification and probabilities of misclassification. *Statistics,* **39**, 3, 183-205.

Mirvalev, M. and Narkhuzhaev, A. A. (1990). On a chi-square test for homogeneit Markov chains. *Izvestiya Akademii Nauk. USSR,* Ser. Fiz.-Mat., **1**, 8-32.

Molina, I., Morales, D., Pardo, L. and Vajda, I. (2002). On size increase goodness c fit tests when observations are positively dependent. *Statistics and Decisions,* **20** 399-414.

Moore, D. S. (1982). The effect of dependence on chi-squared tests of fit. *Annals c Statistics,* **10,** 1163-1171.

Morales, D. and Pardo, L. (2001). Some approximations to power functions of ϕ divergence tests in parametric models. *Test,* **9,** 249-269.

Morales, D., Pardo, L. and Pardo, M. C. (2001). Likelihood divergence statistics fo testing hypotheses about multiple population. *Communications in Statistics (Sim ulation and Computation),* **30**, 867-884.

Morales, D., Pardo, L. and Vajda, I. (1995). Asymptotic divergence of estimates c discrete distributions. *Journal of Statistical Planning and Inference,* **48**, 347-369

Morales, D., Pardo, L. and Vajda, I. (1996). Uncertainty of discrete stochastic systems General theory and statistical inference. *Transactions of IEEE on Systems, Mar and Cybernetics,* **26,** 681-697.

Morales, D., Pardo, L. and Vajda, I. (1997). Some new statistics for testing hypothese in parametric models. *Journal of Multivariate Analysis,* **62,** 1, 137-168.

Morales, D., Pardo, L. and Vajda, I. (2003). Asymptotic laws for disparity statistics i product multinomial models. *Journal of Multivariate Analysis,* **85,** 335-360.

Morales, D., Pardo, L. and Vajda, I. (2005). On efficient in continuous models base on finitely quantized observations. *Submitted.*

Morales, D., Pardo, L., Salicrú, M. and Menéndez, M.L. (1994). Asymptotic propertie of divergence statistics in a stratified random sampling and its applications to tes statistical hypotheses. *Journal of Statistical Planning and Inference,* **38**, 201-221

Morales, D., Pardo, L. and Zografos, K. (1998). Informational distances and relate statistics in mixed continuous and categorical variables. *Journal of Statistic Planning and Inference*, **75**, 47-63.

Morris, C. (1975). Central limit theorems for multinomial sums. *Annals of Statistic* **3**, 165-188.

Mote, V. L. and Anderson, R. L. (1965). An investigation of effect of misclassification o the properties of Chi-Squared tests in the analysis of categorical data. *Biometrik* **52**, 95-109.

Murthy, V. K. and Gafarian, A. V. (1970). Limiting distributions of some variations c the chi-square statistic. *Annals of Mathematical Statistics,* **41,** 188-194

Nadarajah, S. and Zografos, K. (2003). Formulas for Rényi information and related measures for univariate distributions. *Information Sciences,* **155,** 119-138.

Nadarajah, S. and Zografos, K. (2005). Expressions for Rényi and Shannon entropies for bivariate distributions. *Information Sciences,* **170,** 173-189.

Nayak, T. K. (1983). Applications of Entropy Functions in Measurement and Analysis of Diversity. Ph.D. Dissertation, University of Pittsburgh. USA.

Nayak, T. K. (1985). On diversity measures based on entropy functions. *Communications in Statistics (Theory and Methods),* **14,** 203-215.

Nayak, T. K. (1986). Sampling distributions in analysis of diversity. *Shankya,* Series B, **48,** 1-9.

Nei, M. (1973). Analysis of gene diversity in subdivided population. *Proceedings of the National Academy of Sciences* (USA), **70,** 3321-3323.

Neyman, J. (1949). Contribution to the theory of the χ^2-test. *Proceedings of the First Symposium on Mathematical Statistics and Probability,* 239-273. University of Berkeley.

Neyman, J. and Pearson, E. S. (1928). On the use and interpretation of certain test criteria for purposes of statistical inference. *Biometrika,* **8,** 175-240 and 263-294.

Oler, J. (1985). Noncentrality parameters in chi-squared goodness-of-fit analyses with an application to log-linear procedures. *Journal of the American Statistical Association,* **80,** 181-189.

Pardo, J. A. (2004). Inference for three-dimensional contingency tables based on phi-divergences, in *Soft Methodology and Random Information Systems* (M. López, M.A. Gil, P. Grzegorzewski, O. Hyrniewicz and J. Lawri, Eds.), Springer_Verlag, New York.

Pardo, J. A., Pardo, L. and Menéndez, M. L. (1992). Unified (r,s)-entropy as an index of diversity. *Journal of the Franklin Institute,* **329,** 907-921.

Pardo, J. A., Pardo, M. C., Vicente, M. L. and Esteban M. D. (1997). A statistical information theory approach to compare the homogeneity of several variances. *Computational Statistics and Data Analysis,* **24,** 411-416.

Pardo, J. A., Pardo, L. and Zografos, K. (2002). Minimum ϕ-divergence estimators with constraints in multinomial populations. *Journal of Statistical Planning and Inference,* **104,** 221-237.

Pardo, L. (1997a). Generalized divergence measures: Statistical applications, in *Encyclopedia of Microcomputers* (A. Kent, J. G. Williams and N. L. Johnson, Eds.), Marcel Dekker, New York.

Pardo, L. (1997b). *Statistical Information Theory (In Spanish).* Hesperides, Spain.

Pardo, L., Calvet, C. and Salicrú, M. (1992). Comparación de medidas de diversidac *Historia Animalium,* **1**, 3-13.

Pardo, L., Morales, D., Salicrú, M. and Menéndez, M. L. (1993a). The φ-divergenc statistic in bivariate multinomial populations including stratification. *Metrika,* **40** 223-235.

Pardo, L., Morales, D., Salicrú, M. and Menéndez, M. L. (1993b). R_{ϕ}^{h}-divergenc statistics in applied categorical data analysis with stratified sampling. *Utilita Mathematica,* **44**, 145-164.

Pardo, L., Morales, D., Salicrú, M. and Menéndez, M. L.(1995). A test for homogeneit of variances based on Shannon's entropy. *Metron,* **LIII**, 135-146.

Pardo, L., Morales, D., Salicrú, M. and Menéndez, M. L. (1997a). Large sample be haviour of entropy measures when parameters are estimated. *Communications i Statistics (Theory and Methods),* **26**, 483-502.

Pardo, L., Morales, D., Salicrú, M. and Menéndez, M. L. (1997b). Divergence mea sures between s populations: Statistical applications in the exponential famil *Communications in Statistics (Theory and Methods),* **26**, 1099-1117.

Pardo, L., Pardo, M. C. and Zografos, K. (1999). Homogeneity for multinomial pop ulations based on ϕ-divergences. *Journal of the Japan Statistical Society,* **29**, 2 213-228.

Pardo, L. and Zografos K. (2000). Goodness of fit tests with misclassified data base on ϕ-divergences. *Biometrical Journal,* **42**, 223–237.

Pardo, L., Pardo, M. C. and Zografos, K. (2001). Minimum ϕ–divergence estimator fo multinomial populations. *Shankhya,* Series A, **63**, 1, 72-92.

Pardo, L. and Pardo, M. C. (2003). Minimum power-divergence estimator in three way contingency tables. *Journal of Statistical Computation and Simulation,* **73** 819-834

Pardo, L. and Pardo, M. C. (2005). Nonadditivity in loglinear models using ϕ-divergenc and MLEs. *Journal of Statistical Planning and Inference,* **127**, 1-2, 237-252.

Pardo, M. C. (1997a). Asymptotic behaviour of an estimator based on Rao's divergence *Kybernetika,* **33**, 5, 489-504.

Pardo, M. C. (1999a). On Burbea-Rao divergence based goodness-of-fit tests for multino mial models. *Journal of Multivariate Analysis,* **69**, 1, 65–87

Pardo, M. C. (1999b). Estimation of parameters for a mixture of normal distribu tions on the basis of the Cressie and Read divergence. *Communications in Statis tics.(Simulation and Computation),* **28**, 1, 115-130.

Pardo, M. C. and Pardo, J. A. (1999). Small-sample comparisons for the Rukhin goodness-of-fit-statistics. *Statistitical Papers,* **40,** 2, 159–174.

Pardo, M. C. and Vajda, I. (1997). About distances of discrete distributions satisfying the data processing theorem of information theory. *IEEE Transactions on Information Theory,* **43,** 1288-1293.

Parr, W.C. (1981). Minimum distance estimation: A bibliography. *Communications in Statistics (Theory and Methods),* **10**, 1205-1224.

Patil, G. P. and Taille, C. (1982). Diversity as a concept and its measurement. *Journal of the American Statistical Association,* **77,** 548-567.

Patnaik, P. B. (1949). The non-central χ^2 and F distributions and their applications. *Biometrika,* **36,** 202-232.

Pearson, K. (1894). Contributions to the mathematical theory of evolution. *Philosophical Transactions of the Royal Society of London,* Series A, **185**, 71-110.

Pearson, K. (1900). On the criterion that a given system of deviations from the probable in the case of a correlated system of variables is such that it can be reasonably supposed to have arisen from random sampling. *Philosophy Magazine,* **50**, 157-172.

Pérez, T. and Pardo, J. A. (2002). Asymptotic normality for the K_ϕ-divergence goodness of fit tests. *Journal of Computational and Applied Mathematics,* **145**, 301-317.

Pérez, T. and Pardo, J. A. (2003a). On choosing a goodness of fit for discrete multivariate data. *Kybernetes,* **32**, 1405-1424.

Pérez, T. and Pardo, J. A. (2003b). Asymptotic approximations for the distributions of the K_ϕ-divergence goodness of fit statistics. *Statistical Papers,* **44**, 349-366.

Pérez, T. and Pardo, J. A. (2003c). Goodness of fit tests based on K_ϕ-divergence. *Kybernetika,* **39**, 739-752.

Pérez, T. and Pardo, J. A. (2004). Minimum K_ϕ-divergence estimator. *Applied Mathematics Letter,* **17**, 367-374.

Pérez, T. and Pardo, J. A. (2005). The K_ϕ-divergence statistic for categorical data problems. *Metrika.* In press.

Pielou, E. C. (1967). The use of information theory in the study of the diversity of biological populations. *Proceedings of the Fifth Berkeley Symposium on Mathematical Statistics and Probability,* **IV,** 163-177.

Pielou, E. C. (1975). *Ecological Diversity.* John Wiley, New York.

Pitman, E. J. G. (1979). *Some Basic Theory for Statistical Inference.* Chapman and Hall, New York.

Pyke, R. (1965). Spacings. *Journal of the Royal Statistical Society*, Series B, **27**, 395-449.

Quade, D. and Salama, I. A. (1975). A note on minimum chi-square statistics in contingency tables. *Biometrics,* **31**, 953-956.

Quandt, R.E. and Ramsey, J.B. (1978). Estimating mixtures of normal distributions and switching distributions. *Journal of the American Statistical Association,* **73**, 730-738

Quine, M. P. and Robinson, J. (1985). Efficiencies of chi-square and likelihood ratio goodness-of-fit tests. *The Annals of Statistics,* **13,** 727-742.

Radlow, R. and Alf, E. F. (1975). An alternate multinomial assessment of the accuracy of the χ^2 test of goodness of fit. *Journal of the American Statistical Association*, **70**, 811-813.

Ranneby, B. (1984). The maximum spacings method. An estimation method related to the maximum likelihood method. *Scandinavian Journal of Statistics*, **11**, 93-112.

Rao, C. R. (1949). On the distance between two populations. *Sankhya,* **9,** 246-248.

Rao, C. R. (1954). On the use and interpretation of distance functions in statistics. *Bulletin of the International Statistical Institute,* **34,** 90.

Rao, C. R. (1973). *Linear Statistical Inference and Its Applications.* John Wiley & Sons, New York.

Rao, C. R. (1982a). Diversity and disimilarity coefficients: A unified approach. *Journal Theoretical Population Biology,* **21**, 24-43.

Rao, C. R. (1982b). Diversity: Its measurement, decomposition, apportionment and analysis. *Sankhya,* Series A, **44,** 1-22.

Rao, C. R. (2002). Karl Pearson chi-square test. The dawn of statistical inference, in *Goodness-of-Fit Tests and Model Validity.* (C. Huber-Carol, N. Balakrishnan, M. S. Nikulin and M. Mesbah Eds.), Birkhäuser, Boston, 9-24.

Rao, J.S. and Sethuramann, J. (1975). Weak convergence of empirical distribution functions of random variables subject to perturbations and scale factors. *The Annals of Statistics*, **3**, 299-313.

Rathie, P.N. and Kannappan, P.L. (1972). A directed-divergence function of type β. *Information and Control,* **20,** 38-45.

Read, T. R. C. (1984a). Closer asymptotic approximations for the distributions of the power divergence goodness-of-fit statistics. *Annals of the Institute of Statistical Mathematics,* **36**, 59-69.

Read, T. R. C. (1984b). Small-sample comparisons for the power divergence goodness-of-fit statistics. *Journal of the American Statistical Association,* **79**, 929-935.

Read, T. R. C. and Cressie, N. A. C. (1988). *Goodness of Fit Statistics for Discrete Multivariate Data.* Springer-Verlag, New York.

Rényi, A. (1961). On measures of entropy and information. *Proceedings of the Fourth Berkeley Symposium on Mathematical Statistics and Probability,* **1**, 547-561.

Robertson, T. (1966). *Conditional Expectation Given a σ-lattice and Estimation of Restricted Densities.* Ph. D. Dissertation, University of Missouri-Columbia.

Robertson, T. (1978). Testing for and against an order restriction on multinomial parameters. *Journal of the American Statistical Association,* **73**, 197-202.

Robertson, T., Wright, F. T. and Dykstra, R. L. (1988). *Order Restricted Statistical Inference.* John Wiley & Sons, New York.

Rodríguez, C. C. and Van Ryzin, J. (1985). Maximum entropy histograms. *Statistics and Probability Letters,* **3**, 117-120.

Roscoe, J. T. and Byars, J. A. (1971). An investigation of the restraints with respect to sample size commonly imposed on the use of the chi-square statistic. *Journal of the American Statistical Association,* **66**, 755-759

Rousas, G. G. (1979). Asymptotic distribution of the log-likelihood function for stochastic process. *Zeitschrift für Warsheinlichkeitstheorie und Verwandte Gebiete,* **47**, 31-46.

Rudas, T. (1986). A Monte Carlo comparison of the small sample behaviour of the Pearson, the likelihood ratio and the Cressie-Read statistics. *Journal of Statistical Computation and Simulation,* **24**, 107-120.

Rukhin, A. L. (1994). Optimal estimator for the mixture parameter by the method of moments and information affinity. *Transactions 12th Prague Conference on Information Theory,* 214-219.

Sahoo, P. K. and Wong, A. K. C. (1988). Generalized Jensen difference based on entropy functions. *Kybernetika,* **24**, 4, 241-250.

Saks, S. (1937). *Theory of the Integral.* Lwów, Warszawa.

Salicrú, M., Menéndez, M. L., Morales, D. and Pardo, L. (1993). Asymptotic distribution of (h, ϕ)-entropies. *Communications in Statistics (Theory and Methods),* **22**, 7, 2015-2031.

Salicrú, M., Morales, D., Menéndez, M. L. and Pardo, L. (1994). On the applications of divergence type measures in testing statistical hypotheses. *Journal of Multivariate Analysis,* **51**, 372-391.

Schorr, B. (1974). On the choice of the class intervals in the application of the chi-square test. *Mathematische Operationsforschung und Statistik,* **5**, 357-377 .

Searle, S. R. (1971). *Linear Models.* John Wiley & Sons, New York.

Sekiya, Y. and Taneichi, N. (2004). Improvement of approximations for the distribution of multinomial goodness-of-fit statistics under nonlocal alternatives. *Journal o Multivariate Analysis,* **91,** 199-223.

Sekiya, Y., Taneichi, N. and Imai, H. (1999). A power approximation for the multino mial goodness-of-fit test based on a normalizing transformation. *Journal of th Japan Statistic Society,* **29,** 79-87.

Sen, P. K. and Singer, J. M. (1993). *Large Sample Methods in Statistics.* Chapman & Hall, New York.

Serfling, R. J. (1980). *Approximations Theorems of Mathematical Statistics.* Joh Wiley, New York.

Sethuraman, J. and Rao, J.S. (1970). Pitman efficiencies of tests based on spacings in *Nonparametric Techniques in Statistical Inferences.* (M.L. Puri, Ed.), 267-27 Cambridge University Press, London.

Shannon, C. E. (1948). The mathematical theory of communication. *The Bell Syster Technical Journal,* **27,** 379-423.

Sharma, B. D. and Mittal, D. P. (1977). New non-additive measures of relative info mation. *Journal of Combinatory Information & Systems Science,* **2,** 122-133.

Siatoni, M. and Fujikoshi, Y. (1984). Asymptotic approximations for the distributions o multinomial goodness-of-fit statistics. *Hiroshima Mathematics Journal,* **14,** 11 124.

Sibson, R. (1969). Information radius. *Zeitschrift für Warsheinlichkeitstheorie un Verwandte Gebiete,* **14,** 149-160.

Silvey, S. D. (1959). The Lagrangian multiplier test. *Annals of Mathematical Statistic* **30**, 389-407.

Simpson, E. H. (1949). Measurement of diversity. *Nature,* **163,** 688.

Slakter, M. J. (1968). Accuracy of an approximation to the power of the chi-squar goodness-of-fit tests with small but equal expected frequencies. *Journal of th American Statistical Association,* **63**, 912-918.

Stephens, M.A. (1986). Tests for the uniform distribution. In *Goodness-of-fit Tech niques* (R.B. D'Agostino and M.A. Stephens Eds.), 331-336. Marcel Dekker, Ne York.

Stuart, A. (1955). A test for homogeneity of the marginal distributions in a two-wa classification. *Biometrika,* **42,** 412-416.

Sturges, H. A. (1926). The choice of a class interval. *Journal of the American Statistic Association,* **21**, 65-66.

Sy, B. K. and Gupta, A. K. (2004). *Information-Statistical Data Mining.* Kluwer Academic, Dordrecht.

Taneichi, N., Sekiya, Y. and Suzukawa, A. (2001a). An asymptotic approximation for the distribution of ϕ-divergence multinomial goodness-of-fit statistics under local alternatives. *Journal Japan Statistical Society,* **31**, 207-224.

Taneichi, N., Sekiya, Y. and Suzukawa, A. (2001b). Asymptotic approximations for the distributions of the multinomial goodness-of-fit statistics under local alternatives. *Journal of Multivariate Analysis,* **81**, 335-359.

Tate, M. W. and Hyer, L. A. (1973). Inaccuracy of the X^2 test of goodness of fit when expected frequencies are small. *Journal of the American Statistical Association,* **68**, 836-841.

Tavaré, S. and Altham, P.M. E. (1983). Serial dependence of observations leading to contingency tables, and corrections to chi-squared statistics. *Biometrika,* **70**, 139-144.

Tenenbein, A. (1970). A double sampling scheme for estimating from binomial data with misclassification. *Journal of the American Statistical Association*, **65**, 1350-1361.

Tenenbein, A. (1971). A double sampling scheme for estimating from binomial data with misclassification: Sample size and determination. *Biometrics*, **27**, 935-944.

Tenenbein, A. (1972). A double sampling scheme for estimating from misclassified multinomial data with applications to sampling inspection. *Technometrics*, **14**, 187–202.

Theil, H. and Fiebig, R. (1981). *Exploiting Continuity: Maximum Entropy Estimation of Continuous Distributions.* Bollinger, Cambridge.

Theil, H. and Kidwai, S. (1981a). Moments of the maximum entropy and symmetric maximum entropy distribution. *Economic Letters*, **7**, 349-353.

Theil, H. and Kidwai, S. (1981b). Another look at maximum entropy coefficient. *Economic Letters*, **8**, 147-152.

Theil, H. and Laitinen, K. (1980). Singular moment matrix in applied econometrics, in *Multivariate Analysis.* (V.P.R. Krishnaiah Ed.), 629-649. North Holland, Amsterdam.

Theil, H. and Lightburn, P. (1981). The positive maximum entropy distribution. *Econometric Letters*, **6**, 231-239.

Theil, H. and O'Brien, P.C. (1980). The median of the maximum entropy distribution. *Econometric Letters*, **5**, 345-347.

Toussaint, G. T. (1974). Some properties of Matusita's measure of affinity of severa distributions. *Annals of the Institute of Statistical Mathematics,* **26**, 389-394.

Toussaint, G. T. (1978). Probability of error, expected divergence, and affinity of severa distributions. *IEEE Transactions on System, Man and Cybernetics,* **8**, 6, 482-48

Turnbull, B.W. and Weiss, L. (1978). A likelihood ratio statistic for testing goodnes of fit with randomly censored data. *Biometrics,* **34**, 412-427.

Vajda, I. (1973). χ^2-divergence and generalized Fisher's information. *Transactions* *6th Prague Conference on Information Theory,* 873-886. Academy of Sciences the Czech Republic. Institute of Information Theory and Automation, Prague.

Vajda, I. (1989). *Theory of Statistical Inference and Information.* Kluwer Academi Dordrecht.

Vajda, I. (1995). Information-theoretic methods in statistics, in *Research Report. Acad emy of Sciences of the Czech Republic.* Institute of Information Theory and Au tomation. Prague.

Vajda, I. and Vasev, K. (1985). Majorization, concave entropies and comparison experiments. *Problems of Control and Information Theory,* **14,** 105-115.

Vajda, I. and Teboulle, M. (1993). Convergence of best phi-entropy estimates. *IEE Transactions on Information Theory,* **39,** 297-301.

Varma, R. S. (1966). Generalizations of Renyi's entropy of order α. *Journal of Mathe matical Sciences,* **1,** 34-48.

Vasicek, O. (1976). A test of normality based on sample entropy. *Journal of the Roy Statistical Society,* Series B, **38**, 54-59.

Verdú, S. (1998). Fifty years of Shannon entropy. *IEEE Transactions on Informatio Theory,* **44,** 6, 2057-2078.

Wedderburn, R. W. M. (1974). Generalized linear models specified in terms of con straints. *Journal of the Royal Statistical Society,* Series B, **36,** 419-454.

Weiss, L. (1974). The asymptotic sufficiency of a relatively small number of orde statistics in goodness of fit. *The Annals of Statistics,* **2**, 795-802.

West, E. N. and Kempthorne, O. (1972). A comparison of the Chi^2 and likelihood rati test for composite alternatives. *Journal of Statistical Computation and Simulation* **1**, 1-33.

Wolfowitz, J. (1957). The minimum distance method. *Annals of Mathematical Statis tics,* **28**, 75-88.

Woodward, W.A., Parr, W.C., Schucany, W.R. and Lindsay, B. G. (1984). A com parison of minimum distance and maximum likelihood estimation of a mixtur proportion. *Journal of American Statistical Association,* **79**, 590-598.

Woodward, W.A., Whitney, P. and Eslinger, P.W. (1995). Minimum Hellinger distance of mixture proportions. *Journal of Statistical Planning and Inference,* **48**, 303-319.

Worchester, J. (1971). The relative odds in the 2^3 contingency table. *American Journal of Epidemiology,* **93,** 145-149.

Yarnold, J. K. (1972). Asymptotic approximations for the probability that a sum of lattice random vectors lies in a convex set. *Annals of Mathematical Statistics,* **43,** 1566-1580.

Zografos, K., Ferentinos, K. and Papaioannou, T. (1990). φ-divergence statistics: Sampling properties, multinomial goodness of fit and divergence tests. *Communications in Statistics (Theory and Methods),* **19,** 5, 1785-1802.

Zografos, K. (1993). Asymptotic properties of ϕ-divergence statistic and applications in contingency tables. *International Journal of Mathematics and Statistical Sciences,* **2**, 5-21.

Zografos, K. (1994). Asymptotic distributions of estimated f-dissimilarity between populations in stratified random sampling. *Statistics and Probability Letters*, **21,** 147-151.

Zografos, K. (1998a). f-dissimilarity of several distributions in testing statistical hypotheses. *Annals of the Institute of Statistical Mathematics*, **50**, 295-310.

Zografos, K. (1998b). On a measure of dependence based on Fisher's information matrix. *Communications in Statistics (Theory and Methods),* **27**, 1715-1728.

Zografos, K. (1999). On maximum entropy characterization of Pearson's type II and VII multivariate distributions. *Journal of Multivariate Analysis,* **71,** 67–75.

Zografos, K. (2000). Measures of multivariate dependence based on a distance between Fisher information matrices. *Journal of Statistical Planning and Inference,* **89**, 91–107.

Zografos, K. and Nadarajah, S. (2005). Expressions for Rényi and Shannon entropies for multivariate distributions. *Statistics and Probability Letters,* **71,** 71-78.

Index

Printed in the United States
by Baker & Taylor Publisher Services

Printed in the United States
by Baker & Taylor Publisher Services